Neurobiology of Sensory Systems

Neurobiology of Sensory Systems

Edited by

R. Naresh Singh

Tata Institute of Fundamental Research
Bombay, India

and

Nicholas J. Strausfeld

The University of Arizona
Tucson, Arizona

Springer Science+Business Media, LLC

Library of Congress Cataloging in Publication Data

International Conference on Neurobiology of Sensory Systems (1st: 1988: Velha Goa, India)
Neurobiology of sensory systems / edited by R. Naresh Singh and Nicholas J. Strausfeld.
p. cm.
"Proceedings of the First International Conference on Neurobiology of Sensory Systems, held September 25-30, 1988, in Goa, India"—T.p. verso.
Includes bibliographical references.
ISBN 978-1-4899-2521-3
1. Senses and sensation—Congresses. 2. Comparative neurobiology—Congresses. I. Singh, R. Naresh. II. Strausfeld, Nicholas James, 1942- . III. Title.
[DNLM: 1. Neurobiology—congresses. 2. Receptors, Sensory—physiology—congresses. 3. Sense Organs—physiology—congresses. WL 700 I5908 1988]
QP431.I57 1988
591.1′82—dc20
DNLM.DLC 89-22960
for Library of Congress CIP

ISBN 978-1-4899-2521-3 ISBN 978-1-4899-2519-0 (eBook)
DOI 10.1007/978-1-4899-2519-0

The cover design is inspired by the hand gestures used to depict the five senses in the "Kathakali" style of Indian Classical Dance. These are, clockwise from the top: vision, smell, touch, taste, and hearing.

Proceedings of the First International Conference on Neurobiology of Sensory Systems, held September 25-30, 1988, in Goa, India

PREFACE

The traveller to India is urged to visit that country's western shore with the Arabian Sea where, about 300 miles to the south of Bombay, an exceedingly lovely coast reaches the peak of its harmony at the erstwhile Portuguese enclave of Goa. The ambience of this alluring province is an exquisite balance of palm trees and rice fields, aged colonial homes - many still elegant and brightly painted - slowly being swallowed up by the exuberant tropical vegetation, incredible blossoms, colorful and courteous people and, deeper inland, some splendid examples of 17th and 18th century Portuguese ecclesiastical architecture. A feast for the eyes by day, and in the evening enough fresh fish and other good food to satisfy the most demanding gourmet. This was the paradisiacal setting for the first International Conference on the Neural Organization of Sensory Systems (ICONOSS for short), sponsored jointly by the International Brain Research Organization (IBRO), the Tata Institute for Fundamental Research at Bombay, the Department of Atomic Energy of the Government of India, and the Department of Science and Technology of the Government of India.

About 100 participants were pleasantly confined at Fort Aguada, a resort cunningly built amongst the ruins of an old Portuguese fort. The conference program achieved an international flavor, recruiting scientists from many nations: India (naturally), Australia, Britain, Canada, Germany, Finland, France, Hungary, Japan, the Netherlands, Sweden, Switzerland and the United States of America. The subjects discussed were as diverse as the countries represented. And although the location alone imbued everyone with an irresistible feeling of collegiality and well-being, what unified the participants was a shared fascination with how sensory systems are organized and operate. This volume summarizes the diverse topics discussed, ranging from mechanosensory receptors of spiders to the mosaic organization of vertebrate photoreceptors, from how crabs view a flat world with eyes on stalks to how bats perceive a three-dimensional world using their mouths and ears, from the perception of speech to the perception of what bits go where in copulating crickets.

A vital function of any international meeting is to dispose of boundaries that humankind in its stupidity has erected between cultures. One great value of scientific pursuit is to provide a common culture through intellectual hybridization. And so it was at this conference: a meeting that introduced scientists from different lands to each other and their research. We departed enriched by the hospitality of our Indian hosts, the beauty of Goa, and by the open and enthusiastic exchange of ideas. For those that were not there, we hope that this volume conveys the spirit as well as the substance of this meeting and whets the appetite for the promise of more to come.

The Editors

FOREWORD AND ACKNOWLEDGEMENTS

We perceive the world around us through our sensory systems, which function as the windows of our brain, and on which learning, memory and the whole richness of experience depend. The senses provide us with the pleasures of life: beauty, aroma, taste, music and touch. And, too, the pressures and the pain, the perception of which, in common with other organisms, is crucial to our survival.

The idea of organizing an International Conference on the neurobiology of sensory systems came to mind about three years ago. Initially, I explored the possibility of holding the conference at the Tata Institute of Fundamental Research (TIFR), Bombay. However, after some exploratory efforts, I realized that the Institute has grown too large and its facilities, like the guest house and the canteen, would not be able to take on the additional load of such an international conference.

We then explored places where we could have the conference. The criteria for selection were (1) living conditions should be good and hygienic and (2) all the participants should be accommodated on the same campus so that there would be ample time and opportunity for fruitful and effective interaction at scientific and personal levels. The younger scientists were more likely to benefit from the more established ones in such an atmosphere. The choice of the Fort Aguada Beach Resort and the Taj Holiday Village as the venue of the Conference was made not only on these criteria but also on the basis of the happy experience of some of the other members of the TIFR, who had organized international conferences there earlier.

A potential disadvantage of holding the Conference away from the home institution and, at a rather luxurious place as well, would have been that not many participants from within the country would be able to bear the cost of attending. Happily we were able to support several of them through local hospitality or reimbursement of travel expenses or both. And I may say that all who had shown a real interest in the conference were accommodated. This was made possible through the kind support of the following:

Tata Institute for Fundamental Research, Bombay
International Brain Research Organization
Department of Atomic Energy, Government of India
Department of Science and Technology, Government of India
Honorable Chief Minister and Government of Goa
Toshniwal Brothers (Bombay) Limited Company

Scientific and Engineering Supply Company, Bombay
The Taj Group of Hotels, India.

I thank all the Members of the International Scientific Advisory Council and the National Organizing Committee for their keen interest and valuable suggestions on many occasions. My appreciation goes to the Members of the Local Organizing Committee for their whole-hearted cooperation. My special thanks are due to M. A. Parelker, the Administrative Secretary of the Conference, for our many discussions about the various organizational aspects, and who provided moral support when it was needed in quite a few anxious moments. Very notable help was rendered by my colleagues Shubha Shanbag, the Scientific Secretary; Seema Deshpande, the Treasurer; Veronica Rodrigues; Chetan Premany; and Kusum Singh who besides being my partner in life has also been an excellent scientific colleague.

For the successful conclusion of any project such as this, one seeks the help of many people at various times and stages. Usually one encounters three kinds of individuals: the first are helpful, the second unhelpful. The former give us happiness and strength to go ahead with the project, while the latter make us more resolute and determined for its success. However, the overwhelming joy of life comes from meeting people of another type -- the third kind -- who came to us on their own accord and offered whatever help they could give. Such were Anindya Sinha, Archana Gayatri, Shashi Acharya, Vishwas Saranghdar, Rasmi Sood, Rohini Balakrishnan, Cheryl Lobo, Joyce Fernandes, Raghu Ram, M. M. Kakeri and H. G. Chauhan.

My special thanks are due to the former Editor Lisa Honski and the present Editor Melanie Yelity, and their colleagues at Plenum Press with whose excellent cooperation and help the publication of the Proceedings of the Conference has been expedited in the shortest possible time. I also thank all the authors for their contributions.

For retyping of certain manuscripts I thank Ms Jennifer Lawrence at the Division of Neurobiology, Arizona Research Laboratories, University of Arizona, and for expert language-editing, Dr. Camilla Strausfeld, at the University of Arizona.

The papers presented in this volume have been broadly arranged according to the subject area: vision -- behavior, systems analysis, development and functional organization; olfaction, taste and receptor channels; auditory, tactile and nociception. I hope this volume "Neurobiology of Sensory Systems" will serve as a ready reference to young and future research workers.

The Convener
R. Naresh Singh

INTERNATIONAL CONFERENCE ON NEUROBIOLOGY OF SENSORY SYSTEMS
GOA 1988

ICONSS

CONTENTS

VISION: Physiology, Functional Organization, Behavior and Developmental Models

OLFACTION AND TASTE: Comparisons with Other Senses, Anatomy, Physiology, Genetic Analysis, Ion Channel Models

ACOUSTIC PERCEPTION, MECANORECEPTION AND NOCICEPTION

PRIMITIVE VISION BASED ON SENSING CHANGE

G.A. Horridge

Centre for Visual Sciences, Research School of Biological Sciences, Australian National University, P O Box 475 Canberra ACT 2601, Australia

An instructive way to discover a new avenue of research is to think about old work with new eyes. If, with a mind full of modern trends and recent theories, one reads through old literature with a discerning eye, it is surprising what crops up. With a bit of luck, you suddenly realise something that the old ones missed. There has been an enormous outburst of research on spatial vision in the past 25 years or so, starting with the controversies about the existence or otherwise of elements in the mammalian visual cortex tuned to particular spatial or temporal frequencies, and arguments about whether visual processing neurons should be considered as matching templates (detecting edges and bars) or as Fourier analysers. This period has also seen the rise of white noise analysis, in which the stimulus is randomly distributed in space and time so that all possible combinations of stimulus pattern have some chance of appearing. Pattern perception has also been intensively studied as a problem in human psychophysics, as a problem in visual behaviour of lower animals, and as a way of giving object recognition to computers.

After all the reviews of it, you would be forgiven for supposing that nothing more can be squeezed out of the old literature. It is not my task to justify old thoughts, to replay old tunes for their ingenuity or repeat old problems to puzzle the young; discovering modern philosophy in Plato, modern science in Aristotle or modern religion in the Buddha, may be a pleasant diversion but essentially it gets us no further forward.

One of the very basic new concepts running through much work on vision is the application of sampling theory to the retina. The receptors are distributed in angular coordinates centred more or less on the optical nodal point of their own region of the eye. This means that they divide up the outside world where they look out from the eye. In an eye with a single large lens, a video-camera or a camera, the receptors cannot overlap and in fact they must be optically distinct in order to keep the image crisp without cross-talk. Photoreceptors, however, cannot be indefinitely small: they must be a certain size in order to catch sufficient light to be sufficiently sensitive in the ambient intensity at which they operate. Sampling theory teaches us that the image is divided up into pixels, which are units of the image space represented by the individual receptors. Within the receptor (or pixel) the origin, direction, colour, polarization etc. of a photon is not reported, the only record passed to the next cell is that the photon is absorbed by a photoreceptor with certain line-labelled absorption features. The cost in having smaller pixels is loss of sensitivity; there is a corresponding gain, increase in spatial resolution, which is the ability to see detail in the image - at least that is the current belief. Now let us turn to some old lost data, never incorporated into these theories.

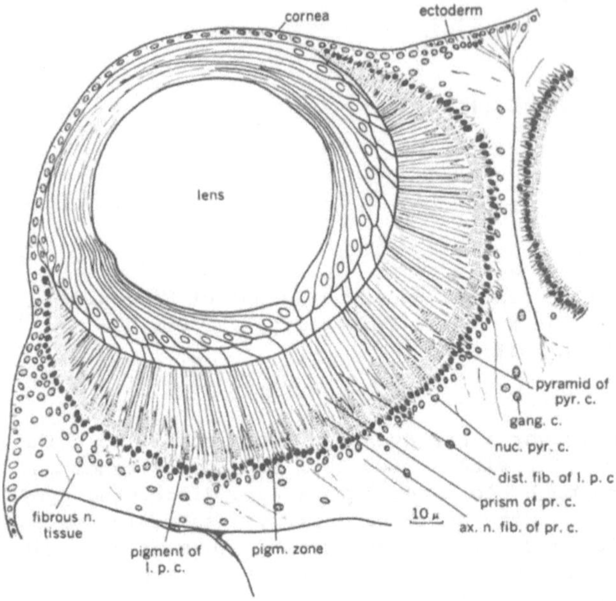

Figure 1.

Eye of Charybdea (Coelenterata: Cubomedusae) Berger:1900

Among the lower Phyla of the Invertebrates are several with scattered examples of remarkably well-developed eyes that have large numbers of small photoreceptors. Good examples of these eyes are found in the Cubomedusae (Coelenterata), the alcyopid worms (Polychaeta), Peripatus (Onychophora), and in errant polychaetes, and such as the common ragworm (Nereis). In none of these examples is the nervous system understood, i.e. no-one has any idea what the sensory neurons are doing, but it is quite obvious from the simplicity of the nervous system and the low level of the behaviour that these animals do not see objects as we do. Their brains are not large enough, there are no signs of spatial projections upon lobes of the brain nor behavioural evidence of visual discriminations of patterns. Among the arthropods also, there are many relatively primitive groups with plenty of receptors in their retinas. Examples are the apterygote insect Machilis, some millipedes, lower crustaceans and even some herbivorous insects with very large numbers of visual receptors but little evidence of pattern discrimination in their behaviour. These animals seem not to be interested in the separation of objects in pictures. The same applies to many molluscs. In fact, it is quite hard to find examples of invertebrates, such as the bee and the octopus, in which it is possible to make behavioural experiments with different responses to different patterns (i.e. pattern discrimination). I think that it is a disgrace that the numerous text-books showing diagrams of invertebrate eyes make no comment on all that spatial resolution apparently going to waste.

The progressive increase in resolution of the eye in many groups of animals over long periods of evolution has gone hand-in-hand with increasing complexity of the brain and behaviour. The eye, the brain and the visual behaviour each need the others as the test bed of successive small improvements. An eye with very many receptors, but without the other developments remains an apparent anomaly. All of these eyes make sense, however, if they are specialised for the detection of motion.

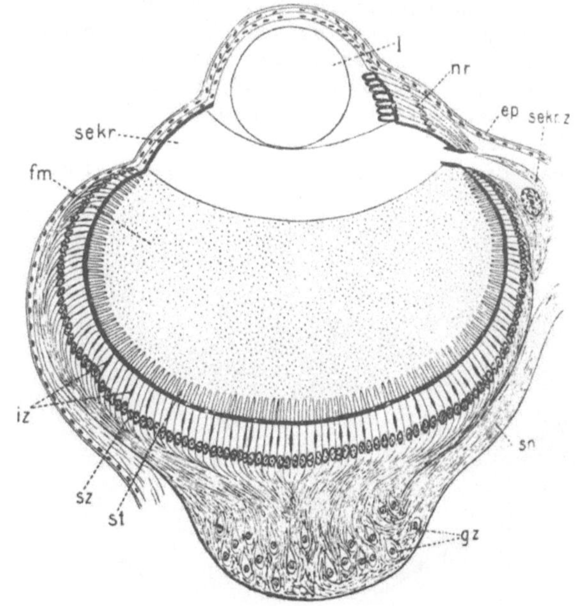

Figure 2.

Eye of <u>Alciopa</u>
(Annelida: Polychaeta)
Demoll: 1909

Vision itself evolved independently in many groups of animals, and most phyla have some representatives with extremely limited vision living in environments full of interesting visual detail that they ignore as pictures. All primitive eyes, when tested, turn out to be sensitive to motion. The best-known very simple examples, such as the barnacle eye, detect the approach of a potential predator that casts its shadow on a few receptors that respond to change in intensity. The next level is the detection of a moving contrast by a <u>stationary</u> eye: some molluscs, e.g. scallops and giant clams, are of this type. Even at this level, there are many types of eyes with large numbers of receptors, and with optical systems that suggest that they approach optimum resolution as limited by diffraction, and have maximum sensitivity to axial rays by utilising long rods of dense visual pigment.

Beyond this stage, several groups of <u>mobile</u> animals have evolved a variety of visual mechanisms which prevent collisions, detect the direction of a distant movement, and pursue prey or mates. Eyes serving this level of complexity have a sampling array of many receptors looking in different directions on adjacent optical axes; an arrangement which can be achieved in an eye like a camera with many light-sensitive grains behind a single lens, or by a compound eye with a separate optical axis for each lens, as in most insects and crustaceans.

On this theory, eyes with arrays of receptors having narrow fields of view were progressively elaborated as sampling arrays with optimum optics in order to detect motion relative to the eye, and not because they could resolve the details in the picture, for at first they lacked the appropriate optic lobe processing mechanisms to do that. In such a theory, the neural mechanisms can be progressively added behind an existing retina that detects motion. All this family of visual systems will therefore depend primarily on detection of motion, or something like motion such as being tuned to successive flicker in adjacent facets. There are many well-studied examples of eyes that depend on motion perception and have additional processing mechanisms based on motion. Examples are jumping spiders, praying mantids and snakes, some of which have diffraction-limited eyes backed by a nervous system that can coordinate a strike at a prey. Eyes with a wide range of structure detect motion of potential prey remarkably well. Possibly the design of these eyes are optimized for motion detection (and therefore for a combination of spatial and temporal resolution), but not for resolving stationary images.

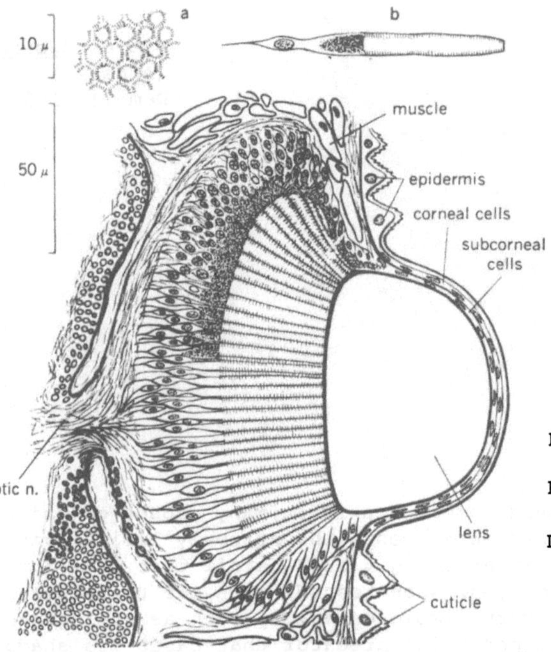

Figure 3.

Eye of Peripatus (Onychophora) Dakin: 1921

Several diverse arthropods that have appropriate behavioural responses (and have been tested) respond to angular movements that are more than an order of magnitude smaller than the interommatidial angle. In fact, low threshold for motion seems to be very widespread in motion perception systems, even in the peripheral vision of vertebrates. In almost all cases it has yet to be demonstrated that motion and not modulation is the stimulus. Besides seeing the direction of a motion, the best examples, such as arthropods, some polychaetes and some molluscs, with brains at this level of complexity, respond also to the direction of a steady light, and sometimes to the direction of a selected colour, but efforts to reveal any sort of form vision have failed except in a few special cases such as the honey bee and the octopus. Nevertheless, most of these animals move freely as if they see separate and discrete objects in a three-dimensional world. There is little evidence that they see objects or patterns and recognize them as belonging to different classes, as we do.

Vision by a moving eye

A freely-moving animal, with eyes that detect the angular bearing and direction of motion of a moving contrast, finds itself with new problems and possibilities introduced by its own motion. When it turns or moves forward on a straight course, all surrounding contrasts move relative to the eye. Because locomotion is irregular and mainly horizontal, most of this induced motion is in the horizontal plane and is not entirely predictable from the motor output. Therefore, to see a movement while it itself moves, the animal must cancel the horizontal motion of the background in at least some of the neurons with fields that respond to motion. But as it moves, the apparent background motion differs at different angles to the line of motion. Therefore, the background must be subtracted region by region. When turning in the horizontal plane, however, it need only subtract the motion of the whole surround together, along the equator. Because of the irregularity of locomotion, the visual system must cancel or ignore all movement of the background in each region relative to the eye, not just that part caused by its own motor output. In agreement with this limitation, we find that eye motion is usually compensated more by visual feedback than by proprioceptive feedback.

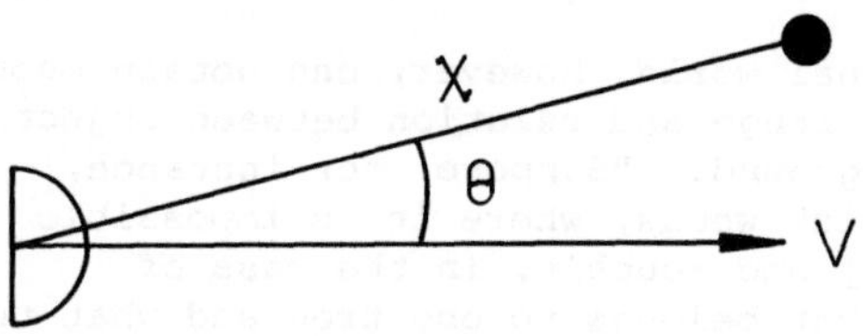

Figure 4.
Motion of an eye generates image motion as an angular velocity in the eye.

As an eye moves sideways (i.e. at right angles to the midline), objects nearby in the visual field generate an image that moves across the retina. The rate at which the image passes the receptors is inversely proportional to the object range. Therefore a moving eye able to make this measurement can tell not only that there is something "out there", but also the direction and the range. That much information is adequate for mobility in a three-dimensional world.

When the eye is moving relative to an object, the rate at which the edges of its image pass the receptors ($d\theta/dt$) is proportional to the sine of their angle θ from the midline, to the eye velocity V, and inversely proportional to the range χ. As shown by Nakayama & Loomis (1974):-

$d\theta/dt = (V \sin \theta)/\chi$ at each instant.......(1)

The time to crash $t = \chi/V = (\sin \theta)/(d\theta/dt)$....(2)

so $\theta/t = d\theta/dt$ when θ is small................(3)

So time to contact, t, depends only on the way that the image moves at each point on the retina. More complex hard-wired detectors of an expanding scene (looming units) would be more effective for measuring the time to contact than simple directional velocity detectors. This type of mechanism has been inferred in flies (Wagner, 1982), in birds (Lee & Reddish, 1981) and man (Lee & Young, 1985) and I can now report that 'looming' neurons have now been found in two groups of insects by my colleagues in Canberra.

Because much insect behaviour, such as the landing response, the dodging response, and the escape response, depends on this kind of measurement by the visual system, and there is little evidence of more complex processing, it would not be unreasonable to think of the insect's visual world as a time-to-contact at each angle on the eye. Whereas we see angle-labelled moving contrasts, the fly would see angle-labelled anticipated collisions, or time-to-reach-goal in each direction.

This idea opens the way towards an understanding of vision which is without perception of objects, a low-level type of vision that enables mobility in a three-dimensional world without actually seeing any <u>thing-as-such</u>. Artificial eyes based on this principle will be useful in all kinds of technological applications, in robot vision and as aids for the blind, and we can imagine a seeing silicon chip that will be used as a universal front-end for a variety of artificial seeing systems.

Measurement of one velocity or many

As described above, measurement of the angular velocity of the image across the retina gives a measure of the range of a contrasting object if the moving eye is informed about its own speed. Secondly, for an eye moving at right angles to the direction of an object, the range is inversely proportional to the angular velocity. For an eye moving directly towards an object, the range is inversely proportional to the rate of angular expansion on the retina, by simple geometry. These are two uses for direct measurement of angular velocity. We shall see later that the main problem is how to make this measurement independent of pattern.

A moving eye in a three-dimensional world, however, can obtain much more information about the existence, range and relation between objects from the relative motion against background. "Suppose, for instance, that a person is standing still in thick woods, where it is impossible for him to distinguish, except vaguely and roughly, in the mass of foliage and branches all around him what belongs to one tree and what to another, or how far apart the separate trees are, etc. But the moment he begins to move forward, everything disentangles itself and immediately he gets an apperception of the material contents of the woods and their relations to each other in space, just as if he were looking at a good stereoscopic view of it" (Helmholtz 1911, transl. Southall 1925, vol.3, p.296).

Neurons involved

Recordings from neurons engaged in visual processing behind eyes have so far been restricted to several insects and other arthropods with compound eyes and some vertebrate species. None of the above eyes with many receptors but little brain or behaviour have been investigated. All that can be said about them is that these animals respond to motion with simple directional behaviour. If the eye is sensitive to motion, not to pictures, clearly the number of receptors can increase without the need for a large brain. All that is required is that motion of the image along the receptor layer is detected and separated into neurons according to its direction in space.

The detection of motion requires a minimum of two adjacent receptors: relative motion requires three, arranged in two pairs. Neurons which detect relative motion are known in vertebrates, where they have a field with the centre sensitive to motion in one direction and a surround inhibited by motion in that direction or excited by motion in the opposite direction (References in Horridge, 1987; Mandl, 1985). Relative motion detectors in insects have fields that are excited by small object motion but inhibited by large object motion. For a freely moving eye, neurons that detect relative motion with medium or small fields are presumably object-detectors.

By progressively increasing the number of neurons sensitive to relative motion, the visual processing system can slowly evolve in complexity from seeing only one outstanding object at a time against total background, to a system that detects the times to contact of different objects in separate directions, by line-labelling them in separate neurons with moderate-sized fields.

Once the neurons that detect relative motion have evolved in conjunction with locomotion, any number of additional hard-wired circuits can be added as outputs of the visual system, to deal with specific elements of behaviour. Examples are alerting neurons, detectors of looming, and neurons for turning towards or away from large discrepancies in the flow field with particularly significant features, such as colour, flicker or contrast. None of this requires an inbuilt memory that would recognize previously familiar objects, although memory circuits can also be progressively superimposed upon the motion system. Such a visual processing system can progressively evolve in complexity without, in the intermediate stages, necessarily recognizing objects with the help of a memory. Bees and other Hymemoptera, however, provide evidence that a memory for patterns and for colours is superimposed upon motion detectors in some insects. Insects have at least two systems; one, colour blind, for motion, and another for colour and shape, in parallel. The comparative study of eyes and optic lobe complexity in a variety of insects and crustaceans is consistent with this progressive elaboration in the number of layers of neurons and their duplication in each layer.

<u>The sampling criteria to see motion</u>

First, as a fundamental principle, if convergence takes place only between adjacent receptors, a small contrast that disappears and reappears too far away on the retina will not excite motion detectors, but it is common (even in insects) for motion to be inferred when the stimulus jump is more than one receptor spacing, possibly by sensitivity to the intensity change alone. In several insect groups the optimum jump distance is one inter-ommatidial spacing, like this:-

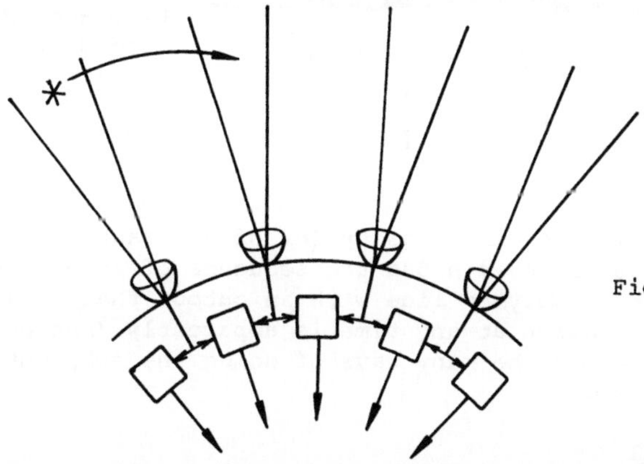

Figure 5.

Secondly, any motion-detecting system must convey the directionality of the signal across the 2-dimensional array of the receptors. We will see that in one model the direction is conveyed by the spatial gradient of the contrast, whereas another employs the directionality of the temporal frequency of passing receptors.

The sampling criterion is illustrated by a row of receptors which have a regular striped pattern projected upon them like this:-

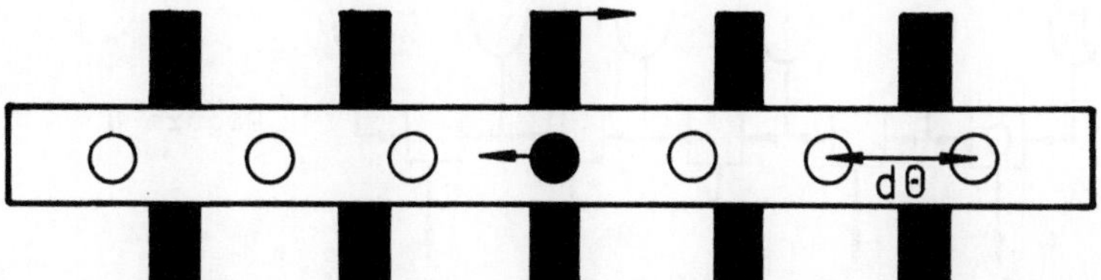

Figure 6.

With the period of the stripes between dθ and 2dθ, the pattern will appear to move in the wrong direction. When the period is dθ, the pattern will never appear to move, and when the period is less than dθ, velocity measurement is not possible. The sampling criterion says that the stimulus period must be more than 2 dθ. Another way to say the same thing is that the periodic pattern can be seen when there is a receptor for each dark and each light band of it.

In so far as any system smooths the output by taking an average over time, it becomes a measure of the drift frequency with which edges or contrast gradients pass the receptors: more edges passing in a giving time give more average output.

If it smoothes the response over time, even the recording gear may be responsible for the apparent measurement of drift frequency irrespective of pattern, because the local response in the eye is event-driven. In other words, an eye which is designed to see objects gives a misleading output when presented with regular stripes or a disrupted pattern.

Measuring velocity across the eye

A contrast moving in the outside world causes a modulation in the receptors, successively across the retina, as in this figure.

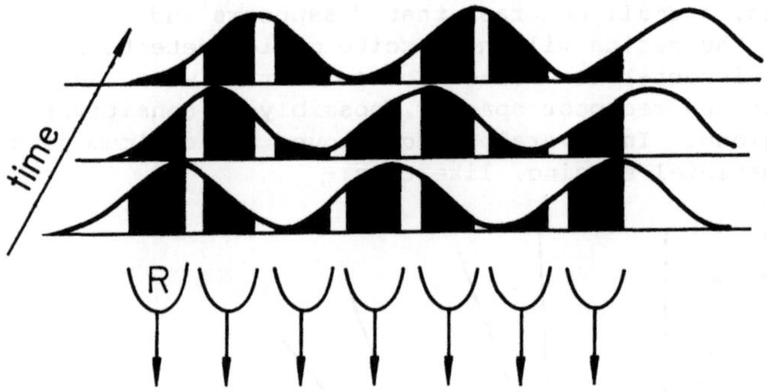

Figure 7.

As the intensity wave travels to the right, each receptor is modulated in time according to the shaded area.

The same modulation of intensity (or of contrast) occurs at successive times as it moves along the sequence of receptors. To measure this velocity continuously on line with an output that is labelled according to its location at any time is apparently (but deceptively) a simple task. There must be many ways of doing the job, but let us examine some of them.

The gradient model

At any moment, the rate of change of the receptor response along the retina is $dR/d\theta$; at any receptor, the rate of change of R with time is dR/dt. Therefore, the rate of change of the angle with time at any point is $d\theta/dt = (dR/dt)/(dR/d\theta)$. Each pair of receptors can therefore give a new output which is theoretically a measure of the velocity over the retina at that point, as in this figure:-

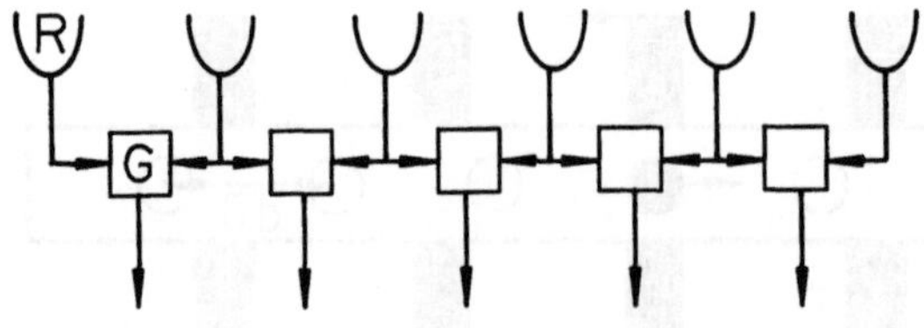

Figure 8

where the square boxes G perform the above mathematical operation. In this model, the spatial resolution of the receptors is preserved and most of the maximum values of the outputs are measures of velocity. The model responds as well to dark as to bright moving objects. Problems with this model are numerous, however. First it is sensitive to pattern, in that a region of the visual field without contrast gives a nonsense answer 0/0; it is little use with a sharp edge or a thin line; it works better with fuzzy images than when much detail is sharply focused. Also, because it looks at small time intervals and involves a division, noise in the input causes large errors. Flicker is disastrous because it causes a large dR/dt but zero dR/dθ. Therefore to make this model work, arbitrary limits must be built into its operation.

In this version of the gradient model the spatial derivative is directional, but the temporal derivative is not. We shall see that in insects the situation is reversed.

The Reichardt model

This well-known model is basically an embodiment of the idea that motion is the autocorrelation of a pattern with itself displaced in time and space (Reichardt, 1961). One receptor picks up a modulation from the pattern; the adjacent receptor receives the same modulation a moment later. Therefore if the first modulation is delayed and then multiplied continuously with the second where they converge, the output reaches a maximum when the velocity of the pattern matches the delay. One version of the model has a high-pass filter behind one receptor and a low-pass behind the other, and the delay depends on the temporal modulation frequency at the receptor. The figure below is a simplified version, the original model also made the correlation in the opposite direction and subtracted one product from the other to eliminate the effect of flicker.

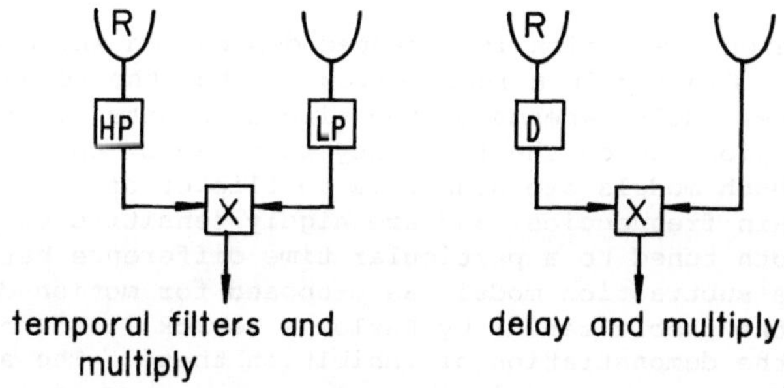

Figure 9

In this model the output is clearly small when the pattern does not move, and is a maximum when the time it takes for the pattern to move from one receptor to the other is the same as the delay in the signal where they converge. Between these values, for a fixed pattern, the output is some measure of the velocity. At greater velocities the response declines, so every value of the response can be caused by two possible inputs (as in all models). Other problems, are numerous. First, the delay where the two arms converge is dependent on the frequency of the modulation at the receptors and therefore to the spatial frequency content of the pattern, but (as in many models) not to the relative phases of the components of the pattern. Therefore the response depends strongly on the spatial frequency, even though changing the components of the pattern relative to each other need have no effect at all. Although one flash should give little response, the flicker of a large source at the frequency which matches the delay gives a large response. This model responds as well to dark moving objects as to light ones, but as soon as we consider its response to a dark-light edge we see that the filters generate a signal that persists in time because an edge contains many frequencies. The multiplication at the convergence therefore is dependent on the slope of the edge and does not give a true measure of its velocity. This model, like all which smooth over time, is more effective in measuring the frequency with which edges pass the eye.

If the environment has a constant texture, or if the constants of the visual processing can be rapidly adjusted for each pattern seen, or if there are many multiplicative unit motion detectors with different time constants, it might be possible to control behaviour sensibly with this system. The optomotor system and some optic lobe neurons of the fly and some other insects behave as if their inputs have multiplicative convergence, as Hassenstein and Reichardt (1953) proposed, and they behave as if they measure drift frequency. The properties, however, are not unique to this model, the tests do not rule out other models, and we have no direct microelectrode demonstration of a mechanism.

Addition or subtraction models

Instead of multiplication, we can have addition at the convergence. In fact, when the receptors respond logarithmically to intensity, the difference is at first sight small. So long as the delay is retained in one arm, we can also have subtraction or inhibition at the convergence, as follows:-

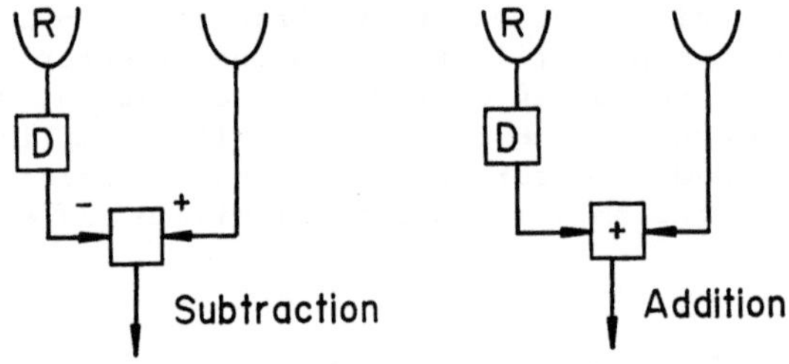

Figure 10

The direction in which the motion is detected depends on which arm has the delayed signal. Clearly in a real nervous system the actual performance of these models depends on the time constants of the synaptic summation or inhibition and on the frequency response of the delay (if that can exist). Both models are sensitive to flicker of a large source, especially at certain frequencies, and are highly sensitive to pattern because they are both tuned to a particular time difference between the two receptors. The subtraction model was proposed for motion-detecting ganglion cells of the rabbit retina by Barlow & Levick (1965) and had some support from the demonstration of inhibition there. The additive and subtractive schemes measure velocity only over a very small range, even for a fixed pattern.

Angular velocity measurement

We have seen that measurement of motion requires convergence between receptors, but that simple systems, even if non-linear, are likely to be influenced by the pattern's complexity, sharpness, brightness or contrast. It is possible to devise a system as follows:-

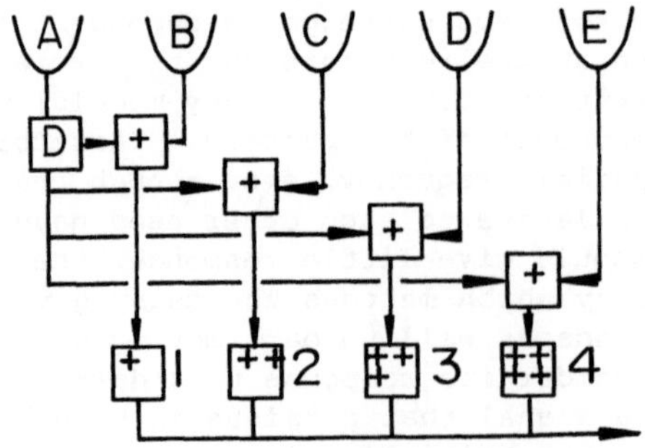

Figure 11

so that a coincidence between A and B results from low velocities, and between A and E for the highest velocities. The final common path measures the total local velocity according to the weighting of paths leading into it. A nervous system can readily generate parallel processing of this complexity. This particular circuit throws away spatial resolution in favour of measurement of velocity, but it is especially sensitive to certain temporal frequencies.

The essential feature resulting from the delay in the temporal filter is the directionality after the convergence. If directionality were not essential, contrast frequency can be measured directly with the same spatial resolution.

Other models

I begin to think that as long as there is convergence between receptors and some kind of measure of time, such as the slopes of the

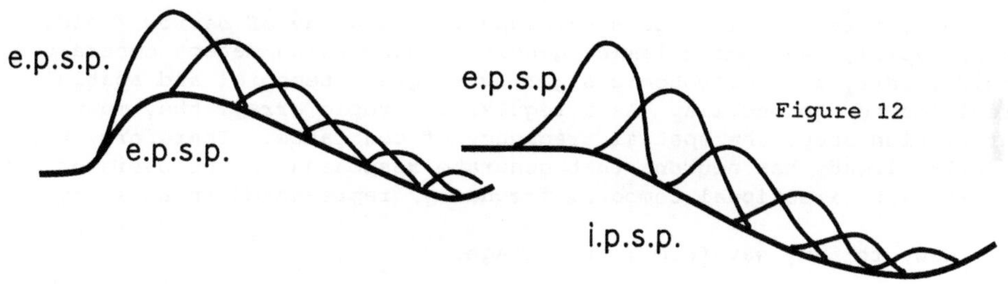

Figure 12

synaptic potentials drawn above, then any interaction at the convergence will suffice to detect motion. Directional motion detection requires an asymmetry in the two arms, but the direction of the motion is not necessarily significant. It is possible that directional motion detection evolved because when the two arms are symmetrical, the system is inevitably equally as sensitive to flicker as to motion.

The interaction at the convergence can be a logical "If X and not Y" or any variant on that: it can be a non-linear addition or inhibition; another possibility is that whenever the input crosses a certain value it starts a clock which runs until the same value arrives down the other arm. A possibility is that the signal is differentiated with respect to time in each arm. Values of dI/dt are then amplified and all values above a mean value are treated as saturated for a certain time while values below mean are taken as zero. The overlap of the resulting waveforms at the convergence is a measure of velocity. As before, however, the system is sensitive to pattern.

There is room for a comparison of the different types of motion-detecting systems when exposed to a wide range of common stimulus situations, so that distinguishing tests can be devised.

Each of the classes of model which detect motion by convergence has an infinite number of members because at the convergence an alteration to one synapse can be compensated by an appropriate change to the other. When we record after the convergence the original presynaptic excitation cannot be recovered.

More lateral spread beyond adjacent receptors is possible, and also there can be successive layers of motion detectors, one below the other. Any system based on these various connections presumably will carry the same basic fault of being pattern sensitive. We see that all simple arrangments to detect and measure unidirectional motion with 2 photoreceptors are unable to measure velocity independently of intensity, contrast and spatial frequency. The most effective is the gradient model but the total response over a short time still depends on the number of edges passing, i.e. on drift frequency.

Reconsideration of the design

To guide the next step, let us consider what insects reveal about structure and processing.

In common insects like flies and butterflies there are many large optic lobe neurons which respond to unidirectional motion as if they measure drift frequency (= contrast frequency) i.e. the rate at which stripes pass each photoreceptor. These neurons are usually wide-field and in flies they appear to be closely related to the optomotor response, but they alone are not sufficient to control behaviour because they do not measure angular velocity on the eye.

A feature of all insects examined is the array of narrow field, high gain, rapidly responding lamina ganglion cells which detect contrast and rapidly carry an electrotonic signal with great temporal and spatial precision to the medulla. As a regular isotropic array, they must carry information about the spatial frequency of contrasts. Therefore, the medulla already has neurons that generate responses on the basis of both spatial and directional temporal frequency, represented in an array.

Now, for any waveform in the image,

Velocity = (drift frequency)/(spatial frequency)....(4)

which is closely related to the gradient model because the drift frequency is proportional to the temporal derivative and the spatial frequency is proportional to the spatial derivative. Insects, however do not convey a directional spatial frequency by the LMC array, which explains why the contrast frequency is directional in insects.

Therefore if the neurons carrying the spatial frequency in their array have the effect of reducing the effect of those measuring directional drift frequency, we have the relation that we seek. Spatial and drift frequencies from the same points in the image are brought together, with the necessary finest possible resolution in space and time.

There are many possible arrangments in which greater local spatial frequency inhibits or shunts out the effect of the drift frequency. For example

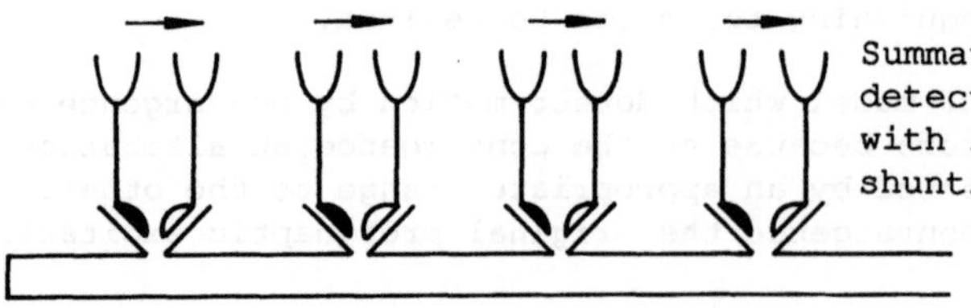

Figure 13.
Summation of unit motion detectors on an interneuron with local interaction by shunting.

In this neuron model, pairs of photoreceptors feed into unit motion detector synapses on a horizontal dendrite. The greater the spatial excitation, the more the second-order cell is reduced in responsiveness by shunting that spreads electrotonically. The similarity to the starburst amacrine cell of the vertebrate retina is obvious.

In insects, the retinal receptors project with great accuracy to the lamina monopolar cells (LMC's), which respond to changes in contrast with great speed. In fact, the LMC cells seem especially designed to feed forward non-directional spatial frequency as rapidly as possible to the medulla. The spatial frequency could be measured as follows by alternating excitation and inhibition in fields of different sizes generated in neurons connected to different numbers of LMC cells, because a single LMC cell cannot measure non-directional spatial frequency irrespective of pattern and velocity even over the limited range required (Ator, 1963). The interneurons (I) would be non-directional on-off small and medium-field units, as described by Osorio (1987). The LMC outputs would be of 2 types (perhaps LMC1 and LMC2).

Figure 14

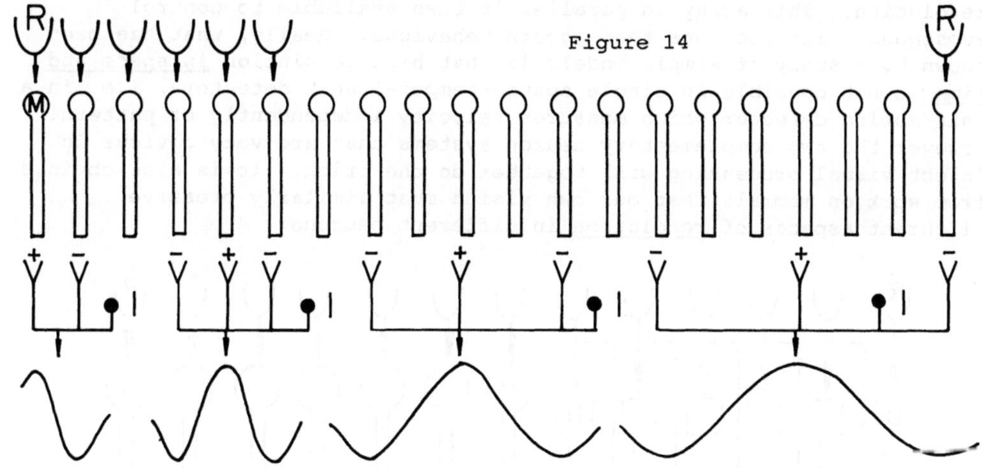

We could imagine that the non-directional LMC outputs reduce the effectiveness of the outputs of the directional contrast frequency detectors in the limited range of spatial frequencies over which motion is useful measured, possibly by presynaptic inhibition as follows:-

Figure 15.

D delay
M lamina monopolar cell
R receptor
UMD unit motion detector

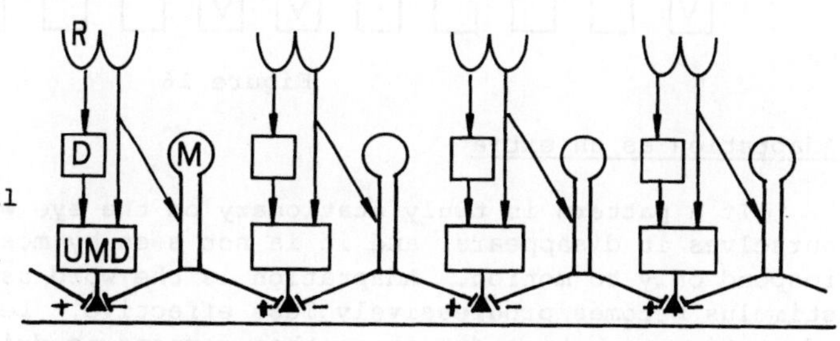

Although the measurement of angular velocity on the retina is obviously of great importance in the behaviour of freely moving animals, especially if they catch or avoid moving objects, so far we do not know of a neuronal processing mechanism that actually achieves it in any animal. Most motion-detecting neurons have not been tested in the appropriate critical way at the motor outputs of the visual processing neurons.

The value of a theory, so long as it is compatible with what is known, is that it leads to new experiments which make the appropriately controlled tests. The problem when many circuits lie in parallel is that suitable experiments are difficult to devise at the best of times, and impossible without a theory.

We can put all this together to make two systems in parallel, as below. The system on the left measures directional contrast frequency (cf), the other on the right measures spatial frequency (sp). The two have opposite effects upon neurons that control behaviour with reference to velocity of motion across the eye (bottom row, V).

These models mean that later outputs to the ventral nerve cord or muscles can control behaviour in a way that depends on velocity by responding to simultaneous inputs. One set of inputs (cf) which detects directional motion have moderate temporal resolution combined with the best possible spatial resolution (from pairs of receptors). The other (sp) set has local loss of spatial resolution (to signal the average local spatial frequency of contrast) but with highly accurate temporal resolution. This array in parallel is then available to control subsequent interneurons that govern behaviour. Really, what has been shown by a study of simple models is that high resolution in space and time is not possible in single spatio-temporal unit detectors, i.e. in a unit motion detector which measures velocity independently of pattern. However the two complementary neuron systems that are very obvious in insect visual processing will together do the trick. It is also obvious from work on mammals that our own vision must similarly preserve different aspects of resolution in different neurons.

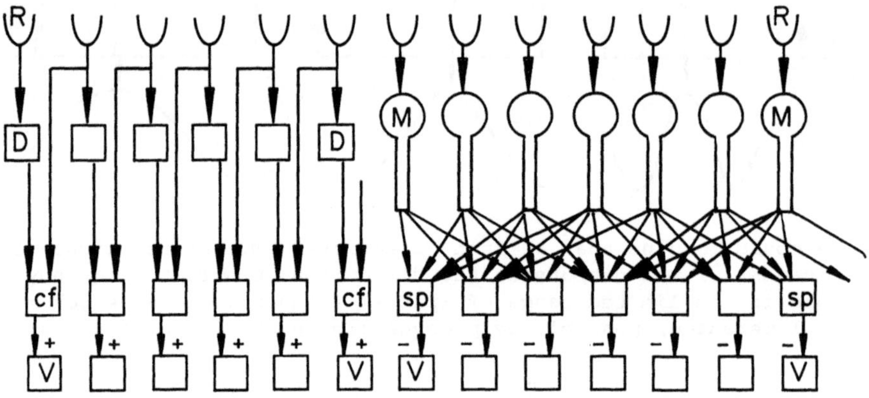

Figure 16

Adaptation as an extra

If a pattern is truly stationary on the eye we know that in ourselves it disappears, and it is not seen by most animals, which respond only to motion. Adaptation is the word used when a constant stimulus becomes progressively less effective. Let us apply the idea of adaptation to motion detectors that respond to drift frequency, i.e. the rate at which contrasts pass each receptor.

For a given pattern in the visual world, the drift frequency is a measure of velocity. The problem is that the pattern is unknown. For an insect flying at constant speed along the side of a wall, the stimulus to one eye is large, to the other much less, but as long as they are constant they remain uninteresting. Maddess and Laughlin (1985) found that the motion perception neurons adapt to the drift frequency. A change in the pattern, or a change in velocity caused by a nearby object, causes a change in the drift frequency. Either cause means that there is something "out there". If the insect adapts in flight, it can avoid crash by avoiding suddenly increasing drift frequencies. For a moving eye, change in drift frequency on any eye region means a change in the flow field or else a change of pattern, either of which could indicate the edges of an object. The local rate of change of drift frequency is a measure of the local image acceleration, either in time measured at one place on the eye, or in motion across the eye, but this is insufficient for free safe motion of the eye through the 3-dimensional natural world.

If we make a flash at one receptor then the other, and assume some interaction times, the response of a model can be calculated as a function of the flash delay between the two receptors, but the spatial properties are fixed by the anatomy. This limitation suggests that the time constants should be under a feed-back control which is dependent on the temporal frequency encountered (shortening the time constant for high temporal frequencies in the stimulus and vice versa). This has been observed experimentally in the H1 neuron of the fly (Maddess & Laughlin, 1985), but such adaptation in a multiplicative system is not sufficient to guide behaviour, because the spatial frequency is unknown.

Seeing objects

From the visual systems of animals without large brains we learn that vision by relative motion alone can be adequate for mobility without the capacity to recognize or even discriminate shapes. In an earlier paper I stressed that higher animals such as birds or man store great numbers of memories of the forms of objects that have been frequently inspected from all angles or handled (Horridge, 1987). When we "see" something we take the form already endowed with meaning from a rich memory from which it has been evoked by a minimum of clues. Much of what we think we see is created by updating our own unique memory pictures as we go along: we don't have to get through our eyes every moment all that detail, form and 3-dimensional structure that we are seeing. Flat pictures, in particular, evoke memories, so that we see 3-dimensional objects in them. Visual illusions are not a peculiar rarity: we see them all the time in normal vision.

Motion perception by animals with large brains also becomes locked into the rich visual memory. Playing fast games has to be learnt. Moving objects appear sharp, and as we move our eye, objects which emerge from behind another object never have a fuzzy edge in conscious vision. This continuous extraction of "cleaned-up" images from memory is the only way to overcome the problem of simultaneous perception of motion, detail, form and location on the retina. As said above preserving the receptor resolution is not compatible with visual processing in a single neuron because the more that neurons integrate, the larger must be their fields. Within the field of an integrating neuron it is not possible to recover the direction (or any other component) of the stimulus. We seem to get around this problem by assembling memories that can be brought into conscious view from large numbers of simultaneously active neurons which carry different aspects of the resolution.

The broadest aspects of category formation

The visual system is therefore similar to the auditory system, which recognizes spoken words in context however they are pronounced, so that words come into consciousness already endowed with meaning. A child learns its native tongue from its mother and slowly builds up a huge memory, combining the meanings of single words, then combinations of them and then familiarity with metaphors. Different languages present us with perplexing differences in the basic categories employed in different cultures, so that languages can never be perfectly translated and another culture can never be absorbed as one's own. As babies develop, we first observe how they learn to see objects by handling their own hands, and by watching things move as coherent objects, and by slowly recognizing things as they associate them as hard, wet, cold or flexible. Later the child learns to speak and understand language, then construct sentences, and ultimate read and write in a long process of learning. The effects of localized brain damage teach us that many of these functions of the brain that appear so integrated and smoothly performed are in fact composed of fragments in the unconscious mechanism that reveal themselves as strange disabilities.

Meaning depends on categories, which in turn depend on associations and a long process of learning. The process of mixing, filtering and abstracting pattern from sense data starts at the eye or ear. The process of recognition of categories is done with the aid of an enormous recall memory before the level of consciousness. This is the only way to explain how thoughts and perceptions arrive in consciousness already coloured by everything which memory recalls for the occasion. The visual or auditory categorization is an unconscious action of the brain, learned in childhood, clearly cultural but also related to the empirical world against which it is continually rechecked.

References

Ator, J.T., 1963, Image-velocity sensing with parallel-slit reticules, J. Opt. Soc. Amer. 53:1416-1422.

Barlow, H.B. and Levick, W.R., 1965, The mechanism of directionally selective units in rabbit's retina, J. Physiol., 178:477-504.

Berger, E.W., 1900, Physiology and histology of the Cubomedusae including Dr. F.S. Connant's notes on the physiology, Mem. Biol. Lab. Johns Hopk. Univ. 4(4):1084.

Dakin, W.J., 1921, The eye of Peripatus, Quart. J. micr. Sci., 66:409-417.

Demoll, R., 1909, Die Augen von Alciopa Cantrainii, Zool. Jb. (Anat), 27:651-686.

Hassenstein, B. and Reichardt, W., 1953, Der Schluss von Reiz-Reaktions Funktionen auf Systemstrukturen, Z. Naturforsch, 8b:518-524.

Horridge, G.A., 1987, The evolution of visual processing and the construction of seeing systems, Proc. R. Soc. Lond.B., 230:279-292.

Lee, D.N. and Reddish, P.E., 1981, Plummeting gannets: a paradigm of ecological optics, Nature Lond., 293:293-294.

Lee, D.N. and Young, D.S., 1985, Visual timing of interceptive action. In "Brain mechanisms and spatial vision," D.J. Ingle et al., ed., 1-30, Nato ASI Series, No.21. Dordrecht: Martinus Nijhoff.

Maddess, T. and Laughlin, S., 1985, Adaptation of the motion-sensitive neuron H1 is generated locally and governed by contrast frequency, Proc. R. Soc. Lond. B, 225:251-275.

Mandl, G., 1985, Responses of visual cells in cat superior colliculus to relative pattern movement, Vision Res., 25:267-281.

Nakayama, K. and Loomis, J.M., 1974, Optical velocity patterns, velocity sensitive neurons and space perception: a hypothesis, Perception, 3:63-80.

Osorio, D., 1987, The temporal properties of non-linear transient cells in the locust medulla, J. comp. Physiol. 161:431-440.

Reichardt, W., 1961, Autocorrelation, a principle for the evaluation of sensory information by the central nervous system, In "Principles of sensory communication, W.A. Rosenblith, ed., 303-317, New York, Wiley.

Southall, J.P.C. (ed), 1925, "Helmholtz's treatise on physiological optics," reprinted edition 1962, London Dover Publications.

Wagner, H., 1982, Flow-field variables trigger landing in flies, Nature Lond., 297:147-148.

PUPIL CONTROL IN COMPOUND EYES: MORE THAN ONE MECHANISM IN MOTHS

D.-E. Nilsson, I. Henrekson and A.-C. Järemo

Department of Zoology, University of Lund
Helgonavägen 3, S-223 62 Lund, Sweden

INTRODUCTION

In many arthropod compound eyes, migrating granules of screening pigment serve as a light-regulating pupil. Mobile pigments are found both in the photoreceptor cells and in specialized pigment cells. Recently there have been several conflicting reports on how the migrations in the pigment cells are controlled in the eyes of moths. Some investigations indicate that the pigment cells are controlled by the receptor cells (White et al., 1983; Bernard et al., 1984; Weyrauther, 1986), whereas other investigations show clearly that the pigment cells are independent and triggered by a light sensitivity of their own (Hamdorf et al., 1986; Land, 1987). We here present evidence for two coexisting mechanisms that control migrations in the pigment cells of some moths. Depending on the experimental design one of these mechanisms may take preference over the other, thus explaining the previous contradictory results.

The number, size and position of cells with migrating pigment is a matter of great variation between species, especially comparing eyes of different optical type. It is therefore obvious that we may encounter variations also in the mechanisms that control the migrations. A short review of previous findings in insect eyes is given below.

Apposition Eyes

The insect ommatidium contains pigment in primary and secondary pigment cells, and in most cases also in the photoreceptor cells. The three insect orders Diptera, Lepidoptera, and Hymenoptera are extensively studied concerning the control of pigment migration in apposition eyes. In these animals the main pigment migrations occur in the photoreceptor cells. In the dark-adapted state pigment granules are dispersed in the cell cytoplasm, and during light adaptation they aggregate close to the rhabdom, and attenuate the light flux. Due to the short migration distance of only a few μm, these migrations are relatively fast: light-adaptation times of less than 10 s are common.

From experiments on flies, butterflies, wasps and bees, it is well established that these migrations are controlled directly by the cell's own visual response (Kirschfeld and Franceschini, 1969; Kolb and Autrum, 1974; Stavenga and Kuiper, 1977: Stavenga et al., 1977; Stavenga, 1979). This is elegantly demonstrated by the fact that only illuminated ommatidia close their pupil, and if the eye is exposed to monochromatic light, pupil closure is seen preferably in the cells that are most sensitive to the particular wavelength. The triggering of pigment migration seems to be initiated by the cell depolarization which causes an influx of calcium ions, and these in turn stimulate the aggregation of pigment granules around the rhabdom (Kirschfeld and Vogt, 1980; Howard, 1984).

Superposition Eyes

The eyes of moths, caddis flies, neuropteran flies and many beetles are of the superposition type where light entering through many facets superimpose to form a single erect image on a deep lying retina (see Nilsson, 1989). In these eyes the most noticeable pigment migrations occur in the primary and secondary pigment cells. In the dark-adapted state the pigment granules are aggregated in the spaces between the crystalline cones, and during light adaptation the pigment moves proximally into the eye so that the superimposition of rays is reduced. Since the pigment granules here have to move a considerable distance, the mechanism is much slower (3-30 min) than the radial migrations in apposition eyes. The photoreceptor cells do in some cases contain a smaller amount of mobile pigment, but in other cases they lack screening pigment altogether (Horridge and Giddings, 1971; Welsch, 1977; Meinecke, 1981).

We shall here focus on the pigment cells of moths because this is the only insect group with superposition eyes where anything is known about the triggering of pigment migrations. A central question has been whether the primary and secondary pigment cells are autonomous or if they receive a controlling input from the photoreceptor cells.

One way of testing this is to compare the spectral sensitivity of the retina's electrical response with an action spectrum of the pigment migration. Such experiments have been carried out on two species of pyralid moth, *Amyelois* and *Ephestia* (Bernard et al., 1984; Weyrauther, 1986), and the sphingid moth *Manduca* (White et al., 1983). In these three species the retina and pigment migration respond with so similar spectra that a retinal control of the pigment cells seems most likely.

In contrast, Hamdorf and Höglund (1981) and Hamdorf et al. (1986) found, in an other sphingid moth, *Deilephila*, that the pigment migration mechanism responds primarily to blue and UV-light, whereas the retinal sensitivity peaks in the green. It was also found (Hamdorf et al., 1986) that the pigment migration can stil be triggered in a preparation where the retina has been removed.

Using non-invasive ophthalmoscopic techniques, Land (1987) obtained conclusive evidence for a control mechanism independent of the retina. The method relies on the fact that different images can be projected simultaneously on the cornea and on the retina. If, for example, a horizontal line of light is projected on the cornea and a vertical line is presented to the retina, then the eye will reveal where the triggering is located by closing the pupil along either a horizontal or a vertical line. In eyes which have a reflecting tapetum below the retina, the result can be observed directly as a fading of the glow. When Land (1987) applied this technique to a sphingid moth (*Theretra*) the eye responded according to the line of light on the cornea, thus proving that the light sensitivity which controls the pigment position is located distally in the eye, far away from the retina.

To further investigate the triggering mechanism we have here used a similar optical technique which is detailed below.

OPTICAL SET-UP

We used an ophthalmoscope based on a compound microscope with an epi-illumination attachment. A horizontal illumination beam combines with the microscope beam-path via a half-silvered mirror at the rear of the front objective (Fig. 1). The eye can thus be viewed and illuminated simultaneously and from the same direction. Illumination was provided by a 50 W halogen lamp with a heat absorbing filter and a gray wedge to control the intensity. The optics of the illumination beam path was designed such that one diaphragm was imaged at the eye's surface (spatial plane) and another at the retina (angular plane). For further details see Land (1984). Both diaphragms could be varied in size and moved laterally. Normally the microscope was focused at the surface of the eye, but a Bertrand lens could also be slided in to view the back focal plane of the objective,

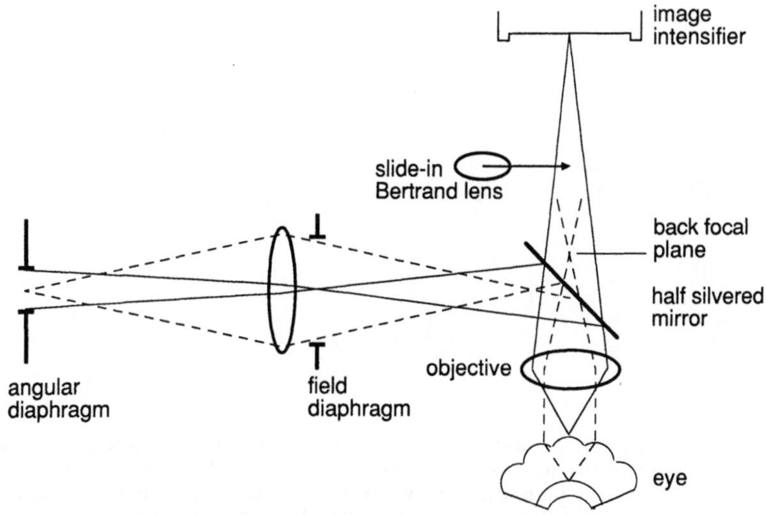

Fig. 1. Ray diagram of the ophthalmoscope. Dashed lines are rays imaging the corneal plane. Rays imaging the retinal (angular) plane are drawn as solid lines.

where the retina is imaged. The microscope was fitted with an image intensifier (Hamamatsu 1366P), making photography possible at low intensities.

EXPERIMENTAL RESULTS

For most of the experiments we used a locally abundant noctuid moth, *Agrotis exclamationis*. In the dark-adapted state (after 1 h in darkness) the eye displayed a large glow upon incident illumination. When exposed to intense light, this glow faded in about 4 min. During the experiments described below, the light intensity was reduced to a level where the pupil did not close entirely. The angular diaphragm was stopped down to produce a nearly parallel beam (3°) which, when passed through the eye's optics, was focused onto a small patch of rhabdoms belonging to the central facets of the glow. Also the spatial diaphragm was closed down so that light entered the eye through only a small number of facets. It was thus possible to illuminate a small stationary spot on the retina through any selected facets within the superposition aperture. With this type of illumination the eye still displayed a large glow (see Fig. 3a) which did not move or change appearance when the illuminated spot on the eye's surface was moved within the glow. If, however, the illuminated spot on the cornea was moved outside the glow, no light reached the retina and the glow was no longer visible.

Initially a group of facets at the periphery of the glow was illuminated. This implies that the eye received light through ommatidial facets that were well to the side of the ommatidia whose rhabdoms were illuminated. There are several possible outcomes of such an experiment (Fig. 2): A pigment response could be evoked in (1) the ommatidia whose rhabdoms are illuminated, or (2) the ommatidia whose facets receive light, or (3) the entire eye with no correlation to the pattern of illumination. The actual result of this experiment, when performed on *Agrotis*, is shown in Fig. 3. The photograph in Fig. 3b is taken 10 min after onset of illumination and it shows clearly that there is a local response in the illuminated facets (equivalent to the result of Land 1987), but there is also a pronounced fading over the entire glow. This was repeated several times with the same result.

The experimental conditions were then altered by moving the corneal spot of light into the center of the eye glow. This seemingly minor alteration caused an unexpected

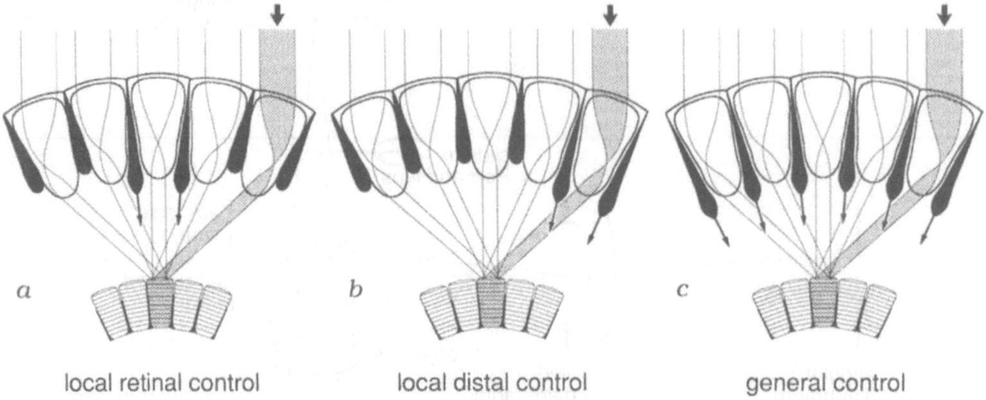

Fig. 2. Schematic diagram of three possible responses of the eye's pupil to illumination through the periphery of the superposition aperture. In *a*, the photoreceptor cells control the pigment cells of their own ommatidium, whereas in *b*, the pigment cells are controlled locally by light sensitive elements in the distal part of the eye. In the third alternative (*c*), pupil closure is independent of where the eye is illuminated. This of course requires a menchanism that mediates the response to pigment cells in the entire eye.

change to the results: there was no longer any trace of local response at the illuminated spot on the cornea, but the general response over the entire glow was the same as in the previous experiment. To make sure that a local response was not just hard to see when it was in the center of the glow, the eye was turned around after the complete experiment so that the glow shifted to other facets. This did not only confirm the observation that there was no local response but it also revealed that the general response had occurred evenly over the entire eye. A check of the contralateral eye showed that it was left unaffected.

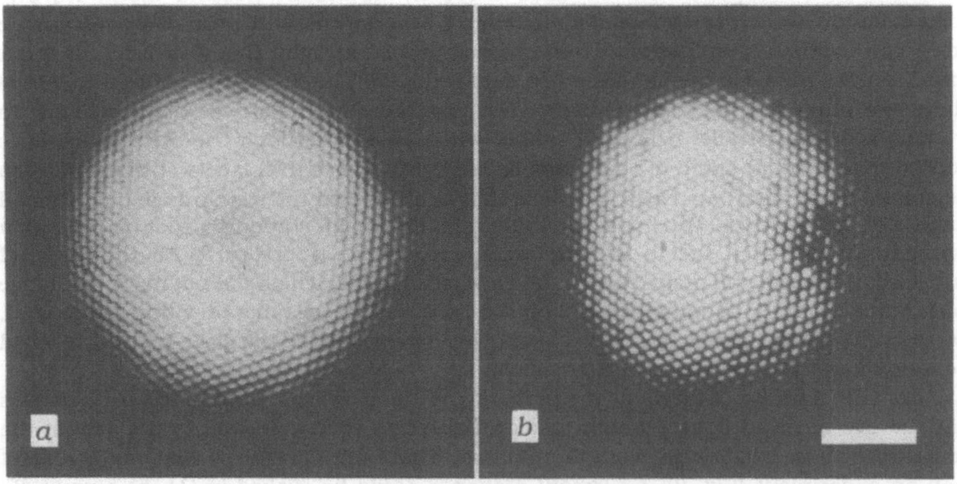

Fig. 3. Eye glow of *Agrotis exclamationis* during an experim ent with partial illumination of both the retina and the cornea. The rhabdoms that receive light are in the center of the glow, and in *a* these are illuminated through a patch of facets at the pariphery of the glow. The illuminated patch on the cornea is seen as a faint circular disc, some eight facets in diameter, at the right edge of the glow. The effect after 10 min stimulation is seen in *b*. During photography in *b* the whole corneal surface was illuminated. Pupil closure is seen over the entire glow but particularly in the corneal region where light entered during stimulation. The coarse hexagonal pattern that is visible in *a* but not in *b* is an artifact resulting from oversaturation of the micro-channel plate in the image intensifier that was used for the photography. Scale bar: 200 μm.

The experiment was again altered, this time by moving the illuminated spot on the cornea to a position just outside the glow. At this position no light reaches the retina and the glow was thus not visible during stimulation. After a 10 min stimulation the corneal diaphragm was removed and the eye was turned to examine the effect. The result was now opposite to the previous experiment: a strong local reaction at the illuminated corneal spot but no general response.

Continued experiments with different positions of the corneal spot revealed that the general response, which occurred evenly and simultaneously in the entire eye, was evoked when the corneal spot was anywhere within the glow, but not when it was moved outside. The light sensitivity that triggers this general response must therefore be in the retina. In contrast, the local response was evoked when the corneal spot was at the periphery of the glow, and even when it was slightly outside. The light sensitivity controlling the local response must thus be located distally in the eye where the crystalline cones and main bodies of pigment cells are. The reason that no local response was seen when the light stimulus was centered in the glow must be that the light sensitive trigger is located peripherally in each ommatidium and thus not exposed by paraxial light.

To summarize the results, there are two independent mechanisms that trigger adaptational response in the pigment cells. One is controlled by the retina, and since it occurs without correlation to the pattern of illumination, it must be hormonally or neurally mediated to all the pigment cells (Fig. 2c). However, if it is a hormonal control, the active substance must be confined within the eye because the effect does not spread to the contralateral side. The other controlling mechanism, which is equivalent to that found by Land (1987), is obviously very local in the distal layers of the eye, and completely independent of the retina (Fig. 2b).

These results now offer a likely explanation to the conflicting results of White et al. (1983), Bernard et al. (1984) and Weyrauther (1986) on the one hand, and Hamdorf et al. (1986) and Land (1987) on the other. Their conclusions are probably all correct, but they may have studied two different mechanisms. We here also have to consider that the reports above are on different species of moth, and it is not unlikely that the two mechanisms can have different relative sensitivity, or one of the mechanisms may be absent in a certain species.

To investigate the generality of our results, we made preliminary tests on a few other species of moth: the privet hawk *Sphinx ligustri* (Sphingidae), the chinese oak moth *Antheraea pernyi* (Saturnidae) and the meal moth *Ephestia kühniella* (Pyralidae). In the sphingid and saturnid species we were able to demonstrate both a general retinal control and a local distal control, just as in *Agrotis*. In *Ephestia*, however, there was no sign of any local distal control nor were we able to clearly demonstrate a general control as in the other moths. Instead, it seems that migration is triggered locally by the retina, but with a pronounced lateral spread of the reaction, extending some 5 to 10 ommatidia away from the illuminated rhabdom. A single experiment on the mysid shrimp *Neomysis integer* gave results similar to those of *Ephestia* but with much less lateral spread of the response.

BIOLOGICAL SIGNIFICANCE OF THE CONTROL MECHANISM

From our present experiments it is clear that the pupil control is not alike in all superposition eyes. There are at least three different mechanisms, two of which may coexist in the same eye. What then could the reason be for this diversity? There is an obvious disadvantage with a local retinal control (Fig. 2a): a motionless visual scene will be imprinted in the eye as a 'pigment position image'. If this happens to an animal at rest, it will suffer from a serious afterimage when it again becomes active. This disturbance will remain until the pupil has evened out by the moving image. Such afterimages would be of little problem if the pupil is fast, as in apposition eyes. But with the slow pupil of superposition eyes the afterimage would impair vision for at least several minutes after the animal has been at rest.

The distal control of pigment position (Fig. 2b) eliminates this problem because the triggering site is far away from the retinal image. Also the general retinal control described here (Fig. 2c), assures that pigment position is adjusted to the average light intensity irrespective of the spatial distribution of light on the retina. It remains unclear, however, why sphingids, saturnids and noctuids possess two different mechanisms producing apparently the same effect. The lateral spread of the response in *Ephestia* partly evens out the pigment position over the eye, but this must of course be less efficient than a true general control or a distal triggering.

The local retinal control in the superposition eye of the shrimp *Neomysis* may be explained by the fact that in these pelagic animals the retinal image rarely stays motionless. It can also be argued that the visual world of a shrimp is generally of such low contrast that any strong local adaptation would never occur.

Animals that would benefit the most from a general control or a local distal control are those that have a slow pupil and frequently reside motionless in a visual world of high contrasts, and, certainly, many moth species qualify to these requirements. It thus seems that the pupil speed and the animal's life-style are sufficient to explain why the pupil is controlled differently in different animals.

REFERENCES

Bernard, G. D., Owens, E. D., and Voss Hurley, A., 1984, Intracellular optical physiology of the eye of the pyralid moth *Amyelois*, *J. Exp. Zool.*, 229:173-187.

Hamdorf, K., and Höglund, G., 1981, Light induced screening pigment migration independent of visual cell activity, *J. Comp. Physiol.*, 143:305-309.

Hamdorf, K., Höglund, G., and Juse, A., 1986, Ultra-violet and blue induced migration of screening pigment in the retina of the moth *Deilephila elpenor*, *J. Comp. Physiol. A.*, 159:353-362.

Horridge, G. A., and Giddings, C., 1971, The retina of *Ephestia* (Lepidoptera), *Proc. R. Soc. Lond. B.*, 179:87-95.

Howard, J., 1984, Calcium enables photoreceptor pigment migration in a mutant fly, *J. exp. Biol.*, 113:471-475.

Kirschfeld, K., and Franceschini, N., 1969, Ein Mechanismus zur Steurung des Lichtflusses in den Rhabdomeren des Komplexauges von*Musca*, *Kybernetik*, 6: 13-22.

Kirschfeld, K., and Vogt, K., 1980, Calcium ions and pigment migration in fly photoreceptors, *Naturwissenschaften*, 67:516-517.

Kolb, G., and Autrum, H., 1974, Selektive Adaptation und Pigmentwanderung in den Sehzellen des Bienenauges, *J. Comp. Physiol.*, 94:1-6.

Land, M., F., 1984, The resolving power of diurnal superposition eyes measured with an ophthalmoscope, *J. Comp. Physiol. A.*, 154:515-533.

Land, M., F., 1987, Screening pigment migration in a sphingid moth is triggered by light near the cornea, *J. Comp. Physiol. A.*, 160:355-357.

Meinecke, C. C., 1981, The fine structure of the compound eye of the african armyworm moth, *Spodoptera exempta* Walk. (Lepidoptera, Noctuidae), *Cell Tissue Res.*, 216:333-347.

Nilsson, D.-E., 1989, Optics and evolution of the compound eye, *in*: "Facets of vision," D. G. Stavenga and R. Hardie, eds., Springer, Heidelberg, New York.

Stavenga, D. G., 1979, Pseudopupils of compound eyes, *in*: "Handbook of sensory physiology," Vol. VII/6A, H. Autrum, ed., Springer, Berlin, Heidelberg, New York.

Stavenga, D. G., and Kuiper, J. W., 1977, Insect pupil mechanisms: I - On the pigment migration in the retinula cells of Hymenoptera (suborder Apocrita), *J. Comp. Physiol.*,113:55-72.

Stavenga, D. G., Numan, J. A. J., Tinbergen, J., and Kuiper, J. W., 1977, Insect pupil mechanisms: II. Pigment migration in the retinula cells of butterflies, *J. Comp. Physiol.*, 113:73-93.

Welsch, B., 1977, Ultrastruktur und funktionelle Morphologie der Augen des Nachtfalters *Deilephila elpenor* (Lepidoptera, Sphingidae), *Cytobiologie*, 14:378-400.

Weyrauther, E., 1986, Do retinula cells trigger the screening pigment migration in the eye of the moth *Ephestia kühniella* ?, *J. Comp. Physiol. A.*, 159:55-60.

White, R, H., Banister, M. J., and Bennett, R. R., 1983, Spectral sensitivity of screening pigment migration in the compound eye of *Manduca sexta*, *J. Comp. Physiol. A.*, 153:59-66.

EARLY VISUAL PROCESSING IN THE COMPOUND EYE: PHYSIOLOGY AND PHARMACOLOGY OF THE RETINA-LAMINA PROJECTION IN THE FLY

Roger Hardie, Simon Laughlin and Daniel Osorio*

Cambridge University, Department of Zoology
Downing St, Cambridge U.K.

ABSTRACT

The defined physiology and anatomy of the retina and the first optic ganglion of the fly provide an excellent opportunity to study the neural mechanisms responsible for early visual processing. Intracellular recordings from an intact preparation have been used to analyse synaptic transfer from the photoreceptors to a major class of second order neurones, the large monopolar cells (LMCs). Coding is optimised to protect pictorial information from contamination by the noise generated at the photoreceptor synapses. Optimisation involves processes of amplification and antagonism that are matched to the statistical properties of images. Having derived a detailed description of coding, we are investigating the underlying physiological and pharmacological mechanisms. The synaptic transfer function has been determined by precise measurements of the responses of photoreceptors and LMCs to identical stimuli; noise analysis identifies a significant contribution of synaptic noise to the postsynaptic signal. Single-electrode current clamp analysis of the LMCs shows that their membranes are approximately Ohmic, so that voltage sensitive mechanisms play little role in signal-shaping. To a first approximation, LMC "on" responses can be described in terms of a single chloride conductance activated by the photoreceptor neurotransmitter, however, the depolarising transient generated at light "off" involves additional depolarizing mechanisms. Lateral inhibition is associated with a conductance decrease thus suggesting a presynaptic mechanism, and is under dynamic control, developing rapidly at the onset of light adaptation. It may vary in strength between different classes of LMC. Ionophoretic studies indicate that histamine mimics the photoreceptor neurotransmitter and the photoreceptor terminals show histamine-like immunoreactivity. This represents the first case of histaminergic neurotransmission in insects, and a pharmacological profile of the putative histamine receptors indicates that they are of a novel class. Immunocytochemical and ionophoretic studies also indicate the involvement of a number of classical neurotransmitters in other lamina interneurones.

---

*Present address: Centre for Visual Sciences,
RSBS ANU Canberra

INTRODUCTION

Sense organs are at the interface between organism and environment, with sensory receptors performing the vital function of transducing environmental energies into the electrical and chemical signals which are the vocabulary of the nervous system. The joint constraints forced upon the receptors by the environment on the one hand, and the biological hardware on the other, may be expected to restrict the range of transformations attainable by transduction alone. In practice the primary sensory signal is transformed again at the first synapse to provide a robust signal which faithfully codes appropriate environmental parameters with a high signal-to-noise ratio. With regard to early visual processing, one of the major problems is to handle, in an efficient manner, the great variety of environmental intensities with which the animal may be confronted. Even a strictly diurnal animal has to deal with light intensities ranging over 5 or 6 orders of magnitude using neurones which have a dynamic range determined by the reversal potentials of the ions carrying the electrical signals and noise levels ultimately limited by random events associated with such processes as photon absorptions, release of synaptic vesicles and channel openings and closings.

Both environmental and biological constraints have been investigated in detail at the first synapse in the arthropod visual system. Early visual processing has been studied in a variety of arthropod systems: notably Limulus lateral eye (rev: Laughlin, 1981a; the barnacle ocellus (rev: Stuart, 1983); the insect ocellus (Wilson, 1978; Simmons, 1981) and dragonfly compound eye (Laughlin, 1974; Laughlin and Hardie, 1978). However, this article will concentrate on the compound eye of the higher diptera, exemplified by Calliphora, Musca, Lucilia and more recently Eristalis. A combination of historical development and experimental amenability have made this system one of the most thoroughly analysed neuropiles in any sophisticated visual system. Several recent reviews have dealt in detail with the basic anatomy, electrophysiology and coding strategies (Shaw, 1981, 1984; Laughlin, 1981a, 1987; Hardie, 1985; Järvilehto, 1985). In the present article we will concentrate on recent studies which address some long-standing problems regarding the mechanisms involved in coding.

ANATOMY OF THE RETINA-LAMINA PROJECTION

A major advantage of working on the insect visual system, and that of dipterans in particular, is the extremely ordered anatomical lay-out, whereby a precise retinotopic projection is maintained through successive visual neuropiles. In the first visual neuropile, or lamina, every point in the image is represented by a neural module or cartridge which is basically repeated in identical fashion across the whole eye. The almost crystalline precision of this projection has attracted neuroanatomists since the time of Cajal and an unprecedented wealth of anatomical detail is now available for the dipteran lamina (revs: Strausfeld 1976, 1984; Strausfeld and Nässel, 1981; Shaw, 1984).

The lamina is a thin (50 um) sheet of neuropile closely adjoining the retina from which it is separated by a basement membrane and a short "fenestrated layer" which represents the paths of the receptor axons. The receptors project onto a parallel array of basically cylindrical cartridges - one per ommatidium - which penetrate the ca. 50-60 um depth of the lamina. Each cartridge is surrounded by a tightly folded glial sheath composed of at least three distinct classes of glia cells (Saint Marie and Carlson, 1983). Tight junctions between the glial cells restrict both diffusion and electrical current flow between neighbouring cartridges

Table 1. Summary of cell types in the Calliphorid lamina. Synaptic connectivity (from anatomy, listed in order of strength - Shaw, 1984) and putative neurotransmitter are indicated. (med.) indicates connections in the medulla.

Cell	Input from:	Output to:	Putative transmitter[1]
Photoreceptors:			
R1-6	L2,Am,C2/3?Tan1?	L1-3,Am,EGC,	Histamine
R7-8		(med.)	Histamine (R8) GABA? (R7)
second order:			
L1	R1-6,L4,C3,C2	(med.)	
L2	R1-6,L4,C3,C2	R1-6,T1,(med.)	(glut.?)
L3	R1-6,L4,C3,C2	(med.)	
Amacrine	R1-6,Am?	T1,R1-6,EGC,L4,L5,Am?	ACh?
glia (EGC)	Am,R1-6		
third order:			
L4	Am,L4	L1-3,L4,(med.)	?
L5	Am,Tan2	(med.)	?
T1	Am,L2	EGC,(med.)	?
centrifugal:			
C2	(med.)	L1-3,R1-6?	GABA (ACh?)
C3	(med.)	L1-3,R1-6?	?
Tan1	(med.)	L1-3,R1-6?	?
Tan2	(med.)	L5	?
Tan3		neurosecretory[2]	5-HT

1. Evidence mainly from immunocytochemistry (except histamine).
2. Tan3 (LOB-5HT) has no synapses as such but forms neurosecretory terminals in the distal part of the lamina.

(Shaw, 1984). The lamina as a whole is also effectively isolated from the haemolymph and the retina, and the extracellular space is maintained at a potential some 70mV negative to the retina in the dark (Zimmerman, 1978; Shaw, 1984).

Of the eight photoreceptors in each ommatidium, only six (so-called R1-6) synapse in the lamina. Each group of six photoreceptors R1-6, which share the same visual axis project to each lamina cartridge which, in addition, contains a total of 12 interneurones (Table 1). Three of these in particular, the large monopolar cells (LMCs) are directly postsynaptic to the photoreceptors. Serial EM reconstructions in Musca (Nicol and Meinertzhagen, 1982) have shown that each photoreceptor makes about 200 synapses on each of two LMCs (L1 and L2), and rather fewer on the third LMC (L3), whose dendrites are restricted to the distal half of the lamina (Strausfeld, 1976). The synapses are tetradic in nature: two of the four post-synaptic elements are always L1 and L2, whilst the remaining two elements may include L3, amacrines or glia. This tetradic synapse accounts for ca. 90% of all lamina synapses (Shaw, 1984). Other connections are summarised in Table 1.

BASIC ELECTROPHYSIOLOGY OF THE RETINA AND LAMINA

Intracellular recordings may be made routinely from both the photoreceptors and their major postsynaptic elements, the LMCs. The or surgery required is the removal of tiny piece of the cornea, whereupon glass microelectrodes may be lowered into the retina, and, by advancing deeper, into the lamina. With care, the optics of the eye, which can be monitored by a variety of sophisticated optical techniques (Franceschini, 1975; van Hateren, 1986a), remain undamaged, thus allowing results to be quantified in terms of environmentally significant measures. In addition, since single quantum captures can be reliably counted as quantum bumps in both photoreceptors and LMCs, all results can in principle be calibrated in terms of effectively absorbed quanta (e.g. Dubs et al., 1981).

Photoreceptor responses

Fly photoreceptors were amongst the first sensory receptors from which intracellular recordings were made (Kuwabara and Naka, 1959) and their physiology has been extensively investigated (rev. Hardie 1985). Like all arthropod photoreceptors, fly retinula cells respond to light increments with a depolarisation mediated via a conductance increase - primarily to sodium ions (Muijser, 1979) - although a number of other voltage and light sensitive conductances are certainly involved in shaping the final response (see review by Fain and Lisman, 1981 for the better studied case of the Limulus eye). Typically, a maximum response of some 60-70 mV may be obtained in response to saturating light flashes, with mechanisms of light adaptation bringing this down to a plateau level of maximally, some 30-40 mV during prolonged light. Once a plateau level has been reached the photoreceptors again respond sensitively to increments or decrements of light intensity (Fig. 1).

The photoreceptor signals are conveyed passively with little decrement to the axon terminals in the lamina (Zettler and Järvilehto, 1973). The six R1-6 axons converging in each cartridge form extensive gap-junctions with their neighbours (Chi and Carlson, 1976; Ribi, 1978). Correlating with this anatomical finding there is a degree of electrical coupling which appears to vary rapidly with the state of light-adaptation - the cells becoming uncoupled at higher intensities (van Hateren, 1986a).

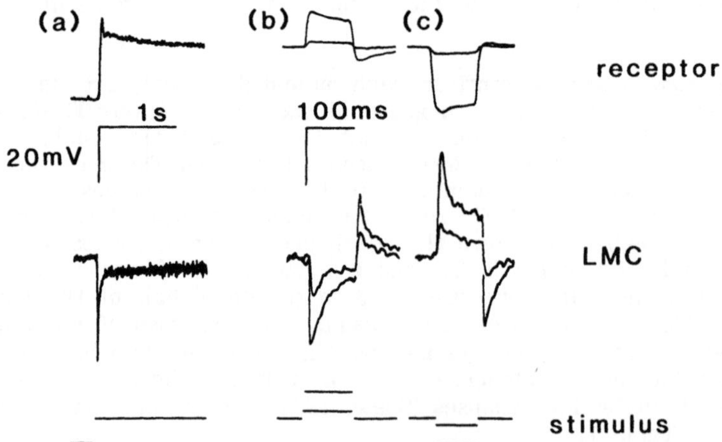

Fig. 1. The responses of a fly photoreceptor and LMC to identical stimuli: a) the onset (from darkness) of an effective background intensity of $5x10^5$ photons s^{-1} per photoreceptor; (b) 100 ms increments, (c) 100 ms decrements about this background. From Laughlin et al. (1987).

LMC Responses

The basic features of the LMC response have been known since the first intracellular recordings from these interneurones (locust, Shaw, 1968; fly, Autrum et al., 1970). The first synapse inverts the signal and LMCs respond to light with a transient hyperpolarization associated with an increase in conductance (Shaw, 1968; Laughlin, 1974).

LMCs have an unusually low resting potential of only some -15 mV (fly: Laughlin and Hardie, 1978; dronefly: Guy and Srinivasan, 1988; crayfish: Wang-Bennett and Glantz, 1987), the precise value being difficult to define because of the variable extracellular potentials encountered. Dark-adapted LMCs in the fly show a maximum hyperpolarising potential to a light flash of ca. 40-50 mV. There is a variable depolarising component at light "off" which increases in size with the state of light-adaptation. The total dynamic range is similar to that of the photoreceptor - ca. 60 mV. Compared to the photoreceptor, the LMC response is very phasic and there is little maintained response to a continuous background light. The LMC will, however, continue to respond sensitively to increments or decrements of intensity about the background. Comparison of LMC and photoreceptor responses (Fig. 1) reveals that the LMC response is amplified, saturating with stimuli which generate only small (<10 mV) receptor signals.

Apart from an "off" spike probably generated in the medulla, the LMCs can be considered as non-spiking interneurones and their signals are probably conducted to the axon terminals in the medulla by passive propagation. Cable modelling, using measured parameters (axon length 500 um, diameter 2 um, 10KOhm cm^2 - Zettler and Järvilehto, 1973) suggest that signals arriving in the medulla terminals retain about 70% of their initial amplitude (van Hateren 1986b). Similar conclusions have also been reached for LMCs in dronefly (Guy and Srinivasan, 1988) and crayfish (Wang-Bennett and Glantz, 1987).

OPERATIONAL PRINCIPLES UNDERLYING SIGNAL TRANSFORMATION

The transformation of light signals by photoreceptors and the photoreceptor-LMC synapses can be described by three simple operations; a log transformation, a subtraction and an amplification (Laughlin and Hardie, 1978). Each of these operations follows a simple design principle that minimises the constraints imposed on coding by the cellular hardware of the eye and the inevitable associated noise.

Log transform: This operation is performed by the photoreceptors. The curve of response versus log intensity is sigmoidal, with a considerable log-linear region. This function is typical of many sensory receptors and results from an unavoidable biophysical constraint, the self-shunting of the EMF as the receptor progressively depolarises towards the response reversal potential (rev. Lipetz, 1971). Once the light intensity is sufficient to drive the response into the log-linear range, adaptation mechanisms hold it there, so achieving a good approximation to the logarithmic transform at high background intensities. Happily for eyes, the log transform produced by self-shunting is pre-adapted for coding visual signals since it automatically generates contrast constancy (rev. Laughlin, 1987).

Subtraction and amplification: The logarithmic transform poses an operational problem for the nervous system. The photoreceptor signal contains two components: a plateau (of up to 30 mV), representing the background illumination; and fluctuations about this plateau (of up to about 10 mV) representing objects of different contrast. This fluctuating component contains all information on contrast and, because of the log transform, is unchanged by the background. Quite sensibly the fly subtracts away the background signal and amplifies the remaining pictorial detail to protect it from noise contamination (Laughlin and Hardie, 1978). The efficient execution of subtraction and amplification raises further operational problems which are familiar to neurophysiologists using oscilloscopes. One requires a quick and efficient system for removing DC-offsets and a gain which takes into account the expected amplitudes of signals. Similarly, in an eye there must be dynamic processes of subtraction to deal with local intensity changes in both space and time, especially when the background signal changes rapidly as the eye moves from an area of sunlight to shade. In the fly lamina the processes of subtraction and amplification are carefully designed to maximise the gain without risking saturation. This design involves a matching of the coding to the properties of incoming signals. The design principles, predictive coding (Srinivasan et al., 1982) and matched amplification, (Laughlin, 1981b) have recently been reviewed (Laughlin, 1987) and only an outline is given here.

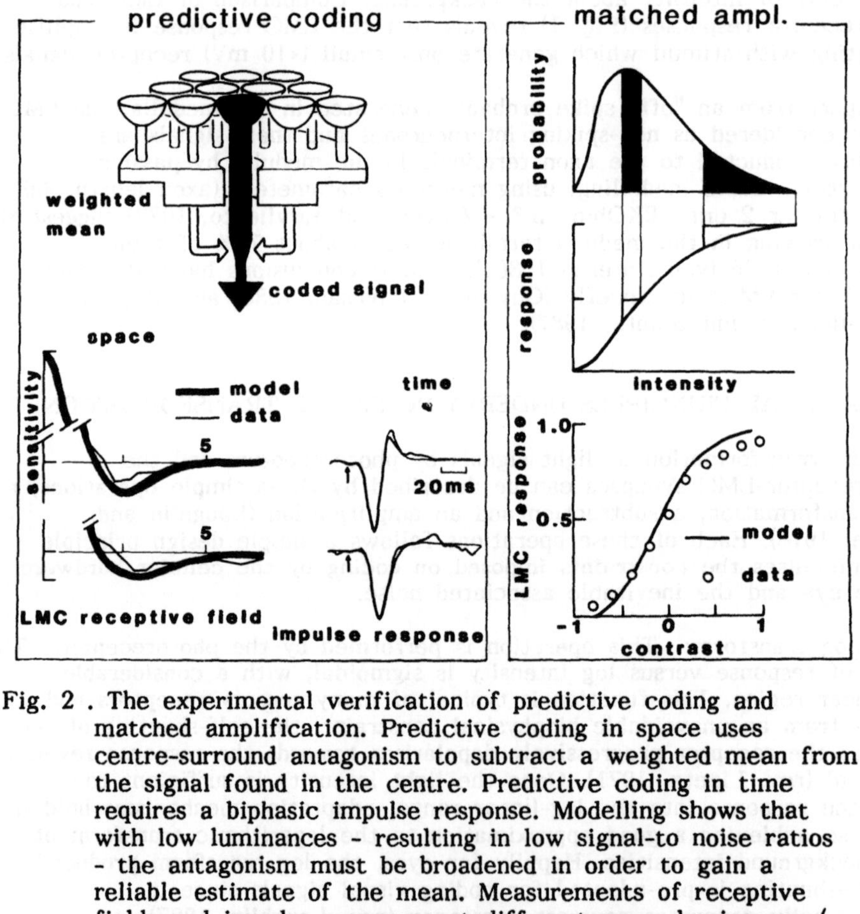

Fig. 2 . The experimental verification of predictive coding and matched amplification. Predictive coding in space uses centre-surround antagonism to subtract a weighted mean from the signal found in the centre. Predictive coding in time requires a biphasic impulse response. Modelling shows that with low luminances - resulting in low signal-to noise ratios - the antagonism must be broadened in order to gain a reliable estimate of the mean. Measurements of receptive fields and impulse responses at different mean luminances (above, high; below, low) are in good agreement with the model. Matched amplification requires a relationship between input intensity and response amplitude in which equal increments in response correspond to equal areas under the probability distribution of input levels. Contrast response functions of LMCs (below) come close to the theoretical expectation. From Laughlin (1987).

The process of subtraction resembles predictive coding, a technique used in digital image compression (rev. Gonzalez and Wintz, 1977). For each point in the image one takes a weighted mean of intensities in surrounding points to provide a statistical estimate of the intensity expected there. This prediction is subtracted from the intensity registered at that point and the remainder coded. Since this coding procedure removes only predictable components, no information is lost. At low light levels prediction is complicated by the unreliability of intensity signals resulting from photon noise. To obtain reliable estimates a larger number of points must be used for the prediction. This intensity dependency was used to show that predictive coding is executed in both time and space by the subtractive processes in the fly lamina (Srinivasan et al., 1982; Fig. 2). As a result, LMCs show lateral antagonism and transient responses that are not observed in the photoreceptors in the retina.

As with an oscilloscope, appropriate choice of amplification requires a knowledge of the expected signal. The first prerequisite therefore is a statistical description (probability distribution) of object contrasts which was obtained by direct photometric measurements of natural scenes (Laughlin, 1981b). Information is optimally coded when all response levels of the cells are used equally often (e.g. Shannon and Weaver, 1949). In other words equal areas under the probability distribution function should correspond to equal increments in response. LMC contrast response functions in fact follow this prediction remarkably closely (Fig. 2), demonstrating that amplification has been matched to signal statistics. By using the full dynamic range, amplification is maximized and this reduces the effects of added synaptic noise (Laughlin et al., 1987).

We have shown that the operating characteristics of the first synapse in the fly visual system can be quantitatively understood on the reasonable assumption that the function of the LMCs is to code contrast signals with optimal fidelity in the face of the unavoidable constraints of limited dynamic range and noise. We now consider in more detail how this is actually achieved, examining the synaptic transfer characteristics, conductance mechanisms in the LMCs and finally the neuropharmacology of synaptic transmission.

THE SYNAPTIC TRANSFER CHARACTERISTIC

It is technically very difficult to perform simultaneous recordings from a photoreceptor and a postsynaptic LMC, and thus synaptic transfer has been investigated by challenging both classes of cells with identical stimuli, the excellent reproducibility of results allowing an accurate reconstruction of the transfer characteristic (Laughlin et al., 1987). The characteristic curve relating presynaptic (photoreceptor) voltage to postsynaptic (LMC) voltage is sigmoidal and can be well fitted by a simple model for synaptic function which assumes an exponential relation between presynaptic depolarization and transmitter release, and a hyperbolic relation between postsynaptic conductance and membrane potential (Fig. 3; Katz and Miledi, 1967; Blight and Linas, 1980). As was also found in the case of the barnacle photoreceptor synapse (Hayashi et al., 1985), the shape of the curve is unaffected by light adaptation (over at least 6 log units of intensity). The maximum slope (a measure of the gain) averages approximately 6 at all intensities. The effect of light adaptation is simply to shift the curve so that it remains centred on the mean level of receptor depolarisation (this is equivalent to the process of subtraction already discussed above). Interestingly, in the completely dark-adapted state the curve starts near the region of maximal slope - so that dim flashes are already transmitted with near maximal gain. This finding implies that even in the dark, there is tonic release of transmitter, a conclusion also reached for ocellar (Simmons, 1982), barnacle (Hayashi et al., 1985) and crayfish photoreceptor synapses (Wang-Bennett and Glantz, 1987) and supported by two independent observations: namely the constant barrage of background noise in the dark and the fact that depolarisations at light "off" are often associated with a conductance decrease to below the dark resting conductance level (Fig. 5). Tonic release of neurotransmitter has been reported in a number of synapses where graded potentials rather than spikes are transmitted (e.g. Simmons, 1981; Burrows, 1979) and may well be a general phenomenon. Functionally, two advantages may be recognized: (i) the synapse may then signal both hyperpolarizations and depolarizations; (ii) tonic release maintains the synapse in the steep (sensitive) part of its operating range (see above). The high constant gain of the photoreceptor LMC synapse corresponds to an e-fold conductance increase in the LMC being generated by a 1.5-1.85 mV presynaptic signal, which is amongst the highest synaptic gains reported (see Laughlin et al., 1987).

Dynamics of Synaptic Transfer

The temporal properties of synaptic transfer have been deduced by comparing the waveforms of photoreceptor and LMC responses. Transfer is described by a simple model in which the photoreceptor drives two exponential processes: a rapid hyperpolarisation and a slower depolarisation. The time constant of the hyperpolarising process, 0.5ms, is unaffected by light adaptation. However, the time constant of the depolarisation decreases from 50ms to 1.5ms, and its relative amplitude increases. This change accounts, in part, for the more transient response of the light-adapted LMC. The hyperpolarising component is much faster

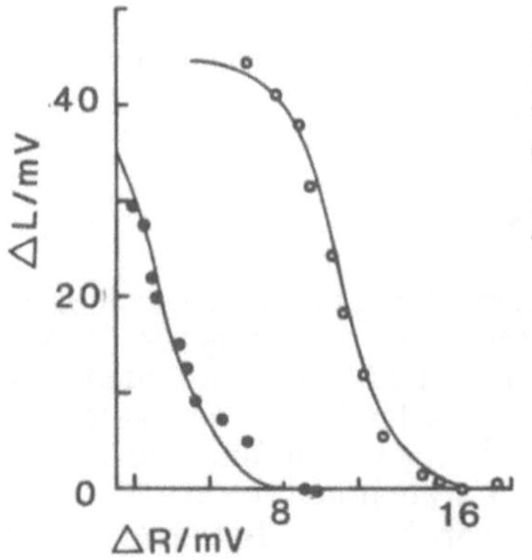

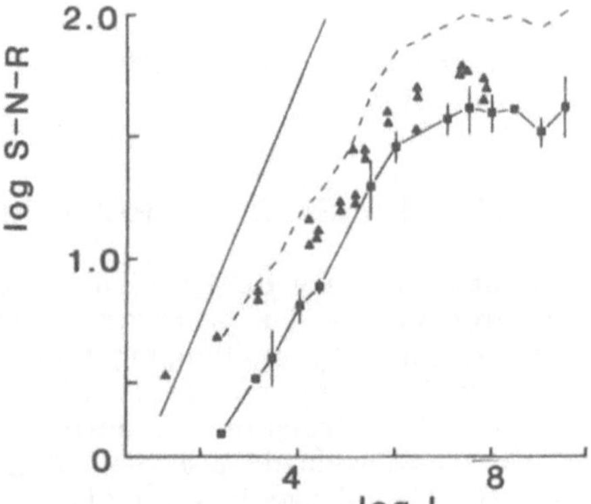

Fig. 3 (left). The voltage characteristic curves measured from the responses of a single photoreceptor and LMC to identical stimuli; (●) dark-adapted and (o) light-adapted. To facilitate comparison L is measured relative to the maximum peak response in the LMC (i.e. L=0 is the peak response). Both sets of data are fitted by the same model synaptic characteristic curve (see text). From Laughlin et al. (1987).

Fig. 4 The S-N-R ratio in photoreceptors (▲) and LMCs (■) measured at different background intensities, I. The dashed line shows the expected LMC S-N-R from the sixfold receptor convergence. The solid line is the relation expected in a purely shot-noise limited system (S-N-R /I). From Laughlin et al.(1987).

than phototransduction and is of the same order as the LMC membrane time constant, consequently synaptic transmission does not limit the fly's ability to detect rapid changes in intensity (Laughlin et al., 1987).

Noise Incurred in Synaptic Transfer

By measuring signal to noise ratios (S-N-R) in receptors and LMCs under identical conditions it is possible to identify the extra noise induced during transmission across the synapse. If no extra noise is generated, then the S-N-R in LMCs should be /6 better than in photoreceptors, because of the convergence of six photoreceptors onto one LMC. At low light intensities, when quantal noise is very large, this is in fact the case (Fig. 4) but with increasing background intensities, the LMC S-N-R falls away from the expected ideal indicating the injection of extra noise (Laughlin et al., 1987). Although Fig. 4 suggests that, in daylight, most of the benefit of receptor convergence is wasted by the extra synaptic noise, in fact the situation is probably not quite so bad: firstly, there is a degree of electrical coupling between the photoreceptors (Dubs, 1982; van Hateren, 1986a) so that the S-N-R measured in the photoreceptors may be an over-estimate. Secondly, power spectral analysis indicates that a significant proportion of the intrinsic noise occurs at high frequencies (roll-off at 400 Hz) which are beyond the high frequeny cut-off of photoreceptor signals (Laughlin et al., 1987). Nevertheless, there can be little doubt that, at high light intensities, synaptic noise is a significant factor in limiting the fidelity of the neural image.

CONDUCTANCE MECHANISMS IN LMCs

The form of the characteristic curve relating LMC response to photoreceptor input, and the observation that the slope and shape of this curve are unaltered by light-adaptation, suggests a simple mechanism for coding. The chemical synapse from receptor to LMCs has the high (constant) gain required for matched amplification. The antagonistic processes responsible for predictive coding then shift this curve to keep it centred on the photoreceptor plateau response. To investigate this further we have recently examined the conductance mechanisms in the LMC membrane using single electrode current clamp. Comparable studies have also recently been performed in the dronefly (Guy and Srinivasan, 1988) and the crayfish (Wang-Bennett and Glantz, 1987).

Passive Properties

In the dark the resting input resistance of the LMCs is only ca. 20 M (this compares with values of 25-40M in the dronefly and 35M in the crayfish). Presumably there are substantial conductances active in the dark since photoreceptor transmitter is released tonically in the dark (see above). Furthermore, despite this, the resting potential of the LMC is relatively depolarized, indicating that a depolarising conductance must also be active. Zimmerman (1978) presented pharmacological evidence suggesting that tonic release of GABA may be responsible, and whilst this has not been confirmed, GABA applied ionophoretically to LMCs does indeed cause a depolarisation (Hardie, 1987).

Injection of current into LMCs reveals no significant voltage sensitive conductances. The I/V curves are essentially linear (Ohmic) over the entire physiological range both in the dark, and when light-adapted (Laughlin, 1974; Laughlin and Osorio in prep.; Wang-Bennett and Glantz, 1987; Guy and Srinivasan, 1988). Voltage sensitive conductances are therefore unlikely to play a significant role in signal-shaping. The only obvious voltage sensitive component is an anode-break spike which is obviously related to the similar spike seen at light "off" (see below).

Light Responses:

"On" Responses. The peak of the LMC response is associated with a resistance decrease of as much as 16 M i.e. ca. 75% of the resting input resistance (Laughlin and Osorio in prep.). The reversal potential of the response is ca. 35 mV negative to the resting potential, which is approximately the same as the amplitude of the maximum hyperpolarising response. The major charge carrier is probably chloride since ionophoretic injection of chloride ions reverses the sign of the on responses (Zettler and Straka, 1987; Hardie, 1987).

We have already argued that the peak of the "on" response is generated by a high gain synapse, and the large conductance change is also related to the great density of synapses (Nicol and Meinertzhagen, 1982). More difficult to explain, however, is the cut back in the "on" response which is a manifestation of the subtractive process we have interpreted in terms of predictive coding.

A novel technique allows conductance changes to be followed with high temporal resolution (van Hateren, 1986a; Laughlin and Osorio, in prep.), whereby it is obvious that the transient voltage response is associated with an equally rapid transient resistance change (Fig. 5; Wang-Bennett and Glantz, 1987; Laughlin and Osorio in prep.). Were the transient to be caused by a depolarising conductance superimposed on the hyperpolarising one, then the cut-back in the on response would be expected to be associated with an additional conductance increase. This result thus suggests that the cut-back in the transient is generated primarily by a reduction in the chloride conductance.

One possible mechanism for this would be a desensitization of receptors on the LMCs: this however would not account for the continued sensitivity of light-adapted LMCs to further light increments; further, ionopohoretic application of the putative photoreceptor neurotransmitter (histamine) induces a maintained hyperpolarisation (Hardie, 1987; see below). Presumably, therefore, there is a reduction in photoreceptor transmitter release for which at least three types of mechanisms can be envisaged: 1. a negative feed-back circuit (pre-synaptic inhibition); 2. a mechanism intrinsic to the photoreceptor axon; and 3. a subtraction of the voltage signal in the receptor axons by extracellular fields (Laughlin, 1974; Shaw, 1975; rev. 1984).

It is unlikely that the transient is generated by a negative feedback circuit from the LMC since current injection into LMCs induces only passive polarization of the membrane and also because the "on" response is still transient even after the LMC has been hyperpolarised beyond its reversal potential (which would be expected to break or reverse any such feedback loop). However, alternative feedback circuits, for example photoreceptor - amacrine - photoreceptor (Table 1) could play a role in shaping the transient (Shaw, 1981). Hayashi et al. (1985) have suggested that a similar "on" transient in barnacle second order neurones (I-neurones) may be generated by calcium activated potassium conductances antagonising photoreceptor depolarisation in the axon terminals. Presently there is no evidence for such a mechanism in fly photoreceptors.

Still the most promising candidate for the generation of the transient response, however, is the extracellular field potential (Shaw, 1975; rev. 1984). Current from the photoreceptor axons flowing across the resistance barriers, set up by the glial sheaths, results in a depolarisation of the extracellular space. This can be expected to effectively reduce the transmembrane potential at the photoreceptor terminals and thus reduce transmitter release. The glial cells are strategically placed to modulate this current flow, and hence the strength of the antagonism, for example under different conditions of light-adaptation (Shaw, 1984). Suggestively, the glial cells are interconnected by gap-junctions and receive significant synaptic input from photoreceptors and amacrine cells (rev. Shaw, 1984).

Depolarising Mechanisms. Although most features of the LMC response can be interpreted in terms of the modulation of a single conductance mechanism (i.e. the chloride conductance activated by the photoreceptor neurotransmitter), there is no doubt that other mechanisms are also involved. If the response were only generated by this one conductance, then voltage responses should faithfully follow the underlying resistance changes. In fact such ideal behaviour is only approximated in light-adapted cells responding to small contrast increments or decrements (Fig. 5a). For example, during the plateau phase of the LMC response to a moderately bright light step, the conductance often remains stable or actually increases whilst the voltage is returning to the baseline (Fig. 5b). This can be most simply interpreted in terms of the slow activation of a depolarising conductance.

"Off" Responses. The LMC response to light "off" (or dimming) is a transient depolarisation which can be rather variable, particularly when dark-adapted; at least three components can be identified. Initially the depolarising "off" transient is associated with a conductance decrease - presumably to the photoreceptor activated chloride conductance. The conductance can often decrease to a level below the resting conductance level suggesting that there must have been a tonic release of transmitter in the dark. Except for low contrast signals in light-adapted cells, this phase is usually followed by a transient conductance increase, presumably to ions with a more positive reversal potential since it is still

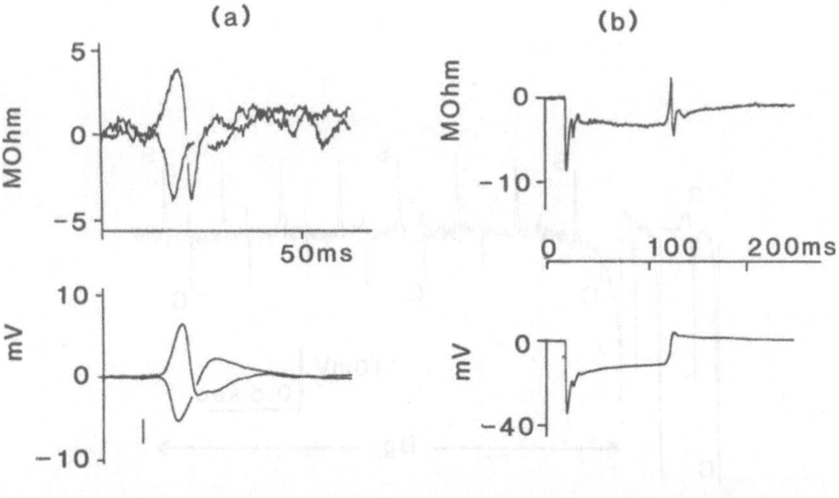

Fig. 5. a) The change in LMC input resistance accompanying brief increments and decrements about a constant background (upper traces); lower traces: voltage responses to same stimuli follow the resistance changes quite closely. b) resistance and voltage responses to a saturating 100ms light pulse. During the plateau phase the voltage declines whilst the resistance decreases; and the "off" transient is accompanied by a rapid resistance increase followed by a decrease. From Laughlin and Osorio (in prep.).

associated with a depolarisation in the voltage signal (Fig. 5). Whether this depolarising conductance is the same as that described above for the "on" response is unclear.

The third component, an "off" spike is only seen following bright stimuli. Since it is more prominent in recording locations near the medulla, it is probably generated in the medulla terminal (Zettler and Järvilehto, 1973; Laughlin and Hardie, 1978; Guy and Srinivasan, 1988; Laughlin and Osorio in prep.). Its role in coding is unclear, but a similar spike in cockroach ocellar neurones has been interpreted as a calcium action potential in the axon terminal (Mitzunami et al., 1987).

Lateral Antagonism

As the stimulus is moved away from the excitatory centre of the LMC's visual field, the maximum hyperpolarising response becomes smaller and antagonistic depolarising components become more pronounced (e.g. Laughlin and Hardie, 1978; Laughlin and Osorio in prep.). Mechanisms underlying this manifestation of lateral inhibition have also been investigated under current clamp conditions. In order to test the effect of light adaptation we used a broad adapting field in addition to a simple centre-surround stimulus. This allowed a striking confirmation of the conclusions of Dubs (1982) and Srinivasan et al. (1982), namely that lateral inhibition is sharpened by light adaptation. Thus, although centre and surround stimuli both give rise to hyperpolarizing responses in the dark, in the presence of a broad adapting field, the response to the surround often becomes depolarizing (Fig. 6). The depolarizing response is associated with a marked conductance decrease, suggesting a presynaptic mechanism (field potential or otherwise, see above), and also rules out the possibility that this effect is an extracellularly recorded artefact. Interestingly this effect was not obviously seen in all cells, and dye injections suggested that a prominent off-axis depolarization of this sort may be a specific property of L2 cells. Although further confirmation is required,

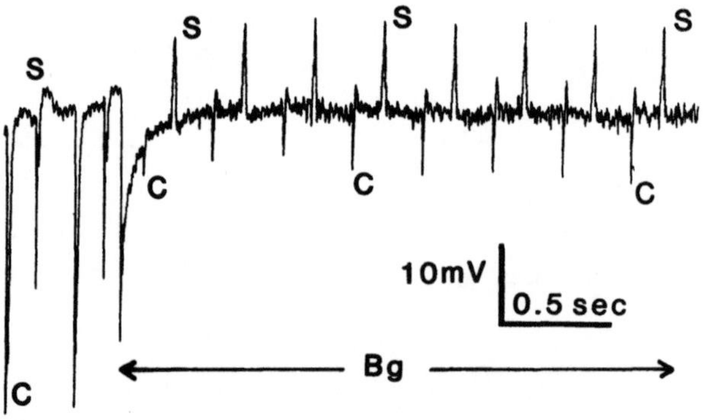

Fig. 6. The time course of the onset of intra-cartridge and lateral antagonism. The LMC is alternately stimulated with 5ms pulses delivered to the centre (C) and surround (S). Following two cycles of presentation in the dark, a broad-field background adapting light (Bg) is turned on. The polarity of the response to the surround is immediately reversed, but the centre response develops larger "on" and "off" transients over a period of seconds. From Laughlin and Osorio (in prep.).

this is the first time that any physiological difference between different classes of LMCs has been reported. It is difficult to envisage how presynaptic inhibition mediated via extracellular fields can differentiate between different classes of LMCs. If this finding is confirmed it would indicate additional synaptic mechanisms involved in lateral antagonism.

The time course of the effect of light-adaptation can also be followed by presenting rapidly repeated centre and surround flashes as the adapting field is turned on (Fig. 6). For the surround the hyperpolarizing response converts to a depolarizing response within 100 ms whilst the "on" and "off" transients to the centre adjust more slowly. These different time-scales support the conclusion that lateral antagonism and self-inhibition represent distinct mechanisms.

NEUROPHARMACOLOGY

A complete description of synaptic transfer requires a knowledge of the neurotransmitter(s) involved. In addition such knowledge can be of considerable value in helping to unravel the circuitry.

Histamine is the putative photoreceptor transmitter

Several substances have been suggested as the photoreceptor neurotransmitter, however, there is now substantial evidence to support the hypothesis that it is histamine (Hardie 1987; rev. 1988a). (i) Histamine is present in the retinae of various insects (moth, cockroach and locust) at high concentrations and enzymatic pathways exist for its synthesis (from histidine) and inactivation (via n-acetylation) (Maxwell et al., 1978; Elias and Evans, 1983). (ii) Ionophoretic studies have shown that, out of a wide range of potential transmitter candidates, only histamine mimics the action of the natural neurotransmitter. The maximum hyperpolarisation induced by histamine is similar to that which can be induced by light (Fig. 7), and is also largely mediated by an increase in conductance to chloride ions (Hardie, 1987). In the locust ocellar

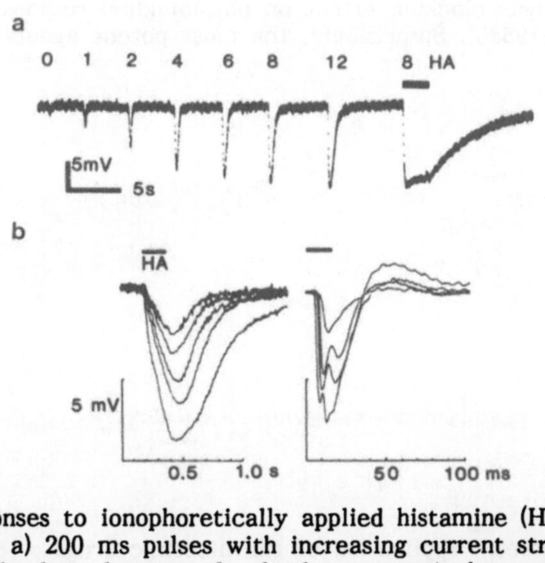

Fig. 7. Responses to ionophoretically applied histamine (HA) in an LMC. a) 200 ms pulses with increasing current strengths (in na), the last, longer pulse leads to a tonic hyperpolarization; b) responses to a similar regime of histamine pulses (left) compared to a series of light flashes (right) on a faster time scale - both saturate at similar amplitudes.

preparation it has also been shown that the reversal potential for light and histamine-induced responses are the same (Simmons and Hardie, 1988). The receptors for histamine are presumably situated on the LMCs themselves since cobalt applied in sufficient concentration to block synaptic activity does not affect the response to exogenously applied histamine (Hardie, 1987; Simmons and Hardie, 1988). Even more convincing is the observation that histamine induces chloride currents in whole-cell recordings of LMCs isolated from blowfly laminae (Hardie unpubl.). (iii) Ionophoretic studies in the housefly (Musca) have also shown that the pharmacology of light and histamine induced responses are similar. Thus a range of antagonists could be identified which block responses to both light and histamine (Fig 8; Hardie 1988b). (iv) An immunocytochemical study (Nässel et al., 1988) reveals strong histamine-like immunoreactivity localised in the terminals of all R1-6 and also R8 photoreceptors. Apart from a few fibres in the lobula plate, the optic lobes are otherwise devoid of histamine immunoreactivity.

In addition to the photoreceptors of a wide range of insects (lepidopterans, orthopterans, blattodeans and dipterans) recent evidence has suggested that histamine is the neurotransmitter of barnacle photoreceptors (Stuart and Callaway, 1988). It may well turn out that histamine is the neurotransmitter of most arthropod photoreceptors.

Histamine Receptors

The photoreceptor-LMC synapse is the first case in which histamine has been implicated as a neurotransmitter in any insect, and indeed histamine is a rather rarely encountered transmitter in any animal, only having been definitely established in some identified neurones in Aplysia (e.g. McCaman and Weinreich, 1985). In vertebrate brain, histamine is suspected of playing a neuromodulatory function - possibly related to arousal states (Pollard and Schwartz, 1987) - and three classes of histamine receptors (H1, H2 and H3) have been identified (rev. Schwartz et al., 1986).

In order to get a preliminary pharmacological characterization of insect histamine receptors, a range of antagonists was applied via ionophoresis and their blocking effect on physiological responses of LMCs observed (Hardie, 1988b). Surprisingly, the most potent agents tested were

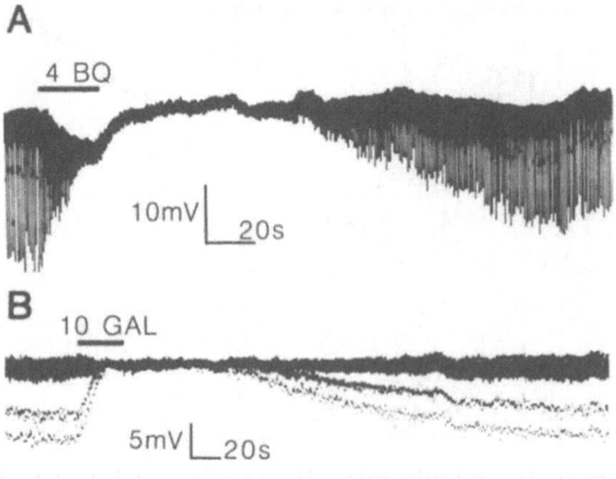

Fig. 8. Ionophoretic application of the nicotinic drugs benzoquinonium (BQ) and gallamine (GAL) both completely block an LMC's responses to light flashes (the vertical deflections at ca. 1s intervals) within 10-15s. The cells recover completely after one or two minutes. From Hardie (1988b).

agents classically considered as nicotinic acetyl choline antagonists (gallamine and benzoquinonium). The most effective of the anti-histamine agents tested were all of the H2 variety (ranitidine and cimetidine). Even without this pharmacological evidence, the rapid synaptic time constant (see above) would strongly suggest that the receptors are directly coupled to ionic channels and hence unlikely to be related to the vertebrate receptor classes which all operate via second messengers. Patch-clamp recordings of isolated outside-out patches from dissociated LMCs have revealed a histamine-sensitive chloride channel, (Hardie unpubl.).

There is also some circumstantial evidence for a second class of histamine receptor in the lamina (Hardie, 1988b). From a variety of histaminergic drugs which antagonised the light response, one (an H1 agonist, 2-thiazolyl-ethylamine) had a qualitatively distinct effect. Although it barely affected responses to bright light flashes, responses to very dim flashes were obviously attenuated. In addition the level of tonic noise in the dark was reduced suggesting that this drug interferes with the tonic release of histamine by the photoreceptors (Fig. 9). This would shift synaptic activity to the start of the sigmoidal operating characteristic, where the gain is much lower (Fig. 3). The implication is that 2-TE inhibits tonic release of photoreceptor transmitter by interacting with a distinct class of histamine receptor, either on feedback neurones (amacrines would be the most likely candidates), or with histamine autoreceptors on the photoreceptor terminals themselves (cf. Sarantis et al., 1988 for a discussion of autoreceptors on vertebrate photoreceptors).

Other Lamina Neurotransmitters

Despite the increasing availability of antibodies to putative transmitters or enzymes involved in their metabolism only a few of the lamina interneurones have been tentatively assigned a transmitter (see Table 1; rev. Hardie, 1988a). Most other information regarding

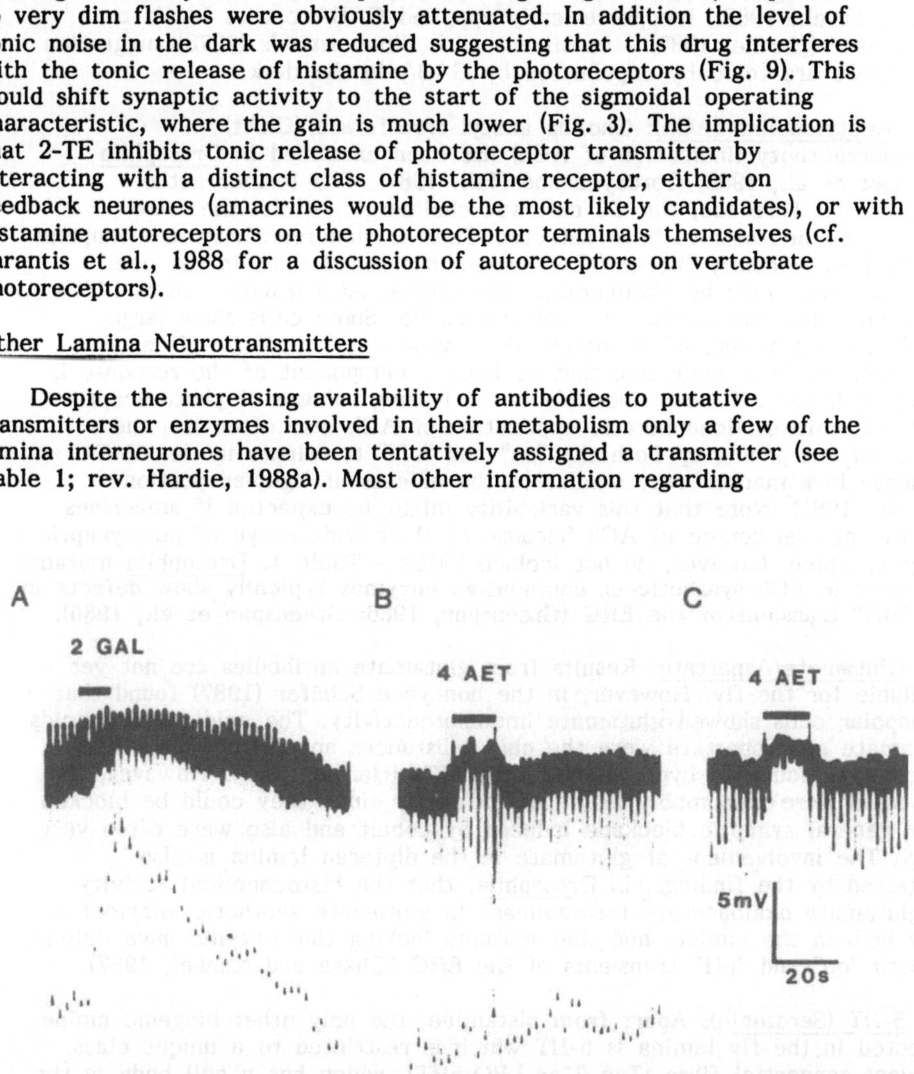

Fig. 9. The H1 agonist 2-thiazolyl-ethylamine (AET) blocks the response to weak light flashes (C) but not to bright (sub-saturating) ones (B), which are, however, blocked by gallamine (GAL) applied from a neighbouring barrel of the same ionophoretic electrode (A). From Hardie (1988b).

neurotransmitter identities and actions comes either from ionophoretic studies (Hardie, 1987, 1988b) or from studies of Drosophila mutants deficient in certain neurotransmitters (rev. Hall, 1982). It is probably premature to make specific hypotheses regarding neurotransmitter circuitry in the lamina, but gradually sufficient information is becoming available to allow a pharmacological dissection of signal processing in this neuropile.

-amino-butyric acid (GABA): Antibodies raised against GABA or GAD (an enzyme required for GABA synthesis) stain one class of centrifugal fibre, the so-called C2 cell (Meyer et al., 1986; Datum et al., 1986). Datum et al. also report that R7 is GABAergic. Although this was not confirmed by Meyer et al, it appears, at least, that R7 and R8 are likely to contain different neurotransmitters since R8 (and not R7) show histamine immunoreactivity. This would suggest that R7 and R8 have different (antagonistic?) effects on postsynaptic neurones in the medulla (Nässel et al., 1988). Ionophoretically appplied GABA causes LMCs to depolarise (Hardie, 1987) - consistent with Zimmerman's (1978) suggestion that LMCs are tonically depolarised by GABA in the dark.

Acetylcholine (ACh): Choline acetyl transferase (ChAt) immunoreactivity (indicative of ACh) has been examined in Drosophila (Buchner et al., 1986; Gorczyca and Hall, 1987). The former authors suggest that here too, the C2 cell was cholinergic, whilst the latter authors claimed also to have detected ChAt immunoreactivity in monopolar cell bodies. Recently Bicker (pers. comm) has found that, in the bee, amacrine cells may be cholinergic. Like GABA, ACh usually causes LMCs to depolarize, but the results are rather variable. Some cells show large, rapid depolarizations, whilst others show slow noisy depolarizations or hyperpolarizations suggesting that at least a component of the response is being mediated indirectly (possibly by activating release of photoreceptor neurotransmitter). Ionophoretic application of ACh also often has the effect of sharpening up both the "on" and "off" transients in the LMC response in a manner very similar to the effects of light-adaptation (Hardie, 1987). Note that this variability might be expected if amacrines are the natural source of ACh because of their wide range of postsynaptic targets, which, however, do not include LMCs - Table 1. Drosophila mutants deficient in ACh synthetic or degradative enzymes typically show defects in the "off" transient of the ERG (Greenspan, 1980; Greenspan et al., 1980).

Glutamate/Aspartate: Results from glutamate antibodies are not yet available for the fly. However, in the honeybee Schäfer (1987) found that monopolar cells showed glutamate immunoreactivity. The acidic amino-acids glutamate and aspartate were the only substances apart from histamine which were found to hyperpolarise the LMCs (Hardie, 1987). However, the responses were presumably mediated indirectly since they could be blocked by a general synaptic blockade induced by cobalt and also were often very noisy. The involvement of glutamate in the dipteran lamina is also suggested by the findings, in Drosophila, that the histochemical activity of glutamate oxaloacetate transaminase (a glutamate synthetic enzyme) is very high in the lamina, and that mutants lacking this enzyme have defects in both "on" and "off" transients of the ERG (Chase and Kankel, 1987).

5-HT (Serotonin): Apart from histamine, the only other biogenic amine detected in the fly lamina is 5-HT which is restricted to a unique class of giant tangential fibre (Tan 3 or LBO-5HT), which has a cell body in the protocerebrum (Nässel et al., 1983; rev. Nässel, 1987). No information is yet available regarding the function of 5-HT in the lamina, but the structure of these unique neurones (there is only one, bilaterally symmetrical pair of these in each animal, each with ramifications in every optic lobe on both sides of the brain) suggests a role such as overall sensitivity setting, perhaps related to circadian rhythms.

CONCLUSIONS

A major function of the lamina is to act as an interface between photoreceptors and neurons. This interface takes into account properties of both the photoreceptor signals and neurons. Antagonistic mechanisms in the lamina strip away the redundant signal components from the receptor response so that the amplified signal can be better accommodated within the limited dynamic response range of the post-synaptic neurons. It appears that these antagonistic mechanisms operate upon transmitter release from the photoreceptor terminal but the underlying set of mechanisms has yet to be established. Nor do we know the mechanisms responsible that compute the redundant component by averaging over space and time, and the way in which this averaging is adjusted to take account of the background light intensity. The remaining receptor signal is amplified at the photoreceptor-LMC synapses to protect it from synaptic noise contamination. This process is well designed in a number of respects. The amplification of signal is matched to the expected amplitude. Our analysis of mechanisms suggests that this matching requires that the photoreceptor LMC synapse operates with the correct gain. It remains an open question as to whether this gain is determined genetically or is set by visual experience.

With so many mechanisms unresolved what is the prospect for further research on the fly lamina? Current hypotheses suggest that the apparent simplicity of the lamina belies a sophisticated range of mechanisms including interneuronal circuits, the regulation of neuronal milieu by activity (via field potentials) and by modulation of glial activity, and processing within single cells resulting from the interactions between different type of channels. With such a complexity of mechanisms, further progress will require the implementation of diverse and novel experimental approaches as exemplified by the recent introduction of neuropharmacological current-clamp, and patch-clamp analysis. The problems facing the fly lamina are common to nervous systems everywhere, and it is to be expected that the continued analysis of this exceptionally well-defined system will reveal further principles and mechanisms of general significance.

REFERENCES

Autrum, H., Zettler, F. and Järvilehto, M. 1970. Postsynaptic potentiuals from a single neuron of the ganglion opticum I of the blowfly Calliphora erythrocephala. Z. vergl. Physiol., 48:357-384.

Blight, A.R. and Llinas, R., 1980. The non-impulsive stretch-receptor complex of the crab: a study of depolarization-relesase coupling at a tonic sensorimotor synapse. Phil. Trans. R. Soc. Lond. B, 212:1-34

Buchner, E., Buchner, S., Crawford, G., Mason, W.T., Salvaterra, P.M. and Sattelle, D.B. 1986. Choline acetyltransferase-like immunoreactivity in the brain of Drosophila melanogaster. Cell Tissue Res., 246: 57-62.

Burrows, M. 1979. Synaptic potentials effect the release of transmitter from locust non-spiking interneurons. Science, 204:81-83

Chase, B.A. and Kankel, D.R. 1987. A genetic analysis of glutamatergic function in Drosophila. J. Neurobiol., 18:15-41.

Chi, C. and Carlson, S.D. 1976, Close apposition of photoreceptor axons in the housefly. J. Insect. Physiol. 22:1153-1156..

Datum, K-H., Weiler, R. and Zettler, F. 1986. Immunocytochemical demonstration of -amino butyric acid and glutamic acid decarboxylase in R7 photoreceptors and C2 centrifugal fibres in the blowfly visual system. J. comp. Physiol., 159:241-249.

Dubs, A. 1982. The spatial integration of signals in the retina and

lamina of the fly compound eye under different conditions of luminance. J. comp. Physiol. 146:321-343.
Dubs, A., Laughlin, S.B. and Srinivasan, M.V. 1981. Single photon signals in fly photoreceptors and first order interneurones at behavioural threshold. J. Physiol., 317:317-334.
Elias, M.S. and Evans, P.D. 1983. Histamine in the insect nervous system: distribution, synthesis and metabolism. J. Neurochem., 41:562-568.
Fain, G.L. and Lisman, J.E. 1981. Membrane conductances of photoreceptors. Prog. Biophys. molec. Biol., 37:91-147.
Franceschini, N. 1975. Sampling of the visual environment by the compound eye of the fly: Fundamentals and applications, pp. 98-125. In: Photoreceptor Optics. Snyder, A.W. and Menzel, R. eds. Springer, Berlin Heidelberg New York.
Gonzalez. R.C. and Wintz, P. 1977. Digital image processing. Addison Wesley, Reading Mass
Gorczyca, M.G. and Hall, J.C. 1987. Immunohistochemical localization of choline acetyltransferase during development and in Chats mutants of Drosophila melanogaster. J. Neurosci., 7:1361-1369.
Greenspan, R.J. 1980. Mutations of choline acetyltransferase and associated neural defects in Drosophila melanogaster. J. comp. Physiol. 137:83-92.
Greenspan, R.J., Finn, J.A. and Hall, J.C. 1980. Acetylcholinesterase mutants in Drosophila and their effects on the structure and function of the central nervous system. J. comp. Neurol., 189:741-774.
Guy, R.G. and Srinivasan, M.V. 1988. Integrative properties of second-order visual neurons: a study of large monopolar cells in the dronefly Eristalis. J. Comp. Physiol. A, 162:317-332.
Hall, J.C. 1982. Genetics of the nervous system in Drosophila. Q. Rev. Biophys., 15:223-479.
Hardie, R.C. 1985. Functional organization of the fly retina, pp. 1-79. In: Ottoson, D. ed. Prog. sensory physiology Vol. 5. Springer, Berlin, Heidelberg, New York, Toronto
Hardie, R.C. 1987. Is histamine a neurotransmitter in insect photoreceptors? J. comp. Physiol., 161:201-213.
Hardie, R.C. 1988a. Neurotransmitters in compound eyes, pp. 235-256. In: Facets of Vision. Stavenga D.G., Hardie R.C. (eds.). Springer, Berlin Heidelberg New York Toronto.
Hardie, R.C. 1988b. Effects of antagonists on putative histamine receptors in the first visual neuropile of the housefly (Musca domestica). J. Exp. Biol., 138:221-241.
Hateren, J.H. van 1986a. Electrical coupling of neuro-ommatidial photoreceptor cells in the blowfly. J. comp. Physiol. 159:795-811.
Hateren, J.H. van 1986b. An efficient algorithm for cable theory, applied to blowfly photoreceptor cells and LMCs. Biol. Cybern., 54:301-311.
Hayashi, J.H., Moore, J.W. and Stuart, A.E. 1985. Adaptation in the input-output relation of the synapse made by the barnacle photoreceptor. J. Physiol., 368:179-195.
Järvilehto M. 1985. The eye: vision and perception, pp. 355-429. In: Comprehensive insect physiology, biochemistry and pharmacology. Kerkut, G.A. and Gilbert, L.I. eds. Pergamon Oxford
Katz, B. and Miledi R. 1967. A study of synaptiuc transmission in the absence of nerve impulses. J. Physiol. Lond., 224:655-699
Kuwabara, M. and Naka K. 1959. Responses of a single retinula cell to polarized light. Nature, 184:455-456
Laughlin, S.B. 1974. Resistance changes associated with the response of insect monopolar neurons. Z. Naturforsch., 29c:449-450
Laughlin, S.B. 1981a. Neural principles in the peripheral visual systems of invertebrates In: Handbook of sensory physiology Vol VII/6b. Autrum, H. ed pp. 133- 280 Springer, Berlin Heidelberg NewYork.
Laughlin, S.B. 1981b. A simple coding procedure enhances a neuron's information capacity. Z. Naturforsch. 36c:910-912

Laughlin, S.B. 1987. Form and function in retinal processing. Trends Neurosci. 10:478-483.
Laughlin, S.B. and Hardie, R.C. 1978. Common strategies for light adaptation in the peripheral visual systems of fly and dragonfly. J. Comp. Physiol. 128:319-340.
Laughlin, S.B. and Osorio, D. 1989. Mechanisms for neural signal enhancement in the blowfly compound eye. (submitted).
Laughlin, S.B., Howard, J. and Blakeslee, B. 1987. Synaptic limitations to contrast coding in the retina of the blowfly Calliphora. Proc. Roy. Soc. Lond. B, 231:437-467.
Lipetz, L.E. 1971. The relation of physiological and psychological aspects of sensory intensity. In: Handbook of sensory physiology Vol. I (Loewenstein W.R. ed.) pp. 191-225. Springer Berlin Heidelberg New York
McCaman, R.E. and Weinreich, D. 1985. Histaminergic synaptic transmission in the cerebral ganglion of Aplysia. J. Neurophysiol. 53:1016-1037.
Maxwell, G.D., Tait J.F. and Hildebrand J.G. 1978. Regional synthesis of neurotransmitter candidates of the moth Manduca sexta. Comp. Biochem. Physiol. 61C:109-119.
Meyer, E.P., Matute, C., Streit, P. and Nässel, D.R. 1986. Insect optic lobe neurons identifiable with monoclonal antibodies to GABA. Histochem. 84:207-216.
Mitzunami, M., Yamashita, S. and Tatea, H. 1987. Calcium-dependent action potentials in the second-order neurones of cockroach ocelli. J. Exp. Biol., 130:259-274.
Muijser, H. 1979. The receptor potential of retinular cells of the blowfly Calliphora: The role of sodium, potassium and calcium ions. J. Comp. Physiol. 132:87-95.
Nässel, D.R. 1987. Serotonin and serotonin-immunoreactive neurons in the nervous system of insects. Prog. Neuroblol., 30:1 85.
Nässel, D.R., Ohlsson, L. and Sivasubramanian P. 1983. A new possibly serotonergic neuron in the lamina of the blowfly optic lobe: an immunocytochemical and Golgi-EM study. Brain Res., 280:361-367.
Nässel, D.R., Holmqvist, M.H., Hardie, R.C., Hakånson. R. and Sundler, F. 1988. Histamine-like immunoreactivity in photoreceptors of the compound eyes and ocelli of flies. Cell Tissue Res., 253:639-646.
Nicol, D. and Meinertzhagen, I.A. 1982. An analysis of the number and composition of the synaptic populations formed by photoreceptors of the fly. J. Comp. Neurol., 207:29-44.
Pollard, H. and Schwartz, J-C. 1987. Histamine neuronal pathways and their functions. Trends Neurosci., 10:86-89
Prell, G.D. and Green, J.P. 1986. Histamine as a neuroregulator. Ann. Rev. Neurosci., 9:209-254.
Ribi, W.A. 1978. Gap junctions coupling photoreceptor axons in the first optic ganglion of the fly. Cell Tiss Res., 195:299-308.
Saint Marie, R.L. and Carlson, S.D. 1983. The fine structure of neuroglia in the lamina ganglionaris of the housefly, Musca domestica L. J. Neurocytol., 12:213-241.
Sarantis, M., Everett, K. and Attwell, D. 1988. A presynaptic action of glutamate at the cone output synapse. Nature, 332:451--453.
Schäfer, S. 1987. PhD Thesis Free University Berlin
Schwartz, J-C., Arrang, J-M., Garbarg, M. and Korner, M. 1986. Properties and roles of the three subclasses of histamine receptors in brain. J. Exp. Biol., 124:203-224.
Shannon, C.E. and Weaver, W. 1949. The mathematical theory of communication. University of Illinois Press Urbana.
Shaw, S.R. 1968. Organization of the locust retina. Symp. Zool. Soc. Lond., 23:135-163.
Shaw, S.R. 1975. Retinal resistance barriers and electrical lateral inhibition. Nature, 255:480-483.
Shaw, S.R. 1981. Anatomy and physiology of the identified non-spiking cells in the photoreceptor-lamina complex ofthe compound eye of insects, especially Diptera, pp.61-116. In: Neurones without

impulses. Roberts, A. and Bush, B.M.H. Soc. Exp.Biol. Seminar series 6.
Shaw, S.R. 1984. Early visual processing in insects. J. exp. Biol. 112:225-251.
Simmons, P.J. 1981. Synaptic transmission between second and third-order neurones of a locust ocellus. J. comp. Physiol 145:265-276.
Simmons, P.J. 1982. The operation of connexions between photoreceptors and large second-order neurons in dragonfly ocelli. J. comp. Physiol., 149:389-398.
Simmons, P.J. and Hardie, R.C. 1988. Evidence that histamine is a neurotransmitter in the locust ocellus. J. Exp. Biol., 138:205-219
Srinivasan, M.V., Laughlin, S.B. and Dubs, A. 1982. Predictive coding: a fresh view of inhibition in the retina. Proc. Roy. Soc. Lond. B, 216:427-459.
Strausfeld, N.J. 1976. Atlas of an insect brain. Springer, Berlin Heidelberg New York.
Strausfeld, N.J. 1984. Functional neuroanatomy of the blowfly's visual system, pp. 483-522. In: Photoreception and vision in invertebrates. Ali, M.A. ed. Plenum Press, New York London.
Strausfeld, N.J. and Nassel, D.R. 1981, Neuroarchitectures serving compound eyes of Crustacea and Insects. In:Handbook of Sensory Physiology, Vol VII/6b, Autrum, H. ed. Springer Berlin Heidelberg New York.
Stuart, A.E. 1983, Vision in barnacles. Trends Neurosci., 6:137-140.
Stuart, A.E. and Callaway, J.C. 1988, Histamine is synthesized by barnacle ocelli and affects second-order visual cells. Invest. Ophthalm. Vis. Sci., 29:223 (abstr.)
Wang-Bennett, L.T. and Glantz, R.M. 1987, The functional organization of the crayfish lamina ganglionaris. I. Non-spiking monopolar cells. J. Comp. Physiol. A 161:131-145.
Wilson, M. 1978, Generation of graded potential signals in the second order cells of locust ocellus. J. comp. Physiol., 124:317-331.
Zettler, F. and Järvilehto, M. 1973. Active and passive axonal propagation of non-spike signals in the retina of Calliphora., J. Comp. Physiol., 85:89-104.
Zettler, F. and Straka, H. 1987. Synaptic chloride channels generating hyperpolarising responses in monopolar neurones of the blowfly visual system. J. Exp. Biol. 131: 435-438.
Zimmerman, R.P. 1978. Field potential analysis and the physiology of second-order neurons in the visual system of the fly. J. Comp. Physiol. A , 126:297-317.

ANALYSIS OF SENSORY SPIKE TRAINS

Hiroko M. Sakai

National Institute for Basic Biology
Okazaki 444 Japan

INTRODUCTION

The neuron network of the vertebrate retina processes signals by means of graded potentials, whose complex end products are encoded into spike trains by ganglion cells. Spike discharges are the means of long-distance communication from the retina, and are ubiquitous means of cell-to-cell communication in the central nervous system. Generation of an action potential is a highly nonlinear process, which depends upon the threshold in all-or-none fashion. Biophysical properties underlying the generation of an action potential have been a fundamental problem of membrane physiology; in the last few decades, great strides have been made in this field.

From the standpoint of information transmission, however, generation of one spike does not mean anything; it is the train of spikes, or array of spikes that carries information. Whereas graded potentials carry information in analog forms, a spike generation occurs from one point to another; it is a point process. One of the most fundamental questions of neurophysiology, the importance of which is unfortunately not yet fully recognized, is a problem of signal transformation from analog to point process: whether or not the dynamic properties of information change during the voltage-spike, or analog-point transformation.

In the retina, ganglion cells produce spikes as final products of complex signal processing carried out by means of graded potentials among retinal interneurons, whereas in other central nervous systems, cell-to-cell communication is generally assumed to be by means of spike trains. Even in the latter case, the spike trains are transformed into graded potentials at synaptic sites, which are again encoded into spike trains at the spike generation site of postsynaptic cells. Transformation of analog-point signals is a basic problem common in the central nervous system.

In this article, the problem is approached through the Wiener analysis of spike trains and slow potentials of ganglion cells, together with results of analysis performed on the preganglionic cells in the catfish retina (Sakuranaga and Naka, 1985abc; Sakai and Naka, 1987ab). A conclusion is that no major information modifications result from the analog-point transformation in the catfish ganglion cells. In addition, a simple analytical explanation is presented here to account for such information conservation.

MATERIALS AND METHODS

Eyecup preparation of channel catfish, *Ictalurus punctatus* was used. Intracellular recordings were made either from a ganglion cell or a preganglionic cell, and extracellular spike discharges were recorded from a nearby ganglion cell. Responses were evoked by flash of light given in the dark, and by Gaussian white-noise modulated light, $30\mu W/cm^2$ in mean luminance, covering the entire retinal surface. Then a white-noise modulated current, with a standard deviation of 2 to 5 nA, was injected into a cell through an intracellular glass electrode and the resulting current-evoked responses were recorded by an extracellular platinum-coated tungsten electrode. The power spectrum of the white-noise modulated current was flat from 1 to 100 Hz, and that of the white-nose modulated light was from 1 to 50 Hz. First- and second-order kernels were computed by cross-correlation between white-noise modulated input, either light or current, and the resulting cellular responses.

Analyses were performed off-line through a software system, STAR, running on a combination of VAX-11/780 computer and an AP-120B array processor. Extracellular spike discharges were transformed into unitary pulses of 1 ms duration. Time resolution of analyses was 1 ms for the current-evoked and 2 ms for the light-evoked kernels. Details of the digitization procedure and data processing are found in Sakai and Naka (1988a).

RESULTS AND CONCLUSIONS

Generation of spikes

Responses of a retinal ganglion cell consist of slow potentials and spike discharges. Figure 1 shows responses of an off-center ganglion cell in the catfish retina (B) evoked by a Gaussian white-noise modulated light (A). By using an algorithm described by Sakuranaga et al. (1987), ganglion-cell responses can be segregated into slow potentials (C) and unitized spike trains (D). The slow potential can be analyzed by means of the standard white-noise analysis without any contamination from the spike discharges. Spike discharges, on the other hand, have to be converted into an analog form, to compute kernels. Two problems are involved in analyzing the dynamics of the spike trains: 1) the trains are a point process with a maximum firing frequency of <1 kHz to limit the amount of information carried by the trains; 2) the firing of a spike discharge is a stochastic process with a jitter in the spike occurrence. The former indicates that signals are transmitted only intermittently, and the latter indicates that no two spike trains produced by two identical stimuli are identical. Several attempts were made to analyze the dynamics of spike trains. The most straightforward approach is the trigger or reverse correlation in which the average (optimal) input waveform to trigger a spike discharge is sought (de Boer and Kuyper, 1968). Sakuranaga et al. (1987) have shown that a direct cross-correlation between analog white-noise against the resulting spike discharges (transformed into unitary pulses) produced first- and second-order Wiener kernels similar to those obtained by a cross-correlation of the analog white-noise input and the resulting intracellular slow potentials. The first-order kernel computed by this method is identical to that produced by the trigger reverse correlation by de Boer and Kuyper, P.(1968), although time runs in the opposite direction.

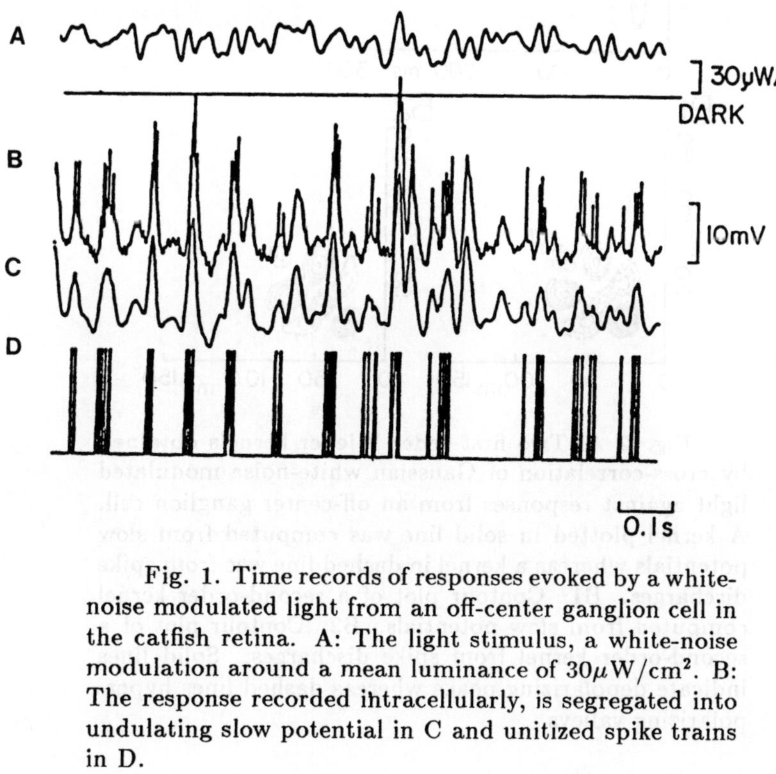

Fig. 1. Time records of responses evoked by a white-noise modulated light from an off-center ganglion cell in the catfish retina. A: The light stimulus, a white-noise modulation around a mean luminance of $30\mu W/cm^2$. B: The response recorded intracellularly, is segregated into undulating slow potential in C and unitized spike trains in D.

Figure 2 shows first-order and second-order Wiener kernels thus produced from the ganglion-cell responses shown in Fig. 1: the kernel plotted as a solid line was from the extracted slow potential in Fig. 1C, and the kernel plotted as a dashed line was from spike discharges transformed into unitary pulses in Fig. 1D. While there are some minor differences, the first-order kernels computed from both slow (analog) and unitized spike (point process) potentials are similar, although the spike kernel (dashed line) is more differentiating with a slightly shorter peak response time. Moreover, both the slow potential (Fig. 2B1) and unitized spike discharge (Fig. 2B2) produced similar second-order kernels. The considerable similarity can better be appreciated by comparing these kernels with the second-order kernels from various cell types shown in Sakai and Naka (1987a, 1988a).

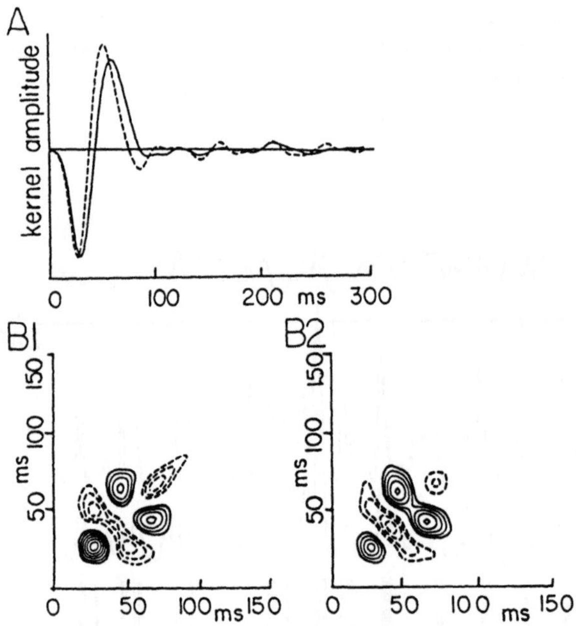

Fig. 2 A: Two first-order Wiener kernels obtained by cross-correlation of Gaussian white-noise modulated light against responses from an off-center ganglion cell. A kernel plotted in solid line was computed from slow potentials whereas a kernel in dashed line was from spike discharges. B1: Contour plot of a second-order kernel computed from slow potentials. B2: Contour plot of a second-order kernel from spike discharges. Solid lines indicate depolarizing peaks whereas dashed lines hyperpolarizing valleys.

Sakuranaga et al. (1987) analyzed a considerable number of ganglion-cell responses by decomposing them into slow potentials and spike trains, and found that the first-, and second-order kernels computed from the two different type of signals were similar. Sakai and Naka (1987a) compared the time course and structure of first- and second-order Wiener kernels computed from extracellular spike discharges with those from intracellular responses of ganglion cells, and concluded that no major transformation of kernel structure takes place when the signal is transformed from analog signal (PSPs) to spike trains. This conclusion can be verified if the voltage-spike generation, a highly nonlinear operation, is nearly a static function.

Figure 3A shows an example of simultaneous recordings made from a single ganglion cell. The upper trace shows intracellular activity recorded with an intracellular glass electrode whereas the lower trace shows extracellular spike discharges recorded by an extracellular tungsten electrode. The bar, LIGHT, indicates the approximate timing of the presentation of light stimulus. The cell is an on-center ganglion cell. Exact synchrony of the spike discharges indicates that recordings were made from a single ganglion cell. Figure 3B shows a first-order Wiener kernel produced by cross-correlation of white-noise modulated current injected into a ganglion cell through the intracellular electrode and the resulting spike discharges recorded through the extracellular electrode. The kernel is mainly depolarizing, with a peak response time of 1 ms, the limit of our time resolution because spikes were transformed into unitary pulses of 1 ms in duration. This kernel waveform approximates an impulse function, or static, suggesting that the transformation of slow potential into spike train does not appear to involve a long memory for slow potential past value.

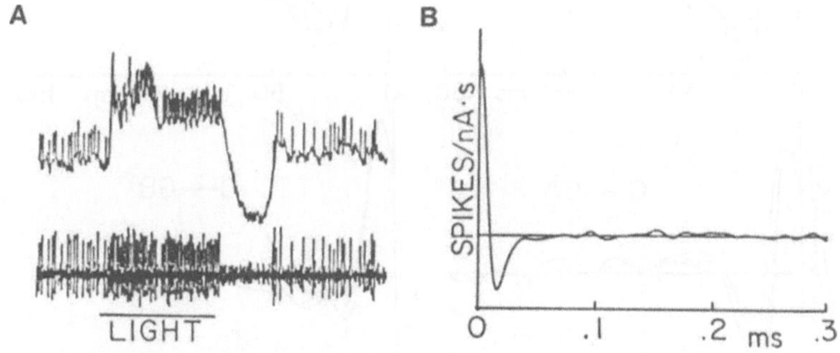

Fig. 3 A: Simultaneous recordings with an intracellular electrode (upper record) and an extracellular electrode. Exact synchrony of spike discharges indicates that recordings were made from a single ganglion cell. The bar indicates the approximate timing of a flash of light. B: First-order Wiener kernel computed from white-noise modulated current injected intracellularly into the ganglion cell and the resulting extracellular spike discharges from the same ganglion cell.

Origin of Linear and Nonlinear Components Computed from Ganglion-cell Slow Potentials

Well-defined first- and second-order kernels are obtained from ganglion-cell slow potentials. An example is shown in Fig. 2, A and B1. A question is whether these kernels are unique only to ganglion cells, or are they transmitted from preganglionic cells without major transformation? In the former case, inputs to ganglion cells must undergo some transformations, whereas in the latter case, transmission from preganglionic-cell to ganglion-cell PSPs must be impulse-like.

Sakai and Naka (1987a), and Sakuranaga et al. (1987) found that first-order kernels computed from ganglion-cell slow potentials are similar to those computed from bipolar and amacrine cells, and that the second-order kernels are similar to those computed from amacrine cells: there are no first- and second-order kernels unique to ganglion cells. The first- and second-order kernels shown in Fig. 2, A and B1, for example, are similar to those obtained from NB amacrine (off amacrine) cells. This similarity suggests that signals of preganglionic cells are transmitted to ganglion-cell PSPs without major dynamic transformation.

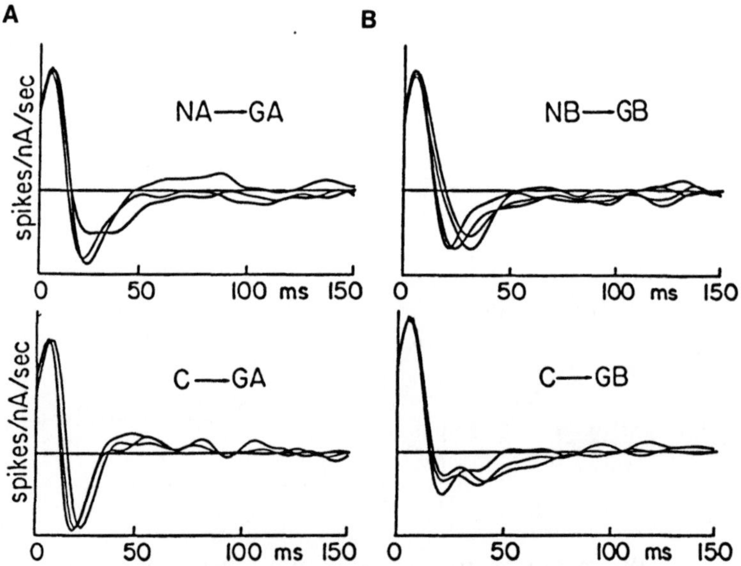

Fig. 4. First-order kernels produced from white-noise current injected into an amacrine cell, either NA (on), NB(off) or C (on-off) type, and the resulting extracellular spike discharges recorded from a neighboring GA (on) or GB (off) ganglion cell.

Figure 4 shows first-order kernels produced from white-noise current injected into an amacrine cell and the resulting extracellular spike discharges recorded from a neighboring ganglion cell. "NA", "NB" and "C" indicate on, off, and on-off amacrine cells, while "GA" and "GB" are on-center and off-center ganglion cells respectively (Sakai and Naka, 1988a). In all transmissions, the first-order kernels are mainly depolarization with peak response time of 2–4 ms. The kernel waveforms are nearly an impulse function. Fourier transformation of the kernels show that the transfer functions from amacrine cell to ganglion cell spike discharges are constant gain, lowpass with cutoff frequency of 30–40 Hz (Sakai and Naka, 1988a).

Anatomical evidence is that ganglion cells receive monosynaptic inputs from bipolar and amacrine cells (Dowling and Boycott, 1966; Dowling and Werblin, 1969; Sakai et al., 1986). In addition, ganglion-cell dendrites are presynaptic to other ganglion cells as well as to bipolar and amacrine cells in the catfish retina (Sakai et al., 1986). With the two-electrode current injection experiments, physiological evidence is provided that these are indeed so in the catfish retina. First-order kernels produced for transmission from bipolar to ganglion cell, as well as that from one ganglion to another ganglion cell are similar to those shown in Fig. 4. Kernel waveforms are mainly depolarization with short peak response time and the transmission was approximated by the first-order kernel; transmission to a ganglion cell is linear or quasi-linear (Sakai and Naka, 1988ab).

Figure 5 summarizes the findings of the current injection experiments. Bipolar cells, either on-center (BA) or off-center (BB), respond linearly to input modulation (Sakuranaga and Naka, 1985a; Sakai and Naka, 1987a). Bipolar

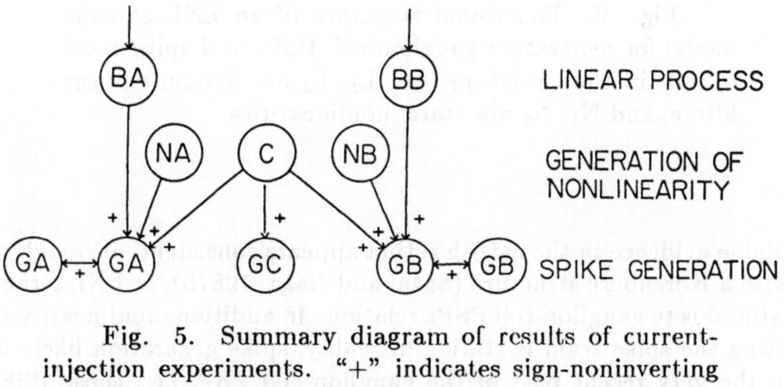

Fig. 5. Summary diagram of results of current-injection experiments. "+" indicates sign-noninverting transmission with linear lowpass filter. Virtually, all the transmission to ganglion cells are marked "+", indicating no major dynamic transformation taking place. BA (on-center bipolar), BB (off-center bipolar), NA (on amacrine), C (on-off amacrine), NB (off amacrine), GA (on-center ganglion), GC (on-off ganglion) and GB (off-center ganglion cell).

signals are linear. Second- and third-order nonlinearities are generated only in amacrine cells (Sakuranaga and Naka, 1985bc; Sakai and Naka, 1987ab). Linear signals of bipolar cells and nonlinear signals of amacrine cells are transmitted to ganglion cells and encoded into spike trains of ganglion cells without major dynamic transformation. Moreover, GA cells are interconnected with neighboring GA cells, while GB cells are interconnected with neighboring GB cells. The transfer function is constant gain, lowpass, with almost no phase shift at the lower frequency regions.

A simple analytical explanation is provided here for the strong similarity between light-evoked kernels computed from slow potentials and those computed from spike discharges. Suppose, as in Fig. 6, that the signal processing from the light stimulus, l(t), leading to the ganglion cell postsynaptic potentials (PSP) can be represented by a Korenberg structure, which comprises a first dynamic linear system (L_1) followed by a static nonlinearity (N_1), then followed by a second dynamic linear system (L_2). Suppose further that the process involved in generating the spike train is a static (i.e. no-memory) nonlinear function of the ganglion cell PSP. This means the decision whether or not to generate a spike at a given "instant" of time depends theoretically only upon the value of the PSP at that instant.

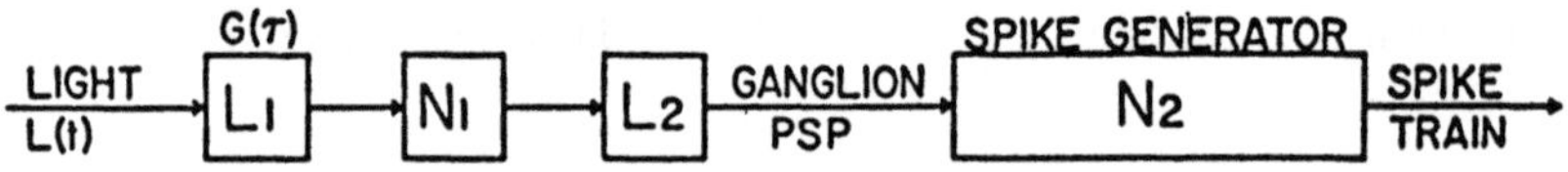

Fig. 6. Functional structure of an LNL cascade model for generating ganglion-cell PSPs and spike train output in the catfish retina. L_1, L_2 are dynamic linear filters, and N_1, N_2 are static nonlinearities.

Available evidence in the catfish retina appears consistent, although not conclusive, with a Korenberg structure (Sakai and Naka, 1987b), or LNL structure for the light stimulus to ganglion-cell PSPs relation. In addition, nonlinearity involved in generating the spike train is static. In reality, spike generation likely depends also upon the very recent past of the ganglion cell PSP, i.e. those PSP values immediately preceding the spike. However, the assumption that spike generation is very nearly a static function of the PSP is supported by Fig. 3B.

In this article, experimental results and heuristic explanations were presented that the neural circuitry leading from the light stimulus to the generation of ganglion cell spike discharges can be represented by a Korenberg structure followed by a highly nonlinear static process of spike generation. A similar cascade structure to that shown in Fig. 6 was used by Marmarelis et. al. (1986), in which a Wiener structure, or LN structure was followed by spike generation in the form of a threshold-trigger. They pointed out that the cascade of the static nonlinearity with the threshold-trigger was equivalent to a single static nonlinearity. A LNL structure was adopted by Victor and Shapley (1979) to describe the nonlinear pathway of the Y-cell in the cat retina. They suggested that the nonlinearity was originated in the amacrine cell. In two retinas, that of cat and catfish, the process leading to ganglion-cell spike generation are remarkably similar. Accurate cascade identification of the process leading from light to ganglion-cell PSPs is presently underway in the catfish retina.

REFERENCES

de Boer, E. and Kuyper, P. 1968, Triggered correlation. IEEE. Trans. Biomed. Eng. 15: 169–179.

Dowling, J. E. and Boycott, B.B. 1966, Organization of the primate retina: electron microscopy. Proc. R. Soc. Lond. B.Biol. Sci. 166: 80–111.

Dowling, J. E. and Werblin, F.S. 1969, Organization of retina of the mudpuppy *Necturus maculosus*. I. Synaptic structure. J. Neurophysiol. 32: 315–338.

Marmarelis, V. Z., Citron, M. C., and Vivo, C. 1986, Minimum-order Wiener modeling of spike-ooutput system. Biol. Cybern. 54: 115–124.

Sakai, H. M. and Naka, K.-I. 1987a, Signal transmission in the catfish retina. IV.transmission to ganglion cells. J. Neurophysiol. 58: 1307–1328.

Sakai, H. M. and Naka, K.-I. 1987b, Signal transmission in the catfish retina. V. Sensitivity and circuit. J. Neurophysiol. 58: 1329–1350.

Sakai, H. M. and Naka, K.-I. 1988a, Dissection of the neuron network in the catfish inner retina. I. Transmission to ganglion cells. J. Neurophysiol. 60: 1549–1567.

Sakai, H. M. and Naka, K.-I. 1988b, Dissection of the neuron network in the catfish inner retina. II. Interactions between ganglion cells. J. Neurophysiol. 60: 1568–1583.

Sakai, H.M., Naka, K.-I. and Dowling, J. E. 1986, Ganglion cell dendrites are presynaptic in catfish retina. Nature Lond. 319: 495–497.

Sakuranaga, M. and Naka, K.-I. 1985a, Signal transmission in catfish retina. I. Transmission in the outer retina. J. Neurophysiol. 53: 373–389.

Sakuranaga, M. and Naka, K.-I. 1985b, Signal transmission in catfish retina. II. Transmission to type-N cells. J. Neurophysiol. 53: 390–410.

Sakuranaga, M. and Naka, K.-I. 1985c, Signal transmission in catfish retina. III. Transmission to type-C cells. J. Neurophysiol. 53: 411–428.

Sakuranaga, M., Ando, Y.-I. and Naka, K.-I. 1987, Dynamics of ganglion-cell response in the catfish and frog retina. J. Gen. Physiol. 90: 229–259.

Victor, J. D. and Shapley, R. M. 1979, The nonlinear pathway of Y ganglion cells in the cat retina. J. Gen. Physiol. 74: 671–687.

SIGNAL CODING AND SENSORY PROCESSING IN THE PERIPHERAL RETINA OF THE COMPOUND EYE

Matti Järvilehto, *Matti Weckström and *Eero Kouvalainen

Department of Zoology and *Department of Physiology
University of Oulu
90570 Oulu, Finland

ABSTRACT

The structure of the dipteran compound eye provides an extensive neuronal network for sensory processing of visual signals. The coding of impinging photons as transmembrane voltage fluctuations provides the means for visual neurons to extract relevant information from the visual surround.

The simultaneous processing of these signals, by lateral feed-forward and feed-back interactions amongst parallel channels, functions as a mechanism to reduce information. The spatial resolution is limited by the optics i.e. the arrangement and the number of ommatidia. The channel capacity of information transfer in all neuronal elements limits the temporal and spatial resolution of the visual system.

In this paper we present an evaluation of signal coding and processing by means of linear systems analysis using different stimulation approaches and intracellular transmembrane voltage recordings. In summary, our results show that: (1) Transformation of light stimuli into photoreceptor signals is essentially a linear process; (2) response summation and adaptation at high degrees of stimulus modulation is increasingly non-linear; (3) the direct pathway from receptor axons to second order interneurons shows a frequency dependent amplification that can be modelled quantitatively by fractional differentiation; (4) lateral interaction is already present between the photoreceptor terminals in the first synaptic ganglion where the mechanism acts like a capacitive coupling. This finding also explains certain time domain characteristics of the receptive field configuration in second order neurons.

INTRODUCTION

There are three major approaches to the functional analysis of the compound eye and underlying visual system. These are: (1) Structural: anatomical and chemical analysis of neurons (the latter directed at elucidating putatative transmitters and modulatory peptides); (2) Functional: physiological, biophysical and electrical properties; (3) Systems analytical: information properties. In this contribution we will present an analysis of the visual system's input-output properties in relation to the structural and functional properties of the cells. The analysis is based on white noise stimulation approach for linear systems analysis. First we will consider the known functional units in the insect retina and the different possibilities for their analysis, and afterwards present a short synopsis of the theoretical background to linear systems analysis finally presenting some important results of its application to insect vision.

Functional units in the fly retina: A functional unit comprises a defined group of cells which perform an arbitrarily given function (Järvilehto 1985). The structure of the compound eye and underlying neuropils suggests that both spatial and temporal information processing is performed by a variety of hypothetical functional architectures and the large variety of morphological elements suggest all manner of cellular configurations that could be considered as elementary functional units. For a description of the general anatomy see Strausfeld (1979, 1984) and Shaw (1984).

A familiar example of separate functional units is in the eye's periphery where different groups of photoreceptor (R1 - R6 and R7 + R8) provide an obvious basis for a system having the properties of high versus low sensitivity on the one hand, and low versus high acuity on the other. The difference between a high sensitivity system (HSS) and a high acuity system (HAS) simply defines them as functional units. This is also supported by the anatomical/functional observation that information conducted by the R1 to R6 cell axons is transmitted and released to the second order cells (LMCs) in the optic cartridges of the first optic neuropile (the lamina) whereas R7 and R8 bypass this synaptic neuropil to converge on second order cells in the deeper optic neuropile (the medulla). It seems that the peripheral organization of functional units is rigorously reorganized so that much irrelevant visual information is filtered out. The following two anatomically overlapping pathways are suggested to mediate neural information processing in the peripheral levels of the visual system.

1. Direct pathway: The photoreceptor-LMC synapse. Anatomically, six photoreceptor axons converge to three lamina monopolar cells (LMCs) L1-L3, forming approximately 50 μm-long regions of multiple synaptic contacts. All the evidence so far supports the idea that L1 - L3 share similar electrical activity, but through their different anatomical connections each of them may individually perform different functions. Synaptic contact between receptors and LMCs is chemical and probably histaminergic in nature (Hardie, 1987; also this volume). Other second order neurons (e.g. L4, L5), and a fourth type of second order output neuron (beta, T1 or "basket" cells) may carry the information directly to the medulla, but their functional role is not established at present.

2. Indirect pathway: Lateral interaction. A network of interconnections amongst retinotopically organized neurons in the lamina provides an extensive anatomical basis for as yet functionally obscure lateral interactions between either receptor cell terminals and/or second order neurons (for review: Shaw, 1984).

Lateral inhibition as a means of spatial information processing is well documented from the hyperpolarizing responses of the large monopolar neurons (L1/L2) in flies and other insects (Zettler and Järvilehto, 1972; Laughlin, 1974; Dubs, 1982). Its effect is to enhance spatial contrast sensitivity. Typically, second order relay neurons from the lamina have much smaller receptive fields than the photoreceptor cells (R1-6) presynaptic to them. This finding comes from recording receptive field organization of receptor cells and comparing these with receptive fields recorded by the stimulation of postsynaptic neurons by presynaptic elements and comparing their respective maximum response amplitudes to single stimuli, though without taking the time course of the responses into account.

The cellular elements mediating this inhibition have remained unidentified, although anatomical evidence (Strausfeld, 1984; Shaw, 1984) suggests the likely candidates to be the L4 monopolar cell and the alpha cell (amacrine neuron), the latter having an intriguing role as a mediator between photoreceptor terminals and T1, and L4 neurons (for review see Strausfeld 1984). Recently, it has been suggested that the mechanism for lateral interaction is not necessarily through neuronal intermediaries, but could also be carried out by electrical current flowing through some low resistance channels or junctions between cells (Shaw 1984). Barriers that restrict ion-diffusion between single optic cartridges have been shown to exist in the lamina (Shaw, 1975; 1984; Järvilehto, et al., 1985) and these may provide a physical basis for interaction between photoreceptor terminals in adjacent cartridges using the principle of capacitive coupling. In the following sections, therefore, we have based our analysis on different transfer functions, recorded in comparable stimulus conditions.

Electrical responses

Quantum bumps - units of the receptor potentials. Information in retinal cells is carried in the form of electric voltage fluctuations over the membrane. The absorption of a photon is followed by a discrete membrane depolarization with an amplitude of a fraction of a millivolt. These unitary events are called bumps. The summation of these quantal events produces the actual receptor potential. The recorded amplitude of the smooth voltage wave is a function of time and stimulus intensity. The time course of the summated response, the receptor potential, seems be dependent on both the time course of the individual bumps as well as their latency.

Response analysis. We can considered the response analysis as looking at the same problem in two interchangeable ways, with respect to the time and frequency domains. If spatial distribution is taken into account the modal domain is also needed for a complete description of the system and no information is lost in changing from one domain to another. In the present context we shall be considering the time and frequency domain.

The time domain. This is the traditional way of observing recorded signals. The stimulus function can almost be of any kind e.g. impulse, triangle, sawtooth, sinusoid, ramp, exponentially increasing, random etc., but the step function has proven to be the most commonly used. A typical step response and its associated parameters are illustrated in Fig. 1.

Several characteristics of the step response make it universally acceptable. It is easy to generate, and it is easily modelled by differential equations, which are used to predict the system's time domain response, and a variety of determining factors in the performance of the system can be derived from using step-function stimuli. In particular, the peak value is commonly measured to characterize the response of the system.

In the case of a linear system the impulse response is the weighting function of the system, with which responses to all kinds of inputs can be generated, via an appropriate convolution. Its Laplace (or Fourier) transform fully characterizes the system in the frequency domain. In the linear systems analysis, the step response and impulse response can be easily calculated from each other. In biological systems the impulse stimulus is in practice often set at a constant level and the response level is additionally influenced by some additional complex, but controllable adaptational non-linearities.

All amplitude parameters of recorded biological signals are strongly related to time which alone is not a parameter, but in the response measured may show important biological relevance. Measurement of different kinds of time-dependent parameters indicate the system's reaction time (see e.g. Weckström and Järvilehto, 1986). Thus, it is possible to decompose the complex time record into its components and consider them with respect to an additional third axis, the frequency.

The frequency domain. This represents the recorded signal as a set of sine waves with different amplitudes and phases. Using this representation results neither in a gain nor loss of information, but merely represents the signal differently. The frequency domain analysis has several advantages over the time domain analysis. It allows us to measure the frequency response function, which represents the ratio of the system's output-to-input in the frequency domain, and thus fully characterizes a stable and linear time-invariant physical system, which is assumed but not present in the real world. Using frequency responses it is also easier to predict responses to various kinds of stimuli than it would be if only time domain (i.e. convolution) were used.

White noise analysis has become one of the standard methods in the characterization of a control system (Marmarelis and Marmarelis, 1978) and this method can also be applied to sensory systems. Both linear and non-linear system analysis and identification with Wiener kernels can be applied to systems subjected to a randomly fluctuating input signal. When frequency response functions are calculated using cross-spectra, another function - the coherence function - can be obtained as a by-product. The coherence function is a normalized measure of linear correlation (as a function of frequency) between the system's input and output signals. Both system non-linearities and inherent noise diminish the coherence from a maximum value of 1 towards a minimum of 0, and their presence is thus indicated by a non-unity coherence.

Generally, when the input signal is periodic, time domain averaging can be used to increase the signal-to-noise ratio and to discriminate between noise and non-linearity. As pointed out by French (1980a), time domain averaging can also be used when identical sequences of pseudorandom noise are applied as stimuli.

Utilizing this concept we have developed a system for the measurement of visual transfer function of photoreceptors and higher order neurons. Because averaging suppresses only the non-correlated noise, but not the "harmonic noise" due to nonlinearities, the coherence function obtained using time domain averaging is more indicative of the system's non-linearity than the signal-to-noise ratio. The white noise stimulation method has also the advantage that it mimics natural conditions allowing at the same time quite accurate control of the system's adaptational state.

Noise. Every system is subject to variable sized random inputs, "noise", which behave as if they consisted of periodic inputs of all frequencies and phases. All levels of the visual signal processing are contaminated by noise from different sources, including noise in photoreceptors (e.g. photon noise, bumps), transmitter release noise in synapses and neuronal membrane noise. All these obscure the real nature of the signals when system analytical methods are used. One should bear in mind that the linearity is assessed by means of a coherence function, which is decreased by noise as well as by non-linearities.

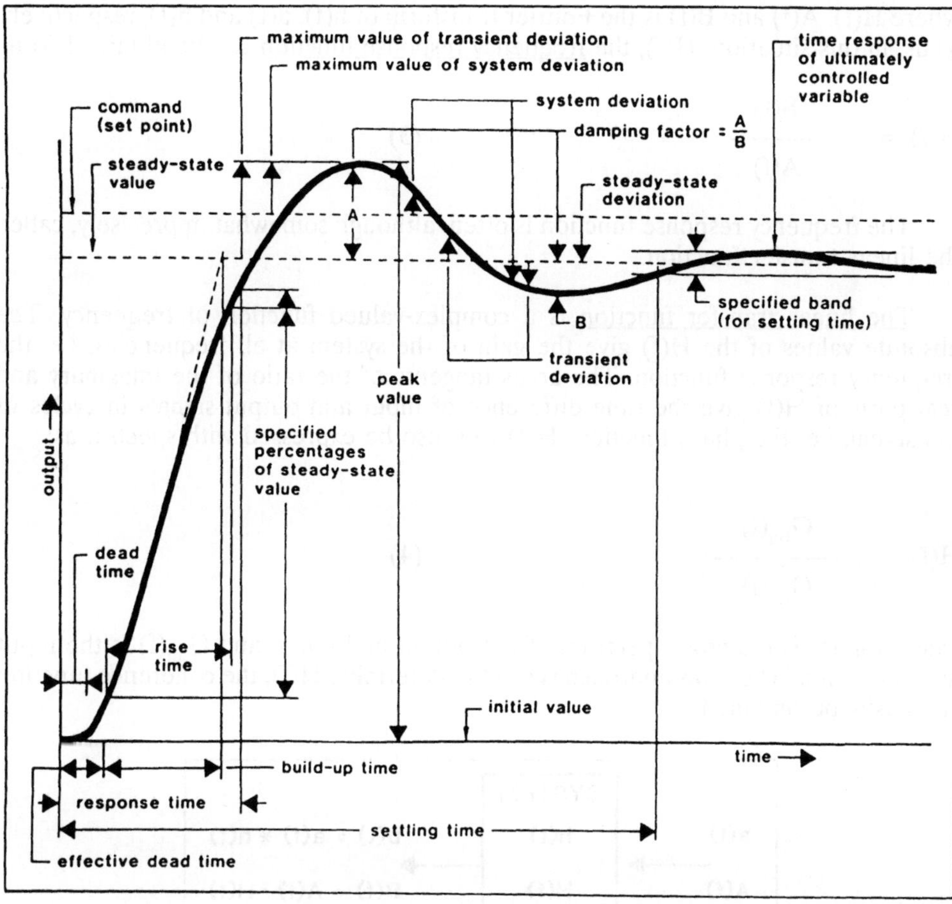

Fig. 1. Generalized time response of a system to a step increase of input. (Modified from the American National Standard ANSI MC85.1M-1981. Terminology for Automatic Control)

THEORY AND METHODS

Analytical approach

To help in understanding the meaning of the frequency response function H(t), Fig. 2 shows the relationships between the input signal a(t) and output signal b(t) for a stable, linear, time-invariant system ("ideal" system) in the absence of noise.

The system is characterized by its impulse response h(t) and as described in connection with following equation, the output signal b(t) is the convolution (denoted "*") of a(t) with h(t), thus:

$$b(t) = a(t) * h(t) \qquad (1)$$

By the convolution theorem, it follows that

$$B(f) = A(f) \cdot H(f) \qquad (2)$$

where H(f), A(f) and B(f) is the Fourier transform of h(t), a(t) and b(t), respectively. Thus, in this situation, H(f), the frequency response function can be obtained from:

$$H(f) = \frac{B(f)}{A(f)} \qquad (3)$$

The frequency response function is often, although somewhat imprecisely, called the linear transfer function.

The linear transfer function is a complex-valued function of frequency. The absolute values of the H(f) give the gain of the system at all frequencies, i.e. the frequency response function. The arcus tangents of the ratio of the imaginary and real parts of H(f) give the time difference of input and output signals in cycles of the signal, i.e. the phase function. H(f) can also be expressed with spectra, as:

$$H(f) = \frac{G_{ba}(f)}{G_{aa}(f)} \qquad (4)$$

where $G_{ba}(f)$ is the cross-spectrum of the output and input, and $G_{aa}(f)$ is the input autospectrum. When this approach is used to determine H(f), the coherence function can easily be obtained.

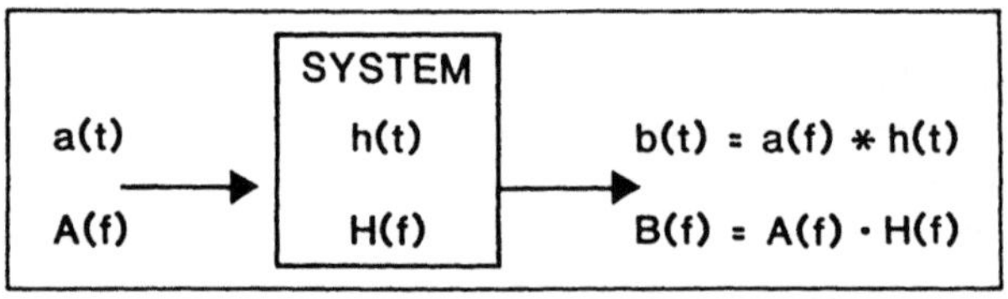

Fig. 2. Input - output relations for a linear system shown as a block scheme.

The coherence function is defined by

$$\gamma^2(f) = \frac{G_{ba}(f) \cdot G^*_{ba}}{G_{bb}(f) \cdot G_{aa}(f)}$$

The value of the coherence function is unity in a linear noise-free case. Non-linearities and measurement of noise diminish this value. The coherence function is thus a measure of the linearity and noisiness of the system. If non-linearities are present, complete identification of the system requires the use of more sophisticated methods such as Wiener kernel analysis. Nevertheless, the non-linearities can be seen in time the domain in certain cases. They are also seen as harmonic distortions, where the higher order non-linearities are seen as successively higher harmonics of the fundamental (stimulating) frequency.

Large amounts of noise in intracellular voltage recording from LMCs is generated in the synapses themselves between photoreceptors and the neurons. It can be estimated that about 50% of the total noise in LMCs originates in the synapse (Laughlin et al., 1987). Thus the time domain averaging should increase the values of the coherence much more prominently in case of LMCs than in photoreceptor responses.

Experimental approach

Stimulation. Pseudorandom noise (spectrum flat within 0.1-200 Hz) from a shift register type generator was led to a light emitting diode (LED, Siemens LG 5411, green, with peak emission at 555 nm) via a controlled current source in order to stimulate the receptor cell. The current passing through the LED was recorded to monitor the instantaneous luminous output of the diode. Impulse stimuli (white, duration approx. 10 us with 2.2×10^{12} quanta/cm^2, given to a dark adapted eye), were produced by a xenon flash tube, the intensity of the pulses being attenuated over four decades with neutral density filters.

Experimental animal and the recording techniques. The intracellular voltage responses were recorded from the retinular cell bodies in the retina or retinular cell axons or LMC's in the lamina of the wild-type blowfly *Calliphora erythrocephala*. Recordings were made with glass capillary microelectrodes filled with 3 M KCl, having resistances of 100 to 300 megaohms as measured in the tissue. The stimulus and response were sampled at 1 kHz and input, output, and cross spectra were calculated by standard FFT techniques. The transfer and coherence functions were subsequently calculated. A more detailed account of the methods is given elsewhere (Weckström et al., 1988a; Kouvalainen et al., 1988).

RESULTS

Responses of the receptor cells in the time domain. We have recorded the impulse responses of receptor cells at different adaptational conditions. Fig. 3A shows a set of responses in dark adapted conditions. The time domain signals are usually analyzed with respect to their maximum amplitude. This is shown in Fig. 3B. The resulting function is called the operating function of the cell. One should bear in mind that this resulting function is characteristic for the particular cell only in respect to the analyzed voltage parameter. Many other time response parameters may be analyzed, but usually only the latency and rise time are shown. The linearity of the system cannot be assessed from these functions, although it can be made to appear as such if fitting dimensions for axes are chosen.

Responses in frequency domain. A typical transfer function of a photoreceptor cell obtained with white noise analysis is shown in Fig. 4 where the luminous intensity of the LED stimulus is about 50 mcd mean and 0-100 mcd peak-to-peak. It shows underdamped behaviour with a small resonance at approximately 30 Hz. The slope of the high frequency asymptote is about 30 dB/octave. This indicates a five pole (6 dB/octave indicates a pole) system, which can be treated as a cascade of linear low-pass filters. The mean corner frequency is at 63 Hz (SD +/- 12 Hz, n=13).

The phase curve should display a high frequency asymptote to -450°, if the system is a minimum phase system, i.e. contains no delay elements. Instead, the phase curve decreases beyond this point until the variance of its estimate grows too high. The coherence curve shows remarkably high value between the frequencies from 0.5 Hz to 100 Hz and remains over 0.5 until 120 Hz (mean 121 Hz, SD +/- 32 Hz, n=13). This indicates the high signal-to-noise ratio as well as the high degree of linearity of the system under these stimulation conditions.

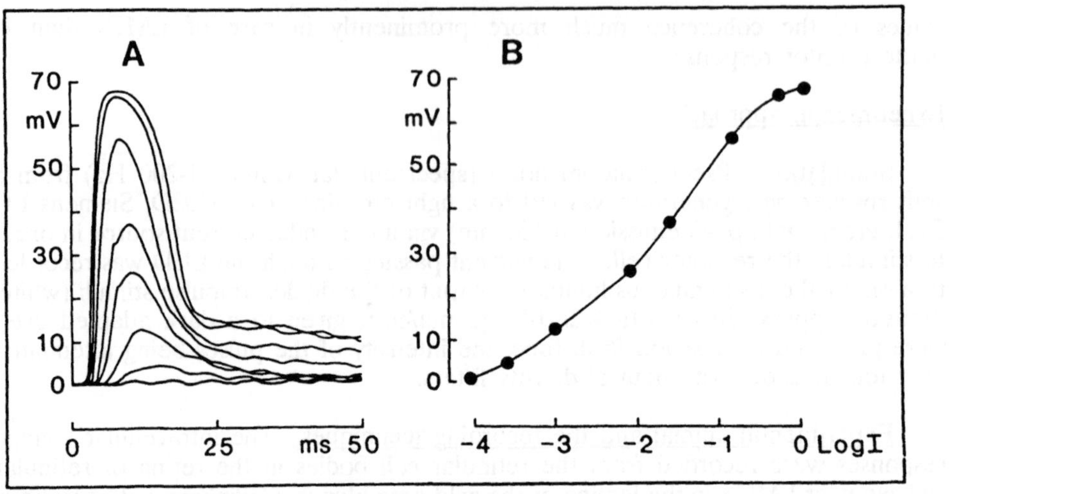

Fig. 3A. Intracellularly recorded impulse responses elicited by a 30 μs white flash in a set of different stimulus intensities, in relative values 100, 49, 14, 1.5, 0.21, 0.023 and 0.008. B. The corresponding V_{max} - Log I function of the recorded receptor cell response.

A recording is shown from a green sensitive cell (probably R1-6) with two different adaptational stages and with the same intensity modulation, but at a different mean intensity level (Fig. 5). The gain functions are normalized to the same dB-scale. The differences in gain functions are found in high frequencies. The light adaptation decreases the coherence values, especially in frequencies lower than 20 Hz. When the signal-to-noise ratio can be regarded as constant, as in these experiments, the decrease of coherence values indicates an increase of non-linearities, or some non-coherent input influencing the recorded receptor response. The phase-lead caused by light adaptation (Fig. 5. phase plot) is found between 0.5 to 5 Hz in these stimulus conditions. It can be also be clearly seen stimulating with a 1 Hz sine-modulated light from 0 to 100 mcd of intensity. The amplitude maximum lead (time difference) in that case is about 47 ms, which gives a corresponding phase lead of 17°.

Responses in frequency domain: Model for the first synapse. The second order relay neurons (L1, L2 and L3) receive inputs from the terminals of R1-R6 in the lamina. Neuronal voltage response to presynaptic depolarization is a hyperpolarization. This is seen also in the phase plot of the transfer function as a 180° phase shift. The monopolar response amplitude is sensitive to the rise of the presynaptic response. This is also seen in the comparison of the peak times of the response maximum amplitudes. The synaptic transformation of the signals conveyed by the photoreceptors has been proposed to be like that made by a high-pass filter (French and Järvilehto, 1978).

The linear transfer function of the synapse is obtained by subtracting the computed photoreceptor transfer function from that of a monopolar cell response. It shows high-pass characteristics, which cannot be accounted for with conventional passive models, as the low-frequency asymptote has a slope less than 20 dB/decade. We propose here a model -- a fractional order differentiator -- which contains only two parameters. The output of such a system is described by:

$$y(t) = \frac{d^k x}{dt^k} \qquad (6)$$

where x(t) and y(t) are input and output signals as functions of time and k, the order of differentiation, is not necessarily an integer. In the frequency domain, k can be easily calculated from experimental data as:

$$\log G(f) = k \log f + A \qquad (7)$$

where G(f) is the gain of the synaptie transfer function and A is a constant defining the gain at 1 Hz.

The transmission of noise from photoreceptors to the monopolar neurons is affected by the described transfer characteristics of the synapse between them. These characteristics might indeed be one major reason for the large values of noise variance obtained from the LMC recordings.

Responses in frequency domain: Lateral interaction. The photoreceptors in the same ommatidium share the same optical apparatus but their optical axes diverge by about 2°. The axon terminals from six different ommatidia having cells with the same optical axes converge to form the primary elements of a lamina subunit, called a neurommatidium or optic cartridge. Conversely, the receptor axons from the same ommatidia are distributed to six different neurommatidia. The receptive fields of fly photoreceptor cells have a half-width of about 2° to 4°.In two of the second order neurons, L1 and L2, which are postsynaptic to photoreceptors and respond to by phasic hyperpolarization to incremental light increase, their receptive field is about 1° to 2° narrower. This finding is based on the analysis of the maximum amplitudes of responses to stimulation with short light pulses.

As shown previously, the synapse between the receptor and LMC acts like a frequency dependent amplifier (Järvilehto and Zettler, 1971). Thus all changes in the time course of the presynaptic response with constant maximum amplitude will be reflected in an amplification of the postsynaptic response. Should the presynaptic responses have no mutual interaction displayed in response to the time course, the response time course should stay invariant with respect to all intensity compensated changes in the stimulus location within the receptive field. This is also true if the recording is made from the receptor cell body, but the difference is easily seen even in the pulse responses recorded from the axon terminals in equivalent stimulus conditions. This is the reason for analyzing the lateral interaction in the frequency domain.

The lateral interaction between the receptor terminals in different neurommatidia was studied using both on-axis, 2° off-axis and 4° displaced off-axis stimuli, i.e. stimulation from the periphery of the receptive field of the recorded cell. To avoid the effects of changing the operating point or the adaptational state of the cell, it was important to adjust the intensity of the light pulses as well as the intensity of the white noise modulated light to elicit voltage responses of equal amplitude modulations in both the on-axis and off-axis recordings (mean: 3 mV, modulation: 5 mV peak-to-peak). The recordings originating from axon terminals of the short R1-R6 photoreceptor as identified by their response properties and by iontophoretic dye injection. The stimulation was applied at the centre of the receptive field to obtain transfer functions of the fly photoreceptor cells (as shown in Figs. 4 and 5) as well as at 2° and 4° displaced from the center (Fig. 7).

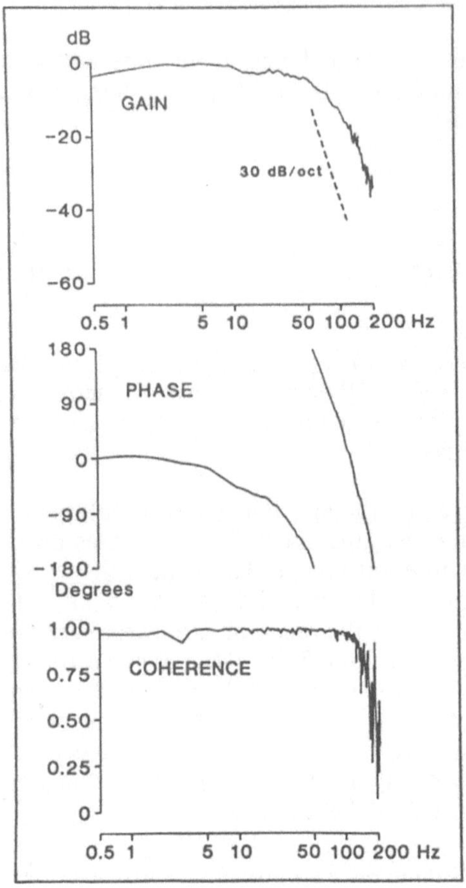

Fig. 4.

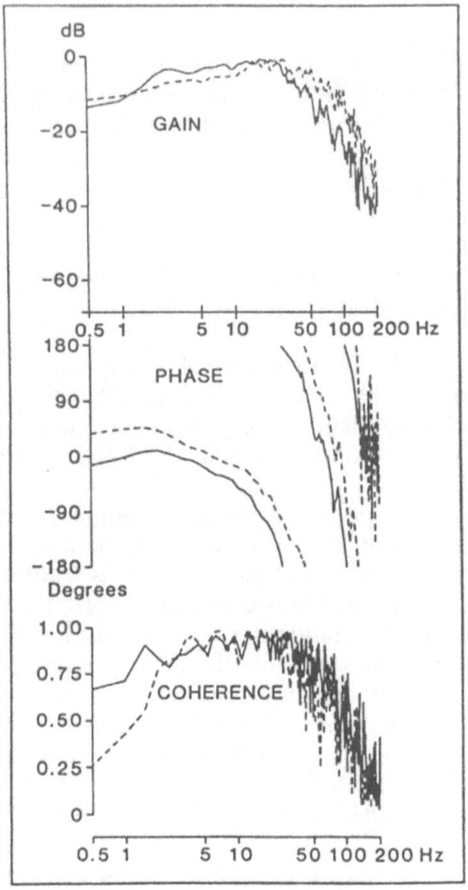

Fig. 5.

Fig. 4. The transfer function of a photoreceptor cell with the gain, phase and coherence functions. The broken line with a slope of 30db/octave is shown as a reference. The record is obtained with seven time averages (and with an appropriate segment average), and with luminous intensity modulation (as LED output) from 0 to 50 mcd. The arrow denotes the small resonance frequency. The coherence function shown as a measure of the linearity of the system.

Fig. 5. The transfer functions of a photoreceptor cell (an other cell than in Fig.4) at two different adaptational states. The continuous line: Low stimulus intensity, modulation from 0 to 50 mcd, the mean intensity 25 mcd. The broken line: High stimulus intensity, modulation from 50 to 100 mcd, the mean intensity 75 mcd.

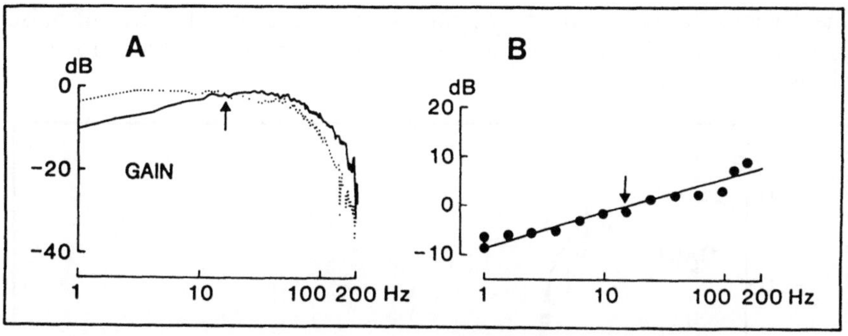

Fig. 6. A. The gain of the linear transfer functions of a photo-receptor (dotted line) and a photoreceptor-monopolar neuron system (solid line) obtained by using white noise as stimulus with time-averaging. B. The difference function calculated from the normalized transfer functions above showing the transfer function of the multiple site synapses between photoreceptor and monopolar neuron. The line denotes a fit of the present model with Log G(f) = 0.36 log (f) + 0.43, whereby 0.36 is the estimate of the differentiation order in question with 0.43 as only a relative value of the normalized functions. The correlation coefficient of the line is 0.97.

The transfer functions of the photoreceptors have a corner frequency averaging about 63 Hz +/- 12 Hz, n=14). As seen in Fig. 7 the transfer function obtained with 4°-displaced stimulation differs from the on-axis one, and shows a decrease in corner frequency averaging about 21 Hz SD +/- 3Hz, n=7, $p<0.001$, with Student's paired t-test), whereas the transfer function does not decrease significantly with 2°-displaced stimulation. The phase plot shows increasing phase-lag in the case of 4° off-axis-induced responses (the difference averages 25°, SD +/- 5° at 15 Hz, $p< 0.001$). The coherence function in our recordings ranges from >0.5 to as high as about 120 Hz.

The attenuation of voltage responses at higher frequencies is also seen in the impulse responses obtained with equal response amplitudes, whereby the time-to-peak is delayed by an average of 15% (SD +/- 8%, $p<0.001$), which means 2.3 ms in responses of about 10 mv of amplitude.

DISCUSSION

Systems analysis originated from problems arising from man-made control systems, where the principles of the construction are well known. In contrast, in biological systems, such as the insect's compound eye and optic lobes, the design principles are not known beforehand. In this study, the function of photoreceptors and postsynaptic second order cells, the LMCs, were studied by systems analysis. This approach to the experimentally acquired data can, in the best of all possible worlds, serve only as a model to envelope the observed phenomena. The best model is the object itself, because it shows its intrinsic predictability. The worst model is an internally contradictive one. Since the goal of analysis and modelling is to avoid such contradictions, a good model with known limitations exhibits a high predictability of the observed phenomena.

One obvious limitation to linear systems analysis is the presence of noise from various sources. By definition, uncontrolled noise is non-coherent and appears in the analysis as a non-linearity of the system.

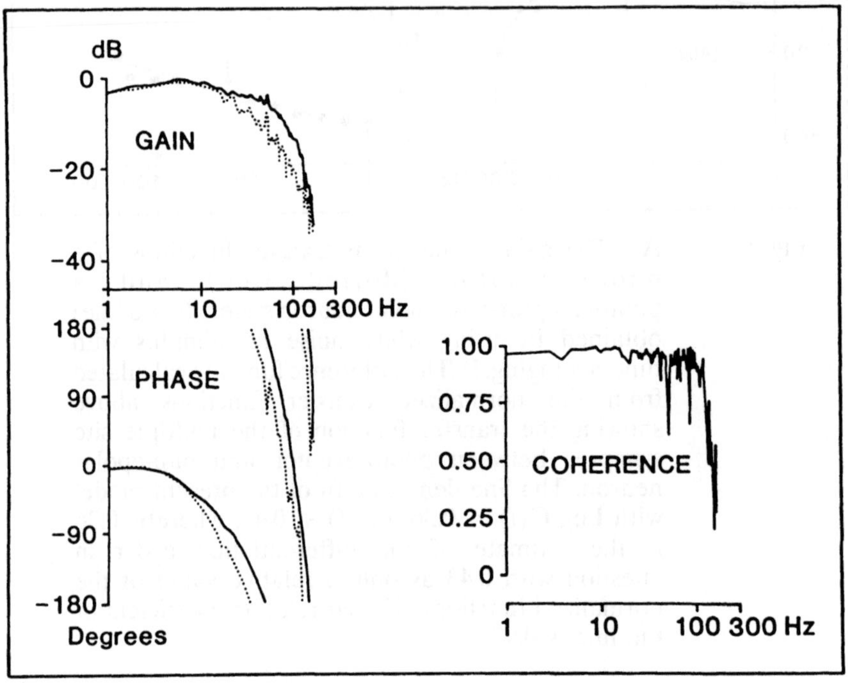

Fig. 7. Gain and phase plots of the transfer function of the blowfly photoreceptor cell, obtained by stimulation from the centre of the receptive field (continuous line) and from the periphery, stimulus displaced 4° (dotted line). The responses are adjusted by the stimulus intensity to match the amplitude at different angles of incident light. Note the shifts in gain and phase plots towards the lower frequencies.

However, time averaging is a powerful method to reduce noise. Even if, as here, pseudorandom input signals are used for the analysis of the system, the identical time sequence can be repeated and averaged. In this manner noise from external sources is reduced (compare in Fig. 4 with the coherence function in Fig. 5). The high coherence value in Fig. 4 shows that the system is more linear than previously argued. The non-coherent inputs from various other sources e.g. random arrival of photons, amplifying transducer mechanism (intrinsic noise), opening and closing of ionic channels, additional random release of transmitters, and many other as yet uncharacterized processes are important system properties. Depending on the time course of the processes, they may become identifiable during the analysis. One of the adaptation processes is the pupil mechanism, which contributes to non-linearities, has a time constant of several seconds and evidently does not affect the transfer function in our recording time (Franceschini, 1972; Leutscher-Hazelhoff, 1975).

Transfer function. Several internal mechanisms of phototransduction contribute to the transfer function. We may surmise that these include the following: conformational changes in the photopigment molecule; numerous enzyme activated reactions in the microvillar membrane; diffusion; opening of sodium channels in the photoreceptor membrane with the concomittant production of rapid inward sodium currents manifested as a receptor potential. The system's order, as determined here, is based on the maximum (asymptotic) slope of the gain plot with the consequence that the order may be higher than actually observed because additional very fast events are not detected by the necessarily limited bandwith of the analysis.

The mean lifetime of the light-activated channels seems to be of the order of 16 ms, which corresponds to a frequency of about 10 Hz. The lifetime of the channels is a limiting factor in that it essentially performs low-pass filtering to the signals. Thus, the first pole seen in the transfer function, i.e. the first downward turn of the curve, is argued to arise from the channel kinetics (Heimonen et al., 1988).

The resonance in the transfer function has been attributed to the dynamic control processes leading to release of the internal transmitter or, more specifically, to feedback inside the transduction chain (French, 1980a,b). This hypothesis is supported by the evident accentuation of the resonance in more light-adapted conditions. When a larger degree of modulation is used, the system's behaviour progressively changes to a non-linear one in the low frequency end of the spectrum, caused by the adaptive regulation. However, there seems to also be another source of resonance behaviour. The photoreceptor membrane may contain voltage dependent conductances, which can cause oscillations in the voltage response, and which can also be activated with depolarizing current injections. Both mechanisms contribute to the transfer function. These properties of the frequency response function are even more accentuated in the first synapse because of its differentiator function. The mean level of illumination seems not to be a very important signal in the information processing of the compound eye, and thus it is removed (Laughlin & Hardie, 1978).

The first synapse. The major information transmission channel from the retina to the deeper optic lobe neuropil -- the medulla -- is via the first synapse between R1-R6 and LMCs. This pathway is parallel to a direct route from the compound eye to the medulla, carried by the R7 and R8 receptors. The signal transformation in the R1 to R6-LMC synapse can be simply modelled by fractional differentiation, which is a form of frequency dependent amplification as previously discussed by Weckström et al. (1988b). Fractional differentiation is a linear operator and results, as predicted, in the near-unity coherence values, if noise is avoided with time-domain averaging. This modelling applies equally as well to dark- as to feebly light-adapted systems.

The low-modulation white-noise stimulation used in this study can be argued to represent quite normal stimulation conditions of the fly photoreceptors, as the contrast of natural signals is usually very restricted (Laughlin, 1981). However, the fly is a diurnal animal and point-like flashes or dim flickering light by no means comprise the whole scale of the functions of its compound eye. It ca be argued that if the system is more light-adapted, many mechanisms controlling the total sensitivity (to avoid saturation of the system) and lateral information transfer (to process spatial as well as temporal imaging) are activated. This implies additional synaptic inputs to both photoreceptors and LMCs in the synaptic region of the lamina. Laughlin et al. (1987) found that the non-linearities in LMC signals (in light adapted conditions) are of the order of 17 dB lower than the linear part of the response (which corresponds to a coherence value of 0.86 in totally noise-free, i.e. ideally averaged cases). The observed non-linearity seems to be associated with the responses of the LMCs to cessation or decrease of light stimulation, the so-called "off-response" (Djupsund et al., 1987; Laughlin and Osorio, in preparation; Weckström et al., 1988a; Hardie et al., this volume). A component of the off-response has been attributed to inputs other than photoreceptor synapses. Thus it seems that in light adapted-conditions the function of the peripheral compound eye progressively increases in complexity, and non-linearities become more pronounced. The basic signal transformation procedures, the linear low-pass filtering in photoreceptors and linear frequency-dependent amplification in the photoreceptor-LMC synapse, do not disappear but are probably not as dominating.

The additional inputs into photoreceptors and LMCs also narrow the receptive fields of LMCs, thus sharpening the neural image. Our results indicate that at least part of this lateral information processing takes place presynaptically. Specifically, it has the form of capacitive interaction, by which high-frequency components of the photoreceptor signals are attenuated. When just those high-frequency signals are relatively amplified in the process of photoreceptor light-adaptation, and the first synapse acts a differentiator, this pre-synaptic phenomenon becomes very effective in reducing the receptive fields of LMCs (Zettler and Järvilehto, 1972). Although this high-frequency inhibition may result from capacitive interaction via glial cell barriers, it is surely not the only form of lateral information processing in lamina. Preliminary reports of the different conductances activated during the LMC voltage response suggest that additional synaptically mediated mechanisms may play a role in lateral interactions (Hardie et al., this volume), something that is not surprising considering the wide variety of lateral synaptic contacts found in lamina (Strausfeld, 1984).

The source of the fractional order differentiator function of the first synapse is somewhat enigmatic. It seems that the characteristic curve of mere amplitude transmission is similar to previously studied graded chemical synaptic transmission (French and Järvilehto, 1978; Laughlin et al. 1987). Thus, it would appear that this frequency dependent transmission cannot be attributed to the photoreceptor-LMC synapse itself, but must result from some interactions inside the lamina cartridges. This suggests specific hypotheses on which further investigations can be based. (1) One such mechanism could be the proposed voltage dependent (calcium?) channels in the photoreceptor terminals. This would tend to accentuate high frequencies. However, since these would also saturate the first synapse very easily, they are not very suitable candidates. They can, however, participate in producing the oscillatory behaviour of the impulse response. (2) Various functions have been attributed to the lamina depolarization. However, its generation is far from clear. If it is simply restricted to within the cartridges (neurommatidia), it could play a part in generating the transient characteristics of the LMCs. The lamina depolarization can be proposed to be subtracted from voltage signal of photoreceptor terminals, albeit with a time delay, thus differentiating the postsynaptic response. However, the LMC responses in the lamina (and ocellar neuropil) of other species are very similar to those in *Calliphora*. If the action of lamina depolarization seemingly requires a very special kind of anatomy typical of higher dipterans, employing resistive glial barriers, the stated lamina hyperpolarization hypothesis is considerably weakened. (3) Perhaps the best candidate is the control of the LMC voltage by some indirect pathway, e.g. via amacrine cells, where the simplest model could then include one excitatory and one inhibitory synapse having different frequency characteristics, thus producing a frequency dependent attenuation of the LMC voltage. Of course, this kind of synaptic pathway is highly speculative, and more powerful stimulation techniques are needed to corroborate or falsify any of these hypotheses.

Lateral inhibition. The lateral inhibitory interaction, if present, is summed with the normal photoreceptor responses (Fig. 8), and the transfer functions should be interpreted accordingly. What we have determined, therefore, is the transfer function of the photoreceptor G_{Tx} (i.e. with on-axis stimulation, and equally the off-axis stimulated photoreceptor, $G_{Tx'}$). In this case the inhibitory element can be shown to have a transfer function G_i equal to

$$G_i = 1 - \frac{G_{Tx'}}{G_{Tx}} \qquad (8)$$

The Gaussian distribution of intensities due to the optics of the ommatidial receptive field causes reduction of stimulation intensity at a particular point. This can be compensated by normalizing the transfer functions. If the transfer functions are identical no frequency dependent inhibitory element (G_i) is present. The results show that a band of higher frequencies in the frequency response functions is reduced relative to other frequencies. This indicates that the photoreceptor terminals are mutually inhibited. Inhibition becomes significant, not in the responses obtained with a 2°-displacement of the stimulus, but with 4° off-axis stimulation cells. As the pattern of the photoreceptor axon projections to lamina cartridges indicates, this means that inhibition is effective when neighboring neurommatidial cartridges are sufficiently activated. These results are also supported by the finding that if the transfer functions are measured distally in the photoreceptor cell bodies, where absorbed quanta are univariant and the increase of stimulus intensities in any location of the receptive field produces no variance in measured transfer functions. This also explains the negative finding of any effect in comparison of field and point stimulation recorded from the cell bodies as reported earlier (French and Järvilehto, 1978).

The lateral inhibition shows capacitive characteristics, in the sense that the frequency response is attenuated only in higher frequencies. A morphological substrate for capacitive lateral inhibition is found in the neurommatidia (cartridges) pof the lamina, where photoreceptor axons synapse onto large monopolar neurons. The cartridges are isolated from each other by epithelial glial cells (EGCs: Shaw, 1975; 1984). If this isolation forms a high-enough resistance barrier, it will act like a capacitive element between neighboring cartridges. This is shown in the electric analogue of Fig. 9. Although by no means complete, the model demonstrates how such a high-pass coupling may be achieved in practice. The current flowing through the EGC can be shown (in this model) to have a frequency response (G_i) equal to:

$$G_i = \frac{j\omega C(R_2R_4) + R_2}{j\omega C(R_2R_4 + R_3R_4) + R_2 + R_3 + R_4} \qquad (9)$$

From this two time constants:

$\tau_2 = R_4C$ and $\tau_1 = \tau_2(1/[1+R4/(R_2+R_3)])$

can be derived, which define the corner frequencies of the start and leveling-off of the inhibitory influence. If the values of the mediating RC- element (R_4 and C in the model) are evaluated, we found that if the capacitance has values of 1 to 10 $\mu F/cm^2$, then the resistance is 10 to 1 kohms x cm^2. These match the range of values for glial cell membranes.

Capacitive lateral inhibition, as reported here, represents a new type of spatial information processing in the retina. Capacitive coupling is a very suitable means for spatial information processing. By emphasizing the differentiator function of the synapse, inhibition indirectly reduces the amplitude of the hyperpolarizing responses in the large monopolar neurons, as shown by the decrease in the width of the receptive field of these neurons. Contrast sensitivity is thereby enhanced. Because the structural basis for capacitive coupling is not complex, it could possibly be one general method of spatial information processing in tightly packed neural subsystems. It is also conceivable that the inhibiting capacitive coupling may be mediated by a neural network which would enable integrative properties of synapses, for example, to cause frequency-dependent lateral inhibition.

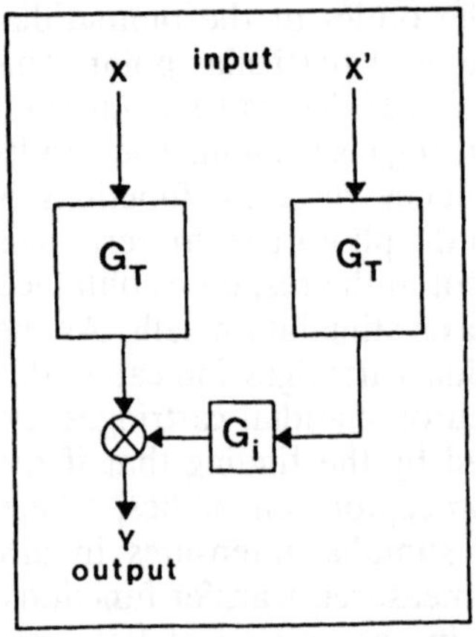

Fig. 8.

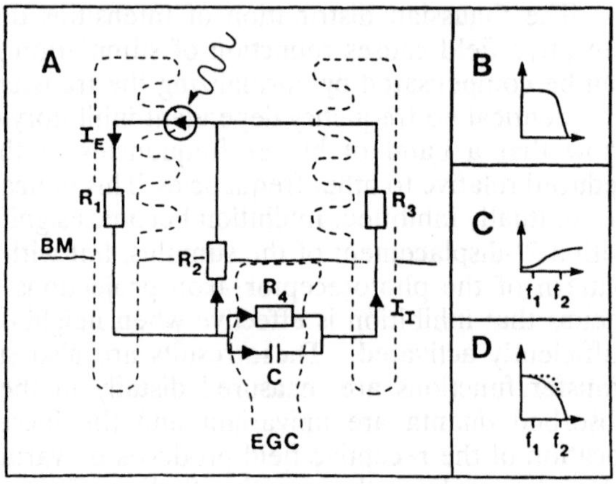

Fig. 9.

Fig. 8. Block diagram of the elements involved in lateral inhibition. X, X', and Y are the recorded on-axis input, off-axis input, and output signals, respectively. G_T is the LTF of transduction in photoreceptors, and G_i the transfer function of the element mediating the inhibition from neighboring cartridges.

Fig. 9. A. Two-dimensional electrical analogue of photoreceptor cells coupled to each other via an EGC barrier, which forms the high resistance and consequently the capacitive coupling element. The capacitances of membranes other than those of the EGCs have been omitted for clarity, but this does not affect the analysis in principle. B. Photoreceptor transfer function. C. Transfer function of the element mediating the inhibition. D. Effect of inhibition on the photoreceptor transfer function.

Conclusion. The analysis in the frequency domain demonstrates that peripherally in the insect visual system there exist several powerful information-reduction and processing mechanisms. An interesting example is a capacitive-like interaction in the presynaptic sites amongst different lamina cartridges. As shown here, the synapses between photoreceptors and neurons possess properties of a frequency dependent gain control. The output of these neurons (the responses of which are very phasic) can be regulated efficiently by selective changes of the presynaptic frequency response function as performed by the proposed interaction. Although this kind of information processing turns out to be especially suitable for contrasting objects moving at a certain frequency band, it would still allow all additional variations of synapse-based information processing at different levels of the visual system. Fly behaviour suggests that its vision is highly frequency dependent and thus matches the accuracy of processing in accordance with variations of the light intensities in the animal's habitat.

Acknowledgements: We would like express our warmest thank to our collaborators J. Korkalainen and K. Heimonen for valuable suggestions and discussions as well as A. Rautio for technical help during completing this work.

REFERENCES

Djupsund K, Kouvalainen E and Weckström M (1987) Light-elicited responses of monopolar neurones in the compound eye in the blowfly studied with a current clamp method. J Physiol 390:165P.

Dubs A (1982) The spatial integration of signals in the retina and lamina of the fly under different conditions of luminance. J Comp Physiol 146:321-343.

Franceschini N (1972) Pupil and pseudopupil in the compound eye of *Drosophila*, In: Wehner R (ed) Information processing in the visual system of arthropods. Springer, Berlin, pp 25-82

French A S (1980a) Phototransduction in the fly compound eye exhibits temporal resonances and a pure time delay. Nature 283:200-202.

French A S (1980b) The linear dynamic properties of the phototransduction in the fly compound eye. J Physiol 308:385-401.

French A S and Järvilehto M (1978) The transmission of information by first and second order neurons in the fly visual system. J Comp Physiol 126:87-96.

Hardie R C (1985) Functional organization of the fly retina. In: Ottoson D (ed) Progress in sensory physiology. Springer, Berlin, pp 1-79

Hardie R C (1987) Is histamine a neurotransmitter in insect photoreceptors? J Comp Physiol 161:201-213.

Hateren van J H (1987) Neural superposition and oscillations in the eye of the blowfly. J Comp Physiol A 161:849-855.

Heimonen K, Weckström M, Kouvalainen E, Korkalainen J, Järvilehto M (1988) Quantum bumps and membrane noise in fly photoreceptors. Acta Physiol Scand Suppl.

Järvilehto M (1985) The eye: Vision and perception. In: Kerkut G A, Gilbert LI (eds) Comprehensive insect physiology, biochemistry and pharmacology, Pergamon Press, Oxford

Järvilehto M, Meinertzhagen I A and Shaw S R (1985) Diffusional restriction and dye coupling in insect brain slices. Soc Neurosci Abstr 11/1:240.

Järvilehto M and Zettler F (1971) Localized intracellular potentials from pre- and postsynaptic components in the external plexiform layer of an insect retina. Z Vergl Physiol 75:422-440.

Kouvalainen E, Weckström M, Korkalainen J, Heimonen K and Järvilehto M (1988) Measurement of visual transfer function with pseudorandom stimulus using time domain averaging. In: Malmivuo J, Nousiainen J. (eds) Proc VI Natl Meet Biophys & Med Eng, Helsinki, pp 189-192

Laughlin S B (1974) Neural integration in the first optic neuropile of dragonflies. III. The transfer of angular information. J Comp Physiol 92:377-396.

Laughlin S B (1981) Neural principles in the peripheral visual systems of invertebrates. In: Autrum H (ed) Handbook of sensory physiology, Vol VII/6B. Springer, Berlin, pp 135-280

Laughlin S B (1987) Form and function in retinal processing. TINS 10:478-483.

Laughlin S B and Hardie R C (1978) Common strategies for light adaptation in the peripheral visual systems. J Comp Physiol 128:319-340.

Laughlin S B, Howard J and Blakeslee B (1987) Synaptic limitations to contrast coding in the retina of the blowfly. Proc R Soc Lond B 231:437-467.

Leutscher-Hazelhoff J T (1975) Linear and non-linear performance of transducer and pupil in *Calliphora* retinula cells. J Physiol 246:333-350.

Marmarelis P Z and Marmarelis V Z (1978) Analysis of physiological systems: The white noise approach. Plenum, New York.

Shaw S R (1975) Retinal resistance barriers and electrical lateral inhibition. Nature 255:480-483.
Shaw S R (1984) Early visual processing in insects. J Exp Biol 112: 225-251.
Strausfeld N J (1979) The representation of a receptor map within retinotopic neuropil of the fly. Verh Dtsch Zool Ges 167-179.
Strausfeld N J (1984) Functional neuroanatomy of the blowfly's visual system. In: Ali M A (ed) Photoreception and vision in invertebrates. Plenum, New York, pp. 483-522
Weckström M and Järvilehto M (1986) Photoreceptor latency: Analysis and definition. Exp Biol 45:45-54.
Weckström M, Kouvalainen E and Järvilehto M (1988a) Non-linearities in response properties of insect visual cells: an analysis in time and frequency domain. Acta Physiol Scand 132: 103-113.
Weckström M, Korkalainen J, Kouvalainen E, Heimonen K, Torkkeli P and Järvilehto M (1988b) A model for information transmission across the first synapse in the fly visual system. In: Malmivuo J, Nousiainen J. (eds) Proc VI Natl Meet Biophys & Med Eng, Helsinki, pp 185 - 188p 185-188
Zettler F, Järvilehto M (1972) Lateral inhibition in an insect eye. Z Vergl Physiol 76:233-244.

DYNAMICS OF SECOND-ORDER NEURONS OF COCKROACH OCELLI

Makoto Mizunami

Department of Biology, Faculty of Science
Kyushu University, Fukuoka 812, Japan

ABSTRACT

The incremental responses from second-order ocellar neurons of the cockroach, Periplaneta americana, were measured. The responses consisted of two components, graded potentials and spikes.

Dynamics of the graded potential responses were studied using white-noise-modulated light with various mean illuminances. I found that (a) white-noise-evoked responses were linear, (b) the responses were an exact Weber function, that is the contrast sensitivity remained unchanged over a mean illuminance range of 4 log units, and (c) the dynamics of the responses remained unchanged over the same range of the mean illuminance. I conclude that (a) the graded potentials linearly encode contrast of fluctuation inputs and (b) the signal processing in the cockroach ocellus is different from that in other visual systems, including vertebrate retinas and insect compound eyes, in which the system's dynamics depend on the mean illuminance.

The dynamic relationship between graded potentials and spikes of second-order neurons was studied using sinusoidally-modulated light with various mean illuminances. A solitary spike was generated at the depolarizing phase of the modulation response. Analysis of the relationship between the amplitude/frequency of voltage modulation and the rate of spike generation showed that (a) the spike initiation process was bandpass at about 0.5-5 Hz, (b) the process contained a dynamic linearity and a static non-linearity, (c) the spike threshold at optimal frequencies (0.5-5 Hz) remained unchanged over a mean illuminance range of 3.6 log units. I conclude that (a) the spikes of second-order neurons encode decremental inputs whose contrast exceeded a fixed level, (b) the spike initiation process may be modeled by an integrate-and-fire generator and (c) the role of ocellar second-order neurons is to produce two kinds of signals, linear (graded) signals and nonlinear (spike) signals.

INTRODUCTION

Much knowledge has been accumulated about signal processing in visual systems, but we are still far from the full understanding of it. An approach to elucidate the visual processing is to analyze the relationship between the light input and the resulting responses of visual neurons. White-noise and sine-wave analysis, the most powerful means to analyze such input/output relationship, have been applied to neurons of vertebrate retinas (e.g. Victor and Shapley, 1979: Naka et al., 1979, 1987) and of insect compound eyes (e.g. Pinter, 1972: French and Jarvilehto, 1978). These analysis provided fruitful results but also showed a difficulty in understanding these complex visual systems.

One of the practical way to overcome this difficulty is to examine simple visual systems, like cockroach ocellus, in which information processing is limited. Consider that the photic inputs which animals receive consist of three parameters, i.e., spatial, spectral and temporal parameters, and thus, the visual systems analyze these parameters to subtract some features of the external world. In cockroach ocellus, the photoreceptors consist of a single spectral type (green receptors; Goldsmith and Ruck, 1958), and all of the second-order neurons has identical receptive field (Mizunami et al., 1982). Therefore, the ocellus analyzes neither color nor spatial information, or more strictly, changes of color or spatial distribution of the light stimulus are translated into temporal changes of the stimulus intensity. The simplicity of the cockroach ocellus allows us to concentrate on the problem: how does the system process temporal changes of stimulus intensity?.

This report summarizes my works on the dynamics of incremental responses of cockroach ocellar second-order neurons. White-noise-modulated or sinusoidally-modulated light was used to analyze the responses. The responses consisted of two components, slow potential and spikes. I first describe dynamics of the slow potential response, and then discuss the dynamic relationship between the slow potential and spikes.

MATERIALS AND METHODS

Preparation, stimulus and recording

Adult males of the cockroach, *Periplaneta americana*, were studied. The whole animal was mounted, dorsal side up, on a Lucite stage and fixed with bee's wax. The compound eyes and one of two ocelli were shielded from light by bee's wax mixed with carbon black. The dorsal part of the head capsule was removed and the dorsal surface of the brain was exposed.

Intracellular recordings were made from large ocellar second-order neurons, L-neurons, using glass microelectrodes filled with 2 M potassium acetate. The recordings were made in the ocellar nerve, to measure the dynamics of the slow potential response. Since the spikes of the L-neurons initiate in the ocellar tract of the brain (Mizunami et al., 1987), the recordings were made in the ocellar tract to measure the relationship between the slow potential and spikes. Stable recordings of over 60 minutes were feasible. These neurons were identified to be L-neurons, following criteria described previously (Mizunami et al., 1987).

A light emitting diode, LED, was used as a light source. The LED had a spectral peak at 560 nm. The LED was driven by a sinusoidal or white-

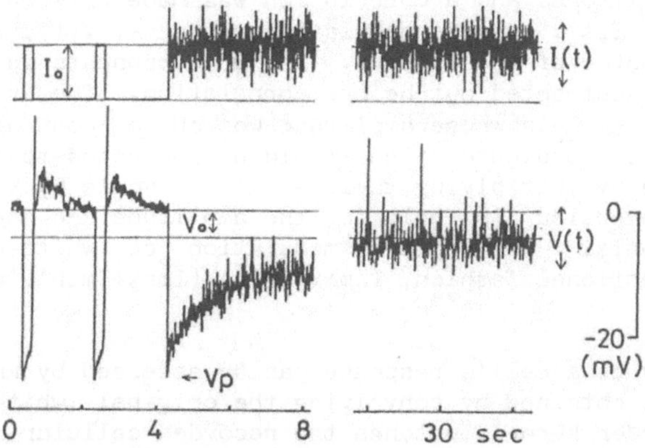

Fig. 1. Responses from an L-neuron evoked by steps of light given in the dark or by white-noise-modulated light. The relationship between Io and Vp or Vo is the cell's DC (static) sensitivity and the relationship between I(t) and V(t) is the incremental sensitivity. Spike potentials are seen at the offset of step stimulation as well as during white-noise stimulation. (From Mizunami et al., 1986).

noise current provided by a function oscillator or a random signal generator. The electrodes were connected to a preamplifier which is equipped so that a constant current can be passed through an active bridge circuit. The stimulus light was monitored by a photodiode. Light stimulus and cellular response were observed on an oscilloscope and stored on an analog tape. White-noise analysis were made off-line on a VAX 11/780 computer with an array processor.

White-noise analysis of the slow potential

The light stimulus the cockrtoach received daily or nightly consisted of two parts, one with a steady mean, Io, and the other with a modulation around the mean, I(t), as shown in Fig. 1. The mean illuminance, Io, changes slowly but covers a large range. The modulation depth of the fluctuation around the mean illuminance, however, is moderate and should remain roughly constant. The response evoked, therefore, consists of two components, the steady mean, Vo, and modulation response, V(t), the former being related to Io and the latter to I(t). In a white-noise analysis, the relationship between I(t) and V(t) is represented by kernels obtained by cross-correlating the white-noise input with the resulting cellular response. The results of first-order cross-correlation, weighted by the power of the stimulus, are the first-order kernels. The first-order kernel is the linear part of the cell's response to an impulse input superposed on a mean illuminance. If a cell's response is linear or quasi-linear, the amplitude and waveform of first-order kernels are, therefore, the comprehensive measure of a cell's incremental sensitivity and the response dynamics. If a cell's response contains second-order nonlinear components, the first- and second-order kernels represent the linear and nonlinear components of the cell's incremental response, and their amplitudes and waveforms represent the cell's incremental sensitivity and response dynamics.

In actual experiments, the light signal was monitored before it was attenuated by ND filters, and a correlation was made between the monitored light signals, $10^n \cdot I(t)$, and the modulation response, V(t), where n is the log attenuation factor of the filters. The DC components in both signals, Io and Vo, were subtracted out before correlation. The results of the correlation were kernels whose amplitude was on a contrast sensitivity scale. The kernel's ordinate values could be converted to an incremental sensitivity scale by multiplying their amplitude scale by the attenuation factor, 10^n. Conversion is only for the amplitude and does not affect waveform of kernels. The depth of modulation of white-noise stimulus defined in a conventional fashion, (Imax-Imin)/(Imax+Imin), was about 0.7-0.9 (70-90 %).

The linearity of a cell's response can be assessed by knowing how well the linear model, obtained by convolving the original white-noise signal with the first-order kernel, matches the recorded cellular response. The degree of accuracy is the mean square error (MSE).

Sine-wave analysis of slow-spike conversion process

The sinusoidal light stimulus consisted of two components, a steady mean, Io, and a dynamic component, I(f). I(f) was defined by the modulation frequency (Hz) and depth of modulation (%). The resulting response of L-neurons consisted of a sinusoidal slow potential modulation and spikes, as shown in Fig. 4. The slow potential response contained three components, a steady mean potential, Vo, a voltage noise, Vn, and a modulation response, V(f). Vo and Vn was related to the mean illuminance, Io, and V(f) was related to light modulation, I(f). V(f) was defined as Vpeak-Vbottom, where Vpeak was the potential at the peak and Vbottom is the potential at the bottom of the voltage modulation. The magnitude of spike response, S(f), could be defined as a probability of spike generation for one voltage cycle, as will be discussed in the Results. The L-neuron had no maintained discharge under steady illumination or in the dark: the spike response had no steady component. I analysed the manner in which the spike response, S(f), is related to parameters of the slow potential response, V(f), Vn and Vo.

RESULTS

White-noise analysis of slow potential responses

The cockroach ocellus contains about 10,000 photoreceptors, and they converge on four large second-order neurons, called L-neurons (Weber and Renner, 1976; Toh and Sagara, 1984). The axon of the L-neuron exits the ocellus and projects into the ocellar tract of the brain, through the ocellar nerve (Mizunami et al., 1982: Toh and Hara, 1984).

Fig. 1 shows the responses of an L-neuron evoked by steps and white-noise modulated stimuli. Brief steps of light given in the dark produced step-like responses with a peak, Vp. A spike is seen at the off-set of the stimulus. At the beginning of white-noise stimulation, a transient peak similar to the one produced by steps of light was seen. With continued white-noise stimulation, the membrane potential reached a steady level, Vo, within 30-40 sec. The steady-level was maintained as long as the stimulus was continued, i.e., the L-neuron reached a dynamic steady-state. The kernels were computed by cross-correlating the slow potentials against the white-noise inputs during the dynamic steady state. Spike potentials were

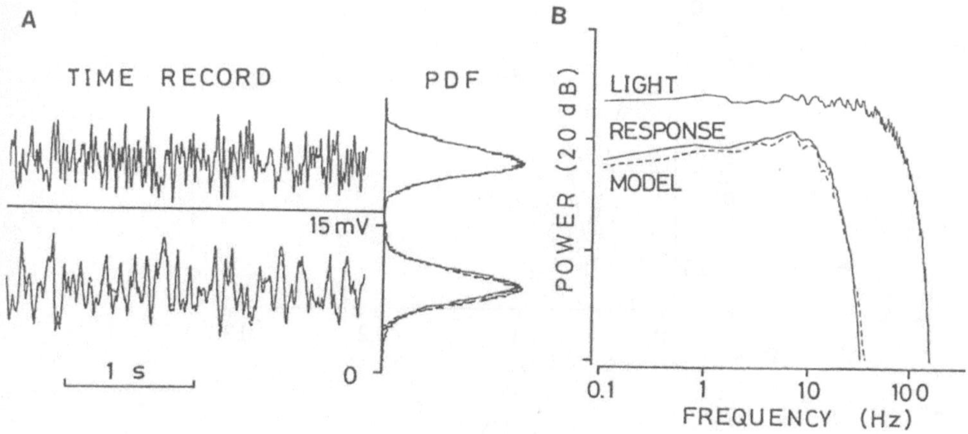

Fig. 2. (A) Time records of a pair of white-noise stimulus and the resulting cellular response (continuous line). Superposed on the response trace is the linear model (broken line). PDFs for the light stimulus and the recorded response are also shown. The light PDF is also superposed on the response PDF. (B) Power spectra of the light stimulus, response (continued line) and model (Broken line). The mean illuminance of the stimulus is 20 $\mu W \cdot cm^{-2}$. (From Mizunami et al., 1986).

removed with a low-pass filter (0.1 to 50 Hz) for slow potential analysis.

If the response of a cell is linearly related to the stimulus modulation, the cell's response to the modulation should be predictable from the first-order kernels with a fair degree of accuracy. Fig. 2A shows the time records of white-noise stimulus (upper trace) and resulting response (lower traces in a continuous line). Superposed on the response trace is the model response predicted by the first-order kernel (broken line). Although there are occasional deviations, the two traces matched well, which shows that the response could be predicted from the first-order kernel fairly accurately. Indeed, the averaged mean square error (MSE) computed from five L-neurons was 11.1% with a standard deviation of 2.1%. Fig. 2A also shows that the probability-density-functions (PDFs) of the light stimulus and of the response PDF. The light stimulus PDF is also superimposed on the response. Although there is a minor deviation between the two PDFs near the mean, they were in good agreement. Fig. 2B shows the power spectra of the light stimulus, response, and linear model. The power spectrum of the response (continuous line) matched well with that of the model (broken line). Both had a slight band-pass filtering property as seen by the lower power for the low-frequency region. A similar analysis made on the responses at a mean illuminance level of 20-0.002 $\mu W \cdot cm^{-2}$ showed that the responses produced by white-noise modulation were linear.

Fig. 3A shows the kernels obtained at five mean illuminance levels, plotted on a contrst sensitivity scale. The kernels were hyperpolarizing and monophasic (integrating). The waveforms were identical, with constant peak response times of about 50 ms, and the amplitudes differed only by 30%. This is remarkable because the mean illuminance for which the kernels were computed covered a range of 1:10,000. Stimuli dimmer than -4 log units produced no reliable results. The results in 3A shows that (a) the response is an exact Weber function, that is the contrast sensitivity is independent of the mean illuminance over a 4-log range of mean illuminance (in other words, the incremental sensitivity decreases by a factor of ten

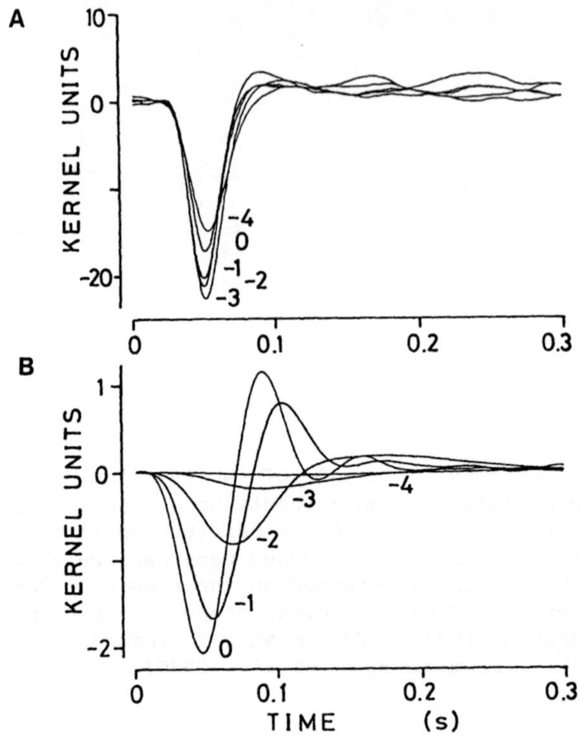

Fig. 3. (A) First order kernels, plotted on a contrast sensitivity scale, obtained at five mean levels. The first-order kernels were calculated by cross-correlating the white-noise light stimuli with the recorded responses. Kernels are labeled 0 through -4 to indicate log density of filters interposed. The mean illuminance at 0 log units is 20 $\mu W \cdot cm^{-2}$. (B) shows turtle horizontal cell kernels plotted as in A. The peak response times, waveforms and amplitudes differed for different levels of mean illuminance. Kernel units are in $mV/(\mu W/cm^{-2}) \cdot sec$. (From Mizunami et al., 1986).

for a tenfold increase in the mean illuminance), and (b) the response dynamics remained unchanged over the same range of mean illuminance. The mean illuminance controlled only the scaling of incremental sensitivity, but not response dynamics.

For comparison, kernels from a horizontal cell of the turtle, Pseudemys scripta elegans, obtained under comparable conditions, are shown in Fig. 3B. Note that the turtle's cellular response could be predicted from the first-order kernels with MSEs of less than 10% (Chappell et al., 1985). In the turtle's horizontal cell, the amplitude of kernels on a contrast sensitivity scale decreased as the mean illuminance decreased. As the mean illuminance was increased, the peak response times became shorter from 100 to 50 msec and the waveform became more biphasic (differential). Thus, the response dynamics depend on the mean illuminance. Sets of impulse responses or kernels similar to the one shown in Fig. 3B have been obtained in the human visual system (Kelly, 1971) and lower vertebrate horizontal cells (Naka et al., 1979; Chappell et al., 1985).

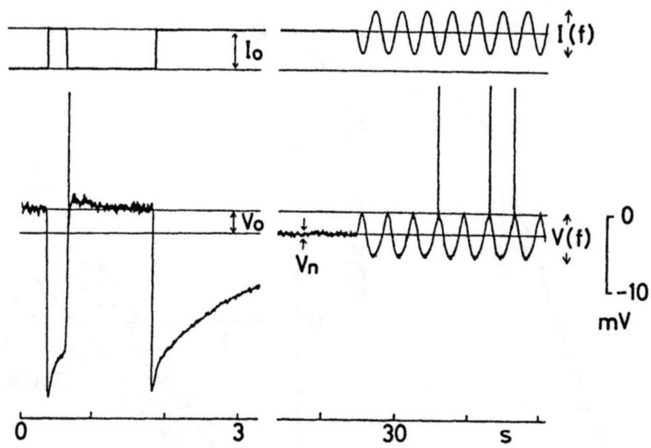

Fig. 4. Responses of an L-neuron evoked either by a step-light stimulus given in the dark or by a sinusoidally-modulated light stimulus. The L-neuron responded to the sinusoidal stimulus with a sinusoidal voltage modulation, V(f), around a mean voltage, Vo. The response contained noise, Vn. Spikes were seen at the offset of step stimulation and at the peak of the voltage modulation. The mean illuminance of the stimulus, Io, was 20 $\mu W \cdot cm^{-2}$, modulation frequency, f, was 2 Hz, and the depth of modulation of the stimulus, I(f), was 60%. (From Mizunami and Tateda., 1988a).

Dynamics of slow-spike conversion process

Fig. 4 shows typical records of responses from an ocellar L-neuron to sinusoidal light modulation. Response of the L-neuron consisted of two components, graded voltage fluctuation and the spikes. The waveform of the slow potential response to sinusoidal light was roughly sinusoidal, thereby confirming a quasi-linear response. A solitary spike was evoked at the peak or depolarizing phase of the modulation response. One modulation cycle of stimulus usually evoked only one spike, even when a stimulus with a large modulation depth (75%) was applied. I defined the magnitude of spike response as the rate of spike generation for one cycle of modulation stimulus. The rate of spike generation was measured as follows. The responses to 30-80 cycles of stimulus were recorded, and the rate of spikes was calculated by dividing the number of voltage modulation cycles in which spikes were induced by the total number of the cycles.

Fig. 5A shows the relationship between the peak-to-peak amplitude of the slow potential response and the rate of spike generation at a modulation frequency of 1 Hz. There was a dead-zone, in which spikes were not generated. Beyond that zone, the spike rate increased with the increase in the amplitude of the slow response. The plot is almost sigmoidal, with a linear part covering the spike rate ranging from 10 to 90%. Similar sinusoidal or quasi-linear relationships between the amplitude of voltage modulation and the spike rate were obtained over a frequency range of 0.1-20 Hz and over a mean illuminance range of 3.6 log units.

Fig. 5B shows the relationship between the peak-to-peak amplitude of

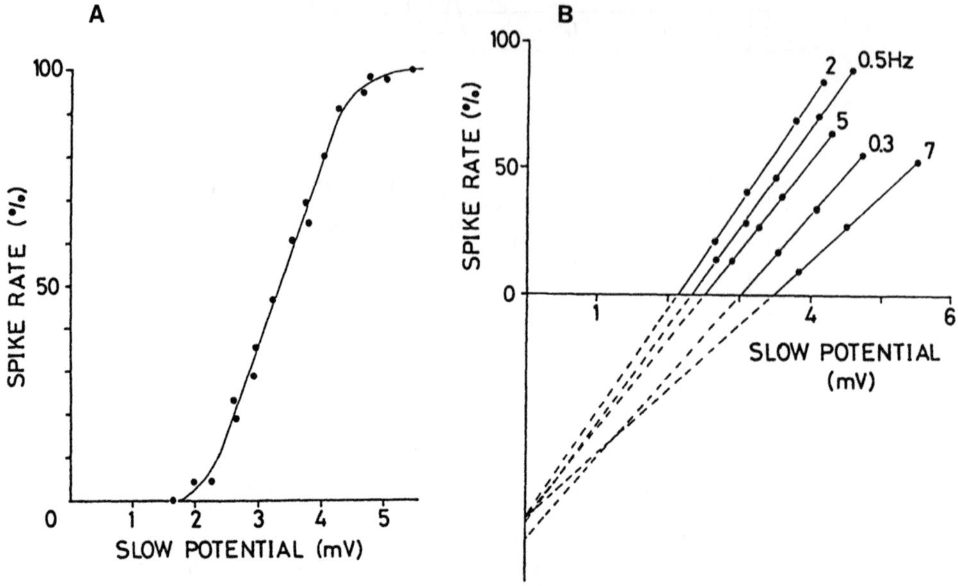

Fig. 5. (A) The rate of spike generation plotted against the peak-to-peak amplitude of the slow potential response, obtained at a frequency of 1 Hz. The form of the curve was sigmoidal, with the linear part covering the range of spike rate of about 10-90%. (B) The spike rate plotted against the amplitude of slow response, obtained at five different frequencies. The extrapolated dashed lines are the regression lines for each frequency. These lines cross the vertical axis, at almost the same point. The stimulus had a mean illuminance of 2 $\mu W \cdot cm^{-2}$. (From Mizunami and Tateda, 1988a).

the slow potential response and the spike rate obtained at different frequencies. The results at a spike rate of between 10 and 90% are shown. The extrapolated straight lines are regression lines for each frequency. The lines cross the vertical axis, at almost the same point. This suggests that the nonlinear threshold is frequency independent: the nonlinearity of the spike initiation process is static. On the other hand, the slope of the lines changes with the frequency, which indicates that the spike initiation process contains a dynamic linearity. In short, the spike initiation process contains a dynamic linearity and a static non-linearity. A simple model for spike initiation process will be proposed, based on these observations (see Fig. 8).

Here I define 50% threshold as the peak-to-peak amplitude of the slow response at a spike rate of 50%. Further analysis was made using the 50% threshold. Fig. 6A shows the relationship between the 50% threshold and the modulation frequency, obtained at a mean illuminance range of 3.6 log units. The 50% threshold was smallest at frequencies of about 0.5-5 Hz: the slow-spike conversion process was bandpass. The 50% threshold at optimal frequencies, where the 50% threshold was the smallest (about 0.5-5 Hz), was unchanged over a mean illuminance range of 3.6 log units. However, the 50% threshold at frequencies of less than 0.5 Hz did change depending on the mean illuminance levels. The 50% threshold was lower at dimmer mean illuminance. Fig. 6B shows the phase characteristics of spike generation, measured from the peak of the potential modulation. Measurements were done from the records of responses in which the spike

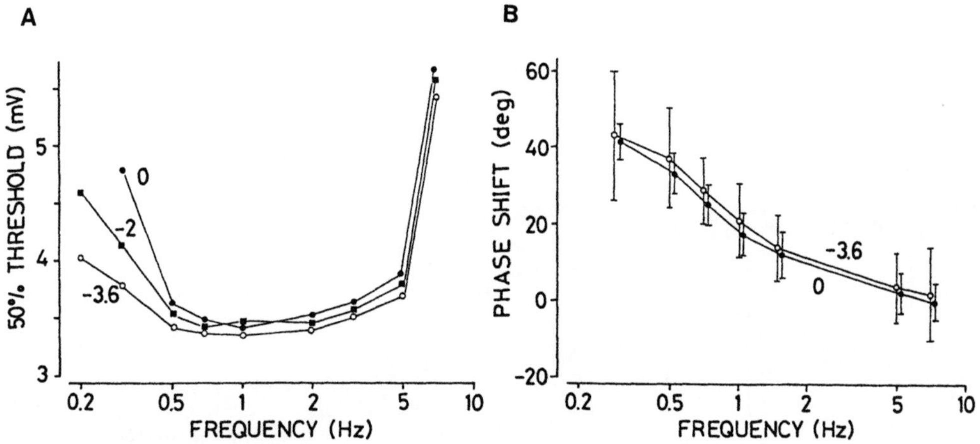

Fig. 6. (A) 50% threshold of spike response, defined as peak-to-peak amplitude of the potential modulation at a spike rate of 50%, plotted against the modulation frequency. The plots are from the responses of an L-neuron to sinusoidal lights with a mean illuminance of 0.005 (-3.6 log), 0.2 (-2 log), and 20 $\mu W \cdot cm^{-2}$ (0 log). (B) Phase characteristics of the spike response measured from the peak of the potential modulation. Measurements were done from the records of responses in which the spike rate was about 50% (40-60%), and the results at a mean illuminance of 0 log and -3.6 log are shown. (A) and (B) are from the same L-neuron. (From Mizunami and Tateda, 1988a).

rate was about 50%. The phase of spike generation progressively led that of the slow potential with decreases in the frequency, which suggests that the process has a differential or change-sensitive nature. The variance of the phase was larger at a dimmer mean illuminance. The variance of the phase at -3.6 log was about twice of that at 0 log units.

The mean potential level and the noise magnitude of the L-neurons depended on the mean illuminance levels (Fig. 7A). The membrane potential was more negative and the voltage noise was less prominent under brighter illumination. In Fig. 7B, steady or noise current was injected into an L-neuron during light stimulation, and the 50% threshold was measured. The mean illuminance of the stimulus was 0 log units, at which the mean membrane potential was about 3 mV negative to the dark level. There, the noise was slight. A steady depolarizing current of 4 nA, estimated to depolarize the neuron about 3 mV based on input resistance data (Mizunami and Tateda, 1988a), had no significant effects on the 50% threshold. The amplitude of the noise did not change by the injection of the DC-current. I conclude that the mean potential level does not affect the 50% threshold. This phenomenon can be explained by a steady-state inactivation of spike initiation (Mizunami and Tateda, 1988a). When a noise current was injected, the 50% threshold at low modulation frequencies became smaller (Fig. 7B, open circles). The noise current had a peak-to-peak intensity of about 4 nA, and the power spectrum of which was similar to that of the actual noise. The curve of the plot in the presence of noise current was similar to that observed under dim illumination (see Fig. 6A). In addition to the change in the 50% threshold, I observed that the noise current produced an increase in the variance of the phase of spike generation. This finding was similar to that observed with a dim mean illuminance. I conclude that the decrease in spike threshold at a low-frequency potential

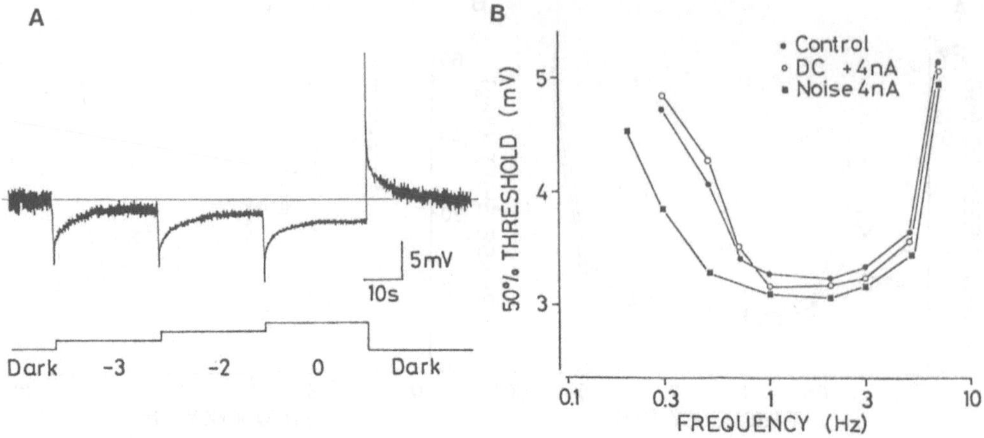

Fig. 7. (A) Responses of an L-neuron to prolonged illuminations. The light intensities are indicated as log attenuation (0 log units = 20 $\mu W \cdot cm^{-2}$). (B) Effects of noise current and steady current injection on the 50% threshold of an L-neuron. The mean illuminance was 0 log unit. The 50% threshold at a low-frequency range decreased when noise current was injected. The steady depolarizing current had little effect on the 50% threshold. (From Mizunami and Tateda, 1988a).

modulation by the decrease in the mean illuminance was due to the increase in the noise: the noise had a facilitatory effect on the initiation of the spike.

DISCUSSION

Dynamics of slow potential response

I used white-noise-modulated light to evoke responses from cockroach ocellar neurons. Cockroaches in their natural environment do not experience a flashing spot of light in the dark. Their photic inputs fluctuate around a mean illuminance, and their visual systems including ocelli must be developed to appreciate changes around a mean illuminance, not a sudden flash in darkness. I found that: (a) the incremental responses of the L-neuron were linear with MSEs of about 10%, (b) the cell's incremental sensitivity was an exact Weber function over a mean illuminance range of 4-log units, and (c) the response dynamics remain unchanged in the same range of mean illuminance. These observations indicate that the levels of mean illuminance controlled the amplitude scaling of the incremental response but not its dynamics. This is a remarkable finding because the response dynamics, as well as incremental sensitivity of all the visual systems so far studied, depend upon the levels of mean illuminance. This is the case with *Limulus* photoreceptors (Fuortes and Hodgkin, 1964), photoreceptors of insects compound eyes (Pinter, 1972; Dubs, 1981), vertebrate cone (Baylor and Hodgkin, 1973) and the second-order neurons (Naka, 1979; Tranchina et al., 1983; Chappell et al., 1985), and the human visual system (Kelly, 1971). Such a coupling has been one of the principle features of models of visual systems (Fuortes and Hodgkin, 1964; Kelly, 1971). The present results show that the couplings

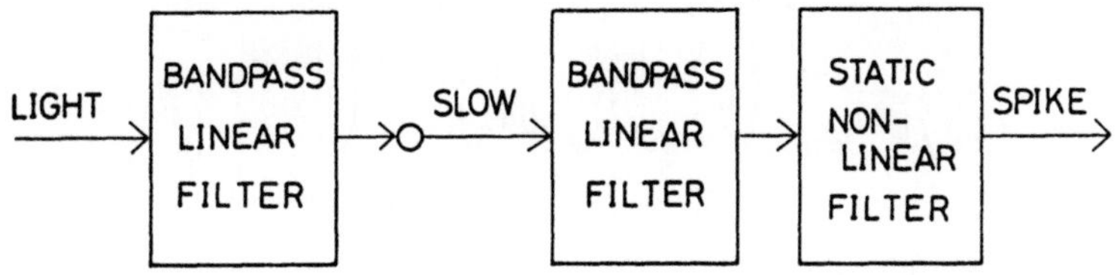

Fig. 8. A model for signal processing in cockroach ocellus. Light signals are passed through a bandpass linear filter and produce a slow potential response in L-neurons. The slow potential is further passed through a linear/nonlinear cascade and produces a spike discharge. The linear filter is bandpass, and the non-linear filter is a static threshold. (Modified from Mizunami and Tateda, 1988a).

of sensitivity and dynamics are not necessarily ubiquitous characteristics of the visual system.

Linear nature of the modulation response have been noted from photoreceptors of vertebrate retina (Baylor and Hodgkin, 1973), insect compound eyes (Pinter, 1972; Dubs, 1981) and *Limulus* eyes (Fuortes and Hodgkin, 1964). Also linear response have been found from second-order neurons of vertebrate retina (Tranchina et al., 1983; Chappell et al., 1985) and of insect compound eyes (French and Järvilehto, 1978: Laughlin and Hardie, 1978). Apparently, linear encoding of contrast is a general principle of peripheral visual processing.

Dynamics of slow-spike conversion process

Cockroach ocellar L-neurons do not discharge spontaneously, and a single cycle of voltage modulation produces a single spike or very few spikes. The cockroach L-neurons probably encode signals in the form of a single spike or very few spikes. This is different from neurons in other visual systems, in which signals are, in most cases, encoded in the frequency of spike trains.

I examined the relationship between the slow potential and spikes of cockroach ocellar L-neurons, using a sinusoidally-modulated light stimulus. I found that (a) the spike initiation process of cockroach ocellar L-neuron is probabilistic (stochastic): the process has an internal noise, (b) the process has bandpass filtering properties, and (c) the process contains a dynamic linearity and a static nonlinearity. The observations suggest that a simple model, such as the integrate-and-fire model developed by Knight (1972a,b), may represent actual spike initiation process. The final form of his model consists of a 'forgetful integration' process, which is frequency-dependent (dynamic) and linear, followed by a 'stochastic firing' process which is frequency-independent (static) and nonlinear. Bryant and Segundo (1976) concluded that a similar model was quite accurate in predicting experimentally observed spike discharges of *Aplysia* neurons. Sakuranaga et al. (1987) examined the spike discharge of catfish retinal ganglion cells and concluded that it can be modeled by an integrate-and-fire generator. These similarities among different preparations suggest that the integrate-and-fire model does represent the actual spike initiation process.

Knight (1972a,b) concluded that the spike encoding in *Limulus*

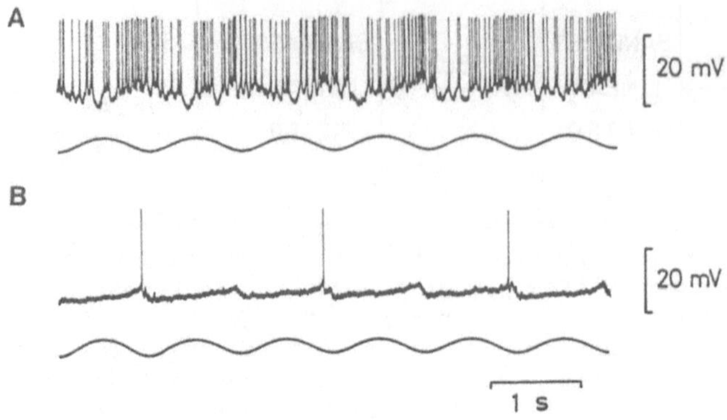

Fig. 9. Typical responses of two types of third-order neurons to sinusoidal light stimulation. One type of third-order neuron, OL-I neuron, showed sinusoidal modulation of spike frequency (A), whereas the other type, D-I neuron, generated solitary spikes at the decremental phase of light modulation (B). Lower traces indicate stimulus light, monitored by a photodiode. The stimuli had a modulation frequency of 1 Hz and modulation depth of 50%. The mean illuminance was 2 uW cm^{-2}. (From Mizunami and Tateda, 1988a).

eccentric cells allows for a firing rate that is a replica of the shape of the stimulus. Sakuranaga et al. (1987) also concluded that signals contained in slow potentials remain substantially unchanged in the spike trains in catfish ganglion cells. In these neurons, little signal modification occurs during the spike initiation process. In the cockroach L-neuron, however, there are fundamental differences between the slow potential response and the spike response: (a) the slow potential response is linear whereas the spike response is nonlinear and (b) the dynamics of the slow response are independent of the mean illuminance levels while dynamics of the spike response depend on the mean illuminance. Therefore, I conclude that the spike initiation in the L-neuron is an important step in visual signal processing. The function of the spike initiation process in the L-neuron is to filter out specific features from the slow potential signals, dimmings from the mean level, rather than to produce a replica of the slow potential.

<u>Signal processing in cockroach ocellar system</u>

In conclusion, I propose a simple model for signal processing in the cockroach ocellus (Fig. 8). Light signals that enter the ocellus are passed through a bandpass linear filter and produce a slow potential response in L-neurons. The linear filter consists of photoreceptors and synapses between photoreceptors and L-neurons. The slow potential is passed through a linear/nonlinear cascade and produces a spike discharge. The linear filter is bandpass, containing both a differential and an integrative nature. The nonlinear filter is a static threshold with a sigmoidal (probabilistic) input/output relationship. The model will be a good base to further extend our analysis of signal processing in cockroach ocellar system.

The ocellar L-neurons make output synapses onto several types of third-order neurons in the ocellar tract (Toh and Hara, 1984; Mizunami and Tateda, 1986). An important question to be solved is whether the slow potential signals or the spike signals of L-neurons, or both, are encoded in the third-order neurons. Fig. 9 shows typical responses of two types of third-order neurons, called OL-I neuron and D-I neuron (Mizunami and Tateda, 1986), to sinusoidal light stimulation. One type of third-order neuron, the OL-I neuron, had a spontaneous spike discharge and exhibited a modulation of spike frequency around a mean (Fig. 9A). The pattern of the response was similar to, although not the same as, that of the slow potential response of the L-neurons. The other type, the D-I neuron, had no spontaneous spike activity and exhibited single spikes at the decremental phase of light modulation (Fig. 9B). The pattern of the response was similar to that of spike response of L-neurons. The observations suggest that both slow potential signals and spike signals in L-neurons are encoded into the third-order neurons. I conclude that the role of ocellar L-neurons is to produce two kinds of signals, linear (slow potential) signals and nonlinear (spike) signals. The signal processing in third-order neurons will a major subject of our future study (Mizunami and Tateda, in press; Mizunami, in preparation).

ACKNOWLEDGEMENTS

I wish to thank Drs. Hideki Tateda and Ken-Ichi Naka, for their valuable cooperation. This study was supported in part by grants from ministry of education of Japan.

REFERENCES

Baylor D A, Hodgkin A L (1973) Detection and resolusion of visual stimuli by turtle photoreceptors. J Physiol 234: 163-198.

Bryant H L, Segundo J P (1976) Spike initiation by transmembrane current: a white-noise analysis. J Physiol 260: 279-314.

Chappell R L, Naka K -I, Sakuranaga M (1985) Dynamics of turtle horizontal cell response. J Gen Physiol 86: 423-453.

Dubs A (1981) Non-linearity and light adaptation in the fly photoreceptors. J Comp Physiol 144: 53-59.

French A S, Järvilehto M (1978) The transmission of information by first and second order neurons in the fly visual system. J Comp Physiol 126: 87-96.

Fuortes M G F, Hodgkin A L (1964) Changes in time scale and sensitivity in the ommatidia of Limulus. J Physiol 172: 239-263.

Goldsmith T, Ruck P (1958) The spectral sensitivity of the dorsal ocelli of cockroaches and honeybees. J Gen Physiol 41: 1171-1185.

Kelly D H (1971) Theory of flicker and transient responses. I. Uniform fields. J Opt Soc Amer 61: 537-546.

Knight B W (1972a) Dynamics of Encoding in a population of neurons. J Gen Physiol 59: 734-766.

Knight B W (1972b) The relationship between the firing rate of a single neuron and the level of activity in a population of neurons. J Gen Physiol 59: 767-778.

Laughlin S B, Hardie R C (1978) Common strategies for light adaptation in the peripheral visual systems of fly and dragonfly. J Comp Physiol 128: 319-340.

Mizunami M, Tateda H (1986) Classification of ocellar interneurones in the cockroach brain. J Exp Biol 125: 57-70.

Mizunami M, Tateda H (1988a) Dynamic relationship between the slow potential and spikes in cockroach ocellar neurons. J Gen Physiol 91: 703-723.

Mizunami M, Tateda H (1988b) Synaptic transmission between second- and third-order neurones of cockroach ocelli. J Exp Biol (in press).

Mizunami M, Tateda H, Naka K -I (1986) Dynamics of cockroach ocellar neurons. J Gen Physiol 88: 275-292.

Mizunami M, Yamashita S, Tateda H (1982) Intracellular stainings of the large ocellar second order neurons in the cockroach. J Comp Physiol 149: 215-219.

Mizunami M, Yamashita S, Tateda H (1987) Calcium-dependent action potentials in the second-order neurones of cockroach ocelli. J Exp Biol 130: 259-274.

Naka K -I, Chan R Y, Yasui S (1979) Adaptation in catfish retina. J Neurophysiol 42: 441-454.

Naka K -I, Itoh M -A, Chappell R L (1987) Dynamics of turtle cones. J Gen Physiol 89: 321-337.

Pinter R B (1972) Frequency and time domain properties of retinular cells of the desert locust (Schistocerca gregalia) and the house cricket (Acheta domesticus). J Comp Physiol 77: 383-397.

Sakuranaga M, Ando Y -I, Naka K -I (1987) Dynamics of ganglion cell response in the catfish and frog retina. J Gen Physiol 90: 229-259.

Toh Y, Hara S (1984) Dorsal ocellar system of the American cockroach. II. Structure of the ocellar tract. J Ultrastruct Res 86: 135-148.

Toh Y, Sagara H (1984) Dorsal ocellar system of the American cockroach. I. Structure of the ocellus and ocellar nerve. J Ultrastruct Res 86: 119-134.

Tranchina D, Gordon J, Shapley R (1983) Spatial and temporal properties of luminosity horizontal cells in the turtle retina. J Gen Physiol 82: 573-598.

Victor J D, Shapley R M (1979) Receptive field mechanisms of cat X and Y retinal ganglion cells. J Gen Physiol 74: 275-298.

Weber G, Renner M (1976) The ocellus of the cockroach, Periplaneta americana (Blattariae). Receptory area. Cell Tissue Res 168: 209-222.

PROCESSING OF MOVEMENT INFORMATION IN THE FLY'S LANDING SYSTEM: A BEHAVIORAL ANALYSIS

Alexander Borst and Susanne Bahde*

Max-Planck-Institut für biologische Kybernetik
Spemannstrasse 38
7400 Tübingen, FRG

* present address:
Max-Planck-Institut für Limnologie
August-Thienemann-Strasse 2
2320 Ploen-Holstein, FRG

ABSTRACT

When approaching a landing site flies extend their legs in order to prevent crash-landing (Goodman 1960). This reflex has been analysed with respect to the underlying release mechanism. Pattern expansion in front of a tethered fly can mimic an approach towards a landing site. Under these conditions landing is a rather stereotyped motor pattern. Only the latency of the onset of the landing response varies with the stimulus strength (Borst 1986). Quantitative studies of the stimulus-latency relationship led to the formulation of a simple model which describes the way movement information at the fly's retina is processed in order to release landing (Borst and Bahde 1986, 1988b). We propose that the output of local movement detectors sensitive for front-to-back motion in each eye are pooled and subsequently processed by a leaky integrator. Whenever the level of the leaky integrator reaches a fixed threshold landing is released.

INTRODUCTION

Velocity-distance control and the detection of obstacles are common visual information processing problems in biological and technical systems (e.g. plunge-dives of sea birds, emergency-breaking in human car drivers, computer vision). In order to initiate the appropriate response in time the information about an impending collision has to be extracted from the retinal velocity field. This problem has been investigated using the the landing response of houseflies as an experimental paradigm. As it will be shown in this article the landing response proved to be an advantageous behavior where the principal steps of information processing from the retinal input up to the motor output could be studied by a quantitative behavioral analysis. All the experiments were done under well defined conditions with tethered flying flies. This has the advantage that one has control over the retinal input, while kinematic response analysis can be done 'in situ'. Landing was released by a periodic pattern of vertical stripes which moved from the front to the back on both sides of the animal. This apparent pattern expansion mimicked the fly's approach towards a landing site. Since the flies are fixed, they are prevented from actually landing at some place. They even do not stop flying, as they only do so, if their feet touch ground. Therefore, only the initial part of the landing routine, i.e. the typical extension of the legs, is taken into account. It is this pre-landing behavior which we exploited to study the processing of visual information in the fly's landing system.

MOTOR OUTPUT

We studied the leg movements which occur during landing by taking a single picture of the fly every 10 milliseconds after pattern movement has been started. In this way the complete time-course of the response could be derived. Fig. 1 gives an example of a housefly *Musca domestica* in flight and in the final landing position. The coordinates of the legs were digitized so that the responses of many flies could be quantitatively evaluated and

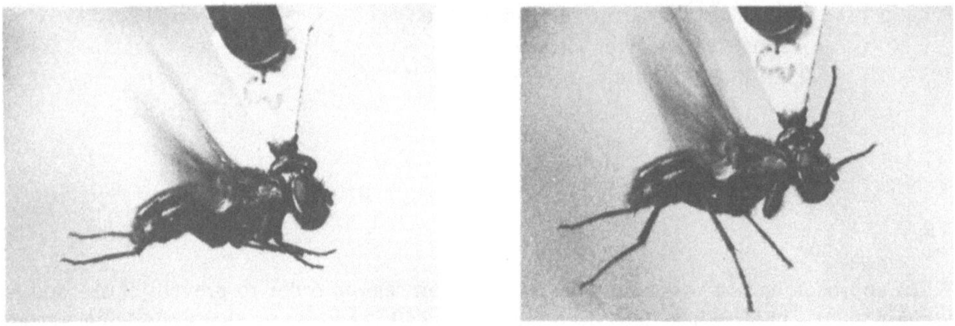

Fig. 1.
Fly with its legs in flight (left) and in the final landing position (right). During flight, the first two pairs of legs are folded underneath the body, while the hindlegs are stretched backwards to help the fly stearing. In the landing position the forelegs are lifted above the head, the midlegs are lowered and the hindlegs have the same position as during flight.

averaged. The reaction turned out to be highly stereotyped with only little variability from fly to fly. The forelegs which during flight are folded underneath the body are being stretched and lifted upwards for 4 mm within less than 30 ms. The main component of this movement comes from the opening of a single joint. The midlegs which, in flight position, are also held close to the body first are moved sidewards and then become lowered. This is also due to the opening of just one joint. The hindlegs participate strongly in stearing and therefore are held stretched backwards during flight. During the landing response these legs do not show any consistent movements. Sometimes they become lowered, sometimes they do a little kick upwards, some flies do not move them at all. In the average the response of the hindlegs is zero. Taking all occuring leg movements together the response represents a fixed action pattern with constant amplitude and time-course of its trajectories (Borst 1986). The only variable parameter of this motor program is the latency of its onset. This can be shown by using different stimuli to release landing (Fig. 2).

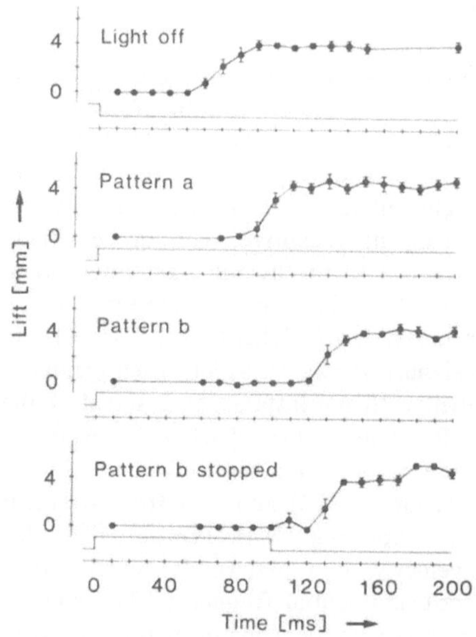

Fig. 2.
Time course of the lift of prothoracic tarsi in response to different stimuli: a reaction starts 50 ms after light has been turned off, b pattern is expanding with 240°/s, the latency is 80 ms, c pattern is expanding with 144°/s, the latency is 120 ms, d same as c except that the pattern is stopped after 100 ms. Data are the mean ± SEM of 6 flies (from Borst 1986).

This fact offers a great methodological advantage because the system has only one output variable, i.e.the latency. It can be easily measured by means of a light-gate in front of the fly which becomes interrupted whenever the fly lifts its forelegs (Fig. 3). Furthermore, the latency of the response can serve as a measure for the stimulus strength. Its quantitative dependence on any stimuli can now be used to study the underlying processing of these stimuli.

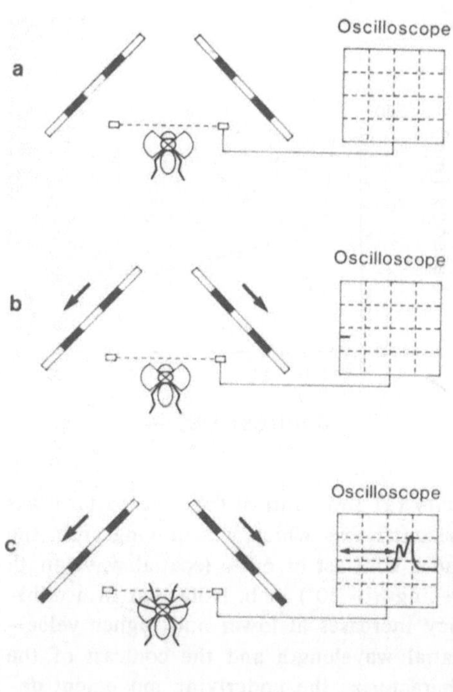

Fig. 3.
Experimental set-up: the fly is suspended with a cardboard triangle glued to its thorax (not shown) between the projection screens of two CRTs. The pattern consisting of bright and dark vertical bars is produced by an image synthesizer (Picasso, Innisfree Inc.) Its intensity is sinusoidally modulated along its horizontal axis with adjustable wavelength and contrast. A light gate monitors lifting of the forelegs. The output of the detector unit is displayed on an oscilloscope. The three phases of an experiment are shown from top to bottom: a the pattern is at rest, b the pattern has been started moving from the front to the back on either side and the oscilloscope time base is triggered simultaneously, c the fly performs the landing response and interrupts the light-beam with its forelegs. This leads to a deflection seen on the oscilloscope. Latency is the time between the onset of the visual stimulus and the deflection signal (from Borst and Bahde 1986).

UNDERLYING MOVEMENT DETECTION SYSTEM

Landing is released by pattern expansion in front of the fly (Braitenberg and Taddei-Ferretti 1966, Fischbach 1981, Wehrhahn et al. 1981). Thus, the evaluation of movement information must play a decisive role in this system and will therefore be considered first. It might be noteworthy that movement detection is not at all a trivial problem for the visual system. A single photoreceptor is just measuring light intensities and, therefore, cannot discriminate whether there is a moving object crossing its receptive field or somebody dimming the light. To decide between motion and a change in the overall light intensity at least two photoreceptors are necessary the outputs of which have to be connected in an appropriate way. One scheme which has been proposed by Reichardt and Hassenstein consists in a correlation-like interaction between the signals of two adjacent visual elements: one of the signals is delayed and subsequently multiplied with the instantaneous signal of the neighbouring one (Hassenstein and Reichardt 1956, Reichardt 1961). Such a 'correlation-type' movement detector has been demonstrated for the optomotor system of various insect species (Hassenstein and Reichardt 1956, Kunze 1961, Götz 1972) and also for movement perception of vertebrates including man (van Santen and Sperling 1984). The output of this movement detector does not invariably reflect the velocity of a pattern. It depends in a characteristic way on pattern properties such as contrast and spatial frequency. The landing response was investigated accordingly. Fig. 4 shows how latency depends on pattern velocity and on pattern contrast at a given spatial wavelength. Numerous of similar measurements were necessary to characterize the underlying movement detection system. We studied in some detail the dependence of optimum pattern velocity, i.e. the one which leads to a minimum latency, on both contrast and spatial wavelength of the pattern. Our results are in accordance with the specific properties of a correlation-type movement detector (Borst and Bahde 1986). Thus, the first principle step of signal processing in the landing system, the extraction of movement information from the light flux in the retina, has been elucidated by these findings.

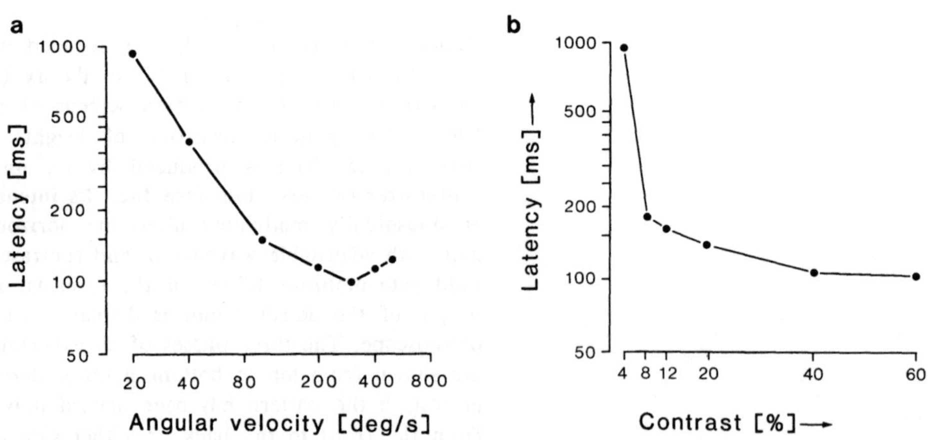

Fig. 4.

Dependence of the response latency on pattern velocity (a) and contrast (b). The pattern was a periodic grating modulated sinusoidally in the horizontal axis which was moving from the front to the back on either side of the animal. It had a contrast of 60 % (spatial wavelength = 20°) in a and was moved with 340°/s (spatial wavelength = 30°) in b. Note that in a minimum latency is obtained at about 300°/s. The latency increases at lower and higher velocities. The minimum latency also depends on the spatial wavelength and the contrast of the pattern in a systematic way. This can be used to characterize the underlying movement detection system. Data are the mean ± SEM of 15 (a) and 10 (b) flies tested 10 times per stimulus.

FURTHER SIGNAL PROCESSING

A comparison between the dynamic response properties of movement detectors (Egelhaaf and Reichardt 1987, Egelhaaf and Borst 1989) and the latencies obtained for the landing response showed that at least two additional steps of signal processing were necessary between movement detection and the final release of the landing response. These will be discussed separately.

1) Temporal Integration

To explain the considerably long latencies of up to seconds one has to assume that the movement detectors' output signals accumulate in time before the sum is large enough to trigger a landing response. Temporal integration followed by a threshold provides a simple mechanism for the conversion of stimulus strength into response latencies. The higher the input to the integrator, the faster its level increases and the earlier threshold is reached. Accordingly, response latencies, in a first approximation, should be inversely proportional to the movement detectors' output signals. This assumption is in qualitative agreement with all

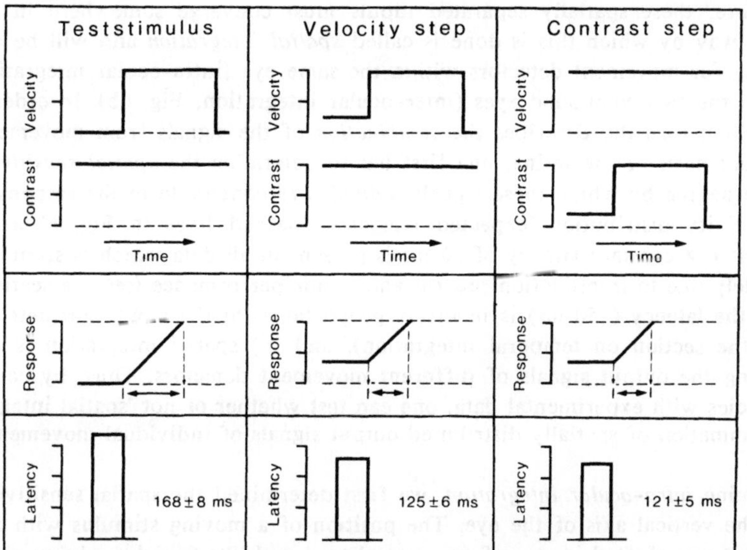

Fig. 5.

The prestimulus experiment provides supporting evidence for the temporal integration process. The top row illustrates the different stimuli. Left side: A pattern (spatial wavelength = 30°, contrast = 0.40) moving for 1 s at 60°/s (test stimulus) elicits a landing response after about 168 ms latency. The effect of prestimulation was tested, independently, in two situations: In the experiment shown in the center, the test stimulus was given at the end of a time period of 5 s of reduced pattern velocity (3°/s instead of 60°/s), whereas in the experiment shown on the right the test stimulus was given at the end of a 5 s period of reduced contrast (0.01 instead of 0.40). Prestimulus velocity as well as prestimulus contrast were too low to release landing during the 5 seconds of their presentation. The expected effect of such a prestimulation is illustrated in the second row: without a prestimulus, the integrator starts from zero level and the signal takes a certain time to reach the threshold for landing. Loading the integrator from a prestimulation level requires less time to reach this threshold. The bottom trace gives the averaged latencies (and their standard errors of the mean) of 20 flies each tested 10 times per stimulus. In both situations the latency of the response to the test stimulus is significantly shorter than without a prestimulation (from Borst and Bahde 1988b).

data obtained so far (Borst and Bahde 1986). An additional way to prove the concept of temporal integration is to precede the actual test stimulus by a subthreshold stimulus. If there is a temporal integration process within the landing system, then subthreshold stimulation preceding the test stimulus should shorten the latency as compared to the situation without prestimulation (the integrator is already 'loaded' to some degree). Conversely, if no temporal integration takes place but the instantaneous signal determines response latency, prestimulation should not affect the response latency to the test stimulus. Figure 5 shows the result of such an experiment. The latency in response to a test stimulus was significantly shortened when it was preceded for 5 seconds by pattern movement of reduced efficiency due to either low speed or low contrast. These results suggest that the signal releasing landing is not an instantaneous output of the movement detectors but rather a temporal accumulation of such outputs.

2) Spatial Integration

Movement detection is thought to be a local process and there is much evidence that a 2-dimensional array of movement detectors covers the whole visual field of the fly (Buchner 1976). The landing response can be released by movement stimuli in different parts of the eye. Therefore, these spatially separated inputs must converge somewhere in the nervous system. The way by which this is done is called *spatial integration* and will be examined in the following for movement detectors within the same eye (intra-ocular integration, Fig. 6a) and between the two compound eyes (inter-ocular integration, Fig. 6b). In order to find an algorithm which formally describes the *combination* of the signals from movement detectors from different parts of the retina, one first has to determine the *spatial sensitivity distribution*, i.e. the degree by which these signals *individually* contribute to the response. From the spatial sensitivity distribution 'expected latencies' (dashed lines in Fig. 6) are calculated assuming that i) a constant latency of 50 ms is present in all data which is stimulus independent and solely due to transduction process and motor performance (see the section on motor output), ii) the latency (-50 ms) is inversely proportional to the movement detectors' output signal (see the section on temporal integration), and iii) spatial integration is achieved by simply adding the output signals of different movement detectors. Thus, by comparing expected latencies with experimental data, one can test whether or not 'spatial integration' is an algebraic summation of spatially distributed output signals of individual movement detectors.

To examine *intra-ocular integration*, we first determined the spatial sensitivity distribution along the vertical axis of the eye. The position of a moving stimulus with only a small vertical extent was shifted in steps from ventral to dorsal. We found a minimum latency at a position ventrally to the horizontal plane through the fly's eye (Fig. 6a, left diagram). When the height of the stimulated area is increased the latencies become increasingly smaller (Fig. 6a, right diagram). A comparison between these data and the expected latencies shows that for high contrast patterns the response to the combined areas fits well the algebraic sum of the responses to the single constituents. In this respect the landing response seems to differ from other movement dependent behaviors like e.g. figure-ground detection by relative motion and also from spatial-integration properties of large-field motion-sensitive neurons in the fly brain: they all usually show a smaller response to the sum of pattern elements than is expected from the sum of the responses to the single ones (Reichardt et al. 1983, Hausen 1984). For low contrast patterns, the landing response to the combined areas is even stronger than expected in case of algebraic summation (compare solid with dashed line in Fig. 6a for C = 0.16). As will be shown at the end of this section, this deviation can be accounted for by the characteristics of the temporal integration process.

To study *inter-ocular integration* we compared the responses to monocular front-to-back stimulation with the responses to the same stimulus presented to both eyes. This was done at two different pattern velocities. At the higher velocity, the latency of the response to binocular stimulation is exactly what is expected from the sum of the responses to the monocular stimuli (Fig. 6b, lower trace). With the low pattern velocity, the latency is significantly shorter, i.e. the response to binocular stimulation is stronger than expected from the sum of the responses to monocular stimuli (Fig. 6b, upper trace). By using stimuli which are close to threshold, an even stronger inter-ocular enhancement has been found (Eckert and Hamdorf 1983). The weaker the stimulus, the stronger is the inter-ocular enhancement. Do we have to assume facilitation between movement detectors on either side of the visual system?

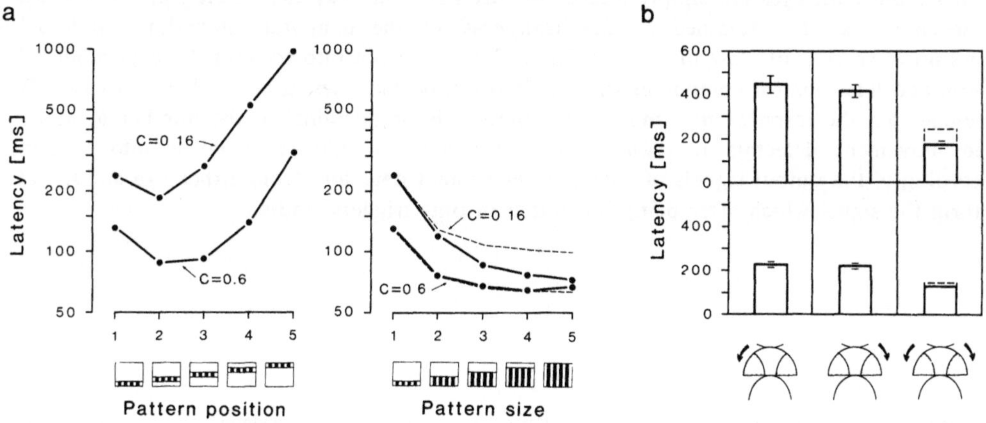

Fig. 6. Spatial integration properties of the landing system.

6a) Intra-ocular integration: to investigate the spatial sensitivity distribution the movement stimulus was presented in different regions of the eye (left diagram). The stimulated area had a vertical extent of 13° and its position was shifted from -26° to +26° with respect to the horizontal plane of the fly's eye. Optimum response (minimum latency) was obtained by stimulation in position 2 which is ventral to the horizontal plane. When the vertical extent of the pattern is increased from 13° to 64° in steps of 13° (right diagram) the response latencies decrease gradually. For high pattern contrast (C = 0.60) the response fits the latencies expected for algebraic summation of the contributions of the individual detectors. However, for low contrast (C = 0.16) the responses to the combined stimuli are significantly faster than expected in case of algebraic summation (expected latencies represented by dashed lines). The pattern had a spatial wavelength of 20° and was moved at 40°/s. Data are the means of the latencies of 15 flies each testet 10 times per stimulus.

6b) Inter-ocular integration: latencies of the responses to monocular and binocular stimuli are compared. Using a comparatively weak movement stimulus (upper row, pattern velocity 60°/s) the binocular stimulus is more effective than the response derived from the sum of the monocular constituents (expected latencies indicated by dashed bars). At a pattern velocity of 150°/s the latency to the binocular stimulus is as long as the response derived from the sum of the monocular constituents (bottom row). The pattern had a spatial wavelength of 30° and a contrast of 0.4. Data are the means and standard errors of the means of the latencies of 20 flies, each tested 10 times per stimulus (from Borst and Bahde 1988b).

In case of both intra- and inter-ocular integration experiments the response to a sufficiently weak compound stimulus is stronger than the sum of the responses to its single constituents. The difference decreases with increasing stimulus strength. This phenomenon can be explained by the temporal integration of the signals. For several reasons physiological integrators have to be leaky in order to prevent long-term accumulation of a triggering signal. Assuming that the leak-induced decay of the accumulated signal is proportional to the level of this signal, the response latency and input strength must be inversely proportional to each other only as long as the input is sufficiently high with respect to threshold. For input signals which are close to threshold, latencies become prolonged because the integrated signal already starts to bend off to its steady state level before threshold is reached. Fig. 7 illustrates how this leads to the observed inter-ocular enhancement although the signals coming from both eyes are simply added. In just the same way the finding on intra-ocular enhancement can be explained by the 'leakyness' of the temporal integrator which only shows up at sufficiently low input intensities. If this is taken into account the experimentally determined latencies for compound stimuli fit the expected ones rather well in *all* cases. We conclude that the response to a movement stimulus is proportional to the number of stimulated movement detectors if their position-dependent weightage is taken into account. Accordingly, the output signals of different movement detectors from either eye simply add to form the signal which after being integrated in time triggers landing.

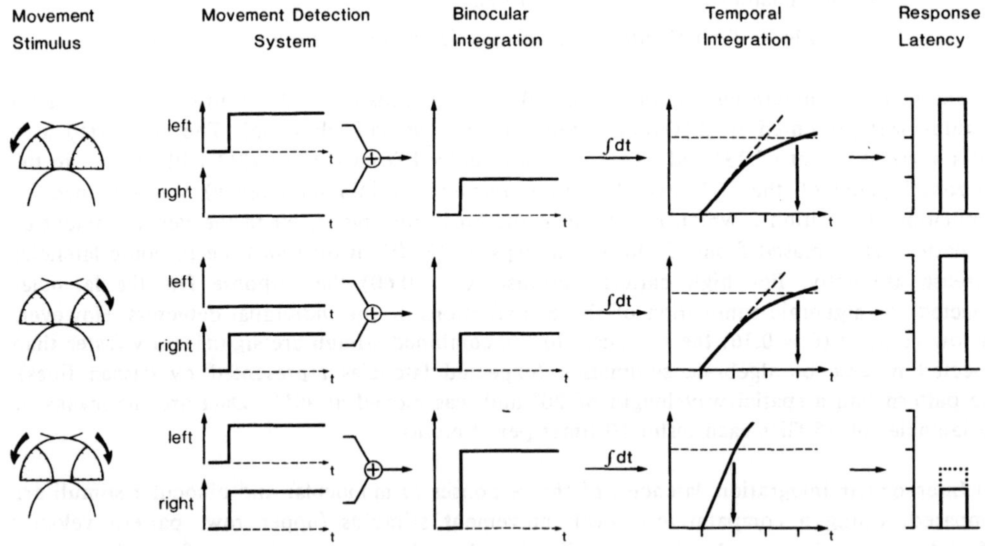

Fig. 7.

The leaky integrator provides a simple explanation for the phenomena shown in Fig. 5. Given high input amplitudes response latencies are inversely proportional to these amplitudes. If the input signal is small response latencies become prolonged because the integrated signal has almost adopted its steady state level before reaching threshold. This leads to the effect illustrated above: if the latency to a weak monocular stimulus is determined first then the latency found for the corresponding binocular stimulus will be shorter than predicted (dashed bar) although the signals from both eyes interact in a linear way.

CONCLUSIONS AND PROSPECTS

All behavioral experiments on the landing response were done with the objective to elucidate the principle according to which a movement stimulus leads to the respective motor output. The model derived from these experiments (Fig. 8) is strikingly simple, but, nevertheless, provides an explanation for apparently complicated phenomena. The model can serve for a further analysis of the landing response. On one hand it predicts, quantitatively, the behavior under natural conditions. For example when approaching a landing site the distance at which the fly initiates landing should depend on the velocity of the fly as well as on the size, contrast and structure of the target. Such experiments have been performed using moving discs of various size and structure to release landing in tethered flies. The results confirmed the predictions in every respect (Borst and Bahde 1988a) and are in agreement with similar studies done previously (Eckert and Hamdorf 1980, Tinbergen 1987). However, when comparing these experiments on tethered animals with free-flight studies, there remains a contradiction to be solved: in contrast to tethered flight no dependence of landing distance on the target size was detectable in free-flight (Wagner 1982). On the other hand the present results facilitate the investigation of the landing response on the neuronal level. Behavioral experiments can be used to determine whether or not any recorded neuron is part of the network controlling the landing response. In a similar way the neuronal network underlying other visual information processing tasks has already been successfully studied and the analysis of behavior always proved to be an indispensable tool (Egelhaaf 1985, Egelhaaf et al. 1988). Preliminary studies on neurons connecting the fly's visual center in the head with motor centers in the thorax have revealed that, indeed, there exist neurons which not only have the same preferred directions as the landing response, but also show a time course which could reflect temporal integration of the movement detector output signals as postulated by the model. Those neurons, thus, are likely candidates to release a landing response in response to visual stimulation (Borst, in prep.). In addition, using anatomical methods, the pathway is currently being traced backwards from the foreleg tibia extensor muscle (Borst and Buchstäber, in prep.). With a combination of behavioral, electrophysiological and anatomical methods along with computer modelling it may, thus, be possible to unravel the neural circuit underlying the landing response.

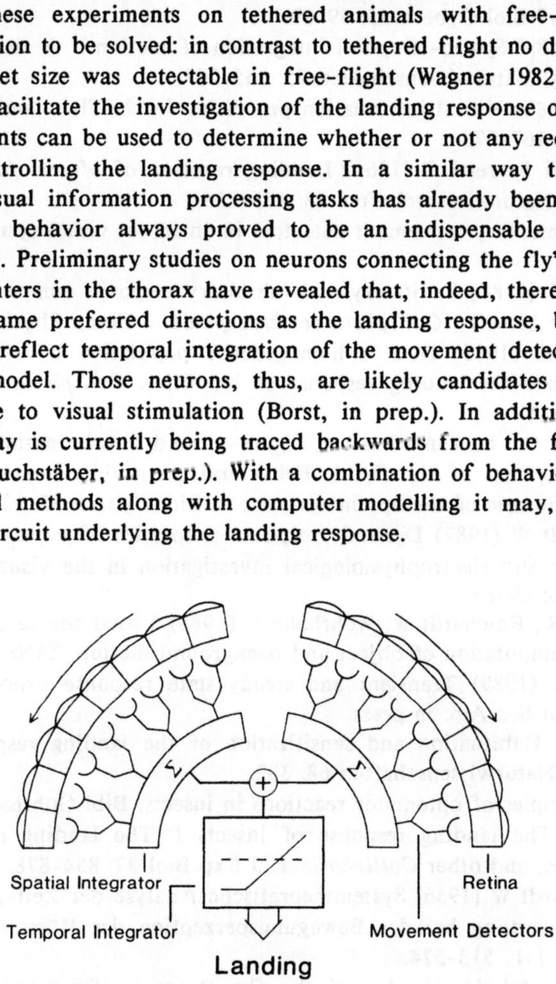

Fig. 8.
Model proposed to underly the landing response: the output of correlation movement detectors sensitive for front-to-back motion are pooled from each eye. These large-field units feed into a temporal leaky integrator. Whenever its level reaches a fixed threshold landing is released and a preprogrammed leg motor sequence is performed (from Borst and Bahde 1988b).

ACKNOWLEDGEMENTS

We are grateful to Prof. Dr. K.G. Götz and to Dr. M. Egelhaaf for their enthusiastic support of our experiments, for many discussions and for critically reading the manuscript. Special thanks to the students and the faculty of the MBL summer course 'Neural Systems and Behavior' for useful and creative suggestions.

REFERENCES

Borst A (1986) Time course of the houseflies' landing response. Biol Cybern 54: 379-383.

Borst A, Bahde S (1986) What kind of movement detector is triggering the landing response of the housefly? Biol Cybern 55: 59-69.

Borst A, Bahde S (1988a) Spatio-temporal integration of motion - A simple strategy of safe landing in flies. Naturwissenschaften 75: 265-267.

Borst A, Bahde S (1988b) Visual information processing in the fly's landing system. J Comp Physiol A 163: 167-173.

Braitenberg V, Taddei Ferretti C (1966) Landing reaction of *Musca domestica* induced by visual stimuli. Naturwissenschaften 53: 155-156.

Buchner E (1976) Elementary movement detectors in an insect visual system. Biol Cybern 24: 85-101.

Eckert H, Hamdorf K (1980) Excitatory and inhibitory response components in the landing response of the blowfly, *Calliphora erythrocephala*. J Comp Physiol 138: 253-264.

Eckert H, Hamdorf K (1983) Does a homogeneous population of elementary movement detectors activate the landing response of blowflies, *Calliphora erythrocephala*? Biol Cybern 48: 11-18.

Egelhaaf M (1985) On the neuronal basis of figure-ground discrimination by relative motion in the visual system of the fly. I. Behavioral constraints imposed on the neuronal network and the role of the optomotor system. Biol Cybern 52: 123-140.

Egelhaaf M, Reichardt W (1987) Dynamic response properties of movement detectors: theoretical analysis and electrophysiological investigation in the visual system of the fly. Biol Cybern 56: 69-87.

Egelhaaf M, Hausen K, Reichardt W, Wehrhahn C (1988) Visual course control in flies relies on neuronal computation of object and background motion. TINS 11: 351-358.

Egelhaaf M, Borst A (1989) Transient and steady-state response properties of movement detectors. J Opt Soc Am: in press.

Fischbach KF (1981) Habituation and sensitization of the landing response of *Drosophila melanogaster*. Naturwissenschaften 68: 332.

Götz KG (1972) Principles of optomotor reactions in insects. Bibl Ophthal 82: 251-259.

Goodman LJ (1960) The landing response of insects I. The landing response of the fly, *Lucilla sericata*, and other *Calliphorinae*. J Exp Biol 37: 854-878.

Hassenstein B, Reichardt W (1956) Systemtheoretische Analyse der Zeit-, Reihenfolgen- und Vorzeichenauswertung bei der Bewegungsperzeption des Rüsselkäfers *Chlorophanus*. Z Naturforsch 11b: 513-524.

Hausen K (1984) The lobula complex of the fly: structure, function and significance in visual behavior. In: Ali MA (ed) Photoreception and vision in invertebrates. Plenum Press, New York London, pp 523-559.

Kunze P (1961) Untersuchung des Bewegungssehens fixiert fliegender Bienen. Z Vergl Physiol 44: 656-684.

Reichardt W (1961) Autocorrelation, a principle for the evaluation of sensory information by the central nervous system. In: Rosenblith WA (ed) Sensory communication. MIT Press, Wiley and Sons Inc, New York London, pp 303-317.

Reichardt W, Poggio T, Hausen K (1983) Figure-ground discrimination by relative movement in the visual system of the fly. II. Towards the neural circuitry. Biol Cybern 46 (Suppl): 1-30.

van Santen JPH, Sperling G (1984) Temporal covariance model of human motion perception. J Opt Soc Am A 1: 451-473.

Tinbergen J (1987) Photoreceptor metabolism and visually guided landing behaviour of flies. Thesis, University of Groningen, The Netherlands.

Wagner H (1982) Flow-field variable triggers landing in flies. Nature 297: 147.

Wehrhahn C, Hausen K, Zanker J (1981) Is the landing response of the housefly driven by motion of a flowfield? Biol Cybern 41: 91-99.

MOTION SENSITIVITY IN INSECT VISION: ROLES AND NEURAL MECHANISMS

M.V. Srinivasan

Centre for Visual Sciences, Research School of Biological Sciences, Australian National University
P.O. Box 475, Canberra, A.C.T. 2601, Australia

ABSTRACT

Over the past thirty or so years, motion processing in invertebrates has been studied primarily through the "optomotor response", a turning response evoked by the movement of a large-field visual pattern. The first part of this paper reviews the major insights that have been gained through this approach. More recently, however, evidence is accumulating to suggest that, in addition to the optomotor pathway, there are other pathways which use motion information in subtler ways. When an insect moves in a stationary environment, the resulting flow field (the apparent motion of various elements of the scene) is rich in range information that can be exploited to estimate the range of an object, or to distinguish an object from the background. This paper summarizes recent behavioural studies in our laboratory, investigating how honeybees accomplish such tasks.

INTRODUCTION

The visual systems of insects are exquisitely sensitive to motion. When placed in the middle of a rotating striped drum, most insects exhibit a tendency to turn in the direction of the drum's rotation. This response, termed the optomotor response, serves to stabilize the insect's orientation with respect to the visual environment, helping it maintain a straight course and simplifying the detection of objects moving within the environment. Behavioural studies of the optomotor response have shown that flies can detect and respond to moving patterns whose contrast is as low as 2 %, or whose mean luminance is so low that each photoreceptor absorbs only 2 photons/sec, on average (Dubs et al, 1981). Locusts show behavioural responses to patterns moving at speeds as low as 40 deg/day (Thorson, 1966). Contemporary research is focussed on understanding how such sensitivities are achieved, and discovering the ways in which they are utilized by the visual system.

Directionally-Selective Movement Detection

In neural terms, directionally-selective movement detection is carried out by comparing the visual signal arising from a small patch of visual space, with a delayed (or otherwise filtered) version of the signal from a neighbouring patch (rev. Reichardt, 1961). The comparison itself takes the form of a nonlinear interaction which functionally resembles multiplication, but which in neural terms probably involves shunting inhibition (Barlow & Levick, 1965; Torre & Poggio, 1978). Such an "elementary movement detector" (EMD) will produce the strongest response when a visual pattern moves in a specific direction (termed the "preferred" direction), and the weakest response when the pattern moves in the opposite direction (termed the "null" direction).

The lobula complex of many insects features neurones with large visual fields, typically covering one entire eye. These neurones can be roughly grouped into four functional classes, according to preferred direction: (i) Horizontal inward (ii) Horizontal outward (iii) Vertical upward and (iv) Vertical downward (Hausen, 1976). The evidence suggests that each of these large-field neurones spatially summates the responses of a large number of EMDs, each with the appropriate preferred direction. Taken together, the four groups of large-field neurones can detect and encode any motion of the visual environment in terms of its horizontal and vertical components.

At levels of ambient illumination approaching daylight, each EMD compares visual signals arising from neighbouring ommatidia. At lower levels of illumination, however, the movement-computing interactions extend over larger distances; this may reflect a strategy for coping with poorer signal/noise ratios (Srinivasan & Dvorak, 1980). Sensitivity to movement is also enhanced by spatially filtering the visual signal in special ways, prior to detecting movement. For example, it appears that neurones inputting to horizontal EMD's have receptive fields with excitatory centres flanked by inhibitory zones that are confined to the horizontal plane. This type of spatial prefiltering suppresses horizontally-oriented features in the visual scene (which provide no cues with regard to horizontal movement), and emphasizes vertically-oriented features (which deliver the most useful cues with regard to horizontal movement).

Other Uses for Motion Information

It is now becoming evident that, in addition to the optomotor pathway, there are other pathways which use motion information in subtler ways. Owing to the small interocular separation (in comparison with humans and other vertebrates), most insects cannot rely on stereo vision to measure the range of objects at distances greater than ca. 1 cm; they need to exploit motion cues (Horridge, 1977; Collett & Harkness, 1982). Recently it has been proposed that locomoting insects perceive the distances of objects in terms of the speeds of their images on the retina, a higher image speed being associated with a smaller range (Collett & Harkness, 1982; Horridge, 1986, 1987). Indeed, the "peering" head motions displayed by locusts, grasshoppers and mantids prior to jumping, or reaching out for a nearby twig, appear to be a means of judging the distance of objects in the visual environment by inducing motion of their images on the retina (Wallace, 1959; Collett, 1978; Eriksson, 1980; Horridge, 1986). The rest of this paper reviews behavioural research in our laboratory, where we have recently been able to demonstrate for the first time, that flying insects -- honeybees -- use retinal image motion to judge object distance. This work was carried out in collaboration with M. Lehrer of the University of Zurich, W. Kirchner of the University of Wurzburg, S.W. Zhang of the Academia Sinica, Beijing, and G.A. Horridge of the Australian National University, Canberra.

EXPERIMENTS, RESULTS AND CONCLUSIONS

Five questions concerning distance perception are posed below, and in each case the experimental procedure, results, and conclusions are summarized. In all of these experiments the procedure consisted of marking groups of ca. 6 bees, and training them to fly into a laboratory, enter an experimental apparatus, perform a specific visual task, receive a reward of sugar solution, and return to their hive. On a warm summer's day, a trained bee will continue to visit the apparatus every 5 min throughout the day.

(1) Do Bees Use Image Motion to Estimate Range ?

When bees are made to fly through a small opening (e.g. in a window), they tend to fly through its centre, balancing the distances to the left and the right boundaries of the opening. How do they gauge and balance the distances on the two sides ? One possibility is that they balance the speeds of image motion on the two sides. To investigate this, we trained bees to enter an apparatus which offered sugar solution at the end of a tunnel formed by two walls (Fig. 1). Each wall carried a pattern consisting of a vertical black-and-white grating. The grating on one of the walls could be moved horizontally at any desired speed, either toward the reward or away from it. After the bees had received ca. 60 rewards with the gratings stationary, the flight trajectories of the bees were filmed from above, as they flew along the tunnel either toward the reward, or homeward after having been rewarded.

When both gratings were stationary, the bees tended to fly primarily along the midline of the tunnel, i.e. equidistant from the two walls (Fig. 1a). However, when one of the gratings was moved at constant speed in the direction of the bees' flight -- thereby reducing the velocity of retinal image motion on that eye relative to the other eye -- the bees veered toward the side of the moving grating (Fig. 1b). Conversely, when one of the gratings was moved in a direction opposite to that of the bees' flight -- thereby increasing the speed of retinal image motion on one eye relative to the other -- the bees veered away from the side of the moving grating (Fig. 1c). There is a highly significant difference in the average locations of the flight trajectories, measured along the width of the tunnel, between the experimental conditions corresponding to Figs. 1a and 1b, between 1a and 1c, and between 1b and 1c (t-test, $p < 0.001$).

These experiments and others (Srinivasan & Kirchner, unpublished results) demonstrate that, when the gratings were stationary, the bees maintained equidistance from the walls by balancing the speeds of the retinal images on the two eyes. A lower retinal image speed on one eye was evidently taken to imply that the grating on that side was further away, and caused the bee to veer toward it; a higher retinal image speed, on the other hand, had the opposite effect. Motion perception under these conditions appears to be independent of the structure (e.g. spatial frequency) of the gratings (Srinivasan & Kirchner, unpublished results). We believe that this experiment probes an innate behaviour that many locomoting insects might use to avoid collisions, especially while passing through narrow gaps.

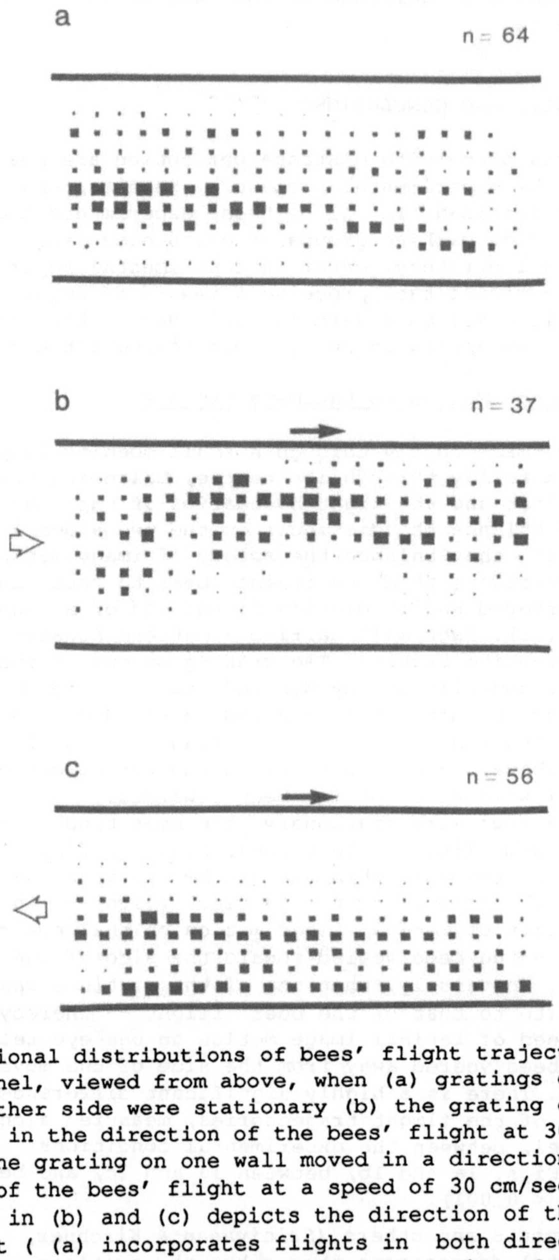

Fig. 1 Positional distributions of bees' flight trajectories within a tunnel, viewed from above, when (a) gratings on the walls on either side were stationary (b) the grating on one wall moved in the direction of the bees' flight at 30 cm/sec and (c) the grating on one wall moved in a direction opposite to that of the bees' flight at a speed of 30 cm/sec. The large arrow in (b) and (c) depicts the direction of the bees' flight ((a) incorporates flights in both directions). The small arrow depicts the direction of motion of the grating. The postional distributions were measured by digitizing a number (n) of flight trajectories, dividing the floor (plan) of the tunnel into a matrix of 20 x 10 cells, and counting the number of times a bee occupied each cell. The size (linear dimension) of each filled square is proportional to the count in the associated cell. The tunnel was 40 cm long, 20 cm tall and 12 cm wide. The height of each grating was 12 cm, and its spatial period 3.5 cm

(2) Can Bees be Trained to Discriminate Between Various Distances ?

We investigated this question by training bees to visit an artifical "meadow" and collect a reward (a droplet of sugar water) from a "flower", raised 70 mm above the floor of the meadow, and presented along with six other flowers, offering no reward, placed on the floor of the meadow (Fig. 3a). The flowers were black, the floor of the meadow was white and the meadow was surrounded by a wall which encouraged the bees to fly above the constellation of flowers, viewing the flowers ventrally before landing on one of them. The rewarded flower was changed frequently to exclude odour cues, and the sizes and positions of all of the flowers were randomly varied between rewards, so as to ensure that the bees were trained to associate only the height of the flower (or, more accurately, its ventral distance from the eye), and not its position, or angular subtense, with the reward (Lehrer et al, 1988).

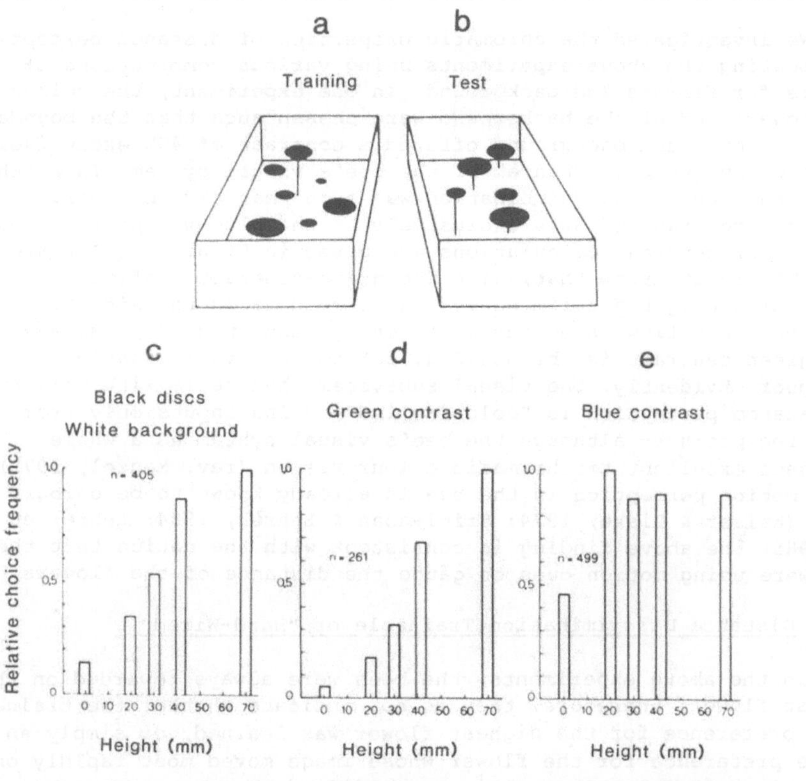

Fig. 2 (a) Training configuration consists of one "high" flower, carried on a 70 mm high stalk, offering a reward of sugar water. In addition six "low" flowers, bearing no reward, are present on the floor. Flower diameters range from 20 mm to 70 mm. (b) Test configuration consists of five flowers of different diameters (ranging from 20 mm to 70 mm) and heights (ranging from 0 mm to 70 mm). In both configurations, the meadow is 30 cm x 40 cm and is surrounded by a wall 12 cm in height. The bar charts show the relative choice frequencies with respect to the flowers at various heights (normalized to a maximum value of unity) for (c) black flowers on a white background; (d) the green-contrast colour combination ; and (e) the blue-contrast colour combination. n = total number of landings counted.

After receiving ca. 30 rewards, the trained bees were tested using a constellation of five flowers of different sizes, each placed at a different height, ranging from the training height (70 mm) to the floor level (0 mm; Fig. 2b). During tests, no reward was offered and the number of landings on each of the five flowers was counted. The sizes and positions of the flowers were varied randomly from one test to another. The choice frequencies with respect to the five test flowers are compared in Fig. 2c. The bees showed the highest preference for the flower at the training height, and a progressively decreasing preference for flowers at lower heights. Thus, bees are capable of distinguishing between flowers at various ventral distances. Since the training procedure eliminated size and positional cues, this finding suggests that the bees were using the only remaining visual cue that was available to them, namely absolute angular velocity of the flowers' images on the retina.

3) What is the Role of Colour in this Form of Distance Perception ?

We investigated the chromatic properties of distance perception by repeating the above experiments using various combinations of colours for flowers and background. In one experiment, the colour of the flowers and of the background were chosen such that the boundary between flower and background offered a contrast of 40% exclusively to the green-receptor channel of the bee's visual system. In another experiment, the colour combination was such that this boundary offered a contrast of 56 % exclusively to the blue-receptor channel. (Details of contrast calculations are given in Srinivasan & Lehrer, 1984).The reults show that, with the green-contrast configuration (Fig. 2d), height discrimination is as good as in the situation with the black flowers on the white background (Fig. 2c). However, when green contrast is absent (Fig. 2e) height discrimination is very poor. Evidently, the visual subsystem that deals with this type of distance perception is "colour-blind", using inputs only from green receptors -- although the bee's visual system as a whole possesses excellent trichromatic colour vision (rev. Menzel, 1979). Since motion perception in the bee is already known to be colour-blind (Kaiser & Liske, 1974; Srinivasan & Lehrer, 1984; Lehrer et al, 1985) the above finding is consistent with the notion that the bees were using motion cues to gauge the distance of the flowers.

4) Is Distance Discrimination Trainable or "Hard-Wired" ?

In the above experiments, the bees were always rewarded on the highest flower; therefore, they do not indicate whether the trained bees' preference for the highest flower was learned, or simply an innate preference for the flower whose image moved most rapidly on the retina. To investigate this, we modified the above setup by using a transparent perspex sheet, raised 5 cm above the floor. Three discs, representing "flowers" were used, each of which could be presented either "low" (on the floor) or "high" (affixed to the underside of the perspex sheet). With this arrangement, it was possible to reward either a "high" (rapidly-moving) flower or a "low" (slowly-moving) flower -- by placing a drop of sugar water on the perspex sheet, within the area projected by the flower on the sheet -- whilst ensuring that the bees always viewed all of the flowers ventrally, i.e. that they never flew at heights below that of the "high" flower (Lehrer et al, in press). As before, the sizes and positions of the flowers were varied randomly between rewards as well as between tests. In tests, commenced after 30 rewards, we recorded the landings on the area projected by each flower on the perspex sheet above it.

Bees rewarded on a "high" flower (with two additional flowers placed on the floor) showed a strong preference for the "high" flower in subsequent tests: 73.0 % of the landings occurred on the "high" flower (n = 488). On the other hand, bees rewarded on a "low" flower (with two additional flowers affixed to the perspex sheet) subsequently showed a strong preference for the "low" flower, which attracted 84.4 % of the landings (n = 243). In a experiment we used two perspex sheets, one 5 cm, the other 10 cm above the floor to present each of the three flowers at a different level. The bees were rewarded on the intermediate-level flower, by placing a drop of sugar water above it on the uppermost perspex sheet (details in Lehrer et al, in press). Again, the bees learned this task well, choosing the rewarded flower 68.6 % of the time (n = 461), and demonstrating that they were capable of discriminating an "intermediate" distance from nearer and farther distances. We conclude that distance preference in the honeybee is a flexible, trainable process, like colour preference (rev. Menzel,1979).

5) Can Bees Segregate an "Object" from a "Background" Solely on the Basis of Motion Cues ?

In all of the above experiments, the "flowers" were readily visible to the bees, since they presented a strong contrast -- in luminance or colour -- against the background. We wondered whether bees are capable of detecting an object when the luminance or colour contrast that is normally present between object and background is replaced by "motion contrast", i.e. relative motion between the two.

This question was tackled by examining whether bees could be trained to find a textured figure when it was presented raised over a background of the same texture (Srinivasan & Lehrer, unpublished results). The figure was a disc, bearing a random black-and-white texture, carried on the underside of a transparent perspex sheet that could be placed at any desired height (h) above the background (Fig. 3a). During training, the bees were rewarded by a droplet of sugar water placed on the perspex sheet above the disc. The position of the disc on the perspex sheet was frequently changed, so as to exclude the possibility of the bees' using positional cues to find the disc. The texture of the disc was always oriented parallel to that of the background, so as to exclude the possibility of the bees' using orientational cues to detect the disc.

Bees, trained on this configuration with the disc raised 5 cm above the floor, were subsequently tested at various disc heights with the reward removed (Fig. 3b). With h = 5 cm, the bees experienced no difficulty in locating the disc: 82 % of the landings occurred within it (Fig. 3b). As the disc was lowered, it became progressively more difficult to locate. Nevertheless, for all heights ranging between 5 cm and 1 cm (including 1 cm), the percentage of landings occurring within the disc continued to remain significantly above that expected if the bees were landing at random anywhere within the area of the background ($p < 0.0005$, Chi-squared test) -- the random-choice level being 2.3 %, the ratio between the areas of the disc and the background . When the height of the disc is reduced to 0 cm (i.e. it is placed directly on the background), the bees are no longer able to detect it: the average frequency of landings occurring within the disc (4.4 %, Fig. 3b) is no longer significantly different from the random-choice level ($p > 0.05$, Chi-squared test). The threshold height for detection of the disc is thus between 0 cm and 1 cm.

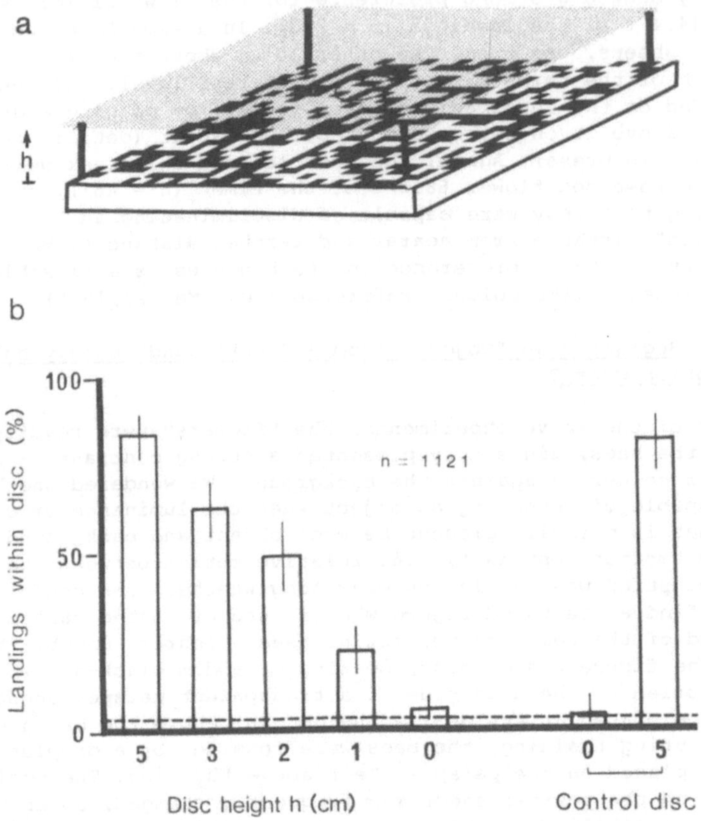

Fig. 3 (a) Apparatus for training and testing discrimination of a textured disc (diameter 6 cm) presented at a height of h cm over a similarly-textured background (size 42 cm x 30 cm, texture pixel size 5 mm x 5 mm). (b) Results of tests using apparatus of (a). The bars on the left show the percentage of landings occurring within the disc, for various heights (h) of the disc above the background. The two bars on the right depict the results of control experiments in which the disc was replaced by one of the same size, but carrying a texture of one-third the original density. Dashed horizontal line depicts the level of random choice (2.3 %).

Control experiments, using discs of altered textural density (Srinivasan & Lehrer, unpublished results), assured us that the bees were not using textural differences (i.e. the difference in apparent textural density between the disc and the lower background) as a visual cue in detecting the disc.

Video films of the bees' flight trajectories showed that, when the disc was visible to the bees, they did not land at random on it: rather, they landed primarily near the boundary of the disc. This preference for the boundary was independent of the shape of the raised figure: it persisted when bees trained on the textured disc were offered a textured square, or a triangle, instead (Srinivasan & Lehrer, unpublished results).

These experiments demonstrate that bees, like flies (Reichardt & Poggio, 1979) are capable of detecting an object against a background by using as a cue the apparent motion of the object relative to the background. They also reveal that the boundary between the object and background -- where the abrupt change of range causes an abrupt change in the speed of motion of the image on the retina -- is of special visual significance. The discrimination of objects through motion discontinuities is likely to be important when an insect attempts to land on a leaf of a shrub: a situation where it may be difficult to identify individual leaves, and to establish which leaf is nearest, when motion cues are absent. This difficulty is not confined to insect vision: for example, it is vividly experienced by anyone attempting to distinguish individual trees in a dense forest (von Helmholtz, 1866). Further experiments, investigating the significance of various kinds of motion contrast at object boundaries, are currently in progress.

REFERENCES

Barlow, H.B. and Levick, W.R., 1965, The mechanism of directionally selective units in rabbit's retina. J. Physiol, Lond. 178:477-504.

Collett, T.S., 1978, Peering - a locust behaviour pattern for obtaining motion parallax information. J. Exp. Biol. 76: 237-241.

Collett, T.S. and Harkness, L.I.K., 1982, Depth vision in animals, in Analysis of Visual Behavior, eds. D.J. Ingle, M.A. Goodale and R.J.W. Mansfield, Boston: MIT Press, 1982, ch. 4, 111-176.

Dubs, A., Laughlin, S.B. and Srinivasan, M.V., 1981, Single photon signals in fly photoreceptors and first order interneurons at behavioural threshold. J. Physiol., Lond. 317: 317-334.

Eriksson, E.S., 1980, Movement parallax and distance perception in the grasshopper (Phaulacridium vitatum (Sjostedt)). J. exp. Biol. 86: 337-340.

Hausen, K., 1976, Functional characterization and anatomical identification of motion-sensitive neurons in the lobula plate of the blowfly Calliphora erythrocephala. Z. Naturf. 31c: 629-633.

von Helmholtz, H., 1866, Handbuch der Physiologischen Optik, Hamburg: Voss (Transl. J.P.C. Southall, New York: 1924, repr. Dover: 1962).

Horridge, G.A., 1977, "Insects which turn and look", Endeavour N.S. 1: 7-17.

Horridge, G.A., 1986, A theory of insect vision: velocity parallax. Proc. R. Soc. Lond. B 229: 13-27.

Horridge, G.A., 1987, The evolution of visual processing and the construction of seeing systems. Proc. R. Soc. Lond. B 230: 279-292.

Kaiser, W. and Liske, E., 1974, Die optomotorischen Reaktionen von fixiert fliegenden Bienen bei Reizung mit Spektrallichtern. J. Comp. Physiol. 89: 391-408.

Lehrer, M., Srinivasan, M.V., Zhang, S.W. and Horridge, G.A., 1988, Motion cues provide the bees' visual world with a third dimension. Nature 332: 356-357.

Lehrer, M., Wehner, R. and Srinivasan, M.V., 1985, Visual scanning behaviour in honeybees. J. Comp. Physiol. A 157: 405-415.

Menzel, R., 1979, Spectral sensitivity and colour vision in vertebrates. In: Vision in Invertebrates, H. Autrum (ed), Handbook of Sensory Physiology, vol 7/6A, Berlin: Springer, pp. 503-580.

Reichardt, W., 1961, Autocorrelation, a principle for the evaluation of sensory information by the central nervous system. In: Sensory Communication, ed. Rosenblith, W. pp. 465-493. Cambridge: MIT Press and Wiley: New York.

Reichardt, W. and Poggio, T., 1979, Figure-ground discrimination by relative movement in the visual system of the fly. Part I. Experimental results. Biol. Cybern. 35: 81-100.

Srinivasan, M.V. and Dvorak, D.R., 1980, Spatial processing of visual information in the movement-detecting pathway of the fly. J. Comp. Physiol. 140: 1-23.

Srinivasan, M.V. and Lehrer, M., 1984, Temporal acuity of honeybee vision: behavioural studies using moving stimuli. J. Comp. Physiol. A 155: 297-312.

Thorson, J., 1966, Small-signal analysis of a visual reflex in the locust. Kybernetik 3: 41-66.

Torre, V. and Poggio, T., 1978, A synaptic mechanism possibly underlying directional selectivity to motion. Proc. R. Soc. Lond. B 202: 409-416.

Wallace, G.K., 1959, Visual scanning in the desert locust Schistocerca gregaria Forskal. J. Exp. Biol. 36: 512-525.

RESPONSE BEHAVIOUR OF ELEMENTARY MOVEMENT DETECTORS IN THE VISUAL SYSTEM OF THE BLOWFLY

R Bult, F.H. Schuling and H.A.K. Mastebroek

Laboratory for General Physics, Biophysics Department
University of Groningen, Westersingel 34
9718 CM Groningen, The Netherlands

ABSTRACT

In this paper we present experiments, involving sequential micro-stimulation of two (or more) adjacent neuro-ommatidia in the compound eye of the blowfly, *Calliphora erythrocephala (Meig.).* Experiments, using brief flashes of 3 ms duration with low intensities in the order of $2.4 \cdot 10^{-2}$ Cd/m², were performed to isolate the individual response contributions of single Elementary Movement Detectors (EMD's) in the input micro-circuitry of the motion-sensitive directionally-selective H1 neuron. A two-dimensional mapping of single EMD contributions to the overall response will be presented for the dark-adapted eye. It is concluded that under such low illumination levels (when compared to normal daylight situations, where illumination typically varies between 1 and 200 Cd/m²), contributions from EMD's with sampling bases up to 8 $\Delta\phi_h$, oriented along the horizontal sensitivity axis of the neuron, contribute most to the response of the neuron.

Supplementary experiments indicate that, when in addition to an ongoing sequence in (for instance) the null direction, thus inhibiting the activity of the neuron, a second sequence is presented somewhere in the receptive field of the H1 neuron, the total response is a non-linear combination of both individual responses. Interpretation in terms of a pooling correlation scheme, which, over a limited target region, sums the activities of the EMD's in a highly non-linear fashion, seems to provide a qualitative explanation.

INTRODUCTION

In the fly visual system, all optic ganglia (lamina, medulla, lobula and lobula plate) are organized as retinotopically ordered 2-dimensional sets of neural columns. The highest order ganglion, the lobula plate, contains wide-field movement-sensitive neurons which have large receptive fields (Eckert, 1981; Hausen, 1981, 1982a, 1982b; Hengstenberg, 1982; Hengstenberg *et al.*, 1982) These neurons perform a spatial summation of the activity evoked in the columns of the lobula plate and they transform the total amount of electrical activity into spike trains. There is strong evidence that these large neurons are primarily involved in providing input information for the motorcontrol of course stabilization (Hausen and Wehrhahn, 1983; Hausen, 1984). One of these wide-field neurons, classified as H1 by Hausen (1976), is highly sensitive to movement in the horizontal direction: spike rates up to 300 spikes/s can be recorded from this neuron in the case of visual stimuli moving horizontally inward, i.e. from back to front, in the visual field (preferred direction), whereas movement in the reverse (null) direction suppresses the activity of the neuron.

In order to obtain more insight in some fundamental spatial properties of the neuronal micro-circuitry underlying movement detection, we recorded the nerve activity of the H1 neuron under micro-stimulation of separate input channels (at lamina level) of Elementary Movement Detectors (EMD's). These EMD's form the input units of the H1 neuron and it is believed that they are distributed uniformly over the entire eye. Also it is assumed that the output signals of all EMD's in the receptive field of the H1 neuron are summed by this neuron. Here we are mainly concerned with the following questions: 1. How do the input channels of the EMD's relate to the ommatidial lattice and how are the outputs of EMD's with different sampling base and orientation combined to give the output of the large tangential H1 cell? Do the output signals combine linearly under certain conditions, and if so, what are the weighing factors of individual contributions to the total response? 2. Do our results justify an interpretation in terms of a spatial neural pooling correlation scheme, as modelled by Poggio, Reichardt and Hausen (Poggio and Reichardt, 1976; Poggio *et al.*, 1981; Poggio, 1983; Reichardt *et al.*, 1983)? In this model large-field binocular pool cells sum the output of a retinotopic array of small-field EMD's over a large part of the visual field of the two compound eyes. These pool cells inhibit, via shunting inhibition, the signals provided by the EMD's, irrespective of their preferred direction. After inhibition of each channel, all signals are fed into a large-field output cell. This cell has two important properties: a) it signals relative motion of a moving object superimposed on a stationary background of the same texture as the object itself, and b) its output is independent of the size of the moving object. This model has proven to agree very well with behavioural data from the fly. What adjustments can be concluded from our electro-physiological data? Are EMD signals summed equally or by means of some weighing process? Non-linear summation effects will be examined for a few special stimuli (transparency phenomena).

MATERIALS AND METHODS

Preparation of the fly, recording and data-analysis: Adult (i.e. 1-2 weeks post emergence) female wild-type blowflies *C. erythrocephala (Meig.)*, reared from outdoor caught specimens, were used for the experiments. After the wings were immobilized with wax, the animal was put into a little plastic cylinder, that could be mounted in a perimeter.

In order to be able to penetrate the lobula plate with a tungsten micro-electrode (tip diameter 1-5 μm) the fly's head was bent forward and its genae were attached to the thorax. In addition the antennae were attached with their base to the head. On the rear of the head, on the contralateral side, a piece of the integument was cut away to expose the lobula plate. Also on the ipsilateral side, a small piece of cuticle was cut away to be able to place a perspex light guide (necessary for the antidromic illumination during the initial phase of the experiment) into one of the air sacks (Fig. 1, part I). A thin silver wire (diameter 0.1 mm) was used as reference electrode. The electrical activity of the H1 neuron, recorded extracellularly, was stored on-line by a NOVA 1200 micro-computer on disk as digitized inter-spike intervals together with a 1 bit stimulus synchronization signal, generated by a 68K micro-processor (FORCE SYS68K, CPU1) based device that controls the real-time display of the visual stimuli (see Schuling, 1988 and Fig. 1, part III). Further data processing details can be found in Zaagman (1977). Off-line data analysis was performed on a general purpose MV/4000 mini-computer (Fig. 1, part II). When a single pair of visual input elements is stimulated, the response of the H1 neuron will be rather small when compared with the situation in which a moving grating stimulates simultaneously several columns of inputs. Therefore, a certain stimulus has to be presented repeatedly for about 200-300 times, in order to obtain final averaged responses with relatively small (i.e. 10 %) statistical errors. We identified the H1 neuron by its very characteristic (and well-

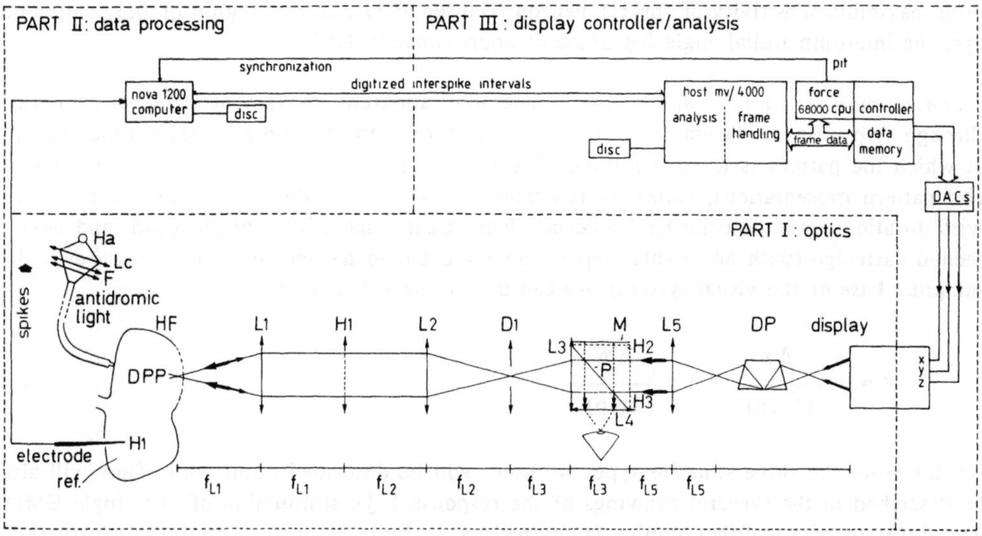

Fig. 1 Schematic drawing of the stimulus generating system, optical setup and data acquisition for micro-stimulation of single EMD's. Part I: Optical setup: HA: halogen light source, LC: light condensor, F: heat filter and broad-bandpass filter (575 nm), D_1: diaphragm, L_5: lens for projection of the display spots, DP: Dove prism, L_1 microscope objective, L_4: eyepiece for viewing the images in the plane H_3, H_1, H_2, H_3: planes in which the far field can be seen, P: pellicle (beamsplitter), M: mirror, L_2, L_3,: lenses (25 mm), electrode: tungsten micro-electrode, ref: silver reference electrode, HF: head of the fly, DPP: Deep Pseudo Pupil. Part II: Data processing setup: the analogue output generated by the display controller system is used to synchronize the digitization of the measured action potentials. After the experiment is finished, data are transferred to the host for off-line analysis. Part III: Display controller setup: Frame data can be downloaded from the host into 256 kB of DRAM, from which the contents can be directly addressed via DMA. P.I.T.: Programmable Interval Timer.

specified) responses to wide-field movement in both preferred and null direction (Hausen, 1976, 1981). Using a Dove prism (DP in Fig. 1, part I), a reliable inspection of the directional selectivity of any neuron, whose electrical activity was picked up by the electrode, was alleviated by rotation of a standard wide-field motion stimulus.

Generation of the visual stimuli and optical setup: The stimuli were generated on a high-intensity, high-resolution display (HP 1332A, aluminized phosphor P31) by means of the previously mentioned real-time display controller system (Fig. 1, part III). Unless mentioned otherwise, each individual spot had a diameter of about 0.1 $\Delta\phi$ ($\Delta\phi$ is the inter-ommatidial angle) in the projection plane of the stimuli. The stimuli were displayed with a repetition frequency of 1 kHz.

A precise optical imaging technique (Fig. 1, part I), described first by Franceschini (1975), enables us to view the regular hexagonal mosaic of the optical axes of the ommatidia in the intact eye of a living blowfly using antidromic illumination. The far field radiation pattern (FFRP) thus generated is projected onto the plane H_2 (using the lenses L_1 through L_3 and a diaphragm D_1). In this plane also the stimuli generated on the HP 1332A display are imaged using a Dove prism (DP) and lens L_5. Both stimulus pattern and FFRP can be seen simultaneously by means of a pellicle P, lens L_4 and a mirror M.

In all experiments, stimuli were projected on neuro-ommatidia in the frontal region of the eye slightly above the equator and lateral of the vertical symmetry plane, i.e. the region with maximum sensitivity of the H1 neuron (Hausen 1981). In this region of the compound eye, the interommatidial angle $\Delta\phi$ amounts approximately 1.6°.

Stepwise moving stimuli, definitions: Generally, apparent movement can be generated through step wise displacement of a stimulus pattern between two successive time instants at which the pattern is to be displayed. The interjump time, i.e. the time interval between two pattern presentations, generally is defined as Δt. If we use brief display times, each with duration p and activate first a certain lamina cartridge with a bright flash, and next a second cartridge (with an angular separation Δx equal to $n \cdot \Delta\phi$ (n=1,...), where $\Delta\phi$ is the sampling base of the visual system), we can define the velocity by:

$$v = \frac{\Delta x}{(\Delta t + p)} = n \cdot \frac{\Delta\phi}{(\Delta t + p)}$$

At this point the three stimulus types we used, will be described point wise (They will also be described in the serveral drawings of the responses). 1): stimulation of one single EMD. The duration of the full stimulus is defined as T, built up out of an initial delay time (typically 45 ms), the first pulse with duration p (typically 3 ms), the interjump time Δt (typically 78 ms), the second pulse (also with duration p), and a long recovery time meant to minimize memory effects. (Mastebroek *et al.* 1982). 2): sequential stimulation of 8 cartridges. Duration p of each flash was 3 ms and the interjump time Δt was in all cases 78 ms (as for type 1 flashes). A sequence of adjacent cartridges, each separated by an angle of exactly 2 times $\Delta\phi$, was illuminated, first in the preferred direction and later on in the reverse direction. Also, these antagonistic sequences (both subsequently illuminating 8 adjacent cartridges) were projected simultaneously to the same row of neuro-ommatidia, both starting at opposite ends of the row. Similarly, experiments were performed, in which two rows were stimulated at the same time in the preferred direction, and also one row of cartridges in the opposite, i.e. null direction, and *vice versa.* In these trials the vertical distance between the separately stimulated rows varied from $\Delta\phi_V$ to 3 times $\Delta\phi_V$. 3): A row of 7 cartridges was stimulated in the null direction with identical temporal parameters as in the type 2 experiments. In conjunction with the illumination of elements 5 and 6 in this sequence, an additional single correlator (type 1) was activated in its preferred direction. In this case we varied the distance between the positions of the flashes from the ongoing sequence at this moment and the additional single EMD stimulus from $\Delta\phi_V$ to 3 times $\Delta\phi_V$.

Sequential illumination of single EMD's, nomenclature: A certain EMD is denoted according to the H(horizontal)/V(vertical) system (Fig. 2). For instance, an EMD with a horizontal separation between both cartridges of $4\Delta\phi_h$, and a vertical separation of $2\Delta\phi_V$, is defined as a $C_{4,2}$ correlator, when the preferred direction is meant and correspondingly as $C_{-4,-2}$ when the null direction is meant.

Typical responses to type 1 stimuli are shown in the Post Stimulus Time Histograms (PSTH's) of the Figs. 3a and 3b. The intensity response is defined as the number of spikes per second per stimulus presentation, being elicited by the intensity step, projected onto the first cartridge of a certain EMD. The second peak in the PSTH is a combined response to the intensity step, projected onto the second cartridge, and the movement induced excitatory or inhibitory response. Each EMD is stimulated twice, first in its preferred direction and later in its null direction, or *vice-versa.* Then, both the intensity responses of the two cartridges are known, as well as the two compound (intensity and movement) responses. The pure movement response $RR_{mov,pref/null}$ for the preferred/ null direction is defined as the difference between the combinatory response, obtained when a sequence of two cartridges is stimulated in that preferred/null direction, and the intensity response of the cartridge, which is illuminated second in that sequence. In all cases we subtracted the stationary background activities.

RESULTS

Experiment 1. Movement response dependence on flash luminance, flash duration and interjump times: In order to determine the dependence of the responses on both luminance level and flash duration, single $C_{2,0}$ correlators were activated both in the preferred direction as well as in the null direction for a range of luminances and pulse durations. Fig. 4 shows the movement responses of a single $C_{2,0}$ EMD as a function of luminance level for 4 different pulse durations of e.g. 1, 3, 6 and 9 ms. The interjump time between both pulses is 78 ms for all cases. Except for the shortest pulse duration, the movement response curves are rather similar. Furthermore, no significant movement responses appear to be elicited by luminance levels of the flashes below about $3 \cdot 10^8$ photons/s/cm² (corresponding to $2.5 \cdot 10^{-4}$ Cd/m²). This value is almost identical to the luminance threshold for the turning response in *Musca*, reported by Pick and Buchner (1979). In addition, the data shown here confirm earlier results obtained by Lenting (1985), indicating that there are no saturation effects for brief flashes with a pulse duration of 3 ms and a luminance level of about $9.5 \cdot 10^9$ photons/s/cm² (i.e. $2.4 \cdot 10^{-2}$ Cd/m²; this is well below the threshold of screening pigment migration in R_{1-6}, which is in the order of 1 Cd/m² (Heisenberg and Buchner, 1977)).

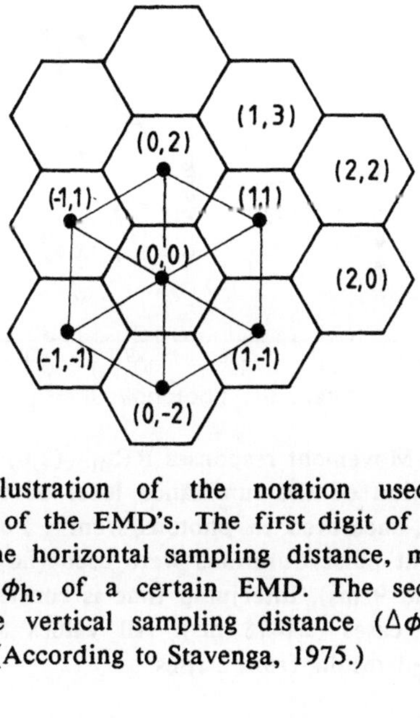

Fig. 2. Illustration of the notation used for the description of the EMD's. The first digit of each index indicates the horizontal sampling distance, measured in units of $\Delta\phi_h$, of a certain EMD. The second index denotes the vertical sampling distance ($\Delta\phi_v$) of that correlator. (According to Stavenga, 1975.)

For a variety of EMD's, normalized movement responses as a function of the interjump time Δt, are shown in Fig. 5. It may be concluded that for these stimulus conditions responses for apparent movement in the preferred direction are about optimal for interjump times between 20 and 100 ms, irrespective of the type of correlator activated. Responses in the null direction are about best for an even broader range of interjump times: between 20 and 250 ms. These times implicate an optimal motion-detection for angular velocities up to 250°/s (that is, when a $C_{1,3}$ correlator is stimulated with an interjump time of 20 ms). The shapes of the response curves indicate that the detection of movement by this system is confined to a sharply defined range of angular velocities, but within this range all EMD's seem to respond about equally, irrespective of their sampling distance.

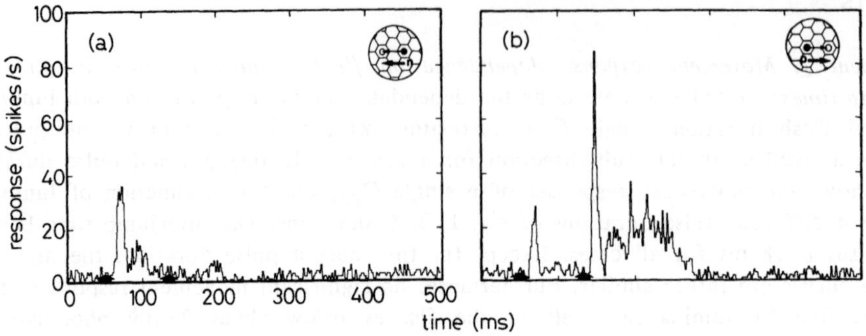

Fig. 3. PSTH's of the responses of the H1 neuron for stimulation of a single $C_{2,0}$ correlator. a) PSTH for a type 1 stimulus in the preferred direction b) PSTH for the same stimulus sequence in the null direction.

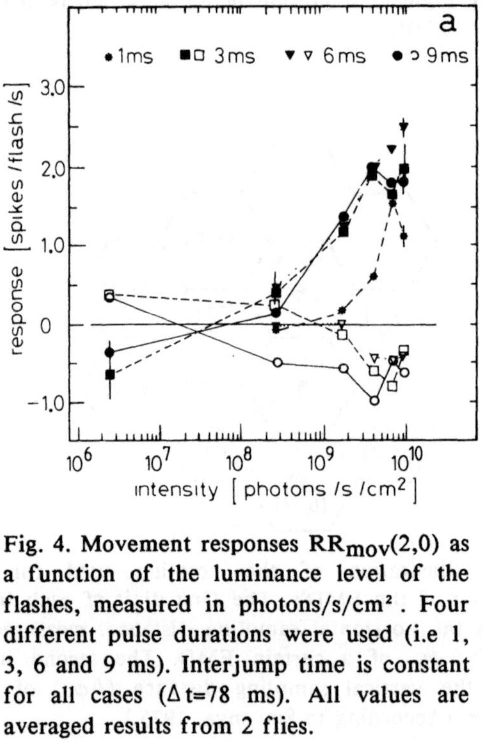

Fig. 4. Movement responses $RR_{mov}(2,0)$ as a function of the luminance level of the flashes, measured in photons/s/cm². Four different pulse durations were used (i.e 1, 3, 6 and 9 ms). Interjump time is constant for all cases (Δt=78 ms). All values are averaged results from 2 flies.

Experiment 2. Relative contributions of Elementary Movement Detectors to the total response of the H1 neuron in the dark-adapted eye: Responses of the H1 neuron to stimulation of single EMD's from which the visual input elements in some cases were separated more than 20 times the inter-ommatidial base $\Delta\phi_h$ (along the horizontal axis), were measured by us for the dark-adapted eye. A two-dimensional mapping of contributions of single EMD's to the overall response of the neuron could be obtained this way.

Luminance levels of the flashes, evoking the apparent movement responses, measured in the projection plane of those flashes, were roughly equal to $2.4 \cdot 10^{-2}$ Cd/m². Individual absolute response values, defined as $RR_{mov,pref/null}$ (in spikes/flash/s), which relate to sampling bases $2n \cdot \Delta\phi_h$ with n=1,...,10 for EMD's along the horizontal axis, are shown in Fig. 6. The stimulus spot size was 0.1 $\Delta\phi$ and the interjump time between the two flashes

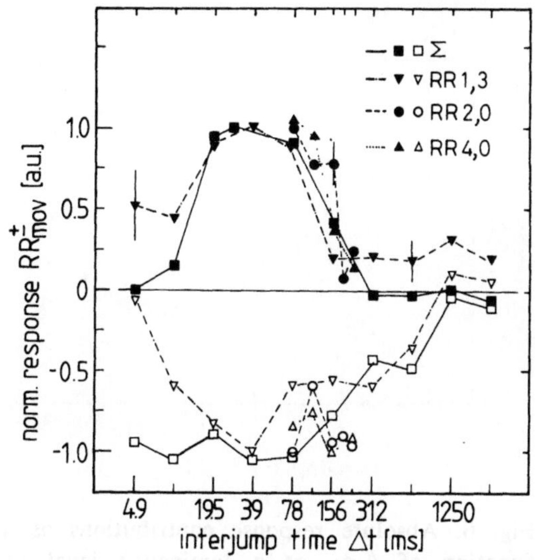

Fig. 5. Normalized movement responses of serveral EMD's as a function of the interjump time Δ t between the flashes. Three different EMD's have been stimulated. The responses for the $C_{2,0}$ and the $C_{4,0}$ correlator pairs are average values for three flies. The responses for the $C_{1,3}$ and the sum over 16 correlators are reconstructed from Lenting (1985). Luminance level for all stimuli was about $2.4 \cdot 10^{-2}$ Cd/m²

(3 ms duration each) was kept constant at 78 ms. This limits the range of angular velocity magnitudes to values between 30°/s and 300°/s. Values are averages over 3-8 flies. For the preferred direction the first three response coefficients are approximately equal (respectively 1.00, 1.06 and 0.91). With increasing sampling base the coefficients fall off gradually (respectively 0.54, 0.34 and 0.38) and can not significantly be distinguished from zero after the contribution, relating to a sampling base of 12 times $\Delta\phi_h$ (i.e. remaining normalized coefficients: 0.06, 0.05, 0.34, 0.06). For the null direction the first 4 (i.e. up to 8 times $\Delta\phi_h$) responses are small, but roughly equal (i.e. coefficients normalized to the largest preferred value are: -0.19, -0.20, -0.31, -0.27). For larger sampling bases the responses appear to reverse in sign (normalized values: 0.20, 0.19) and can be neglected starting with the coefficient relating to 14 times $\Delta\phi_h$ (i.e. next values are: 0.04, 0.02, -0.16, 0.03). Negative numbers do not bear any physiological meaning, but rather indicate suppression of the neural activity below the spontaneous dark level. It is clear however that after the first four coefficients, the difference between the preferred and null direction has disappeared completely. $\Delta\phi_h$ is approximately 1.25° in the frontal equatorial zone of the eye. This implicates that for the preferred direction under the present stimulus conditions the motion-sensitive system is able to measure angular velocities in a range from at least 30°/s up to 125°/s.

Fig. 7 shows two-dimensional maps of relative response contributions, measured for 42 to 44 EMD's, drawn in a regular hexagonal coordinate system. A correlator $C_{x,y}$ is defined here by the position of the second cartridge with respect to the first cartridge, as measured in the H,V coordinate system. The first (i.e. Horizontal) index represents the number of columns that separates both lamina cartridges (within an EMD) and the second (Vertical)

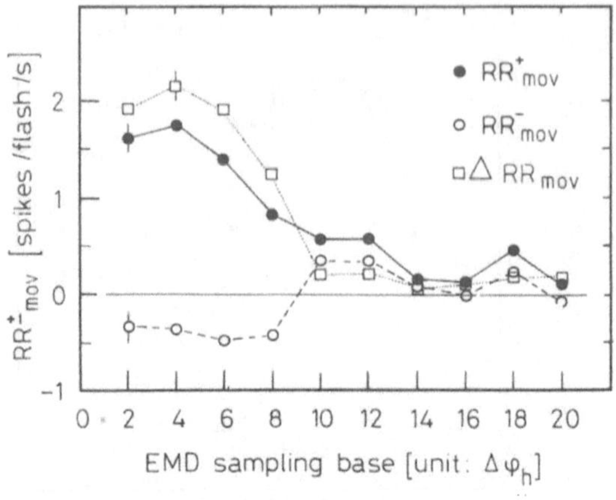

Fig. 6. Absolute response contributions as a function of $\Delta\phi_h$ at a luminance level of approximately $2.4 \cdot 10^{-2}$ Cd/m² for the dark-adapted eye. Results are averages over 3-8 flies. Typical standard errors are shown.

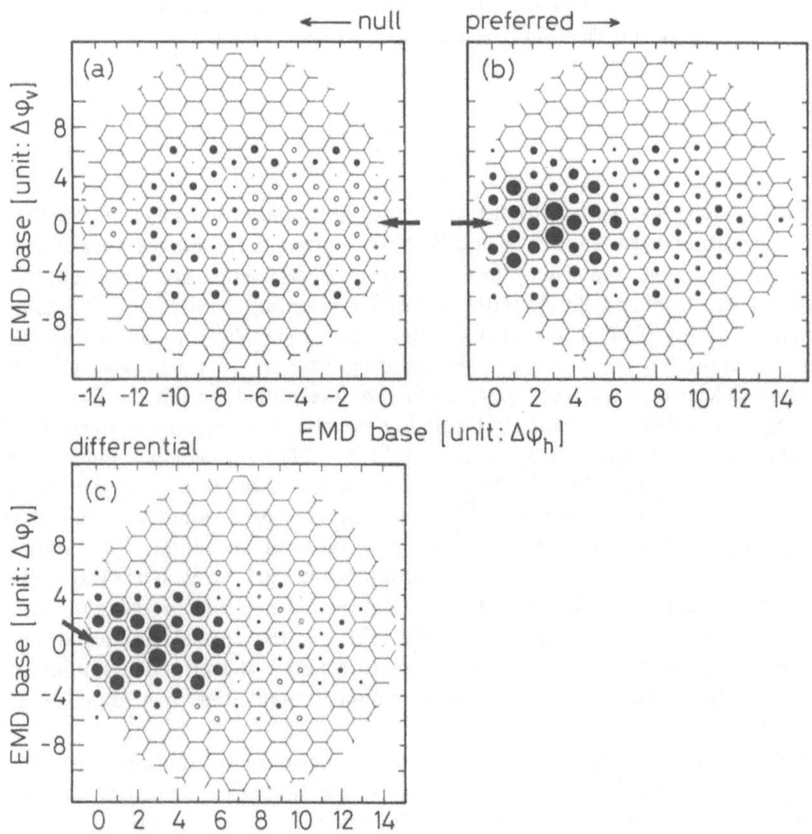

index the number of rows. In all maps, besides the correlators $C_{x,y}$ with (x,y=0,1,...), also the correlators $C_{x,-y}$ (with x,y=0,1,...), which lie symmetrically with respect to the horizontal axis, are drawn. It is presumed that these homologous pairs respond in an identical way. The size of the circles within each thus defined hexagon represents the relative response contribution for that specific EMD, normalized to the strongest (positive) contribution within this map. Positive (excitatory) contributions are drawn as black circles, negative (inhibitory) contributions as open circles. The correlator map for the null direction of the H1 neuron is shown in Fig. 7a, and the map for the preferred direction in Fig. 7b. Although the negative response contributions are smaller than the positive coefficients, a similar pattern as in Fig. 7b emerges from inspection of Fig. 7a, i.e: the nearest input channels up to 6 times $\Delta\phi_h$ and 2 times $\Delta\phi_v$ contribute about equally to the total response of the H1 neuron. Peculiar is the center-surround zone in Fig. 7a at larger distances with small, but definitely positive response coefficients. Fig. 7c is a differential map. It shows a clear, distinct zone of positive interaction coefficients, surrounded by small, sometimes negative contributions. From these maps it is evident that EMD's, with sampling bases of 2, 4 and 6 times $\Delta\phi_h$, preferentially oriented along the horizontal axis, contribute about equally to the response of the H1 neuron.

Experiment 3. Pooling phenomena for interactions between two moving stimuli: The weighed pooling mechanism, mentioned in the introduction, can be examined by apparent movement experiments using more prolonged sequences of flashes. These experiments have been described as type 2 and 3 before: A spot is displaced step wise with interjump times of 78 ms along a row of adjacent cartridges (separations equal to $\Delta\phi$ or 2 times $\Delta\phi_h$), eliciting movement responses in EMD's distributed over a large angular region in the receptive field of the H1 neuron. In concert with this movement, another (wide-field or small-field) sequence is presented and the response to the combined movement is interpreted in terms of the pooling scheme as proposed by Reichardt *et al* (1983).

Figs. 8a and 8b show typical PSTH's, recorded in response to apparent movement in the preferred and the null direction, respectively, when a sequence of 8 adjacent cartridges is activated. The PSTH in Fig. 8a shows a strong facilitation of the activity of the neuron, whereas Fig. 8b shows inhibition of all activity after the initial intensity response. It is clear that the sharp peaks in Fig. 8a are corresponding to the responses of single cartridges to the intensity steps of 3 ms duration, each followed by a vigorous movement dependent response, which partly coincides with the subsequent intensity peak. Also evident is the peculiar (at first sight) envelope of the responses. A simple explanation is: the fact that the step size is in the order of 2.5°, results in an angular extension of the entire movement of about 20°. According to Hausen (1981), the sensitivity of the H1 neuron is about maximal in the frontal, equatorial region, but falls off to 70-80 % of the maximum value within 20°. Therefore the envelope of the individual responses resembles the angular sensitivity profile of the H1 neuron in the frontal region of the eye. Adaptation of the neuron to the ongoing pattern motion is a second phenomenon, which may account for the decrease in response to the last few movement steps. The total number of spikes, registered in the interval between 0 and 800 ms, divided by the number of stimulus presentations and normalized per second, is denoted in Fig. 9 by an open circle, marked with *pref* on the x-

Fig. 7. a) Correlator map for the individual relative contributions of 44 EMD's to the total response of the H1 neuron for the null direction in the dark adapted compound eye. b) Correlator map, with individual contributions of 42 EMD's obtained for the preferred direction of the H1 neuron. c) Differential map. Results obtained for flashes with a duration of 3 ms, interjump time of 78 ms and intensity of $2.4 \cdot 10^{-2}$ Cd/m². Most values are averaged results over 3-8 flies. Only a few were measured for only one fly.

axis. This number is meant as a reference value for the experiments to be described next, and has been set to 1.0 for matters of convenience. Correspondingly, the null response is very small. The PSTH in Fig. 8c results if over exactly one and the same row of visual input elements simultaneously two oppositely directed sequences are moved step wise. It is shown here that the inhibition of activity dominates when both apparent movement events

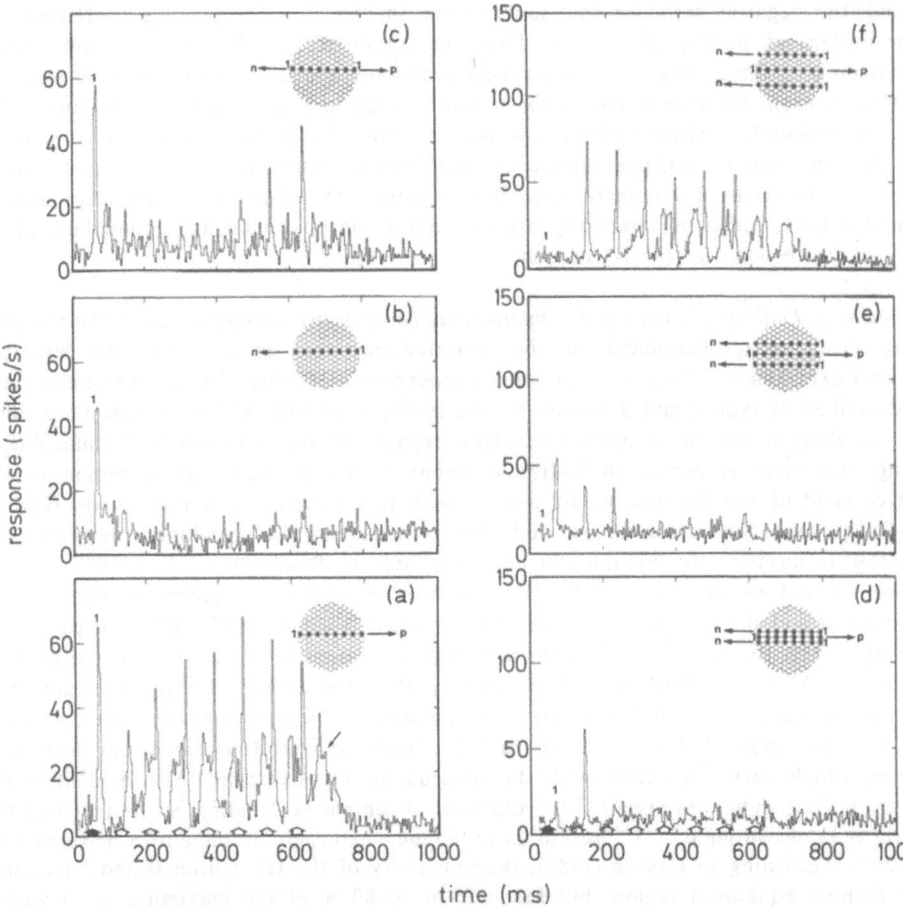

Fig. 8. PSTH's of the responses of the H1 neuron for the type 2 stimuli. The inset in each subFig. shows the spatial organization of the presented stimuli. Black arrows indicate the first flash of 3 ms duration. The intensity peak, marked "1", was elicited by the first flash, projected onto the cartridge, also denoted by a "1" in the inset. Open arrows indicate subsequent flashes. Interjump time was in all cases 78 ms. All six PSTH's were recorded for the same fly. Spot sizes were in all cases 0.1 $\Delta\phi$ and all flashes had equal intensity (i.e. $2.4 \cdot 10^{-2}$ Cd/m²). The PSTH's result from 300 averages. Note the different vertical scales. a) PSTH, recorded in response to ongoing step wise movement in the preferred direction. b) PSTH, measured for the null direction. c) PSTH resulting from simultaneous movement in the preferred and the null direction. Only one row of cartridges is illuminated. d) Simultaneously, the central row is stimulated in the preferred direction and the two directly adjacent rows in the null direction (i.e. separation equal to $\Delta\phi_V$). e) As in d, but the vertical separation between two rows is enlarged to 2 times $\Delta\phi_V$ f) As in d and e, but vertical separation is now 3 times $\Delta\phi_V$.

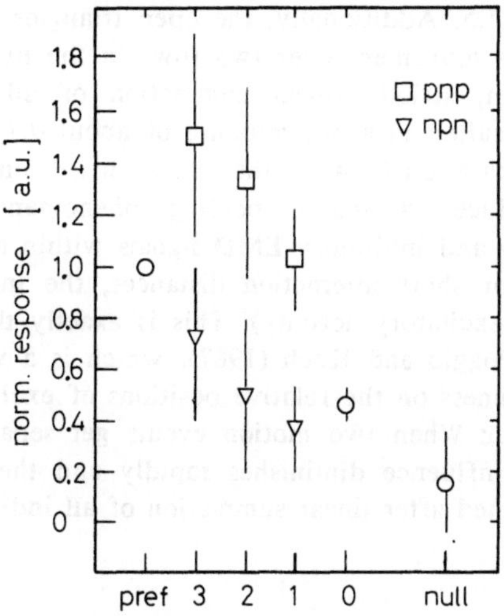

Fig. 9. Graphs of the total number of spikes registered between 0 and 800 ms in the PSTH's of Fig. 8, all normalized to the value obtained for motion in the preferred direction only (like the PSTH in Fig. 8a). The open circles, marked with *pref*, *null* and *0*, respectively, are the estimates obtained from the PSTH's in Figs. 8a through 8c. The open squares result for simultaneous motion over two rows in the preferred and one row in the null direction. Similarly, the open triangles are measured for simultaneous apparent movement over two rows in the null direction and over one row in the preferred direction. All results are averages over three flies.

occur within a few inter-ommatidial distances from each other, but at the end of the sequence, when both stimuli are moving further apart, their mutual interaction becomes less and less and as can be seen in the PSTH, the facilitation of the preferred direction regains somewhat of its initial strength. The open circle marked with *0* on the abscissa in Fig. 9 represents the normalized response for this situation. Figs. 8d through 8f are PSTH's, recorded in response to oppositely directed stimuli with increasing vertical separation, as shown in the insets. From those PSTH's it is clear that the larger the angular separation between the two events, the smaller their mutual influence. From the results it is argued that, whenever moving objects come close enough for visual interaction, the corresponding changes in activity (due to facilitation or inhibition) are pooled in a highly non-linear fashion. The open squares in Fig. 9 are estimated responses for the case that two rows are stimulated in the preferred direction and one row in the null direction, for three different arrangements. The non-linear behaviour is evident: When the two events are closest, which is denoted by the square named *1* on the x-axis, the total response is about equal to 1.0, whereas in case of linear summation of all the spikes, if recorded in three independent experiments in which all three apparent movement stimuli (two in the preferred direction and one in the null direction) would have been presented separately, would have amounted approximately 1.8. In Fig. 9 it is demonstrated that the larger the separation between the two movement events, the larger the net response, until it eventually (denoted by *3* on the x-axis) reaches the value 1.5. Additionally, the open triangles indicate the total response, measured for simultaneous movement over two rows in the null direction and one in the preferred direction. Again, strictly linear summation of all excitatory and inhibitory activities would have resulted in a net response of about 0.7, which is indeed the value reached for the largest separation between the two events. From the data we can conclude that the interaction induces a spatial pooling phenomenon, which sums weighed contributions of excitatory and inhibitory EMD signals within a well defined target region in such a manner that for short interaction distances, the inhibition predominates over facilitation (i.e. it vetoes excitatory activity). This is exactly the effect induced by silent inhibition, described by Poggio and Koch (1987), which is a very localized phenomenon, dependent for its effectiveness on the relative positions of excitation and inhibition in the dendritic tree of a neuron: When two motion events get separated beyond a few inter-ommatidial angles, their influence diminishes rapidly and the net response "at infinity" resembles the result, obtained after linear summation of all individual movement responses.

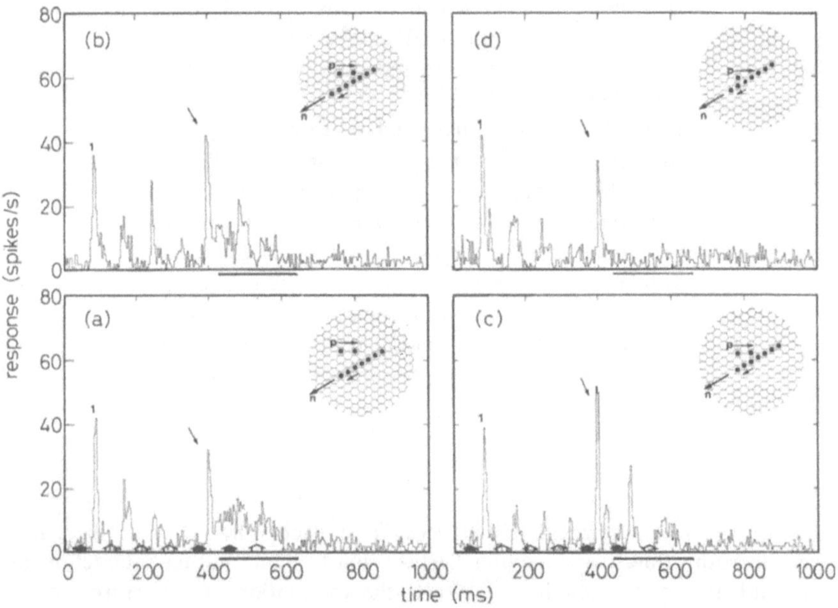

Fig. 10. PSTH's of the responses of the H1 neuron for the type 3 stimuli. The inset in each subFig. shows the spatial arrangement of the presented stimuli for that case. Black arrows indicate the first presentation of a flash of 3 ms duration. Open arrows indicate the following flashes. Interjump time was in all cases 78 ms. All four PSTH's were recorded for the same fly. Spot sizes were in all cases were 0.1 $\Delta\phi$ and all flashes had equal intensity of $2.4 \cdot 10^{-2}$ Cd/m^2). PSTH's are the averages over approximately 300 trials. For all four PSTH's the adapting motion was along the 30° axis in the null direction. a) PSTH, recorded when an additional single $C_{2,0}$ EMD was stimulated in the preferred direction. Timing of the extra flashes is indicated by the black arrows around 400 ms. The horizontal bar, stretching from 450-650 ms, indicates the time window, used for the graph in Fig. 11. Separation between extra flashes and initial flashes is more than 2 times $\Delta\phi_V$. b) As in a, but separation is smaller, approximately 2 times $\Delta\phi_V$. c) As before, separation about equal to $\Delta\phi_V$. d) As before, separation less than $\Delta\phi_V$.

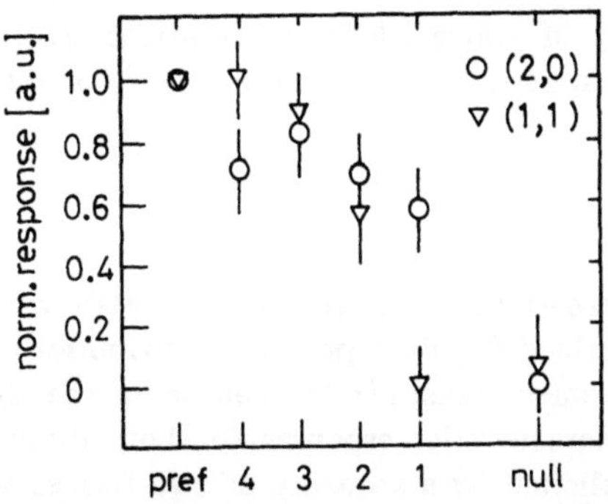

Fig. 11. Graphs of the total number of spikes registered between 450 and 650 ms in the PSTH's of Fig. 10, all normalized to the number of spikes obtained for the PSTH measured in response to activation of the preferred direction (not shown here). The open triangles are the estimates for the PSTH's in Figs. 10^a through 10^d. The open circles are values obtained when a row of $C_{2,0}$ correlators is stimulated in its null direction. All the results are averages over two flies.

We conclude that the shunting inhibition phenomenon is restricted to target regions, not extending a few interommatidial angles.

Fig. 10 shows four typical PSTH's, we obtained for type 3 experiments. In these experiments a row of adjacent $C_{1,1}$ EMD's are sequentially illuminated in the null direction. When, in concert with a certain EMD in this sequence, an extra $C_{2,0}$ EMD is stimulated with identical parameters, but in its preferred direction, it is shown to depend on their mutual separation, whether the total response will reflect linearly the presence of the second motion event. It has to be mentioned that more than just those EMD's are activated, since the first flash on the extra EMD in fact activates also an EMD with a larger sampling base. From the results presented earlier, however, it seems justified to neglect the contributions of those EMD's to the total response. The PSTH's demonstrate again that a highly non-linear summation of individual EMD contributions is performed when the two events come closer. For short distances, the inhibition of the pursued $C_{1,1}$ stimulation in the null direction highly prevails over the additional $C_{2,0}$ response. Only the extra intensity peak, induced by the short flash at the first input of the extra apparent movement stimulus, is not masked by that inhibition (this intensity peak is omitted form the calculations for Fig. 11.). The responses, normalized to the increase in spike rate for the preferred direction (not shown here) measured in the 200 ms interval, are plotted in Fig. 11. Both a $C_{2,0}$ and a $C_{1,1}$ type EMD from two different flies have been examined twice for this type of experiment. From linear summation of both stimuli, one would expect an estimate of about 0.8-1.0 for both types of EMD (both are equally strong in the dark adapted state - see Fig. 6). As is evident from Fig. 11, this value is only attained for the spatial arrangement of Fig. 10^a, denoted by the circle and triangle at position *4* in Fig. 11. The closer the two apparent movements become, the stronger the weighing of the inhibitory process. These results are complete in agreement with the theoretical predictions by Poggio and Koch (1987). Once again we conclude that for wide-field and for small-field motion, weighed pooling of single EMD-contributions resolves a highly non-linear summation of the elicited neural activity with the prevailence of inhibition over excitation.

DISCUSSION

Using a sophisticated micro-stimulation technique, EMD's with sampling bases up to 20 times $\Delta\phi_h$, have been examined for their possible contributions to the total response of the horizontally-selective movement sensitive H1 neuron in the dark-adapted eye of the blowfly *C. erythrocephala.* Characteristic responses for both directions of movement (i.e. preferred and null) could be elicited by a sequence of two flashes with time intervals from as small as approximately 20 ms up to 150 ms. Such a time span definitely sets a limit to the range of velocities that can be seized by the movement-sensitive system. From the data we can conclude that under scotopic luminance levels, i.e. mean irradiance of the flashes of about $2.4 \cdot 10^{-2}$ Cd/m², the spatial sampling pattern underlying visual movement detection is dominated by nearest neighbour interactions between pairs of cartridges, with sampling bases up to 6 times $\Delta\phi_h$, preferentially oriented along the horizontal sensitivity axis of the neuron. In accordance, Tinbergen (1987) demonstrated that movement detectors with sampling bases of 2 and 3 times the interommatidial angle are involved in the movement response. Such a coupling of EMD output signals, restricts the theoretically discriminable range of velocities to values between 14°/s and 750°/s. A plausible explanation for such a mechanism at these low light levels, is given by a neural *pooling* scheme in which signals from a variety of neighbouring and wide-angle movement detectors are combined through some kind of weighing process. In such a pooling scheme a gain in absolute light sensitivity is provided, however, at the cost of spatial resolution.

An extensive two-dimensional mapping of relevant contributions originating from isolated EMD's has been constructed. It is worthwhile to notice that in these maps for each single correlator the relative response coefficients have been determined for both directions of motion: from the data it can be concluded that these detectors are bi-directional. A uni-directional detector would either cause a spike increase for the preferred direction or suppression of the neural activity when stimulated in its null direction: it certainly would not be able to express both types of response. The correlators in our experiments expressed both types of characteristic behaviour. As a consequence, for each correlator, a differential coefficient can be deduced, defined as the difference between its preferred and its null contribution. Inspection of Fig. 7^c reveals that such a system of bi-directional detectors, is very apt to discriminate visual movement under low light conditions for a broad range of velocities. Interesting with this respect is the observed center-surround organization of the differential map. Beyond the nearest neighbour interactions, a sharp reversal of the sign of the coefficients is seen. In the optomotor course control behaviour, similarly, EMD's with sampling base identical to and larger than the interommatidial angle contribute to the perception of horizontal movement (Buchner, 1976). In the turning response of *Musca*, *positive* interactions occurred at apparent movement of stripes separated by angles up to 8 ° (i.e. 6 to 8 times $\Delta\phi_h$), whereas at larger angles negative interactions with opposite response direction were obtained (Pick and Buchner, 1979). The sign reversal in our data beyond sampling bases of 6 or 8 times $\Delta\phi_h$ seems to underlie this reversal in response magnitude for wide angles.

Moreover the additional experiments we performed, in which wide-angle motion is accompanied by either another wide-field antagonistic motion or a small-field motion indicate that both excitatory and inhibitory output signals from several EMD's are pooled in a non-linear fashion with non-equal weighing factors. These weighing factors are convincingly demonstrated to depend on the mutual distance of the two processes. It is argued that this highly non-linear effect can be explained, at least qualitatively, by a weighed neural pooling correlation scheme, as a modification of the pooling correlation model, initially proposed by Poggio and Reichardt (1976). The process of silent (or shunting) inhibition as described by Poggio and Koch (1987) may underlie such a pooling scheme.

ACKNOWLEDGEMENTS

The authors would like to thank their colleague Dr. W.H. Zaagman for expert techinical assistance in the data processing setup. This work was partly supported financially by the *Netherlands organization for scientific research* (NWO) through the foundation *Stichting voor Biophysica* and *Stichting voor Biologisch Onderzoek.*

REFERENCES

Buchner E (1976) Elementary movement detectors in an insect visual system. Biol Cybern 24: 85-101

Eckert H (1981) The horizontal cells in the lobula plate of the blowfly, Phaenicia sericata. J Comp Physiol 143: 511-526

Franceschini N (1975) Sampling of the visual environment by the compound eye of the fly: Fundamentals and Applications. In: Snyder AW, Menzel R (eds) Photoreceptor Optics. Springer Verlag, Berlin Heidelberg New York, pp 98-125

Hausen K (1976) Functional characterization and anatomical identification of motion sensitive neurons in the lobula plate of the blowfly Calliphora erythrocephala. Z Naturforsch 31c: 629-633

Hausen K (1981) Monocular and binocular computation of motion in the lobula plate of the fly. Verh Dtsch Zool Ges: 49-70

Hausen K (1982a) Motion sensitive interneurons in the optomotor system of the fly. I. The horizontal cells: Structure and signals. Biol Cybern 45: 143-156

Hausen K (1982b) Motion sensitive interneurons in the optomotor system of the fly. II. The horizontal cells: Receptive field organization and response characteristics. Biol Cybern 46: 67-79

Hausen K (1984) The lobula-complex of the fly: Structure, function and significance in visual behaviour. In: Ali MA (ed) Photoreception and vision in invertebrates. Plenum Press, New York London, pp. 523-559

Hausen K and Wehrhahn C (1983) Microsurgical lesion of horizontal cells changes optomotor yaw responses in the blowfly Calliphora erythrocephala. Proc R Soc Lond B 219: 211-216

Heisenberg M and Buchner E (1977) The role of retinula cell types in visual behavior of Drosophila melanogaster. J Comp Physiol 117: 127-162

Hengstenberg R (1982) Common visual response properties of giant vertical cells in the lobula plate of the blowfly Calliphora. J Comp Physiol 149: 179-193

Hengstenberg R, Hausen K and Hengstenberg B (1982) The number and structure of giant vertical (VS) cells in the lobula plate of the blowfly Calliphora erythrocephala. J Comp Physiol 149: 163-177

Lenting BPM (1985) Functional characteristics of a wide-field movement processing neuron in the blowfly visual system. Thesis, University of Groningen.

Mastebroek HAK, Zaagman WH, Lenting BPM (1982) Memory-like effects in fly vision: Spatio-temporal interactions in a wide-field neuron. Biol Cybern 43: 147-155

Pick B and Buchner E (1979) Visual movement detection under light- and dark-adaptation in the fly, Musca domestica. J Comp Physiol 134: 45-54

Poggio T (1983) Visual algorithms. In: Braddick OJ, Sleigh AC (eds) Physical and biological processing of images. Springer Verlag, Berlin Heidelberg New York, pp 128-153

Poggio T and Reichardt W (1976) Visual control of orientation in the fly. Part II. Towards the underlying neural interactions. Q Rev Biophys 9: 377-438

Poggio T, Reichardt W and Hausen K (1981) A neural circuitry for relative movement discrimination by the visual system of the fly. Naturwissenschaften 68: 443-446

Poggio T and Koch C (1987) Synapses that compute motion. Scientific American 255: 46-52

Reichardt W, Poggio T and Hausen K (1983) Figure-ground discrimination by relative movement in the visual system of the fly. Biol Cybern 46 (suppl.): 1-30

Schuling FH, Vorenkamp B and Zaagman WH (1989) Microprocessor-controlled vector scan display system for generation of real-time visual stimuli. Med & Biol Eng & Comput 27 (in press)

Stavenga DG (1975) The neural superposition eye and its optical demands. J Comp Physiol 102: 297-304

Tinbergen J (1987) Photoreceptor metabolism and visually guided landing behaviour of flies. Thesis, University of Groningen

Zaagman WH (1977) Some characteristics of the neural activity of directionally selective movement detectors in the visual system of the blowfly. Thesis, University of Groningen

SPATIAL VISION IN A FLAT WORLD: OPTICAL AND NEURAL ADAPTATIONS IN ARTHROPODS

Jochen Zeil, Gerbera Nalbach (*), and Hans-Ortwin Nalbach (*)

Lehrstuhl für Biokybernetik, Universität Tübingen, Auf der Morgenstelle 28; D-7400 Tübingen 1, FRG and (*) Max Planck Institut für Biologische Kybernetik, Spemannstr 38; D-7400 Tübingen 1, FRG

ABSTRACT

We review evidence to show that in several arthropod families eyes and supporting neural control systems are shaped according to the spatial layout of their environment. Amphibious crabs that live at sandy beaches and mudflats and insects that live above or below the water surface have horizontally aligned acute zones for vertical resolution in those eye regions that look at the horizon. In amphibious crabs acute zones are aligned with the horizon by visual, leg-proprioceptive and statocyst reflexes whereby optokinetic sensitivity to movement around roll and pitch axes reaches a sharp maximum at the eye equator. There is clear evidence of a position dependent mechanism of eye alignment to the horizon in at least two species of flat world crabs. Optokinetic sensitivity to movement around the yaw axis is restricted to the dorsal visual field in flat world crabs and in waterstriders with a maximum just above the eye equator. We discuss the relevance of these specialisations for spatial vision in a flat world.

INTRODUCTION

There were dragonflies flying around in the lower carboniferous, some 250 million years ago. Unless one takes fossils as evidence for frequent crash landings, this means that very complex problems of visual orientation had been solved by that time. In the meantime, evolution must have shaped visual systems under the pressure of specific tasks and specific environments. With the rather short exercises we sofar had with machine vision we are just starting to grasp the significance of such a long experience with the world. For neurobiologists, the general point we would like to make here – by documenting somewhat scurrilious visual systems – is that we need to understand the natural operating conditions if we want to fully understand the organisational principles of visual systems. Beside knowledge of the physical conditions like available light intensity and spectral composition this entails basically two things: a need to analyse the **behaviour** of animals and the **ecological conditions** in which they operate. Here we will focus on the second aspect by presenting evidence of a correlation between the design of visual systems including the supporting neural control systems and the ecological context. Starting point is the observation in a number of very different animals (Fig 1) which, however, live in similar environments, of eye specialisations that can be interpreted as specific adaptations to the needs of spatial vision.

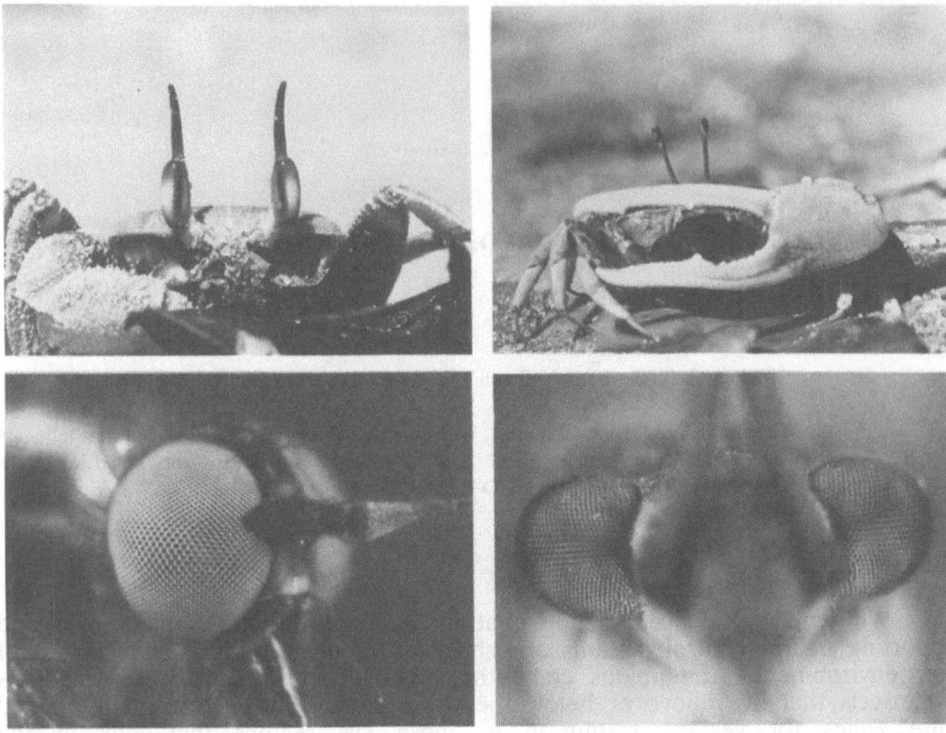

Fig 1 Design of visual systems in the brachyuran crabs *Ocypode ceratophthalmus* (**top left**); *Uca coarctata* (**top right**); the empidid fly *Hilaria sp* (**bottom left**) and the waterstrider *Gerris lacustris* (courtesy of S Heyl) (**bottom right**).

A. EYE SPECIALISATIONS IN A FLAT WORLD

a. Amphibious Crabs: With their eyes on mobile stalks, crabs are unique amongst the arthropods. In terms of evolutionary history this design has – especially in the Brachyura through changes in eye stalk length – allowed modifications of eye separation and eye height above substrate which are otherwise difficult to achieve (cf however the stalk eyed flies (Diopsidae) Burkhardt and laMotte, 1986). In a comparative study of eye design in semi-terrestrial crabs we have shown that the length of eyestalks, eye separation and acute zones for vertical resolving power along the eye equator correlate with the structure of the animals' environment (Zeil et al., 1986). In short, species which live in flat environments like sandy beaches or mudflats have their eyes on long, vertically oriented eye stalks that bring the eyes close together and a horizontally aligned acute zone for vertical resolution; species living in visually more complex environments, like mangrove thickets or rocky coastlines have their eyes on short eye stalks, the eyes lie far apart and lack pronounced acute zones (Fig 2a–d). We suggested that taken together, these design features and their correlation with the structure of the environment have to be interpreted in terms of spatial vision. Variations in eye separation, for instance, have the

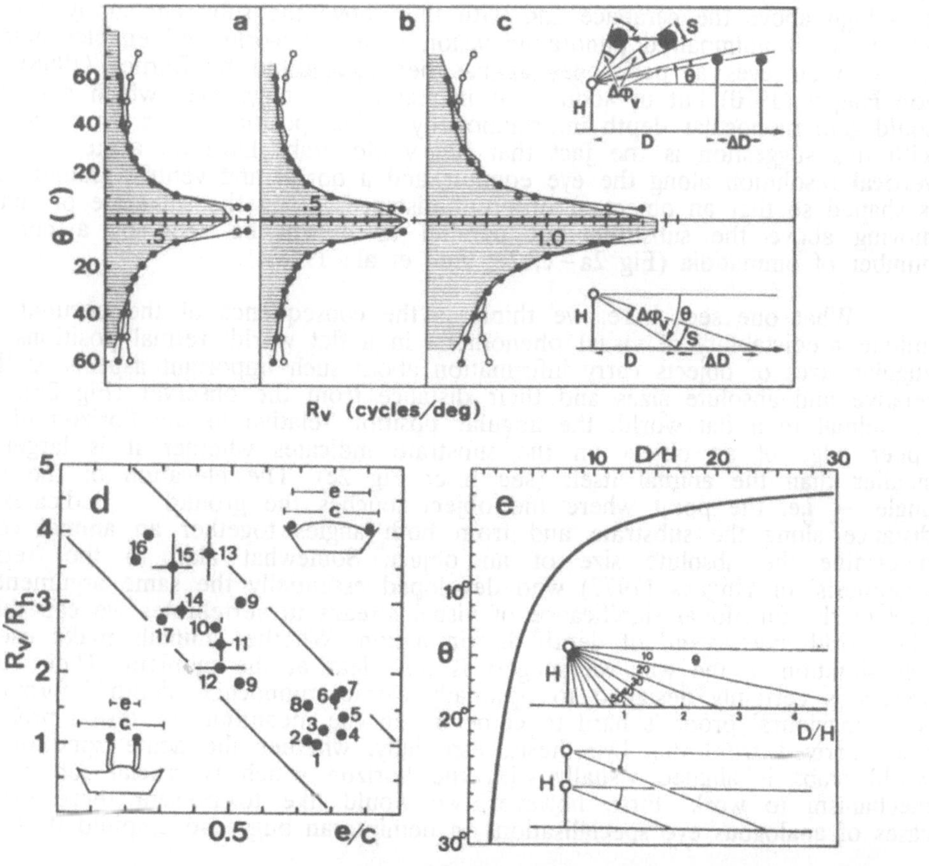

Fig 2 **(a−c)** Profiles of vertical resolving power (**R$_v$**) in the frontal visual field (sagittal plane) of three species of crabs (Ocypodidae) (hatched areas) and profiles which would be expected if an eye was designed to provide size constant information (dots) and relative distance constant information (circles) ($R_v = 1/2\Delta\phi_v$; see Zeil et al. 1986 for details and equations). Abscissa is altitude in degrees with respect to the horizon. Curves are fitted to the profiles found in **(a)** *Macrophthalmus setosus*; **(b)** *Uca sp*; **(c)** *Ocypode ceratophthalmus*. For *M. setosus* the parameter H/S = 12. This means that an object on the ground of the size S = H/12 has the retinal size of one vertical interommatidial angle. For definition of parameters see insets. **(d)** Relationship between eye separation, length of eye stalks and acute zones for 17 species of brachyuran crabs. See Zeil et al. (1986) for details. **e**: eye separation (cm); **c**: carapace width (cm); **R$_v$**: vertical resolving power (cycles/deg); **R$_h$**: horizontal resolving power (cycles/deg). R_v and R_h were measured in the fronto-medial plane at the eye equator. Points in the graph represent means with standard deviation when more than 3 animals were measured. **(e)** Schematic representation of the geometry of vision in a flat world (cf Hughes, 1977; Collett and Harkness, 1982). The curve shows the intersection of visual directions below the horizon (**θ**) with the substrate. D/H is the distance (**D**) along the substrate normalised to eye height (**H**) above the substrate (see insets). The projection of a flat terrain on an eye at height (**H**) above the substrate contains information about the three-dimensional layout of the terrain since $D = H/\tan\theta$ (All figures from Zeil et al. 1986)

specific visual consequence to make binocular disparities – which are the critical limiting factor for binocular stereopsis – larger or smaller. To raise the eyes high above the carapace and with this above the substrate surely has the advantage of unimpaired panoramic vision in a flat world and enables animals to use their eyes as periscopes as has been suggested by Barnes (1968) and von Hagen (1970) but in addition it increases the range over which ani – mals could gain monocular depth information by retinal position alone. In agreement with this suggestion is the fact that flat world crabs have an acute zone for **vertical** resolution along the eye equator and a dorsal and ventral gradient that is shaped so that an object at different distances along the substrate or that is moving above the substrate but parallel to it will be seen by a constant number of ommatidia (Fig 2a – c, 2e; Zeil et al., 1986).

What one sees here, we think, is the consequence of the peculiar and unique predictability of visual phenomena in a flat world: retinal positions and angular sizes of objects carry information about such important aspects as their relative and absolute sizes and their distance from the observer (Fig 2e). For an animal in a flat world, the angular position relative to the horizon of the upper edge of an object on the substrate indicates whether it is larger or smaller than the animal itself (see inset Fig 2c). The elevation of the base angle – i.e. the point where the object touches the ground – indicates its distance along the substrate and from both angles together an animal could determine the absolute size of an object. Somewhat akin to the 'terrain hypothesis' of Hughes (1977) who developed essentially the same argument to explain the functional significance of visual streaks in vertebrates we call this a 'flat world mechanism' of depth discrimination. Whether animals make use of this situation in the way we suggest is not clear at the moment. Their visual system is certainly designed to optimally extract monocular depth information but behavioural proof is hard to come by. In the meantime we tested predic – tions derived from this hypothesis, especially, whether the acute zone of flat world crabs is aligned visually with the horizon which is crucial for such a mechanism to work. First, however, we would like to present three further cases of analogous eye specialisations in hemipteran bugs and empidid flies.

b. Notonecta: The flattest of all worlds is the water surface of puddles and ponds. Through the work of Schwind (1978, 1980) we know that the visual system of *Notonecta glauca* a bug which spends long times suspended from the water surface hunting for insect prey that gets trapped on the surface, shows specific adaptations to the optical conditions of this narrow zone of life. Special to *Notonecta* is that its vertical resolution peaks sharply where in its normal posture hanging from the water surface, it views the boundary between the zone of total internal reflection and the region where it sees the world above the water (Fig 3). Of particular interest to our argument here is the fact that in the eye of *Notonecta*, like in those of shore crabs, there is a gradient of increasing resolution towards the horizon. Schwind (1978), in addition, has recorded from a visual interneuron which represents space constant information: it responds maximally to an object of a quarter of the animal's own size moving in the plane of the water surface, regardless of its angular size and elevation in the visual field. From arthropods this is the only proof sofar, that they make use of environmental geometry for depth and size discrimination.

c. Waterstriders: Waterstriders, on the other hand, live on the airy side of the watersurface. They use landmarks to visually compensate for drift (Junger and Dahmen, 1988) and locate prey by means of watersurface waves (Murphey, 1971). Although they are able to determine the angular position and the distance of a source of watersurface waves even when blinded, their readiness to respond and the accuracy of their attack is higher when they are seeing (Dahmen, personal communication). The eye of *Gerris lacustris* has been mapped by Dahmen und Junger (1988) and one can clearly see from Fig 4 to what extreme degree visual axes of ommatidia are concentrated close to the horizon. The gradient of vertical resolution above and below the eye equator is

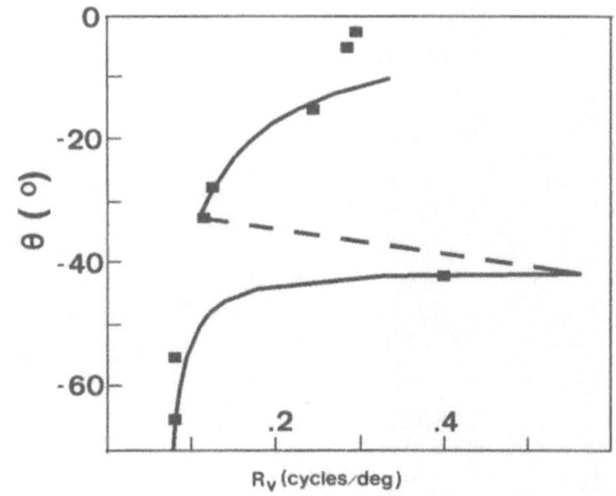

Fig 3 Specialisations in the eye of *Notonecta glauca*. Data from Schwind (1980). The gradient in the zone of the eye that looks at the 'substrate', i.e. at the water surface from below (from the animal's horizon ($\theta = 0°$) to the limit angle for total internal reflection ($\theta = -41.25°$) has been modelled by assuming that it provides the animal with size constant information with $R_V = 1/(2\arctan((S/H)\sin\theta))$, $S/H = 0.15$. See inset Fig 2c for definition of variables and Zeil et al. 1986 for details. The part of the visual field where light reaches the eye from above the water surface (from directly below the animal ($\theta = -90°$) to the limit angle of total internal reflection ($\theta = -41.25°$)) has been separately modelled by $R_V = n\sin\theta/(2d\alpha\sqrt{1-n^2\cos^2\theta})$ with the refractive index of water $n = 1.33$ and the angular size ($d\alpha$) to be kept constant in air ($d\alpha - 9°$). Otherwise conventions as in Fig 2a – c.

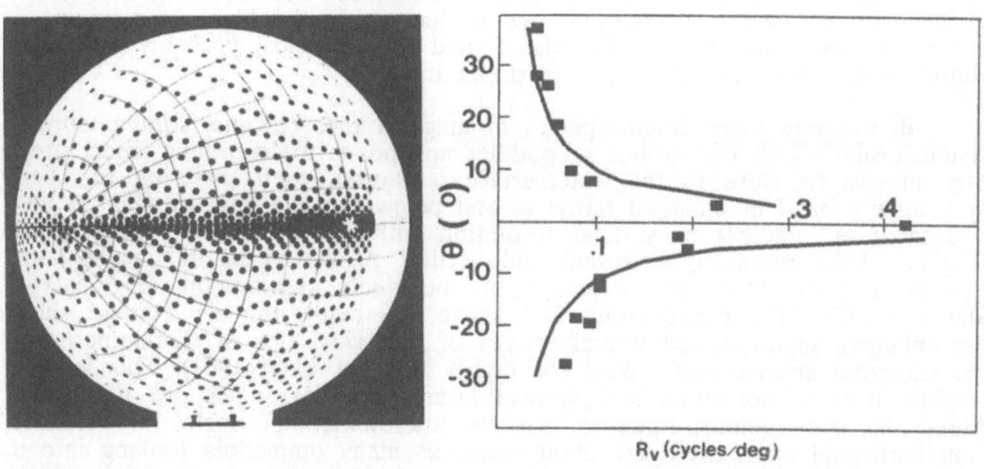

Fig 4 Specialisations in the eye of the waterstrider *Gerris lacustris*. From Dahmen und Junger (1988) and Dahmen (in preparation). **Left**: The pattern of ommatidial visual axes for the right eye projected onto the unit sphere. Star marks forward direction. **Right**: The gradient of vertical resolution along one vertical plane. Curves are fitted to the data by assuming that the gradient provides the animal with size constant information ($R_V = 1/(2\arctan((S/H)\sin\theta))$; dorsal visual field $S/H = 0.5$, ventral $S/H = 0.45$)). Conventions as in Fig 2a – c.

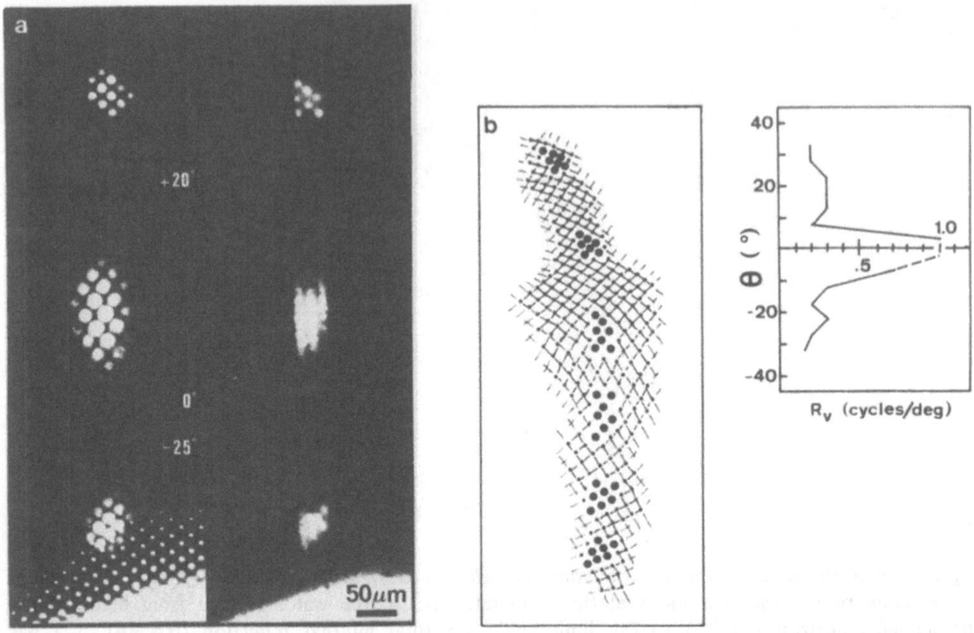

Fig 5 Specialisations in the eye of the fly *Rhamphomyia tephraea* (Empididae). (**a**) Corneal (**left**) and deep (**right**) pseudopupils at different elevations during antidromic illumination. (**b**) The variation of the facetlens pattern in the vertical direction across the eye and the distortion of the arrangement of those ommatidia which are likely to contribute to one neuro-ommatidium. (**c**) The gradient of vertical resolution in the empidid eye. Conventions as in Fig 2a-c.

very similar to the ones we found in amphibious crabs (cf Fig 2a-c). A special property of the waterstrider eye is that within the high acuity zone - but not in the dorsal and ventral visual field - they most probably possess a neural superposition eye (Dahmen, in preparation).

d. Empidid Flies: Some species of empidid flies (*Hilaria, Rhamphomyia*) systematically search the surface of puddles and ponds for drowning insect prey. The animals fly close to the watersurface and often even skate on it. Their eyes show a band of enlarged facets at and below the eye equator (Fig 1) and a pronounced gradient of vertical resolution with a maximum at the equator (Fig 5c). Differences in horizontal and vertical resolution in this acute zone can be inferred from the structure of the deep pseudopupil (Fig 5a; cf Stavenga 1979). The pseudopupils 20° above and below the eye equator show the enlarged, superimposed virtual images of the distal tips of rhabdomeres in the trapezoid arrangement typical for higher flies (cf Land 1981). This is only possible because horizontal and vertical interommatidial angles are the same. Along the eye equator, however, vertical interommatidial angles are smaller than horizontal ones, there are about twice as many ommatidia looking into a given solid angle in vertical than in horizontal directions. At a depth which corresponds to the local horizontal eye radius where the visual axes of horizontally neighbouring ommatidia converge one sees a pattern of three vertical rows of images. There are vertical strings of images since vertical interommatidial angles are smaller than horizontal ones: virtual images of distal rhabdomere tips superimpose exactly in horizontal direction, but in vertical direction the rhabdomere images from those ommatidia that contribute to the deep pseudopupil overlap only partially. The fact that discrete images of rhabdomere tips form at all at this depth (the local horizontal eye radius), shows that horizontal interommatidial angles must be multiples of vertical ones. In the empidid eye we may see an example where – for the purpose of increasing vertical resolution in a restricted part of the eye – the facet pattern becomes extremely distorted while the optical conditions for neural superposition are preserved (Fig 5b, cf Stavenga 1979). A full account of the empidid eye will appear elsewhere (Zeil and Dahmen, in preparation).

B. FUNCTIONAL ADAPTATIONS: STABILISING SYSTEMS

One prerequisite for a flat world mechanism of depth discrimination to work is that the acute zone should be aligned with the horizon since the horizon and the animal's own height above ground serve as the critical reference system. It is compensatory eye movements around roll and pitch axes, therefore, that are relevant for this purpose. We would predict that in flat world animals this alignment is under predominant visual control. Unfortunately there is little work on this in crabs, although compensatory eye movements around the yaw axis have been analysed extensively (cf Neil, 1982). Our own pilot experiments on flat world crabs (*Mictyris, Heloecius, Macrophthalmus*) showed that for roll and pitch stimulation, visual input – even if it was restricted to a 1° wide horizontal black stripe – could override conflicting input from statocysts and legs (Fig 6). This lead to a more extensive comparative study of compensatory eye movements in crabs from different habitats and life-styles (Nalbach et al., a, b, in preparation).

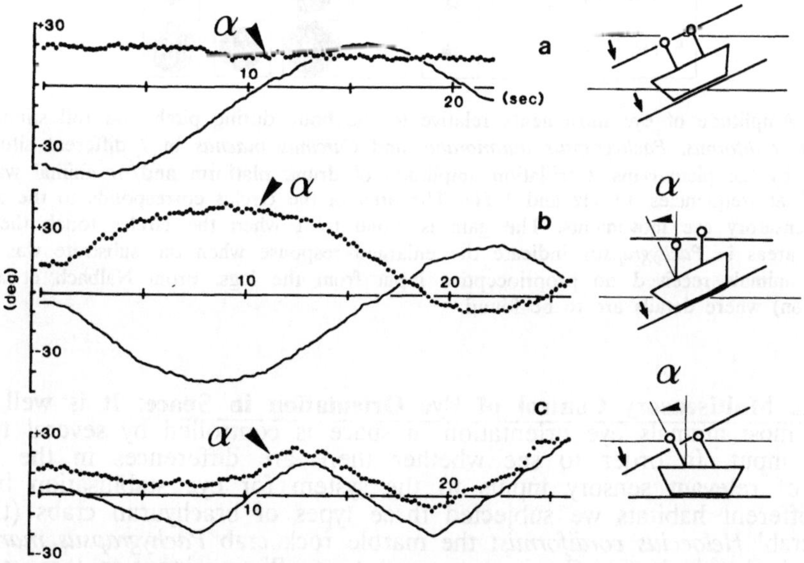

Fig 6 Sample record of compensatory eye movements around the roll axis in *Heloecius cordiformis*. Angular position (α) of one eye stalk relative to the vertical axis of the animal (dotted line) and angular position of the carapace relative to external coordinate system. (**a**) When the animal and the stripe are oscillated in phase, the eye stalks do not move relative to the body. The eyes are locked to the visual input, the vestibular input is suppressed; (**b**) The animal is oscillated while the visual pattern remains stationary. The eye stalks move through equal and opposite angles as the body and through this fully compensate for body roll; (**c**) the animal is stationary while the stripe is oscillated. The eyes follow the pattern movement. From Nalbach et al. (b, in preparation).

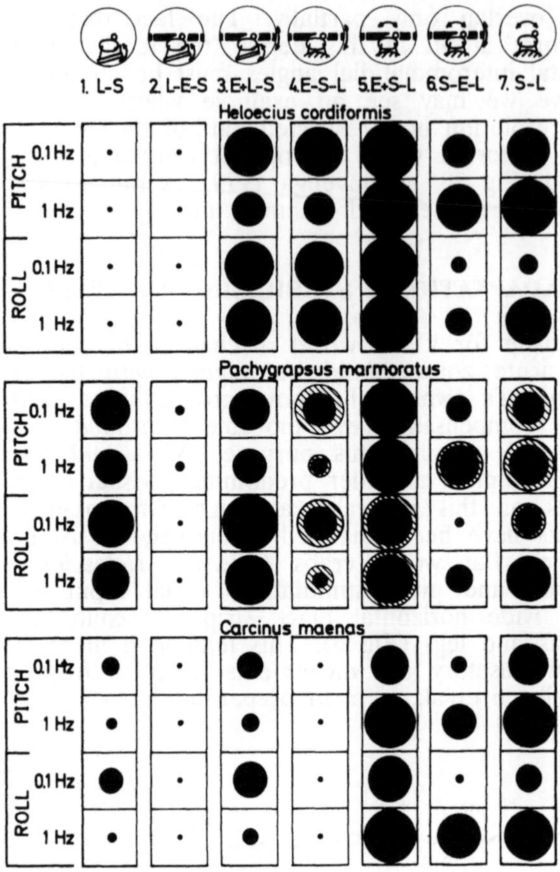

Fig 7 Amplitude of eye movements relative to the body during pitch and roll stimulation of *Heloecius cordiformis, Pachygrapsus marmoratus* and *Carcinus maenas* in 7 different situations as indicated by the pictograms. Oscillation amplitude of drum, platform and/or animal was always $U = \pm 10^\circ$ at frequencies 0.1 Hz and 1 Hz. The area of the circles corresponds to the amplitude of compensatory eye movements. The gain is equal to 1 when the circles touch their boxes. Hatched areas in *Pachygrapsus* indicate the enlarged response when the substrate was removed and the animals received no proprioceptive input from the legs. From Nalbach et al. (b, in preparation) where details are to be found.

a. Multisensory Control of Eye Orientation in Space: It is well known that in most animals eye orientation in space is controlled by several types of sensory input. In order to see whether there are differences in the relative weight of relevant sensory inputs to the system for eye stabilisation in crabs from different habitats we subjected three types of brachyuran crabs (the 'flat world crab' *Heloecius cordiformis*; the marble rock crab *Pachygrapsus marmoratus*; the 'swimming' crab *Carcinus maenas*) to conflicting input to legs, statocysts and eyes and determined their relative contributions to compensatory eye movements around roll and pitch axes (Nalbach et al., b, in preparation). During leg stimulation the animals were suspended above an oscillating platform with which they had leg contact. For statocyst stimulation the whole body of animals and the platform were oscillated together and finally visual input was provided by a surrounding drum with a horizontal black stripe that extended vertically 5° and was positioned at the eye equator of the crab. Platform, drum and animal could be oscillated independently or in various coupling combinations. The summarized results of this analysis (Fig 7) show that in the three species we studied, the different sensory inputs driving compensatory eye movements are given different weights. We suggest that this has to do with the availability of cues in their environment: *Heloecius* for instance, which inhabits relatively flat terrain at the fringes of mangroves relies much more heavily on visual input to stabilise eye orientation in space than does *Carcinus*. To verify this, note the differences we found when we provided visual stimulation with conflicting input from statocysts and legs (4th column in Fig 7). In turn, leg input has a much stronger influence on eye orientation in the rock crab *Pachygrapsus* (1st column in Fig 7) than in the other species while the swimming crab *Carcinus* relies mainly on statocyst input.

b. Distribution of Optokinetic Sensitivity: A closer look at visually driven compensatory eye movements which could help to align acute zones with the horizon revealed that *Heloecius* – our representative of flat world crabs – and the rock crab *Pachygrapsus* both show a maximum of sensitivity to vertical movement close to the eye equator (Fig 8a; Nalbach et al., a, in preparation). A single black stripe that is oscillated with an amplitude of ±10° at different mean positions around the pitch or roll axis of a crab elicits the strongest eye movements when presented close to the horizon.

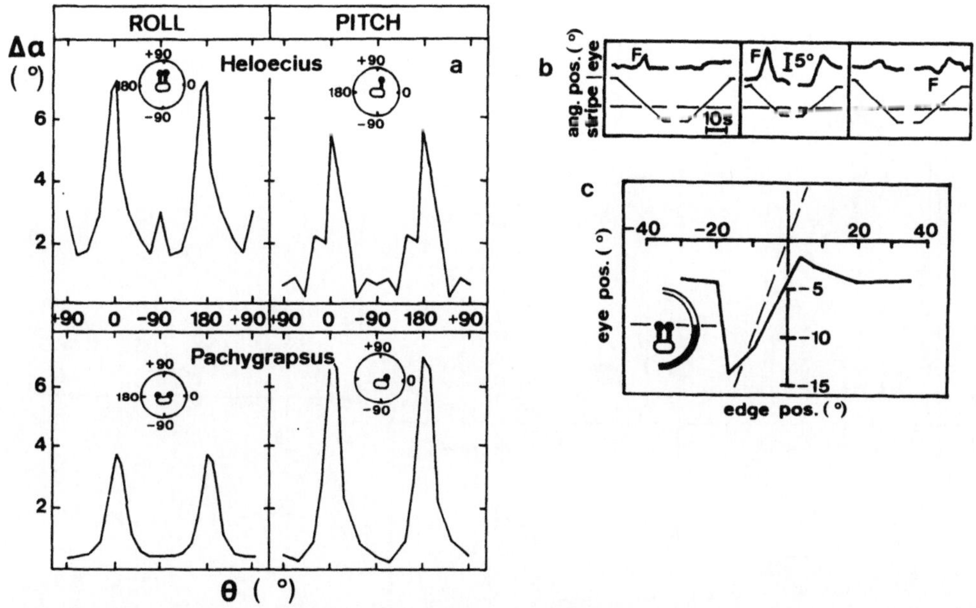

Fig 8 (**a**) Regional distribution of sensitivity to vertical motion around roll (**left**) and pitch (**right**) axis in *Heloecius cordiformis* (**top**) and *Pachygrapsus marmoratus* (**bottom**). Animals were stimulated by a horizontal black stripe of 5° vertical extent that oscillated through ±10° and 0.1 Hz at different elevations (**θ**) in the visual field. **α**: Amplitude of eye movements. (**b**) Sample records of eye stalk movement in *Mictyris longicarpus* (thick lines) when stimulated by a black stripe of 5° or 10° vertical extent moving up and down across the equator of the eye. **F** marks instances of fixating eye movements. (**c**) Relationship between eye stalk orientation and the orientation of a stationary edge in *Mictyris longicarpus*. With a top white edge, the eye fixates the edge within its acute zone when the edge has an elevation between 0° and −15°. Equality is marked by the dashed line. From Nalbach et al., a, in preparation.

The analysis showed further that this property is not exclusive to flat world crabs. This particular distribution of motion sensitivity prevents the crab from compensating the image flow produced by contrast on the ground during translational movements. But it will help to stabilise the eyes against dynamical pitch and roll perturbations during locomotion since the crab is sensitive only to motion across the horizon. However, for a flat world mechanism of depth discrimination to work we would expect not only dynamic alignment of acute zones but a static one in resting animals. In the soldier crab *Mictyris longicarpus*, and to a lesser degree in *Heloecius* we indeed found such a static component in roll and pitch stabilisation (Nalbach et al., a, in preparation): soldier crabs actually fixate a horizontal edge when it is positioned close to the eye equator (Fig 8b,c). The fact that this is a static, position dependent reaction is the more surprising as it is the first fixating eye movement described for any crab and was sofar known only for Stomatopods (Schaller 1953).

There is another striking correlation between the ecology of animals and the particular distribution of optokinetic sensitivity in their visual field. It is not, however, directly related to the eye specialisations and their role in depth

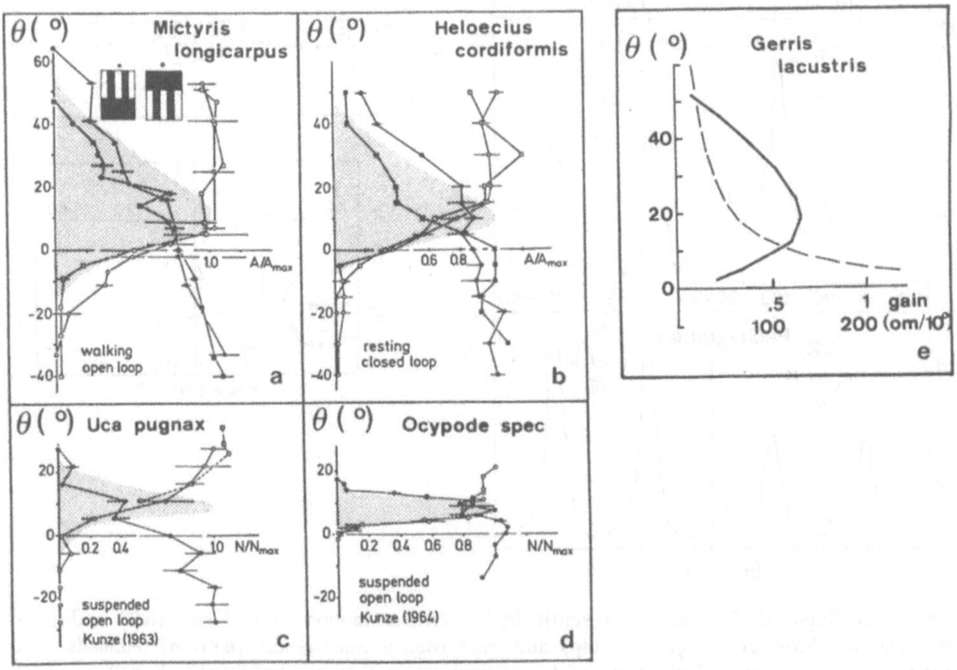

Fig 9 Vertical distribution of optokinetic sensitivity to yaw movement in four species of flat world crabs (hatched areas in **a**–**d**; from Nalbach and Nalbach, 1987 and Kunze, 1963, 1964) and (**e**) in the waterstrider *Gerris lacustris*. Density distribution of ommatidia is shown as dashed line (from Dahmen and Junger, 1988). Response amplitudes are normalised to maximal values of amplitude of eye movements (A/A_{max}) or nystagmus frequency (N/N_{max}). Animals were stimulated with moving stripe patterns which were successively covered from above or from below. For details see the original articles. Compensatory head movements in waterstriders were elicited by a horizontal stripe pattern with 10° vertical extent at different elevations in the visual field. For details see Dahmen and Junger, 1988.

vision that we ascribe to them. In flat world crabs compensatory eye movements around the yaw axis can only be elicited by pattern motion in a narrow zone above the eye equator (Fig 9a–d; Kunze, 1963, 1964; Nalbach and Nalbach, 1987). The rock crab *Pachygrapsus* as a representative for species that inhabit visually more complex environments, in contrast, has its optomotor sensitivity not restricted to a small dorsal area of the visual field (Nalbach and Nalbach, 1987).

Dahmen and Junger (1988), interestingly enough, have found that optokinetic sensitivity in waterstriders is similarily restricted to the visual field above the horizon (Fig 9e). Acute zones in the eyes of waterstriders and ocypodid crabs – although containing up to 40% of all ommatidia of the eye (Dahmen and Junger, 1988) – do not significantly contribute to movement detection in the service of compensatory eye movements around the yaw axis. The special 'topography' of optokinetic sensitivity in these animals is different from the density distribution of visual axes. It must, therefore, reflect the neural organisation at higher stages of visual processing.

Nalbach and Nalbach (1987) have suggested that this specialisation is a simple mechanism for animals living in a flat world to separate the rotational and the translational components of optic flow. Large field image motion at the eye of an animal is always due to the animal's own movement. Image motion at the eye of a translating animal depends on the distance of objects and on their angular position relative to the direction of locomotion; while image motion produced by a turning animal, in contrast, is the same for all objects, regardless of their distance from the observer. In a flat environment, objects that appear above the horizon are likely to be far away. Their angular positions in the visual field of a **translating** animal would not change much. However, they will heavily contribute to image flow once the animal **rotates**. In a flat world, large field movement above the horizon rather reliably tells an animal that it has changed course or that its eyes have rotated. The pure translational component of optic flow may now provide them with unimpaired **motion parallax cues** to the distance of nearby objects on the ground. By restricting their yaw optokinetic sensitivity to a small region just above the horizon, flat world animals can spare more complicated neural computations to separate translational from rotational flow field components. They make pragmatic use of the geometry of their environment.

DISCUSSION

a. Optics of 'Astigmatic' Compound Eyes: There are optical problems with astigmatic compound eyes. When there are large (up to 4 fold) differences between vertical and horizontal resolution, neighbourhood relationships between ommatidia change in visual space and 'optical' requirements are in conflict with 'sampling' requirements. This is exemplified for instance in the eye of *Ocypode ceratophthalmus* where interommatidial angles in vertical directions are 0.3° and in horizontal directions 0.8°. Now, sampling theory (cf Horridge, 1978; Land, 1981) would want the acceptance functions of those ommatidia that look into neighbouring visual directions to overlap at 50% halfwidth, i.e. to be matched to the angular distance between neighbouring visual directions. This is not possible in astigmatic eyes: if acceptance functions were matched to the best vertical resolution there would be gaps between acceptance functions of ommatidia looking into neighbouring horizontal directions and if acceptance functions were matched to horizontal resolution optimal sampling in vertical direction would be impaired (Fig 10a). This essentially means that animals could not make use of their high vertical resolving power. One solution would be to have oval acceptance functions in the astigmatic eye regions, the other one to fill gaps in horizontal directions by oscillating or 'scanning' eye movements (cf Sandeman, 1978). Little is known about the rhabdomere cross-sections in the acute zone of ocypodid crabs. In waterstriders, rhabdomere cross

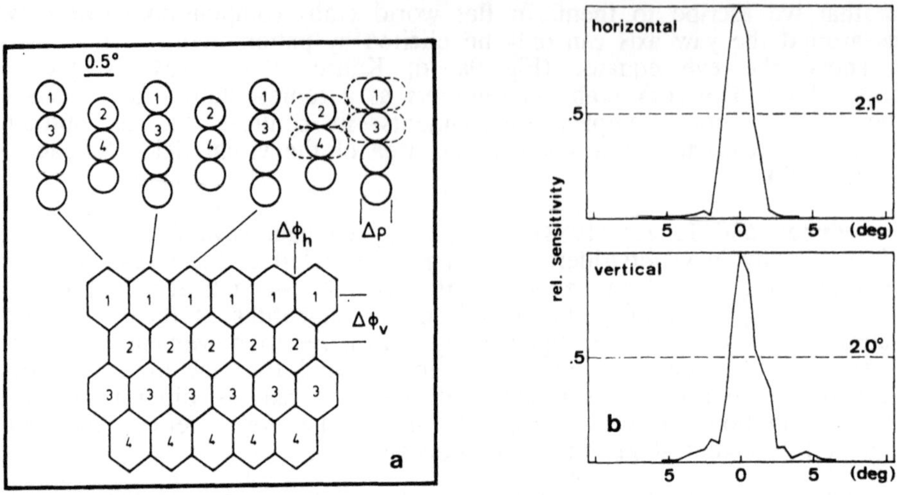

Fig 10 (**a**) Neighbourhood relationships between ommatidia change in visual space when there are large differences between horizontal ($\Delta\phi_h$) and vertical ($\Delta\phi_v$) interommatidial angles. Circles represent the angular extent of acceptance angles (ρ) of ommatidia in visual space. The situation is shown for an eye with $\Delta\phi_v = 0.25°$, $\Delta\phi_h = 0.85°$. Because neighbourhood relationships in vertical directions change, the acceptance angle ($\Delta\rho$) can be $2\Delta\phi_v$ (with $\Delta\phi_v = \lambda/2A$ this gives $\Delta\rho = \lambda/A$, which is the Airy disc at half width) without leading to extensive overlap. In an ideal eye, where the acceptance angles are matched to both vertical and horizontal interommatidial angles, they would have to have an oval shape (dashed lines) (from Zeil et al., 1986). (**b**) Vertical and horizontal angular sensitivity of a photoreceptor within the acute zone in *Gerris lacustris* at a position 87° lateral and 4.1° above the equator. $\Delta\phi_h = 2.2°$ and $\Delta\phi_v = 1.1°$ at this location. The recording was made by K. Bartsch, University of Tübingen.

sections are elongated in horizontal directions in the ventral eye, in the equatorial zone, however, they are round (Schneider and Langer, 1969). The change from one type of rhabdomeres to the next occurs abruptly between two horizontally aligned facet rows but it is unclear at present where this transition comes to lie in visual space. Bartsch (personal communication) has recorded from receptors in the waterstrider's eye but sofar has not encountered asymmetrical acceptance functions. Close to the horizon, acceptance functions are symmetrical in horizontal and vertical directions and have a halfwidth of about 2° which is closer to the horizontal (2.2°) than to the vertical (1.1°) interommatidial angle (Fig 10b).

b. Retinal Elevation as a Cue to Depth: the Question of Reliability: A somewhat worrying aspect of a flat world mechanism of depth perception is the question of accuracy and reliability on surfaces other than the watersurface that are always to some degree irregular. As a rule, if an animal relies solely on the base angle to determine the distance of an object, it will over– or underestimate it by 50% when the relief contains valleys and hills of half the height of the animal itself. On the other hand, ocean beaches and mudflats can be surprisingly level and small remaining irregularities could maybe be tolerated. In case this is not so, however, animals have some further cues they could rely on: sofar, we have only considered the static situation with a resting crab and a still object of interest; the situation changes when either one or both are moving. The image of the top edge of an approaching object that has the same (vertical) size as the crab will oscillate across the equatorial acute zone and will **on average** not change its retinal position. With a smaller object, the top egde will be seen at an increasing angular position below the horizon during the approach. With a larger object, the average position of its top edge will come to lie at increasing angular positions in the dorsal visual field above the horizon.

Quantitatively, the relation between change in angular position ($d\theta$) and change in distance (dD) is given as

$$d\theta/dD = -\sin(2\theta)/2D.$$

The ratio of average angular velocity ($\dot{\theta}$) and translational velocity ($\dot{D}$), therefore, is a monocular measure of distance. The translational velocity of an approaching object can be determined by measuring the average angular velocity ($\dot{\theta}_B$) and the retinal position (θ_B) of the base angle:

$$\dot{D} = -H\dot{\theta}_B/\sin^2(\theta_B)$$

Alternatively, whenever a crab moves itself, it can have a kinesthetic measure of its own velocity.

The cues that become available during locomotion or when an object of interest approaches are to a large degree independent of substrate contours since signals can be averaged over some period of time. To what degree they will be available to an animal will depend on eye design in a similar way as the static 'flat world mechanism' we have introduced above. Both static and dynamic situations, however, require that there is a defined orientation of eyes and acute zones relative to the substrate.

This brings us to the second problem, namely what the reference system for a flat world mechanism should be in case surfaces are slanted. The simplest possibility would be that animals align their acute zones with the surface and not with the true horizon line or gravity. Ghost crab individuals that have their burrows on slanted parts of the beach indeed hold their eye stalks oriented perpendicular to the **surface** (up to about 20° slant) and not to the true horizon or parallel to gravity (Zeil, unpublished observation). This is further evidence for static alignment of acute zones with the local horizon. It may help here to point out that infinity (the visual horizon) for an animal is determined by its height above ground (H) and its vertical resolution ($\Delta\phi_v$) at the eye equator:

$$D_\infty \leq H/\tan(\Delta\phi_v/2)$$

For an adult ghost crab with 5cm eye height and a vertical resolution of $\Delta\phi_{vmax} = 0.3°$ this is about 19m.

c. Behavioural Proof: We have not carried out experiments specifically designed to test whether flat world animals make depth and size discriminations on the basis of retinal elevation and angular size. However, for crabs, a wealth of observations reported in the literature at least shows how important it is for flat world animals to make these discriminations. Some ocypodid crabs for instance choose partners for mating by their size – and we would add by their height above the substrate (e.g. Lighter, 1974; Yamaguchi et al., 1979; Salmon and Hyatt, 1983; Brooke, 1981). Some even rely on symbols of size as is shown by the sand pyramids built by male ghost crabs to attract females to their mating burrows (Linsenmair, 1968). In other animals there is direct evidence for a flat world mechanism of depth discrimination: Collett and Udin (1988) have recently shown that frogs use retinal elevation as a cue to depth and together with electrophysiological evidence in burrowing owls (Cooper and Pettigrew, 1979) and *Notonecta* (Schwind, 1978), this is encouraging.

CONCLUSION

A full understanding of the design of visual systems requires a detailed knowledge of their natural operating conditions. The evolution of eye specialisations in animals shows that to a large degree visual systems are special purpose built. We reviewed work on eye design and the control of eye-head movements in insects and crabs to demonstrate that often visual systems are adapted to the three-dimensional structure of the environment. It seems clear that - although somewhat out of fashion - a comparative approach to the neurobiology of sensory and neural control systems is needed and can be successfully applied to gain insight into their natural operating conditions. Beyond optics, where the basic constraints are rather well understood, we are still quite ignorant, however, what in detail shaped the evolution of visual systems. We lack insight into the specific requirements of the visual tasks that animals are faced with and into the prevailing environmental conditions that might determine the way in which they can be met. For animals living in a flat world, at least, the environmental geometry relevant for spatial vision can be described fairly easily; yet, to find a behavioural test for the flat world mechanism of depth perception remains a challenge. Furthermore, especially crabs may in addition use binocular cues to depth and if they do, it would be interesting to find out whether - depending on habitat - some do more so than others.

Acknowledgements: Financial support came from the DFG (SFB 307) to J.Z. and H.-O. N. Part of the work was carried out during a visit to the School of Zoology, University of New South Wales, Sydney and supported by an ARGS grant to D.C. Sandeman, grants from the Studienstiftung des Deutschen Volkes to G. and H.-O. N., by a Visiting Lectureship from the University of New South Wales and a Feodor-Lynen-Grant, Humboldt Stiftung, Bonn to J.Z. We are grateful to S. Niesiolowski, Lodz, for identifying empidid flies. We thank K. Bartsch and H.-J. Dahmen for allowing us to use some of their unpublished work and U. Henique for her help in preparing the figures.

REFERENCES

Barnes, R. S. K. 1968, On the evolution of elongated ocular peduncles in the Brachyura, System. Zool., 17:182-187

Brooke, M. de L.; 1981, Size as a factor influencing the ownership of copulation burrows by the ghost crab (*Ocypode ceratophthalmus*), Z. Tierpsychol., 55:63-78

Burkhardt, D., de la Motte, I. 1983, How stalk-eyed flies eye stalk-eyed flies: Observations and measurements of the eyes of *Cyrtodiopsis whitei* (Diopsidae, Diptera), J. Comp. Physiol., 151:407-421

Collett, T. S., Harkness, L. I. K. 1982, Depth vision in animals, in: "Analysis of Visual Behavior," D. J. Ingle, M. A. Goodale, R. J. W. Mansfield, eds., M.I.T. Press, Cambridge Mass, London,pp111-177

Collett, T. S., Udin, S. B., 1988, Frogs use retinal elevation as a cue to distance, J. Comp. Physiol., A163:677-683

Cooper, M. L., Pettigrew, J. D., 1979, A neurophysiological determination of the vertical horopter in the cat and owl, J. Comp. Neurol., 184:1-26

Dahmen, H.-J., The compound eye of waterstriders (*Gerris lacustris*, Hemiptera), J. Comp. Physiol., (To be submitted)

Dahmen, H.-J., Junger, W. 1988, Adaptation to the watersurface: Structural and functional specialisation of the gerrid eye, in: "Sense Organs" (Proc. of the 16th Göttingen Neurobiology Conference)," N. Elsner, F. G. Barth, eds., G. Thieme, Stuttgart New York, p233

Hagen, H. O. v., 1970, Zur Deutung langstieliger und gehörnter Augen bei Ocypodiden (Decapoda, Brachyura), forma et functio, 2:13-57

Horridge, G. A., 1978, The separation of visual axes in apposition compound eyes, Phil. Trans. Roy. Soc. Lond., B285:1-59

Hughes, A., 1977, The topography of vision in mammals, in: Handb. Sens. Physiol., Vol. VII/5," F. Crescitelli, ed., Springer, Berlin Heidelberg New York, pp613 – 756

Junger, W., Dahmen, H. – J., 1988, Waterstriders (Gerridae) use visual landmarks to compensate for drift on a moving water surface, in: "Sense Organs" (Proc. of the 16th Göttingen Neurobiology Conference)," N. Elsner, F. G. Barth, eds., G. Thieme, Stuttgart New York, p35

Kunze, P., 1963, Der Einfluβ der Gröβe bewegter Felder auf den optokinetischen Augennystagmus der Winkerkrabbe (*Uca pugnax*), Ergeb. Biol., 26:55 – 62

Kunze, P., 1964, Eye – stalk reactions of the ghost crab *Ocypode*, in: "Neural theory and modelling," R. F. Reiss, ed., Stanford Univ Press, Stanford California, pp293 – 305

Land, M. F., 1981, Optics and vision in invertebrates, in: "Comparative physiology and evolution of vision in invertebrates. (Handb of Sens Physiol, Vol. VII/6B)," H. Autrum, ed., Springer, Berlin Heidelberg New York, pp 471 – 592

Lighter, F. J., 1974, A note on the behavioural spacing mechanism of the ghost crab *Ocypode ceratophthalmus* (Pallas) (Decapoda, Family Ocypodidae), Crustaceana, 27:312 – 314

Linsenmair, K. E., 1967, Konstruktion und Signalfunktion der Sandpyramide der Reiterkrabbe *Ocypode saratan* Forsk. (Decapoda, Brachyura, Ocypodidae), Z. Tierpsychol, 24:403 – 456

Murphey, R. K., 1971, Sensory aspects of the control of orientation to prey by the waterstrider, *Gerris remigis*, Z. vergl. Physiol. 72:168 – 185

Nalbach, H. – O., Nalbach, G., 1987, Distribution of optokinetic sensitivity over the eye of crabs: its relation to habitat and possible role in flow – field analysis, J. Comp. Physiol., A160:127 – 135

Nalbach, H. – O., Nalbach, G., Forzin, L., (a), Multisensory control of eye stalk orientation in space. II. Visual system and eye design, J. Comp. Physiol.,(in preparation)

Nalbach, H. – O., Zeil, J., Forzin, L., (b), Multisensory control of eye stalk orientation in space. I. Crabs from different habitats rely on different senses, J. Comp. Physiol., (in preparation)

Neil, D. M., 1982, Compensatory eye movements, in: "Biology of the Crustacea, Vol. IV," H. L. Atwood, D. C. Sandeman, eds., Academic Press, NewYork, pp133 – 163

Sandeman, D. C., 1978, Eye – scanning during walking in the crab *Leptograpsus variegatus*, J. Comp. Physiol., 124:249 – 257

Salmon, M., Hyatt, G. W., 1983, Communication, in: "The biology of Crustacea Vol. 7: Behavior and ecology," F. J. Vernberg, W. G. Vernberg, eds., Academic Press, New York London, pp1 – 40

Schaller, F., 1953, Verhaltens – und sinnesphysiologische Beobachtungen an *Squilla mantis*, Z. Tierpsychol., 10:1 – 12

Schneider, L., Langer, H., 1969, Die Struktur des Rhabdoms im 'Doppelauge' des Wasserläufers *Gerris lacustris*, Z. Zellforsch., 99:538 – 559

Schwind, R., 1978, Visual system of *Notonecta glauca*: A neuron sensitive to movement in the binocular visual field, J. Comp. Physiol., 123:315 – 328

Schwind, R., 1980, Geometrical optics of the *Notonecta* eye: Adaptations to optical environment and way of life, J. Comp. Physiol., 140:59 – 68

Stavenga, D. G., 1979, Pseudopupils of compound eyes, in: "Handb. Sens. Physiol. Vol. VII/6A," H. Autrum, ed., Springer, Berlin Heidelberg New York, pp357 – 439

Yamaguchi, T., Noguchi, Y., Ogawara, N., 1979, Studies of the courtship behavior and copulation of the sand bubbler crab, *Scopimera globosa*, Publ. Amakusa. Mar. Biol. Lab., 5:31 – 44

Zeil, J., Nalbach, G., Nalbach, H. – O., 1986, Eyes, eye stalks and the visual world of semi – terrestrial crabs, J. Comp. Physiol., A159:801 – 811

SEARCH AND CHOICE IN *DROSOPHILA*

Karl G. Goetz

Max-Planck-Institut für biologische Kybernetik
Spemannstrasse 38
D-7400 Tübingen, FRG

ABSTRACT

Search behaviour of the fruitfly in a choice between inaccessible targets, or 'figures', depends on the spatial and temporal context of the visual input. An account of the attempted flights and runs towards these figures was derived from the wingbeat control responses during tethered flight in a simulator, or obtained by automatic tracking during free walk in an arena. The present results demonstrate new aspects of search strategies in insects.

(1) At least three pairs of flight control muscles respond to displacements of the retinal images of *figure* and *ground*. Each pair contributes to fixation and tracking of a figure, and to stabilization of course and altitude with respect to the ground. Two pairs support a rigid strategy of *'instructional' fixation.* The third pair engages in a flexible strategy of *'operant' fixation* suitable to cope with artificially inverted displacements. Only 'instructional' fixation is found in the corresponding muscles of the mutant *'small optic lobes'*.

(2) Relative movement between figure and ground indicates the proximity of a figure, and seems to serve for long-range *depth perception* during free flight. Tethered flies prefer the nearest figure which might be attractive as a landing site. The attraction subsides gradually whenever the figure is moving in conjunction with the ground. Fixation of a figure requires up to 5 course-control manoeuvres/s, and can be maintained for 24 h of tethered flight in a simulator.

(3) Comparison of the number of 'runs' towards each of two inaccessible figures in the periphery of the arena has revealed a remarkable equidistribution of the frequencies of approach. The results suggest lack of preference even in a choice between two extremely different figures. The temporal sequence of the runs shows, however, that the equidistribution is achieved by *sustained spontaneous alternation* of the preferred target. In search for rewarding sites this strategy disengages the fly from continuous fixation of the nearest figure. The alternation of preference in *Drosophila* is reminiscent of sustained spontaneous depth reversal in human perception which helps to overcome one-sided interpretations of ambiguous figures such as the Necker cube. Alternation of preference may have concealed the ability for pattern recognition and learning in numerous experiments on vertebrates or invertebrates. To exclude this possibility one has to prove the statistical independence of successive choices.

(4) Area-covering search can be observed in the absence of sensory landmarks. Random-walk approximation of the search trajectories has revealed a reversible increase of the persistence of direction, or the 'mean free path', with the time spent under unfavourable condi-

tions. This strategy gradually extends the radius of search until the fly hits upon favourable sensory signals. Ether narcosis irreversibly increases the mean free path, and blocks its control for the rest of the life.

INTRODUCTION

The retinal image of the resting environment is an important reference system for the control of locomotion in *Drosophila melanogaster*. Displacements of this image are resolved, for instance, into a horizontal component which elicits course-control responses of the wings or legs, and a vertical component which elicits altitude-control responses of the wings. The properties of the responses have been investigated in numerous experiments on stationarily flying or walking flies (Bausenwein et al. 1986; Buchner 1976; Buchner et al. 1978; Bülthoff 1981; Bülthoff and Götz 1979; Bülthoff et al. 1982; Fischbach and Heisenberg 1981; Götz 1964, 1968, 1970, 1975, 1977, 1980, 1983a and b, 1985, 1987a and b; Götz and Biesinger 1985a and b; Götz and Buchner 1978; Götz et al. 1979; Götz and Wandel 1984; Götz and Wenking 1973; Heide 1983; Heide et al. 1985; Heisenberg and Buchner 1977; Heisenberg and Götz 1975; Heisenberg and Wolf 1984; Heisenberg and Wolf 1988; Mayer et al. 1988; Zanker 1988).

Fig. 1 shows *Drosophila* during free flight in a resting environment. The cylindrical surface illustrates the visual field of the eyes. Deviation from a straight course is accompanied by a displacement of the retinal image in the opposite direction. To counteract involuntary deviations, the fly tries to follow the displacements of the retinal image. This is done by an increase of the wingbeat amplitude on the outer side, and a decrease of the wingbeat amplitude on the inner side, of the intended turn (Götz 1968, 1983a and b). Course-control is extremely efficient: The fly responds to displacements as slow as 1 revolution/h, or as fast as 50 revolutions/s (Götz 1964). However, stabilization of the course with respect to the *ground* is only one aspect of flight control. Actually, the fly has to detect, approach and explore visual landmarks in search of rewarding sites. This requires *figure*-induced orientation (Götz 1983b; Heisenberg and Wolf 1984). The use of visual flight control for *fixation* and *tracking* of a figure, and for selection of a target according to its distance will be treated in the following two sections. Another section describes a strategy for the efficient exploration of competing targets on the ground. The last section covers search in the absence of sensory landmarks.

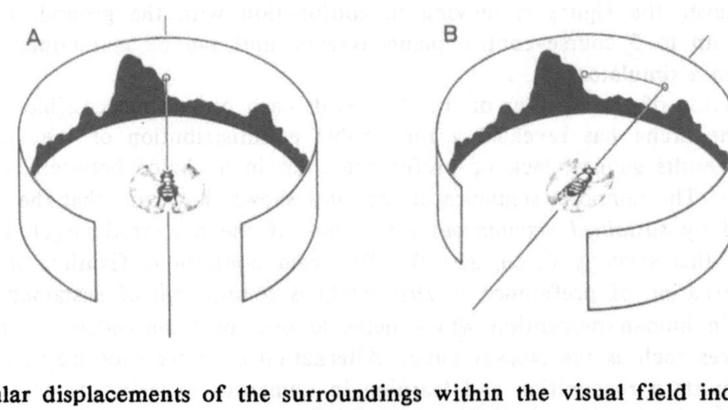

Fig. 1. Angular displacements of the surroundings within the visual field indicate a change in the direction of flight. The figure illustrates the retinal images received before (A), and after (B), a turn to the right. *Drosophila* is capable of stabilizing the retinal image in order to maintain a straight course. (From Götz 1975).

FIXATION OF A FIGURE: TWO STRATEGIES OF APPROACH

Flight control is achieved by up to 17 very small *control muscles* on either side of the bulky *power muscles* in the thorax of the fly. Fig. 2 shows four of these muscles which have been investigated in cooperation with G. Heide. Three muscles, the *basalar*, the *sternobasaler* and the *1st axillary muscle*, respond to visual stimulation and contribute to the control of both course and altitude. These muscles either increase or decrease the wingbeat amplitude of the fly (Götz 1983a and b, 1987b; Heide 1983; Heide et al. 1985).

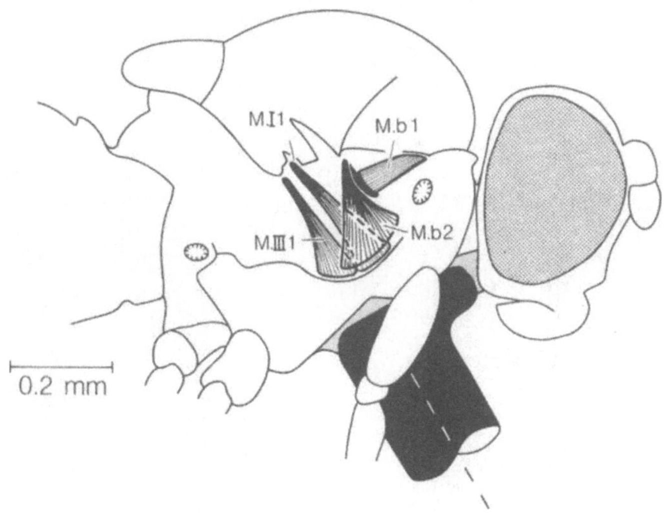

Fig. 2. Flight control muscles on the right side of the thorax of *Drosophila*. M.b1 basalar muscle, M.b2 sternobasalar muscle, M.I1 1st axillary muscle, M.III1 3rd axillary muscle. Redrawn from Zalokar 1947; notation according to Heide 1971. Head and thorax are affixed to the fly-holder (dark colour) of a flight simulator which allows to record the movements of the wings, and the activities of selected control muscles, under flight-induced visual stimulation.

Fig. 3A illustrates the tethered flight of *Drosophila* in a *flight simulator*. An infrared spotlight on either side casts a shadow of the beating wing onto the contralateral mask opening of an optoelectric device. This device measures the wingbeat amplitudes on either side. The difference of the wingbeat amplitudes represents the intended turn. The signal controls, by rotation of a circular transparency in a projector on top of the simulator, the angular speed of a bar on the cylindrical screen around the fly. One half of the screen is removed to show the details. The simulator allows the fly to manoeuvre the figure into arbitrary angular positions (Götz 1987a, b).

Fig. 3B gives an example of figure-induced orientation during 12 h of tethered flight in the simulator. The historgrams represent, in temporal sequence from front to back, the time spent by the figure in different angular positions between 180° to the left and 180° to the right of the fly's forward direction. A maximum at the centre indicates *fixation* of the figure in the frontal area of the visual field. Actually, this experiment has been continued for another 20 h, and fixation is encountered around the clock (Götz 1987a). Fixation requires continuous course control manoeuvres, like driving in heavy traffic. As an example, a succession of course-control manoeuvres, recorded in a time period of 5 s, is given in Fig. 4. The upper traces show the beat amplitudes of the left and right wing, respectively. The lower traces the simultaneously recorded activities of the sternobasalar muscle on the left and on the right. Muscular activity is accompanied by an increase of the *ipsilateral* wingbeat amplitude. At least one pair of antagonists, the 1st axillary muscles, accounts for the simultaneous decrease of the *contralateral* wingbeat amplitude.

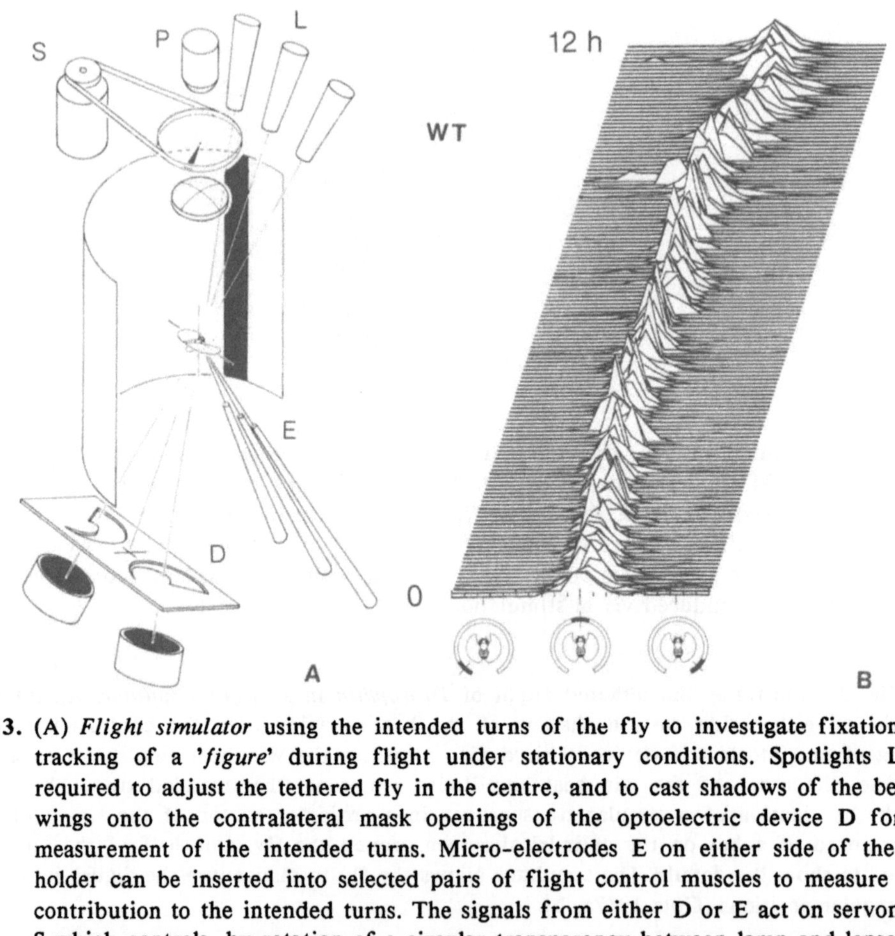

Fig. 3. (A) *Flight simulator* using the intended turns of the fly to investigate fixation and tracking of a '*figure*' during flight under stationary conditions. Spotlights L are required to adjust the tethered fly in the centre, and to cast shadows of the beating wings onto the contralateral mask openings of the optoelectric device D for the measurement of the intended turns. Micro-electrodes E on either side of the fly-holder can be inserted into selected pairs of flight control muscles to measure their contribution to the intended turns. The signals from either D or E act on servomotor S which controls, by rotation of a circular transparency between lamp and lense of a projector P, the angular speed of a vertical bar on a cylindrical screen. The diagram shows half of the screen on the left side of the fly. - (B) Figure-induced orientation during the first 12 h of an experiment in which tethered flight was sustained for 32 h with only short interruptions during feeding. The difference of the wingbeat amplitudes on either side was used to control the angular speed of the bar. The histograms represent, in temporal sequence from front to back, the time spent by the figure in different positions to the centre of the visual field. The maxima near the midline of the histograms demonstrate '*fixation*' of the figure in front of the fly.

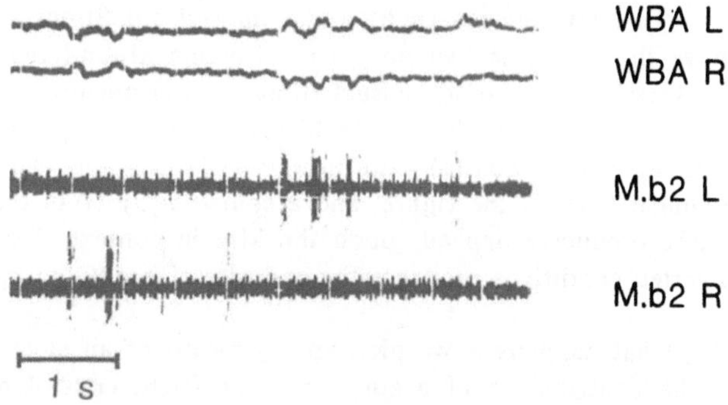

Fig. 4. Time course of both the wing beat amplitude (WBA), and the spike activity of the sternobasalar flight control muscle (M.b2), recorded simultaneously on either side of the thorax (L, R). The example shows a succession of 6 course-correcting manoeuvres during fixation of a figure in the flight simulator. Contraction of a sternobasalar múscle seems to increase the ipsilateral wingbeat amplitude. Simultaneous decrease of the contralateral wingbeat amplitude intensifies the manoeuvre.

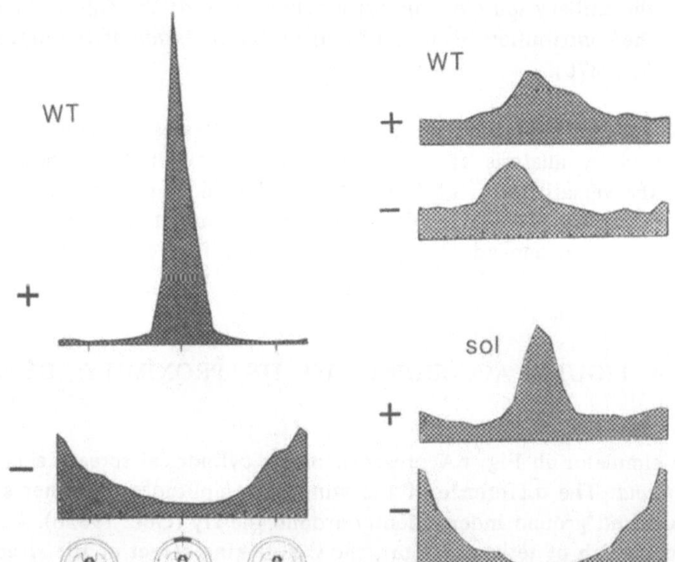

Fig. 5. Figure-induced orientation in the flight simulator. The histograms show the relative time spent by the figure in different positions with respect to the centre of the visual field. The results were obtained by continuous alternation between '*normal*' conditions (+) where the figure is expectedly moving in the direction opposite to the intended turn, and '*inverted*' conditions (-) where the figure is unexpectedly moving in the direction of the intended turn. - (A) *Non-adaptive* strategies of 'instructional' fixation prevail if the wingbeat signal is used to control the movements of the figure: the wild-type fly (WT) alternates between a course *towards* the figure under normal conditions, and a course *away* from the figure under inverted conditions. Flight time 1 h. - (B) The contribution of the pair of axillary muscles (M.I1) to the intended turns of the wild-type (WT; 6 flies, 10 h) revealed the existence of *adaptive* strategies of 'operant' fixation which are not confused by the inversion of the perceived movement (upper histograms). The corresponding muscular subsystem of the mutant '*small optic lobes*' (*sol*; 2 flies, 5h) shows a relapse to *non-adaptive* strategies of 'instructional' fixation (lower histograms).

To simulate free flight under *normal* conditions, the wingbeat signal must be used to move the figure in 'counter-turn' direction around the fly. *Inversion* of the signal changes the *kinetic context* of the visual stimulus, and creates a situation which never occurred in the history of insect evolution. The histograms in Fig. 5A show the results of continuous alternation between *normal* (+), and *inverted* (-) conditions. The fly alternates, accordingly, between a course *towards* the figure, and a course *away* from the figure. Conclusion: Fixation of the figure obviously depends upon the kinetic context. The attempt to fixate the figure under inverted conditions produces the opposite effect (Götz 1983b; Heisenberg and Wolf 1984).

What happens if we pick up, by means of an electrode on either side of the fly in Fig. 2, the contribution of a single pair of flight control muscles, and use only this signal to move the figure around the fly? Of the three pairs of muscles investigated, so far, one was outstanding because of its unexpected properties: The pair of *1st axillary* muscles. The histograms in Fig. 5B show these properties: In '*small optic lobes*' (*sol*), a mutant with reduced number of nerve cells in the 1400 columns of the visual neuropil (Fischbach and Heisenberg 1981; Heisenberg and Wolf 1984), the axillary muscles support a course *towards* the figure under *normal* (+) conditions, and a course *away* from the figure under *inverted* (-) conditions. The contribution of these muscles depends on the kinetic context. However, in the wild-type (WT), the axillary muscles support a course *towards* the figure, independent of the kinetic context. The contribution is not confused by the *inversion* of the perceived movement (Götz 1983b, 1985, 1987b).

The co-existence of non-adaptive and adaptive strategies of fixation in the wild-type has been confirmed by analysis of the torque response (Wolf and Heisenberg 1986). The results illustrate the versatility of the flight control system. It is tempting to speculate about the function of the nerve cells which are missing in the mutant and which do not seem to be required for the ordinary non-adaptive optomotor control of course and altitude in *Drosophila.*

CHOICE OF A FIGURE ACCORDING TO ITS PROXIMITY: DYNAMIC DEPTH PERCEPTION

The flight simulator in Fig. 6A presents, on its cylindrical screen, a bar-shaped *figure* and a dotted *ground*. The difference of the wingbeat amplitudes on either side can be used for moving figure and ground independently around the fly (Götz 1983b). The results in Fig. 6B, obtained during 5 h of tethered flight, show a striking effect of the *structural context* of the visual stimulus on the orientation of a fly which is allowed to control the angular speed either of the figure ('FIG.'), or of figure and ground simultaneously ('FIG. + GRD.'). Fixation is found as long as the figure is moving with respect to the ground. Almost no fixation is found as long as the figure is moving in conjunction with the ground. Two alternative explanations can be given under these conditions:

(1) Without relative movement between figure and ground, the figure will be *invisible*.
(2) Without relative movement between figure and ground, the figure will be *ignored*.

The movement-induced separation of figure and ground in the fly, and the corresponding transition from invisibility to visibility, is now well understood from the work of W. Reichardt and his group (Egelhaaf et al. 1988). However, the slow fading of fixation in the intermediate time period of the present experiment, and the absence of fading in a corresponding experiment on an optomotor-blind mutant (Götz 1983b), suggest that, in the present experiment, the conspicuous figure is still visible but actively ignored.

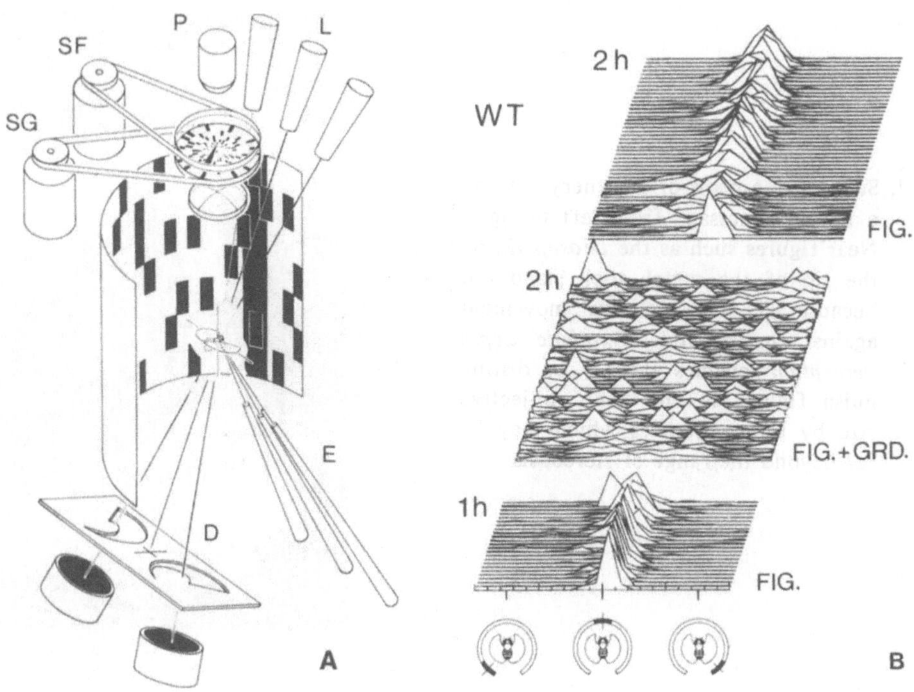

Fig. 6. (A) *Flight simulator* using the intended turns of a tethered fly to control, independently by servomotors SF and SG, the angular speed of a vertical bar (FIG.), and a random-dot pattern (GRD.). Technical details as described in Fig. 3A. - (B) Series of histograms showing, in temporal sequence from front to back, the preferred position of the vertical bar in the visual field of a wild-type fly during 5 h of flight in the simulator. Course towards the *figure* prevails if the *ground* is at rest (FIG.). The attraction of the visual object subsides gradually if the ground is moving in conjunction with the figure (FIG. + GRD.). Relative movement between figure and ground indicates proximity of the figure, and seems to be used for '*dynamic depth perception*' during free flight.

A tentative explanation of this phenomenon is given in Fig. 7. The two frames illustrate the displacement of figures in successive retinal images, received on a passage from left to right. *Near figures*, such as the *Drosophila* on the tip of the match-stick, are likely to become preferred targets because of the relative movement against the ground. *Remote figures* appear to be embedded in the ground and, therefore, do not elicit the attention of the fly.

Evaluation of the *structural context* enables the fly to distinguish figures of similar subjective size by their distance. This facilitates the exploration of visual landmarks and is likely to increase the limited range of *depth perception* in *Drosophila* from 'millimeters' to 'meters'.

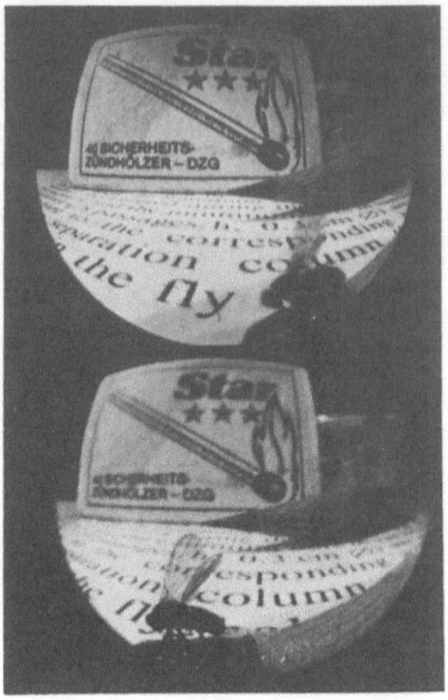

Fig. 7. Successive aspects of a scenery seen by a fly on a passage from left to right. Near figures such as the *Drosophila* on the tip of the match-stick stand out because of their relative movement against the ground. '*Dynamic depth perception*' enables the fly to distinguish figures of the same subjective size by their distances which may be far beyond the range of stereopsis.

SEARCH AMONG COMPETING FIGURES: SPONTANEOUS ALTERATION OF THE PREFERRED TARGET

The continuity of the fixation response during the first 12 h of the experiment in Fig. 3B suggests that *Drosophila* would easily disclose its preference in a choice between two different figures. In Fig. 8 the fly is affixed to a sled, and is kept walking for about 12 h on a tread compensator (Götz and Wenking 1973; Götz 1977), a servo-controlled ball to hold the freely walking fly in a fixed position and orientation to the dotted pattern on either side. Preference, in this experiment, is given by the average curvature of the track in revolutions/pathlength towards the preferred pattern. In a choice between the one-dot-pattern and the multi-dot-pattern of the same luminous area we obtained, with 90 flies, a total pathlength of 3800 m. The average curvature was about 1 rev./2 m pathlength towards the multi-dot-pattern. This preference lasts throughout the experiment, is invariant to the reversal of contrast and amounts to only 3 percent of the optomotor course-control response obtained under similar conditions. Scaled to our bodily dimensions the average curvature would be 1 rev./km pathlength. The effect is significant, but clearly too small to justify the notion 'preference'.

It is conceivable that flies do not easily detect different numbers of elements in the two patterns. We have a similar problem with the *Minsky-Papert spirals* shown in Fig. 9. Our perception fails to inform us about the number of elements in these spirals. We need scrutiny to identify a single element in the spiral on the left, and two disjunct elements in the spiral on the right.

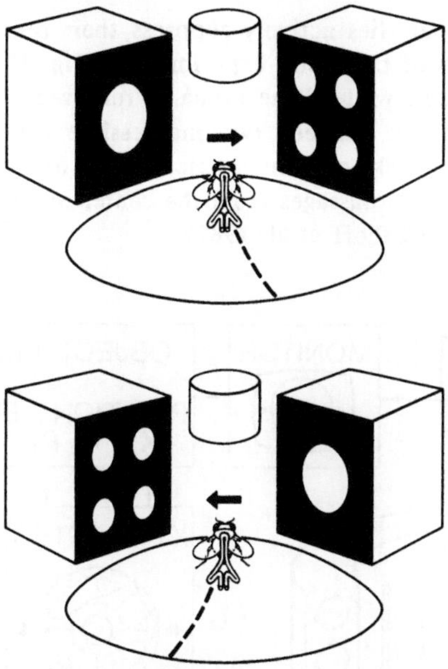

Fig. 8. Choice between differently dotted patterns of the same luminous area. The freely walking fly is held, for several hours, in a state of spontaneous fluctuation around a fixed position and orientation to the patterns on either side of a *tread compensator*. The cylindrical transducer on top of the compensator measures the translatory and rotatory displacements of a tiny sled of paramagnetic wire which is drawn by the fly. A sensor-controlled servo-system (not shown in the diagram) counteracts the displacements by appropriate rotation of a mercury-filled ball on which the fly is walking. Broken lines illustrate the track of the fly. The average curvature in revolutions/meter pathlength is read from counters on the shaft the servo-motors. To diminish the influence of locomotor bias the patterns were exchanged every 5 minutes. Arrows indicate the surprisingly weak preference of the multi-dot pattern.

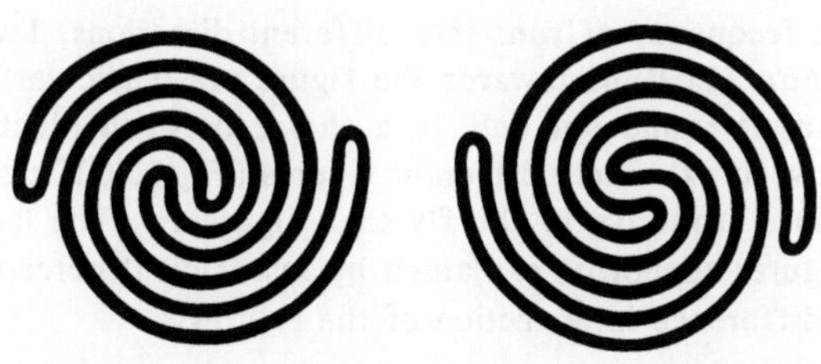

Fig. 9. The *Minsky-Papert spirals* illustrate a difficulty *Drosophila* may have with the discrimination of dotted patterns in the previous experiment. Scrutiny of a human observer is required to detect the actual number of unconnected picture elements: one in the spiral on the left, and two in the spiral on the right. (From Minsky and Papert 1969).

It is also conceivable that flies actively suppress their preferences in a choice between different figures. The results of the arena-experiment in Fig. 10 support this conjecture. The frontal half of the illuminated wall of the arena is removed to show the circular disk, on which the fly is freely walking between two inaccessible figures. Shortened wings, and a water-filled moat around the disk, prevent its escape. An optical scanner is used to plot the track of the fly and to record its passages from the center of the disk to the different sectors in the periphery (Götz 1980; Bülthoff et al. 1982).

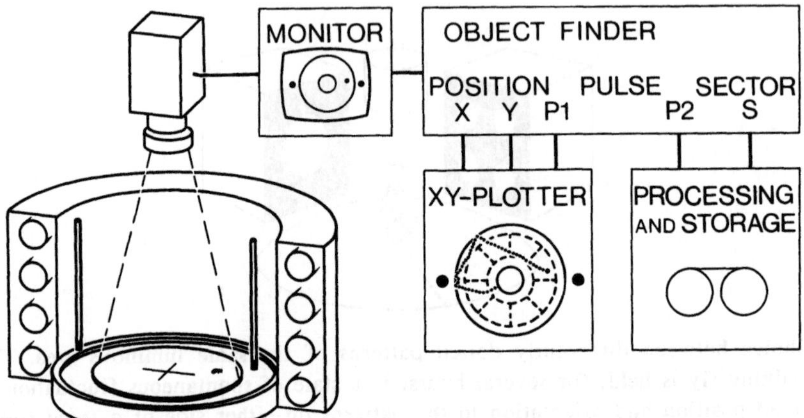

Fig. 10. *'Dipterodrome'* with opto-electrical scanner, used for the investigation of search strategies on the ground. The frontal half of the arena is cut away to show an elevated disk of 88 mm diameter on which the fly is freely walking between two inaccessible figures. Shortened wings, and a water-filled moat around the disk, prevent its escape. To plot subsequent XY-positions along the track of the fly, the 'object finder' delivers a regular sequence of plot pulses P1 as long as the fly is moving. The distances from the centre of the disk are averaged to obtain the mean radial distance R of the track. A pulse P2 indicates the passage from the centre to one of the eight sectors of a ring-shaped area in the periphery of the disk. This pulse activates the storage of the corresponding sector number S. (From Bülthoff et al. 1982).

Fig. 11A shows the frequency of runs into different directions. Two results characterize the behavior of *Drosophila*: (1) Runs towards the figure are particularly frequent. (2) Equidistribution of these runs is found not only in a choice between identical figures (F1, F1), but also in a choice between extremely different figures (F1, F2): If tested separately, figure F2 is about 10 times more attractive to the fly than figure F1. The lack of preference in a choice between these figures cannot be explained by insufficient perception: the fly evidently compensates for the difference in attraction of the figures.

Fig. 11B shows this strategy in action. Each of the diagrams represents the track of a fly which was running for 200 s on a circular disk, either between the identical figures F1 and F1, or between the extremely different figures F1 and F2. Near the center, the trajectories are blanked for technical reasons. The diagrams show that the equidistribution of the choices is achieved by *sustained spontaneous alteration* of the preferred target: the fly keeps running to and fro, between the figures, sometimes for hours. In one of the experiments a fly scored, within 7 hours of walk on the disk, 2500 alterations on a path of 220 m length (Bülthoff et al. 1982). The loop on the right of the upper diagram is one of the rare exceptions where a figure has been visited twice in succession. The results suggest two general conclusions:

(1) A strategy based on sustained spontaneous alteration of target facilitates the exploration of the visual environment in search of rewarding sites. The strategy prevents the fly from being captured by the nearest object. The effect of this strategy can be compared with sustained spontaneous depth reversal in human perception which helps to overcome one-sided interpretions of ambiguous figures such as the classical *Necker cube* shown in Fig. 12.

(2) Sustained spontaneous alteration of target may have concealed the ability for pattern discrimination and learning in numerous apparently 'unsuccessful' experiments on invertebrates as well as vertebrates. To exclude this possibility, one has to prove the statistical independence of successive choices. This has rarely been done in the past.

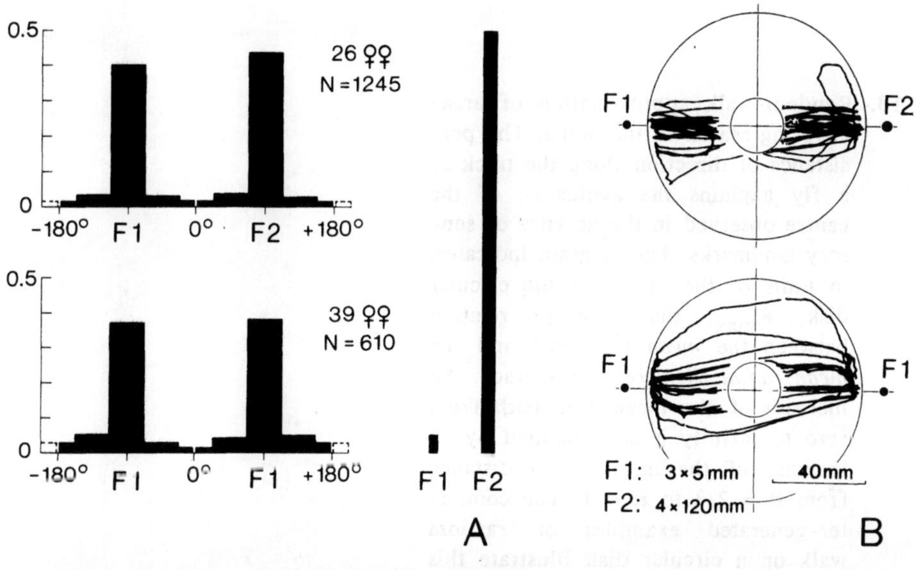

Fig. 11. Choices between inaccessible figures which are either identical (F1, F1, lower diagrams), or extremely different (F1, F2, upper diagrams). - (A) Relative frequencies of runs in different directions, obtained by evaluation of N passages from the centre to the periphery of the circular disk in the arena. Note the apparent lack of preference in the choice between different figures. - (B) Tracks of two flies which were running for 200 s on the circular disk. The scanning device obscures the tracks near the centre or at velocities below 1 mm/s. Equidistribution of the choices is achieved by *sustained spontaneous alteration* of the preferred target.

Fig. 12. The *Necker cube* is a classical example of sustained spontaneous depth reversal in human perception. The ambiguous figure is immediately seen in 3D. Spontaneous alteration between the aspect with the lower left square in front, and the aspect with the upper right square in front, prevents a pre-occupation of the observer. Spontaneous alteration of the preferred target during search in the arena may serve similar purposes in *Drosophila*.

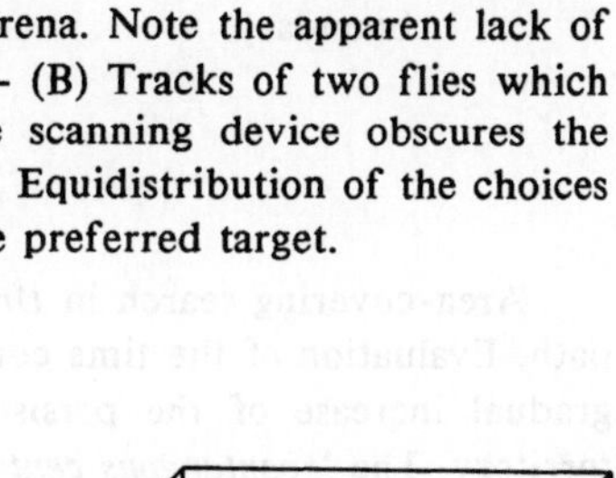

AREA-COVERING SEARCH: ADAPTATION OF THE 'MEAN FREE PATH'

Removal of sensory cues near the fly elicits area-covering search behaviour. The persistence of direction along the track can be described by the *mean free path* of a random-walk approximation. A fly on a circular platform is compelled to run, often for some time, along the outer boundary if the mean free path is sufficiently large to divert locomotor activity from the center to the surround. This effect is likely to account for the avoidance of the centre in experiments with *Drosophila* (Götz and Biesinger 1985a). Fig. 13 illustrates the mean radial distances R of three simulated random-walk tracks from the centre of the platform. The actual distance of a fly in an arena experiment can be used to calculate the corresponding mean free path (Götz and Biesinger 1985b).

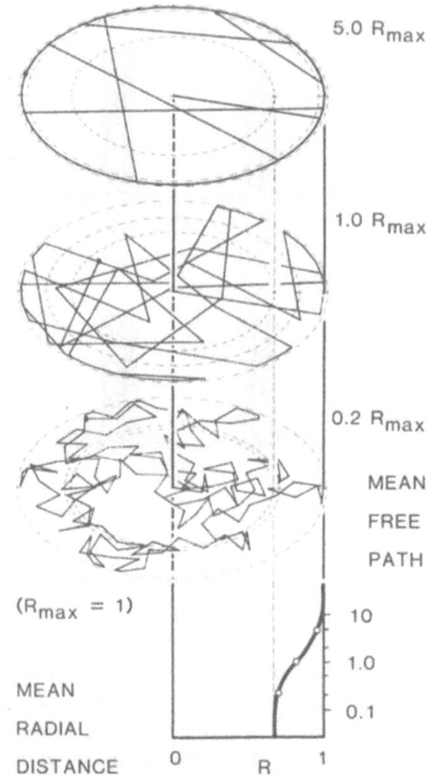

Fig. 13. Random-walk approximation of area-covering search in the arena. The persistence of direction along the track of a fly explains the avoidance of the centre observed in the absence of sensory landmarks. The diagram indicates, in units of the radius of the circular disk, R_{max}, the expected relation between the *mean free path* and the *mean radial distance* of the track. An increase of the mean free path from zero to infinity is accompanied by an increase of the mean radial distance from R = 2/3 to R = 1. The computer-generated examples of random walk on a circular disk illustrate this transition: increasing portions of the tracks of 30 units total length are forced into orbits along the outer boundary.

Area-covering search in *Drosophila* is obviously controlled by variation of the mean free path. Evaluation of the time course of centre avoidance in the arena has revealed a reversible gradual increase of the persistence of direction during accomodation to a non-rewarding territory. The '*spontaneous centrophobia*' of the fly extends its radius of search, most probably in order to facilitate the detection of favourable districts in a natural environment. The tracks of a completely blind mutant '*sine oculis*' (*so*) on the circular disk of the arena shown in Fig. 10 illustrate the transition from a comparatively low mean free path of about 17 mm at the beginning to an increased mean free path of about 180 mm after 20 min of area-covering search in the absence of visual cues (Fig. 14).

The control of the persistence of direction in the tracks of *Drosophila* can be disturbed by '*ether-induced centrophobia*'. A few seconds of exposure to ether vapour are sufficient to increase the initial mean free path by a factor of two, and to block the control mechanism irreversibly for the rest of the life of a fly. Fig. 15 shows the different tracks of a control fly and its sibling which received a very short ether narcosis 4 d before the experiment. Banana odour released from the inner compartment of the arena fails to hold a fly near the

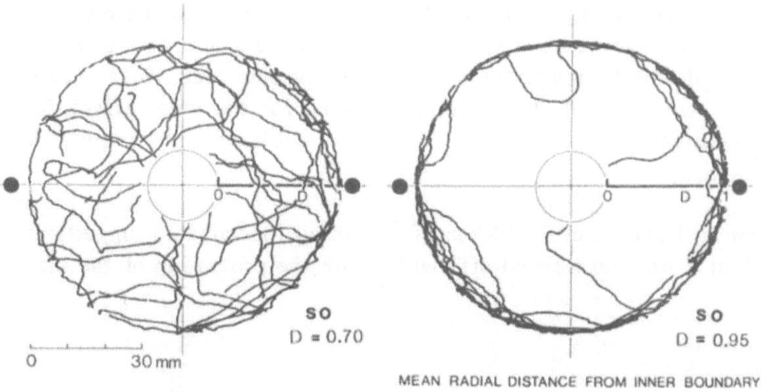

Fig. 14. *'Spontaneous centrophobia'*, explained by an increase of the 'mean free path' during area-covering search in the arena. The diagrams show different tracks of a completely blind mutant *'sine oculis'* (*so*) which was freely walking between two inaccessible figures of 4 x 120 mm size. Each of the tracks represents 200 s of search, either at the beginning (diagram on the left) or at the end (diagram on the right) of a 20 min period of continuous locomotor activity on the circular disk. The blindness allows unnoticed removal of scent marks deposited by the fly. The scanning device obscures tracks near the centre or at velocities below 1 mm/s. Unlike the example in Fig. 13, the mean radial distance (D) of the fly is measured from the inner boundary of the scanning area. The 'mean free path' derived from these distances amounts to 17 mm at the beginning, and 180 mm at the end of the experiment. Increase of the mean free path facilitates the escape from unfavourable districts in a natural environment.

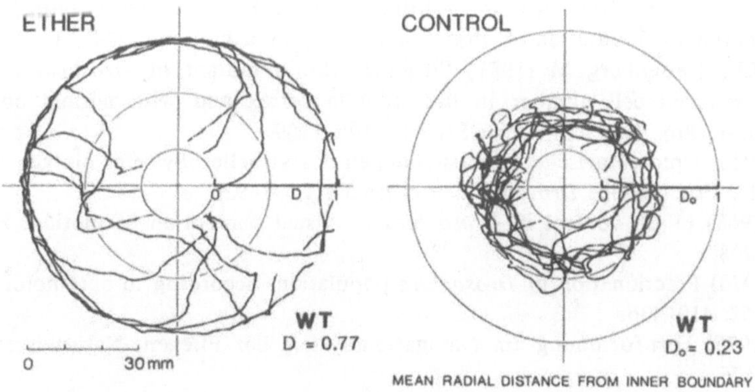

Fig. 15. *'Ether-induced centrophobia'*, a permanent change in area-covering search due to a mild ether narcosis 4 d before the experiment. The concentric circles indicate the outer boundary of the disk in the arena, and the invisible borders of a ringshaped inner compartment below the air-permeable floor from which banana odour is released at a constant rate of 2 ml/s. The diagrams represent the tracks of two 7 d old female siblings of a wild-type (WT) strain which have been reared under identical conditions. Each of the flies was running for 200 s on the disk. Tracks near the centre or at velocities below 1 mm/s were obscured. The mean radial distances of the test fly (D), and of the control fly (D_0), are measured from the inner boundary of the scanning area. Irreversible post-narcotic increase of the persistence of direction along the track seems to account for the translocation of search from the source of odour to the periphery of the disk. The translocation, confirmed in 806 trials, can not be explained by olfactory desensitization. (From Götz and Biesinger 1985a).

centre if the mean free path is sufficiently large compared with the radius of the disk. The olfactory cue and its substitutes, a visual attractant at the centre or a thermal repellent at the periphery, facilitate the demonstration of both '*spontaneous*' and '*ether-induced*' avoidance of the center. However, none of these cues is essential for the ether effect. Search control by variation of persistence in the track of a fly is compatible with observations in other insects.

Acknowledgement. I am grateful to Michael Dickinson for valuable suggestions, and to Ulrike Flaiz and Rainer Zorn who assisted efficiently with the processing of the manuscript and the drawings.

REFERENCES

Bausenwein B, Wolf R, Heisenberg M (1986) Genetic dissection of optomotor behavior in *Drosophila melanogaster*. Studies on wild-type and the mutant *optomotor-blind*^{H31}. J Neurogenetics 3: 87-109.

Buchner E (1976) Elementary movement detectors in an insect visual system. Biol Cybern 24: 85-101.

Buchner E, Götz KG, Straub C (1978) Elementary detectors for vertical movement in the visual system of *Drosophila*. Biol Cybern 31: 235-242.

Bülthoff H (1981) Figure-ground discrimination in the visual system of *Drosophila melanogaster*. Biol Cybern 41: 139-145.

Bülthoff H, Götz KG (1979) Analogous motion illusion in man and fly. Nature 278: 636-638.

Bülthoff H, Götz KG, Herre M (1982) Recurrent inversion of visual orientation in the walking fly, *Drosophila melanogaster*. J Comp Physiol 148: 471-481.

Egelhaaf M, Hausen K, Reichardt W, Wehrhahn C (1988) Visual course control in flies relies on neuronal computation of object and background motion. TINS 11: 351-358.

Fischbach KF, Heisenberg M (1981) Structural brain mutant of *Drosophila melanogaster* with reduced cell number in the medulla cortex and with normal optomotor yaw response. Proc Nat Acad Sci USA 78. 1105-1109.

Götz KG (1964) Optomotorische Untersuchungen des visuellen Systems einiger Augenmutanten der Fruchtfliege *Drosophila*. Kybernetik 2: 77-92.

Götz KG (1968) Flight control in *Drosophila* by visual perception of motion. Kybernetik 4: 199-208.

Götz KG (1970) Fractionation of *Drosophila* populations according to optomotor traits. J Exp Biol 52: 419-436.

Götz KG (1975) Hirnforschung am Navigationssystem der Fliegen. Naturwissenschaften 62: 468-475.

Götz KG (1977) Sehen, Abbilden, Erkennen - Verhaltensforschung am visuellen System der Fruchtfliege *Drosophila*. Verh Schweiz Naturforsch Ges 1975, pp 10-33.

Götz KG (1980) Visual guidance in *Drosophila*. In: Siddiqi O, Babu P, Hall L, Hall J (eds) Development and Neurobiology of *Drosophila*. Plenum, New York London, pp 391-407.

Götz KG (1983a) Bewegungssehen und Flugsteuerung bei der Fliege *Drosophila*. In: Nachtigall W (ed) BIONA report 2, Akad Wiss Mainz. Fischer, Stuttgart New York, pp 21-33.

Götz KG (1983b) Genetic defects of visual orientation in *Drosophila*. Verh Dtsch Zool Ges 76: 83-99.

Götz KG (1985) Loss of flexibility in an optomotor flight control system of the *Drosophila* mutant 'small optic lobes'. Biol Chem Hoppe-Seyler 366: 116-117.

Götz KG (1987a) Course-control, metabolism and wing interference during ultralong tethered flight in *Drosophila melanogaster*. J Exp Biol 128: 35-46.

Götz KG (1987b) Relapse to 'preprogrammed' visual flight-control in a muscular sub-system of the *Drosophila* mutant 'small optic lobes'. J Neurogenetics 4: 133-135.

Götz KG, Biesinger R (1985a) Centrophobism in *Drosophila melanogaster*. I. Behavioral modification induced by ether. J Comp Physiol 156: 319-327.

Götz KG, Biesinger R (1985b) Centrophobism in *Drosophila melanogaster*. II. Physiological approach to search and search control. J Comp Physiol 156: 329-337.

Götz KG, Buchner E (1978) Evidence for one-way movement detection in the visual system of *Drosophila*. Biol Cybern 31: 243-248.

Götz KG, Hengstenberg B, Biesinger R (1979) Optomotor control of wing beat and body posture in *Drosophila*. Biol Cybern 35: 101-112.

Götz KG, Wandel U (1984) Optomotor control of the force of flight in *Drosophila* and *Musca*. II. Covariance of lift and thrust in still air. Biol Cybern 51: 135-139.

Götz KG, Wenking H (1973) Visual control of locomotion in the walking fruitfly *Drosophila*. J Comp Physiol 85: 235-266.

Heide G (1971) Die Funktion der nicht-fibrillären Flugmuskeln von *Calliphora*. I. Lage, Insertionsstellen und Innvervierungsmuster der Muskeln. Zool Jahrb, Abt Allg Zool Physiol Tiere 76: 87-98.

Heide G (1983) Neural mechanisms of flight control in Diptera. In: Nachtigall W (ed) BIONA report 2, Akad Wiss Mainz. Fischer, Stuttgart New York, pp 35-52.

Heide G, Spüler M, Götz KG, Kamper K (1985) Neural control of asynchronous flight muscles in flies during induced flight manoeuvres. In: Gewecke M, Wendler G (eds) Insect Locomotion. Paul Parey, Berlin Hamburg, pp 215-222.

Heisenberg M, Buchner E (1977) The role of retinula cell types in visual behaviour of *Drosophila melanogaster*. J Comp Physiol 117: 127-162.

Heisenberg M, Götz KG (1975) The use of mutations for the partial degradation of vision in *Drosophila melanogaster*. J Comp Physiol 98: 217-241.

Heisenberg M, Wolf R (1984) Vision in *Drosophila*. Studies of Brain Function, vol. 12. Springer, Berlin Heidelberg New York Tokyo.

Heisenberg M, Wolf R (1988) Reafferent control of optomotor yaw torque in *Drosophila melanogaster*. J Comp Physiol 163: 373-388.

Mayer M, Vogtmann K, Bausenwein B, Wolf R, Heisenberg M (1988) Flight control during free yaw turns in *Drosophila melanogaster*. J Comp Physiol 163: 389-399.

Minsky M, Papert S (1969). Perceptrons. Cambridge Mass., M.I.T. Press.

Wolf R, Heisenberg M (1986) Visual orientation in motion-blind flies is an operant behaviour. Nature 323: 154-156.

Zalokar M (1947) Anatomie du thorax de *Drosophila melanogaster*. Rev Suisse Zool 54: 17-53.

Zanker J (1988) How does lateral abdomen deflection contribute to flight control of *Drosophila melanogaster*? J Comp Physiol 162: 581-588.

THE EVOLUTION OF THE TIERED PRINCIPAL RETINAE OF JUMPING SPIDERS (ARANEAE: SALTICIDAE)

A. David Blest[1] and David O'Carroll[2]

[1]Developmental Neurobiology Group, Research School of Biological Sciences, P.O.Box 475, Canberra, A.C.T. 2601 Australia; [2]Department of Biological Sciences, Flinders University, Bedford Park, South Australia 5042

A unique sensory structure with an unusual modus operandi always poses a challenge to a biologist concerned with the fine-print of Evolution. The principal eyes of Salticid spiders offer just such a challenge. This short Review will identify its nature, summarise the outcome of new findings achieved in the last eight years, and outline some new ones.

Land (1969a,b) provided the first functionally informative account of the anatomy of the principal eyes of jumping spiders. In addition to many of their unique features listed below, he showed that it is through the principal eyes alone that key visual discriminations are mediated. The secondary eyes merely generate turning movements of the spiders that bring the principal eyes to bear on small targets in object space of potential significance to the spiders. The movements of the principal retinae that allow a target to be fixated, tracked and scanned are described by Land (1969b) and in various subsequent reviews and will not be specified here. The anatomical account of the principal retina of Phidippus by Land (1969a) has recently been revised by Blest et al.,(1988): their update of the earlier results will be taken for granted in what follows. A major conclusion from the re-examination of Phidippus was that the principal retinae of all phylogenetically advanced jumping spiders can be supposed to be organised along functionally equivalent lines.

The remarkable features of Salticid principal retinae are these:

(i) A retina is supplied with images by a miniature Galilean telescope (Williams and McIntyre, 1980), schematised by Fig. 1.

(ii) The diverging component of the telephoto system is provided by an interface between a dense glial matrix in which the receptive segments of the photoreceptors are embedded, and an effectively fluid-filled tube anterior to it. Apically, the interface takes the form of a tiny, hemispherical pit.

(iii) The receptive segments are tiered. At the fovea (which corresponds to the mid-point of the receptor mosaics) there are four Layers of receptive segments. Proximally in phylogenetically advanced Salticids, Layer I consists of long rhabdomeres constructed as light guides. Layers II-IV, lying distally, consist of much shorter rhabdomeres (Fig. 2). The overall consensus of evidence urges the conclusion that throughout the Salticidae it is parsimonious to assume that Layers I and II consist of receptors with peak responsiveness to green light, and

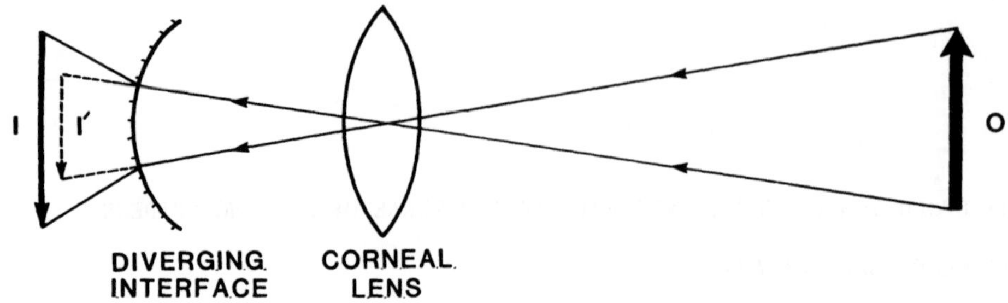

Figure 1. Summary of the optical design of the principal eye of a jumping spider. An image, I, of an object, O, is projected onto four tiered receptor mosaics by a corneal lens. A diverging optical component is formed by the interface between a dense retinal matrix and fluid filling a retinal tube anterior to it; together, corneal lens and interface provide telephoto optics. I' indicates the size and approximate position of the image that would be formed if the diverging component were absent. (From Blest and Sigmund, 1985).

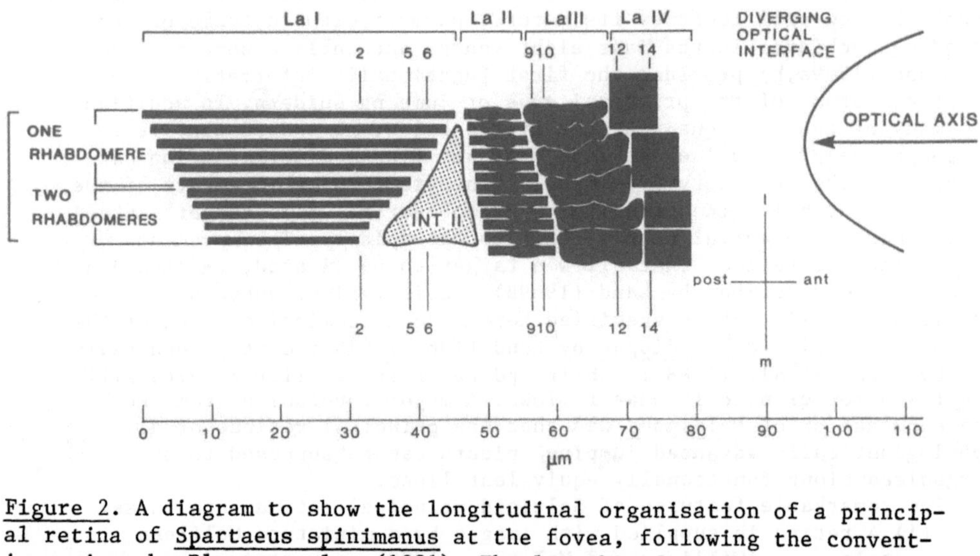

Figure 2. A diagram to show the longitudinal organisation of a principal retina of Spartaeus spinimanus at the fovea, following the conventions given by Blest et al., (1981). The lengths of receptive segments are approximations. The exact horizontal extents of Layers II-IV are not given, but each is wider than Layer I; the distance between Layer IV and the diverging optical interface is correctly represented, as is the horizontal 'staircase' offered by the tips of the rhabdomeres of Layer I. Spartaeus is the type genus of the relatively primitive subfamily Spartaeinae (Wanless, 1984); Layer I retains two rhabdomeres in each receptive segment on the medial side. 1, outer lateral; m, medial; post, posterior; ant, anterior. The small numerals 2-14 refer to relevant half-tone figures of transverse sections taken at those levels, in the original account. INT II, region occupied by intermediate segments of Layer II receptors. (From Blest and Sigmund, 1985).

that Layer IV contains receptive segments maximally responsive in the ultra-violet; the spectral properties of Layer III have not been determined (de Voe, 1975; Blest et al., 1981). The conclusion of Land (1969a) that retinal tiering compensates for the chromatic aberration of the dioptrics is supported in modified form by the results of the last decade.

(iv) The receptor mosaics are disposed as extremely narrow, vertical strips with boomerang-like profiles in transverse sections (Fig. 3). The retinae of other spiders amount to roughly hemispherical cups of receptive segments.

(v) Finally, we have shown that the principal retinae of large Salticids can resolve as little as 2.4 arc min at the fovea, such resolution depending upon the mosaic and light guiding properties of foveal Layer I receptive segments. This conclusion from geometrical optics (Williams and McIntyre, 1980; Blest et al., 1981; Blest, 1985) is supported by an ethological study (Jackson and Blest, 1982a).

The Salticidae are an extremely successful group of predatory arthropods, with ca. 4500 described species distributed world-wide. They present major problems for taxonomists: there is little reliable evidence that might tell us to which contemporary Families they could be allied. The retinae are so unlike those of other spiders as to make Salticid affinities even more enigmatic. A recent review by Blest (1987) noted that an examination of further extant species would be unlikely to resolve the evolutionary history of jumping spiders.

That pessimistic comment was written in 1985. Subsequently, one of us (A.D.B.) has examined the post-embryonic morphogenesis of a Salticid principal retina, and argued that receptor tiering is achieved by a sequence of events that approximately recapitulates the evolutionary events that led to it (Blest and Carter, 1987; Blest, 1988). D.O'C. has examined the principal retinae of a misumenine Thomisid and some Lycosids, and secondary retinae of members of the same Families. The following account attempts an integration of the data provided by ourselves and others, and comes to the comforting conclusion that the evolution of jumping spiders is readily comprehensible without the invocation of Magic, and that it is possible to suggest their systematic affinities rather more realistically than hitherto.

THE PRINCIPAL EYES OF SPIDERS: ANATOMICAL PRECEDENTS

The ultrastructural retinal anatomy of spiders has been summarised by Blest (1985), following extensive surveys by Homann (1951, 1971) conducted by light microscopy. To summarise what was known around 1985 about the anatomy of principal retinae:

(i) Principal retinae of all spiders appeared to be motile (Land, 1985), although other than in Salticids, movements of the twin retinae were not seen to be conjugate; they amount to little more than unco-ordinated tremors.

(ii) As far as was known, all principal retinae comprised a hemispherical sheet of receptive segments that constituted a monolayer. The receptive segments of such a hemispherical cup each contain two rhabdomeres which may be contiguous with the rhabdomeres of adjacent receptive segments, and which are not obviously organised as light guides.

The complex and ethologically effective Salticid principal retinae seemed to have no plausible precedents.

Conventional evolutionary theory requires that the gradual development of a functionally complex organ should proceed <u>via</u> a series of stages, each of which must be effective and viable. The enormous gap between the principal eyes of Salticids and those of any other spider

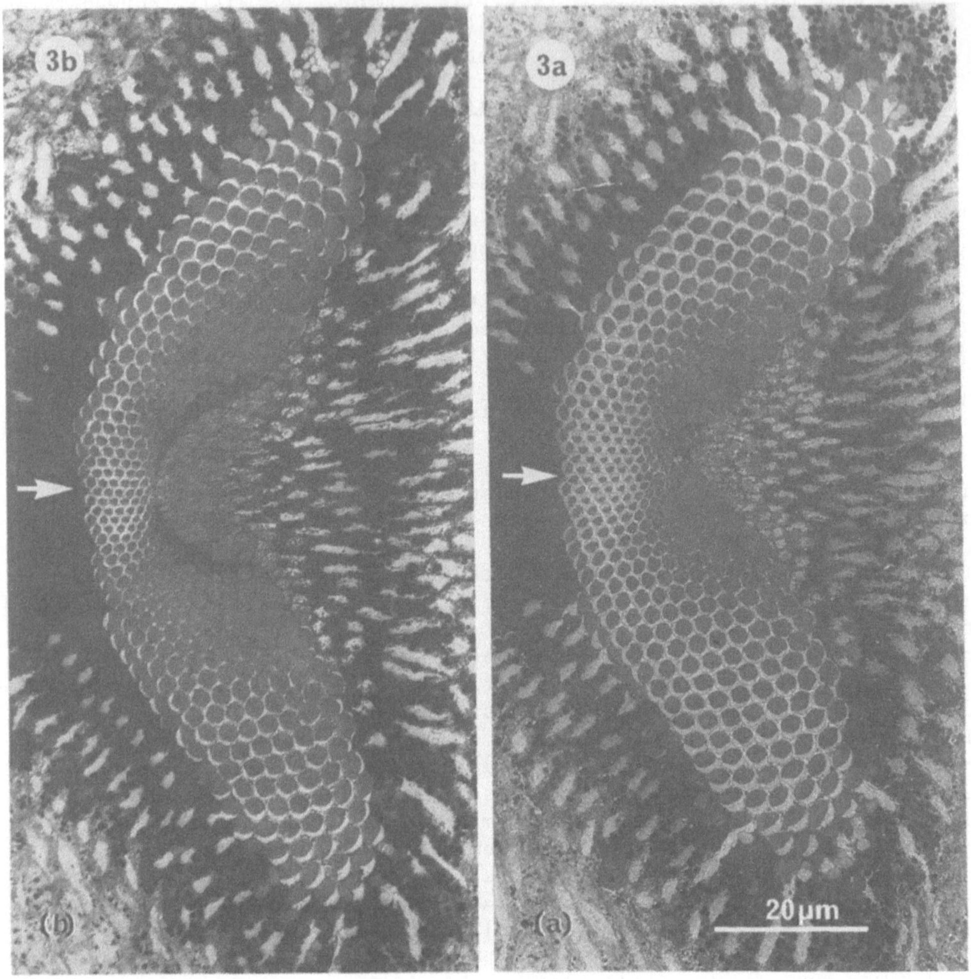

Figure 3a,b. Transverse sections through Layer I receptive segment fields of Amycus spiralifer. The approximate levels of section are indicated schematically by Figure 2: (a) corresponds to level 2, and (b) to level 5, near to the top of the 'staircase'. Land (1969b) showed that when a target is fixated by the retinal movements he described, its inverted image is brought to bear on the foveal region of the retina where inter-rhabdomeral intervals are least and resolution optimal. The fovea, which is also the mid-point of the retinal boomerang is arrowed. Scale bar on Figure 3a.

has presented formidable problems to taxonomists. Much of Salticid anatomy has clearly been determined by their visual organisation. Is there any way in which we can make sense of their evolution?

AN APPROACH TO THE ORIGIN OF THE SALTICID PRINCIPAL RETINA VIA THE COURSE OF ITS POST-EMBRYONIC MORPHOGENESIS

In the restricted sense accepted by De Beer (1958) and, for example, Medawar and Medawar (1983), Haeckel's 'Law of Recapitulation' can be useful when we come to consider phylogenetic enigmas. Given a number of caveats, it is possible look at a morphogenetic sequence, assess the extent to which it may reflect a prior evolutionary history, and erect a phylogenetic model that hopefully can be tested.

Blest and Carter (1987), Blest (1988) and Blest and Carter (1988) examined the morphogenesis of the principal eye of Plexippus validus. The eggs of spiders eclode to provide a post-embryo whose major organ systems are yet to differentiate.

Briefly, during the first three days of post-embryonic development (at 20°C) a primordial principal retina consists of an aggregate of undifferentiated cells that originates from a blastodermal invagination (Homann, 1971). By the end of Day 3, the future retina has differentiated to a distinct hemispherical layer of nascent receptive segments which do not yet bear rhabdomeral microvilli. Around Days 4-5, the hemisphere suffers a dramatic change of shape: it is 'squeezed' horizontally. The sequence of changes generated by squeezing, and by the progressive differentiation and elongation of receptive segments is shown below by Fig. 4. The end-result is that receptive segments that originally lay at the margins of the hemisphere - the retinal 'equator' - come to overlie receptive segments closer to the retinal 'pole', thus generating tiering. Blest (1988) notes that the motive force of 'squeezing' probably derives from differential growth of the glial matrix; full realisation of adult tiering requires not merely growth of the matrix, but the local involution of some parts of it. Such local involution has been observed directly in foveal Layer I after the post-embryos have moulted to the second instar (Blest and Carter, 1988).

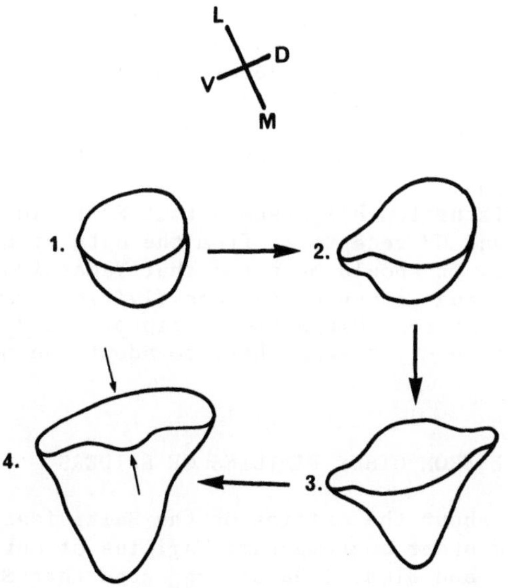

Figure 4. A schematic diagram of the conformational changes undergone by a principal retina between Day 4 and Day 11 of post-embryonic development, displayed clockwise. The horizontal line that defines the future foveal retina is indicated by small arrows. Orientation of the diagrams is given at top: L, lateral; M, medial; D, dorsal; V, ventral. Redrawn and modified from Blest and Carter (1987).

Thus, receptive segments of Layer IV derive from the primordial retinal equator, and those of Layer I from the pole. Layer IV receptors are known to have peak responses in the UV, and those of Layers I and II in the green (Blest et al., 1981). How best can we account for the positions of green and UV receptors both in terms of embryology and of their roles in compensating for the chromatic aberration of the imperfect dioptrics?

Intuitively, two different models can be proposed as extreme cases, each of which assumes that retinae of an evolutionary precursor of the Salticidae contained green and UV receptors only:

(i) Initially, the two spectral classes of receptors were scattered randomly over a retinal hemisphere. If 'squeezing' is assumed to have operated during evolution just as it does now during morphogenesis, it would seem necessary to postulate that receptive segments of the two spectral types underwent differential longitudinal migration parallel to the optical axis.

(ii) Initially, UV receptive segments lay at the equator of the retinal hemisphere; the migration of receptive segments imposed by 'squeezing' moved them to their contemporary distal position as Layer IV.

In merely ultrastructural terms, there is no real evidence that the first model is plausible. It is not obvious why a primordial retina would consist of receptors of two spectral types, randomly distributed. Blest (1988) summarises some evidence that suggests that, at least in Agelenids, UV receptors occupy a retinal equator, and why they may do so. The second model is compatible with the evidence from all sources.

Nevertheless, it clearly cannot be more than a crude outline of events that we can only hypothesise. A mere piling up of receptive segments to create a tiered retina as focal lengths of a dioptric apparatus increased would have been of little effect unless the receptive segments of all four Layers were 'fine-tuned' with respect to an optical system that was progressively becoming more complex. Blest and Carter (1988) present evidence from the morphogenesis of a principal retina that suggests how such fine-tailoring may have taken place, and Blest and Carter (in preparation) examine the problem in in the broader context of the micro-evolution of the Salticidae, and of their visual apparatus.

The foregoing discussion has assumed that Salticid principal retinae contain only green and UV receptors, from the data of De Voe (1975) and Blest et al., (1981). It should be noted that Yamashita and Tateda (1976) described a broader range of spectral sensitivities from a very small number of intracellular recordings from receptors of Menemerus. Until that genus is re-examined, it seems best to adopt the more conservative case.

COMPARATIVE EVIDENCE FROM OTHER FAMILIES OF SPIDERS

Some hypotheses about the origins of the Salticidae have attempted to dissociate them from other contemporary Families of cursorial spiders. For example, Jackson and Blest (1982b) suggested that Salticids may have arisen from web-building spiders as a consequence of selection pressures imposed by the evolution of the strategy of predatory web-invasion. Nevertheless, it is not unreasonable to look closely at members of the Lycosidae and Thomisidae, because their predatory life-styles best resemble those of jumping spiders amongst the advanced spiders. With hindsight, it can be recognised that the most persuasive evidence about the origin of the jumping spider principal retina would be the demonstration of a contemporary retina with receptive segments differentiated as four morphological types in a single Layer.

The Australian misumenine Thomisid genus Hedana proves to match this speculative paradigm. The retinal mosaic is differentiated into four morphological receptor types; the receptive segments exhibit some resemblance to those of Salticids; and, there is a modest degree of tiering of the receptive segments. Also, the range of movements of the retinae is large (see below) compared to those of any other Family

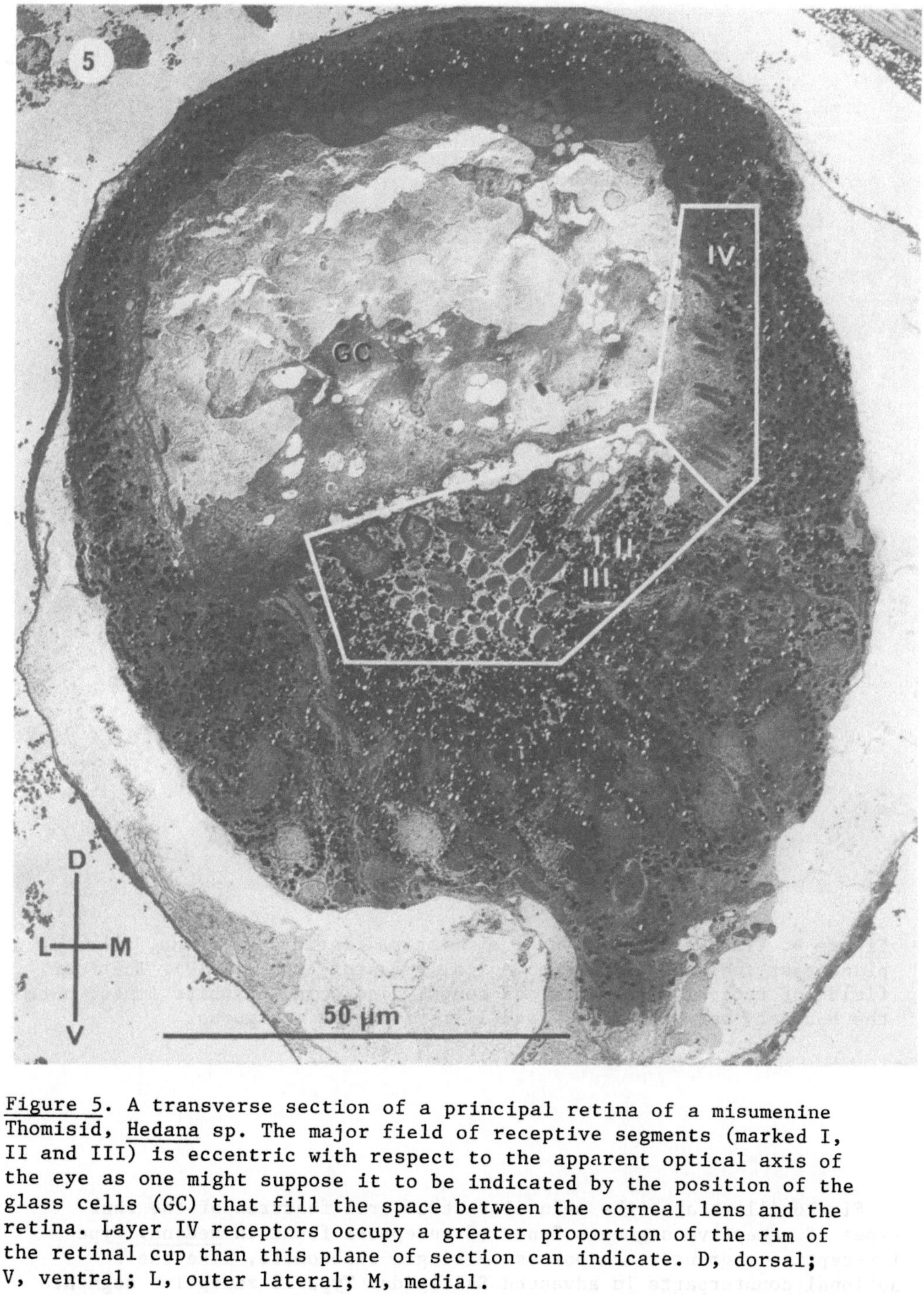

Figure 5. A transverse section of a principal retina of a misumenine Thomisid, Hedana sp. The major field of receptive segments (marked I, II and III) is eccentric with respect to the apparent optical axis of the eye as one might suppose it to be indicated by the position of the glass cells (GC) that fill the space between the corneal lens and the retina. Layer IV receptors occupy a greater proportion of the rim of the retinal cup than this plane of section can indicate. D, dorsal; V, ventral; L, outer lateral; M, medial.

of spiders, other than possibly the Palpimanidae (A.D.B., unpublished observations made in Panama).

The basic anatomy of the principal retina of Hedana is displayed in Figs. 5–7. Receptive segments are coded in terms of the receptive segments of Salticid principal eyes that they most resemble. Since there are no grounds that can establish real homologies, these assignations should not be taken too seriously; they may well prove to be wrong.

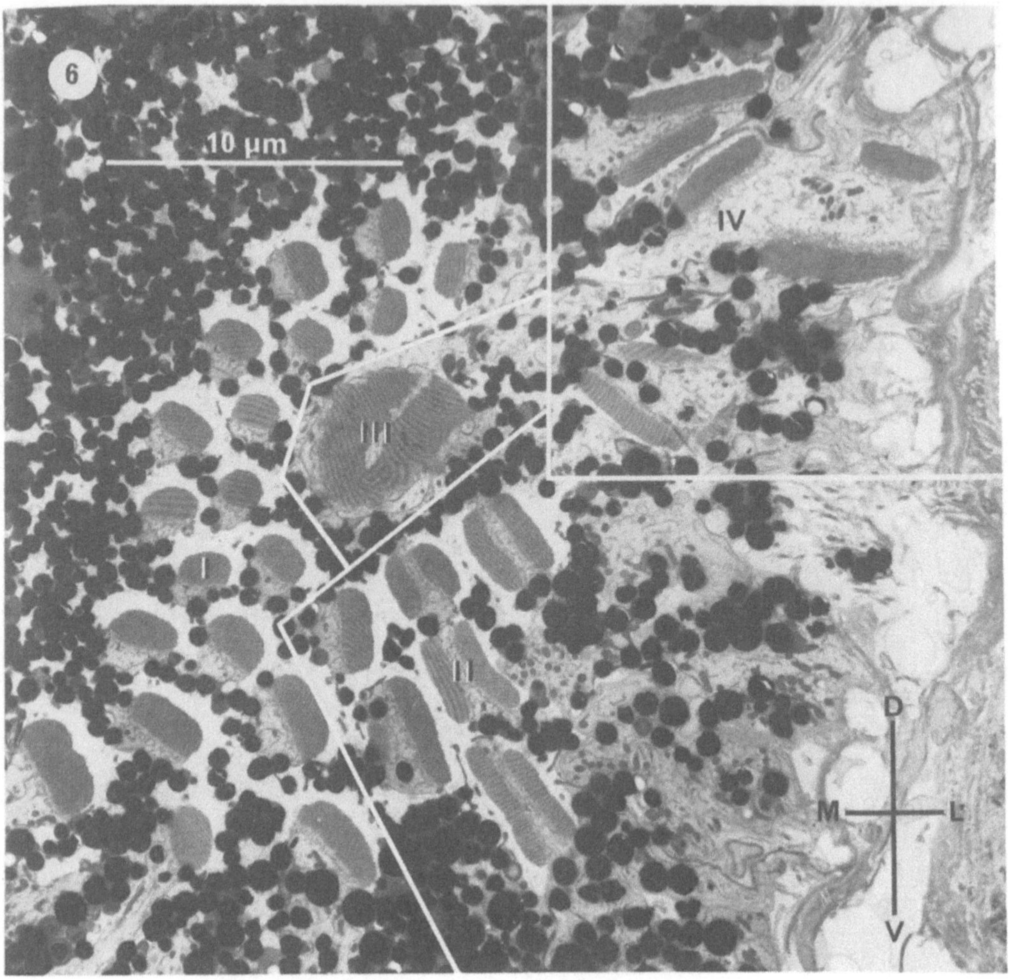

Figure 6. Transverse section of a principal retina of Hedana, taken at a plane that includes all morphological receptor types (I-IV). The four fields of receptive segments are roughly indicated by white lines; note the boundary between Types I and II is somewhat ambiguous.

Fig. 6 illustrates the main ultrastructural features of the four types of receptive segments Types of receptors found in Hedana. Type I receptive segments each possess a single rhabdomere, like their notional counterparts in advanced Salticids. Type II receptive segments each bear two rhabdomeres. The status of the Type III rhabdome is enigmatic. This large, deep structure is probably derived from rhabdomeres contributed by several receptors, but we have not been able to determine the real situation. Type IV receptive segments remarkably resemble Layer IV (4a) receptive segments of Salticids: each contains two rhabdomeres presenting a narrow rectangle in transverse section. Unlike the 4a receptors of Salticids, which form a single strip with the rhabdomeres of adjacent segments tightly contiguous, Type IV receptive segments of Hedana are well separated by pigmented glia.

A detailed assessment of the modest degree of receptor tiering in Hedana will be given by O'Carroll (in preparation) in the context of a full description of the retina. For present purposes, we will pretend that a principal retina of Hedana offers the receptor types of Salticidae dispersed in one plane around an approximate hemisphere. The model is close to reality, except for the extraordinary disposition of the Type III rhabdom - whose relative depth in the retina is illustrated by Fig. 7, below.

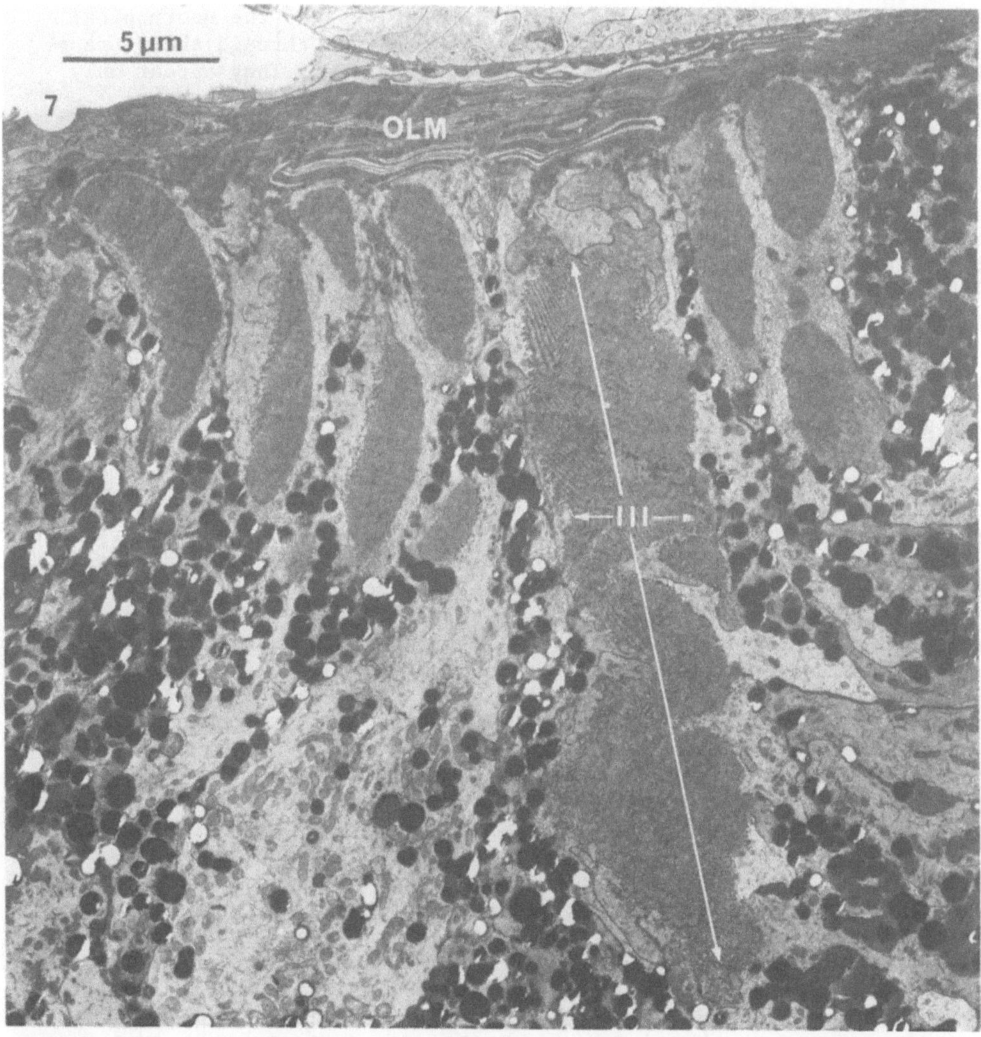

Figure 7. A longitudinal section of a principal retina of Hedana, taken at a plane passing through the whole depth of the Type III receptive segment. The presumptively composite Type III rhabdom is flanked by rhabdomeres attributable to Types I or II receptors that exhibit some tiering. OLM, outer limiting membrane of the retina.

We do not know how this extraordinary retina works. We can, however, consider two aspects of its performance in terms of what we know about the Salticid principal retina: (i) The extent of retinal movements and how they are organised; (ii) The presence or absence of light guides.

MOVEMENTS OF THE PRINCIPAL RETINAE OF SPIDERS

The extensive movements of the principal retinae of advanced jumping spiders during tracking and scanning of targets in object space (Land, 1969b) are not universal amongst spiders: one of us (D.O'C.) has looked at the retinal movements of several Families and has found them to be of no greater extent than ca. 2°, and to amount to little more than uncoordinated tremors. In contrast, the excursions performed by retinae of Hedana amount to ca. 30°. Movements of the two principal retinae are neither conjugate nor appreciably co-ordinated.

Within the Salticidae, the retinal movements of what are believed to be primitive Groups (for example the Lyssomaninae and the neotropical genera allied to Thiodina) can be readily observed through their translucent carapaces (A.D.B., unpublished observations). They appear only infrequently and almost accidentally conjugate, although they are individually of similar extent to the movements of retinae of advanced Salticids. The principal retinae of Portia fimbriata (which achieves a remarkable spatial acuity of 2.4 arc min at the fovea (Williams and McIntyre, 1980)) can be seen by ophthalmoscopy to perform highly conjugate movements. The Spartaeinae within which Portia belongs are considered by Wanless (1984) to be primitive by a number of anatomical criteria.

LIGHT GUIDING BY RHABDOMERES, AND VISUAL ACUITIES

Williams and McIntyre (1980) and Blest et al., (1981) showed that the high spatial acuities indicated by the visual responses of Salticids can only be mediated through the foveal receptor mosaics of Layer I, farthest from the dioptrics. The more distal Layers are composed of less regular and/or fine mosaics. Furthermore, foveal Layer I rhabdomeres are constructed as long (as much as 90 micron) light guides which transmit light focussed on their tips by internal reflection. Within contemporary Salticids there is a progression from a presumptively primitive state at which foveal Layer I rhabdomeres are short and apparently not able to act as light guides, to the condition found in Portia and, for example, in Phiale magnifica (Blest, 1985) both of which have been estimated from geometrical optics to have visual acuities of ca. 2.4 arc mins at the foveae. The fine details of that evolutionary succession will be described elsewhere (A.D.Blest and M. Carter, in preparation). Here, it is sensible to ask whether any aspect of light guiding is anticipated by the principal retina of Hedana?

That the receptive segments that we tentatively equate with Layer I in Salticids do not act as light guides is indicated by their isolation from each other by processes of pigmented glia, and by their relatively electron-dense cytoplasm (Fig. 6). Moreover, the focal length to aperture ratio in principal eyes of Hedana is ca. 1.1, compared to 2.0 - 2.5 in Salticids. Thus, in Hedana, marginal rays would enter the photoreceptors at angles too large for entrapment by total internal reflection at refractive index boundaries between rhabdomeres and adjacent glial processes (Land, 1981).

Despite the short focal length and small diameter (less than 100 microns) of the corneal lens, the small Type I photoreceptors are estimated by geometrical optics to have inter-receptor angles of only just over 1°. This is substantially better than estimates for the principal eyes of other Families of spiders, which typically lie between 4 - 6° (Land, 1985).

Thus, although the diversity of receptor types in Hedana resembles that in Salticids, the performance of the retina is far inferior.

TIERED PRINCIPAL RETINAE IN LYCOSIDAE AND OXYOPIDAE

Homann (1971) in a survey of spider retinae by light microscopy, considered the misumenine Thomisids to belong to a large, monophyletic group that includes the Lycosidae, Pisauridae, Oxyopidae, Ctenidae, Zoropsidae and a number of small Families. On an assumption that jumping spiders may relate to this complex of Families, it is reasonable to examine principal eyes of its members for hints of tiering. It has been found in Geolycosa godeffroyi in the Lycosidae (but not in the type genus, Lycosa): some local regions of the retina have receptive segments each with two rhabdomeres arranged as two layers. A similar arrangement is found in the principal retina of an Australian Oxyopes sp. Both are illustrated in Fig.8, below.

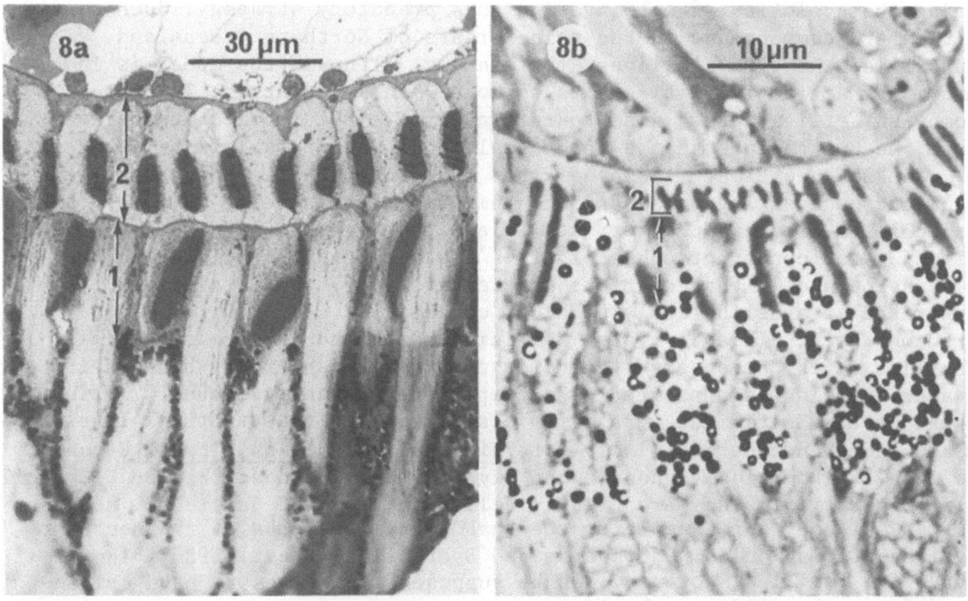

Figure 8. (a) A longitudinal section of a principal retina of Geolycosa godeffroyi (Lycosidae), showing two tiers of receptive segments (1 and 2). Those of Layer 1 have larger, less regular rhabdomeres than those of Layer 2. (b) A similar longitudinal section of a principal retina of Oxyopes sp., to show that there are two tiers of receptive segments. Resin sections one micron in thickness, stained with toluidine blue.

DISCUSSION

The relationships that may pertain between retinae of Salticids, the misumenine Thomisid Hedana, and other more cursorial hunting spiders need to be discussed in two distinct contexts:

(i) The systematic affinities of the Salticidae, and the speculative model for the evolution of jumping spiders first offered by Jackson and Blest (1982b), and elaborated by Blest and Carter (1987).

(ii) Tentative models that might account for the development of a 'differentiated' principal retina in the first place, and then for the conformational changes that must have been imposed on it to generate the elaborately tiered, boomerang-shaped retinae of Salticids.

The Systematics and Evolution of the Salticidae

The taxonomic distance between the Salticidae and other Families of spiders is primarily determined by their unique principal retinae and their extraordinary modus operandi, and secondarily by the many modifications of body structure that stem from them, as we noted above.

Jackson and Blest (1982b) described the complex ethology of Portia fimbriata, a Salticid in the Spartaeinae considered to be primitive (Wanless, 1984). Portia to a far greater degree than Salticids of other subfamilies is web-dependent. Its major predatory strategy is to invade the webs of members of other Families of spiders in order to consume their occupants, deploying a remarkable range of visual and kinaesthetic skills to do so. Jackson and Blest (1982b) proposed that web-dependency within the Spartaeinae conserved some components of the behaviour of their ancestors. The phylogenetic model assumed that the Salticidae arose from web-builders that occupied habitats in which their own webs were closely juxtaposed to the webs of other groups of spiders, facilitating the evolution of web-invasion as a predatory strategy. Such contiguities are commonplace in the rain-forests of Northern Queensland, Australia, where Portia lives, for example. Although the model is in many ways attractive, it amounts to more than speculation. The only component of it that we wish to retain here is the implication derived from the Spartaeinae that a 'proto-salticid' was probably web-dependent.

Jackson and Blest (1982b) did not suppose that a web-dependent proto-salticid necessarily related to contemporary web-building families: the cribellates, for example, or, still less the Argyopids and their allies, all of which construct orb-webs. The 'proto-salticid' implicit in their discussion and in that of Blest and Carter (1987) could fairly be described as a deliberately nebulous concept.

Our present alliance of the Salticidae with the various cursorial Families which commonsense taxonomists have always perceived as their nearest cousins is supported by a comparison of their secondary retinae. Those of the Salticidae were first described at ultrastructural level by Eakin and Brandenburger (1971) for Phidippus and Metaphidippus. Secondary retinae of Thomisids, Lycosids and Oxyopids will be described by O'Carroll (in preparation). They are clearly allied to those of Salticids, and bear no resemblance, for example to the rhabdomeral networks of the secondary retinae of Clubionids, Gnaphosids and Sparassids (Blest, 1985). The cursorial Families, however, can be supposed themselves to have originated from web-builders; the issues become a matter of when the Families diverged, and at what point in their phylogenetic history the Salticidae originated.

The Evolution of Retinal Movements

Homann (1934, 1975) noted the inhomogeneity of the receptors of the principal eyes of Thomisids, and that they mostly lie within a glial cup that he described as the 'pigment body'. He also (Homann, 1934) assessed the distances at which Misumena (Misumeninae) responded to natural prey, assumed that the movements of the principal eyes that he observed were concerned with the fixation of images of prey in object space, and described the two muscles attached to a principal retina responsible for them. He described the movements of the two principal retinae as 'unco-ordinated', commenting that very infrequent conjugate movements were probably no more than transitory accidents.

The beautifully conjugate movements of the principal retinae of advanced Salticids analysed by Land (1969b) seem not to be anticipated in other Families of spiders. Land noted that conjugate movements are periodically uncoupled. Our own observations by ophthalmoscopy of the retinal movements of Portia and Plexippus confirm this. The original ophthalmoscopic analysis of principal eye optics by Land (1969a) depended upon the principal retinae of a spider at rest, confronting no specific target in object space, coming to a standstill. The species that Land studied (in the genera Phidippus and Metaphidippus) are active hunters. Our own attempts at an ophthalmoscopy of Portia were made difficult because their principal retinae are continuously active (D.S. Williams and M.F.Land, unpublished observations). We assume that their uninterrupted movements relate to the lifestyle of Portia, which invade the webs of other species, remaining motionless at their peripheries whilst, presumably, they evaluate occupants at considerable distances on their visual horizons (Jackson and Blest, 1982b).

It would be nice to find a bridge between the non-conjugate but relatively extensive movements of Thomisid retinae and the sophisticated conjugate movements displayed by advanced Salticids. One of us (A.D.B.) has examined movements of the principal retinae of Lyssomanine Salticids in Panama, and also those of some neotropical species in the genera Thiodina, Cotinusa and Scopocira. In all cases, the carapaces are sufficiently translucent for movements of the principal retinae to be observed through them. These species were compared with Itata completa, an advanced Salticid also with a conveniently translucent carapace, and the movements of whose retinae prove to be predominantly conjugate.

The retinal movements of the Lyssomaninae are mostly non-conjugate. Occasional conjugation seems to be more than 'accidental' - as Homann described the infrequent conjugate movements he observed in Thomisids. The primitive status of Lyssomanines is broadly accepted by taxonomists. A relatively primitive status for Thiodina, Cotinusa, Scopocira and some other genera will be justified elsewhere (Blest and Carter, in preparation), largely on grounds of retinal anatomy. The present qualitative observations support a case that only 'advanced' Salticids have acquired conjugate movements of their principal retinae, whilst notionally 'primitive' Salticidae may offer a bridge to a condition exhibited by misumenine Thomisids.

The Evolution of Retinal Tiering

We have shown that a misumenine Thomisid, Hedana, has a complex retina with four morphologically distinct classes of photoreceptive segments, arranged as indicated by Fig. 9(B) above. Despite a hint of tiering in

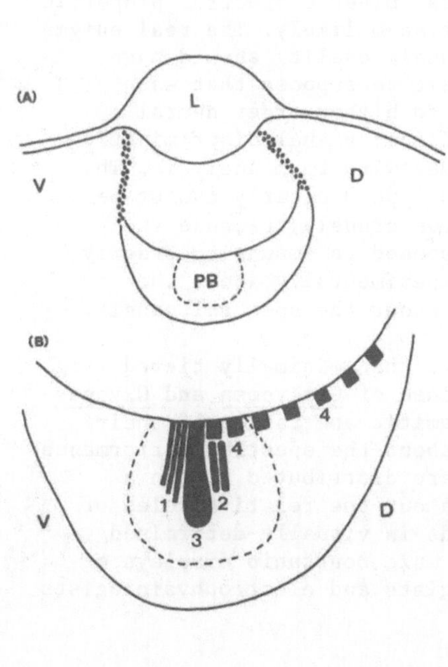

Figure 9 (left). The principal eye of Hedana. Fig. 9(A). A longitudinal profile of an eye that shows the relationships between the corneal lens, L, and the retinal cup that includes the pigment body (PB) described by Homann. Fig. 9(B). A retina, expanded from the longitudinal plane of section shown by Fig. 9B. The region indicated by a dashed line encloses the 'pigment body of Homann. Classes of receptive segments are indicated by 1-4. Note the element of tiering represented by the relationship between rhabdomeres 2 and 4, and the depth of the probably composite rhabdome 3.

the principal retinae of Geolycosa and Oxyopes, retinae of Hedana offer a simple model for a precursor of Salticid principal retinae that is more pertinent.

The transformation of a single hemispherical layer of receptive segments into a complex tiered retina of the Salticid type, by the conformational changes summarised by Fig.4, must have proceeded from an initial state that favoured it. There are some three initial states that are plausible:

(i) Photoreceptors whose receptive segments lay around a hemisphere were morphologically and spectrally undifferentiated. During phylogeny, the progressive conformational change that led to tiering was imposed by increase in the focal length of the dioptrics, and a consequent need to correct for chromatic aberration. This model appears unacceptably naive. For example, it supposes that some differentiation of the spectral responses of the receptors happened after retinae started to elongate. Given the evidence for cells of more than one spectral type in the principal eyes of various Families of spider (reviewed by Blest, 1985, 1988), an early retina that consisted merely of a homogeneous hemispherical monolayer seems unappealing.

(ii) Photoreceptors disposed as a hemispherical monolayer were differentiated into at least two spectral types, green and UV receptors. Blest (1988) reviews the evidence supporting a case that UV receptive segments originally lay at the 'equator' of a primordial retina, so that conformational changes would have caused them to occupy the position of Layer IV receptive segments in contemporary Salticids. This model is rather more attractive than the simplistic assumptions implicit in (i), and was offered by Blest (1988).

(iii) Photoreceptors were initially differentiated into a variety of morphological and spectral types with some correspondence to those found in contemporary Salticids. The principal retina of Hedana exhibits just that kind of pattern. We should regard it as currently a 'best model' for a precursor of Salticid retinae; as a future focus of discussion, it should replace the speculations of Blest (1988).

It must be emphasised that the misumenine Thomisids exemplified by Hedana cannot be supposed to be the linear ancestors of the Salticids by any taxonomic criteria. The principal retina of Hedana suggests a starting point: a virtual monolayer of receptive segments was differentiated to provide a spectrum of morphological types. Diverse spectral properties of the four morphological types of receptor seem likely. The real enigma is why such a complex retina with a poor mosaic quality should have evolved in the first place. It seems simplest to suppose that each receptor type is individually 'hard-wired' to higher-order neural circuits each of which serves a unique, critical visual discrimination. Visual tasks may not necessarily be concerned with form analysis. The single giant rhabdome that we categorise as Type 3 clearly cannot be concerned with the evaluation of shapes, even crudely, because the erratic movements of a retina cannot be supposed to amount to orderly scanning. Unfortunately, little is known experimentally about the visual behaviour of Thomisids, and nothing about the spectral sensitivities of their photoreceptors.

The principal retina of Hedana is no more than minimally tiered (Fig. 9). The local tiering of the principal retinae of Geolycosa and Oxyopes may have evolved to compensate for the chromatic aberration of their dioptrics. Again, we do not know anything about the spectral performance of their receptors and how spectral types are distributed within a principal eye. We have little information about the relative roles of the principal and secondary eyes of Lycosids in visually-determined behaviours, even though some Lycosids have male courtship displays of considerable complexity. Hopefully, ethologists and electrophysiologists will collaborate in the future to elucidate the functional roles of these eyes.

It might seem that much effort has been devoted to a protracted study of a small and specialised corner of arthropod vision. However strange Salticid principal eyes may appear to be, they are superbly effective. That the construct evolved but once is only surprising if one forgets the peculiar route that probably led to it. The modus operandi of these eyes may offer lessons relevant to the design of special purpose sensors.

ACKNOWLEDGEMENTS

We are indebted to Professor M.F.Land for many valuable comments at all stages of the work. The field work of one of us (A.D.B.) in the Republic of Panama has been generously funded by the National Geographic Society, most recently in 1987 (Grant No. 3392-86); he is grateful to the Smithsonian Tropical Research Institute for providing facilities on Barro Colorado Island, and to RENARE (Republic of Panama) for granting the relevant permits. Drs Peter McIntyre and David S.Williams contributed much to the early stages of this project, and Dr R.R.Jackson (University of Canterbury, New Zealand) has provided important ethological information about Salticids. Dr F.R.Wanless (British Museum, Natural History) has been equally informative in the context of Salticid taxonomy. Margrit Carter has provided superb and informed histological assistance throughout our recent research, and we owe much to the services provided by George Weston and the ANU Transmission Electron Microscope Unit.

REFERENCES

Blest, A.D. (1985) The fine structure of spider photoreceptors in relation to function. In The Neurobiology of Arachnids. (F.G.Barth, ed.), pp. 79-102, Springer-Verlag, Berlin-Heidelberg-New York-Tokyo.

Blest, A.D. (1987) Comparative aspects of the retinal mosaics of jumping spiders. In Arthropod Brain: Its Evolution, Development, Structure and Functions. (A.P.Gupta, ed.), pp. 203-229, John Wiley & Sons, New York.

Blest, A.D. (1988) Post-embryonic development of the principal retina of a jumping spider. I. The establishment of receptor tiering by conformational changes. Phil. Trans. R. Soc. Lond. B 320, 489-504.

Blest, A.D. and Carter, M. (1987) Morphogenesis of a tiered principal retina and the evolution of jumping spiders. Nature Lond. 328, 152-155.

Blest, A.D. and Carter, M. (1988) Post-embryonic development of the principal retina of a jumping spider. II. The acquisition and reorganisation of rhabdomeres and growth of the glial matrix. Phil. Trans. R. Soc. Lond. B 320, 505-515.

Blest, A.D., Hardie, R.C., McIntyre, P. and Williams, D.S. (1981) The spectral sensitivities of identified receptors and the function of retinal tiering in the principal eye of a jumping spider. J. comp. Physiol. A 145, 227-239.

Blest, A.D. and Sigmund, C. (1985) Retinal mosaics of a primitive jumping spider, Spartaeus (Araneae: Salticidae: Spartaeinae): a phylogenetic transition between low and high visual acuities. Protoplasma 125, 129-139.

De Beer, G.R. (1958) Embryos and Ancestors (3rd ed.). Oxford University Press.

De Voe, R.D. (1975) Ultraviolet and green receptors in principal eyes of jumping spiders. J. gen Physiol. 66, 193-208.

Eakin, R.W. and Brandenburger, J.L. (1971) Fine structure of the eyes of jumping spiders. J. Ultrastruct. Res. 37, 618-663.

Homann, H. (1934) Das Sehvermögen der Thomisiden. Z. vergl. Physiol. 20, 420-429.

Homann, H. (1951) Die Nebenaugen der Araneen. Zool. Jahrb. Anat. 71, 56-144.

Homann, H. (1971) Die Augen der Araneae. Anatomie, Ontogenie und Bedeutung für die Systematik (Chelicerata, Araneae). Z. Morph. Tiere 69, 201-272.

Homann, H. (1975) Die Stellung der Thomisiden und der Philodromidae in System der Araneae (Chelicerata, Arachnida). Z. Morph. Tiere 80, 181-202.

Jackson, R.R. and Blest, A.D. (1982a) The distances at which a primitive jumping spider makes visual discriminations. J. exp. Biol. 97, 441-445.

Jackson, R.R. and Blest, A.D. (1982b) The biology of Portia fimbriata, a web-building jumping spider (Araneae: Salticidae) from Queensland: utilisation of webs and predatory versatility. J. Zool. (Lond.) 196, 255-293.

Land, M.F. (1969a) Structure of the retinae of jumping spiders (Salticidae: Dendryphantinae) in relation to visual optics. J. exp. Biol. 51, 443-470.

Land, M.F. (1969b) Movements of the retinae of jumping spiders (Salticidae: Dendryphantinae) in response to visual stimuli. J. exp. Biol. 51, 471-493.

Land, M.F. (1981) Optics and vision in invertebrates. In Handbook of Sensory Physiology VII/6B (H. Autrum, ed.), pp. 471-592. Springer-Verlag, Berlin-Heidelberg-New York.

Land, M.F. (1985) The morphology and optics of spider eyes. In Neurobiology of Arachnids (F.Barth ed.), pp. 53-78. Springer-Verlag, Berlin-Heidelberg-New York-Tokyo.

Medawar, P.B. and Medawar, J. (1983) Aristotle to Zoos. Harvard University Press.

Wanless, F.R. (1984) A review of the spider sub-family Spartaeinae nom. nov. (Araneae: Salticidae) with descriptions of six new genera. Bull. Br. Mus. nat. Hist. D 46, 135-205.

Williams, D.S. and McIntyre, P. (1980) The principal eyes of a jumping spider have a telephoto component. Nature Lond. 288, 578-580.

Yamashita, S. and Tateda, H. (1976) Spectral sensitivities of jumping spider eyes. J. comp. Physiol. 105, 1-8.

DEVELOPMENTAL STUDIES ON THE OPTIC LOBE OF *DROSOPHILA MELANOGASTER* USING STRUCTURAL BRAIN MUTANTS

K.-F. Fischbach, F. Barleben, U. Boschert, A.P.M. Dittrich, B. Gschwander, B. Houbé, R. Jäger, E. Kaltenbach, R.G.P. Ramos, and G. Schlosser

Institut für Biologie III
Schänzlestr.1.
D-7800 Freiburg i. Brsg.
Federal Republic of Germany

SUMMARY

On the background of a detailed analysis of wildtype structure, mutants of the visual system of *Drosophila melanogaster* are being used for identification of genetic and epigenetic factors in the development of the compound eye and optic lobe. Mutant analysis reveals the role the larval visual system plays in the development of the adult optic lobe, and that visual fibres are able to reach their retinotopic destination via ectopic pathways. By use of different mutations and their multiple combinations, the visual system can be simplified by drastically reducing the number of cell types in the optic lobe. Simplification is limited by the response of remaining neurons which show sprouting and compensatory innervation. It is expected that the molecular characterization of the neurological genes will yield information about mechanisms of neuronal function, neuronal diversification, axonal pathfinding, and target recognition.

INTRODUCTION

The development of higher organisms can be compared to a game, the outcome of which is determined by the starting conditions and the operating rules. The application of the operating rules invariably leads to certain (the essential and characteristic) features of the organism. Although the developmental (epigenetic) rules are not explicitly implemented in the genome (Stent, 1981), genes and their products play the essential part. Brain morphology, neuronal functions and behavioural abilities are under tight genetic control (Fischbach and Heisenberg, 1984).

Neurogenetics is the application of single gene analysis to neurobiological problems. At least three, overlapping areas of research can be distinguished:

Developmental neurogenetics in the narrow sense is concerned with the questions of which genes are involved in the implementation of the epigenetic rules of neural development, how many they are and how they function. Well known examples of such genes are the *neurogenic loci* which constitute a developmental switch between the fate of equipotent cells (Lehmann et al.,1981). Less stringently defined, developmental genetics comprises in addition studies which use the phenotypes of neurological mutants for the study of epigenetics (e.g. Power, 1943; Fischbach, 1983). The epigenetic rules investigated using both approaches apply e.g. to axonal guidance, target recognition, cell death, cell differentiation etc.

Biochemical neurogenetics is concerned with the question of which genes are involved in neuronal function, how many they are and what they are good for. Special attention has recently been paid to genes coding for ion channel proteins (Tempel et al., 1987, 1988; Pongs et al., 1988) and enzymes in transmitter metabolism (Livingstone and Tempel, 1983; Tempel et al., 1984) or second messenger metabolism (Livingstone et al., 1984; Livingstone, 1985; Dudai and Zvi, 1985).

Behavioural neurogenetics is primarily concerned with genes which are essential for certain behaviours or sensory abilities, e.g. learning, phototaxis, geotaxis, fixation, olfaction, courtship etc. (for reviews see Fischbach and Heisenberg, 1984; Hall, 1986). It is clear that at another level of analysis most of these genes will be found to primarily affect neuronal function or neural development. However, as important selection factors operate at the level of behaviour, it is interesting to learn how behavioural organization is reflected at the genomic level.

It is now commonly accepted that no other organ requires the expression of as many specific genes as the brain (e.g. John and Miklos 1988). Especially the brain of vertebrates displays a wealth of transcripts (up to 150,000) which is unmatched in other organs (Chikaraishi, 1988). Invertebrates are much simpler at the genetic level, possibly due to the smaller diversity of cell types. However, their individual neurons are comparable in function and structural complexity to their vertebrate counterparts. In *Drosophila melanogaster* RNA complexity studies have yielded an estimate of only about 10,000 different transcripts in the head (Levy and Manning, 1981). It is therefore hoped that invertebrates, and especially *Drosophila*, can be used as relatively simple model systems for neurogenetic studies.

The present paper is written to illustrate our approach to the neurogenetics of *Drosophila* using structural brain mutants. Although it may eventually turn out that some of the neurological phenotypes described are not produced by mutations in brain specific genes and are rather due to hypomorphic mutations in genes with more general functions, it will be demonstrated that the mutant *phenotypes* can be used for the study of epigenetic processes. Furthermore, we shall stress our confidence that several of the identified genes will turn out to be involved in important neural functions.

EXPERIMENTAL STRATEGIES FOR THE ISOLATION OF NEUROLOGICAL GENES IN *DROSOPHILA*

Experimental procedures for the isolation of neurological genes starting from DNA clones or antibodies require *a posteriori* the production of mutants or the correlation with an already known mutation. In the worst case it may turn out that elimination of the gene's function does not lead to a noticeable phenotype. Mutant analysis, therefore, remains the strategy we prefer to start with. One can make use of a host of pre-existing mutants with defects in the nervous system (Hall 1982) and of elaborated techniques for the isolation of new ones. Sophisticated as well as simple techniques have been invented to screen the offspring of mutagenized flies for behavioural defects (Benzer, 1967; 1971; Heisenberg and Götz, 1975; Bülthoff, 1982; Thomas and Wyman, 1982) and time saving methods for studying the brain phenotype have been worked out (Heisenberg and Böhl, 1979; Jäger and Fischbach, 1987). These histological methods even allow to screen for new mutants using brain morphology as the only selection criterion. Heisenberg and Böhl (1979; first chromosome), Heisenberg and Fischbach (unpublished; first and second chromosome) and Fischbach et al. (1987a,b; first chromosome) isolated a wealth of such mutants most of which are not yet characterized in any detail. These mutants can be used to study brain development at the level of epigenesis and as starting points for the identification of neurologically active molecules.

Several ways are paved to isolate and characterize mutated genes at the molecular level. After precise mapping of a gene by standard techniques one can take advantage of the large polytene chromosomes in the salivary glands of the *Drosophila* larvae: single bands can be excised to construct a DNA minilibrary of a small genomic region (Pirrotta et al., 1983a,b). After deficiency mapping such clones can be used as starting points for a characterization of the chromosomal region (e.g. Miklos et al., 1988). Alternatively, DNA clones already on the *Drosophila* market (Merriam, 1984) with well defined positions on the cytological or even molecular maps can be used as entry points for chromosomal walking near precisely mapped genes. A straightforward approach to the isolation of a gene is transposon tagging (e.g. Bingham, 1981; Rubin et al., 1982). This is done by inducing stable insertions of a transposon (like a P-element) into the genome. This may result in the inactivation of genes. We have successfully used P/M mutagenesis for the production of new neurological mutants (Fischbach et al., 1987a,b). The P-element sequences can be used as probes to pick the genes from genomic libraries.

Recently, transposons have been designed that derive from a P-element, but additionally contain bacterial reporter genes and sequences needed for plasmid replication and selection (O'Kane and Gehring, 1987). One can be confident that the isolation and molecular characterization of future neurological mutants will become even more straightforward by the use of these modern techniques.

MUTANTS USED FOR THE STUDY OF EPIGENESIS

introvert : Eye-brain differentiation is independent from head eversion

In a screen for the phenotype of lethal mutants near the *sol/slgA* region at the base of the X-chromosome (Miklos et al., 1987) we examined the phenotype of the pupal lethal Q56 (which was kindly given to us by G.L.G. Miklos). Homo- and hemizygous Q56 pupae do not eclose, because they fail to evert their adult head after puparium formation (Fig.1A). The compound eye, the antenna, the ocelli, and the proboscis develop inside the thorax and the brain undergoes final differentiation in the abdomen where it lies upside down on the gut, above the thoracic ganglion (Fig.1B). We therefore named this mutant *introvert* (*intro*). The remarkable feature of its phenotype is that the brain, including the visual neuropiles, differentiates normally, despite its ectopic position. The optic lobes are innervated by the retinula cells of the compound eye.

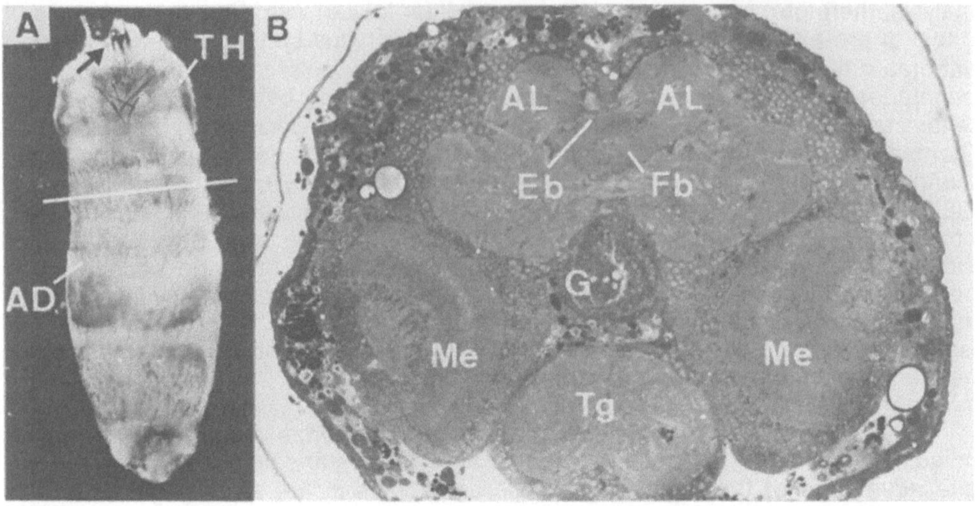

Fig.1. Phenotype of *introvert* mutant. A: External view of *intro* fly carefully prepared from the pupal case. Please note the apparent absence of a head, the small thorax (TH) and the huge abdomen (AD). The arrow points to the larval mouthhooks. Plane of section for B indicated by white bar in A. B: Semithin section through the anterior part of the abdomen. The thoracic ganglion (Tg), the gut (G), and the brain with the antennal lobes (AL), the ellipsoid body (Eb), the fan shaped body (Fb), and the medulla (Me) are seen. Please note that the orientation of the brain is upside down.

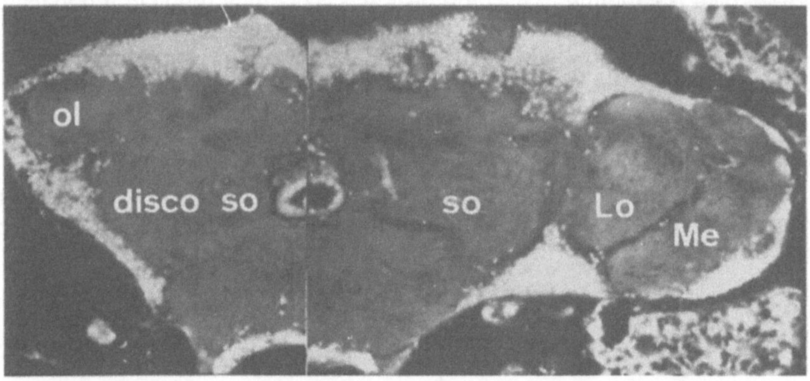

Fig.2. Montage of left half of the brain of an eyeless *disco*^{1} *so*^{1} fly and of right half of the brain of an eyeless *so*^{1} fly. Compare the tiny optic lobe rudiment (ol) of the double mutant with that of the *so*^{1} fly. The latter is still clearly subdivided into medulla (Me), lobula (Lo) and lobula plate (not visible in this plane of section). The size of the optic lobe rudiment in the double mutant is the same as that in *disco*^{1}, i.e. *disco*^{1} is epistatic over *so*^{1} with regard to the optic lobe phenotype.

disconnected : The adult optic lobe is built on larval elements

The connection between the developing compound eyes and the optic lobe is formed well before the failure of head eversion becomes a problem in *intro* mutants. Ingrowth of retinal fibres into the brain hemisphere via the optic stalk takes place already in the 3rd instar larva. It has been shown previously that this innervation of the optic lobe by retinula cells of the compound eye is crucial for its proper development (Power, 1943; Meyerowitz and Kankel, 1978; Fischbach, 1983; Nässel and Sivasubramanian, 1983). In its absence, massive cell degeneration occurs after axon formation (Fischbach and Technau, 1984). Interestingly, in adult *disconnected* (*disco*) mutant flies the optic lobe, which is not connected to the compound eye

(Steller et al., 1987), is much smaller than in *disco*+ flies without eyes (Fischbach, 1983a) and in contrast to the eyeless situation it lacks medullar columnar cell types completely. How can this phenotype of the *disco* optic lobe be explained?

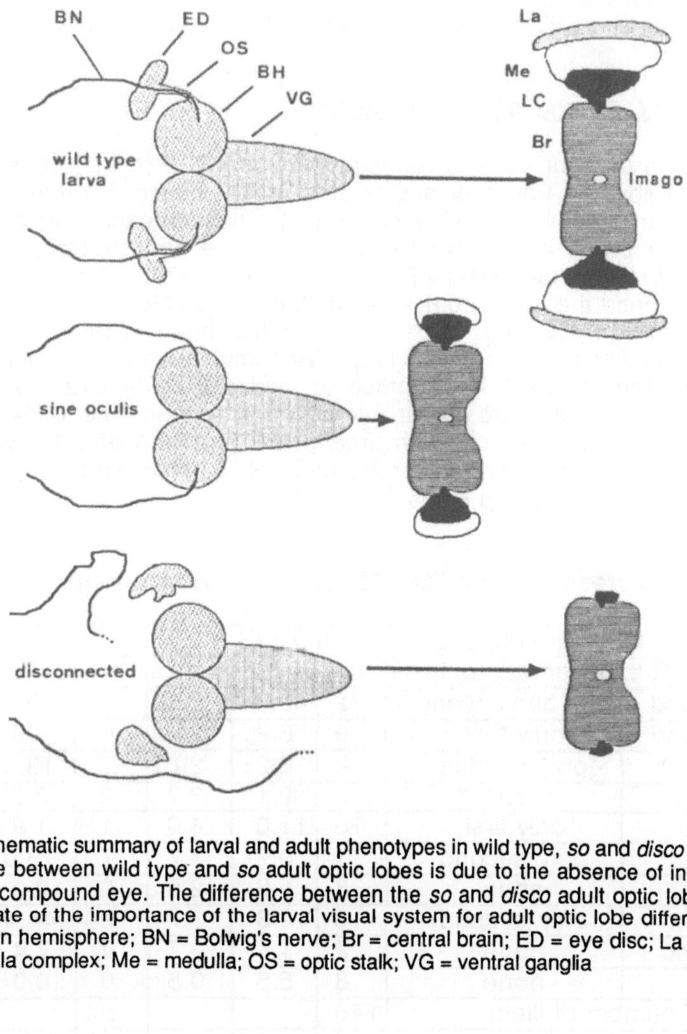

Fig.3. Schematic summary of larval and adult phenotypes in wild type, *so* and *disco* flies. The difference between wild type and *so* adult optic lobes is due to the absence of innervation from the compound eye. The difference between the *so* and *disco* adult optic lobes gives an estimate of the importance of the larval visual system for adult optic lobe differentiation. BH = brain hemisphere; BN = Bolwig's nerve; Br = central brain; ED = eye disc; La = lamina; LC = lobula complex; Me = medulla; OS = optic stalk; VG = ventral ganglia

In *disco* an early developmental defect is that the larval visual nerve (Bolwig's nerve) has failed to connect to its target cells (Steller et al., 1987). Genetic mosaic analysis has shown that the subsequent cell degeneration in the optic lobe of *disco* pupae is most likely a consequence of the misrouting of Bolwig's nerve and not due to an independent effect of the *disco* mutation on the optic lobe (*ibid.*). The comparison between the size of the optic lobe rudiments of eyeless *disco* ;*so* double mutant flies and that of *so* -flies shows that the *disco* mutation is epistatic over *so* with regard to the optic lobe phenotype (Fig.2). The larval visual system seems to serve as a central core for the differentiation of the adult visual system, i.e. Bolwig's nerve is not only responsible for the formation of the optic stalk and proper innervation of the optic lobe by retinula cell axons, it also seems to have some direct function for adult optic lobe development (Fig.3).

It has been suggested recently that one function of Bolwig's nerve might be the stabilization of optic lobe pioneer cells (OLPs), early differentiating neurons of the larval brain (Tix and Technau, 1988). These cells are missing in the 3rd instar *disco* larva of the unconnected phenotype (Tix and Technau, 1987). The absence of innervation by the larval photoreceptors could be the cause for the non-formation or degeneration of the OLPs in *disco* larva (Fischbach and Technau, 1987). The phenomenon is reminiscent of the non-formation of the imaginal lamina in the absence of retinal innervation (Meinertzhagen, 1973; Hofbauer, 1979).

irregular chiasms C : Axonal pathfinding defects

Mutations in the four genes *irregular chiasmsA-D* (*irreA-D*) cause abnormal growth of visual fibres in the inner and/or outer optic chiasm. Here we present the data on *irreC* as an example, since it is presently the best characterized mutant of this group at the phenotypic level. Two *irreC* alleles have been obtained so far, one resulting from P/M mutagenesis (*UB883*; Fischbach et al., 1987b; Boschert and Fischbach, 1987) and the other from X-ray mutagenesis (*1R34*; Jäger, 1988). Their detailed genetic and cytological characterization has been initiated as a first step towards the molecular analysis of the locus. Recombination and deletion mapping place *irreC* near the tip of the X-chromosome, and cytological examination of both alleles has given results that are compatible with this location: The *1R34* mutation is associated with a small inversion, with breakpoints in 1F-2A and 3C, while *in situ* hybridization of a cloned P-element probe to *UB883* chromosomes revealed the presence of P-element insertions in the 2E-3C area.

Table 1. Expressivity of the UB883 and 1R34 alleles of *irreC*

disorders in chiasms of		UB883			1R34		
one optic lobe	other optic lobe	n	%	calc.	n	%	calc.
first and second	first and second	72	49.3	28.9	9	16.7	8.2
first and second	only first	9	6.2	21.4	0	0.0	2.7
first and second	only second	9	6.2	20.7	7	13.0	32.9
first and second	none	3	2.1	7.7	2	3.7	5.3
only first	only first	16	11.0	4.0	1	1.9	0.2
only first	only second	1	0.7	7.7	7	13.0	5.3
only first	none	8	5.5	2.8	0	0.0	0.9
only second	only second	18	12.3	3.7	24	44.4	33.0
only second	none	2	1.4	2.7	4	7.4	10.7
none	none	8	5.5	0.5	0	0.0	0.9
total number of flies		146	-	-	54	-	-
defective first chiasms		215	73.6	-	36	33.3	-
defective second chiasms		213	72.9	-	93	86.1	-

146 *UB883* and 54 *1R34* flies of both sexes were sectioned using the mass histology procedure described in Jäger and Fischbach (1987) and were inspected for disorders of the first and second optic chiasms. As no sexual dimorphism was observed, only pooled data are shown. Calculated values (calc.) are derived from the hypothesis that expression of the mutant phenotype of an optic chiasm is random and independent from the phenotype of any other optic chiasm. Comparison of the actual values (%) shows, however, that symmetric phenotypes are much more frequent than expected, i.e. there is a bilateral correlation between the expression of mutant phenotypes. On the other hand, an interdependence between first and second optic chiasmic defects cannot be detected.

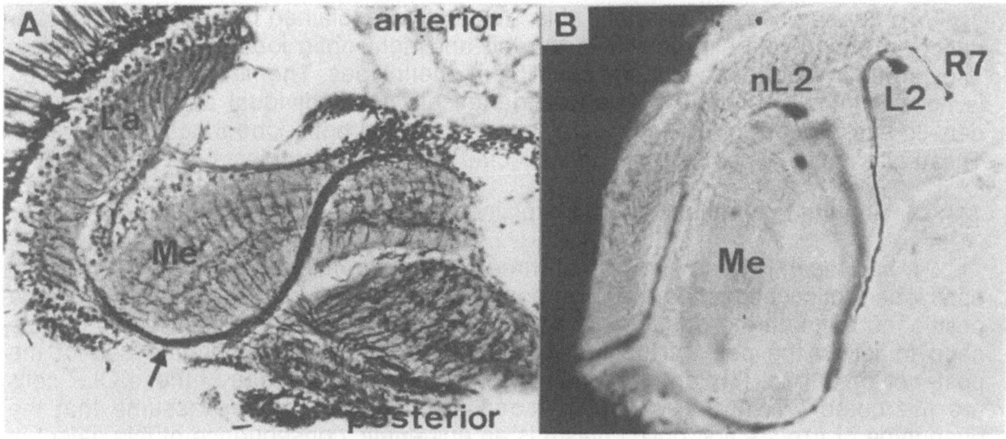

Fig.4. Phenotype of *irreC*^{UB883} : A. A horizontally oriented silver stained section through the optic lobe of the mutant is shown. Please note the ectopic fibre bundle (arrow) which projects from the posterior lamina to the anterior medulla. B. Golgi-impregnated fibres in the ectopic fibre bundle. La = lamina; Me = medulla; nL2 = normal L2 terminal; L2 , R7 = characteristic L2 and R7 terminals of fibres in the ectopic bundle.

Both *irreC* mutant alleles were originally isolated due to a severe disorder of the second optic chiasm. Fibre tracts from the medulla to the lobula plate penetrate the lobula neuropile instead of running in the second optic chiasm. In some severe cases the neuropiles of medulla, lobula and lobula plate seem to be partially fused. Most spectacular in our eyes, however, is a much more subtle, albeit related disorder of the first optic chiasm. One or sometimes several horizontally oriented fibre bundles from the equatorial, most posterior part of the lamina are misrouted on their path to their normal target area in the anterior medulla. They take a long detour around the posterior medulla neuropile. Then they penetrate the medulla neuropile anteriorly from the inner to the outer face where the fibres turn around again and form their normal characteristic terminals in retinotopic positions (Fig.4A,B). It should be noted that the exact position of medulla penetration varies, but in all cases the fibres find their appropriate target regions in the anterior medulla.

The mutant phenotypes are presently used to ask the following questions:

1. Are the disorders of the first and second optic chiasm of one optic lobe epigenetically connected or are they due to *irreC* expression in different cell populations?

The expressivity of the *irreC* phenotype is variable in both alleles. This variability can be used to test the hypothesis whether the disorders of the second optic chiasm are the epigenetic consequences of the disorders of the first chiasm or *vice versa*. The expressivities of the defects in the first and second optic chiasms of one optic lobe are not correlated (see Table 1). This result argues in favour of epigenetic independence of the defects. The conclusion is also supported by the specificity of the two alleles. While allele *UB883* causes defects in the first and in the second optic chiasm with about the same frequency ($p_1 = 0.74$; $p_2 = 0.73$), allele *1R34* affects the optic chiasms differently ($p_1 = 0.33$; $p_2 = 0.86$).

2. Are the disorders of the optic chiasms of one optic lobe dependent on the phenotype of the contralateral optic lobe?

To our surprise, the data of Table 1 cannot be explained by assuming independence of phenotypic expression in the left and right optic lobes of individual flies. They argue in favour of a bilateral coupling of phenotypes. The nature of this coupling is not yet known. It could be caused genetically (by individual differences in the genetic background) or epigenetically, e.g. by bilateral optic lobe tangential neurons. These alternative hypothesis will be subjected to experimental tests.

3. Can the *irreC* phenotype be explained by displaced larval pioneers?

It has recently been shown that three early differentiating neurons of the larval optic lobe (optic lobe pioneers; OLPs) persist into the adult stage where they become positioned beneath the posterior lamina in an equatorial plane. Their axons can be seen to follow the path of the first optic chiasm before they become part of the posterior optic tract (Tix and Technau; 1988). In *irreC* the axons of these OLP cells are misrouted (Tix and Technau, personal communication). We assume that the phenotype of *irreC* 's first optic chiasm is an epigenetic consequence of this defect. It is tempting to speculate that the defect in the second optic chiasm could be caused also by displacement of pioneer neurons. Such neurons, however, have yet to be identified.

4. Why are only fibres from the posterior lamina misrouted?

The differentiation of the retina and lamina proceeds along a posterior - anterior axis (Meinertzhagen, 1973). In *irreC* only the first outgrowing equatorial axons connecting the lamina and the medulla are misrouted. This fact suggests that the first axons use a different set of cues for pathfinding than later ones. The first retinal and laminar axons innervating the medulla may trigger its further differentiation which might provide new information for growth cones coming later.

5. By what mechanism(s) can misrouted fibres find their right target neurons?

One of the amazing features of the *irreC* phenotype is that the misrouted fibres eventually terminate in what seem to be the correct retinotopic loci. They do not find their appropriate target neurons after a random walk. The axonal paths are not erratic. Especially telling is that in mutant flies L3 neurons using ectopic bundles to reach their retinotopic destination in the anterior medulla are able to elaborate their terminal specializations in the right depth of the medulla, although the axons enter this layer normally from the opposite direction. We cannot yet explain target recognition and establishment of retinotopy in *irreC* in molecular terms, but we are fascinated by the apparent similarities to vertebrate systems at the phenotypic level (Stuermer 1988).

6. Is the *irreC* phenotype related to the *bypass fibres* found in the flesh-fly *Boettcherisca peregrina*?

Mimura (1987) described so-called *bypass fibres* in the optic lobe of the flesh-fly. These retinula fibres terminate in the medulla like normal *R7* and *R8* axons, but run a considerable roundabout course (Fig.8 in Mimura, 1987), which is very similar to the course of the ectopic fibre bundles in *irreC*. Furthermore, in the flesh-fly population being tested by Mimura, this phenotype was much more frequent in males than in females (17.6% and 4.7% respectively). This frequency distribution is perfectly compatible with the assumption of a recessive X-chromosomal inheritance. We therefore suggest that the X-chromosome of fleshflies also carries a gene of *irreC* - like function.

SIMPLIFYING THE OPTIC LOBE BY REDUCING THE NUMBER OF CELL TYPES

One of the major structural features of the dipteran optic lobe is the high number of cell types which participate in the formation of each of the repetitive columns, especially inside the medulla (Campos-Ortega and Strausfeld, 1972; Fischbach and Dittrich, in press). We can ask: Is it possible to reduce this complexity by genetic means, i.e. can partially blind mutants be isolated which affect only certain cell types while the organization of the remaining optic lobe and its connections to the retina are undisturbed?

Several mutants have been isolated which partially fulfil the above criteria. Their optic lobes may be regarded as functional, albeit simplified versions of the original. These mutants are listed in Table 2 together with the other mutants discussed in this paper. Not included in the table are the loci *Vam* (Coombe and Heisenberg, 1986), *elav* (Campos et al., 1985), and *l(1)ogre* (Lipshitz and Kankel, 1985) which also affect the optic lobe.

The effect of mutations on fibres in the anterior optic tract (AOT) is well suited to demonstrate that it is possible to gradually dissect brain structures by the use of mutations (Fischbach and Lyly-Hünerberg, 1983).

Table 2. Some structural mutants of the optic lobe

genotype	all.	phenotype of optic lobe	Ref.	Fig.
lop	1	columnar neurons of lobula plate missing	1	5,7-11
omb	1	HS and VS neurons in lobula plate missing	2	5
so	2	lamina and many columnar neurons missing	1,3	2,5
rol	1	<40% reduction of medulla and lobula complex	4	5,6
sol	14	>40% reduction of medulla and lobula complex	1,5,6	5,6
mnb	2	>60% reduction of medulla and lobula complex	4,7	5,6
disco	4	no lamina, lack of medullar columnar neurons	8	2,5
irreC	2	optic chiasms disordered	9	4
flo	1	medulla fibre loops	1	12

References: [1]Fischbach, 1983b; [2]Heisenberg et al., 1978; [3]Fischbach and Technau, 1984; [4]Fischbach and Heisenberg, 1984; [5]Fischbach and Heisenberg, 1981; [6]Fischbach, 1985; [7]Fischbach et al., 1987a; [8]Steller et al., 1987; [9]Boschert and Fischbach, 1987. all. = mutant alleles

We have extended the previous electron microscopic investigations and obtained quantitative data on the number of fibres in the AOT of individuals of 15 different *Drosophila* strains (Fig.5). For the evaluation of the phenotype of multiple mutants it is noteworthy that more than 100 non-visual fibres are running in the AOT (Fischbach and Lyly-Hünerberg, 1983), i.e. that more than 90% of visual fibres have been eliminated in the AOTs of the multiple mutants with the exception of *rol so*. The AOT of this double mutant contains only about 100 fibres less than the single mutants *rol* and *so* which both have lost about 400 fibres. This result means that most fibres missing in the AOT of *rol* are also missing in *so*. On the other hand, Fischbach and Lyly-Hünerberg have shown that *sol* and *so* eliminate nearly exclusive sets of fibres in the AOT. Similar to the double mutant *sol so* the double mutant *rol sol*, therefore, should have also only few fibres left. This is indeed the case (Fig.5). The interesting conclusion that the visual mutants *rol* and *sol* eliminate different sets of neurons in the optic lobe is also supported by planimetric studies (Fig.6A). In contrast to the central brain (Fig.6B), the medulla and lobula complex of *rol sol* are extremely reduced.

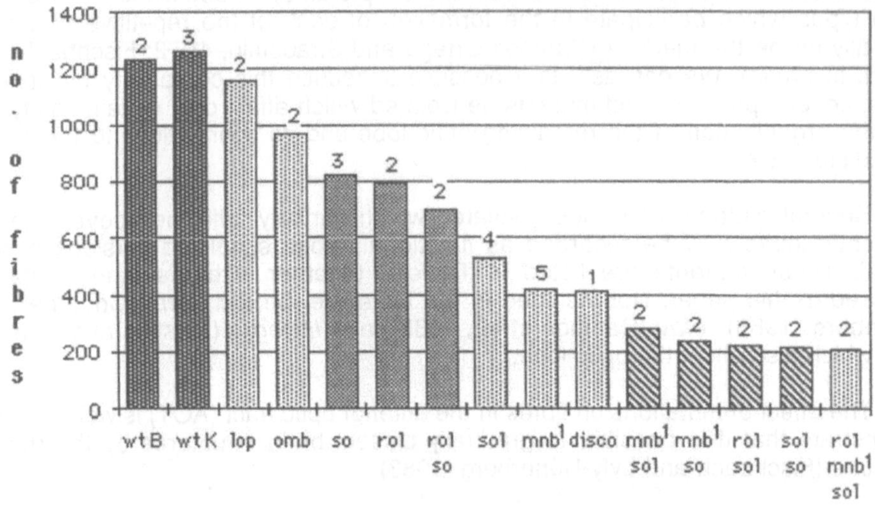

Fig.5. Mean numbers of axonal cross-sections counted in the AOT of male flies of various genotypes.The numbers on top of each bar denote the number of flies sectioned. Different shading supports discussion (see text). wt B = wild type Berlin (Jakob et al., 1977); wt K = wild type Kapelle (Heisenberg and Buchner, 1977); lop = *lobula plate-less*; omb = *optomotor blind* ; so = *sine oculis*; rol = *reduced optic lobe*; mnb = *minibrain*; disco = *disconnected.*

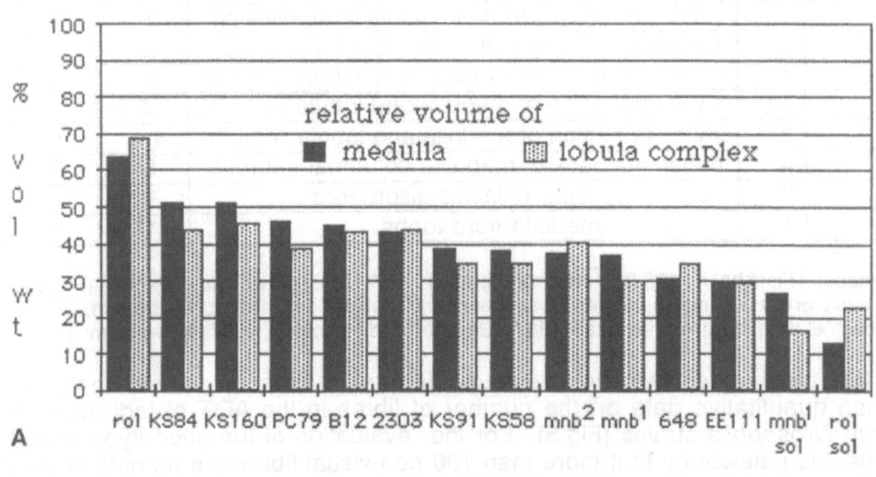

A

Fig.6. Relative volume of neuropiles (in % wild type volume) derived from planimetric measurements of complete series of sections through the optic lobe (A) and central brain (B, next page) of male mutant flies. The autosomes of all mutants had been isogenized for at least 10 generations using back crossing with females of an attXX (Berlin) strain, the males of which were used as wild type controls. The *small optic lobes* alleles used are: *KS160, KS84, PC79, KS58, KS91, EE111* (Fischbach and Heisenberg, 1981); *B12* (Bülthoff, 1982); *2303, 648* (Fischbach, 1985). The allele used for the construction of the double mutants was *KS58.*

The single mutants *rol, sol, mnb, lop* and the multiple mutants offer themselves for a functional analysis. It should be possible to detect visual functions which are not using the missing cell types (e.g. in the AOT). Also, the abolishment of sophisticated visual behaviour may uncover more simpler forms of behaviour. Recently, Wolf and Heisenberg (1986) have shown that the double mutant *rol sol* is completely motion blind. Orientation towards visual stimuli is nevertheless retained to a certain degree. This residual behaviour could be shown to be operant (*ibid.*).

lobula plate-less

The *lobula plate-less* (*lop*) mutant was isolated by M. Heisenberg in a screen for structural brain mutants on the second chromosome. Some of our results concerning the structural phenotype of *lop* have been reviewed in Heisenberg and Wolf (1984). Its phenotype as seen in silver stained preparations (Fig.7) displays several remarkable features. Most obvious in horizontal serial sections is the absence of the lobula plate neuropile (The insert in Fig.7 shows the wild type optic lobe for comparison). Only a very small dorsal rudiment of this neuropile can be seen in the mutant (Fig.7A). Nearly all cell bodies of the columnar neurons of the lobula plate, normally situated posterior to the neuropile, are also absent. These cell bodies belong to *T4, T5,* and *Tlp* cells (Fischbach and Dittrich, in press). Their absence in adult flies is due to extensive cell death in the lobula plate cortex during the first half of the pupal stage (data not shown). Tangential elements of the mutant's lobula plate, however, do not degenerate. The giant fibres of the vertical system (VS; Heisenberg et al., 1978), e.g., can be followed from their terminal arborizations in the posterior central brain via a dorsal ectopic fibre tract of the second optic chiasm (Fig.7B) into the medulla. In this neuropile, the dendrites turn ventrally and can be followed down the neuropile due to their extreme thickness (small white arrows in Fig.7). This projection pattern is reminiscent of the normal arborization pattern of the VS neurons in the lobula plate.

We have been interested in the question, how well the organization of the remaining optic lobe of *lop* flies is conserved. The answer to this question is important for any functional analysis. It is also of interest to learn how surviving cells react to the absence of columnar neurons in the lobula complex. The results of our Golgi studies are as follows:

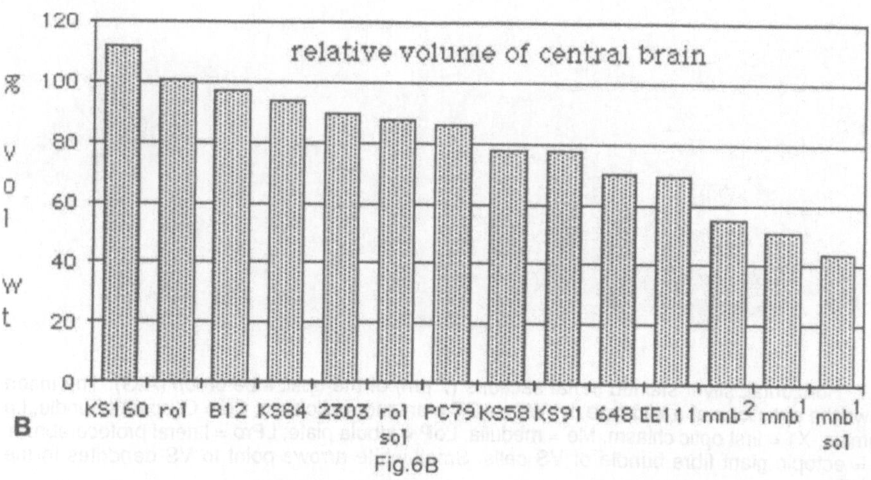

Fig.6B

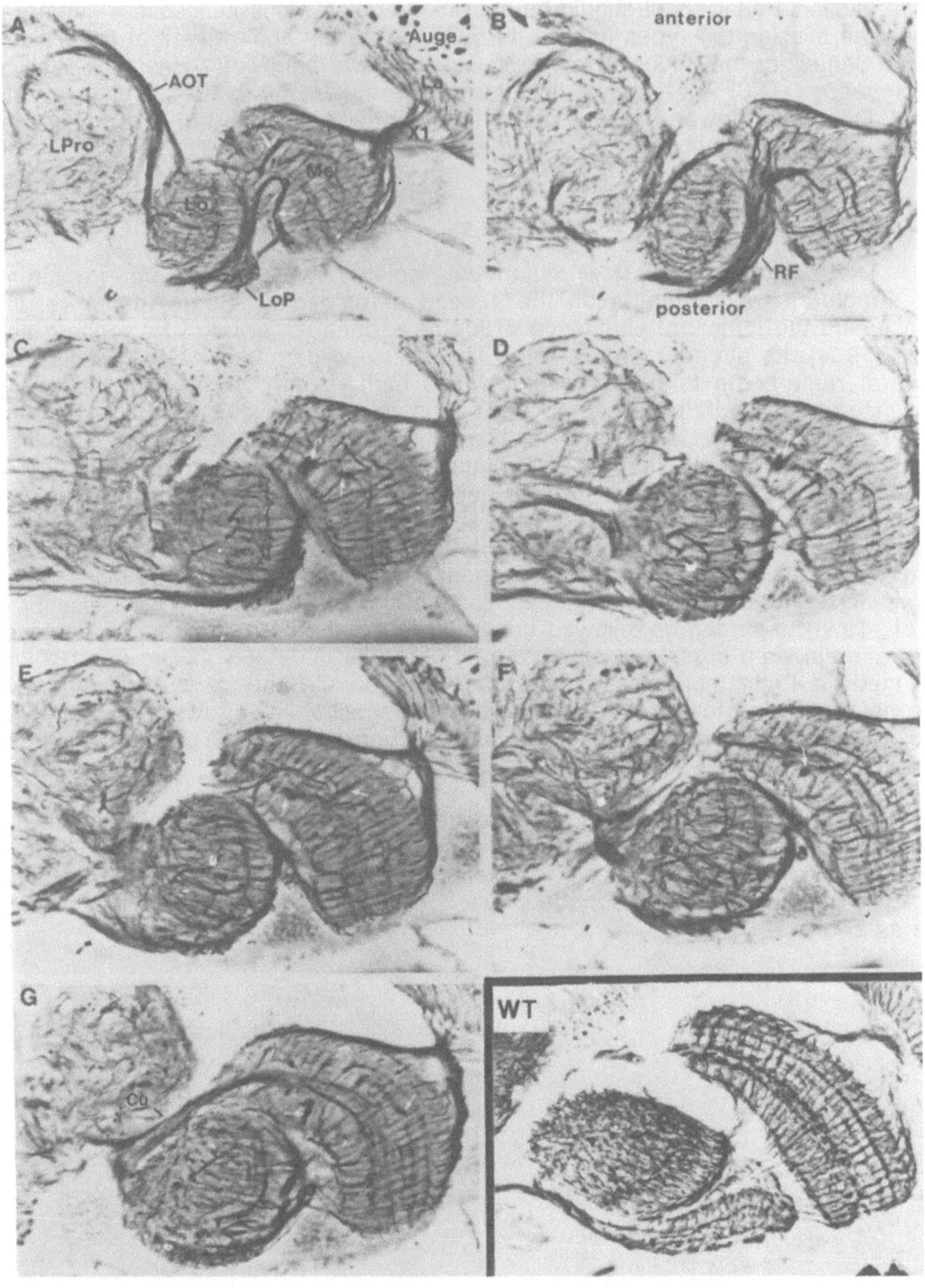

Fig.7. Horizontal, silver stained serial sections (7 µm) of the optic lobe of *lop* (A-G). The insert shows the optic lobe of a wild type fly (WT). AOT = anterior optic tract; Cu = Cuccatti's bundle, La = lamina, X1 = first optic chiasm, Me = medulla, LoP = lobula plate, LPro = lateral protocerebrum, RF = ectopic giant fibre bundle of VS cells. Small white arrows point to VS dendrites in the medulla.

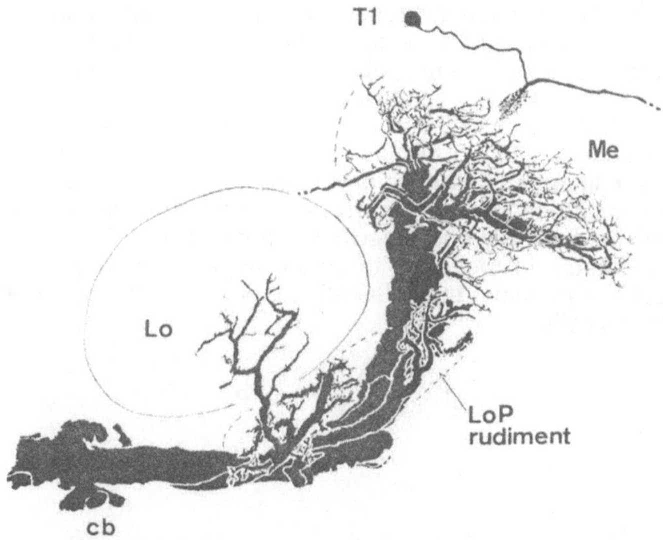

Fig.8. Camera lucida drawing of Golgi-impregnated lobula plate giant fibre(s) sprouting into the medulla of *lop*. Arborizations can also be seen in the rudiment of the lobula plate and in the lobula. It cannot be excluded that the drawing shows the superimposed shapes of several giants. The T1 neuron has been included to demonstrate that the arborizations of the ectopic tangential neurons do not penetrate the most superficial layers of the medulla. cb = cell body.

Tangential neurons: The medulla of *lop* contains its normal tangential neurons which enter the neuropile via Cuccatti's bundle (Fig.7G). The arborizations of these neurons inside the medulla are normal. However, they frequently send branches into the lobula (data not shown). This phenomenon might represent compensatory innervation of this neuropile which lacks several small field neurons (see below). As inferred from silver stained sections, the medulla of *lop* also contains the ectopic arborizations of lobula plate tangential neurons. Their Golgi impregnation (Fig.8) supports this conclusions. However, the Golgi preparations also show arborizations of the giants in the lobula plate rudiment and in the lobula.

Columnar neurons: All columnar cell types establishing connections between the lamina and medulla (*R7, R8, L1-5, T1, C2, C3, La wf*, see Fischbach and Dittrich, in press) are present in *lop* and well differentiated (data not shown). *Tm* neurons are well stratified inside the medulla neuropile and most of them can easily be homologized to their counterparts in wild type flies (Figs.9,10). The *Tm* and at least four types of *TmY* neurons project retinotopically into the lobula, where their arborizations look fairly normal. The *TmY* neurons send a branch into the lobula plate rudiment (Fig.10), i.e. their differentiation is normal and it seems unlikely that these cell types are a primary focus of the *lop* gene action. *T2* and *T2a* neurons (see Fischbach and Dittrich, in press) are often impregnated in *lop* and well differentiated inside the medulla neuropile. Their terminals in the lobula are also of normal shape and extension (Fig.11). It is interesting that, even though most *T4* and *T5* neurons are missing in adult *lop* flies, rarely neurons were impregnated which have to be homologized with these cell types, albeit their dendritic arborizations are more widespread than normal (Fig.11). In our view this means that *lop* does not affect the program for the differentiation of these cell types *per se*. The reason for their own or their precursors' degeneration in the lobula plate cortex during the pupal stage remains obscure. *Tlp* neurons (Fischbach and Dittrich, in press) were never seen in *lop*. The same is true for *T3* neurons, although frequently impregnated in wild type flies. Their cell bodies are normally positioned near those of the *T2* -, *C2*- and *C3*- cells (Fischbach and Dittrich, in press). It therefore might be that *T3* neurons, which normally connect the proximal medulla with the lobula, are missing in addition to most *T4, T5*, and *Tlp* cells.

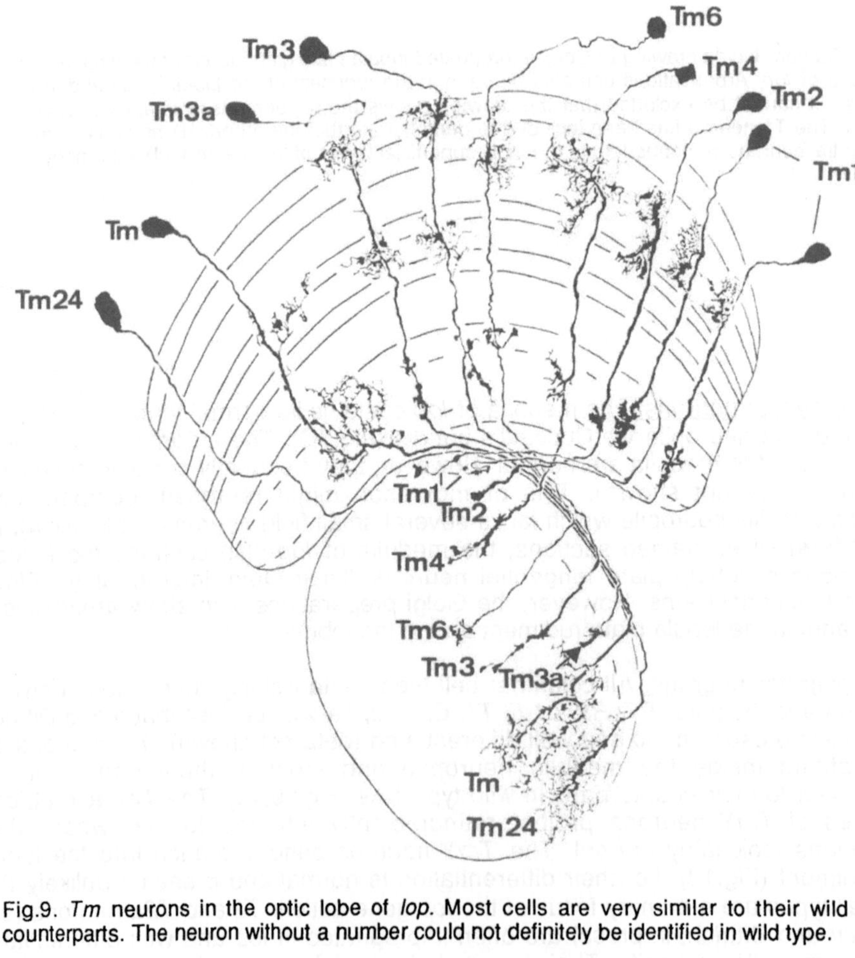

Fig.9. *Tm* neurons in the optic lobe of *lop*. Most cells are very similar to their wild type counterparts. The neuron without a number could not definitely be identified in wild type.

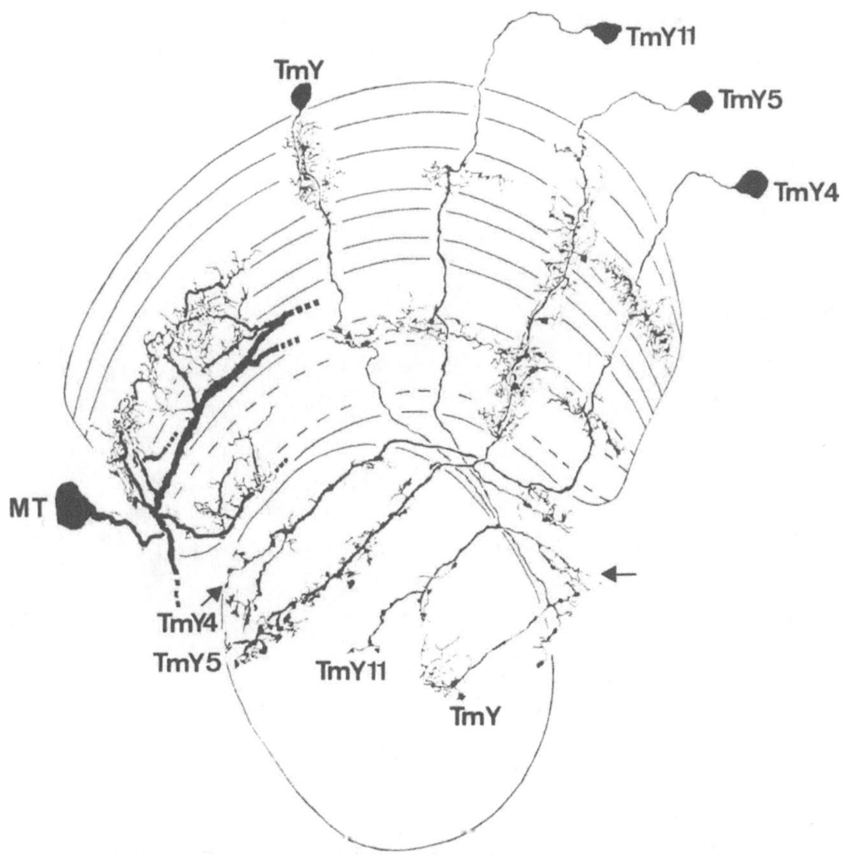

Fig.10. A fragment of a medulla tangential neuron (*MT*) and *TmY* neurons. Most cells are very similar to their wild type counterparts at least inside the medulla (Fischbach and Dittrich, in press). The *TmY* neuron without a number could not be identified in wild type. Please note that *TmY* neurons branch into the small lobula plate rudiment (arrow).

In conclusion, our Golgi impregnations of neurons in the optic lobe of *lop* show that the phenotype of *lop* can satisfactorily be explained by degeneration of lobula plate columnar neurons (possibly with the inclusion of *T3* neurons), although without genetic mosaic analysis it cannot be excluded that this degeneration is due to other primary defects, e.g. Heisenberg and Wolf (1984) found unusually large cell bodies in the medulla cortex of *lop,* the nature of which, however, is not clear. Most cell types in the optic lobe of *lop* are either unaffected by the mutation or their altered shape can be understood as a direct reaction to the absence of the cells mentioned, e.g. the abnormal growth of the VS dendrites into the medulla is most likely sprouting in response to an insufficient synaptic input inside the lobula plate rudiment. The phenotype of other mutants supports this interpretation (see below). It is doubtful whether the sprouting of the VS cells can functionally compensate for the loss of normal input neurons. In fact pitch and roll optomotor responses are severely diminished in mutant flies (Heisenberg and Wolf, 1984). Interestingly, yaw optomotor responses can still be elicited (*ibid.*). This might be explained by the finding of Paschma (1982)

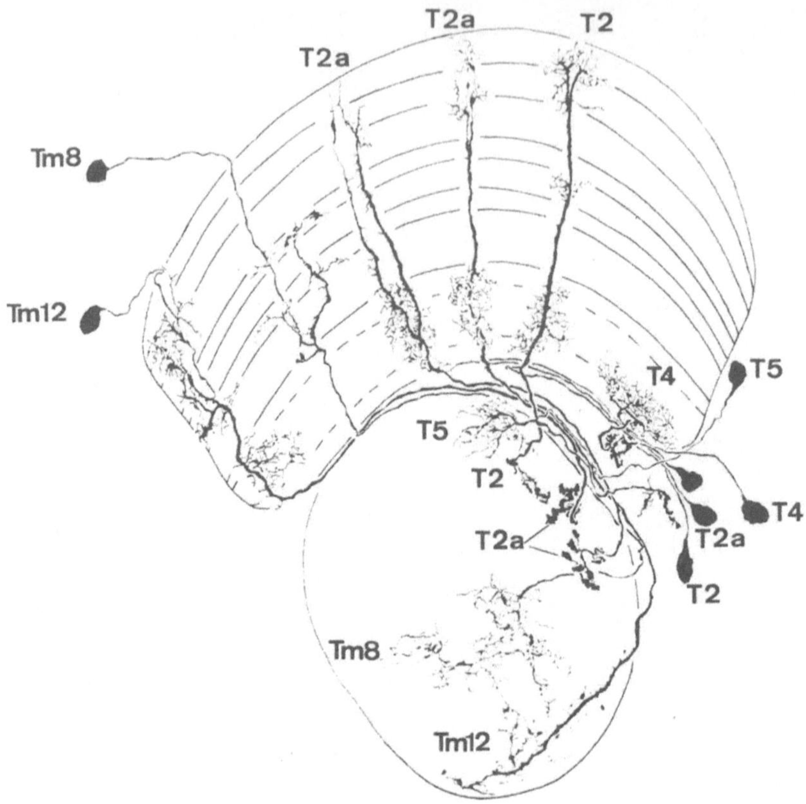

Fig.11. *Tm* and *T*-cells in the optic lobe of *lop*. The lobula terminals of *Tm8* and *Tm12* appear to be more extended than in wild type. Both forms of *T2* neurons (Fischbach and Dittrich, in press) exist in the mutant. *T3* -cells were never seen. In contrast to the situation in wild type, only very few *T4* - and *T5* -cells were impregnated. Typically the dendritic bushes of *T4* in the proximal medulla and of *T5* in the lobula are larger than normal and seem to arborize in several columns.

that, in contrast to the VS cells, the HS neurons terminate in the rudiment of the lobula plate.

The search for other visual functions which are not affected by the *lop* mutation should be successful. Bülthoff and Buchner (1985) have shown that deoxyglucose mapping of nervous activity in the medulla of *lop* yields wild type-like results. The abnormal activity distribution in the lobula is expected due to the absence of *T5, Tlp* and *T3* cells which normally innervate the lobula. Visual pathways not using these neurons should still be functional.

small optic lobes

small optic lobes (*sol*) mutants have been studied in some detail (Fischbach and Heisenberg, 1981; Bülthoff, 1982; Fischbach and Lyly-Hünerberg, 1983; Götz, 1983; Heisenberg and Wolf, 1984; Miklos et al., 1987). In *sol* mutants tissue autonomous cell degeneration in the medulla cortex and possibly secondary degeneration in the lobula complex (Fischbach and Technau, 1984) lead to the absence of many columnar cell types in adult flies. In spite of the severe reduction in optic lobe volume (Fig.6A), optic lobe organization is still intact, e.g. the number of columns is not reduced (Fischbach and Heisenberg, 1981). While visual acuity, optomotor responses, and colour vision are normal (Fischbach, 1983b), other visual functions as pattern discrimination, object-ground discrimination (*ibid.*), and visual plasticity (Götz, 1983; Heisenberg and Wolf, 1984) are defective.

Together with the group of G.L.G. Miklos in Canberra cloning of the genomic *sol/slgA* area in 19F4 on the X-chromosome (Miklos et al., 1987) has just been completed (Barleben et al., 1988). The initiated molecular characterization of the potential *sol* gene product will hopefully explain, why cell degeneration is occurring in the mutant's optic lobe.

minibrain

In a study of the phenotype of the mnb^2 mutant (short cut for mnb^{UB913}, Fischbach et al., 1987b) the following results were obtained (Kaltenbach, 1988):

In mnb^2 embryos no defects of the peripheral or central nervous system could be detected. At the stage of the white pupa only a small reduction of brain size in mnb^2 individuals is apparent. During the first half of pupal development the deficit in brain size increases steadily until the volume of the *mnb* optic lobe is smaller by about 60% and that of the central brain by about 50% as compared to wild-type controls (Fig.6). At 30% of pupal development many differentiated neurons with cell bodies in the lobula-plate cortex and axons in the inner optic chiasm degenerate. No degeneration takes place in the central brain. The reduction of brain size (Fig.6B) which is correlated with a reduction in cell number (Fig.5; and fibre counts in the cervical connective; Houbé, unpublished), is thus due to defective proliferation.

Eclosion of adult flies from the pupal case is delayed. While this process takes only 30-180 seconds in wild type controls, it may take up to several hours in the mnb^2 mutant. It is due to poor leg movements normally required for eclosion. Heidenreich (1982) recovered a mnb^1 strain from a cross with a *y cv v f car* marker strain. Individuals of this mnb^1 strain were unable to leave the pupal case without human help, and the strain was called *cesarean cut*. Therefore, eclosion behaviour is affected in the mnb^1 allele also, and this phenotype can be accentuated by genetic background effects.

The mnb^2 mutation is due to the insertion of a P-element in 16E (Barleben, 1987). The insertion has been localized inside the maternal effect region (16E2-16F2) of the *Shaker* gene complex (ShC; 16E2-16F8) by restriction analysis (Heisenberg, personal communication). This precise mapping in a genomic region which is essentially cloned (Pongs et al., 1988; Kamb et al., 1987; Tempel et al., 1987) makes it likely that the molecular nature of the *mnb* gene product will be known soon.

medulla fibre loops

Mutations that modify properties of only few neurons are certainly of importance when the evolution of brain and behaviour is considered. The *medulla fibre loops* (*flo*) mutant might be an example of such a mutation.

flo was isolated during a screen for structural brain mutants on the second chromosome (Heisenberg and Fischbach, unpublished). It maps 1.6±1.1 Morgan units near *cn* (2-57,5). The only defect known so far are apparent fibre loops at the level of the 2nd medulla layer, near the top of each medullar column, as seen in silver stained sections (Fig.12). These loops are formed by lamina monopolar neurons (most probably by *L2* terminals). The behavioural phenotype of *flo* flies has so far not been characterized. It would be interesting, however, to learn whether structural specificity is correlated with equally specific behavioural defects.

flo might be well suited as a neuropile marker in genetic mosaic experiments. We have used it to demonstrate that the fibres forming the loops are still present in each medulla column of sol^{KS58} *flo* double mutants.

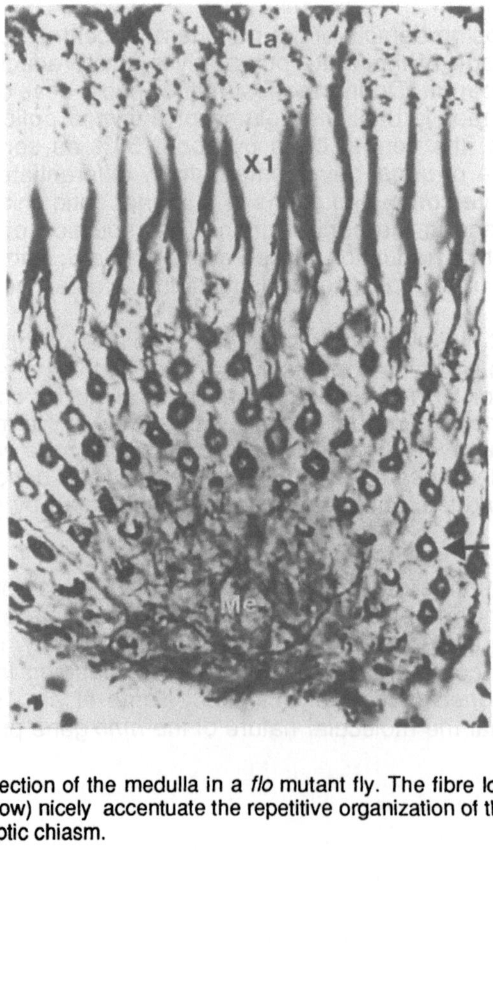

Fig.12. Tangential section of the medulla in a *flo* mutant fly. The fibre loops at the top of each medullar column (arrow) nicely accentuate the repetitive organization of this neuropile. X1 = fibre bundles in the first optic chiasm.

Some limits of optic lobe simplification

The data on the fibre numbers in the AOT (Fig.5) and the planimetric measurements (Fig.6) suggest that the combined actions of mutations in the neurological genes *rol*, *sol*, *so, mnb* on cell numbers can be understood by a simple additive model. Synergistic gene actions are not apparent at that level of analysis. Certainly, this can be used to some degree to engineer wanted phenotypes. Using the right argumentation, i.e. listing still normal instead of defective behavioural subroutines, multiple mutants might yield important information about the function of the optic lobe. However, one should be aware that simplification of the optic lobe by reducing the number of cells is limited by sprouting and compensatory innervation of the remaining neurons.

In the case of the giant VS neurons in *lop* we suspected that their sprouting into the medulla is an epigenetic consequence of the absence of small field input neurons at the level of the lobula plate. In the *sol* ^{KS58} mutant the giant fibres reside still mainly inside the lobula plate, but may send minor branches into the medulla (Fischbach and Heisenberg, 1981). This phenomenon has not been seen in *mnb*^{1} flies. However, in the double mutant *sol mnb* the number of small field input neurons innervating the lobula plate is obviously lowered beyond threshold. The giant fibres now sprout into the medulla (Fig.13) in a manner reminiscent of the situation in *lop* (Fig.8). Due to the size of the participating neurons this is a spectacular demonstration of a possible epigenetic rule, i.e. that growth of dendrites is only stopped when a sufficient number of presynaptic neurons has been encountered. The phenomenon is not restricted to the giant tangential neurons, it is also reflected in the abnormal shapes of columnar neurons (Fischbach and Dittrich, in preparation). The mutants thus reveal by their altered phenotypes epigenetic mechanisms which are active in normal development.

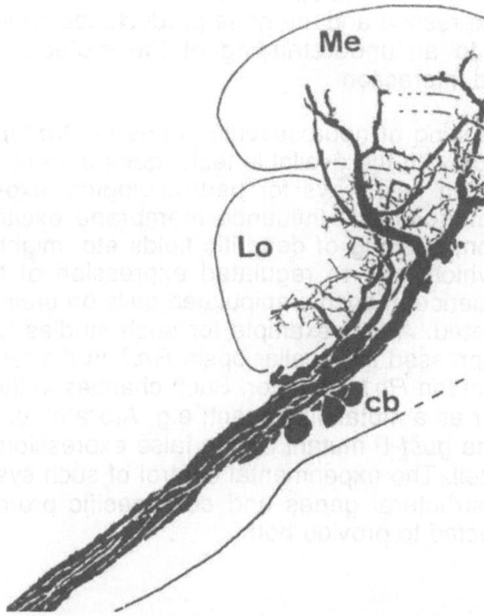

Fig.13. Sprouting of the giant fibres of the lobula plate of the double mutant *sol* ^{KS58} *mnb*^{1} into the medulla.

GENERAL CONCLUSIONS AND PERSPECTIVES

The study presented is based on the analysis of structural brain mutants which have been isolated by screens of adult brains. The procedure necessarily selects for viable alleles. Several of the mutations isolated are therefore expected to be hypomorphic alleles of essential genes. Nevertheless, we have shown that mutant *phenotypes* can be used to ask specific as well as general neurobiological problems, e.g. we addressed questions like:

1. What role plays the larval visual system of *Drosophila* in the development of its adult optic lobe?
2. By what mechanisms are misrouted fibres able to find their targets?
3. What happens to neural populations when others are removed by genetic means?
4. Which neurons are not essential for a given behaviour?
5. What are the limits of engineering simplified brains by multiple mutants?

Not all structural brain mutants are equally well suited to answer anyone of these questions, but the pool of available structural brain mutants is already fairly large and it will be increased further. The first and second chromosomes are not yet saturated for neurological genes and screens on the third and fourth chromosomes have not yet been performed. Therefore, we are confident that the importance of neurological mutants of *Drosophila melanogaster* for the study of neurobiological problems will further increase.

It is likely that several of the genes identified code for important brain specific functions. At least, one would expect that mutations in such genes produce neurological phenotypes. What then are the perspectives of the molecular characterization of the neurological genes presented in this study? It will be interesting to learn which molecules are responsible for the developmental defects of the mutants, and where and when the genes are expressed and the gene products localized. This knowledge will contribute, certainly, to an understanding of the molecular mechanisms of neuronal differentiation and interaction.

The importance of cloning of neurospecific genes in *Drosophila* has an additional aspect. In combination with the available techniques of gene technology in this organism, it is likely to open new ways for neurobiological experimentation, e.g. alleles of genes which autonomously influence membrane excitability, transmitter metabolism, axon formation, shaping of dendritic fields etc. might be fused *in vitro* with specific promotors which lead to regulated expression of the genes in well defined cell types. The influence of such manipulated cells on brain development and function could be investigated. A first example for such studies has been given by Feiler et al. (1988) who expressed the ocellar opsin *Rh2* in the retinula cells *R1-6* of the compound eye by use of the *Rh1* promotor. Such changes in the cell address of a gene sometimes also occur as a mutational event, e.g. Arora et al. (1987) explain the high attractivity of salt in the *gust B* mutant by the false expression of a salt receptor gene in a sugar receptor cell. The experimental control of such systems requires the availability of respective structural genes and cell specific promotors. Cloning of neurological genes is expected to provide both.

ACKNOWLEDGEMENTS

We thank M. Heisenberg for providing the single mutants *omb*^{H31}, *lop*^{N684}, *mnb*^{1}, *rol*^{KS221} and the multiple mutants *rol sol* , *mnb sol*, and *rol mnb sol*. Furthermore, some of our projects were initiated in the lab of M. Heisenberg. We thank B. Bausenwein, U. Baumann and I. Canal for suggestions on the manuscript. Our work is presently being supported by the DFG (grants Fi336/3-1 and Fi336/1-5).

REFERENCES

Arora K., Rodrigues V., Joshi S., Shanbhag S., and Siddiqi O. (1987). A gene affecting the specificity of the chemosensory neurons of *Drosophila. Nature* 330, 5 (1987).

Barleben F. (1987). Genetische und phänotypische Charakterisierung neu isolierter Gehirnstrukturmutanten von *Drosophila melanogaster. Dipl. thesis*, Univ Freiburg.

Barleben F., Baumann U., Davies J., Pirrotta V., Olson J., Hall F., Cotsell J., Delaney S., Hayward D., Schuppler U., Fischbach K.-F., and Miklos G.L.G. (1988). Molecular and genetic analysis of the *small optic lobes/sluggish* region of the X-chromosome of *D. melanogaster. J. Neurogenetics* (in press).

Benzer S. (1967). Behavioral mutants of *Drosophila* isolated by countercurrent distribution. *Proc. Natl. Acad. Sci. USA* 58, 1112-1119.

Benzer S. (1971). From the gene to behaviour. *J. Am. Med. Assoc.* 218, 1015-1022.

Bingham P.M. (1981). Cloning of DNA sequences from the *white* locus of *Drosophila melanogaster* by a novel and general method. *Cell* 25, 693-704.

Boschert U. and Fischbach K.-F. (1987). Mutants with irregular optic chiasms in *Drosophila melanogaster. In* New Frontiers in Brain Research, eds: N. Elsner, O. Creutzfeld. Thieme Verlag Stuttgart.

Bülthoff H. (1982). *Drosophila* mutants disturbed in visual orientation. II. Mutants affected in movement and position computation. *Biol. Cybern.* 41, 71-77.

Bülthoff I. and Buchner E. (1985). Deoxyglucose mapping of nervous activity induced in *Drosophila* brain by visual movement. II. *optomotor blind* ^{H31} and *lobula plate-less*^{N684}, visual mutants. *J. Comp. Physiol. A* 156, 25-34.

Campos-Ortega J.A. and Strausfeld N.J. (1972). Columns and layers in the second synaptic region of the fly's visual system. The case for two superimposed neuronal architectures. *In* Information processing in the visual system of arthropods. Ed. R. Wehner. Springer Verlag. Berlin Heidelberg New York.

Chikaraishi D.M. (1988). Characteristics of Brain Messenger RNAs. *In* From message to mind. Eds. S.S. Easter Jr., K.F. Barald, and B.M. Carlson. Sinauer Associates, Inc. Mass.

Campos A.R., Grossman D., and White K. (1985). Mutant Alleles at the Locus *elav* in *Drosophila melanogaster* lead to Nervous System Defects. A Developmental-Genetic Analysis. *J. Neurogenetics* 2, 197-218.

Coombe P.E. and Heisenberg M. (1986). The structural brain mutant *Vacuolar medulla* of *Drosophila melanogaster* with specific behavioral defects and cell degeneration in the adult. *J. Neurogenetics* 3, 135-158.

Dudai Y. and Zvi S. (1985). Multiple defects in the activity of adenylate cyclase from the *Drosophila* memory mutant *rutabaga. J. Neurochem.* 45, 355-364.

Feiler, R., Harris, W.A., Kirschfeld, K., Wehrhahn, C. and Zuker, C.S. (1988): Targeted misexpression of a *Drosophila* opsin gene leads to altered visual function. *Nature* 333, 737-741.

Fischbach, K.-F. (1983a). Neural cell types surviving congenital sensory deprivation in the optic lobe of *Drosophila melanogaster. Dev. Biol.* 95, 1-18 .

Fischbach, K.-F. (1983b). Neurogenetik am Beispiel des visuellen Systems von *Drosophila melanogaster. Habilitation*, Würzburg.

Fischbach, K.-F. (1985). Neurogenetics of the visual system of *Drosophila melanogaster. Biol. Chem. Hoppe-Seyler* 336 (2), 114-115.

Fischbach K.-F. and Dittrich A.P.M. (1988). The optic lobe of *Drosophila melanogaster.* A Golgi analysis of wild-type structure. *Cell Tissue Res.*, in press.

Fischbach K.-F. and Heisenberg M. (1981). Structural brain mutant of *Drosophila melanogaster* with reduced cell number in the medulla cortex and with normal optomotor yaw response. *Proc. Natl. Acad. Sci. USA* 78: 1105-1109.

Fischbach K.-F. and Heisenberg M. (1984). Neurogenetics and behaviour in Insects. *J. exp. Biol.* 112 : 65-93.

Fischbach K.-F. and Lyly-Hünerberg I. (1983). Genetic dissection of the anterior optic tract. *Cell Tiss. Res.* 231: 551-563.

Fischbach, K.-F. and Technau, G. (1984). Cell degeneration in the developing optic lobes of the *sine oculis* and *small optic lobes* mutants of *Drosophila melanogaster. Dev. Biol.* 104: 219-239.

Fischbach K.-F. and Technau G.M. (1987). Mutant analysis of optic lobe development in *Drosophila. In* New Frontiers in Brain Research. Eds: N. Elsner, O. Creutzfeld. Thieme Verlag Stuttgart.

Fischbach K.-F., Boschert U., Barleben F., Houbé B., and Rau T. (1987a). New alleles of structural brain mutants of *Drosophila melanogaster* derived from a dysgenic cross. *J. Neurogenetics* 4, 126-128.

Fischbach K.-F., Houbé B., Boschert U., Barleben F., and Gschwander B. (1987b). Structural mutants of the visual system of *Drosophila melanogaster* derived from a dysgenic cross. *J. Neurogenetics* 4, 128-130.

Gierer A. (1987). Directional cues for growing axons forming the retinotectal projection. *Development* 101, 479-489.

Götz K.G. (1983). Genetics and ontogeny of behaviour. Genetic defects of visual orientation in *Drosophila. Verh. Dtsch. Zool. Ges.* 1983, 83-99.

Hall J.C. (1982). Genetics of the nervous system in *Drosophila. Quart. Rev.Biophys.* 15, 223-479.

Hall J.C. (1986). Learning and rhythms in courting, mutant *Drosophila. Trends Neurosci.* 9, 414-418.

Heidenreich D. (1982). Die Genetik der Mutante *minibrain* von *Drosophila melanogaster.* Diplomthesis, Würzburg.

Heisenberg M. and Böhl K. (1979). Isolation of anatomical brain mutants of *Drosophila* by histological means. *Z. Naturforsch.* 34, 143-147.

Heisenberg M. and Buchner, E. (1977). The rôle of retinula cell types in visual behavior of *Drosophila melanogaster . J. comp. Physiol.* 117, 127-162.

Heisenberg M. and Götz K.G. (1975). The use of mutations for the partial degradation of vision in *Drosophila melanogaster. J. Comp. Physiol.* 98, 217-241.

Heisenberg M. and Wolf R. (1984). Vision in *Drosophila.* Genetics of microbehavior. Springer Verlag. Berlin. Heidelberg. New York.

Heisenberg M., Wonneberger R., and Wolf, R. (1978). *optomotor blind*^{H31} - A *Drosophila* mutant of the lobula plate giant neurons. *J.Comp.Physiol.A* 124, 287-296.

Hofbauer A. (1979). Die Entwicklung der optischen Ganglien bei *Drosophila melanogaster. Dissertation* Univ Freiburg.

Jacob, K.G., Willmund, R., Folkers, E., Fischbach, K.-F. und Spatz, H.Ch. (1977). T-maze phototaxis of *Drosophila melanogaster* and several mutants in the visual system. *J. Comp. Physiol.* 116, 209-225.

Jäger R. (1988). Isolierung neuer struktureller Gehirnmutanten auf dem X-Chromosom von *Drosophila melanogaster* mit Röntgenmutagenese. *Dipl. Thesis*, Univ Freiburg.

Jäger R. und Fischbach K.-F. (1987). Some improvements of the Heisenberg-Böhl method for mass histology of *Drosophila* heads. *DIS* 66, 162-165.

John B. and Miklos G.L.G. (1988). The eukaryote genome in development and evolution. *Allen & Unwin.* London.

Kaltenbach E. (1988). Phänotypische Charakterisierung der Mutante *minibrain*^{UB913} von *Drosophila melanogaster. Dipl. thesis.* Univ Freiburg.

Kamb A., Iverson L.E., and Tanouye M.A. (1987). Molecular characterization of *Shaker*, a *Drosophila* gene that encodes a potassium channel. *Cell* 50, 405-413.

Lehmann R., Dietrich U., Jiménez F., Campos-Ortega J.A. (1981). Mutations of early mutagenesis in *Drosophila. Wilhelm Roux's Arch* 190, 226-229.

Levy L.S. and Manning J.E. (1981). Messenger RNA sequence complexity and homology in developmental stages of *Drosophila. Dev. Biol.* 85, 141-149.

Lipshitz H.D. and Kankel D.R. (1985). Specificity of gene action during central nervous system development in *Drosophila melanogaster*: Analysis of the *lethal(1) optic ganglion reduced* locus. *Dev. Biol.* 108, 56-77.

Livingston M.S., Sziber P.P., and Quinn W.G. (1984). Loss of calcium/calmodulin responsiveness in adenylate cyclase of *rutabaga*, a *Drosophila* learning mutant. *Cell* 37, 205-215.

Livinstone M.S. (1985). Genetic dissection of *Drosophila* adenylate cyclase. *Proc. Natl. Acad. Sci. USA* 82, 5992-5996.

Livinstone M.S. and Tempel B.L. (1983). Genetic dissection of monoamine transmitter synthesis in *Drosophila. Nature* 303, 67-70.

Meinertzhagen I.A. (1973). Development of the compound eye and optic lobe of insects. pp. 51-104 *in* Developmental neurobiology of arthropods, ed. D. Young. Cambridge University Press.

Merriam, J. (1984). *Drosophila melanogaster* (Cloned DNA). *In* Genetic maps. (Ed. O´Brien, S. J.) *CSH Laboratory* 3, 304-308.

Meyerowitz E.M. and Kankel D.R. (1978). A genetic analysis of visual system development in *Drosophila melanogaster. Dev. Biol.* 62, 112-142..

Miklos G.L.G., Kelly L.E., Coombe P.E., Leeds C., and Lefèvre G. (1987). Localization of the genes *shaking-B, small optic lobes, sluggish-A*, and *stress-sensitive-C* to a well-defined region on the X-chromosome of *Drosophila melanogaster. J. Neurogenetics* 4,1-19.

Miklos G.L.G., Yamamoto, M.T., Davies J., Pirrotta V. (1988). Microcloning reveals a high frequency of repetitive sequences characteristic of chromosome 4 and the ß-heterochromatin of *Drosophila melanogaster. Proc. Natl. Acad. Sci. USA* 85, 2051-2055.

Mimura K. (1987). Two types of very long visual fibers found in the optic lobe of the flesh-fly, *Boettcherisca peregrina. Cell Tissue Res.* 250, 73-78.

Nässel D.R. and Sivasubramanian P. (1983). Neural differentiation in fly CNS transplants cultured *in vivo. J. Exp. Zool.* 225, 301-310.

O'Kane C.J. and Gehring W.J. (1987). Detection *in situ* of genomic regulatory elements in *Drosophila. Proc. Natl. Acad. Sci. USA* 84, 9123-9127.

Paschma R. (1982). Strukturelle und funktionelle Defekte der *Drosophila*- Mutante *lobulaplate-less*^{N684}. *Dipl. Thesis.* Univ Würzburg.

Pirrotta V., Hardfield C., and Pretorius G.H.J. (1983a). Microdissection and cloning of the *white* locus and the 3B1-3C2·region of the *Drosophila* X chromosome. *EMBO J.* 2, 927-934.

Pirrotta V., Jäckle H., and Edstrom J.E. (1983b). Microcloning of microdissected chromosome fragments. p.1-17 *In* Genetic Engineering. Principles and Methods. Vol.5. Eds. Setlow J.K., Hollaender A. Plenum Press New York. London.

Pongs, O., Kecskemethy, N., Müller, R., Krah-Jentgens, I., Baumann, A.,Kiltz, H.H.,

Canal, I., Llamazares, S. and Ferrús, A. (1988). Shaker encodes a family of putative potassium channel proteins in the nervous system of *Drosophila* *EMBO J.* 7, 1087-1096.

Power M.E. (1943). The effect of reduction in numbers of ommatidia upon the brain of *Drosophila melanogaster. J. Exp. Zool.* 94, 33-71.

Rubin G.M., Kidwell M.G., Bingham P.M. (1982). The molecular basis of P-M hybrid dysgenesis: The nature of induced mutations. *Cell* 29, 987-994.

Steller H., Fischbach K.-F., und Rubin G.M. (1987). *Disconnected* : A locus required for neuronal pathway formation in the visual system of *Drosophila* . *Cell* 50, 1139-1153.

Stent G.S. (1981). Strength and weakness of the genetic approach to the development of the nervous system. *Ann. Rev. Neurosci.* 4, 163-194.

Stuermer C.A.O. (1988). The trajectories of regenerating retinal axons in the goldfish. I. A comparison of normal and regenerated axons at late regeneration stages. *J. Comp. Neurol.* 267, 55-68.

Tempel B.L., Jan Y.N., and Jan L.Y. (1988). Cloning of a probable potassium channel gene from mouse brain. *Nature* 332, 837-839.

Tempel B.L., Livingstone M.S. and Quinn W.G. (1984). Mutations in the dopa decarboxylase gene affect learning in *Drosophila. Proc. Natl. Acad. Sci. USA* 81, 3577-3581.

Tempel B.L., Papazian D.M., Schwarz, T.L., Jan, Y.N. and Jan, L.Y. (1987). Sequence of a probable potassium channel component encoded at *Shaker* locus of *Drosophila.Science* 237, 770-775.

Thomas J.B. and Wyman R.J. (1982). A mutation in *Drosophila* alters normal connectivity between two identified neurones. *Nature* 298, 650-651.

Tix S. and Technau G.M. (1987). Pioneer neurones in the optic lobes and imaginal discs of *Drosophila melanogaster. In* New Frontiers in Brain Research, eds: N. Elsner, O. Creutzfeld. Thieme Verlag Stuttgart.

Tix S. and Technau G.M. (1988). Pre-existing neuronal pathways in the developing optic lobes of *Drosophila melanogaster. Development.* (in press).

Wolf R. and Heisenberg M. (1986). Visual orientation in motion-blind flies is an operant behaviour. *Nature,* 323, 154-156.

DEVELOPMENTAL, GENETIC AND MOLECULAR ANALYSES OF LETHAL(1)OGRE, A LOCUS AFFECTING THE POSTEMBRYONIC DEVELOPMENT OF THE NERVOUS SYSTEM IN DROSOPHILA MELANOGASTER

Douglas R. Kankel, Toshiki Watanabe, R. Naresh Singh and Kusum Singh

Department of Biology
Yale University
P.O. Box 6666
260 Whitney Avenue
New Haven, CT 06511-8112
U.S.A.

INTRODUCTION

Among the major sets of problems in developmental neurobiology are the definition of the mechanisms by which developing neurons become different from their non-neuronal brethren and from each other and the elucidation of the molecular bases by which the large numbers of cells in the nervous system find and identify one another, form the correct ensemble of connections and maintain appropriate connectivity for the life of the organism. Our laboratory has chosen to use the developing visual system in ***Drosophila*** as a model; at the current time, we have obtained a fairly extensive description of the overt aspects of development in the visual system of ***Drosophila***, and we have even been able to specify the rules by which various subsets of this system interact with one another during ontogeny. Yet, we have little understanding of the subcellular and molecular phenomena which control these events.

At first glance, there would seem to be a multiplicity of strategies available in the pursuit of the molecular bases for these complex developmental events. First is the classic genetic strategy, i.e. to screen or select for adults or larvae with abnormalities in visually-driven behaviors or in alterations in readily detectable morphological phenotypes. Its major advantages are extreme sensitivity and the fact that it has "function" as a point of departure. However, it makes the assumptions that the loci of interest are capable of mutating to an abnormal but viable condition and further, that in those cases where a viable mutant is obtained, that the mutant is directly disturbing the process under study rather than as a consequence of some secondary or higher order relationship; these assumptions will obviously fail in a number of cases. A second strategy is directly molecular-based and either has been or is being conducted in a variety of laboratories. This strategy can be called molecular-list-making; one assumes, reasonably, that a given gene involved in a particular process is expressed in temporal and spatial proximity to that process and that by making a list of the molecules found in the "right" place and at the "right" time, one will have included those molecules which play a central role in controlling the process. This has led to differential screens for RNAs (*e.g.* Levy *et al.*, 1982) or to the identification of cell- or time-specific antigens (*e.g.* Venkatesh *et al.*, 1985; Zipursky *et al.*, 1984). There are two major problems with this strategy: (a) There is the issue of sensitivity; it is reasonably likely that at least some if not many of the molecules of interest in developmental processes are below the threshold of detectability for the techniques used. (b) There is the problem of sorting the relevant few from among the many. Almost all of the molecular screens of this nature to date have identified a large number of potentially interesting molecules (ones found in interesting places at interesting times). Unfortunately, no easy mechanism exists to determine functional relevance for any particular molecule. Significant effort is required to ascertain which of the identified molecules are relevant to the process of interest and which are not. Although it is clear that these sorts of screens are reasonable, it is equally clear that they are labor intensive. A third strategy takes advantage of having significant information available about the biology and biochemistry of the process of interest. Given enough extant data, one may be able to posit a specific role for a particular molecule of known function or a specific class of molecular function/s in some higher order process. Our lab, for example, has adopted this approach in studies which probe the role of neurotransmitters in the assembly and maintenance of structure in the CNS (Chase & Kankel, 1986, 1987). From experience, we know that this too requires considerable labor and as for the making of molecular lists does not use known function/s as its point of departure.

The basic work to be reported here involves the analysis of a mutant originally identified on the bases of its failure to perform normally in certain visually driven tasks. This is obviously an example of what we have defined in the first class of strategies.

One of the original reasons the visual system in ***Drosophila*** was chosen as a model for the study of the development of the nervous system was based on a belief that it would be relatively straightforward to define a great many, if not most, of the subsets of the genome which had a specific effect on that particular part of the fly's sensory and central processing system (i.e. affected the development and/or the function of the visual system only). This basic strategy was predicated on the assumption that eliminating the function of such genes would lead to significant abnormalities in visual behavior and/or easily observable changes in the structure of the compound eyes or optic lobes without affecting the viability of the fly; it would then be a simple (although perhaps laborious) matter of defining clever schemes of selection or screening to identify relevant mutants and their underlying loci. An inherent assumption in this strategy was the belief that visual-system-specific mutants would be relatively common and that the loci which they represented would play a prominent role in visual system development. Using screens which have identified mutants with abnormalities in both visually driven behaviors (*e.g.* Benzer, 1967; Heisenberg, 1974; Lipshitz & Kankel, 1985) and/or obvious morphological abnormalities in the visual system (*e.g.* Meyerowitz & Kankel, 1978; Heisenberg & Böhl, 1979), in excess of 100-150 loci have been identified with prominent effects on the visual system of the adult. Of these, no more than a handful have been analyzed to any significant extent; nevertheless, I would hazard to categorize them into two broad groups. Those which are, in fact, relatively specific for the visual system [*e.g.* the ninaE locus (O'Tousa *et al.*, 1985) which has been found to encode the rhodopsin for the major class of photoreceptors and the ***sevenless*** locus (Harris *et al.*, 1976) which seems to rather specifically affect the differentiation of a subset of the developing ommatidia]; these tend to function relatively late in the developmental process and are often involved in the terminal differentiation of the specialized functions associated with the particular cell-types in which they operate. The second group is represented by mutants which are probably hypomorphic (leaky) mutations which represent loci that have a broader spatial domain of operation but which can mutate such that specific alleles show a phenotype constrained to some or all of the visual system [*e.g.* the eye mutant ***Glued*** which is in fact a cell lethal (Harte and Kankel, 1983) or the ***l(1)ogre*** locus which was originally thought to be specific to the optic lobes but operates more broadly throughout the central nervous system]. I would take it as significant that no locus has yet been identified which alters in any subtle or regular way the detailed pattern of connectivity seen within the optic lobes; in other than the trivial sense that profound abnormalities in the differentiation of cells of either the compound eye or optic lobe leads to a totally chaotic structure which is demonstrably non-functional. Indeed such mutants (i.e. those showing subtle or regular alteration in connectivity) seem to be rare (or at least have been rarely recovered to date); among the few exceptions reported are the mutants ***passover*** and ***bendless*** found in the laboratory of Robert Wyman (Thomas & Wyman, 1984) in which somewhat subtle abnormalities in the pattern of connectivity on the part of the giant fibers in the flight motor system are seen. Even here, however, more genetic probing has unearthed new alleles at ***passover*** which are clearly lethal and accompanied by a wider range of phenotypic abnormalities (Miklos, personal communication; Wyman, personal communication), and the ***bendless*** locus has yet to be explored more than casually in a genetic sense.

We put forward the hypothesis that the genome of ***Drosophila*** has evolved in a manner which makes use of information for nervous system assembly in a largely redundant fashion; i.e. that the process of ensuring the correct relationships in connectivity in the abdominal ganglion is not significantly different from that needed to ensure appropriate connections in the visual system and that, in fact, the same mechanisms and therefore the same genes are used in both instances. The consequence of this redundancy in function from the standpoint of genetic analysis is that the most interesting genes are apt to be represented by lethality when there is any significant abnormality in their function. This may very well make a screen for such mutants problematic since approximately 90% of the classically defined loci seem capable of mutating to lethality; whether these data for classic loci are representative of the genome as a whole is unknown. Even if it were, what fraction of lethal loci capable of giving rise to readily detectable viable alleles is also unknown. We present below the results of a series of analyses on a locus in which mutations are associated with a broad spectrum of morphological abnormalities in the development of the imaginal nervous system during the larval and pupal stages.

METHODS AND RESULTS

<u>The ***ogre*** locus and its phenotype</u>

This locus was originally identified by the isolation of a single viable allele in a screen for mutants with abnormal performance in a simple test for visual pattern detection (Lipshitz and Kankel, 1985). During the course of the routine genetic localization of the mutant, we became aware of the possibility that it might have been allelic to one of the complementation groups identified by Tom Cline and his co-workers in the vicinity of the ***Sxl*** locus (Nicklas and Cline, 1985). We were able to establish that the lethal locus designated ***jnl3*** (Nicklas and Cline, 1985) and the behavioral mutant we had designated ***cb8*** were allelic. We have named the locus (with Dr. Cline's permission) ***l(1)optic ganglion reduced***. All known alleles at the locus are characterized by severe abnormalities in the structure of both the adult and developing optic lobes. In the single available viable allele, no other abnormalities are seen. In the original lethal allele and in all other lethals studied to date, there are additional and profound structural abnormalities in various subsets of the CNS outside of the optic lobes. Genetic mosaic analyses have suggested that the phenotypes observed in the CNS are at least regionally autonomous; there is some question, however, as to whether there is strict cell autonomy, i.e. we can not eliminate the possibility that there is local non-autonomy at the cellular level, and, in fact, our failure to recover somatic recombinant clones for any ***ogre*** allele makes this seem likely. Our working hypothesis based on the earliest data we have gathered is that ***l(1)ogre*** encodes a function which is potentially specific for the subset of the CNS which is imaginal-specific, i.e. that portion of the adult CNS which arises from the progeny of the giant neuroblasts of the larval brain and ventral ganglion. These latter cells seem to play no functional role in the larval CNS but begin to differentiate either late in the 3rd instar or during metamorphosis and ultimately become part of the functioning adult CNS. The final phenotype of note is a prolonged larval developmental; in the most extreme lethal allele studied, it is often several days longer. The primary lethal phase is at the larval/pupal boundary although a significant number of individuals can pupariate, and there is a low frequency of pharate adults formed with an occasional, and extremely sick, adult escaper found.

<u>Fine structure</u>

A recent electron microscopic analysis using the viable and the most severe lethal alleles has confirmed most of the earlier observations made at the level of the light microscope. It has revealed the following in the CNS: (i) a significant number of cells with many electron-translucent vacuoles which seem to enlarge as a function of developmental time, ultimately fusing together to destroy the cell and forming extracellular channels, (ii) a variety of inclusion bodies and (iii) electron-dense cells (assumed to be degenerating) in the CNS. Males hemizygous for the viable allele ***l(1)ogre***^{vcb8} evidenced a phenotype intermediate to that of the *l(1)ogre*^{ljnl3}/*Y* males and the control *l(1)ogre*^{ljnl3}/+ females with respect to these characters. This analysis, however, does not allow us to distinguish between cells of the imaginal-specific and functional larval nervous systems. We have also occasionally found structures outside the nervous system which evidence similar phenotypes but at substantially reduced frequencies. Whether this is a direct effect of the mutation on those structures or a secondary effect of the abnormalities in the nervous system is unknown. The newly hatched first instar mutant larva is essentially indistinguishable from its wild-type control, and beginning in the second instar the severity of abnormalities in the mutant relative to the wild-type increases.

Extent and pattern of mitotic activity

Because we suspect that *l(1)ogre* has a prominent effect on the imaginal-specific subset of the CNS, we carried out an analysis of the pattern and extent of mitotic activity in mutant and wild-type larvae. As mentioned above male larvae hemizygous for *l(1)ogre*^{ljnl3} grow more slowly and moult later than their heterozygous (*l(1)ogre*^{ljnl3}/+) female siblings. At 18° C the *l(1)ogre*^{ljnl3} male larvae undergo first and second instar moults around 70 and 118 hours, respectively, compared to heterozygous female siblings which moult first at 62 hours and the second time at 106 hours. Autoradiographic detection of [methyl-^{3}H]-thymidine incorporating cells in appropriately staged second and third instar larvae shows that *l(1)ogre*^{ljnl3} males have, on an average, less than one-third of the cells incorporating thymidine in the central nervous system (CNS) than do *l(1)ogre*^{ljnl3}/+ females of an equivalent stage.

Genetics

Our genetic analysis to date has focussed on the more-or-less straightforward tasks of localization and the generation of new mutant material. We have used a series of extant duplications and deficiencies as well as several new rearrangements which we have generated ourselves to localize ***ogre*** on the cytogenetic map. Figure 1 shows a subset of the most relevant data which establishes that the *l(1)ogre* locus falls somewhere within the ***6E1-2*** to ***6E4-5*** interval as does a distal flanking semi-lethal locus *l(1)jnlX* and a proximal flanking lethal locus *l(1)jnl2*. Nicklas and Cline (1983) had previously placed the ***ljnl3*** allele at 18.1 on meiotic recombination map. We have carried out a series of modest scale mutageneses using x-rays, ENU and EMS as mutagens and have recovered an additional 2 and possibly 3 lethal alleles. With the exception of our ENU screen, *l(1)ogre* mutants have been relatively rare; this is consistent with the results of the putative saturation mutagenesis of the region by Nicklas and Cline (1983) in which only a single allele at the locus was recovered. We have also gathered a collection of *X*-linked lethals in the ***6E/F*** region from other laboratories to test for allelism to our known *l(1)ogre* alleles. One such mutant (Eeken *et al*., 1985), provided by Dr. A.P. Schalet, has proven to be a small (10 Kb) deletion within the *l(1)ogre* locus.

Molecular biology

As mentioned above, genetic studies showed that *l(1)ogre* was located between the ***Df (1)HA32*** breakpoint in ***6E4-5*** and the ***Df (1)Sxl***^{bt} breakpoint in ***6E1-2***. In order to clone the *l(1)ogre* locus, all the DNA between these breakpoints (about 70 Kb) was isolated by way of chromosome-walking. The small deficiency mentioned above further localized the ***ogre*** locus as did comparing the pattern of transcripts between the flanking breakpoints among the wild-type and various ***ogre*** alleles. These data guided a series of transformations with genomic fragments, success ultimately being yielded by a 12.5 Kb fragment which completely includes the subset of the genome associated with a 2.9 Kb species of RNA which we believe to be the single transcript of the *l(1)ogre* locus. We have recovered cDNA clones corresponding to the 2.9 Kb transcript; a 2.15 Kb Xho I fragment from one of these was subcloned into a pGEM-7Zf(+) vector and an ^{35}S-labelled antisense probe constructed for use in hybridizing to tissue sections. The pattern of expression as inferred from the hybridization of this probe is both spatially and temporally complex. Substantial expression is seen throughout the embryonic, larval and pupal stages, but in the adult, expression seems limited to the follicle cells of the ovary and to portions of the stomodael

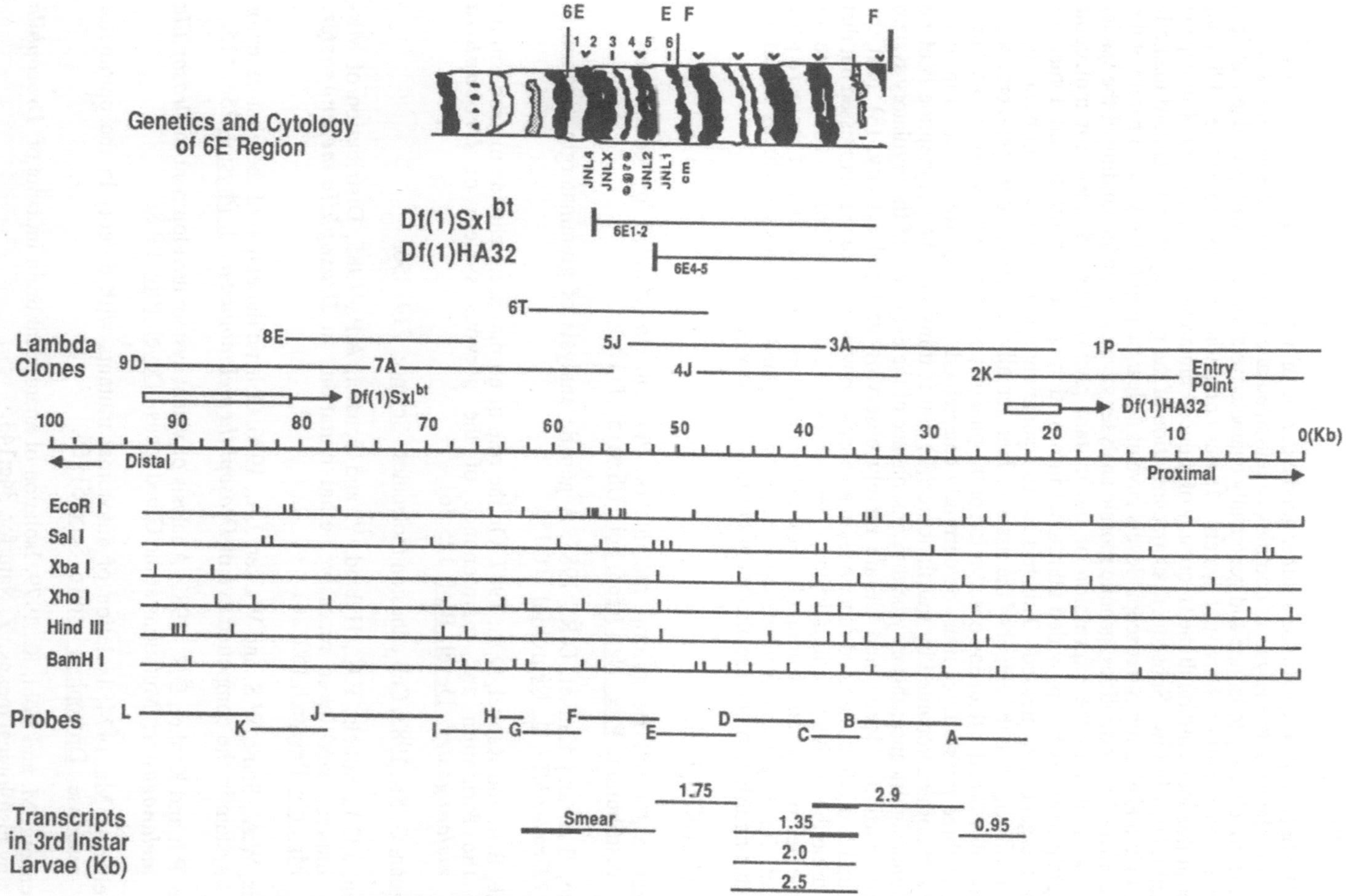

Figure 1. Summary of some basic genetic and molecular data for the ***l(1)ogre*** locus.

valve. We are struck by extensive expression in tissues where we have previously seen no phenotype, even in the small deficiency which lacks detectable transcript.

DISCUSSION

We believe this work to be a rather good example of both the strengths and weaknesses inherent in the genetic analysis of complex developmental events within the nervous system. It was relatively easy to isolate and genetically characterize a mutation with prominent effects on the development of the visual system. It was comparatively easy to isolate additional alleles at this locus although the frequency of induced mutations with a variety of mutagens seemed relatively low. Subsequent characterization of the phenotypes by standard methods at the light and electron microscopic levels gave us a clear picture of the general nature of the developmental abnormalities; genetic mosaic analyses gave us an understanding of the spatial domains critical to the expression of the mutant phenotypes. Subsequent molecular biological analysis has provided additional information concerning the spatial domain in which the gene is expressed. This has led to a somewhat perplexing condition, i.e. it has become obvious that the extent of the tissues which actually express the ***ogre*** gene product is considerably broader than any observation of the phenotype had previously led us to believe. At least three possibilities must be currently entertained: (1) That transcription in many tissues is not accompanied by translation or (2) that if translated the polypeptide is of no functional significance; this may be the consequence of the evolution of the regulatory system for this particular locus, and we are merely observing vestigial activity. (3) That the polypeptide product is indeed functional in all tissues in which the gene is expressed but that there are either active functionally equivalent genes in tissues other than nervous tissues or that the nervous system is uniquely sensitive to the absence or reduction of the ***ogre*** gene product. It is hoped that further molecular analysis of the locus and its product/s will allow us to distinguish among these alternatives or to define new ones.

LITERATURE CITED

Benzer, S., 1967, Behavioral mutants of ***Drosophila*** isolated by countercurrent distribution. Proc. Nat. Acad. Sci. (USA) 58:1112.

Chase, B.A. and Kankel, D.R., 1987, A genetic analysis of glutamatergic function in ***Drosophila***. J. Neurobiol. 18:15.

Chase, B.A. and Kankel, D.R., 1987, On the role of normal acetylcholine metabolism for the formation and maintenance of the nervous system of ***Drosophila melanogaster***. Devel. Biol. 125:361.

Edelman. G. M., 1983, Cell adhesion molecules. Science· 219: 450.

Eeken, J.C.J., Sobels, F.H., Hyland, V. and Schalet, A.P., 1985, Distribution of MR-induced sex-linked recessive lethal mutations in ***Drosophila melanogaster***. Mutation Research 150: 261.

Harris, W.A., Stark, W.S. and Walker, J.A., 1976, Genetic dissection of the photoreceptor system in the compound eye of ***Drosophila melanogaster***. J. Physiol. 256: 415.

Harte, P.J. and Kankel, D.R., 1983, Analysis of visual system development in ***Drosophila melanogaster***: Mutations at the Glued locus. Devel. Biol. 99:88.

Heisenberg, M., 1974, Isolation of autosomal mutants with defects in the optomotor response. Drosophila Inform. Serv. 51: 64.

Heisenberg, M. and Böhl, K., 1979, Isolation of anatomical brain mutants of ***Drosophila*** by histological means. Z. Naturfor. 34c:143.

Levy, L.S., Ganguly, R., Ganguly, N. and Manning, J.E., 1982, The selection, expression and organization of a set of head-specific genes in ***Drosophila***. Devel. Biol. 94:451.

Lipshitz, H.D. and Kankel, D.R., 1985, Specificity of gene action during development in ***Drosophila melanogaster***: Analysis of the ***l(1)optic ganglion reduced*** locus. Devel. Biol. 108:56.

Meyerowitz, E.M. and Kankel, D.R., 1978, A genetic analysis of visual system development in ***Drosophila melanogaster***. Devel. Biol. 62:63.

O'Tousa, J.E., Baehr, W., Martin, R.L., Hirsh, J., Pak, W.L. and Applebury, M.L., 1985, The ***Drosophila*** ninaE gene encodes opsin. Cell 40:839.

Ripoll, P. and Garcia-Bellido, A., 1979, Variability of homozygous deficiencies in somatic cells of ***Drosophila melanogaster***. Genetics 91:443.

Rutishauser, U. , 1984, Developmental biology of a neural cell adhesion molecule. Nature 310:549.

Rutishauser, U., Watanabe, M., Silver, J., Troy, F. A., and Vimr, E. R. , 1985, Specific alteration of N-CAM-mediated cell adhesion by an endoneuraminidase. J. Cell Biol. 101:1842.

Thomas, J.B. and Wyman, R.J., 1984, Mutations altering synaptic connectivity between identified neurons in ***Drosophila***. J. Neurosci. 4:530.

Venkatesh, T.R., Zipursky, S.L. and Benzer, S., 1985, Molecular analysis of the development of the compound eye in ***Drosophila***. Trends Neurosci. 8:251.

White, K., 1980, Defective neural development in ***Drosophila melanogaster*** deficient for the tip of the X chromosome. Devel. Biol. 80:332.

Zipursky, S.L., Venkatesh, T.R., Teplow, D.B. and Benzer, S., 1984, Neuronal development in the ***Drosophila*** retina: monoclonal antibodies as molecular probes. Cell 36:15.

DEVELOPMENT AND FINE STRUCTURE OF THE NERVOUS SYSTEM OF LETHAL(1)OPTIC GANGLION REDUCED VISUAL MUTANTS OF DROSOPHILA MELANOGASTER

R. Naresh Singh*, Kusum Singh* and Douglas R. Kankel

Department of Biology, Yale University, P.O.Box 6666

New Haven, Connecticut 06511, United States of America

ABSTRACT

Drosophila melanogaster male larvae hemizygous for $l(1)ogre^{ljnl3}$ grow more slowly and moult later than their heterozygous ($l(1)ogre^{ljnl3}$/+) female siblings. At 18°C the $l(1)ogre^{ljnl3}$ male larvae undergo first and second instar moults around 70 and 118 hours, respectively, compared to heterozygous female siblings which moult first at 62 hours and the second time at 106 hours. Autoradiographic detection of tritiated[methyl-^{3}H]-thymidine incorporating cells in appropriately staged second and third instar larvae shows that l(1)ogre males have, on an average, less than one-third of the cells incorporating thymidine in the central nervous system (CNS) than do $l(1)ogre^{ljnl3}$/+ females of an equivalent stage.

Transmission electron microscopy of the $l(1)ogre^{ljnl3}$ larvae revealed : (i) a large number of electron-translucent vacuoles which seem to enlarge and ultimately fuse together to form channels, (ii) a variety of inclusion bodies and (iii) electron-dense cells (assumed to be degenerating) in the CNS. Males hemizygous for the viable allele $l(1)ogre^{vcb8}$ evidenced a phenotype intermediate to that of the $l(1)ogre^{ljnl3}$/Y males and the control l(1)ogre/+ females with respect to these characters.

INTRODUCTION

A previous study of mutations at the lethal(1)optic ganglion reduced (l(1)ogre) locus included a light microscopic analysis of the development of the optic lobe proliferation centres and the CNS in general in individuals carrying both viable and lethal alleles. In individuals hemi- or homozygous for any of the alleles, the proliferation centres were morphologically abnormal. Histological preparations showed that the regular repeating architecture characteristic of the wild-type is nearly absent. A clear division into lamina, medulla and lobula complex is rarely detectable. The external and internal chiasmata are missing or highly disarrayed (Lipshitz and Kankel, 1985).

---

* *Permanent address*: Molecular Biology Unit, Tata Institute of Fundamental Research, Homi Bhabha Road, Navy Nagar, Colaba, Bombay 400 005, India.

In addition, sections from the paraffin-embedded individuals either hemi- or homozygous for one of the lethal alleles showed the presence of relatively large extracellular vacuoles in the supra- and suboesophageal ganglia and thoracic nervous system. Defects outside the optic lobes were not seen in comparable sections from individuals (either larvae or adults) carrying the viable allele $l(1)ogre^{vcb8}$)Lipshitz and Kankel, 1985).

By using a temperature sensitive heteroallelic combination, $l(1)ogre^{ljnl3}/l(1)ogre^{vcb8}$, it was found that the lethal phase of the l(1)ogre gene extends from the middle of third instar to late third instar and perhaps into early pupal period (Lipshitz and Kankel, 1985). However, histological data gathered in these same studies clearly showed an effect of l(1)ogre mutations on the optic lobes as early as the second instar.

The analysis of gynandromorphs suggested that the morphological abnormalities were autonomous to the extent that the morphology of regions which were overwhelmingly wild-type genotypically was that of the wild-type, while that of tissues which were overwhelmingly mutant in genotype was that of the mutant. Because there is lack of a clear-cut phenotype at the cellular level, it could not be concluded that the mutations are in fact cell-autonomous (Lipshitz and Kankel, 1985).

Based on the previous results, Lipshitz and Kankel (1985) had suggested the working hypothesis that this locus has a prominent and perhaps rather specific effect on that subset of the larval CNS which is destined for the production of the imaginal-specific portion of the adult CNS. The imaginal-specific subset is defined as that portion of the nervous system which originates by mitosis during the larval period but which plays no known functional role during the larval instars. This includes the entirety of the adult optic lobes and all the progenitors of the division of the so-called giant neuroblasts of the larval brain and ventral ganglia.

The aims of the present study have been : (i) To analyze the mitotically active population of cells in the CNS of l(1)ogre mutants; these are the progenitors of the imaginal-specific subset of the adult CNS. (ii) To examine the fine structure of the CNS of l(1)ogre individuals with the hope of elucidating how the mutant phenotype comes into being. To achieve the first objective the dividing cells of the larval CNS were labelled with [methyl-^{3}H]-thymidine and identified by autoradiography, after fixation and sectioning. For the attainment of the second objective, we used transmission electron microscopy to examine the fine structure of the CNS and a few other tissues and organs during several developmental stages (e.g. imaginal discs and muscles in larvae and retina and muscles in pharate adult or the imago).

MATERIALS AND METHODS

Fly Culture

Drosophila melanogaster were cultured at 18°C on cornmeal-molasses-agar-yeast medium (Doane, 1967) supplemented with fresh yeast.

Genetic Strains Used

The strains used were $l(1)ogre^{ljnl3}$ cm v/Binsnscy;ywl(1)$ogre^{vcb8}$/FM6; y w $l(1)ogre^{vcb8}$/C(1)A,y/Y and wild type, Canton Special (CS).

Larval Growth Measurements

Since hemizygous l(1)ogre males are known to have a prolonged post-embryonic developmental period (Lipshitz and Kankel, 1985), it was necessary for the comparison of mitotically active populations in mutant and wild type to determine with some precision the times at which l(1)ogre homozygous and hemizygous individuals were at comparable developmental stages. Specific stages such as hatching from eggs, first and second larval moults and pupariation may be treated as landmarks of nearest equivalence in the overall ontogeny of individuals of different genotypes with different growth rates.

Females from the $l(1)ogre^{ljnl3}$ cm v/Binsnscy stock were allowed to lay eggs at 18°C. Batches of larvae which hatched during 1 hour interval were collected, and each batch allowed to continue development at 18°C in a small food vial. At varying times after hatching, the larvae from a given vial were collected, sexed, hemi- and homozygous Binsnscy individuals discarded. The remaining ones were examined under a compound microscope at the minimum magnification of 10x12.5 for morphological features such as mouth-hooks and the anterior spiracles for estimating the developmental stage. Transition to the next larval instar was timed by these observations (Bodenstein, 1950).

Autoradiography

Larvae were grown at 18°C, and individuals of appropriate stage and genotype collected. A batch of about 20-30 $l(1)ogre^{ljnl3}$/Y or $l(1)ogre^{ljnl3}$/+ larvae were fed upon 1.5 ml of agar-sucrose-yeast extract-inactivated yeast medium (White and Kankel, 1978) containing 100 uCi/ml [methyl-^{3}H]-thymidine (50 Ci/mole specific activity, New England Nuclear). At the end of a six hour feeding the larvae were etherized, nicked at the posterior and anterior extremities and fixed with alcoholic Bouin's fixative overnight at 5°C. Specimens were dehydrated with a graded series of ethanol, cleared with xylene and embedded in 'Paraplast'. Five µm thick paraffin sections were made on a Sorvall JB4 Porter Blum Microtome equipped with water trough, where specimens were advanced manually by a micrometer for more uniform thickness of the sections in the ribbon. Sections were de-waxed with xylene and the autoradiography essentially done according to the method of White and Kankel (1978), using Kodak NTB2 nuclear emulsion.

Emulsion coated slides were exposed for 2-3 weeks in the dark at 5°C. They were subsequently developed at 15°C with Kodak D19 developer for 4 minutes, rinsed with water for 30 seconds, fixed with Kodak fixer without hardner for 4 minutes and washed with water for 10 minutes. Sections were stained with a solution of 0.01% toluidine blue, 0.05% methylene blue and 0.05% borax ($Na_2B_4O_7.10H_2O$), dehydrated with 70 through 99% ethanol, dried overnight at room temperature then on a warm slide drier at 50°C and finally covered with a layer of 'Cytoseal-60' and glass cover-slip.

In the CNS of larvae, counts of cells with a cluster of silver-grains in the emulsion over their nuclei were made at 312x with a microscope fitted with a square-mesh graticule in the eye-piece. Counts were made from all sections of a given larva. Since sections were 5 µm thick, some cells may span more than one section and consequently may be counted more than once. To obviate this difficulty, we took two approaches. First, grain clusters less than 1 µm in diameter were excluded from the count; the 1 µm cut-off is essentially arbitrary. Second, photomicrographs of serial sections were taken and the outline of tissue and labelled cells traced onto transparent plastic sheets such that the contributions from different sections could be distinguished. Labelled cells which overlap in adjacent sections and the total number of labelled cells were counted; the percentage of labelled cells which overlap in adjacent sections was found to be 8, 13, 20 and 22 in the samples that were examined. We conservatively estimate that the extent of error in our counts is on the order of 25%, i.e., our counts could be as much as 25% higher than the actual number of labelled cells. This should have no impact on our comparison of mutant with wild-type since the same error should obtain in both cases.

The spatial distribution of the tritium labelled cells within a single section of the larval CNS was recorded by tracing the outline of the cells and that of the CNS onto transparent plastic sheets with the aid of a Zeiss camera lucida at a magnification of 580x. Since the signal to background ratio was quite high, there was no difficulty or ambiguity in deciding whether a cell was labelled or not (Fig. 1).

Transmission Electron Microscopy

Newly emerged, second or third instar larvae were used. They were pre-cooled on ice to immobilize them, transferred to a few drops of Karnovsky's fixative (1965) on a glass slide and their posterior end nicked with a sharp razor blade. Such specimens, or ones in which the CNS was recovered by dissection, were transferred to 10 ml of the same fixative and fixation allowed to proceed for 6 hours at room temperature. The specimens were washed with 0.1 M sodium phosphate buffer of pH 7.4 with several changes over a period of 1 hour, and the anterior end of the larva was then nicked. Post-fixation was done with Dalton's chrome-osmium tetroxide fixative (Dalton, 1955) on ice until the entire specimen was coloured brown due to osmium; this was followed by a 1 hour incubation in the same fixative at room temperature. Specimens were washed with buffer, dehydrated with graded series of ethanol solutions and embedded in Spurr's low viscosity embedding mixture, or in Durcupan ACM (Fluka), in flat silicone rubber moulds, through the intermediate dilutions with propylene oxide. Polymerisation was done at 60°C overnight.

Adults were processed in a manner similar to larvae except that the specimens were sufficiently opened while submerged in a few drops of fixative by cutting a horizontal slit in the dorsal posterior region of the head capsule, severing one antenna, part of one eye, the entire proboscis, the cervical connective, appendages like legs and wings and the posterior tip of the abdomen. After fixation with Karnovsky's fixative the abdomen was also removed from the specimen. Pupae were processed like adults after removal of the pupal case.

Table 1 lists the strains and developmental stages used in the fine structure studies. Approximately 60-90 nm thick sections were cut using glass knives, collected on 3 mm diameter copper slots with 2x1 mm window pre-coated with 'Formvar', stained with an aqueous 4% uranyl acetate solution at 60°C for 5 minutes and with Reynolds' lead citrate at room temperature for 5 minutes (Reynolds, 1963). Subsequent examination was with a Philips 300 or a Zeiss 10 A electron microscope. Kodak Ester thick based 4489 film 8.3x10.2 cm sheets were used for photography.

About 700 electron micrographs were made for these studies. Each category of observation is defined on the basis of sections taken from 2-4 individuals. A total of forty-eight individuals have contributed tissue to this fine structure analysis.

RESULTS

During the development of the larva,structures of the anterior spiracles and mouth-hooks were found to be reliable criteria for following the transition to the next larval instar. The results based on these morpho-

Table 1. Developmental stages at which the different strains of <u>Drosophila melanogaster</u> were used for transmission electron microscopy and some identifying characteristics.

	Experimental		Control	
Strain genotype	Larval instar	Adult (ad.) or pharate adult (pad.)	Larval instar	Adult (ad.) or pharate adult (pad.)
$\frac{l(1)ogre^{ljnl3}\ cm\ v}{Binsnscy}$	II: y^+♂ III: y^+♂ Black mouth-hook	(pad.) Orange-eye ♂ Head & Thorax	II: y^+♀ III: y^+♀ Black mouth-hook	(pad.) Bar, red-eye ♀ & White-eye ♂ Head & Thorax
$\frac{y\ w\ l(1)ogre^{ljnl3}}{FM6}$	I y ♂ Brown mouth-hook		I y/y^{31d} ♀ Black (or nearly so) mouth-hook	
$\frac{y\ w\ l(1)ogre^{vcb8}}{C(1)A,\ y/Y}$	II y ♂ Brown Mouth-hook	(ad.) White-eye ♂ Head & Thorax	II y ♀ Brown mouth-hook	(ad.) Normal-eye ♀ Head & Thorax
$\frac{y\ w\ l(1)ogre^{vcb8}}{FM6}$	II y ♂ Brown mouth-hook	(ad.) White-eye ♂ Head& Thorax	II y/y^{31d}♀ Black(or nearly so) mouth-hook	(ad.) Bar-eye ♀ Head & Thorax

The genetic notations are mainly adapted from Lindsley and Grell (1972) and Lipshitz and Kankel (1985).

logical features are given in Fig.1, where the l(1) ogre^{ljnl3}/+ female larvae served as controls for the hemizygous l(1)ogre^{ljnl3} males.

For control females, the first to second instar moult occurs at around 62 hours and the second to third instar moult around 106 hours. Corresponding moults in hemizygous l(1)ogre^{ljnl3} male larvae are 70 and 118 hours, respectively. In addition, the moulting of individuals in the population is spread over a longer time period in l(1)ogre^{ljnl3}/Y males compared to l(1)ogre^{ljnl3}/+ females (Fig. 1).

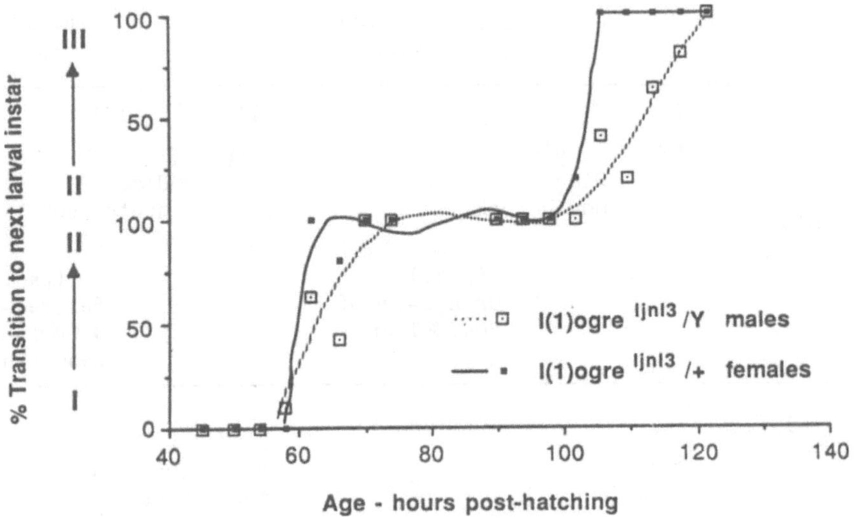

Fig. 1. Development of l(1)ogre^{ljnl3}/Y male and l(1)ogre^{ljnl3}/+ female larvae at 18°C. Male and female larvae were characterized by the presence or absence of testes. The age of the larva is given in hours and is counted from the time of hatching from the egg. Curves were fitted to the data using a higher order polynomial function and the graph produced on a Laser Writer Plus by the presentation graphics software "Cricket Graph", version 1.1 running on a Macintosh Plus computer.

Autoradiography

Actively dividing cells in the early second and third instar larvae were labelled with [methyl-^{3}H]-thymidine as described in Materials and Methods. These stages were chosen because of the precision with which they could be identified, based on the reliability of defining newly moulted animals by the structure of their mouth hooks and spiracles (Bodenstein, 1950). Dividing cells were subsequently detected in serially sectioned material by autoradiography. Cell counts were made directly in a compound microscope, and drawings of the CNS and the positions of dividing cells were made with a camera lucida. It is clear that the number of thymidine-incorporating cells in l(1)ogre^{ljnl3}/Y males is greatly reduced relative to l(1)ogre^{ljnl3}/+ females (Fig. 2). Counts of the labelled cells in such preparations and the average number of cells with labelled nuclei per CNS are given in Table 2 (last column).

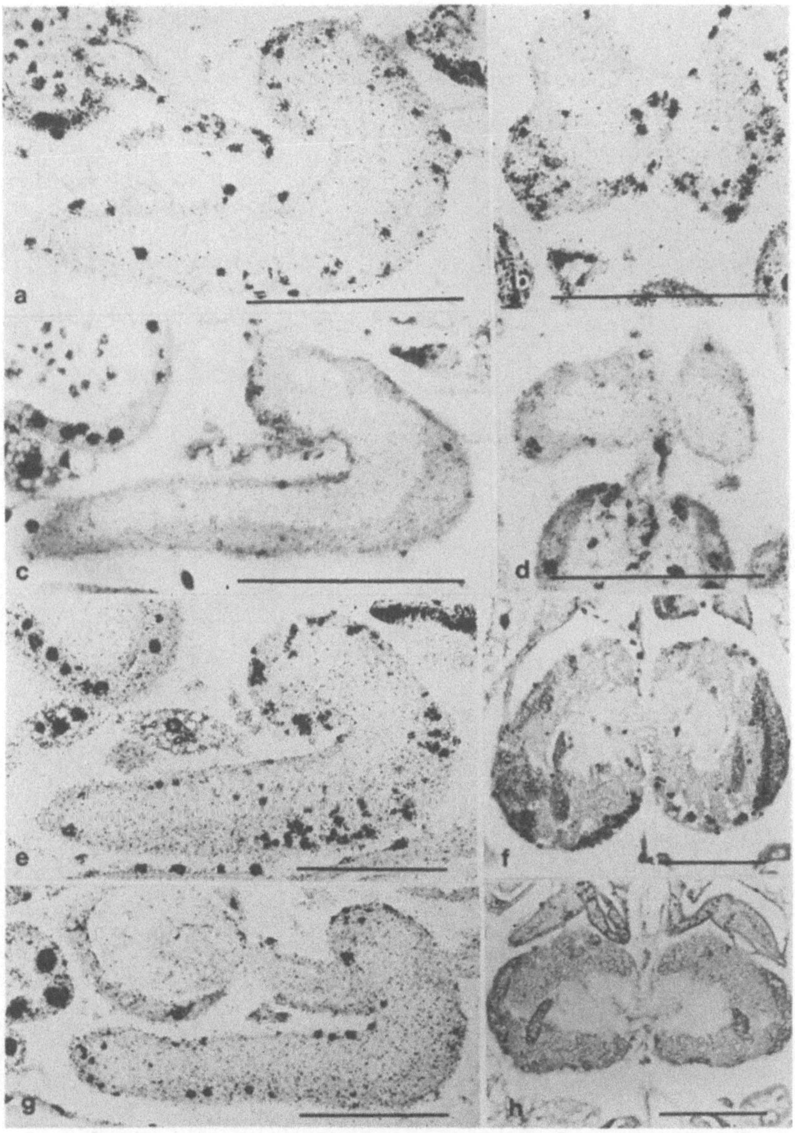

Fig. 2. Autoradiographs of sections of CNS of l(1)ogre[ljnl3] cm v /+ and l(1)ogre cm v/Y larvae grown at 18°C. The female larvae served as controls for males. Larval age is counted from the time of hatching from the egg; the age given in the following is the age at the beginning of a 6-hour period in which the larvae were allowed to feed on medium containing [methyl-^{3}H-thymidine, (a) Vertical section of 62 hour old female larva. (b) Horizontal section of 62 hour old female larva. (c) Vertical section of 70 hour old male larva. (d). Horizontal section of 70 hour old male larva. (e) Vertical section of 106 hour old female larva. Note the non-uniform distribution of grains in the CNS. (f) Horizontal section of third instar female larva (g) Vertical section of 118 hour old male larva. (h) Horizontal section of third instar male larva. Magnification bar = 100 µm.

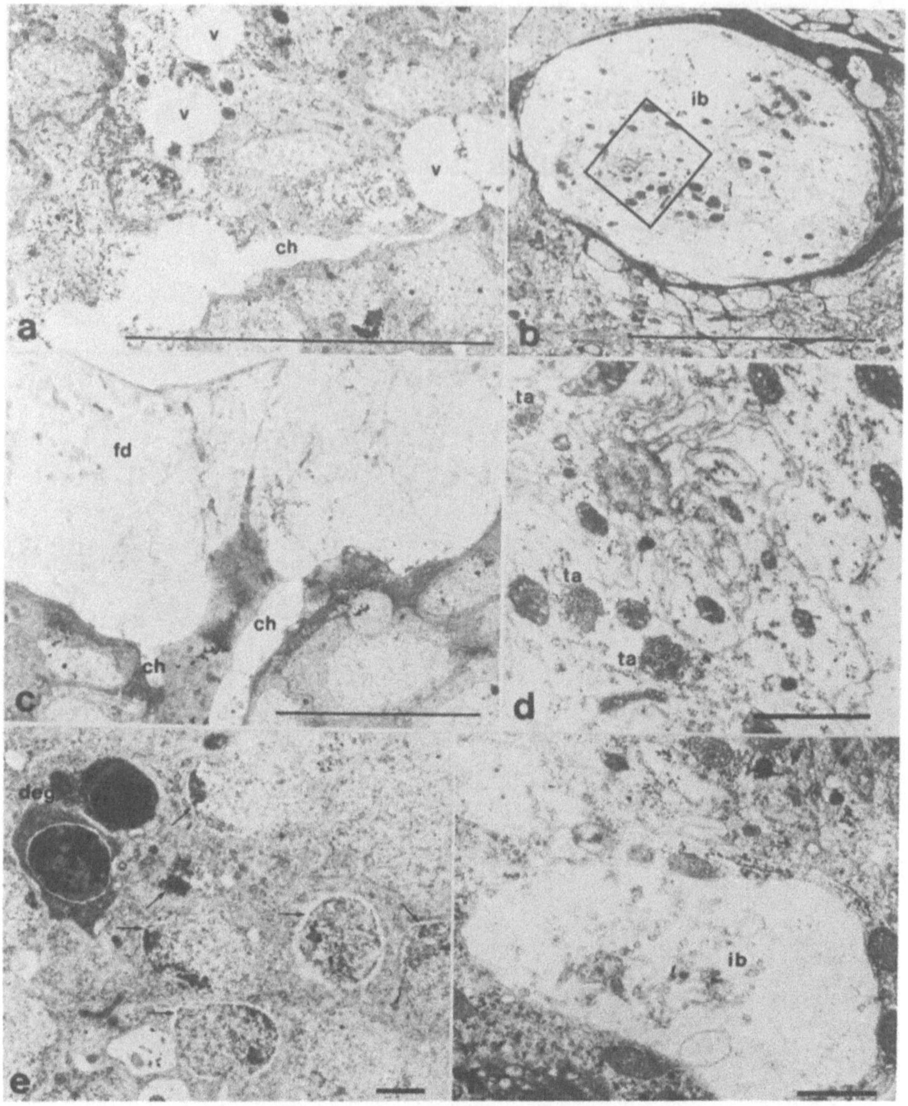

Fig. 3. Electron micrographs of sections through the larval brain hemispheres of l(1)ogre^{1jnl3} cm v/Y individuals. (a) Cells in the cortex with vacuoles v present which are translucent to electrons. Such vacuoles fuse together and form small channels ch. (b) and (f) Two typical inclusion bodies ib containing mitochondria, remnants of membrane and cellular components in the neuropiles of larval brain hemispheres. (c) Deposits of electron-lucent fluid fd around cortex at the dorsal junction of brain hemispheres. Two channels ch contiguous with the deposit are in field of view. (d) A higher magnification of the boxed region in Fig.3b showing cellular components and aggregates of tubule like structures ta. (e) Degenerating cells deg in the cortex of the brain hemisphere. Some of the other cells showing early signs of degeneration are indicated by arrows.

Thick magnification bar = 1 µm

Thin magnification bar = 10 µm.

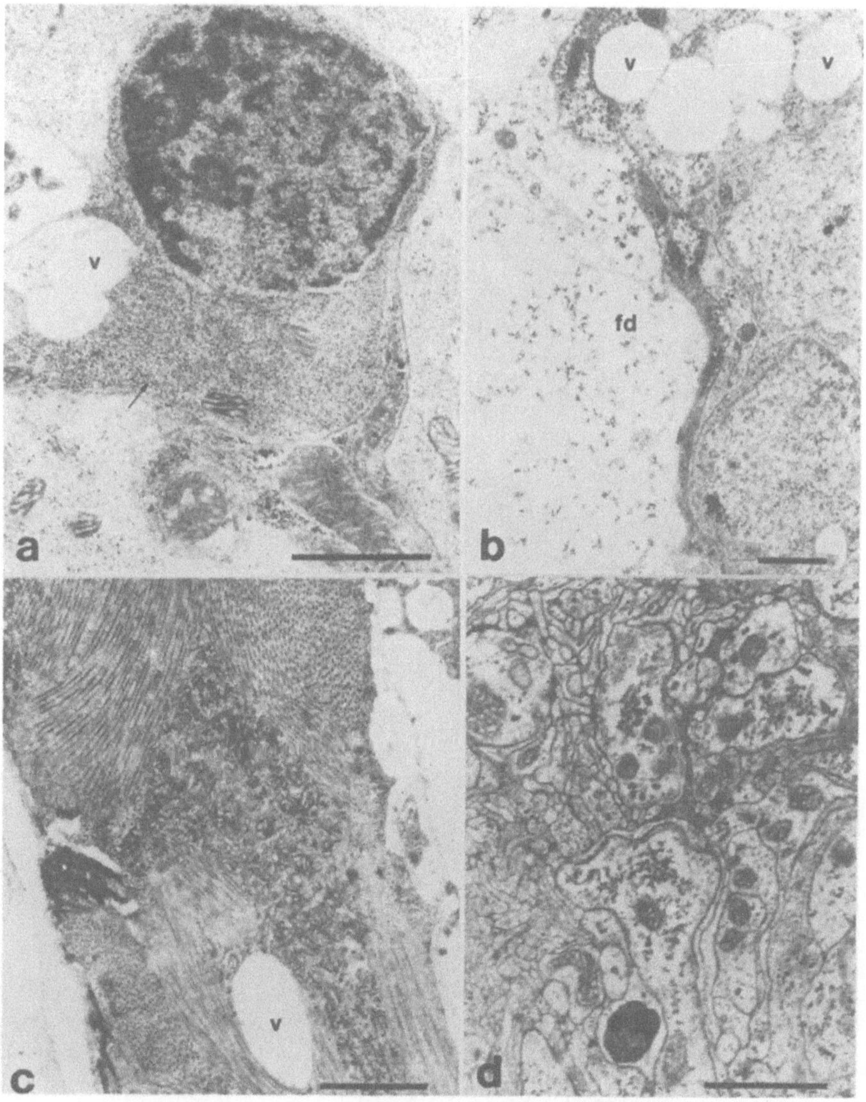

Fig. 4. Electron micrographs of the larval ventral ganglia and muscles of $l(1)ogre^{ljnl3}/Y$ and $l(1)ogre^{vcb8}/Y$ male larvae. (a)-(c) are from $l(1)ogre^{ljnl3}/Y$ individuals. (a) Degenerating cell (arrow) and vacuoles v in the cortex of the ventral ganglia. (b) Vacuoles v in the cortex of the ventral ganglia and accumulation of fluid fd around the cortex (as in brain hemispheres cf. Fig. 3c). (c) A vacuole v in the musculature of the oesophageal wall. (d) Ventral ganglia neuropile of a $y\ w\ l(1)ogre^{vcb8}$ male larva. Note the essentially normal appearance.
Magnification bar = 1 µm.

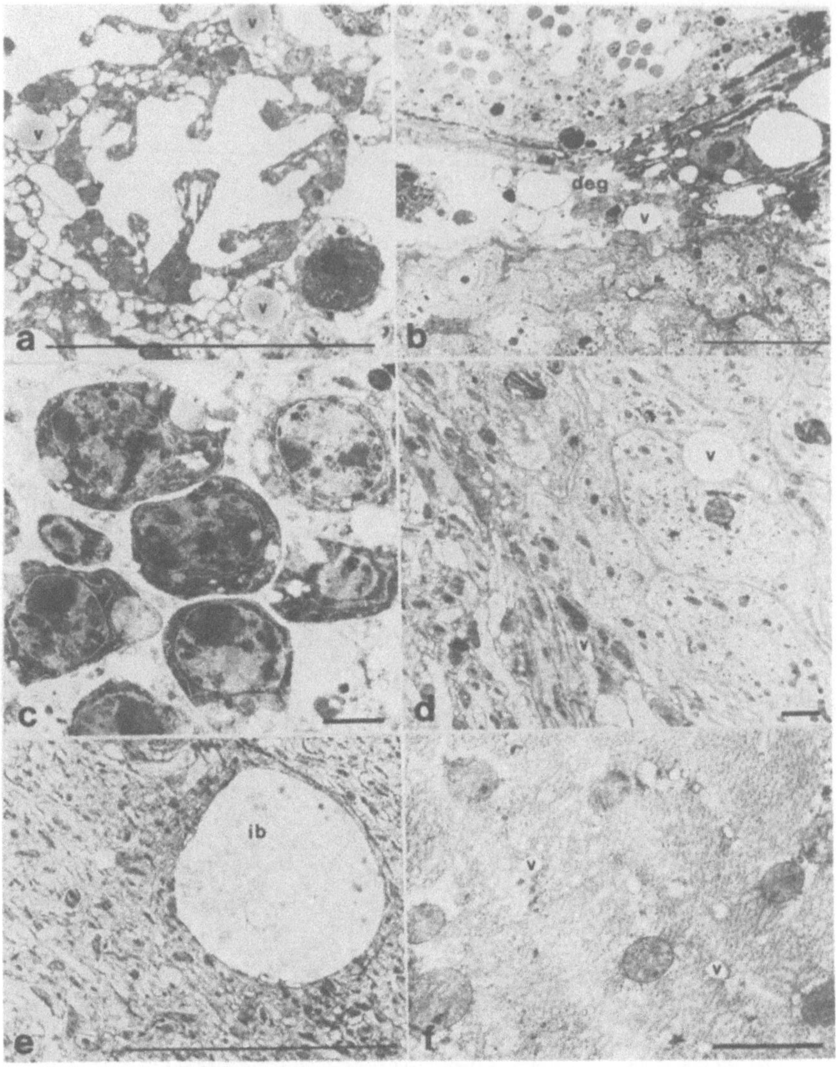

Fig. 5. Electron micrographs of the eye, brain and muscles in the head of pharate adults. (a) Retinula cells of a *l(1)ogre*^{ljnl3}*/Y* male. Vacuoles v are sometimes present. Note abnormal rhabdomeres (cf. Fig. 7a). (b) Vacuoles v and massive degeneration deg in the region between retina and lamina of a *l(1)ogre*^{vcb8}*/Y* male. (c) Degenerating cells in the medulla of a *l(1) ogre*^{ljnl3}*/Y* male. A few knife marks with 1-7 O'clock orientation are artifacts. (d) Cortex of an individual hemizygous for the viable allele *l(1)ogre*^{vcb8}. In such individuals, the frequency of vacuoles v is greatly reduced. (e) Brain neuropile of a *l(1)ogre*^{vcb8}*/+* female with a comparatively rare inclusion body ib containing membrane fragments. (f) Muscles of the head of a *l(1)ogre*^{ljnl3}*/Y* male with vacuoles, some marked v.
Thick magnification bar = 1 μm.
Thin magnification bar = 10 μm.

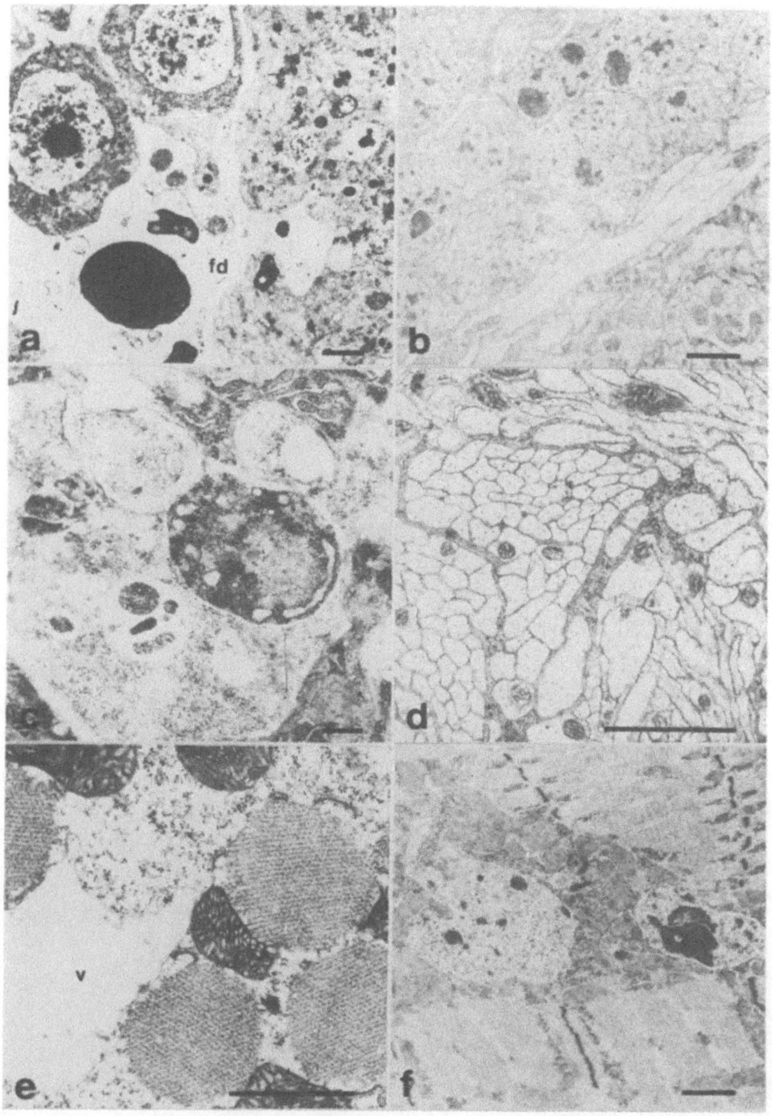

Fig. 6. Thoracic ganglia and muscles of pharate adults. (a) and (c) Cortex of male pupa hemizygous for $\underline{l(1)ogre^{ljnl3}}$ showing extensive degeneration and fluid fd accumulation. (b) Neuropile and cortex boundary of a $\underline{l(1)ogre^{vcb8}/+}$ female pharate adult which appears quite normal. (d) Neuropile of a $\underline{l(1)ogre^{vcb8}/+}$ female pharate adult which appears quite normal. (e) A vacuole v between thoracic muscle-fibres of a $\underline{l(1)ogre^{ljnl3}/Y}$ pharate adult. (f) Thoracic muscles of a $\underline{l(1)ogre^{vcb8}/Y}$ individual with little sign of degeneration.

Magnification bar = 1 µm.

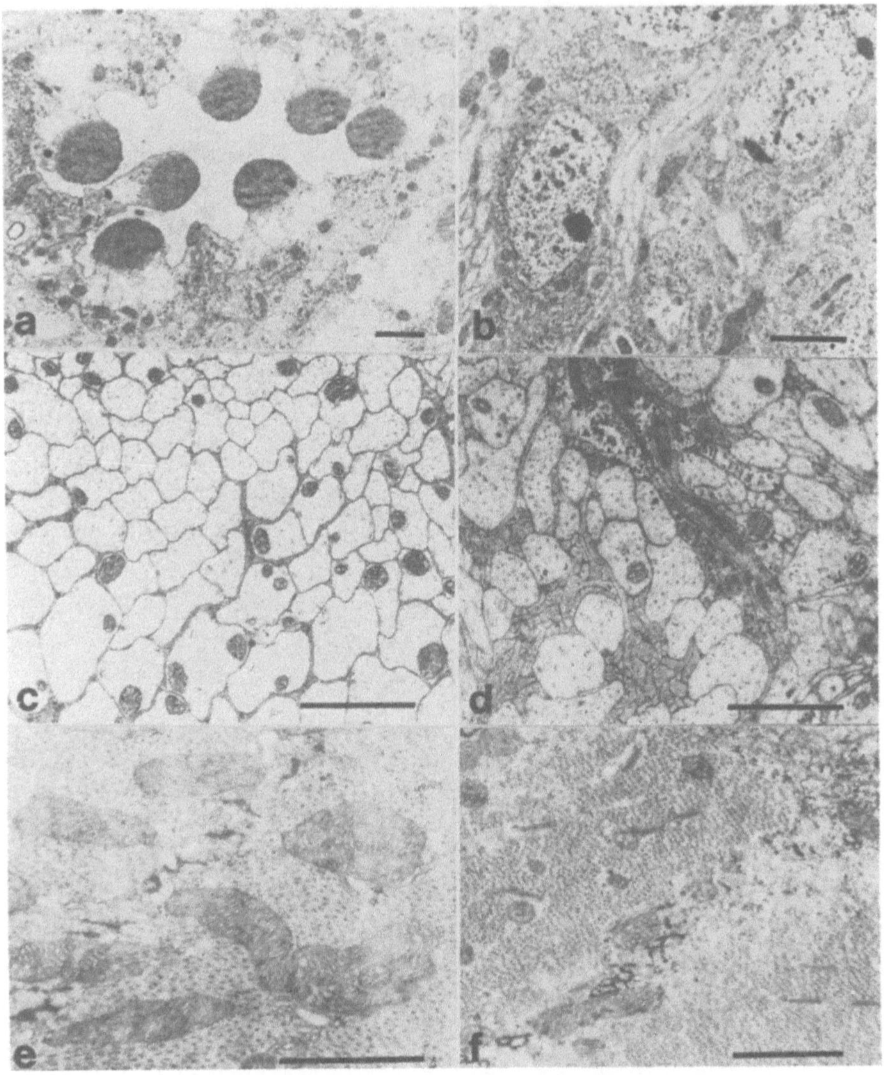

Fig. 7. Electron micrographs of the control female flies (Fig. 7a, c, e) and larvae (fig. 7b, d, f). (a) Retinula cells from a C(1)A, y/Y individual (cf. Fig. 5a). (b) Cortex of a second instar C(1)A, y/Y female (cf. Fig. 5d, e). (c) Brain neuropile of heterozygous l(1)ogre ljnl3/+ female pharate adult. (d) Neuropile of larval brain hemisphere of a C(1)A, y/Y female. (e) Head musculature of a C(1)A, y/Y adult. (f) Larval muscles of a $l(1)ogre^{ljnl3}/+$ female

Magnification bar = 1 μm.

Table 2. Counts of [methyl-^{3}H]-thymidine-incorporating cells in the CNS of second and third instar larvae of Drosophila melanogaster as detected by autoradiography. Female $l(1)ogre^{ljnl3}/+$ larvae grown at 18°C for 62 and 106 hours served as controls for $l(1)ogre^{ljnl3}/Y$ males grown for 70 and 118 hours, respectively. The age of the larvae is counted from the time of emergence from the egg. Note that there is no overlap in the populations of males and their female controls at either the early second (70/62 hrs) or early third (118/106 hrs) instars.

Larval genotype (age hours)	Sex	Counts of thymidine incorporating cells	Arithmetic mean
l(1)ogre/+ (62)	♀	599;810;500;741;833;486;1097; 836;626;535;473;816;816;1044.	730
l(1)ogre/Y (70)	♂	163;232;190;292;225;202;240; 249;267;221;241;316;274;261; 379;258;343;260.	256
l(1)ogre/+ (106)	♀	1769;1626;1482;1537;1760;1319; 1768;1380;1387;1166;1117;983; 1369;1240;1128;1715;1527;1606; 1458;1505.	1442
l(1)ogre/Y (118)	♂	331;367;461;729;415;406;473; 823;595;398;228;411;326;546; 313;387;316;898;472.	468

Analyses showed none of the data (Table 2) are normally distributed [probability plots Gnanadesiken, 1977)], and inspection shows that the counts from populations of mutants never overlap with their heterozygous controls. This makes statistical tests for significance of difference almost unnecessary, but we have nevertheless used two nonparametric tests, the Wilcoxan signed ranks test and the Kolmogorov-Smirnov two-sample test (Hollander and Wolfe, 1973), to establish that data from mutants and heterozygotes are not drawn from the same population.

Fine Structure Analysis

Electron microscopy shows the presence of electron-translucent vacuoles within the cells of the brain hemispheres of $l(1)ogre^{jnl3}/Y$ larvae (Fig. 3a, c). Such vacuoles enlarge and ultimately fuse to form channels which become continuous with the periphery of the brain (Fig. 3c). Brains of $l(1)ogre^{ljnl3}/Y$ larvae were also found to have a variety of inclusion-bodies containing cellular components such as mitochondria and membrane-fragments (Fig. 3b, d, f). In addition, many electron-dense cells are present (Fig. 3e); this appearance is typical of degenerating cells, and we assume it to be indicative of degeneration here as well. Similar abnormalities were found in the ventral ganglia of larvae (Fig. 4a, b, d) as well as in both the brain (Fig. 5a-e) and ventral ganglia of pupae (Fig. 6a,c). Such abnormalities in $l(1)ogre^{ljnl3}/+$ female controls either do not occur or are very rare (Fig. 7a-d). As expected individuals hemizygous for $l(1)ogre^{vcb8}$, the viable allele, have a phenotype intermediate between that of l(1)ogre^{ljnl3}/Y males and l(1)ogre^{ljnl3}/+ females (Fig. 6b,d).

We have carried out a preliminary examination of a small number of very early first instar (1 hour post-hatching) l(1)ogre^{ljnl3}/Y males and l(1)ogre^{ljnl3}/+ females. We see no evidence of the presence of the intracellular vacuoles which become so prevalent at later stages but do see a low frequency of cells which we consider degenerate. The mutant and the heterozygote at this stage are indistinguishable.

In addition to the CNS, other tissues in l(1)ogre^{ljnl3}/Y larvae and pharate adults, such as muscles, salivary glands and imaginal discs, occasionally have vacuoles and degenerating cells (Figs. 4c,5f,6e); once again, such abnormalities are extremely rare in controls l(1)ogre^{ljnl3}/+, Fig. 6f,7e,f and are present, but infrequent, in l(1)ogre^{vcb8}/Y individuals.

DISCUSSION

Results from the measurements of growth, from autoradiography and from electron microscopy reveal the following:

(a). As previously shown by Lipshitz and Kankel (1985), overall development in l(1)ogre^{ljnl3}/Y larvae is slower than in l(1)ogre^{ljnl3}/+ controls; additionally, we show that moults are less synchronous in mutants than in heterozygous controls. The first and second moults in l(1)ogre^{ljnl3}/Y larvae occur around 70 and 118 hours, compared to control larvae, which moult at 62 and 106 hours, respectively.

(b). Actively dividing cells in the CNS of l(1)ogre^{ljnl3}/Y larvae outside of the optic lobes are about one-third as numerous as those in the control larvae of equivalent developmental stages.

(c). A large number of electron-translucent vacuoles were found in the cells of the developing optic lobe and elsewhere in the CNS of l(1)ogre^{ljnl3}/Y individuals. However, outside of the optic lobes, we were unable to distinguish imaginal-specific cells from those of the functioning larval CNS.

(d). A variety of inclusion bodies are seen within cells of both the developing optic lobes and other regions of the CNS of l(1)ogre^{ljnl3}/Y individuals. As in (c) above, we were unable to distinguish imaginal-specific from functioning larval cells.

(e). Electron-dense cells, which we assume to be degenerating, are seen in extremely large numbers in the medulla and lobula regions of pupae and in the posterior end of the ventral ganglia of larvae of the l(1)ogre^{jnl3}/Y genotype. They are also found in other regions of the CNS, but it is our impression that they exist at a somewhat lower frequency.

(f). Individuals hemizygous for the l(1)ogre^{vcb8} allele show the same qualitative phenotypes as l(1)ogre^{ljnl3} hemizygotes. There is always, however, a reduction in the severity of the abnormalities seen as well as a substantial reduction in the frequency of those phenotypes seen outside the optic lobes.

Among the most prominent of the l(1)ogre phenotypes is the pronounced reduction in size and grossly abnormal organization of the optic lobes (Lipshitz and Kankel, 1985). To a significant extent, this may be a consequence of the reduction in size of the optic lobe primordia themselves (Lipshitz and Kankel, 1985). However, we believe that the massive cell degeneration in the developing medulla and lobula complex reported here contribute to these abnormalities as well.

There is also a pronounced effect of mutations at the l(1)ogre locus outside of the optic lobes; this was previously characterized at the light microscope level as the appearance of a large number of (presumably) extracellular vacuoles in various subsets of the CNS in the pharate adults (Lipshitz and Kankel, 1985). Our current results using individuals hemizygous for $l(1)ogre^{jnl3}$ indicate a significant reduction in the number of thymidine-incorporating cells. We think that this leads to a deficit in the number of cells available for the formation of the imaginal-specific portion of the adult CNS and, at least in part, that this explains the profound morphological abnormalities seen. We also draw attention to the presence of many cells containing vacuoles filled with electron-translucent material; these vacuoles seem to coalesce and ultimately lead to the destruction of the cell in which they are found. This also, in part, may explain the profound structural abnormalities seen in the light microscope. On the basis of the fine structure morphology these vacuole-rich cells resemble those seen in patients suffering from a number of disorders in the sphingolipid metabolism (Suzuki, 1979). Most inherited lipid storage diseases are now considered to be consequence of abnormalities in lysosomal hydrolases (Johannessen, 1979). At least 100 inherited metabolic disorders with ocular manifestations are at present known (Kenyon and Green, 1979). To know if mutations at the l(1)ogre locus are similar to any of the ocular disorders in humans would be of considerable interest.

ACKNOWLEDGEMENTS

The work was supported by NIH Grant (NS11788) to D.R.K. We thank Barry Piekos for the courtesy extended to us while using the E.M. Laboratory. We thank Mary Brown, Karen Stark and Toshi Watanabe for reading and commenting on the manuscript. K.S. and R.N.S. thank all members of the laboratory for making their stay pleasant and are particularly grateful to Sara Jones and Gregory Fitzgerald for making them familiar with the setup of the laboratory and help on many occasions. R.N.S. thanks Dr. Rajan Nayyar, Loyola University, for helpful suggestions.

REFERENCES

Bodenstein, D. 1950. The postembryonic development of Drosophila, pp. 275-367. In M.Demerec (ed.) Biology of Drosophila. John Wiley & Sons, Inc., New York.

Dalton, A.J. 1955. A chrome-osmium tetroxide fixative for electron microscopy. Anat. Record 121:281 A.

Doane, W.W. 1967. Drosophila, pp. 219-244. In F.Wilt and N. Wessels (eds.) Methods in Developmental Biology. Crowell, New York.

Gnanadesikan, R. 1977. Methods for Statistical Data Analysis of Multivariate Observations. John Wiley & Sons, New York.

Hollander, M. and Wolfe, D.A. 1973. Nonparametric Statistical Methods. John Wiley & Sons, New York.

Johannessen, J.V. 1979. Electron Microscopy in Human Medicine, Vol. 6, p.207, McGraw-Hill International Book co., London, New York.

Karnovsky, M.J. 1965. A formaldehyde-glutaraldehyde fixative of high osmolarity for use in electron microscopy. J.Cell Biol. 27:137 A.

Kenyon, K.R. and Green, W.R. 1979. Inborn lysosomal diseases, pp. 267-283. InJ.V.Johannessen (ed.) Electron Microscopy in Human Medicine, Vol. 6. McGraw-Hill International Book Co., London, New York.

Lindsley, D.L. and Grell, E.H. 1972. Genetic Variations of Drosophila melanogaster. Carnegie Institution of Washington, Publication No. 627.

Lipshitz, H.D. and Kankel, D.R. 1985. Specificity of gene action during central nervous system development in Drosophila melanogaster: Analysis of the lethal(1)optic ganglion reduced locus. Dev. Biol.108:56-77.

Reynolds, E.S. 1963. The use of lead citrate of high pH as an electron opaque stain in electron microscopy. J. Cell Biol. 17:208-221

Suzuki, K. 1979. Metabolic diseases, pp. 3-53. In J.V.Johannessen (ed.) Electron Microscopy in Human Medicine, Nervous System, Sensory Organs, and Respiratory Tract, Vol.6. McGraw-Hill International Book Co., London, New York.

White, K. and Kankel, D.R. 1978. Patterns of cell division and cell movement in the formation of the imaginal nervous system in *Drosophila melanogaster*. Dev. Biol. 65:296-321.

CRITICAL DEVELOPMENTAL PHASES IN THE ONTOGENY OF HUMAN LATERAL GENICULATE NUCLEUS DURING PRENATAL LIFE

Shashi Wadhwa and Veena Bijlani

Department of Anatomy
All-India Institute of Medical Sciences
New Delhi-110029, India

ABSTRACT

Abnormalities of external and internal environment during prenatal and early postnatal life have been known to result in developmental disorders. It is evident that irreversible microstructural and chemical changes result in permanent functional teratological sequelae and dysfunction of information processing. The present study on the human lateral geniculate nucleus (LGN) which is the first relay station in the visual pathway, was undertaken with a view to highlight the critical developmental periods which render this nucleus susceptible to environmental conditions. Human embryos, fetuses and premature infants ranging in age from 8 to 37 weeks of gestation, belonging to both sexes were obtained by hysterotomy and autopsy. The developing neural substrate was analysed using rapid Golgi, electronmicroscopy, immunohistochemical methods and quantitative morphometry. Cell density in the human LGN increases from 8 to 12 weeks of gestation with a subsequent decline continuously upto 37 weeks. It thus appears that the events of cell proliferation and migration resulting in the formation of LGN occur upto 12 weeks and constitute a critical phase. Events of synaptogenesis and dendritic proliferation begin from 13-14 weeks onwards. At 16-17 weeks the maximum number of retinal fibres have been quantified in the optic nerve, and numerical density and percentage of gamma-aminobutyric acid immunopositive neurons is highest. This probably forms another critical period of susceptibility. Between 21 and 24 weeks, significant regressive event occurs resulting in the loss of optic nerve axons and simultaneous formation of laminae in the LGN. This comprises yet another sensitive phase in the development of LGN. From 24 weeks onwards morphological differentiation of cells is apparent since different varieties of cell types described in the adult become recognisable and there is continuous dendritic proliferation and modelling. During this period many varieties of synaptic contacts are increasingly identified. From these events it is evident that there are multiple sensitive periods of susceptibility in the prenatal development of lateral geniculate nucleus.

INTRODUCTION

Developmental disorders are known to result from alterations in the normal internal and external environment during prenatal and early postnatal life. The developing nervous system demonstrates considerable plasticity which while allowing for adaptability can also be influenced by changes in its environment and adverse environmental factors can disrupt the normal developmental process. The visual system is an extensively studied region in this respect. The concept of a critical period in the visual system development evolved with the visual deprivation studies of Hubel and Wiesel (1970). Using different modes of sensory disruption, such as rearing in the dark (Blake and Di Gianfilippo, 1980; Timney et al., 1978), monocular deprivation (Cynader et al., 1980; Wilkinson, 1980), and surgically produced unilateral convergent or divergent strabismus (Jacobson and Ikeda, 1979; Von Grunau and Singer, 1980) in the cat, it has been observed that between 3 weeks to 6 months postnatal, the experiential alteration has a critical and permanent effect. Similarly in the mcnkey, the critical period concerned with establishment of normal binocular vision is observed upto nine months. In man, the first seven years of life are regarded as a clinically sensitive period during which the visual system is affected by alterations in visual environment (Assaf, 1982). Recent studies have shown that there is not a unitary sensitive period for the whole visual system rather there are different sensitive periods of development for various levels of visual pathway. Behavioural studies have demonstrated that psychophysical functions of rods and cones, monocular spatial vision and binocular spatial vision have separate short sensitive periods of development (Harwerth et al., 1986). There also exists a brief critical period of time during which the corpus callosum interacts with the developing visual system (Elberger, 1984). While the critical periods of visual development influenced by visual experience have been estimated by behavioural studies, analysis of the neurological mechanisms underlying these periods have required invasive experimental laboratory techniques. Although a beginning has been made in the use of non-invasive evoked potential methods to investigate the human, anatomical techniques to analyse structural development and pathological material are helpful in comparing, correlating and extrapolating data obtained from the experimental animals to humans. The influence of visual experience on structural development of the human visual system and the close temporal relationship with functional changes measured experimentally or clinically during postnatal life have been discussed by Hickey (1981) and Garey (1984).

The role of binocular competition is also evident in the development of visual pathways during the prenatal period as demonstrated by the monocular deprivation studies in fetal monkey (Rakic, 1981) and cat (Williams and Chalupa, 1983). Removal of an eye during a particular period in fetal life is observed to result in markedly expanded projections to lateral geniculate nucleus from the intact eye. A number of studies in the recent years have been directed, to understand the events in fetal life which give rise to mature patterns of connections in the mammalian visual system (Rakic, 1977; Shatz, 1983; Williams, 1983; Bunt et al., 1983).

Most of the studies have been conducted on experimental animals as the cat and rat, in whom the visual pathways develop mostly postnatally. Only recently the monkey, a primate closely resembling man, has been used for vision experiments and efforts have been directed to analyse its developmental morphology. The prenatal developmental studies on visual system in man are few. Some aspects of prenatal development of the lateral geniculate nucleus, which is the principal thalamic relay station in the visual pathway, are reported. These include displacement and rotation (Rakic, 1974), time period of laminae formation (Dekaban, 1954) and sequence of cellular maturation using Nissl stained preparations (Hitchcock and Hickey, 1980; Damyanti et al., 1983). However, events like changes in neuronal density, dendritic development, maturation of interneurons and relay neurons, mode of development of afferent innervation, synaptogenesis and neurochemical maturity are not known. The present study attempts to analyse these aspects to know the time schedule of events of proliferation, migration, synaptogenesis, dendritic proliferation and modelling, axonal elimination and lamination with a view to highlight the critical developmental periods which render this nucleus susceptible to environmental influences.

MATERIAL AND METHODS

The present study was conducted on lateral geniculate nuclei (LGN) taken from fifty-five human embryos, fetuses and premature infants. The fetal material ranging in age from 8 to 37 weeks of gestation, belonging to both sexes, was obtained by hysterotomy and autopsy*. The time period elapsed between the death of the fetus and tissue fixation varied from 30 to 45 min in the case of hysterotomy specimens and 90 min to 4 h in the case of autopsies. The age of the fetus was estimated by measurement of crown-rump length (Hamilton et al., 1962), crown-heel length and biparietal diameter.

Pieces of diencephalon with LGN taken from different gestational ages were stained with rapid Golgi technique and blocked in celloidin. Coronal sections of 100-120 μm thickness were cut and camera lucida drawings of neurons with completely stained arbors were made.

Slices of LGN 2-3 mm thick and optic nerves from fetuses of different gestational ages were fixed by immersion in modified Karnovsky's fixative and processed for electronmicroscopy. Semithin sections (0.5-1.0 μm) from LGN's of different age periods were subjected to immunohistochemical localisation of gamma-aminobutyric acid (GABA). Adjacent sections were stained with toluidine blue. Numerical densities of the GABA positive and GABA negative neurons as well as composite cell population were stereologically estimated. Fibre counts of the optic nerves at different gestational ages were made from electronmicrographs.

RESULTS

The mean volumetric cell density of 301×10^3 per cu.mm at 8 weeks rises to 429×10^3 per cu.mm at 12 weeks and at 15-16 weeks it declines to 287×10^3 per cu.mm. Thereafter there is a continuous decline in the neuronal density upto 37 weeks of gestational period (Fig. 1). The precipitous fall between 15 and 17 weeks is due to the glial cells not being counted from the age of 17 weeks onwards.

The GABA immunopositive cells constitute 1% of the cell population at that age. At 17 weeks, however, there is a peak rise in the numerical density of GABA immunopositive cells which form 16% of the total neurons while there is a fall in the overall numerical density of GABA negative neurons. Subsequently the percentage of GABA positive neurons falls (Fig. 2). Details of these observations have been published in a previous paper (Wadhwa et al., 1988).

---

*Consent from patients and ethics committee were taken.

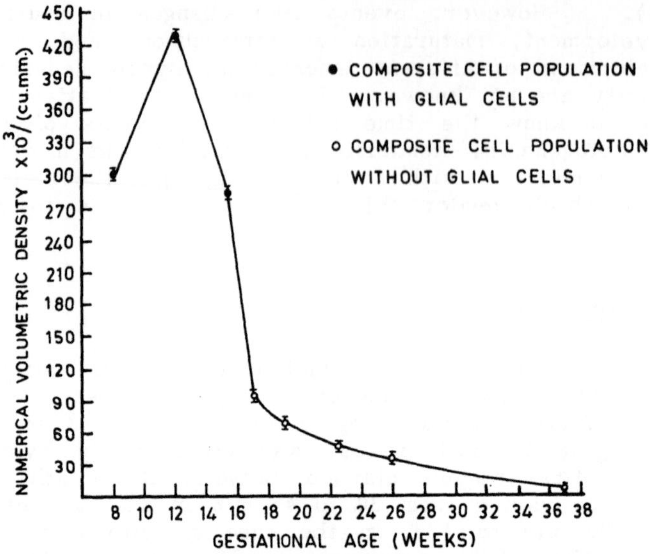

Fig. 1. Graphic representation of the volumetric cell density in the human lateral geniculate nucleus at different gestational ages.

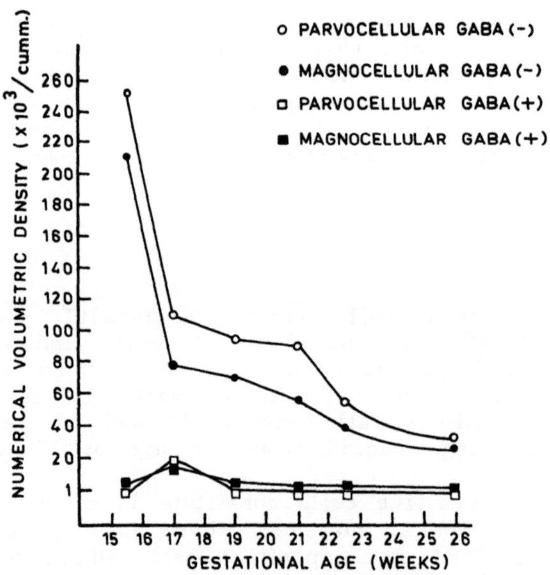

Fig. 2. Graph showing the comparative numerical densities of GABA positive and GABA negative neurons in the parvo- and magno-cellular regions at different gestational ages.

Human optic nerve fibre counts made at different gestational age periods show an increase in the total number of fibres upto 16-17 weeks with the count being 7.24 million. Thereafter progressively lower estimates of fibres are obtained, being 2.2 million at 26 weeks. In the period between 16-17 and 19 weeks, 2.8 million fibres and between 19 and 26 weeks another 2.2 million axons are eliminated (Fig. 3). Around 23-24 weeks lamination is observed in the LGN.

Golgi studies reveal a relatively homogeneous population of bipolar and few multipolar neurons at 15-16 weeks (Fig. 4). From 24 weeks onwards increasing number of multipolar neurons acquire the shape and form of various morphological types of neurons described in the adult (Courten and Garey, 1982) (Fig. 5). Between 27 and 37 weeks neurons undergo modelling with increase in the number of dendrites emanating from the soma and the dendritic branches (Fig. 6). Detailed observations are reported elsewhere (Wadhwa and Bijlani, 1988).

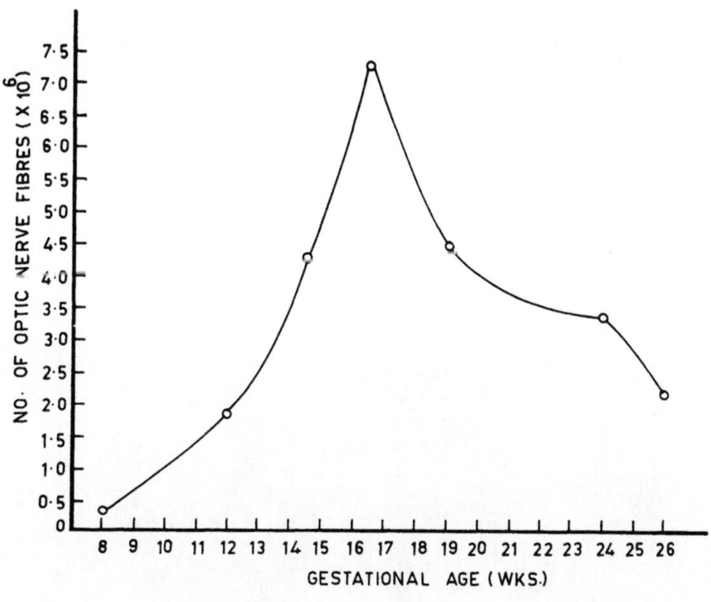

Fig. 3. Graph showing the rise and decline in number of optic nerve fibres with increasing gestational age.

Ultrastructurally synapses are observed from 13-14 weeks onwards, presynaptic terminals of which appear to be of retinal origin. At 17 weeks synaptic terminals of cortical origin are identifiable. From 26 weeks onwards many varieties of synaptic contacts are increasingly identified (Fig. 7).

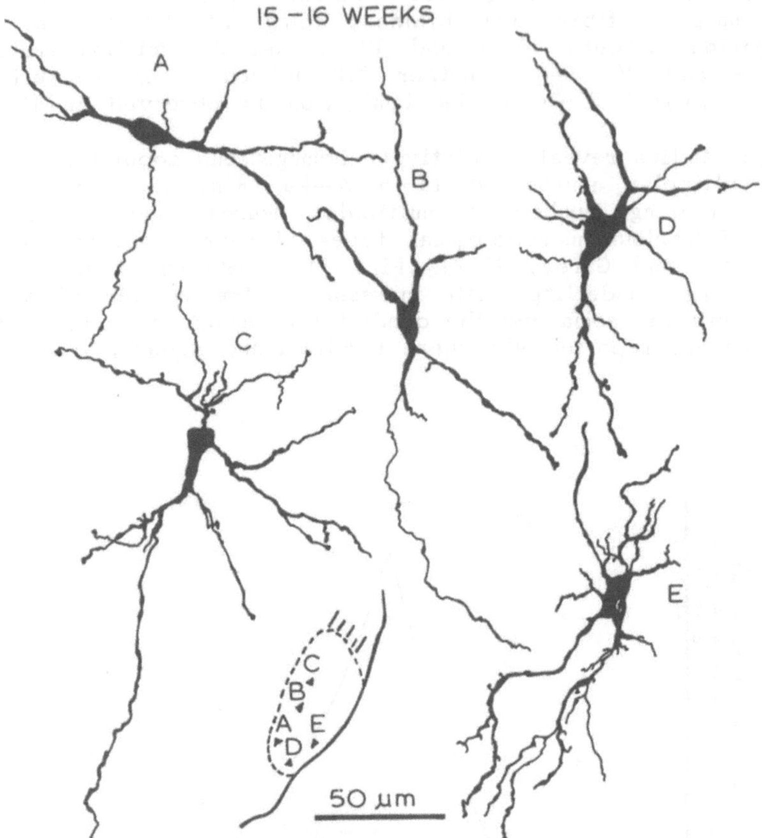

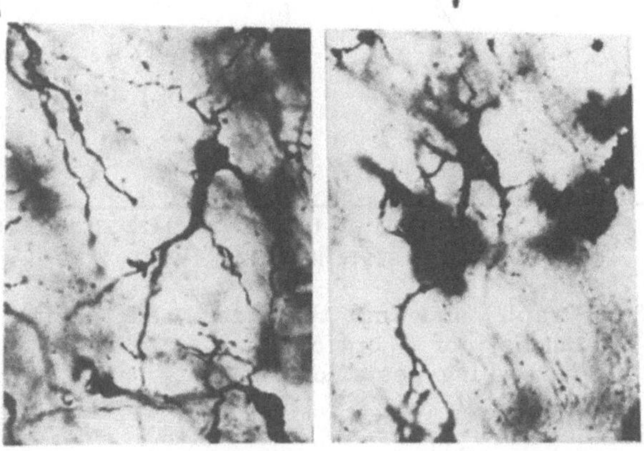

Fig. 4. Camera lucida drawing of LGN cells at 15-16 weeks of gestation. Note the relatively greater maturity of cells D and E in the magnocellular zone as compared to cells A, B, and C of the parvocellular zones. Photomicrographs below show the representative cells C and E from the parvo- and magnocellular zones respectively.

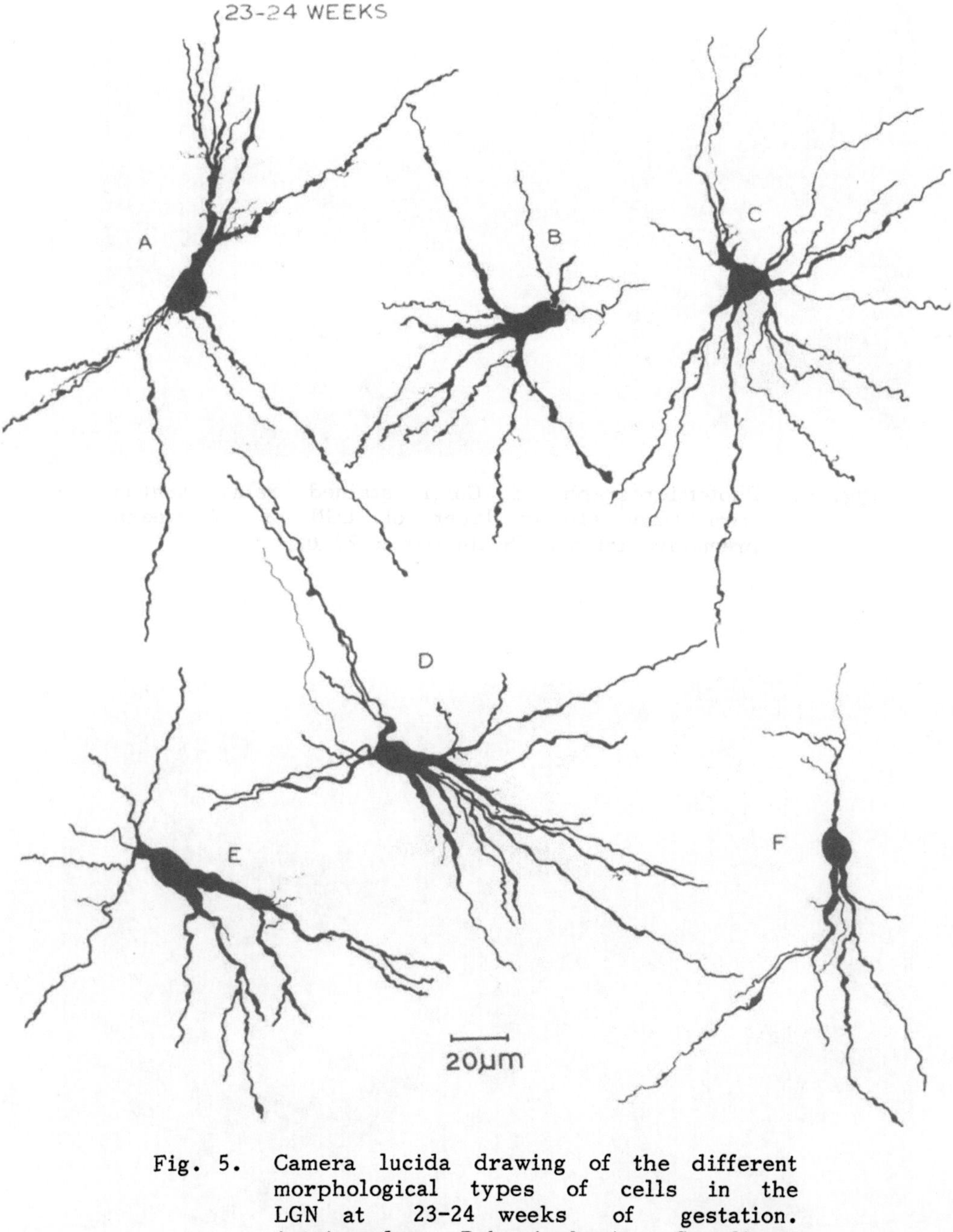

Fig. 5. Camera lucida drawing of the different morphological types of cells in the LGN at 23-24 weeks of gestation. A-triangular, B-hemispheric, C-radiate, D-bitufted, E-capsular, F-presumed interneuron.

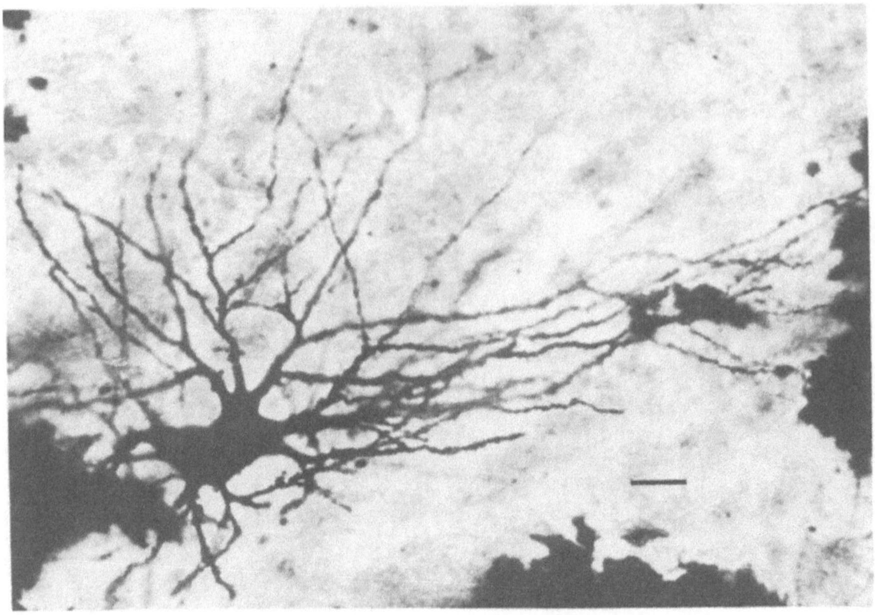

Fig. 6. Photomicrograph of Golgi stained relay neuron from magnocellular layer of LGN of 37 weeks premature infant. Scale bar = 20 µm

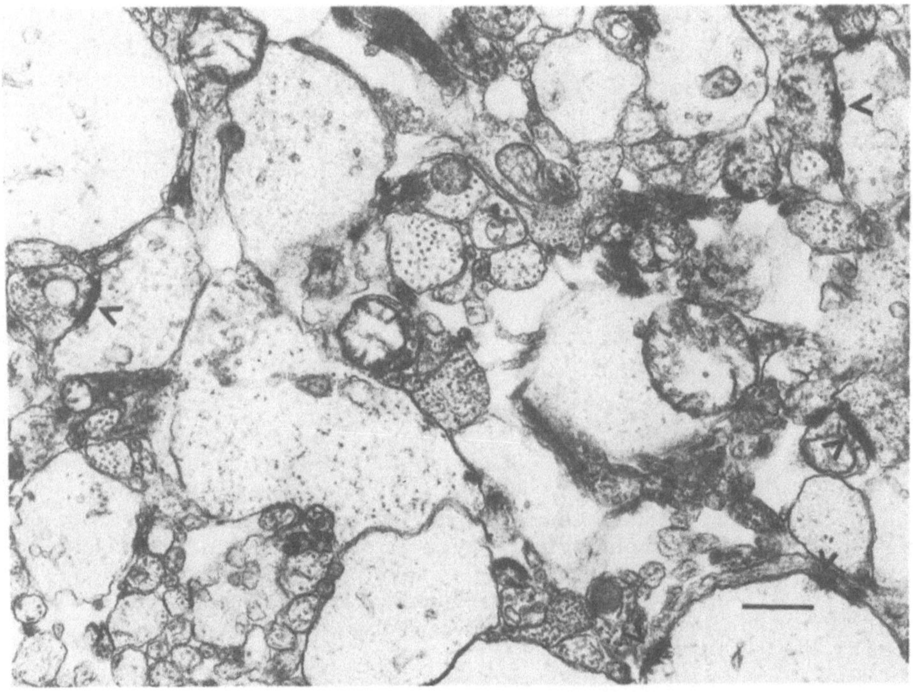

Fig. 7. Electronmicrograph from LGN of 26 weeks gestation showing multiple synapses. Scale bar = 0.5 µm.

DISCUSSION

The present study highlights the time periods during which developmental events of cell proliferation, migration, excessive production of axons and their elimination, synaptogenesis and dendritic proliferation occur in the human lateral geniculate nucleus.

An increase in the numerical density of cells in the human LGN is observed upto 12 weeks of gestation followed by a subsequent decline. In our earlier study (Damyanti et al., 1983) rows of migrating cells have been found extending towards the LGN in a 12 weeks human fetus. Rakic and Sidman (1969) have noted a peak proliferation in the diencephalic wall between 8 and 15 weeks of gestation in the human fetus. Autoradiographic studies of Rakic (1977) in the monkey have shown that the cells of LGN are derived from the proliferative activity of ventricular and subventricular zones of the diencephalon and the cells migrate to LGN. From the above mentioned reports in human fetuses and our own observations on numerical density it appears that the cells destined for the LGN reach it around 12-15 weeks of gestation. Thus the period upto 12 weeks constitutes a critical phase during which the events of cell proliferation and migration result in the formation of LGN.

Analysis of the population kinetics of inhibitory interneurons which are GABA positive shows that the highest percentage of these cells is present at 16-17 weeks. At about the same period the number of optic nerve fibres is the largest. This is followed by a decline in both the parameters. This period probably forms another critical period of vulnerability. A similar pattern of excessive production of optic nerve fibres followed by a loss has been demonstrated to occur in the chick, rat, cat and monkey (Raegar and Raegar, 1978; Lam et al., 1982; Perry et al., 1983; Sefton and Lam, 1984; Crespo et al., 1985; Ng and Stone, 1982; Rakic and Riley, 1983). Recent studies of Provis et al. (1985) on human fetal optic nerves also show a similar pattern. The period of precipitous axonal loss in the optic nerves between 17-19 weeks precedes laminae formation in the LGN. Subsequent gradual loss from 19 weeks onwards overlaps the period of lamination. Both retinogeniculate and corticogeniculate afferent inputs are stated to be involved with the formation of distinct laminae in the LGN (Rakic, 1981; Brunso-Bechtold et al., 1983). During the period of 21 to 24 weeks gestation, lamination is completed and is accompanied by significant regression in the number of optic axons. This perhaps comprises yet another sensitive phase in the development of LGN.

Dendritic development begins from 13-14 weeks onwards such that at 15-16 weeks mostly bipolar but some multipolar cells are also present. From 24 weeks morphological differentiation of cell types is apparent since different varieties of cell types described in the adult become recognisable. Thereafter there is continuous proliferation and modelling of the dendrites. During this period synaptic contacts of different types are also formed. This later period of development is indicative of growth and modelling for consequent achievement of interactive stabilisation between the resident population of neurons and the afferent hard core wiring. This therefore represents another critical period of growth of LGN. Dobbing (1974) pointed to two peaks in the growth of the human brain in general during which the brain is vulnerable to growth restriction:the first between 10-17 weeks during the period of neuronal multiplication; and second, the later period of growth spurt when the processes at risk are dendritic arborisation,synaptic connectivity and myelination.

Our studies demonstrate that there are multiple overlapping sensitive periods during which various vital growth processes are occurring that could render the lateral geniculate nucleus susceptible to environmental influences.

REFERENCES

Assaf, A.A., 1982, The sensitive period:transfer of fixation after occlusion for strabismic amblyopia, Br J Ophthalmol., 66:64.

Blake, R., and DiGianfilippo, A., 1980, Spatial vision in cats with selective neural deficits, J Neurophysiol., 43:1197.

Brunso-Bechtold, J.K., Florence, S.L., and Casagrande, V.A., 1983, The role of retinogeniculate afferents in the development of connections between visual cortex and dorsal lateral geniculate nucleus, Dev Brain Res., 10:33.

Bunt, M.S., Lund, R.D., and Land, P.W., 1983, Prenatal development of the optic projection in albino and hooded rats, Dev Brain Res., 6:149.

Courten, de C., and Garey, L.J., 1982, Morphology of the neurons in the human lateral geniculate nucleus and their normal development, Exp Brain Res., 47:159.

Crespo, D., O'Leary, D.D.M., and Cowan, W.M., 1985, Changes in the number of optic nerve fibres during late prenatal and postnatal development in the albino rat, Dev Brain Res., 5:263.

Cynader, M., Timney, B.N., and Mitchell, D.E., 1980, Periods of susceptibility of kitten visual cortex to the effects of monocular deprivation extends beyond six months as a function of age. Brain Res., 191:545.

Damyanti, N., Wadhwa, S., and Bijlani, V., 1983, Development and maturation of the lateral geniculate body in man, Ind J Med Res., 77:279.

Dekaban, A., 1954, Human Thalamus. An anatomical, developmental and pathological study. II. Development of the human thalamic nuclei, J Comp Neurol., 100:63.

Dobbing, J., 1974, The later development of brain and its vulnerability, in: "Scientific Foundations of Paediatrics," J.A. Davis and J. Dobbing, eds., William Heinmann Medical Books Ltd., London.

Elberger, A.J., 1984, The existence of a separate brief critical period for the corpus callosum to affect visual development, Behav Brain Res., 11:223.

Garey, L.J., 1984, Structural development of the visual system of man, Human Neurobiol., 3:75.

Hamilton, W.J., Boyd, J.D., and Mossman, J.W., 1962, "Human Embryology", Second Edition, Heffer, W. and Sons Ltd.,Cambridge.

Harewerth, R.S., Smith, E.L. III, Duncan, G.C., Crawford, M.L. J., and Von Noorden, G.K., 1986, Multiple sensitive periods in the development of the primate visual system, Science, 232: 235.

Hickey, T.L., 1981, The developing visual system, TINS, 4:41.

Hitchcock, P.F., and Hickey, T.L., 1980, Prenatal development of the human lateral geniculate nucleus, J Comp Neurol., 194:395.

Hubel, D.H., and Weisel, T., 1970, The period of susceptibility to the physiological effects of unilateral eye closure in kittens, J Physiol (London), 206:419.

Jacobson, S.G., and Ikeda, H., 1979, Behavioural studies of spatial vision in cats reared with convergent squint:is amblyopia due to arrest of development, Exp Brain Res., 34:11.

Lam, K., Sefton, A.J., and Bennett, M.R., 1982, Loss of axons from the optic nerve of the rat during development, Dev Brain Res., 3:487.

Ng, A.Y.K., and Stone, J., 1982, The optic nerve of the cat:appearance and loss of axons during normal development, Dev Brain Res., 5:263.

Perry, V.H., Henderson, Z., and Linden, R., 1983, Postnatal changes in retinal ganglion cells and optic axon populations in the pigmented rat, J Comp Neurol., 219:356.

Provis, J.M., Van Driel, D., Billson, F.A., and Russell, P., 1985, Human optic nerve:Overproduction and elimination of retinal axons during development, J Comp Neurol., 238:92.

Raegar, G., and Raegar, U., 1976, Generation and degeneration of retinal ganglion cells in the children, Exp Brain Res., 25:551.

Rakic, P., 1974, Embryonic development of the pulvinar-LP complex in man, in: "The Pulvinar-LP Complex," I.S. Cooper, M. Riklan, and P. Rakic, eds., Thomas, Springfield, Illinois.

Rakic, P., 1977a, Prenatal development of the visual system in rhesus monkey, Phil Trans R Soc Lond Biol., 278:245.

Rakic, P., 1977b, Genesis of the dorsal lateral geniculate nucleus in rhesus monkey:site and time of origin, kinetics of distribution of neurons, J Comp Neurol., 176:23.

Rakic, P., 1981, Development of visual centres in the primate brain depends on binocular competition before birth, Science, 214:928.

Rakic, P., and Riley, K.P., 1983, Overproduction and elimination of retinal axons in the fetal rhesus monkey, Science, 219:1441.

Rakic, P., and Sidman, R.L., 1969, Telencephalic origin of pulvinar neurons in the fetal human brain, Z Anat Entwicklungsgesch., 129:53.

Sefton, A.J., and Lam, K., 1984, Quantitative and morphological studies on developing optic axons in normal and enucleated albino rats, Exp Brain Res., 57:107.

Shatz, C.J., 1983, The prenatal development of the cat's retinogeniculate pathway, J Neurosci., 3:482.

Timney, B., Mitchell, D.E., and Giffin, F., 1978, The development of vision in cats after extended periods of dark rearing, Exp Brain Res., 31:547.

Von Grunau, M.W., and Singer, W., 1980, Functional amblyopia in kittens with unilateral extropia. II. Correspondence between behavioural and electrophysiological assessment, Exp Brain Res., 40:305.

Wadhwa, S., and Bijlani, V., 1988, Cytodifferentiation and developing neuronal circuitry in the human lateral geniculate nucleus, Int J Dev Neurosci., 6:59.

Wadhwa, S., Tákacs, J., Bijlani, V., and Hámori, J., 1988, Numerical estimates of GABA immunoreactive neurons in the human lateral geniculate nucleus in the prenatal period, Human Neurobiol., 6:261.

Williams, R.W., 1983, Prenatal development of the cat's visual system, Thesis, University of California, Davis.

Williams, R.W., and Chalupa, L.M., 1983, Expanded retinogeniculate projections in the cat following prenatal unilateral enucleation: functional and anatomical analysis of an anomalous input, Soc Neurosci Abstr., 9:701.

Wilkinson, F.E., 1980, Reversal of the behavioural effects of monocular deprivation as a function of age in kittens, Behav Brain Res., 1:101.

IN OCULO DIFFERENTIATION OF EMBRYONIC NEOCORTEX INTO RETINA IN ADULT RAT

Gomathy Gopinath, Ashok Kumar Shetty, Ranjita Banerjee and P.N. Tandon

Departments of Anatomy and Neurosurgery
All-India Institute of Medical Sciences
New Delhi-110029, India

ABSTRACT

Embryonic neocortex of 13 day gestation period, when transplanted into the anterior eye chamber of the adult rat differentiated into retina as well as the phenotypical neocortex. The layers of the retina clearly identifiable on the 30th posttransplantation day had close resemblance to those of the adult retina. The few differences observed were (1) absence of an optic fibre layer, (2) low population of cells in the ganglion cell layer, and (3) a narrow outer plexiform layer. Subsequently the ganglion cell layer disappeared followed by the inner nuclear layer. The outer nuclear layer and the photoreceptors continued to be present and were seen in small bits mixed with differentiated neocortical tissue till the end of experimental period, which is 6 months. These findings indicated that early embryonic cortex is capable of differentiating into retina in the anterior chamber of the eye in adult rat, which may have some trophic factor. The continued survival of the retina seems to depend on the connections of the major cell types with the appropriate target sites.

INTRODUCTION

Intraocular transplantation of neural tissue is extensively used for studying the survival and growth of single or multiple grafts and also the interaction of the grafts from related and unrelated areas of the brain (Olson et al., 1984). This technique has certain advantages over the transplantation in the central nervous system. Identification and also the sequential growth and behaviour of the transplant without intervention are easy and possible. This technique so far has yielded a wealth of information in relation to a number of different developing regions of the central nervous system of mammals.

The present study was undertaken to standardize the technique of transplantation in the anterior eye chamber and also to follow up the behaviour of the transplanted neural tissue. The neural tissues chosen for the study were neocortex (13-17 gestation days), substantia nigra (13-20 gestation days) and adrenal medulla of adult rats. Subsequent examination of the transplant grown from the embryonic neocortex of 13 gestation day has yielded some interesting results which are presented here.

MATERIAL AND METHODS

Stock-bred Wistar rats were used in the present study. Donor tissue was collected from embryos on the respective days of gestation determined by sperm-positive vaginal smear in the female rat following mating. The sperm-positive day was taken as the zero day of pregnancy. After anaesthetizing the pregnant rat with intramuscular ketamine the embryos were removed by cesarian section. The embryos were kept in chilled Ringer lactate solution and the brains were removed and dissected free of the meninges and blood vessels under an operation microscope. Thereafter the anterior region of the neocortex was removed and one cubic millimetre of the tissue was taken into a glass capillary tube connected to a Hamilton syringe with a polypropylene tube.

Meanwhile 90 day old adult rats were anesthetized with ketamine and the pupil of the eye, used for transplantation was dilated using atropine ointment. The transplantation was carried out according to the procedure described by Olson et al (1983). After depositing the tissue into the anterior chamber of the eye through a small incision in the cornea the graft was pushed gently towards the corneo-scleral junction with the flat side of an iris forceps. A total of 30 rats were transplanted using the neocortex from embryos of the age group 13 to 17 gestational days. Antibiotics were used for 3 days post-operatively.

The transplants were examined and measured at regular intervals under the operation microscope. The hosts were sacrificed at varying periods between 30th and 180th day after transplantation. The transplanted eye was carefully dissected out without injuring the eye ball. The eye ball was immediately fixed and the following day the transplant was removed from the anterior surface of the iris after dissecting out the cornea. The remaining layers of the eye ball were examined to make sure that these were intact.

The transplants were processed for electron microscopy. Semi-thin and ultra-thin sections were examined and photographed. Ultra-thin sections from specimens of 30 and 40 days were processed for immunogold-labelling after treating with antiserum against Vasoactive-intestinal Polypeptide (VIP) to identify the amacrine cells. The population of amacrine cells containing VIP are comparatively more than the number positive for other neurotransmitters or peptides (Brecha, 1983). The sections were treated with either hydrogen peroxide or sodium periodate before immunocytochemical processing.

RESULTS

Neocortex of the 13 day old embryo was a thin shell of tissue forming the wall of the lateral ventricle. The cells close to the ventricular wall were tall columnar type and were actively dividing. The neuroblasts in the superficial region were rounded in shape. The subpial region had loosely arranged cells (Fig. 1).

27 rats transplanted with neocortical tissue had successful grafts. These rats were free of infection, cloudiness or opacity in the transplanted eye. Vascularization of the grafts was seen on the 3rd post operative day. The size of the graft had reduced during the first week but increased subsequently till the end of the first month. Thereafter the size remained stationary till the end of the study.

Six grafts of a total of eight grown from the neocortical tissue

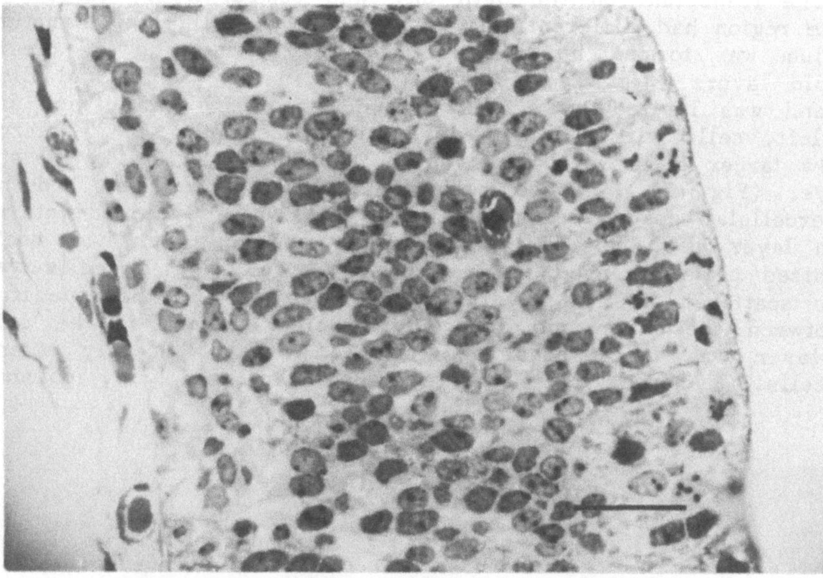

Fig. 1 Section through neocortex of 13th gestation day. Ventricular lining shows mitotic figures. Bar - 25 microns.

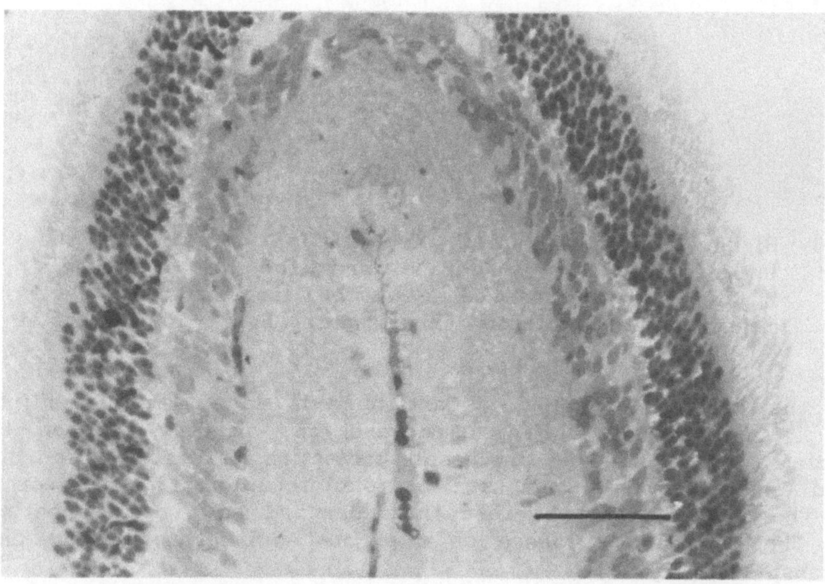

Fig. 2 30 days old transplant. Section through one of the folia showing double fold of retina. Bar - 50 microns.

of 13th embryonic day had patches of tissue showing retinal morphology in addition to the neocortical features on days 30,40,60,90 and 180 days. The transplant on the 30th day was comparatively larger and a sizeable region had differentiated into islands of folia like structures Each folium was formed by a double fold of retina showing clearly identifiable layers (Fig. 2). A small cleft in the centre had blood vessels and was limited by a well-formed membrane. On either side of the cleft, cells with light to dark nuclei measuring 14-25 microns and a few larger cells with clear nuclei containing nucleoli were seen in a layer (Fig. 3). These cells were loosely arranged and had wide intercellular spaces separating them. A well developed internal plexiform layer was seen next. The inner nuclear layer had mostly medium sized cells arranged in 3-4 rows. A few light er cells were also seen scattered in this layer. There was a very thin plexiform layer between the inner and the outer nuclear layers. The outer nuclear layer was thick consisting of 6-8 rows of densely stained small cells. These cells were seen arranged in columns.

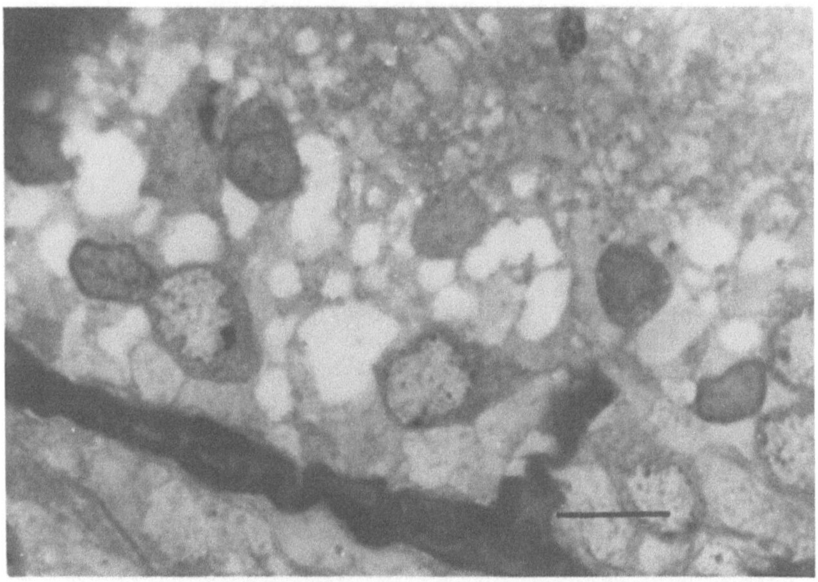

Fig. 3 Higher magnification of the ganglion cell layer, the internal limiting membrane and the blood vessels in the cleft. The cells in the ganglion cell layer measured 14-25 microns in diameter. Bar - 10 microns.

Many of the cells in the outer part of this layer showed mitotic activity. An external limiting membrane was present between the outer nuclear layer and the processes projecting outwards which had inner and outer segments just as the photoreceptors in the normal retina. The outer segments were rather long and had lamellar pattern inside. There was no evidence of a pigment epithelium in any part of the transplant (Fig. 4).

Almost the same features persisted in the specimen obtained on the 40th day including the mitotic activity in the superficial region of the outer nuclear layer. But the large cell type with clear nuclei observed in the ganglion cell layer of the 30th day specimen was not present. Proportionately more of cortical tissue was seen when compared with the earlier graft.

In the grafts obtained subsequently the area occupied by tissue resembling retina gradually dwindled and the graft of 6 months duration had only a few scattered patches of dense cells similar to the ones seen in the outer nuclear layer. At many places in the older graft the outer nuclear layer was seen mixed with the deeper part of the differentiated neocortex (Fig. 5). There was gradual disappearance of ganglion cell layer and thinning of the inner nuclear layer. The outer plexiform layer also became obliterated at places where the cells of the inner and the outer nuclear layers intermingled. Towards the end of the study, on the 180th day in one specimen there were small bits of outer nuclear layer most of which was devoid of it's photoreceptor-like processes. This layer was thrown into folds and in the cleft the photoreceptor-like processes were intact. One other specimen examined on 180th day had only cortical tissue.

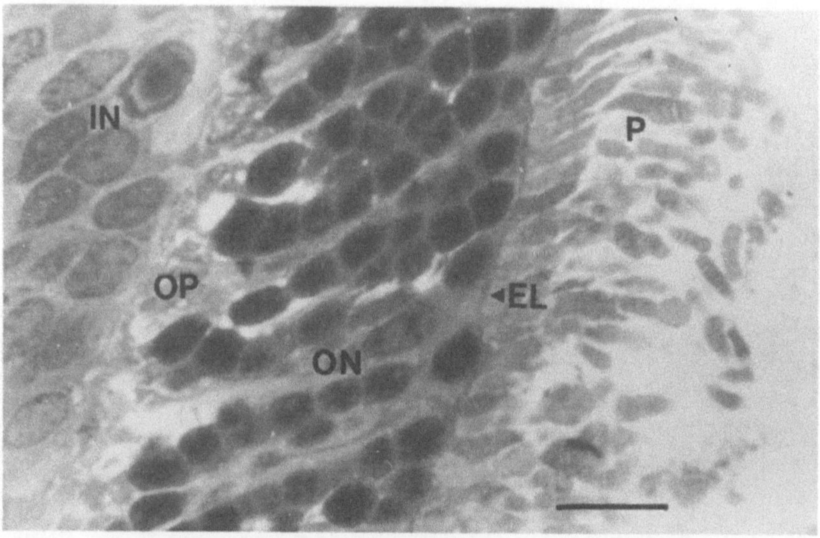

Fig. 4 Higher magnification showing outer segments of the retina. The processes (P) have lamellar pattern. Arrow indicates external limiting membrane. The outer nuclear layer (ON) shows mitotic figures and columnar pattern. Outer plexiform layer (OP) and inner nuclear layer (IN). Bar - 10 microns.

The observations under the light microscope were confirmed by electron microscope. The inner and outer segments of the outer processes of the grafts of 30 to 60 days had the morphological characteristics of the photoreceptors of the normal retina. The outer plexiform layer was only sparsely populated with processes and synapses. The inner plexiform layer was more dense with profiles of axons, dendrites, glial fibres and synapses (Fig. 6). The internal limiting layer was uninterrupted. Immunogold labelling was negative for VIP in the specimens examined.

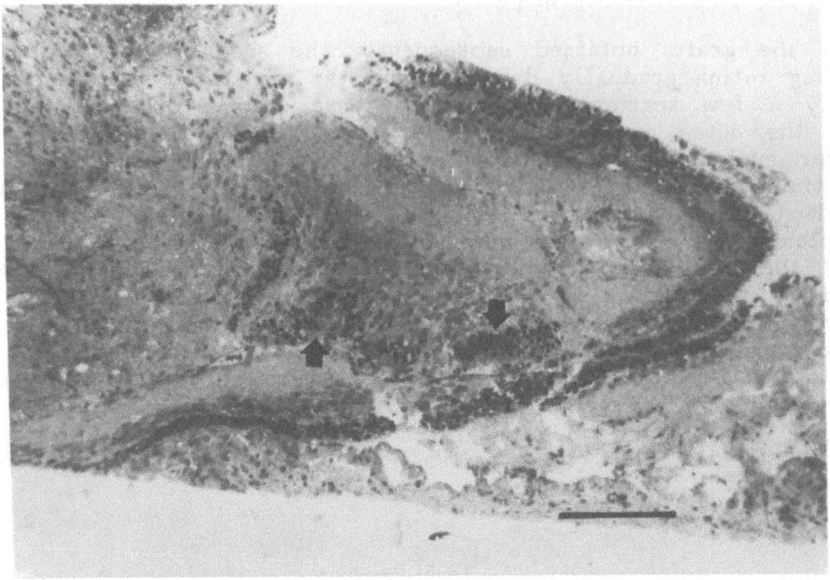

Fig. 5 Mixture of neocortex and retinal layers in the transplant of 90 days. Arrows indicate outer nuclear layer in the deeper aspect of the neocortex. Bar - 100 microns.

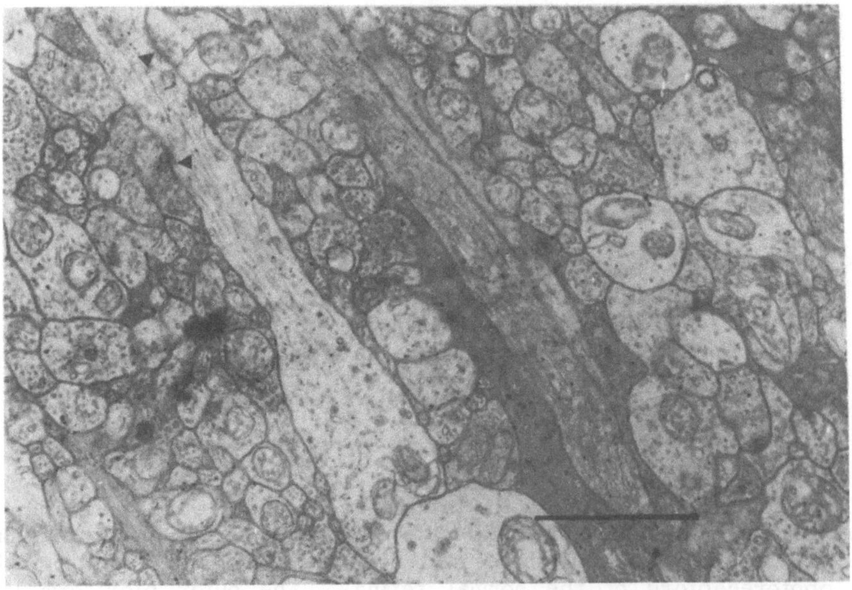

Fig. 6 Electronmicrograph of the internal plexiform layer showing profiles of axons, dendrites and glial processes. Arrows show axo-dendritic synapses. Bar - 2 microns.

None of the other intraocular grafts grown from older embryonic neocortex, substantia nigra or adrenal medulla had shown retina-like morphology.

DISCUSSION

The observation of the present study of differentiation of embryonic neocortex of 13th gestation day into structure similar to retina has been most unexpected. It is significant to note that none of the other transplants grown from older neocortex, substantia nigra or adrenal medulla had any feature remotely resembling adult retina. So it stands to reason that the neocortex of 13th day having a close resemblance to the neural retina of the 16th gestation day (del Cerro et al., 1985) with multiplying neuroepithelial cells and neuroblasts has still retained the potentials to differentiate into retina. Much of the neocortex seems to have given rise to retina at the initial period of growth as seen in the 30th day specimen. At this stage, though many of the features of the retina are of the adult type, there are certain characteristics which require special mention.

There are no fibre outgrowths from the ganglion cell layer. An uninterrupted limiting membrane was present between the retina and the cleft between the two retinal layers. Only blood vessels were seen in the cleft. The ganglion cell layer itself was sparsely populated with amacrine cells identified according to the morphology and size (McLoon and Lund, 1984). The few large cells with clear nuclei and nucleoli seen only on the 30th day appeared to be ganglion cells. It cannot be concluded whether the ganglion cells had not formed in sufficient quantity from the neocortical tissue or whatever number formed had differentiated into amacrine cells later (Hinds and Hinds, 1978). In the normal retina ganglion cells are identifiable around the 15th gestation day from the inner layer of the optic vesicle (Braekvelt and Hollenberg, 1970). The 13th embryonic day neocortex, though has a close resemblance to the 16th day old embryonic retina still in the process of differentiation (del Cerro et al., 1985), may not have the full capabilities to give rise to adequate number of the earliest forming cell component, which is the ganglion cell. The subsequent disappearance of the few ganglion cells formed could be due to the absence of connections with the appropriate target site (Hughes and McLoon, 1979). The lack of formation of sufficient number of ganglion cells or their early disappearance may be responsible for the low population of the later developing amacrine cells. A close developmental relationship has been suggested between amacrine cells and ganglion cells during development (Hirokawa, 1978). In retinal transplants with intact ganglion cells at the time of transplantation, formation and continued survival and immunoreactivity the amacrine cells to different neuropeptides had been recorded (Matthews et al., 1982; McLoon and Lund, 1984; and Personal observations). Immunoreactivity for VIP is reported to appear in the amacrine cells only by the 12th postnatal day in rat (Terubayashi et al., 1982) when amacrine and ganglion cell contacts are being established (Weidman and Kuwabara, 1968). Thus the absence of immunoreactivity for VIP in the amacrine cells on the 30th day and the subsequent disappearance of the amacrine cells in the present study can be explained.

In spite of the paucity of cells in the ganglion cell layer, the inner plexiform layer is better developed with apparently mature neuronal processes and synapses. The few axodendritic contact sites observed during the early posttransplantation periods may be between the bipolar cells and the amacrine cells. The bipolar cell layer persisted longer than the ganglion cell layer, though the formation of the outer plexiform layer was not in proportion to the size of the two nuclear layers. The outer plexiform layer in the normal retina starts appearing after the 6th postnatal day when the processes of the bipolar cells extend horizontally and separate the two nuclear layers (Braekevelt and Hollenberg, 1970). Subsequently this layer widens by the growth and connectivity of the cell processes. Instead of the normal increase in size, the present observation is reduction and subsequent disappearance of this layer showing that the developmental time table of this region is not proceeding according to the same schedule as in the normal retina.

The presence of mitotic figures in the superficial part of the outer nuclear layer indicates cell formation well after the cessation of mitotic activity by the end of first week in the normal retina (Braekevelt and Hollenberg, 1970). The persistence of the outer nuclear layer and the outer segments even in patches till the 180th post transplantation day shows the resilience of this region and the better potential of the neuroepithelial cells of the neocortex to form these cells. The photoreceptors normally differentiate by 5th postnatal day and the mature features are apparent by the second week or later (Weidman and Kuwabara, 1968) which fit in well with the present observations.

The differentiation of the ventricular cells of the neocortex into the various major cell types of retina must be due to the influence of the surrounding environment, which in this case is anterior chamber of the eye. The idea that cells in the central nervous system can differentiate into cells other than the expected phenotypical neurons depending on the kind of environment provided is in agreement with other experimental studies. Le Dourin et al (1975) have shown that embryonic neural crest cells, which normally form adrenergic neurons when grafted into the vagal region differentiated into cholinergic neurons. Neurons even after forming noradrenergic synapses converted to cholinergic type in the presence of nonneuronal supporting cells in culture (Patterson, 1978). Thus it appears that at a critical period in development, the future differentiation of at least certain cell types can be influenced by altering the environment. The optic vesicle which develops from the prosencephalon is in place by the 11th embryonic day. The optic cup formation takes place later and then only the inner neural layer starts actively multiplying and differentiating, a stage at which the neocortex is also starting to form. So it is possible that some of the ventricular cells of the embryonic neocortex of early period can still differentiate into retina in the proper environment. Neocortical transplants from 11-20 day old embryos in other sites like tectum (McLoon and McLoon, 1984) and spinal cord (Bernstein et al., 1985) have differentiated into phenotypical neurons only. Since the older neocortical grafts in the anterior eye chamber had not shown such changes it is reasonable to assume that this property of the ventricular cells is retained only for a short period and can be probably influenced only by environment in the eye. del Cerro et al (1987) have postulated that the adult eye retained trophic capabilities which are lost or drastically diminished in the adult brain.

CONCLUSIONS

Present studies have shown that the early embryonic neocortex has the capability of differentiating into retina in the environment of the anterior chamber of the eye. The developmental sequence more or less resembles that of the normal retina. But the maturation and maintenance of the different cell types and layers seems to depend on the trophic influence exerted by the early forming cells and the establishment of their connections with the target sites.

REFERENCES

Bernstein, J.J., Hoovler, D.W., and Turtil, S., 1985, Initial growth of transplanted E 11 fetal cortex and spinal cord in adult rat spinal cord, Brain Res., 343:336.

Braekevelt, C.R., and Hollenberg, M.J., 1970, The development of the retina of the albino rat, Am. J. Anat., 127:281.

Brecha, N., 1983, Retinal Neurotransmitters: Histochemical and biochemical studies, in: "Chemical Neuroanatomy", P.C. Emson, ed., Raven Press.

del Cerro, M., Gash, D.M., Rao, G.N., Notter, M.F., Weigand, S.J., and Gupta, M., 1985, Intraocular retinal transplants, Invest. Ophthalmol. Vis. Sci., 26:1182.

del Cerro, M., Gash, D.M., Rao, G.N., Notter, M.F., Weigand, S.J., Sathi, S., and del Cerro, C., 1987, Retinal transplants into the anterior chamber of the rat eye, Neuroscience, 21:707.

Hinds, J.W., and Hinds, P.L., 1978, Early development of amacrine cells in the mouse retina: an electron microscopic serial section analysis, J. Comp. Neurol., 179:277.

Hirokawa, N., 1978, Characterisation of various nervous tissues of the chick embryos through responses to chronic application and immunocytochemistry of B-bungarotoxin, J. Comp. Neurol., 180:449.

Hughes, W.F., and McLoon, S.C., 1979, Ganglion cell death during normal retinal development in the chick: Comparisons with cell death induced by early target field destination, Exp. Neurol., 66:587.

Le Douarin, N.M., Renaud, D., Teillet, M., and Le Douarin, G.H., 1975, Cholinergic differentiation of presumptive adrenergic neuroblasts in interspecific chimeras after heterotopic transplantations, Proc. Natl. Acad. Sci. (U.S.A.), 72:728.

Matthews, M.A., West, L.C., and Riccio, R.V., 1982, An ultrastructural analysis of the development of the fetal rat retina transplanted to the occipital cortex, a site lacking appropriate target neurons for optic fibres, J. Neurocytol., 11:533.

McLoon, S.C., and Lund, R.D., 1984, Loss of ganglion cells in fetal retina transplanted to rat cortex, Develop. Brain Res., 12:131.

McLoon, S.C., and McLoon, L.K., 1984, Transplantation of the developing mammalian system, in: Neural Transplant, Development and Function, J.R. Sladek Jr. and D.M. Gash, eds., Plenum Press.

Olson, L., Seiger, A., and Stromberg, I., 1983, Intraocular transplantation in rodents: A detailed account of the procedure and examples of its use in neurobiology with special reference to brain grafting, in: Advances in Cellular Neurobiology, S. Fedroff, and L. Hertz, eds., Academic Press, New York.

Olson, L., Bjorklund, H., and Hoffer, B.J., 1984, Camera bulbi anterior. New vistas on a classical locus for neural tissue transplantation, in: Neural Transplants, Development and Function, J.R. Sladek, Jr., and D.M. Gash, eds., Plenum Press.

Patterson, P.H., 1978, Environmental determination of autonomic neurotransmitter functions, Ann. Rev. Neurosci., 1:1.

Terubayashi, H., Okamura, H., Fujisawa, H., Itoi, M., Yanaihara, N., and Ibata, Y., 1982, Postnatal development of vasoactive-intestinal polypeptide immunoreactive amacrine cells in the rat retina, Neuroscience Lett., 33:259.

Weidman, T.A., and Kuwabara, T., 1968, Postnatal development of the rat retina, Arch. Ophthal., 79:470.

AXONAL OUTGROWTH AND PROCESS PLACEMENT OF SENSORY LUMBAR NEURONS IN THE NEMATODE CAENORHABDITIS ELEGANS

Shahid S. Siddiqui

Laboratory of Molecular Biology
Department of Materials Science
Toyohashi University of Technology
Toyohashi 440, Japan

ABSTRACT

One of the most remarkable features of the brain is the degree of precision with which neurons are interconnected. As the neural circuitry of the nematode Caenorhabditis elegans has been completely established through serial section electron micrographs(White et al., 1986), and because the nematode is amenable to genetic analysis, it is an excellent system to study the genetic basis of axonal outgrowth and process placement in the formation of a neural network. Here, I will describe the identification of 11 genes which affect the axonal outgrowth and guidance of five pairs of lumbar neurons in C. elegans.

A pair of bilaterally symmetric lumbar ganglia are situated in the tail of C. elegans, each consisting of 12 neurons. We have examined the morphology and pattern of axonal processes of three embryonic(PHA, PHB, and PLM), and two postembryonic(PHC and PVN) lumbar neurons in the wild type and existing uncoordinated(unc) mutants, immunocytochemically. We have earlier shown that antibodies against horseradish peroxidase(HRP) stain PHA, PHB, PHC and PVN neurons(S. Siddiqui and J. Culotti, unpublished), and monoclonal antibodies specific to different tubulin isotypes stain the mechanosensory neuron PLM(Siddiqui et al., 1989). A monoclonal antibody TY21, raised against C. elegans crude homogenate is specific for phasmid neurons PHA, and PHB. In the wild type, PHA, PHB, PHC, and PVN send anteriorly directed processes ventrally via the lumbar commissures to targets in the ventral cord; whereas, PLM sends a process anteriorly in a subventral position, sending a branch ventrally to the ventral cord near the vulva. PHA, PHB, PHC, and PLM are bipolar neurons, extend a backward process into the tail spike. Mutations in nine genes(unc-6, unc-13, unc-33, unc-44, unc-51, unc-61, unc-71, unc-73, and unc-98) result in abnormal axonal outgrowth and process placement of PHC, PVN and PLM neurons. Four(unc-6, unc-33, unc-44, and unc-51) of the nine genes identified above(S. Siddiqui and J. Culotti, unpublished), and unc-76 were previously shown to affect the growth of PHA and PHB neurons(Hedgecock et al., 1985). Mutants in unc-53, which are apparently normal in the growth of PHA, PHB, PHC, and PVN neurons, and unc-76, harbor defects in PLM axonal outgrowth and guidance. In summary, we have identified a set of 11 genes, which exert overlapping, but distinct effects on the axonal outgrowth and process placement of embryonic(PHA, PHB, and PLM), and postembryonic(PHC and PVN) lumbar neurons in C. elegans.

INTRODUCTION

How individual neurons recognize and interact with one another during development to form a neural network is a major unsolved problem in neurobiology. Several distinct mechanisms have been implicated in axonal outgrowth and guidance(e.g. reviewed by Eisen, 1988), but these will not be elaborated here. Instead, I will describe the genetic approach to identify the genes which code for or regulate the expression of molecules that are critical for axonal outgrowth and guidance. The soil nematode Caenorhabditis elegans has a number of features which make it an attractive system for a genetic dissection of the structure and function of the nervous system (Brenner, 1973, 1974).

C. elegans is small and transparent. The adult hermaphrodite(1 mm) has precisely 302 neurons, grouped into 118 different classes, based on fiber morphology and synaptic contacts(White et al. 1983). The neural circuitry ('wiring diagram') of each of the 302 neurons has been established through serial section electron micrographs(Albertson and Thomson, 1976; White et al. 1976; Ward et al., 1975; Ware et al., 1975; White et al., 1986; and Hall and Russell, 1988). The cell lineage of all the somatic cells, including the nervous system has been determined from the zygote to the mature adult (Sulston and Horvitz, 1977; Kimble and Hirsh, 1979; and Sulston et al., 1983), making it the only biological system with complete neural circuitry and cell lineage information. Most remarkably, the structure of the nervous system and the entire cell lineage is invariant from one individual to another. Amenability to genetics could provide an alternative method to disrupt the normally determinant program for the establishment of the nematode neural circuitry. Mutants with abnormal axonal guidance can be examined to reveal the molecular nature of cues used by axonal growth cones to form the appropriate synaptic contacts. Mutants could also provide a large number of individuals, for both cellular and molecular analyses.

More than one hundred genes have been identified in C. elegans which affect a variety of behaviors, including chemosensory, mechanosensory, osmotic avoidance, thermosensory, egg laying, male mating, and dauer formation (reviewed in Chalfie and White, 1988); the largest number of mutants are uncoordinated (Unc) mutants with defective locomotory behavior. The mutant phenotypes of unc mutants, first described by Brenner(1973, 1974), range from slightly slow movement on agar plate to those which are completely paralysed. Many kinds of alterations in the nervous system could produce the uncoordinated phenotype. Indeed, about a quarter of unc genes have been shown to harbor defects in the muscle structure(Waterston, 1988), but the majority of unc genes remain uncharacterized at the ultrastructural level as considerable effort is required to reconstruct animals with serial section electron micrographs. Recently, several techniques have been used to stain neurons and their processes in C. elegans. These include formaldehyde induced fluorescence to stain dopaminergic and serotinergic neurons(Sulston et al. 1975; Horvitz et al. 1982), fluorescein dye filling of chemosensory neurons(Hedgecock et al., 1985), and use of plant lectins and antibodies to visualize neurons with immunofluorescence(Okamoto and Thomson, 1985; Siddiqui and Culotti, 1984, 1988; Siddiqui et al., 1988; H. Bhatt and E. Hedgecock; C. Desai, H. Ellis, S. McIntire, J. Yuan, and R. Horvitz; C. Li, and M. Chalfie; H. Takano-Ohmuro, T. Kaminuma, Y. Tabuse, J. Miwa, and S. Siddiqui, unpublished results, and personal communications).

In this paper, I will describe the identification of 11 genes affecting axonal outgrowth and process placement of four sensory(PHA, PHB, PHC, and PLM), and one motor/inter-neuron PVN, by screening existing uncoordinated mutants immunocytochemically, with three antibodies which selectively stain lumbar neurons in the tail nervous system of C. elegans.

MATERIALS AND METHODS

Nematode Strains: Wild type *C. elegans* N2 (var Bristol) was obtained from S. Brenner. Other nematode strains were obtained from the MRC, Laboratory of Molecular Biology, Cambridge, UK, and *Caenorhabditis* Genetics Center, University of Missouri, Columbia, Missouri. Nematodes were cultured on nutrient agar petri plates, seeded with OP50 strain of *E. coli*, as described by Brenner (1974), at 20°C.

Antibodies: Polyclonal antibodies against horseradish peroxidase(HRP) were raised in rabbits, and affinity purified as described(Siddiqui and Culotti, 1988). Monoclonal antibody 6-11B-1 (Piperno and Fuller, 1985) is specific for acetylated alpha tubulin from a variety of species, and was generously given by Dr G. Piperno, Rockefeller University. Another monoclonal antibody 3A5, specific for many alpha tubulin isotypes(Piperno and Fuller, 1985), was a gift from Dr M. Fuller, University of Colorado. TY21, a monoclonal antibody raised against *C. elegans* crude homogenate was kindly provided by Dr T. Kaminuma, Tokyo Metropolitan Institute of Medical Sciences .

Immunofluorescent Staining of Nematodes: Wholemount 'squash' immunofluorescent staining of nematodes has been described(Siddiqui et al., 1988). A Standard Zeiss microscope, equipped with epifluorescence illumination, and filter sets 10, and 15, for FITC and rhodamine, respectively, was used for microscopy. Photographs were taken with MC63 camera, using Kodak Technical Pan film.

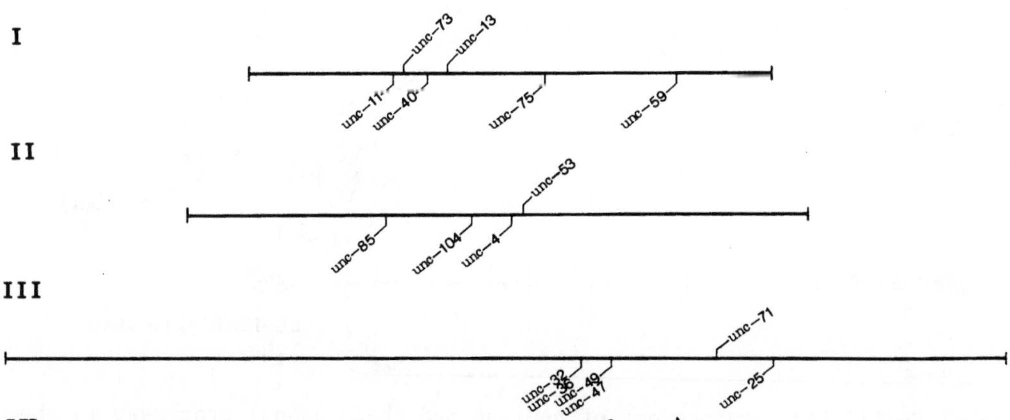

Fig. 1 Genetic map. Map positions of genes affecting axonal outgrowth and process placement described in this paper are shown above the line. Genes affecting nervous system , identified by other techniques including ultrastructural analyis, immunocytochemistry, etc. are shown below the line. (For reference see Swanson et al., 1984).

RESULTS

General Features of the Tail Nervous System

The adult hermaphrodite of *C. elegans* has exactly 40 neurons in the tail nervous system which have been grouped into four distinct ganglia. Fig. 2 shows schematically, the disposition of cell bodies in the tail nervous system. These include a pair of lumbar ganglia, each consisting of 12 neurons, dorsorectal ganglion containing three neurons, and the pre-anal ganglion consisting of 13 neurons.

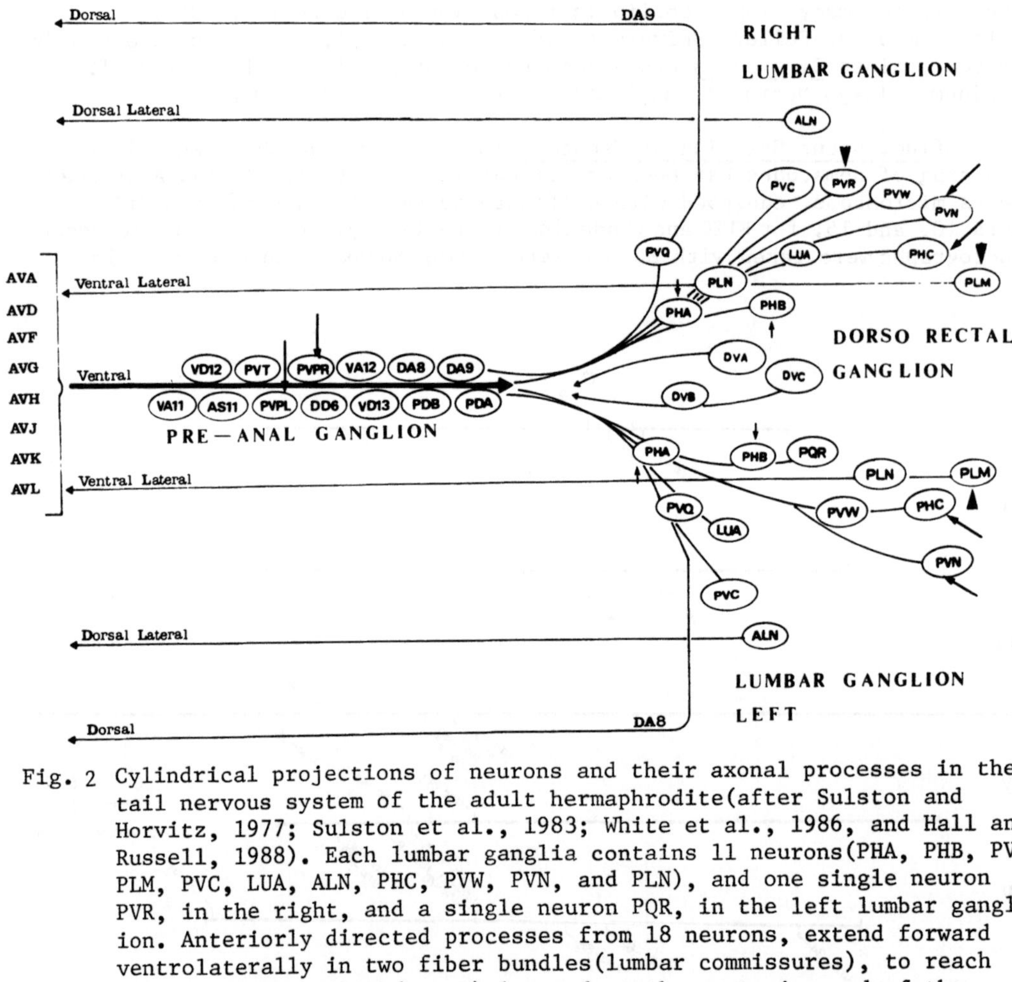

Fig. 2 Cylindrical projections of neurons and their axonal processes in the tail nervous system of the adult hermaphrodite(after Sulston and Horvitz, 1977; Sulston et al., 1983; White et al., 1986, and Hall and Russell, 1988). Each lumbar ganglia contains 11 neurons(PHA, PHB, PVQ, PLM, PVC, LUA, ALN, PHC, PVW, PVN, and PLN), and one single neuron PVR, in the right, and a single neuron PQR, in the left lumbar ganglion. Anteriorly directed processes from 18 neurons, extend forward ventrolaterally in two fiber bundles(lumbar commissures), to reach the pre-anal ganglion(p.a.g) located at the posterior end of the ventral nerve cord. Pre-anal ganglion serves as the major neuropil of the posterior nervous system, as it receives processes from lumbar ganglia, dorsorectal ganglion(DVA,DVB,DVC) , and the posteriorly directed processes of neurons located in the head ganglia(AVA, AVD, AVF, AVG, AVH, AVJ, AVK, and AVL). Six neuron pairs in the lumbar ganglia(PHA, PHB, PHC, PLM, ALN, and PLN), and two single neurons, PVR and PQR, are bipolar, extend a backward directed process towards the tail spike. Lumbar neurons PHA, PHB, PHC, PVN, PLM, and PVR could be labeled with different antibodies using immunofluorescence and are marked with arrows and arrow heads to show the cell body positions.

Staining of PHC and PVN Lumbar Neurons

When adult hermaphrodites are stained with anti-horseradish peroxidase (HRP) antibodies, four pairs of lumbar neurons can be visualized with immunofluorescence. Based on their morphology, fiber disposition, location of cell bodies(determined simultaneously by staining with a nuclear dye, Hoechst 33258), these were identified as two pairs of phasmid neurons PHA and PHB, and two post-embryonic neurons PHC and PVN(Siddiqui and Culotti, 1988). Fig. 3 schematically shows the cell body positions of PHC and PVN neurons and their axonal processes. The staining of phasmid neurons PHA and PHB is faint and usually obscured by the bright staining due to the phasmid sheath cell (not shown in Fig. 3). However, the PHC and PVN pairs could be clearly seen.

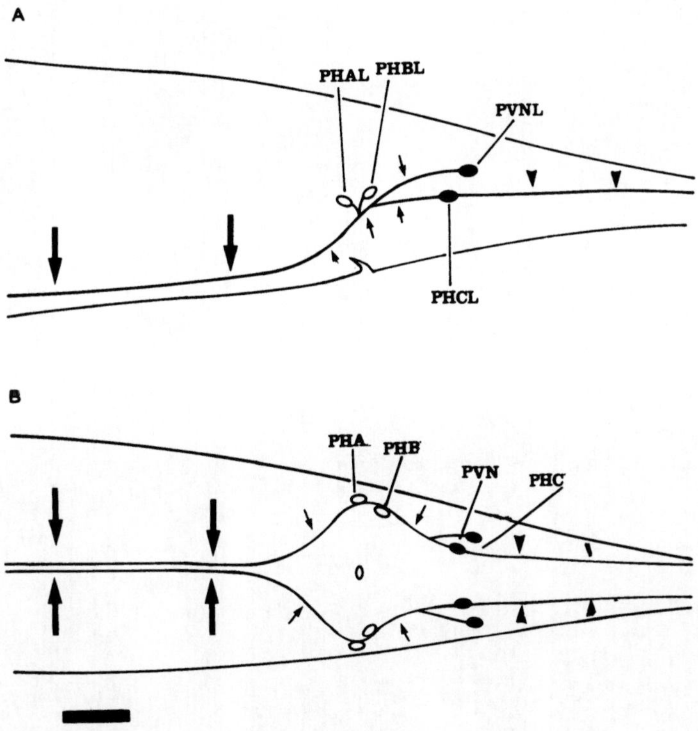

Fig. 3 Schematic diagram of PHC and PVN lumbar neurons(after White et al., 1986, Hall and Russell, 1988). (A) Left lateral view. PHC is a bipolar neuron located in the posterior region of the lumbar ganglia. An anteriorly directed process extends forward into the pre-anal ganglion(big arrows) via the lumbar commissures (small arrows), and terminates in the anterior region of the pre-anal ganglion. A posteriorly directed process extends backward (arrow heads) into the tail spike. The opening in the body is the anus. PVN is a monopolar neuron located dorsally to the PHC cell body. An anteriorly directed process extends forward along the anteriorly directed process of PHC via the lumbar commissures, into the pre-anal ganglion, and extends further into the nerve ring area, along with other ventral cord fibers. This part of PVN axonal process is not shown here, as we could not distinguish it from among other fibers of the ventral cord. PHA and PHB cells (open cell bodies) stain faintly. (B) Ventral view. The scale bar represents 20 microns.

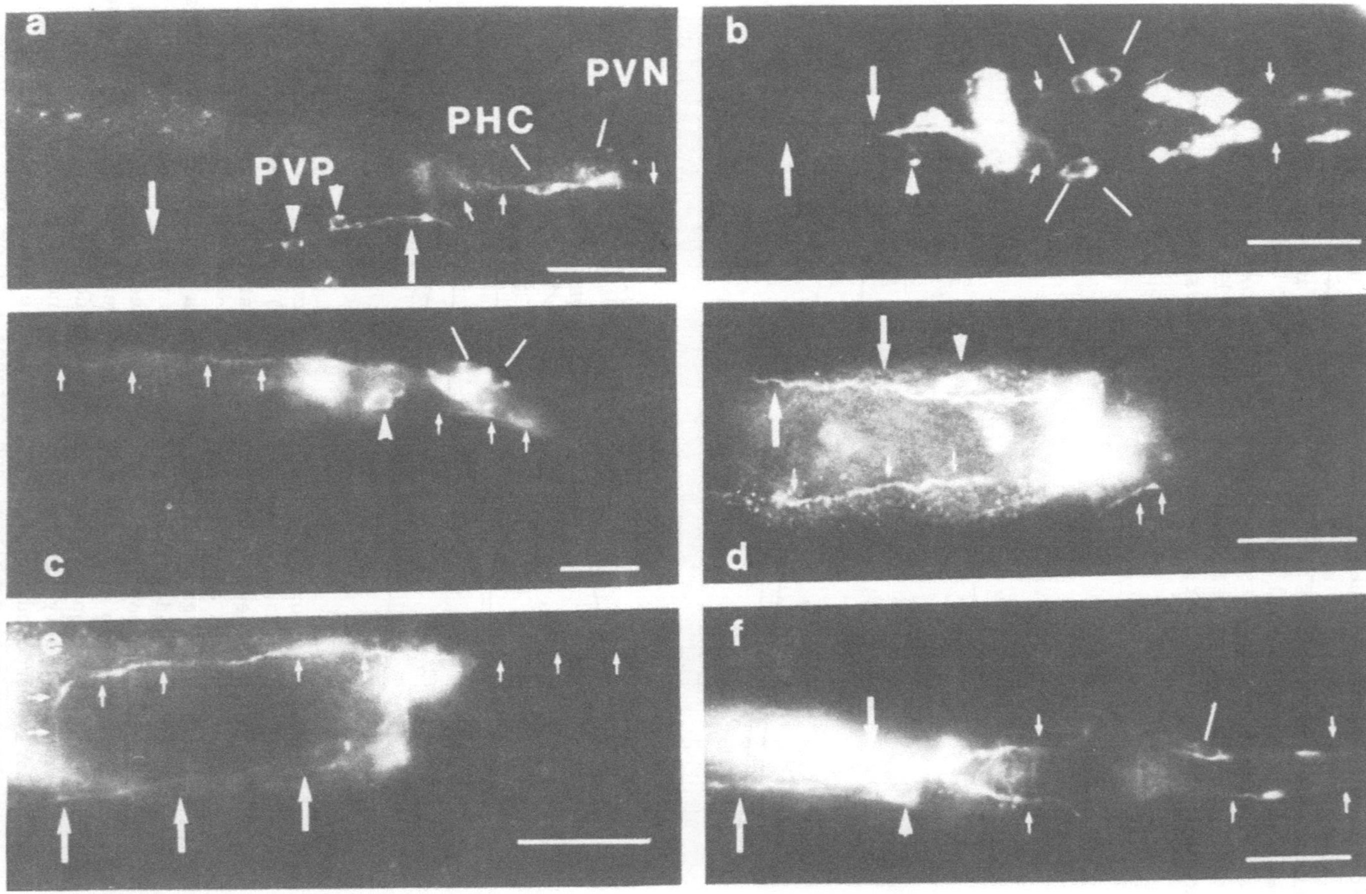

Fig. 4   (See facing page for legend)

Mutants with Abnormal Placement of PHC and PVN axons

We screened uncoordinated mutants representing all existing *unc* genes with anti-HRP antibodies, immunocytochemically. The pattern of axonal outgrowth and process placement for PHC and PVN neurons was found to be abnormal in mutants of nine genes, *unc-6*, *unc-13*, *unc-33*, *unc-44*, *unc-51*, *unc-61*, *unc-71*, *unc-73*, and *unc-98*(Siddiqui and Culotti, 1988). The mutants displayed a complex variety of axonal placement and outgrowth. Most commonly, the anteriorly directed processes of PHC and PVN grew in lateral positions, dorsal to the normal growth along the ventral cord. Mutant axonal processes terminated prematurely, or wandered aimlessly in all directions. Fig. 4 shows immunofluorescent staining of PHC and PVN neurons in the wild type and three mutants(*unc-13*, *unc-33*, and *unc-73*). Mutant axonal processes show growth along the lateral cords. Mutant in *unc-73* often display the abnormal lateral process veering into the ventral cord at a position much further anterior than in wild type, making a 'loop' with the ventral cord. On the other hand, *unc-76* mutants(known to be blocked in the anteriorly directed processes of phasmid PHA and PHB neurons, along the ventral cord; Hedgecock et al. 1985) show normal disposition of PHC and PVN neurons and their axonal processes along the ventral cord (Fig. 4 f). Remarkably, the fraction of mutant animals among the nine genes identified above, displaying the abnormal placement of PHC and PVN axons is much smaller than the uncoordinated phenotype of the animals, i.e. the penetrance of axonal defects ranges from 20-80%, whereas, the locomotory defect is 100% penetrant. Table 1 shows the fraction of animals showing abnormal placement of axons in mutants of the nine genes.

PHC and PVN axons in unc-6 mutants. The fraction of animals showing mutant axonal phenotype is smaller than 25%. The abnormal axons invariably grow much longer. Fig. 6. shows a random selection of mutant individuals with typical placement of PHC and PVN axons. Both left and right pairs of PHC/PVN displayed placement errors. The neuronal cell bodies were frequently mispositioned. Mutants in *unc-6* are known to harbor defects in PHA and PHB axons(Hedgecock et al., 1985, where it has been mentioned as *unc-106*). In rare cases, mutant animals displayed multiple branching of the posteriorly directed process of PHC neuron. We tested three different alleles of *unc-6* (*e78*, *ev 400*, and *e181*), which displayed very similar errors in the process placement (Table 1).

Fig. 4 Immunofluorescent staining of hermaphrodite tail with anti-HRP antibodies. Anterior is to the left. Bar represents 40 microns,

(a) Wild type, left lateral view. Anteriorly directed axons(small arrows) of PHC and PVN neurons, extend forward along the lumbar commissures, into the pre-anal ganglion(big arrows). PVP cell pair is also stained in the pre-anal ganglion. (b) Wild type, ventral view. The posteriorly directed process of PHC(curved arrow) is obscured due to the staining of phasmid sheath cell.

(c) *unc-13* mutant, left lateral aspect. Abnormal PHC/PVN axon is placed in a lateral position(small arrows).
(d) *unc-33* mutant, ventral aspect. Abnormal PHC/PVN anteriorly directed axon is placed in lateral position(small arrows).
(e) *unc-73* mutant, left lateral aspect. Anteriorly directed PHC/PVN axon is placed laterally(small arrows), much dorsal to the ventral nerve cord. The mutant process runs parallel to the ventral cord, projects ventrally by making a right angle turn, joins the ventral cord at a position much further anterior than in the wild type, and making a 'loop'.
(f) *unc-76* mutant, ventral aspect. Normal growth of PHC/PVN axons along the ventral cord.

PHC and PVN axons in unc-33, unc-44, unc-51, and unc-73 mutants. Mutant axons in this group displayed very complex and variable defects. The fraction of mutant animals displaying the errors of process placement is highest for these mutants, as it ranges from 30-74%(Table 1). The anteriorly directed mutant processes showed abnormal branching and bifurcation, premature termination, lateral growth in lateral ventral, lateral dorsal, and even the dorsal cord. Posteriorly extended axonal process of PHC was also abnormal, most frequently in unc-51 and unc-73 animals. Fig. 6 and Fig. 7, schematically show a random selection of typical errors seen in these mutants. In rare animals (2 of 160 animals), PVN had a posteriorly directed process, which is projected into the tail spike. Cell body positions of both PHC and PVN neurons were abnormal, and appeared somewhat smaller in size than in the wild type . Hedgecock and his colleagues(1985) have earlier shown that mutants in unc-33, unc-44, unc-51, and unc-76 have abnormal PHA and PHB phasmid axons, as they stop growing on entering the pre-anal ganglion region of the ventral cord.

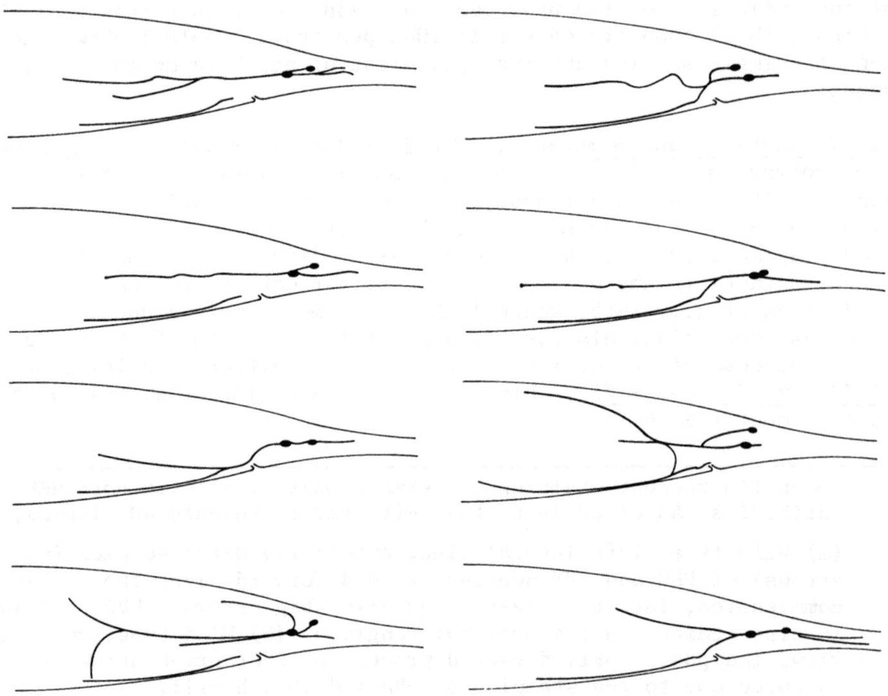

Fig. 5 Schematic free hand drawings of typical guidance errors in PHC and PVN axonal outgrowth and placement. Eight unc-6 mutant animals are shown, selected from a larger set. Filled circles are the cell bodies of the PHC, and PVN(located dorsal to PHC)neurons. Apparently both left and right neuron pairs showed equal frequencies of axonal abnormalities. Most frequntly, the mutant axons grew in lateral positions, and were much longer than in the wild type. The position of neuronal cell bodies is also altered in many individuals. In rare cases, the posteriorly directed process of PHC showed bifurcation, as it extended backward into the tail spike.

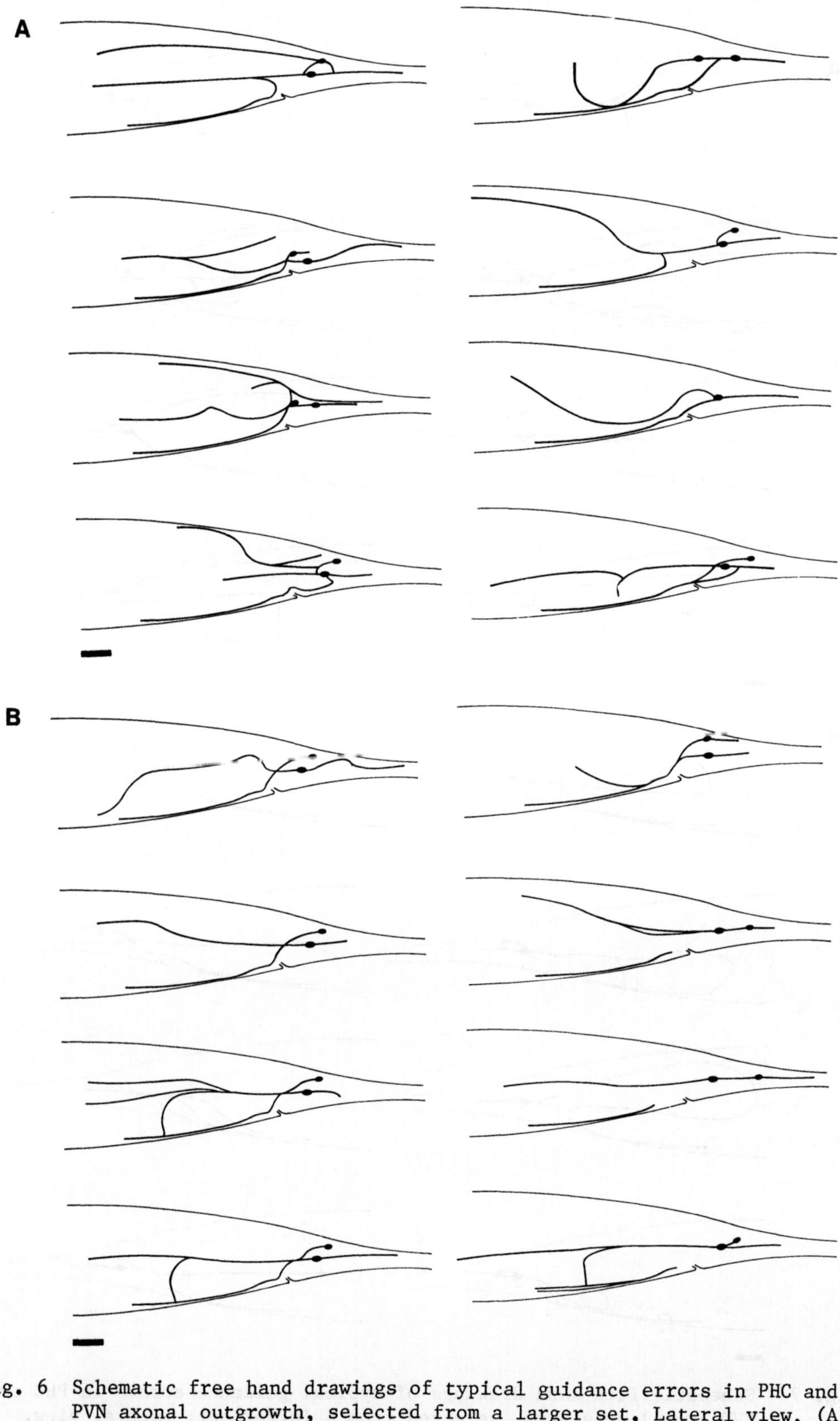

Fig. 6 Schematic free hand drawings of typical guidance errors in PHC and PVN axonal outgrowth, selected from a larger set. Lateral view, (A) unc-33, and (B) unc-44 mutants. Bar represents 20 microns.

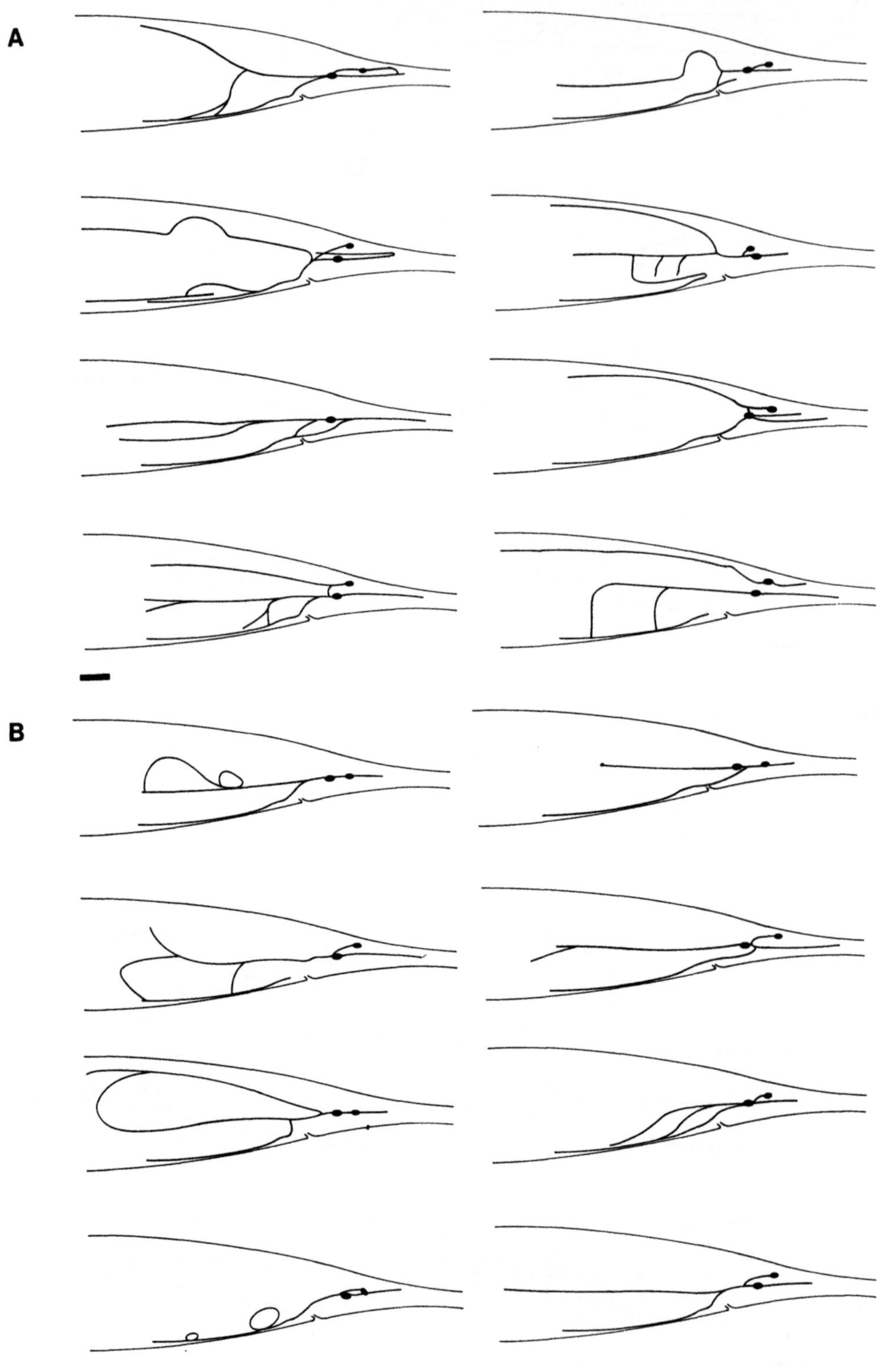

Fig. 7 Schematic free hand drawings of typical guidance errors in PHC and PVN axonal outgrowth, selected from a larger set. Lateral view. (A) unc-51, and (B) unc-73 mutants. Bar represents 20 microns.

Table 1. Summary of PHC and PVN axonal defects in uncoordinated mutants based on immunocytochemical staining with anti-HRP antibodies.

Gene	allele	animals tested for PHC/PVN normal	animals tested for PHC/PVN abnormal	percent abnormal
Wild type		177	1	very small
unc-6 X	e78	272	68	25%
	ev400	149	30	20%
	e181	166	36	22%
unc-33 IV	e204	130	67	51%
	e572	111	45	40%
	e735	120	41	33%
	e1193	164	50	30%
	m7	230	102	44%
unc-44 IV	e362	149	55	36%
	e427	180	85	47%
	e638	166	75	45%
unc-51 V	e369	238	160	67%
	e389	250	186	74%
	e432	135	66	48%
unc-13 I	e51	304	69	23%
	e312	156	44	28%
	e450	192	48	25%
unc-61 V	e228	275	91	33%
unc-71 III	e541	271	86	31%
unc-98 X	su130	388	79	20%
unc-76 V	e911	138	1	very small
unc-53 II	e404	111	1	very small
unc-73 I	e936	240	113	47%

The incidence of abnormal PHC/PVN axons was about the same for the left or right side of the lumbar ganglia.

The data presented here includes the results of Siddiqui and Culotti, 1988.

Only one allele is available for unc-61, unc-71 and unc-76 genes.

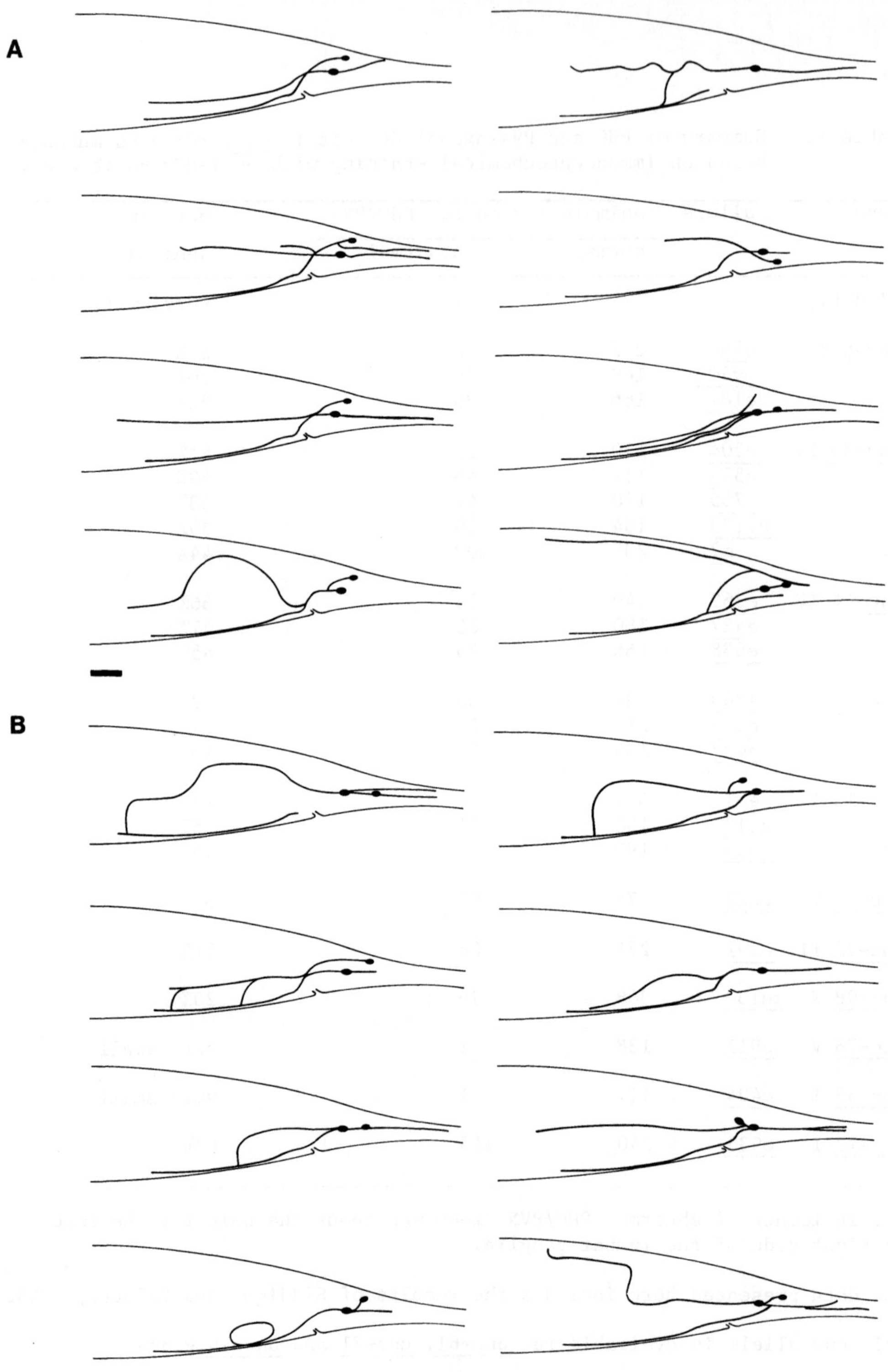

Fig. 8 Schematic free hand drawings of typical guidance errors in PHC and PVN axonal outgrowth, selected from a larger set. Lateral view. (A) unc-13, and (B) unc-61 mutants. Bar represents 20 microns.

A

B

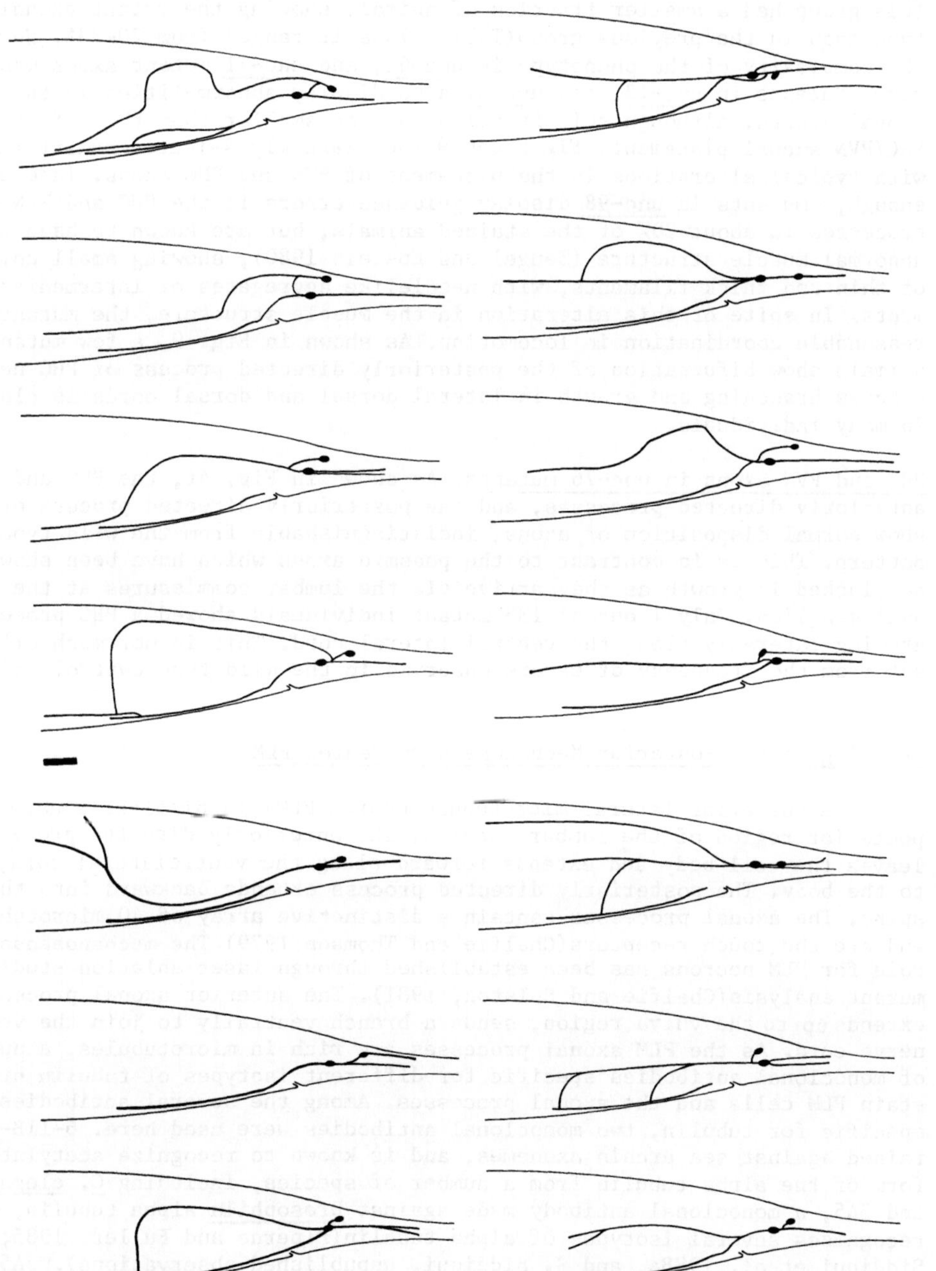

Fig. 9 Schematic free hand drawings of typical guidance errors in PHC and PVN axonal outgrowth, selected from a larger set. Lateral view. (A) unc-71, and (B) unc-98 mutants. Bar represents 20 microns.

PHC and PVN axons in unc-13, unc-61, unc-71, and unc-98 mutants. Mutants in this group had a smaller fraction of animals showing the mutant axonal phenotype than in the previous group(Table 1) as it ranged from 20-33%. However, the complexity of the phenotype in unc-61, and unc-71 mutant axons was the same. Mutants in unc-13, and unc-61 also display abnormalities in the phasmid axonal growth, although this fraction is even smaller than the defect in the PHC/PVN axonal placement. Fig 8 and 9 show randomly selected mutant animals with typical alterations in the placement of PVN and PHC axons. Interestingly enough, mutants in unc-98 display guidance errors in the PHC and PVN axonal processes in about 20% of the stained animals, but are known to have an abnormal muscle structure (Zengel and Epstein 1980), showing small collection of thin and thick filaments, with needlelike aggregates of intermediate filaments. In spite of this alteration in the muscle structure, the mutants show reasonable coordination in locomotion. As shown in Fig. 9, a few mutant animals show bifurcation of the posteriorly directed process of PHC neuron. Process branching and growth in lateral dorsal and dorsal cords is also seen in many individuals.

PHC and PVN axons in unc-76 mutants. As shown in Fig. 4f, the PHC and PVN anteriorly directed processes, and the posteriorly directed process of PHC, show normal disposition of axons, indistinguishable from the wild type pattern. This is in contrast to the phasmid axons which have been shown to be blocked in growth as they arrive via the lumbar commissures at the preanal ganglion. Only 1 out of 138 mutant individuals showed a PHC process growing laterally along the ventral lateral cord. This is not much different from the frequency of errors observed in the wild type control animals.

Staining of the Posterior Mechanosensory Neuron PLM

The posterior lateral microtubule neuron(PLM) is bipolar, located in the posterior region of the lumbar ganglia. The anteriorly directed process, leaves the cell body and extends forward along the ventrolateral cord, apposed to the body. The posteriorly directed process extends backward into the tail spike. The axonal processes contain a distinctive array of 40 microtubules, and are the touch receptors(Chalfie and Thomson 1979). The mechanosensory role for PLM neurons has been established through laser ablation studies and mutant analysis(Chalfie and Sulston, 1981). The anterior axonal process extends up to the vulva region, sends a branch ventrally to join the ventral nerve cord. As the PLM axonal processes are rich in microtubules, a number of monoclonal antibodies specific for different isotypes of tubulin brightly stain PLM cells and the axonal processes. Among the several antibodies specific for tubulin, two monoclonal antibodies were used here. 6-11B-1 was raised against sea urchin axonemes, and is known to recognize acetylated form of the alpha tubulin from a number of species, including *C. elegans*; and 3A5, a monoclonal antibody made against *Drosophila* alpha tubulin, which recognizes several isotypes of alpha tubulin(Piperno and Fuller, 1985; Siddiqui et al. 1988a, and S. Siddiqui, unpublished observations). 3A5 stains cell bodies and axonal processes of the entire set of neurons in *C. elegans*, but 6-11B-1 selectively stains the mechanosensory neurons, including PLM in the lumbar ganglia. The staining of PVM is faint compared to other touch neurons(Siddiqui et al., 1988). In addition, 6-11B-1 stains the unique PVR neuron in the right lumbar ganglion. Fig.10 shows the immunofluorescent staining of PLM neurons with 6-11B-1.

Mutants affected in the process placement of PLM neurons. We screened mutants in all the existing unc genes to examine the placement of PLM axons, with 6-11B-1 antibody, immunocytochemically. Interestingly, mutants in all the nine genes which we had identified using anti-HRP antibodies to study PHC/PVN axons displayed alterations in the axonal outgrowth and guidance of PLM axons. In addition, mutants in unc-76 and unc-53, also showed placement errors, and premature termination of PLM axonal processes.

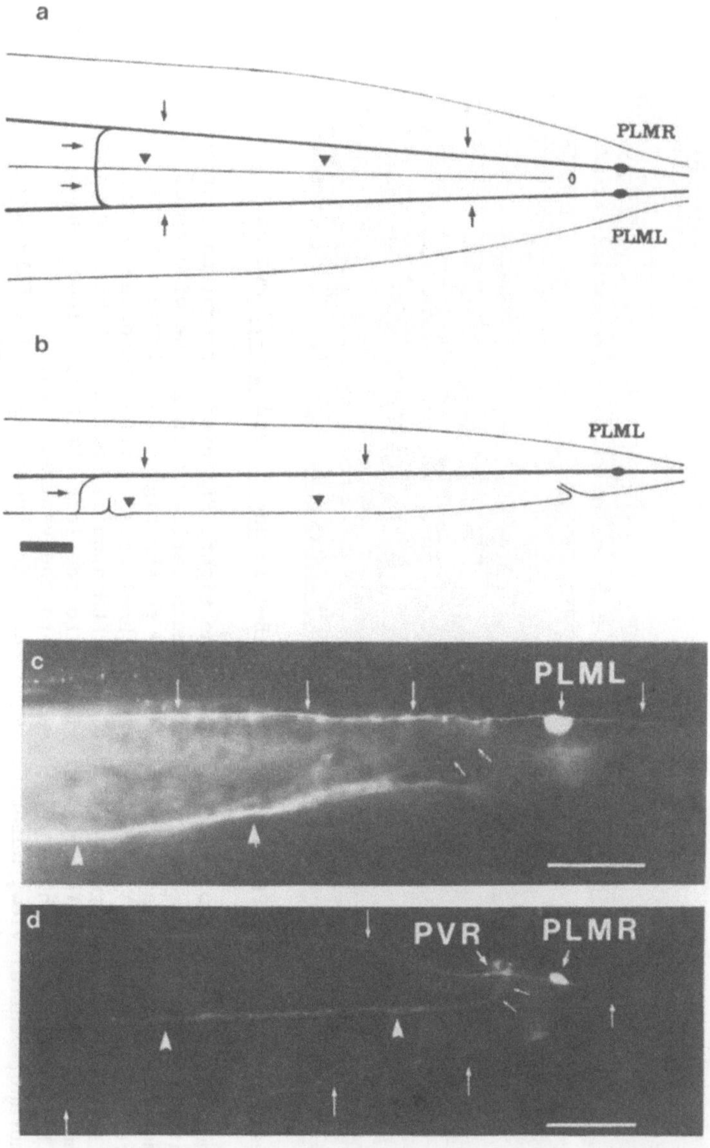

Fig. 10 Schematic drawings of PLM neurons(After Chalfie and Sulston, 1981 and White et al., 1986). Anterior is to the left.
(A) Both anteriorly, and posteriorly directed processes extend forward and backward from the PLM cell body. Ventral view. Anteriorly extended process grows along the ventral lateral cord (vertical arrows), and sends a branch to the ventral cord near the vulva region(horizontal arrows). Ventral cord is marked with the arrow heads. Empty hole at the posterior end of the ventral cord is anus. (B) Left lateral view. Scale bar is 40 microns.

Immunofluorescent staining of PLM neurons with 6-11B-1
(C) Left lateral aspect. Axonal processes of PLM(long arrows) extend anteriorly forward, laterally, and posteriorly backward into the tail spike. Ventral cord(arrow heads) also stain. PVR sends anteriorly directed process from the right lumbar ganglion (small arrows). (D) Ventral view, including PVR. Bar is 50 microns.

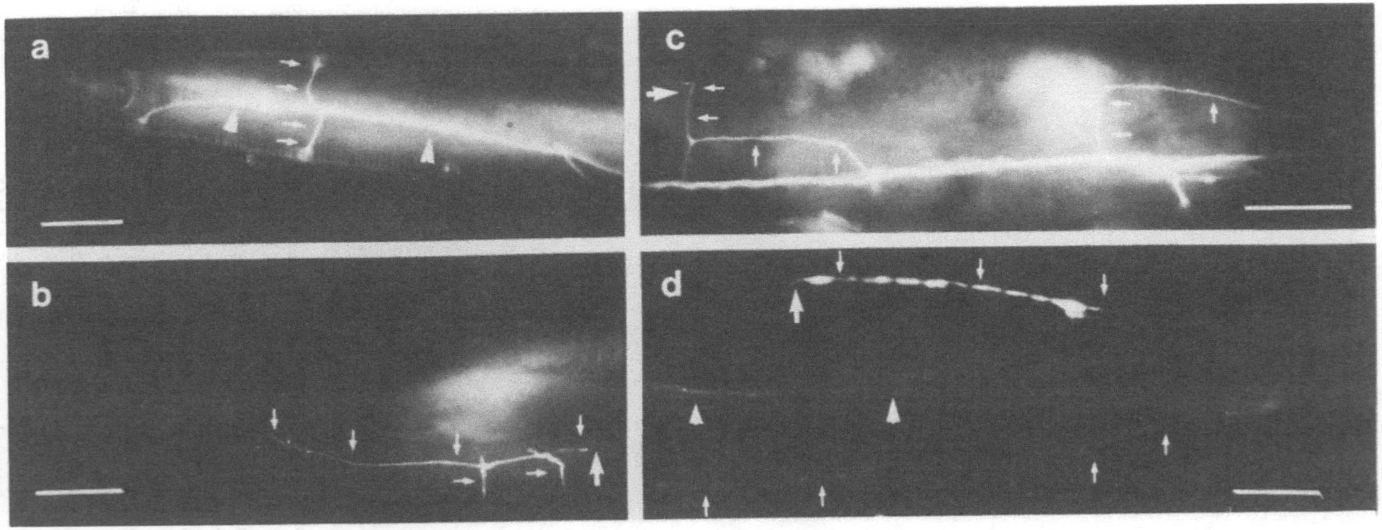

Fig. 11 Immunofluorescent staining of PLM neuron with 6-11B-1 in unc-53 and unc-73 mutants

(a) unc-53 mutant, ventral view, anterior is to the right. PLM axons branch(small arrows) into the ventral cord(arrow heads) at a position much earlier than in the wild type.
(b) unc-53 mutant, right lateral view, anterior is to the right. Anteriorly directed PLM axon(vertical small arrows) projects branches(horizontal arrows) into the ventral cord(not in this plane). Premature termination of PLM axon is marked with big a big arrow.
(c) unc-73 mutant, left lateral view, anterior is to the left. Multiple 'loop' phenotype caused by premature branching of PLM axons into the ventral cord.
(d) unc-73 mutant, ventral view, anterior is to the left. Abnormally thick PLMR axon with prominent varicosities(small arrows). Premature termination of PLMR is marked with a big arrow. Apparently the PLML is normal. Scale bar represents 50 microns.

PLM defects in unc-53 and unc-73 mutants. Among the mutants in 11 genes which we identified with 6-11B-1 and 3A5 antibodies, two mutants unc-53 and unc-73 share a common phenotype in the PLM axonal defects. These are described here, whereas, the detailed analysis of the other nine genes will be described elsewhere(S. Siddiqui, in preparation). Fig. 11 shows the immunofluorescent staining of unc-53, and unc-73 mutants, and Fig. 12 shows the schematic free hand drawings of mutant animals, selected from a larger set of 97 unc-53 animals, and 120 unc-73 animals. The anteriorly directed PLM axons, extend forward along the ventral lateral cord, but send a branch or project ventrally into the ventral cord, at a position more posterior than the wild-type(near the vulva region). Mutants in both unc-53 and unc-73 also display premature termination of PLM axons, multiple branching and multiple commissures into the ventral cord, making several loops with the fibers in the ventral cord. In addition, mutant axons, in many cases, appear much thicker than in the wild type, with prominent varicosities along the process. This phenotype is shown in Fig. 11d, for an unc-73 mutant individual. Interestingly the 'loop' phenotype of the anteriorly directed PLM axons in unc-53 and unc-73 mutants is different from the 'loop' phenotype of PHC/PVN axons in unc-73, in that PLM axons make a loop at somewhat regular intervals, and frequently more than one loop is observed. On the other hand, the loop phenotype of PHC/PVN axons has variable size, and in general, only one loop is seen(Fig. 4e).

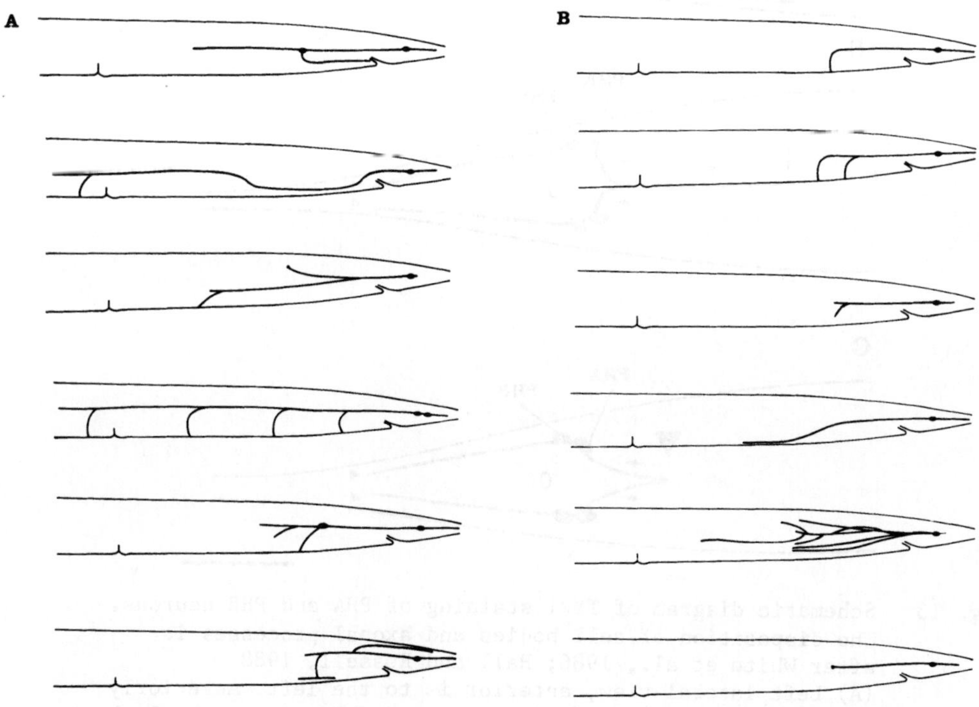

Fig. 12 Schematic free hand drawings of typical abnormalities in PLM axonal outgrowth and process placement in unc-53 and unc-73 mutants, based on immunofluorescent staining with 6-11B-1. Left lateral aspects. Anterior is to the left. (A) unc-53 mutants 51 of 97, showed abnormal axons, including the six individuals shown above. Abnormal branching, 'loop' formation, and premature termination are seen.(B) 44 of 120 animals showed defects in unc-73 Typical guidance errors are shown. Scale bar is 40 microns.

Antibody Staining of PHA and PHB Phasmid Neurons

Two pairs of phasmid neurons (PHA and PHB) are laterally located in the anterior region of the lumbar ganglia. Both PHA and PHB are bipolar neurons. Anteriorly directed axons from the cell bodies extend forward, via the lumbar commissures into the pre-anal ganglion, located at the posterior end of the ventral cord. Posteriorly directed axonal processes extend backward, into the tail spike, up to the opening of the phasmid sensilla (White et al., 1986; Hall and Russell, 1988). A monoclonal antibody TY21 raised against *C. elegans* crude homogenate stains both phasmid neurons and their axonal processes. Fig. 13 shows a schematic diagram of the phasmid neurons, and Fig. 14 shows the immunofluorescent staining due to TY21 in an adult hermaphrodite tail.

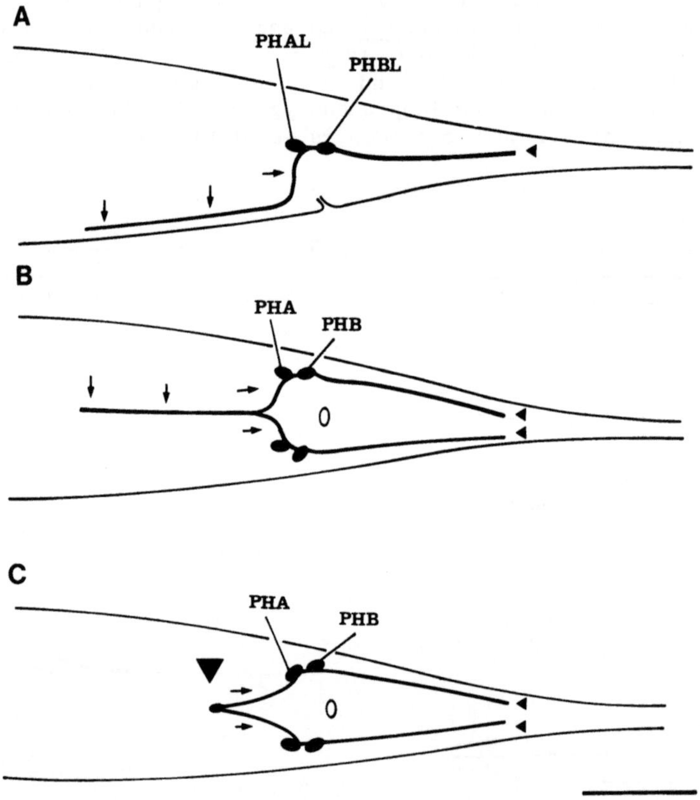

Fig. 13 Schematic diagram of TY21 staining of PHA and PHB neurons. The disposition of cell bodies and axonal processes is after White et al., 1986; Hall and Russell, 1988
(A) Left lateral view, anterior is to the left. Anteriorly directed phasmid axons(arrows) extend forward via the lumbar commissures(horizontal arrow) into the pre-anal ganglion (vertical arrows) and terminate in the ganglion. Posteriorly directed processes extend backward and terminate at the opening of the phasmid sensilla(small arrow head). (B) Ventral view. (C) *unc-73* mutant, ventral aspect. Anteriorly directed axons prematurely terminate as they grow along lumbar commissures to reach the posterior end of the pre-anal ganglion (big arrowhead). Posteriorly directed axons show normal growth.

Mutants with abnormal outgrowth and process placement of phasmid axons

We screened mutants in 11 genes identified previously by staining with anti-HRP and 6-11B-1 antibodies, immunocytochemically with TY21, to study the outgrowth and phasmid axonal phenotypes. Mutants in eight genes(unc-6, unc-13, unc-33, unc-44, unc-51, unc-61, unc-73, and unc-76) showed premature termination and abnormal placement of phasmid axons. Of these, five genes (unc-6, unc-33, unc-44, unc-51, and unc-76) were previously identified to harbor defects in the outgrowth of phasmid axons, by fluorescein dye filling in live animals(Hedgecock et al., 1985). The three new genes identified in the present study, share similar phenotype, albeit the axonal defect is less penetrant than reported for the four genes(unc-33, unc-44, unc-51, and unc-76) by Hedgecock et al. 1985. Fig. 14 b shows immunofluorescent staining of a unc-73 mutant displaying the premature termination of anteriorly directed axons of phasmid neurons. Among the three newly found genes(unc-13, unc-61 and unc-73), the penetrance of phasmid axonal defects were highest in unc-73, and as such could be grouped with the four other genes identified previously by the dye filling technique. The details of phasmid axonal defects in eight genes studied with TY21 will be published elsewhere.

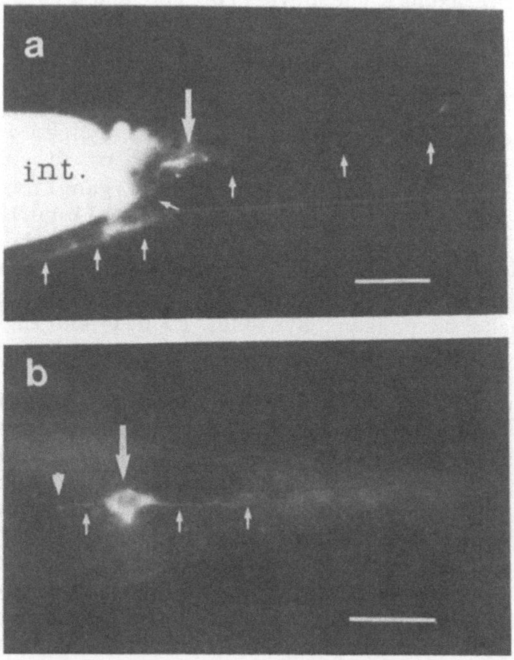

Fig. 14 Immunofluorescent staining of PHA and PHB neurons with TY21 Anterior is to the left. Scale bar represents 40 microns.
(a) Wild type, left lateral view. The cell bodies(big arrow) of phasmid neurons extend anteriorly and posteriorly directed processes(small arrows). Anteriorly directed process enters the pre-anal ganglion via the lumbar commissures(oblique arrow). Bright staining of the intestine(int.) is also visible.
(b) unc-73 mutant, left lateral view. Phasmid neurons(big arrow) extend porcesses(small arrows) anteriorly and posteriorly, but the anteriorly directed process terminates prematurely(arrowhead). The posteriorly directed process is apparently normal.

DISCUSSION

I have described mutants in 11 genes which are affected in the axonal outgrowth and process placement of three embryonic (PHA, PHB, and PLM), and two post-embryonic (PHC and PVN) lumbar neurons. Mutants in these genes could be of use in two important ways. (a) To study the role of sensory neurons PHA, PHB, PHC, and PLM, and the interneuron PVN, in the strucuture and function of the nematode nervous system in general, and in the tail neural circuitry in particular. (b) It is likely, that some of these mutants may be lacking in the molecules critical for the neural specificity and synapse formation, and therefore may reveal the nature of molecules essential for neural circuitry.

It is clear from the results described in this paper, that the mutants have no simple phenotype which could be grouped in a distinct class. Axonal defects are not limited to one type of receptor neuron, or follow a very specific pattern. This is in contrast to the phasmid axonal defects found by fluorescein dye filling, in unc-33, unc-44, unc-51, and unc-76 mutants (Hedgecock et al., 1985). Another remarkable difference is that while the phasmid axonal defect of premature termination is highly penetrant, axonal errors for PHC/PVN and PLM neurons are variable, and in majority of animals, these axons do not show any apparent misplacement. Table 1 summarizes the percentage of abnormal axonal phenotype for PHC and PVN neurons in different mutants.

Neural Circuitry of the Tail Nervous System

The major neuropil of the posterior nervous system is the pre-anal ganglion, essentially a fiber bundle, consisting of fibers from the neurons located in the head, lumbar ganglia, dorsorectal ganglion, and those located in the pre-anal ganglion. In spite of the very small number of neurons in *C. elegans*(302), of which 40 are located in the tail, the morphology and branching pattern of neurons is very simple. Most of these neurons are simple monopolar or bipolar cells. Axonal fibers run parallel to each other in fiber bundles and make *en passant* synapses to neighboring processes(White et al., 1983, 1986; Hall and Russell, 1988).

Sensory receptor neurons. Based on ultrastructural analysis, and geographical location, PHA, PHB, PHC, PVR, PQR, and PLM have been classified as sensory receptors(White et al., 1986, Hall and Russell, 1988). Only in the case of PLM, it has been firmly established, both by laser ablation studies and by mutant analysis, that it serves as the posterior mechanosensory touch receptor(Chalfie and Sulston, 1981, and Chalfie et al., 1985). Phasmid neurons PHA and PHB, are most likely the chemosensory receptors, as they resemble the anterior amphid sensilla, located in the head. PVR may have a mechanosensory role, perhaps in the modulation of the touch response in tail, as it makes synaptic contacts with all of the six touch receptor neurons (Chalfie et al. 1985), and shares the neural antigen recognized by 6-11B-1, which stains the cell bodies of all touch receptor neurons(Siddiqui et al., 1988a and Siddiqui,1988). PHC cells extend a posteriorly directed process backward into the tail spike, very close to the cuticular surface, suggesting a mechanosensory role for these cells(Hall and Russell, 1988).

Interneurons in the tail circuitry. As shown in Fig. 15, two lumbar neurons LUA and PVC, along with DVA in the dorsorectal ganglion, and AVA and AVD whose cell bodies lie in the head ganglia, serve as the major interneurons of the tail circuitry. PVN has a dual role as an interneuron, and as it also synapses onto muscles, as a motor neuron. Remarkably, PVN unlike other cells, is highly branched, especially in the anterior region of the ventral nerve cord.

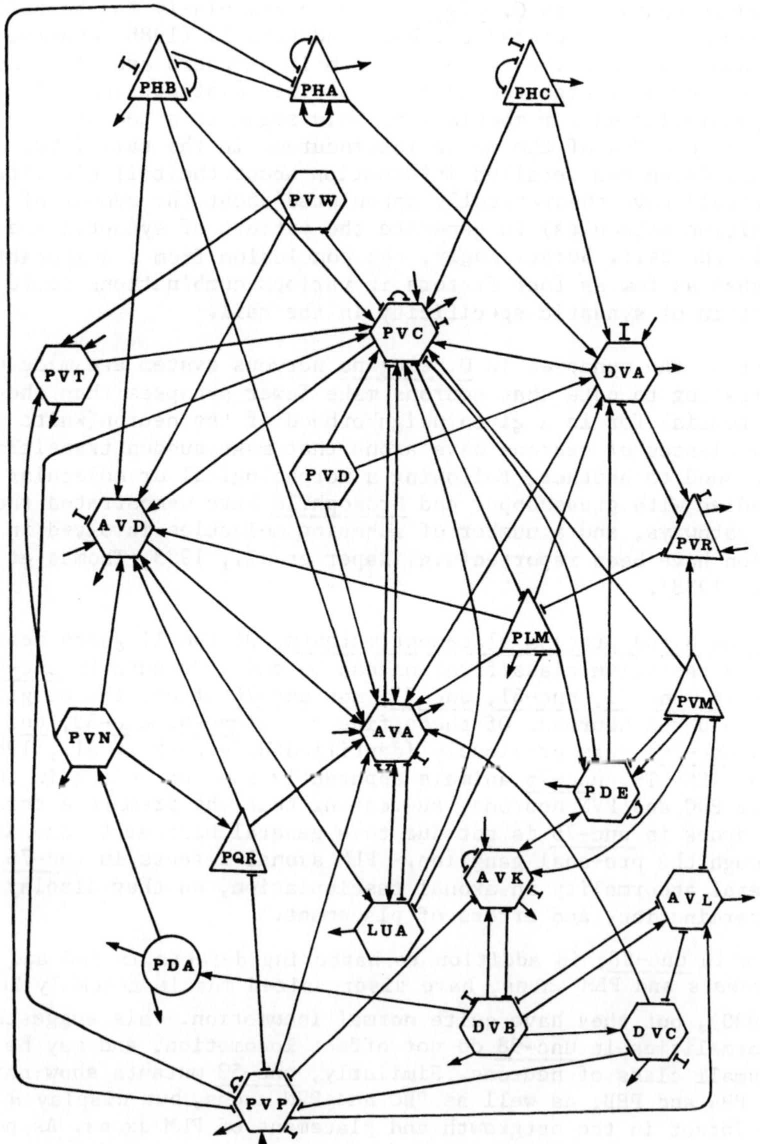

Fig. 15 Neural circuitry of the hermaphrodite tail nervous system (drawn after White et al., 1986, and Hall and Russell, 1988) All receptor neurons(PHA, PHB, PHC,PVR, PLM, PQR) shown by triangles are located in the lumbar ganglia. Motor neurons are shown by circles, and interneurons by hexagons. Chemical synapses are drawn as arrows and the gap junctions are shown by (T) in the diagram. Also see Fig. 2 for the disposition of neuronal cell bodies in the tail nervous system. PVM and PDE, lie in the posterior body, anterior to the pre-anal ganglion and posterior to the vulva. For details of synaptic partners , refer to White et al., 1986, 1988, and Hall and Russell, 1988.

Synapse formation in the tail circuitry. Most of the synapses of the tail nervous system are formed within the pre-anal ganglion. Remarkably, most of the synaptic contacts in C. elegans, and particularly in the tail, are dyadic or triangular in character. Hall and Russell (1988) observe that in the pre-anal ganglion the two post-synaptic partners are never homologous, and therefore could diverge the information into distinct channels, with different possibilities for modulation. This suggests a convergence of information onto a few of the major interneurons in the tail (AVD, AVA, LUA and PVC). Given the detailed information about the tail circuitry, Hall and Russell have theoretically speculated about the number of factors (say recognition molecules) to generate the pattern of synaptic specificity observed in the tail. Surprisingly, the conclusion from a computational study was that as few as four factors in various combinations could account for the pattern of synaptic specificity in the tail.

As most of the synapses in C. elegans nervous system are made en passant, it is interesting to note that neurons make fewer synapses than they have the potential for in a given neighborhood of the neuron(White et al., 1983). Some classes of neurons have axons that make sudden transitions from one neighborhood to another, following a morphological or molecular cue. Elegant studies with grasshopper and Drosophila have demostrated the notion of labeled pathways, and a number of adhesion molecules involved in selective fasciculation have been reported(e.g. Raper et al., 1983, Thomas et al., 1984, Snow et al., 1988).

Axonal outgrowth and process placement mutants. Of the 11 genes described in this paper, a tentative classification can be made. Mutants in unc-6, unc-13, unc-33, unc-44, unc-51, unc-61, unc-73, and unc-76 affect the outgrowth of phasmid PHA and PHB neurons. Of these, five genes(unc-6, unc-33, unc-44, unc-51, and unc-76) were previously identified(Hedgecock et al., 1985). As shown in the RESULTS, unc-76 animals apparently show no defect in the axonal processes of PHC and PVN neurons, suggesting that the premature termination of phasmid axons in unc-76 is not due to a general barrier to axonal outgrowth through the pre-anal ganglion. PLM axonal defects in unc-76 suggest a more general abnormality in axonal fasciculation, as they display both premature termination, and errors of placement.

Mutants in unc-98, in addition to harboring defects in PHC and PVN axonal processes and PLM axons, have disorganized muscle assembly(Zengel and Epstein, 1980), but they have quite normal locomotion. This suggests that axonal abnormalities in unc-98 do not affect locomotion, and may be limited to a very small class of neurons. Similarly, unc-53 mutants show normal phasmidial PHA and PHB, as well as PHC and PVN axons, but display a very stereotype defect in the outgrowth and placement of PLM axons. As pointed out by Hedgecock et al., 1987, the defects may be restricted to the posterior body. The PLM phenotype of unc-53 ('loop formation') by the premature branching of the anteriorly directed axon, into the ventral cord is shared by unc-73.

Molecular nature of defects in the mutants. A component of the cytoskeleton, perhaps microtubules have been implicated in the axonal defects observed in unc-33 and unc-44 mutants(Hedgecock et al., 1985). However, mutants in unc-6 which have an unusual phasmid axonal, and PHC and PVN axonal defects also affect ventral, circumferential growth of sensory axons, and the ventral migration of mesodernal cells(Hedgecok et al., 1987). Recently, Hedgecock and his colleagues have cloned unc-6 gene by transposon tagging and the part of the sequenced gene suggests it is a B2 laminin(Hedgecock, E. unpublished results). As the technique of transposon tagging is widely used in C. elegans, (Greenwald et al., 1985), it is likely that many of the genes identified in this study will be cloned and their molecular identity known(J. Culotti, A. Otsuka, R. Herman, R. Horvitz, and our unpublished results).

Future directions. Axonal growth mutants harbor defects in a number of axonal processes. These defects could be primarily limited to the neurons, muscles or other tissues. Genetic mosaics provide a powerful technique to determine the cellular focus of these mutations, especially since the entire cell lineage is completely known in C. elegans. Recently, Herman(1984) has vastly improved over the technique of generating mosaics by X-irradiation of heterozygotes(Siddiqui and Babu, 1980), and generated mosaics by somatic loss of chromosomal duplications. This technique is now generally used in C. elegans, and can be applied to the axonal growth and guidance mutants.

Laser ablation of specific cells(Sulston and White, 1980), has a great potential to study the role of specific neurons in the neural circuitry. For example, Chalfie and others(1985) have determined the neural circuitry mediating touch response, based on neuroantomical data from serial section electron micrographs and testing specific cells by laser ablations. Such laser ablation studies, combined with mutant analysis could reveal the role of certain cells in the function of nervous system.

Progress has been made in identifying antibodies which recognize specific subsets of neurons and axonal bundles(Okamoto and Thomson, 1985; S. Siddiqui unpublished results). These antibodies can be used to determine the cellular and molecular specificity of the neural antigens, and in the molecular cloning of antigens of interest by direct screening of expression libraries in E. coli (unpublished results). Finally, the techniques of molecular genetics are widely applicable, and have been used to clone genes defined only by mutations(e.g. transposon tagging of lin-12 by Greenwald et al., in C. elegans. Physical mapping of the entire genome of C. elegans(Coulson et al., 1986) will greatly fascilitate molecular cloning of genes, including those affecting nervous system. This, in combination with the possibility of microinjecting molecules of interest in the nematode(Fire 1986), could allow experimental manipulations to establish the role of specific molecules in the structure and function of the nervous system. In conclusion, I have described 11 genes affecting axonal growth and process placement in C. elegans, which could be a good substrate to study the molecules critical for the formation and function of neural networks in a simple model system.

REFERENCES

Albertson, D. G., and Thomson, J. N. 1976. The pharynx of C. elegans. Phil. Trans. R. Soc. B. 275:299-325.
Brenner, S. 1973. The genetics of behaviour. Brit. Med. Bull. 29:269-271.
Brenner, S. 1974. The genetics of Caenorhabditis elegans. Genetics 77:71-94
Chalfie, M. and Thomson, J. N. 1979. Organization of neuronal microtubules in the nematode Caenorhabditis elegans. J. Cell Biol. 82:278-289.
Chalfie, M. and Sulston, J. E. 1981. Developmental genetics of the mechanosensory neurons of Caenorhabditis elegans. Dev. Biol. 82:358-370.
Chalfie, M., Sulston, J. E., White, J. G., Southgate, E., Thomson, J. N., and Brenner, S. 1985. J. Neurosci. 5:956-964.
Chalfie, M., and White, J. G. 1988. Nervous system, pp.337-391, in:Caenorhabditis elegans, W. B. Wood, ed. Cold Spring Harbor Labs. New York.
Coulson, A., Sulston, J., Brenner, S., and Karn, J. 1986. Towards a physical map of the genome of the nematode Caenorhabditis elegans. Proc. Natl. Acad. Sci. 83:7821-7825.
Eisen, J. S. 1988. Growth cone guidance and pathway formation. Trends in Neurosci. 11:333-335.
Fire, A. 1986. Integrative transformation of Caenorhabditis elegans. EMBO J. 5:2673-2680.
Greenwald, I. 1985. lin-12, a nematode homeotic gene, is homologous to a set

of mammalian proteins that includes epidermal growth factor. Cell. 43: 583-590.

Hall, D., and Russell, R. L. 1988. Electron microscopic anatomy of the posterior nervous system of the nematode Caenorhabditis elegans. J. Comp. Neurol. (in press).

Hedgecock, E. M., Culotti, J. G., Hall, D. H., and Stern, B. D. 1987. Genetics of cell and axon migrations in Caenorhabditis elegans. Development. 100:365-382.

Hedgecock, E. M., Culotti, J. G., Thomson, J. N., and Perkins, L. A. 1985. Axonal guidance mutants of Caenorhabditis elegans identified by filling sensory neurons with fluorescein dyes. Dev. Biol. 111:158-170.

Herman, R. K. 1984. Analysis of genetic mosaics of the nematode Caenorhabditis elegans. Genetics. 108:165-180.

Horvitz, H. R., Chalfie, M., Trent, C., Sulston, J. E., and Evans, P. D. 1982. Serotonin and octapamine in the nematode Caenorhabditis elegans. Science. 216:1012-1014.

Kimble, J., and Hirsh, D. 1979. The postembryonic cell lineages of the hermaphrodite and male gonads in Caenorhabditis elegans. Dev. Biol. 70:396-417.

Okamoto, H., and Thomson, N. J. 1985. Monoclonal antibodies which distinguish certain classes of neuronal and supporting cells in the nervous tissue of the nematode Caenorhabditis elegans. J. Neurosci. 5:643-653.

Piperno, G., and Fuller, M. T. 1985. Monoclonal antibodies specific for an acetylated form of alpha tubulin recognize the antigen in cilia and flagella from a variety of organisms. J. Cell Biol. 101:2085-2094.

Raper, J. A., Bastiani, M., and Goodman, C. S. 1983. Pathfinding by neural growth cones in grasshopper embryos. J. Neurosci. 3:31-41.

Siddiqui, S. S., and Babu, P. 1980. Genetic mosaics of Caenorhabditis elegans A tissue specific fluorescent mutant. Science. 210:330-332.

Siddiqui, S. S., and Culotti, J. G. 1984. A neural antigen conserved in different invertebrates. Ann. N. Y. Acad. Sci. 435:341-343.

Siddiqui, S. S., Aamodt, E., Rastinejad, F., and Culotti, J. G. 1988. Molecular and cellular characterization of tubulin isotypes in Caenorhabditis elegans. J. Neurosci. (in press).

Snow, P. M., Zinn, K., Harrelson, A. L., McAllister, L., Schilling, J., Bastiani, M., Makk, G., and Goodman, C. S. 1988. Characterization and cloning of fasciclin I and fasciclin II glycoproteins in the grasshopper. Proc. Natl. Acad. Sci. 85:5291-5295.

Suslton, J. E., Dew, M., and Brenner, S. 1975. Dopaminergic neurons in the nematode Caenorhabditis elegans. J. Comp. Neurol. 163:215-226.

Sulston, J. E., and Horvitz, H. R. 1977. Postembryonic cell lineages of the nematode Caenorhabditis elegans. Dev. Biol. 56:110-156.

Suslton, J. E. and Horvitz, H. R. 1981. Abnormal cell lineages in the mutants of the nematode Caenorhabditis elegans. Dev. Biol. 82:41-55.

Suslton, J. E., Schierenberg, E., White, J. G., and Thomson, J. N. 1983. The embryonic cell lineage of the nematode Caenorhabditis elegans. Dev. Biol. 100:64-119.

Sulston, J. E., and White, J. G. 1980. Regulation and cell autonomy during postembryonic development of Caenorhabditis elegans. Dev. Biol. 78: 577-597.

Swanson, M., Edgely, M. L., and Riddle, D. L. 1984. Caenorhabditis elegans, pp286-299. In: Genetic maps. (ed.) S. O'Brien. Cold Spring Harbor Labs. New York.

Thomas, J. B., Bastiani, M., Bate, M., and Goodman, C. S. 1984. From grasshopper to Drosohila: a common plan for neuronal development. Nature 310:203-207.

Ward, S., Thomson, N. J., White, J. G., and Brenner, S. 1975. Electron microscopical reconstruction of the anterior sensory anatomy of the nematode Caenorhabditis elegans. J. Comp. Neurol. 160:313-337.

Ware, R. W., Clark, D., Crossland, K., and Russell, R. L. 1975. The nerve ring of the nematode Caenorhabditis elegans. J. Comp. Neurol. 162:71-110.

Waterston, R. H. 1988. Muscle, pp281-335. In: The nematode *Caenorhabditis elegans*. (ed.) W. B. Wood. Cold Spring Harbor Lab. New York.

White, J. G., Southgate, E., Thomson, J. N., and Brenner, S. 1976. The structure of the ventral cord of *Caenorhabditis elegans*. *Phil*. *Trans*. *Roy*. *Soc*. *B*. 275:327-348.

White, J. G., Southgate, E., Thomson, J. N., and Brenner, S. 1983. Factors that determine connectivity in the nervous system of *Caenorhabditis elegans*. *Cold Spring Harbor Symp*. *Quant*. *Biol*. 48:633-640.

White, J. G., Southgate, E., Thomson, J. N., and Brenner, S. 1986. The structure of the nervous system of the nematode *Caenorhabditis elegans*. *Phil*. *Trans*. *Roy*. *Soc*. *B*. 314:1-340.

White, J. G., Southgate, E., and Durbin, R. 1988. Neuroanatomy, pp433-455. In: The nematode *Caenorhabditis elegans*. (ed.) W. B. Wood. Cold Spring Harbor Lab. New York.

Zengel, J. M., and Epstein, H. F. 1980. Identification of genetic elements associated with muscle structure in the nematode *Caenorhabditis elegans*. *Cell Motil*. 1:73-97.

ACKNOWLEDGEMENTS

I thank Prof. Y. Hotta for his encouragement and discussions. Part of this work was conducted during my stay at the Northwestern University, Evanston, Illinois. I thank J. Culotti, R. Holmgren, and W. Klein for their support. E. Aamodt, M. Chalfie, J. Culotti, E. Hedgecock, J. Miwa, J. Sulston and J. White, provided valuable suggestions for which I am grateful. The work in Toyohashi was supported by a grant in aid from Ministry of Education Science, and Culture, Japan; NEC Corporation of Japan; and a special research grant from the President of the Toyohashi University of Technology. I thank members of my laboratory for their help with the manuscript.

RECENT PROGRESS IN THE IDENTIFICATION OF NEUROTRANSMITTERS USED BY VERTEBRATE AND INVERTEBRATE PHOTORECEPTORS

P. Vijay Sarthy

Departments of Ophthalmology, Physiology and Biophysics
University of Washington
Seattle, WA 98195, USA

ABSTRACT

In both vertebrates and invertebrates, the synaptic connections and electrophysiological properties of photoreceptors have been well documented. The neurotransmitters that mediate signal transmission from these cells are, however, less well understood. Several lines of evidence strongly suggest that L-glutamate is the transmitter used by cone photoreceptors. L-glutamate and L-aspartate are also likely to be the transmitter employed by rod photoreceptors in vertebrates. Recent biochemical, pharmacological and immunocytochemical data suggest that histamine is likely to be the transmitter used by insect and possibly other invertebrate photoreceptors.

INTRODUCTION

The morphological characteristics and physiological responses of photoreceptor cells have been well studied in both invertebrate and vertebrate retinas (reviewed in Dowling, 1987). It is clear that while invertebrate photoreceptors respond to light stimulus by depolarization, the vertebrate photoreceptors hyperpolarize resulting in suppression of endogenous transmitter release.

Intracellular recordings from second order neurons, the bipolars and horizontal cells, show that for a substance to be considered a putative photoreceptor neurotransmitter, it must: (a) depolarize horizontal cells by increasing their membrane conductance; (b) hyperpolarize the depolarizing bipolar cells by decreasing their membrane conductance and (c) depolarize the hyperpolarizing bipolars by increasing their membrane conductance. This complex set of postsynaptic actions raises the question as to whether the rod and cone photoreceptors use a multiplicity of neurotransmitters, or alternatively whether a single transmitter acts on different receptors each associated with a specific ion channel.

The vertebrate retina contains substances that are known or suspected to be transmitters elsewhere in the CNS (Graham, 1974). Among these substances, the excitatory amino acids, aspartate and glutamate, ACh, GABA and taurine have been proposed as the photoreceptor transmitter.

Glutamate as the cone transmitter

In 1955 Furukawa and Hanawa reported that application of L-aspartate to retina abolished the b-wave of the ERG while the a-wave was unaffected. Subsequently, intracellular recordings from second order neurons showed that application of 0.1 M aspartate or glutamate resulted in depolarization of horizontal cells while photoreceptors were largely unaffected (Cervetto and McNichol, 1972; Dowling and Ripps, 1972; Murakami et al., 1972). The demonstration that exogenously applied L-aspartate and glutamate produced responses of opposite polarity in ON- and OFF-bipolars strongly suggested that aspartate or glutamate could be the photoreceptor neurotransmitter (Murakami et al., 1975; Kaneko and Shimazaki, 1975; Negishi and Drujan, 1979; Kondo and Toyoda, 1980; Shiells et al., 1981; Slaughter and Miller, 1981; Bloomfield and Dowling, 1985). These studies also suggested that a single transmitter released from photoreceptors could influence the response of secondary neurons possibly by acting on different types of receptors (Miller and Slaughter, 1986).

A major limitation of these studies was that extremely high concentrations (mM range) of aspartate and glutamate were required to elicit response in secondary neurons. It was, however, argued that the high concentrations were needed to overcome the diffusion barriers and potent uptake mechanisms that exist in the retina (see Ishida and Fain, 1981). Recently, L-glutamate and its analogs have been shown to be effective at more physiological (μM) concentrations when applied to dissociated horizontal and bipolar cells; at the same concentrations, L- or D-aspartate were less effective (Ishida, 1984; Ishida et al., 1983; Lasater and Dowling, 1982; Lasater et al., 1984)

Kinetic experiments show that the mammalian retina contains a Na^+-dependent, high affinity uptake system with a $K_m \sim 10\ \mu M$ (White and Neal, 1976; Sarthy et al., 1986). Autoradiographic experiments have been carried out to examine whether aspartate and glutamate uptake is localized to the rod and cone photoreceptors. In the monkey retina, while rods accumulate both aspartate and glutamate, cones accumulate only glutamate (Sarthy et al., 1986). This pattern of uptake is seen both in the fovea and in other regions of the retina. Similarly, in the human retina, while rods accumulate both amino acids, cones show heavier labeling with L-glutamate (Lam and Hollyfield, 1980). A preferential uptake of these amino acids has also been noted in other vertebrate retinas (Ehinger, 1981; Marc and Lam, 1981).

L-Aspartate aminotransferase (AAT) is an important enzyme in L-glutamate metabolism that catalyzes the formation of glutamate from α-ketoglutarate (Fonnum, 1984). Cytoplasmic AAT has been suggested as a useful histochemical marker for glutamergic neurons in the CNS (Altschuler et al., 1981). Immunocytochemical studies show that anti-AAT preferentially stains the cone somata, inner segments and synaptic pedicles, while the outer segments are unstained. The rods, however, are not stained (Sarthy et al., 1986). Similar results have also been obtained with other mammalian retinas (Altschuler et al., 1982; Brandon and Lam, 1983). The uptake autoradiography and immunocytochemical data point to the existence of major differences between rods and cones in the uptake and metabolism of L-aspartate and L-glutamate, and are consistent with L-glutamate being a cone transmitter.

In order to establish that L-glutamate is a cone transmitter, it is important to demonstrate either that L-glutamate release occurs from cones in response to depolarization or that hyperpolarization shuts off glutamate release. Although Ca^{2+}-dependent release of glutamate and aspartate has been observed from the retina, it has been rather difficult to prove that the amino acids come from photoreceptors (Neal et al., 1979; Miller and Schwartz, 1983). Recently, two groups of investigators have approached this problem by entirely different techniques. Copenhagen and Jahr (1988) used outside-out patches of membrane from neonatal hippocampal neurons as a source of N-methyl D-aspartate (NMDA) receptors. When these patches were positioned close to solitary rods and cones that were being electrically stimulated, a tenfold increase in the frequency of channel openings was noted. The channel open times and conductance were similar to that reported for NMDA-gated channels. These data suggest that photoreceptor depolarization leads to the efflux of excitatory amino acids. The Ca^{2+}-dependency of the release process remains to be established.

Ayoub et al. (1988) employed a highly sensitive NADH-coupled assay to measure glutamate release from single cones isolated from the spiny lizard retina. They found that K^+-depolarization leads to an increase in L-glutamate release from basal levels to about 1.5 μM glutamate/sec. Moreover, this increase is suppressed in the presence of 1 mM Co^{2+} or 20 mM Mg^{2+}. It was also found that only the principal member of the double cones releases L-glutamate while the accessory cone does not. An unusual feed-forward loop has been proposed for the autoregulation of transmitter release from cones by L-glutamate (Sarantis et al., 1988; see also Kaneko and Tachibana, 1987).

Taken together these lines of evidence show that cone photoreceptors are likely to use L-glutamate as their neurotransmitter. None of the results obtained so far, however, rule out the possibility that in addition to glutamate, aspartate is also be a cone transmitter. It appears that excitatory transmitters, probably aspartate or glutamate, are used by rods in synaptic transmission.

Other transmitter candidates

In addition to the excitatory neurotransmitters, other substances such as ACh, GABA and taurine have also been suggested as photoreceptor transmitters. Turtle cones have been shown to synthesize and accumulate ACh (Lam, 1972; Sarthy and Lam, 1979). Nicotinic and muscarinic receptors have been demonstrated on secondary retinal neurons (Vogel and Nirenberg, 1976; Gerschenfeld and Piccolino, 1977; Yazulla and Schmidt, 1976; Schwartz and Bok, 1979; James and Klein, 1985). The presence of GABA- and GAD-immunostaining in a small population (~25%) of monkey photoreceptors has been proposed as evidence that these cells are GABAergic (Nishimura et al., 1986). Recent, *in situ* hybridization data, however, show that GAD mRNA is absent from monkey photoreceptors (Fu et al., 1988). Finally, large concentrations of taurine are present in photoreceptors although its function is not understood (Sturman and Hayes, 1980).

Histamine as invertebrate photoreceptor transmitter

As with vertebrate photoreceptors, several substances have been suggested as transmitter candidates for invertebrate photoreceptors. These include, Ach, dopamine, aspartate, glutamate, taurine, GABA and histamine (Autrum and Hoffman, 1957; Konopka, 1972; Campos-Ortega, 1974; Langer et al., 1976; Klingman and Chappell, 1978; Hall, 1982; Whitton et al., 1987; Datum et al., 1986; Meyer et al., 1986; Maxwell et al., 1978; Elias and Evans, 1983; Hardie, 1987). Immunocytochemical studies with GABA- and GAD-specific antibodies suggest that R7 photoreceptors and C2 centrifugal fibers in the blowfly visual system might be GABAergic (Datum et al., 1986; Meyer et al., 1986). In octopus retina, synthesis and accumulation of both Ach and dopamine have been reported (Lam et al., 1974).

Among the putative transmitter substances, histamine appears to be a leading transmitter candidate for insect photoreceptors. In three insect species, *Locusta*, *Manduca* and *Periplaneta,* both the retina and the lamina neuropil of the optic lobe contain large amounts of histamine (Elias and Evans, 1983). The tissues can also synthesize and accumulate histamine from exogenous ^3H-histidine (Maxwell et al., 1978; Elias and Evans, 1983). Intracellular recordings from large monopolar neurons in the the lamina of the house fly show that these cells are more sensitive to histamine than to other known transmitter candidates

(Hardie, 1987). In addition, the responses to iontophoretically applied histamine are similar to those induced by light, i.e. they are fast hyperpolarizations and involve membrane conductance changes. Light and histamine-induced responses can be blocked by histamine antagonists. Although both aspartate and glutamate hyperpolarize these neurons, their effect can be blocked by Co^{2+}. Other transmitters such as serotonin, dopamine, glycine, taurine, noradrenaline and octopamine have no effect while GABA and ACh cause depolarization (Hardie, 1987).

Recently, we have shown that large amounts of histamine were synthesized from exogenously supplied histidine by normal *Drosophila* heads (Sarthy, unpublished results). In contrast, heads from the eye-deficient mutant *sine oculis*, synthesized only minute amounts of histamine. Histidine decarboxylase activity was about tenfold higher in extracts of normal heads compared to *sine oculis* heads. Moreover, histidine decarboxylase activity was blocked by α-fluoro methyl histidine, which is known to be a specific inhibitor of the enzyme.

Since the fly eye contains several cell types, it is necessary to demonstrate that histidine is present in the photoreceptors. Immunocytochemical localization studies with a histamine-specific antiserum showed that there is strong staining in photoreceptors (Sarthy, unpublished results). Pre-treatment of the antibody with histamine drastically reduces immunostaining. Finally, ^{3}H-histamine synthesized from ^{3}H-histidine, can be released from normal *Drosophila* heads by a Ca^{2+}-dependent process. These data strongly suggest that histamine is likely to be the neurotransmitter used by photoreceptors in *Drosophila* and possibly by other insects as well. A similar role for histamine has also been proposed for barnacle photoreceptors (Stuart and Callaway, 1988). These cells have been shown to synthesize and contain histamine. Furthermore, histamine mimics the action of the endogenous transmitter. It remains to be seen whether other invertebrate photoreceptors also employ histamine in synaptic transmission.

In summary, there is good evidence for the role of glutamate as a cone transmitter. An excitatory amino acid, either glutamate or aspartate, is also likely to be the transmitter used by rods. It appears that photoreceptors in insects and possibly in other invertebrates as well, employ histamine as a major neurotransmitter. GABA might also be the transmitter for a subpopulation of insect photoreceptors. These findings raise a basic question as to why invertebrate and vetebrate photoreceptors use different transmitters. In addition, whether receptor cells in other sensory systems use excitatory amino acids and histamine as synaptic transmitters remains to be examined.

Acknowledgements. During preparation of this review, the author was supported by NIH grants, EY-03523 and EY-03664. The author thanks his colleagues and collaborators who have contributed to the work presented here and Dan Possin for preparing the manuscript.

REFERENCES

Altschuler, R. A., Neises, G. R., Harmison, G. C., Wenthold, R. J., Fex, J., (1981). Immunocytochemical localization of aspartate aminotransferase immunoreactivity in cochlear nucleus of the guinea pig. Proc Natl Acad Sci 78: 6553-6557.

Altschuler, R. A., Mosinger, J. L., Harmison, G. C., Parakkal, M. H., Wenthold, R. J., (1982). Aspartate aminotransferase-like immunoreactivity as a marker for aspartate/glutamate in guinea pig photoreceptors. Nature 298: 657-659.

Autrum, H., Hoffman, E., (1957). Die Wirkung von Pikrotoxin und Nikotin auf das Retinogramm von Insekten. Z Naturforsch 12b: 752-757.

Ayoub, G. S., Korenbrot, J. I., Copenhagen, D. R., (1988). Glutamate is released from individual photoreceptors. Invest Ophthalmol Vis Sci 29: 273.

Bloomfield, S. A., Dowling, J. E., (1985). Roles of aspartate and glutamate in synaptic transmission in rabbit retina. I. Outer plexiform layer. J Neurophysiol 53: 699-713.

Brandon, C., Lam, D. M. K., (1983). L-Glutamic acid: a neurotransmitter candidate for cone photoreceptors in human and rat retinas. Proc Natl Acad Sci 80: 5117-5121.

Campos-Ortega, J. A., (1974). Autoradiographic localization of ^{3}H-γ-aminobutyric acid uptake in lamina ganglionaris of *Musca* and *Drosophila*. Z Zellforsch 147: 415-431.

Cervetto, L., McNichol, E. F. Jr., (1972). Inactivation of horizontal cells in the turtle by glutamate and aspartate. Science, 178: 767-768.

Copenhagen, D. R., Jahr, C. E., (1988). Release of excitatory amino acids from turtle photoreceptors detected with NMDA receptor-rich membrane patches. Invest Ophthalmol Vis Sci 29: 223.

Datum, K-H., Weiler, R., Zettler, F., (1986). Immunocytochemical demonstration of γ-aminobutyric acid and glutamic acid decarboxylase in R7 photoreceptors and C2 centrifugal fibers in the blowfly visual system. J Comp Physiol A 159: 241-249.

Dowling, J. E., (1987). The Retina. An approachable part of the brain. Belknap Press, Cambridge, MA.

Dowling, J. E., Ripps, H., (1972). Adaptation in the skate photoreceptors. J Gen Physiol 60: 698-719.

Ehinger, B., (1981). [^{3}H]-D-Aspartate accumulation in the retina of pigeon, guinea pig and rabbit. Exp Eye Res 33: 381-391.

Elias, M. S., Evans, P. D., (1983). Histamine in the insect nervous system: distribution, synthesis and metabolism. J Neurochem 41: 562-568.

Fonnum, F., (1984). Glutamate: A neurotransmitter in mammalian brain. J Neurochem 42: 1-11.

Fu, M., Orcutt, J. C., Sarthy, P. V., (1988). Localization of L-glutamic acid decarboxylase mRNA in cat and monkey retinas by *in situ* hybridization. Invest Ophthalmol Vis Sci 29: 204.

Furukawa, T., Hanawa, I., (1955). Effects of some common cations on electroretinogram of the toad. Jap J Physiol 5: 289-300.

Gershenfeld, H. M., Piccolino, M., (1977). Muscarinic antagonists block cone to horizontal cell transmission in turtle retina. Nature 268: 257-259.

Graham, L. T., Jr., (1974). Comparative aspects of neurotransmitters in the retina In: Davson, H., Graham, L. T., (eds.). "The Eye", Vol 6, Academic Press, New York, pp 283-342.

Hall, J. C., (1982). Genetics of the nervous system in *Drosophila*. Q Rev Biophys 15: 223-479.

Hardie, R. C., (1987). Is histamine a neurotransmitter in insect photoreceptors? J Comp Physiol A 161: 201-213.

Ishida, A. T., (1984). Responses of solitary horizontal cells to L-glutamate and kainic acid are antagonised by D-aspartate. Brain Res 298: 5890-5894.

Ishida, A. T., Fain, G. L., (1981). D-Aspartate potentiates the effects of glutamate on horizontal cells in goldfish retina. Proc Natl Acad Sci 78: 5890-5894.

Ishida, A. T., Kaneko, A., Tachibana, M., (1983). Solitary horizontal cells in culture.II. A new tool examining effects of photoreceptor neurotransmitter candidates. Vis. Res. 23: 1217-1220.

James, W. M., Klein, W. L., (1985). α-Bungarotoxin receptors on neurons isolated from turtle retina: Molecular heterogeneity of bipolar cells. J Neurosci 5: 352-361.

Kaneko, A., Shimazaki, H., (1975). Synaptic transmission from photoreceptors to bipolar and horizontal cells in the carp retina. Cold Spr Harb Symp Quant Biol 40: 537-546.

Kaneko, A., Tachibana, M., (1987), Effects of L-glutamate on isolated turtle photoreceptors. Invest Ophthalmol Vis Sci 28: 50.

Klingman, A., Chappell, R. L., (1978). Feedback synaptic interactions in the dragonfly ocellar retina. J Gen Physiol 71: 157-175.

Kondo, H., Toyoda, J. I., (1980). Dual effect of glutamate and aspartate on the on-center bipolar cells in the carp retina. Brain Res 199: 240-243.

Konopka, R. J., (1972). Abnormal concentrations of dopamine in a *Drosophila* mutant. Nature 239: 281-282.

Lam, D. M. K., (1972). Biosynthesis of acetylcholine in turtle photoreceptors. Proc Natl Acad Sci 69: 1987-1991.

Lam, D. M. K., Hollyfield, J. G., (1980). Localization of putative amino acid neurotransmitters in the human retina. Exp Eye Res 31: 729-732.

Lam, D. M. K., Weisel, T. N., Kaneko, A., (1974). Neurotransmitter synthesis in cephalopod retina. Brain Res 82: 365-368.

Langer, H., Lues, I., Rivera, M. E., (1976). Arginine phosphate in compound eyes. J Comp Neurol 107: 179-184.

Lasater, E. M., Dowling, J. E., (1982). Carp horizontal cells in culture respond selectively to L-glutamate and its agonists. Proc Natl Acad Sci 79: 936-940.

Lasater, E. M., Dowling, J. E., Ripps, H., (1984). Pharmacological properties of isolated horizontal and bipolar cells from the skate retina. J Neurosci 4: 1966-1975.

Marc, R. E., Lam, D. M. K., (1981). Uptake of aspartic acid and glutamic acid by photoreceptors in goldfish retina. Proc Natl Acad Sci 78: 7185-7189.

Maxwell, G. D., Tait, J. F., Hildebrand, J. G., (1978). Regional synthesis of neurotransmitter candidates in the CNS of the moth *Manduca sexta*. Comp Biochem Physiol 61C: 109-119.

Meyer, E. P., Matute, C., Streit, P., Nässel, D. R., (1986). Insect optic lobe neurons identifiable with monoclonal antibodies to GABA. Histochem 84: 207-216.

Miller, A. M., Schwartz, E. A., (1983). Evidence for the identification of synaptic transmitters released by photoreceptors of the toad retina. J Physiol 334: 325-349.

Miller, R. F., Slaughter, M. D., (1986). Excitatory amino acid receptors of the retina: diversity of subtypes and conductance mechanisms. Trends Neurosci Res 9: 211-218.

Murakami, M., Ohtsu, K., Ohtsuka, T., (1972). Effects of chemicals on receptors and horizontal cells in the rat retina. J Physiol Lond 227: 899-913.

Murakami, M., Ohtsu, T., Shimazaki, H., (1975). Effects of aspartate and glutamate on the bipolar cells in the carp retina. Vis Res 15: 456-458.

Neal, M. J., Collins, G. G., Massey, S. C., (1979). Inhibition of aspartate release from the retina of the anaesthetized rabbit by stimulation with light flashes. Neurosci Lett 14: 214-245.

Negishi, K., Drujan, B. D., (1979). Effects of some amino acids on horizontal cells in the fish retina. J Neurosci Res 4: 351-363.

Nishimura, Y., Schwartz, M. L., Rakic, P., (1986). GABA and GAD immunoreactivity of photoreceptor terminals in primate retina. Nature 320: 753-756.

Sarantis, M., Everett, K., Attwell, D., (1988). A presynaptic action of glutamate at the cone output synapse. Nature 332: 451-453.

Sarthy, P. V., Lam, D. M. K., (1979). Endogenous levels of neurotransmitter candidates in photoreceptor cells of the turtle retina. J Neurochem 32: 455-461.

Sarthy, P. V., Hendrickson, A. E., Wu, J-Y., (1986). L-Glutamate: A neurotransmitter candidate for cone photoreceptors in the monkey retina. J Neurosci 6: 637-643.

Schwartz, I. R., Bok, D. (1979). Electron microscopic localization of ^{125}I-α-bungarotoxin binding sites in the outer plexiform layer of the goldfish retina. J Neurocytol 8: 53-66

Shiells, R. A., Falk, G., Nagh Shineh, S., (1981). Action of glutamate and aspartate on rod horizontal and bipolar cells. Nature 294: 592-594.

Slaughter, M. D., Miller, R. F., (1981). 2-Amino-4-phosphonobutyric acid: A new pharmacological tool for retina research. Science 211: 182-185.

Stuart, A. E., Callaway, J. C., (1988). Histamine is synthesized by barnacle ocelli and affects second order visual cells. Invest Ophthalmol Vis Sci 29: 223.

Sturman, J. A., Hayes, K. C., (1980). The biology of taurine in nutrition and development. In: Draper, H. H., (ed) "Advances in Nutritional Research" Vol 3. pp. 231-299.

Vogel, Z., Nirenberg, M., (1976). Localization of acetylcholine receptors during synaptogenesis in retina. Proc Natl Acad Sci 73: 1806-1810.

White, R. D., Neal, M. J., (1976). The uptake of L-glutamate by the retina. Brain Res 111: 79-83.

Whitton, P. S., Strang, R. H. C., Nicholson, R. A., (1987). The distribution of taurine in the tissues of some species of insects. Insect Biochem 17: 573-577.

Yazulla, S., Schmidt, J., (1976). Radioautographic localization of ^{125}I-α-bungarotoxin binding sites in the retinas of goldfish and turtle. Vis Res 16: 878-880.

COLOUR VISION AND IMMUNOLOGICALLY IDENTIFIABLE PHOTORECEPTOR SUBTYPES

Àgoston Szèl and Pal Röhlich+

2nd Department of Anatomy, Histology and Embryology

+I. Laboratory of Electron Microscopy

Semmelweis University of Medicine, T zoltó u. 58. Budapest, H-1450 Hungary

ABSTRACT

Colour vision is based on a family of visual pigments localized in the retinal photoreceptor cells. Monoclonal and polyclonal antibodies generated against these proteins were used to discriminate the rods and the cone types of various vertebrate species. The different cell types were demonstrated by light microscopic immunocytochemistry on retinal sections of the turtle, the gecko, and several mammalian species. The human retina was also included. A good correlation could be found between the colour-sensitivities and the immunologic staining pattern of the photoreceptor subpopulations. This novel approach to distinguish the visual cells with immunocytochemistry proved to be useful to map the distribution of the colour-specific elements of the retina.

INTRODUCTION

Considerable amount of data has been accumulated during the past decades on the colour-specificity of retinal photoreceptor cells by using microspectrophotometry (MSP), electrophysiology and selective degeneration or stimulation by monochromatic light (Marc and Sperling, 1977; Mansfield et al., 1984; Nunn et al., 1984; Ahnelt, 1985; Schnapf et al., 1987; a.o.). These methods require living, in most cases dark-adapted retinas and sophisticated instrumentation. Although the most important characteristics of the identified cells can be observed with these methods, the ultrastructure and the distribution of the colour-specific elements are hardly or not detectable, as the quality of the specimens cannot be well preserved after the necessary manipulations.

Recently, we developed a novel approach based on the idea that various visual pigments are responsible for recognizing colours, and the differences in the protein components of the pigments can be detected immunologically (Szèl et al., 1985; 1986a). The visual cells could be distinguished from each other by immunocytochemistry using visual pigment-specific antibodies on the sections of resin-embedded material. Three antibodies produced in our laboratory were especially useful in discriminating the colour-specific photoreceptor cells.

One of them, monoclonal antibody (mAb) COS-1, labelled about two thirds of the cones in the retina of all vertebrates investigated so far, while another one mAb OS-2, recognized practically all photoreceptor cells in the avian and reptilian retina (Szèl et al., 1986a,b). In mammals, however, this latter mAb was bound only to about 10% of all cones (Szèl et al., 1988). The reactivity of the third antibody (Ab RO-1) produced against bovine rhodopsin was confined exclusively to the rod cells of the mammalian retinas (unpublished), in the other two vertebrate classes, a great number of cone cells were recognized by the anti-rhodopsin antibody (Szèl et al., 1985, 1986a).

In the submammalian species striking morphologic differences (single and double cone cells, presence of coloured oil droplets, etc.) are existing between the visual cell types. The colour-specifications of the three antibodies were identified by correlating the spectral and morphologic characteristics of the photoreceptors with the immunological staining pattern in the chicken, the pigeon and the gecko. The mAb COS-1 was found to be specific to the long wavelength-sensitive iodopsin localized in the majority of cones in the chicken and pigeon (Szèl et al., 1986a; Cserhati et al., 1988), and to the green-sensitive photo-receptors of the gecko (Szèl et al., 1986b). The anti-rhodopsin antibody stained the rods together with some single cone types with shorter spectral sensitivity (blue and green). The blue-sensitive cones of the gecko retina were also recognized by Ab RO-1. MAb OS-2 labelled the overwhelming majority of all photoreceptors (rods as well as cones) in all submammalian species mentioned so far (Szèl et al., 1986a).

The spectral specificities of the antibodies to the mammalian cone cells were proven by an indirect way. As the cone cells of the mammalian retina are generally uniform in shape and size, the colour-specific subtypes must first be marked by the method of selective photic damage (Sperling et al., 1980; Szèl et al., 1988). As a result of comparative experiments performed on rabbit, monkey and ground squirrel retinas, mAb OS-2 and COS-1 proved to be specific to the blue-sensitive and to the middle-to-long wavelength-sensitive cones, respectively Szèl and Röhlich, 1988b).

If an anti-visual pigment antibody binds consistently to a definite photoreceptor subtype, characterized by a certain wavelength-specificity, this antibody can be considered as specific to the pigment recognizing the given colour in that species. Our results indicate that this correlation is existing among the species of certain animal classes as well. The question has arisen whether the spectral specificities of the antibodies are exactly the same in phylogenetically distant species. The aim of the present study was to examine and discuss the relationship between the antigenic determinants of spectrally similar photo-pigments of different species.

For this reason, the retinas of three turtle species together with those of several mammalian animals were investigated. This study can be considered as the presentation of preliminary results obtained from a couple of vertebrate species. The detailed comparative description of the retinal photoreceptor mosaic together with quantitative data will be published elsewhere.

MATERIALS AND METHODS

Experimental animals. Two fresh-water turtles (*Pseudemys scripta elegans* and *Emys orbicularis*) and a land-turtle (*Testudo horsefieldi*) were used in this study. The animals were decapitated, the eyes were quickly removed, the posterior halves of the eyes were immersed in ice-cold 0.66-0.1 M phosphate-buffered 1% glutaraldehyde fixative (pH 7,2), and the retina was separated from the underlying pigment epithelium in one minute after death. Following one hour fixation (22 C), the retinas were cut into smaller pieces, washed several times in phosphate buffer, followed by incubation in 0.1 M Tris buffer overnight. After dehydration in ethanol, the retinal pieces were embedded in Araldite (Durcupan ACM, Fluka, Buchs). 0,5-1 μm thick sections were cut on a Reichert OMU-2 ultramicrotome for light microscopic immunocytochemistry. For comparison, gecko (*Tertoscincus scinsus*), chicken and pigeon retinas were also included.

Bovine and pig retinas. The source of these animals was the local slaughter-house. The enucleation followed the death of the animals within 5 minutes. The eyes were cut into two parts in the equatorial plane. The posterior halves were processed as described before.

Rats, mice, hamsters, guinea pigs and rabbits together with ground squirrels (*Citellus citellus*) were also used. Following ether narcosis, the animals were decapitated. The eyes were enucleated within 1 minute after death, and the retinas were processed as above.

The cats and dogs (used for other physiological experiments) were killed by an overdose of intravenous Nembutal. After enucleation, the eyes were carried to the laboratory in ice cold 0.1 M phosphate buffer. Within 20 minutes the eyes were cut into two parts and subsequently immersed in the fixative.

The monkeys (*Cercopithecus aethiops*) were immobilized, narcotized intravenously and exsanguinated. The enucleation was carried out about 5 minutes after death. The fovea was fixed separately.

The retina of a human patient (50 year old woman, suffering from malignant melanoma) was also used. The retina was fixed as described above within 2 minutes after the operative enucleation. As the fovea was completely destroyed as a result of the tumor, only the peripheral parts of the retina could be processed for immunocytochemistry.

The morphology of the single photoreceptor cells was studied on carefully oriented radial sections of the retinas. Tangential sections cut at the outer segment level, were also used to demonstrate the distribution of the photoreceptor cell types.

Antibodies, Immunocytochemistry. The production and characterization of the three antibodies was described elsewhere (Szél et al., 1985, 1986a). Shortly: mice were immunized by a photoreceptor membrane suspension derived from the cone-rich chicken retina. When screening the hybridoma cell lines, only those clones were selected which specifically stained the photoreceptor outer segments. These cell lines were grown up, cloned repeatedly, and were characterized by immunobiochemical methods. Two monoclonals (COS-1 and OS-2) were used in the present study together with the polyclonal anti-rhodopsin serum (Ab RO-1). The latter antibody was produced in rats by immunizing the animals with the excised opsin bands from electrophoretically separated bovine photoreceptor outer segment proteins (Szél et al., 1986a).

The mAbs were diluted 1:10000 for avian and reptilian retinas, and 1:5000 for mammalians. The Ab RO-1 was diluted 1:1000 for sections derived from chicken, turtle and gecko, and 1:2000 for the mammalian retinas.

Prior to the immunoreaction, the embedding resin was removed from the sections by sodium-methoxide (Mayor et al., 1961). After preincubation in 2% bovine serum albumin (BSA), the sections were reacted with the primary antibodies for 1 h at room temperature. The bound antibodies were detected by the avidin-biotin system (Vectastain, Vector, Burlingame). The biotinylated second antibodies were followed by the avidin-peroxidase complex. The immunoreaction was revealed by diaminobenzidine.

Consecutive serial sections reacted with the different antibodies were also used to decide if there were common elements of two photoreceptor subpopulations, and if there were cones which were not recognized by any of the antibodies. The immunocytochemical technique was controlled by the omission of the primary or the secondary antibodies and by immunoreactions carried out on non-retinal tissues of the same animals.

RESULTS

The outer segments of most, but not all visual cells in the retina of the submammalians were recognized by the three anti-visual pigment antibodies. In contrast, the total photoreceptor population was labelled by one of the three antibodies in the mammalian retinas.

All the three antibodies were absolutely specific to the photoreceptor outer segments. No other parts of the photoreceptors, or other cell types of the retinas were labelled (Figs. 1-3 and 5-7). The controls were completely negative (not shown).

Submammalians. The turtle retinas contained the same visual cell types as described in the diurnal birds. The majority of the photoreceptors were cones (double cones together with some single cone types characterized by the colour and position of oil droplets). A relatively small fraction of the visual cells were rods without oil droplets. It must be pointed out, however, that the colours of the oil droplets could only be seen in the native retinas. As a result of the organic solvents used in the embedding procedure, these coloured lipid droplets were lost. Therefore, only the position of these droplets was available for discriminating the single cone types.

The staining patterns of the antibodies in the three turtle species were similar to those demonstrated earlier in the avian retina. MAb OS-2 labelled practically all photoreceptor cells. All double cones, all rods and most, but not all single cones were stained (Fig.1A). The other mAb, COS-1, stained both members of all double cones and one type of single cones having a sclerally located oil droplet. None of the double cones remained unstained (Fig.1B). For comparison, the same immunoreactions performed on chicken and pigeon retinas are shown in Fig.2. Note the structural and immunological similarity between the retinas of the three species.

Similarly to the avian retina, all rod cells together with some single cones were stained by Ab RO-1 in the turtle retinas (Fig.1C). The oil droplets of the anti-rhodopsin positive single cones were located at a slightly more vitreal level compared to those stained by mAb COS-1. With immunocytochemistry carried out in adjacent serial section, it could be observed that the COS-1 and RO-1 antibodies bound to different cell populations (Fig.2B-C). Only the Pseudemys retina is shown, since the light microscopic appearance together with the immunocytochemical staining pattern of the investigated turtle species were identical.

All photoreceptors of the gecko retina were stained by mAb COS-1, and only a small fraction of cones was labelled by Ab RO-1 (Fig.3).

Mammalian retinas. Earlier, only the two main photoreceptor subtypes - the cones and the rods - could be discriminated by morphology, as the cones were indistinguishable from one another on normal retinal sections. Using the three anti-visual pigment antibodies, however, three different photoreceptor cell types could be selectively labelled in the mammalian retinas. The rod cells of all retinas, as expected, were stained by the polyclonal Ab RO-1, but none of the cones bound this antibody. On the contrary, the cones were recognized by the mAbs. While mAb COS-1 labelled the majority of cones in all species, OS-2 was specific to a rarely occurring cone type. Comparing the identical elements on serial section pairs of the ground squirrel retina (Fig.4) and retinas of the monkey and the dog (not shown), it was clearly observable that each cone was stained by either mAb OS-2 or COS-1, and no cones were stained by both of them.

Although all mammalian retinas were similar, with regards to the presence of these three visual cell types, the distribution of the photoreceptors was strikingly different between the primate and non-primate retinas on one hand, and between the diurnal and nocturnal animals on the other. The total area of the subprimate retinas seemed to be uniform, as no special territories with missing or differently sized and shaped cell types could be observed. The three kinds of immunologically different visual cells were distributed rather evenly, albeit irregularly in the whole eye. The retinas of the primates, in contrast, exhibited a relatively well organized array of visual cells, especially in the rod-free area, the fovea centralis which could be identified clearly.

Most of the mammalians, considered as nocturnal animals, possessed rod-dominated retinas. The frequency of the rods, related to the number of all photoreceptors, measured on large areas of tangentially cut retinal pieces, were 80,87,92 and 99% for bovine, pig, dog and rat, respectively (Figs. 5-88). In contrast to these animals, the majority (97%) of all photoreceptors were cones in the retina of the diurnal ground squirrel (Fig.4). While the rod-cone ratio was changing between 3 and 99 percent, the ratio of the OS-2 positive elements to the COS-1 positive cells seemed to be much more constant, not exceeding the values of 6-11% of all cones (6,9,10,11% for bovine, dog, rat and pig, respectively). The estimated frequencies of the two cone types were very similar to these values in the other species as well (cat, rabbit, hamster, guinea pig, mouse, not shown here). In the framework of this communication we could not deal with the interspecies variations of the shape, size and total densities of the photoreceptor cells.

In the fovea centralis of the monkey, the cones were organized in a very regular, hexagonal pattern (Fig.9A and C). The OS-2 positive elements could be found mainly in the centers of hexagonally symmetric areas. Moving away from the fovea centralis, this regularity changed due to the appearing rods.

Around the centrally located cone cells, numerous rods could be found in the periphery, in a more or less regular organization. Each cone was surrounded by several circles of rod cells. As the rod-cone distances were somewhat greater than those between the neighboring rods, the cone cells could be easily identified on tangential sections (Fig.9B and D). It is worth mentioning that, although the ratio of the cones to the rods gradually decreased from 100 to 20, while leaving the fovea towards the periphery, the ratio of the two cone types did not change considerably. The OS-2 positive elements comprised about 6-10% of all cones throughout the whole retina.

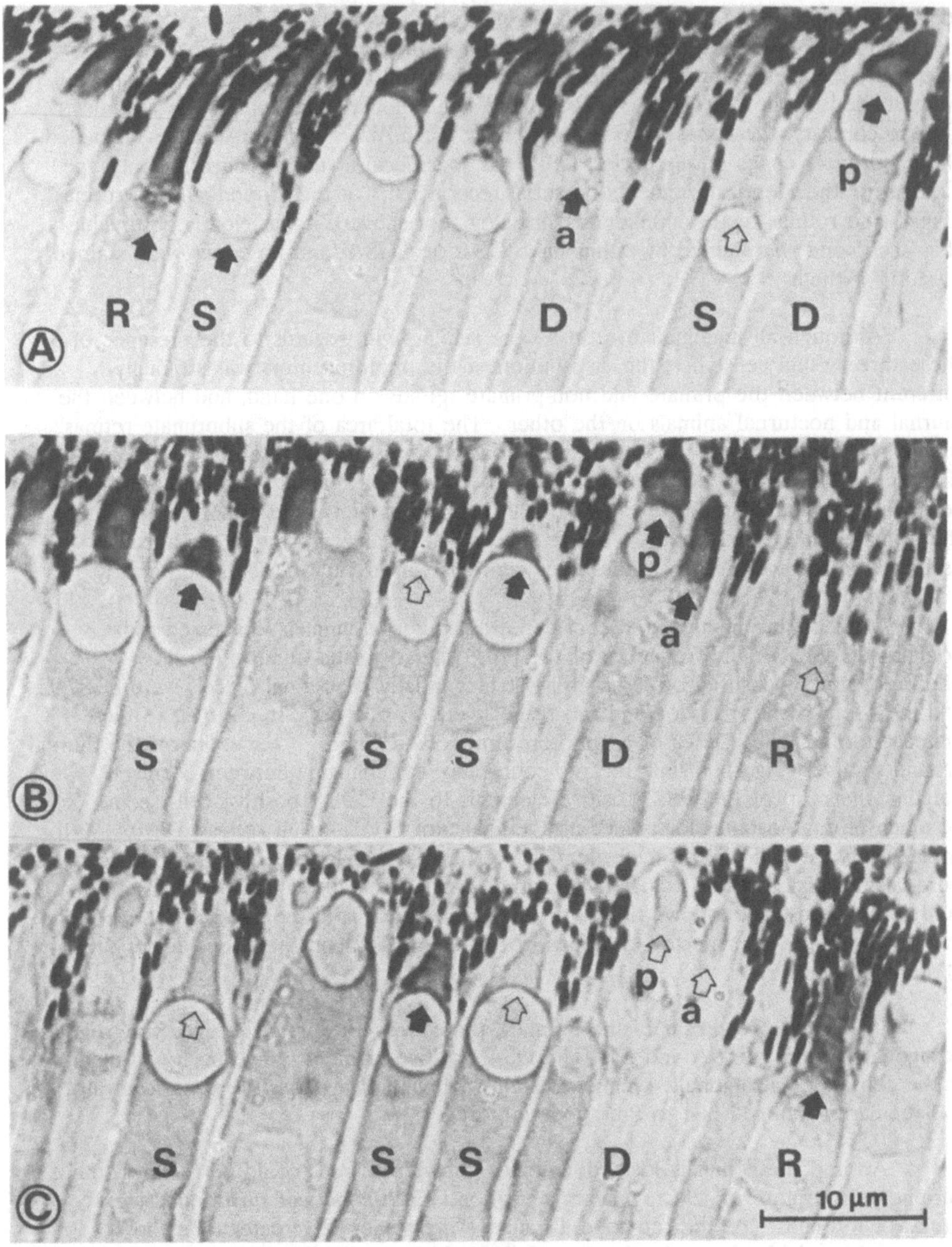

Fig.1. Immunoreactions carried out on *Pseudemys* retina. Section A was reacted with mAb OS-2. Outer segments of nearly all photoreceptors (rods as well as cones, filled arrows) were stained with the exception of a rarely occurring single cone (open arrow). B and C are adjacent serial sections reacted with mAb COS-1 and Ab RO-1, respectively, showing the complementarity of the two antibodies. The principal (p) and the accessory (a) members of the double cones (D) and some single cones (S) were recognized by COS-1, while the rods (R) and another single cone population remained unstained. The COS-1 negative photoreceptors, in turn, were labelled by Ab RO-1. Some identical elements of the two sections were marked with filled and open arrows for the positive and negative outer segments, respectively.

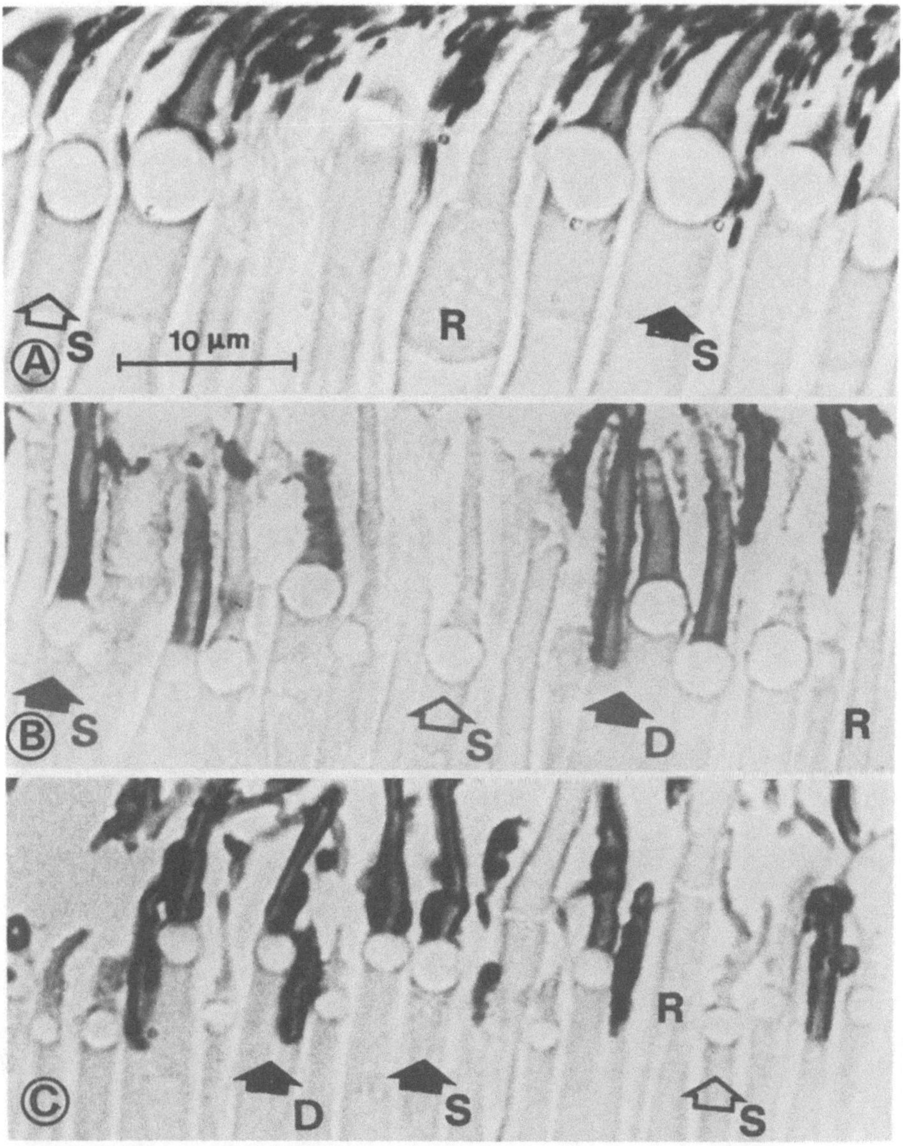

Fig.2. COS-1 immunocytochemistry carried out on *Pseudemys* (A), chicken (B) and pigeon (C) retinas. The same symbols are used as on Fig.1. Note that the majority of cones (both members of double cones together with some single cones) were stained. Although the size and shape of the photoreceptors are different in the three species, the staining pattern is very similar.

Although the fovea of the human retina was not included in the present study, a few conclusions might be drawn from the immunocytochemical results obtained on the peripheral part of the human retina (Fig.10). The immunologically distinguishable photoreceptor types together with their distribution were the same as those described in the monkey retina. Similarly to the other mammalians, the rods comprised the majority (89%) of all photoreceptors, while the ratio of the OS-2 positive cones was 10% of the total cone population.

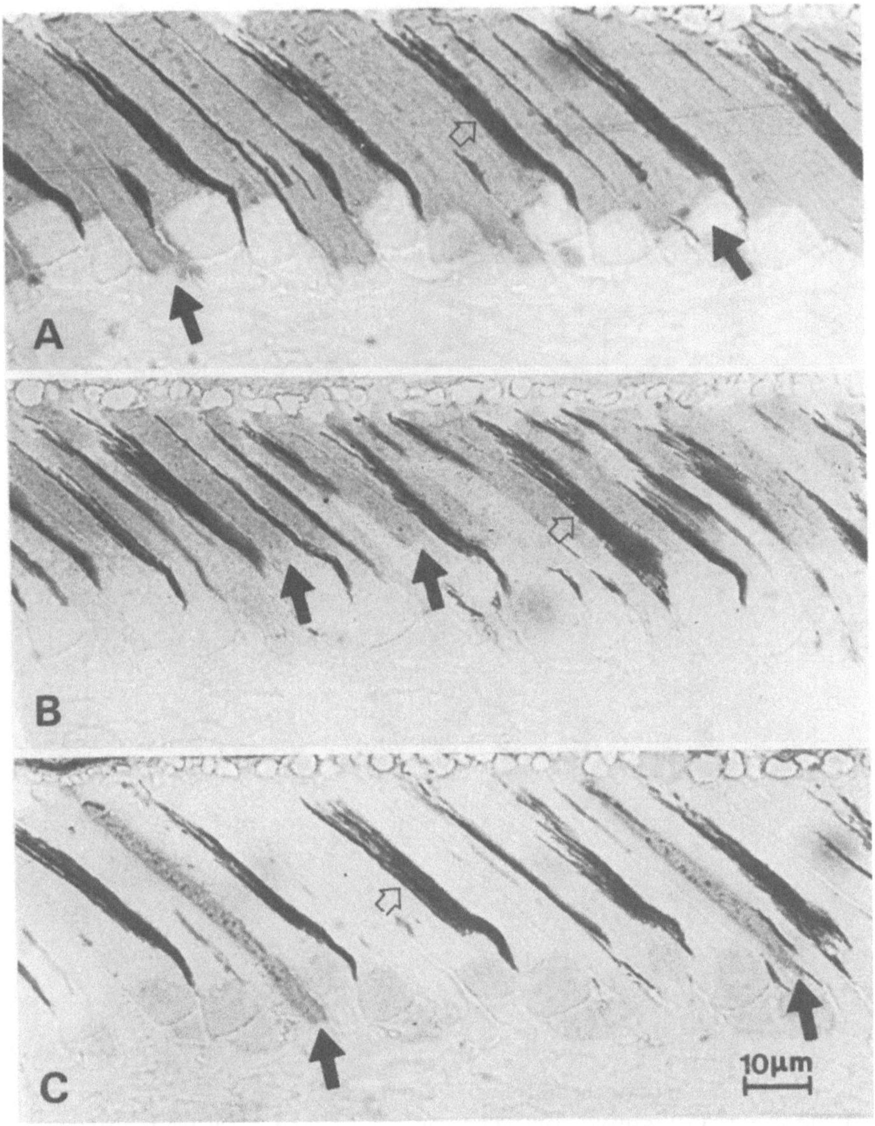

Fig.3. Semithin sections of the gecko retina, reacted with mAb OS-2 (A), mAb COS-1 (B) and Ab RO-1 (C). The outer segments of all cones were stained by OS-2. COS-1 recognized the majority of the cones. A small fraction of the cones were labelled by RO-1. The filled arrows point to the stained elements. The outer segments are separated by the black processes of the pigment epithelial cells (open arrows).

DISCUSSION

The subpopulations of photoreceptors recognized by the described antibodies could be correlated with morphologically and/or spectrally defined cone types. The colour-specificities of the antibodies, however, did not prove to be the same in all individual species.

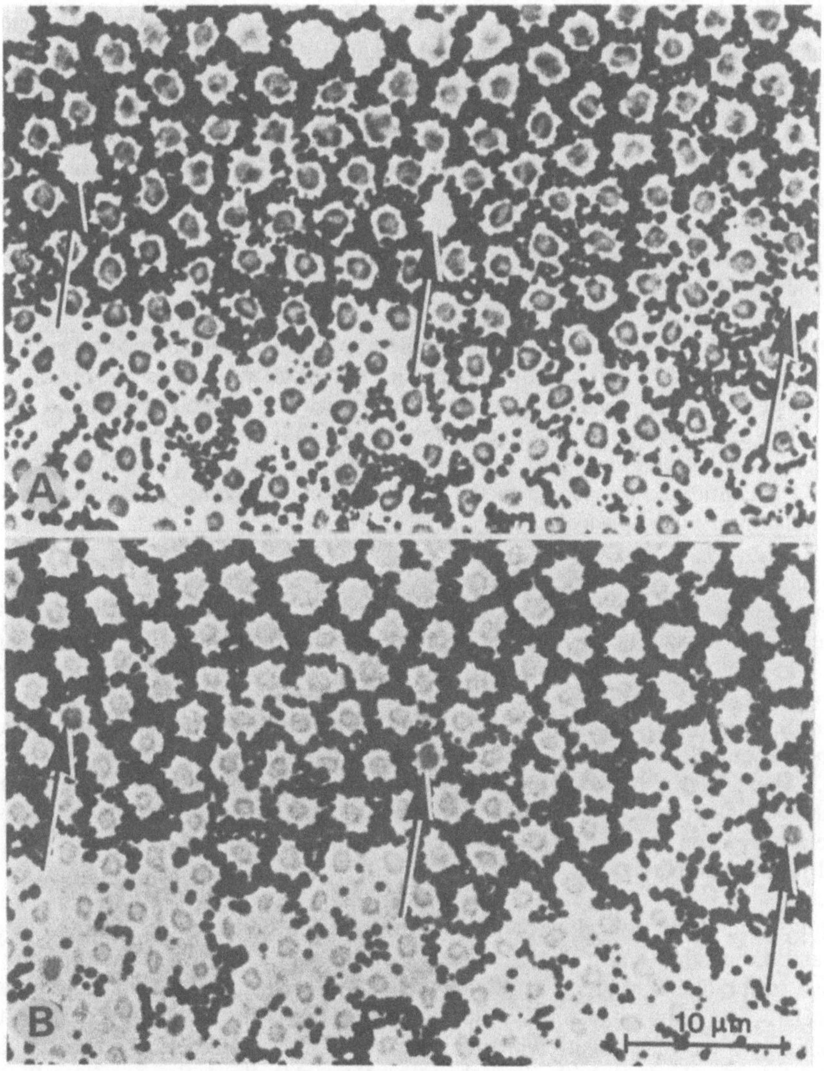

Fig.4. Consecutive serial tangential sections of the ground squirrel retina, reacted with mAb COS-1 (A) and mAb OS-2 (B). Some of the identical elements are marked by arrows. Note that the majority of cones stained by mAb COS-1, were left unlabelled by mAb OS-2. The OS-2 positive ones, in turn, did not bind COS-1. The cones, comprising about 97% of all photoreceptors, are surrounded by the black granules of the pigment cells.

Submammalians. The double cones and some single cones were shown to be sensitive to the red light in different turtle species (Ohtsuka, 1985), chicken (Bowmaker and Knowles, 1977) and pigeon (Bowmaker, 1977). In these animals, mAb COS-1 was specific to both members of the double cones together with one type of single cones having a sclerally localized oil droplet (Szél and Röhlich, 1985; Cserhati et al., 1988). The similar immunologic and spectral features of these elements show the same or very similar visual pigments are present in all these species.

The long wavelength sensitive avian photoreceptors contain the retinal-based pigment, iodopsin (Wald et al., 1955: Liebman, 1972; Szél et al., 1986a). It was shown by MSP that the cones of the land turtles contain the same red-sensitive visual pigment as those of the chicken (562 nm; Liebman and Granda, 1971). In contrast, the longwave-sensitivity of the fresh-water turtles is represented by a different photopigment, having a longer (620 nm; Ohtsuka, 1985) absorption maximum, and called cyanopsin (Wald et al., 1953). The shift to somewhat longer wavelengths can be attributed to the fact that instead of retinal, the chromophore of the visual pigments of the fresh-water turtle is 3-dehydroretinal (Granda and Dvorak, 1977). Our immunocytochemical results point to the identity or strong similarity of the protein components of the two red-sensitive pigments, consequently mAb COS-1 can be considered as specific to both iodopsin and cyanopsin.

MAb COS-1 recognized a morphologically heterogenous, albeit spectrally uniform (absorption maximum: 535 nm) cone population in the gecko retina and the other species discussed so far is remarkable, as the COS-1 positive elements constituted the majority of the total cone population, and in addition, COS-1 recognized the photoreceptors exhibiting the longest wavelength sensitivity of each species. Therefore, mAb COS-1 is considered as specific to the middle-to-long wavelength sensitive pigments of the reptilian and avian species.

Similarly to the chicken and pigeon, Ab RO-1 stained another photoreceptor population. Besides the rods, numerous single cones were labelled in all turtle species. Immunoreactions carried out on adjacent sections unequivocally show that these single cones are different from those stained by mAb COS-1. The morphological, spectral and immunological features of these single cells, containing visual pigments which are thought to be sensitive to the shorter wavelengths of light (blue and green), have not been identified completely (Szél and Röhlich, 1985; Kolb and Jones, 1987; Cserhàti et al., 1988). The similarity between the protein components of the retinal- (rhodopsin) and 3-hydro-retinal- (porphyropsin) based pigments of the birds and freshwater turtles, respectively (van Veen et al., 1986), is evidenced by the simultaneous reactivity of all avian and reptilian retinas to the same antibody.

Quite recently, a monoclonal antibody was reported (Gaur et al., 1988) to label the rods and some single cones of the turtle. This antibody, which was hypothesized to be sensitive to the blue and green sensitive cones, must recognize the same epitope as our Ab RO-1. The only submammalian species where the clear-cut identification of the RO-1 positivity to the blue-sensitive pigment was successful, is the gecko, possessing a relatively simple visual system with altogether two detectable absorption maxima (Szél et al., 1986b).

These findings show that the shorter wavelengths are recognized by a couple of visual pigments being very close relatives of rhodopsin in all submammalians. The degree of molecular similarity between the short wavelength sensitive pigments and rhodopsin, however, deserves more thorough bio-chemical and genetic experiments.

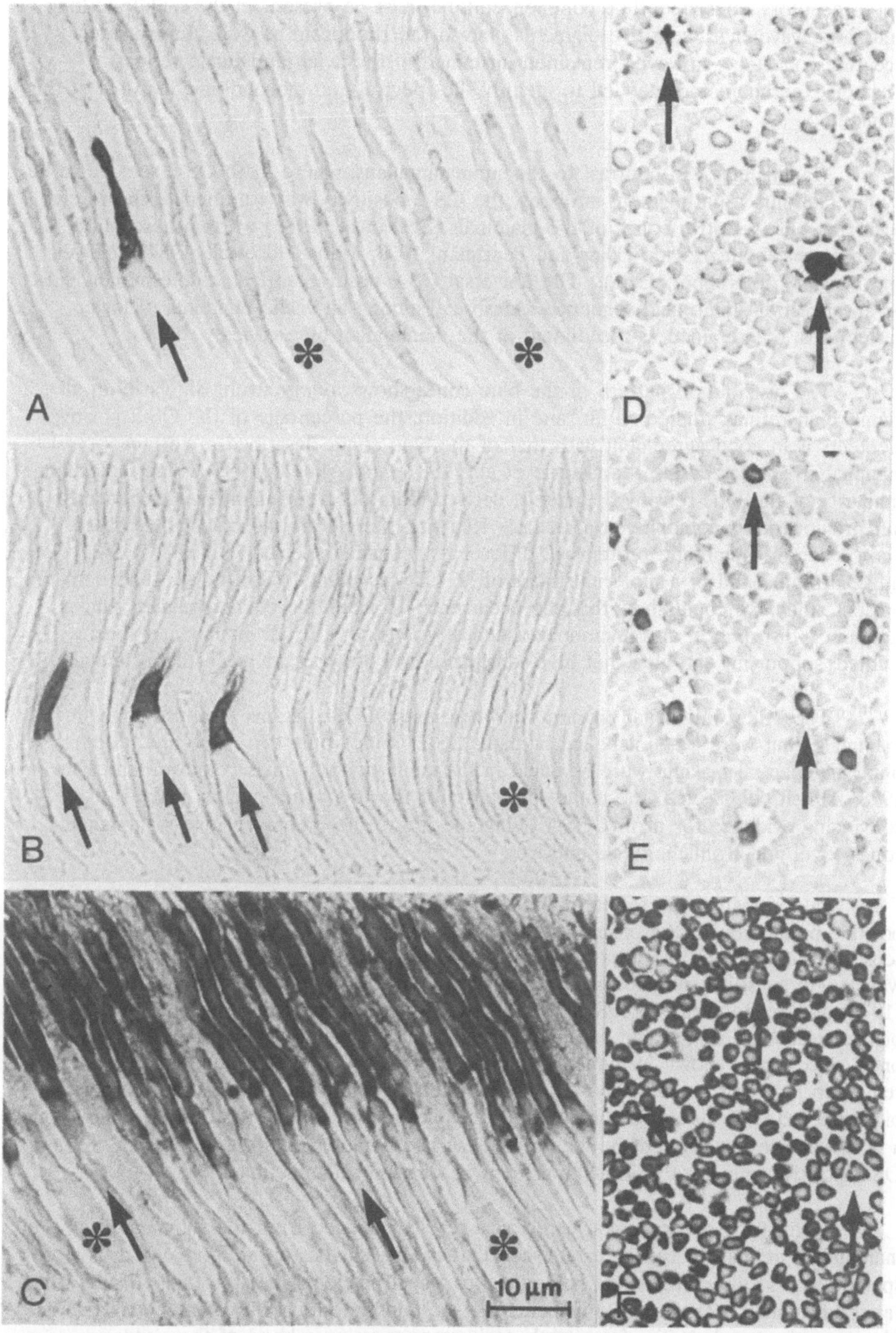

Fig.5. Cross sections (A-C) and tangential sections (D-E) of the bovine retina reacted with mAb OS-2 (A and D), mAb COS-1 (B and E) and Ab RO-1 (C and F). Note, that while a rarely occurring cone type was stained with mAb OS-2, mAb COS-1 labelled almost all cones with a few exceptions. The rods were stained by Ab RO-1. The arrows point to their stained elements, the unlabelled cones are marked with asterisks.

The antibody OS-2 labels nearly the whole series of the submammalian photoreceptors, consequently a common epitope must be present in the middle-to-long and short wavelength sensitive pigments, and in the rhodopsin as well. A very rarely occuring cone type, however, remained unstained in the avian and turtle retinas. Further experiments are needed to identify the spectral and morphologic features of this element.

Mammalians. In contrast to the submammalians, where mAb OS-2 seemed to be a universal anti-visual pigment antibody, the OS-2 positivity was confined to the rarely occurring blue-sensitive cones of the mammals (Szél et al., 1988) with an absorption maximum of about 425 nm (Daw and Pearlman, 1970; Mansfield et al., 1984; Caldwell and Daw, 1978; Nuboer, 1986). The universal OS-2 positive antigenic determinant must be an ancestral amino acid sequence widespread among all visual pigments of lower vertebrates, but retained as a rudiment in the mammalian blue cones.

The distribution pattern of the blue cones shows a very strong similarity in all mammalian retinas studied so far, and in addition, the percentage of the OS-2 positive elements is in good agreement with that of the blue cones described in certain mammalian species by De Monasterio et al., (1981) and Ahnelt (1985). The molecular similarity of the blue-sensitive pigments, derived from different animals, is evidenced by our previous immunobiochemical results. Reacting the electrophoretically separated photoreceptor membrane proteins of different mammalian species with mAb OS-2, one single protein band of a molecular mass of 36 kD was stained (Szél et al., 1988). In the present study, we could show that the occurrence of a definite cone type with similar frequency, distribution and immunocytochemistry is general in all mammalian species studied, supporting the idea that blue-sensitive cones are present in all these animals.

The other anti-visual pigment antibody, mAb COS-1, stains all those cones in the mammals, that were left unlabelled by mAb OS-2. Since only two (blue and green) photopic spectral maxima were revealed in the subprimate animals (Caldwell and Daw, 1978; Ahnelt, 1985; etc.); the complementarity of the two cone-specific mAbs is a convincing evidence for mAb COS-1 being specific to the green-sensitive mammalian cones (Szél and Röhlich, 1988a and b).

In contrast to the retina of the dichromatic subprimate animals, the COS-1 positive population of the primate retinas must represent the green as well as the red-sensitive elements, as in addition to the blue-sensitive cones, two different middle-to-long wavelength sensitive subtypes were demonstrated in the primate retina by Nunn et al., (1978) and Schnapf et al., (1987). It is not surprising that only the blue-sensitive pigment can be distinguished immunologically from the middle-to-longwave sensitive ones, since it was shown in human (Nathans et al., 1986) that the blue and the red/green pigments exhibit only 43% mutual identity, while the red- and green-sensitive pigments are 96% identical. Immunoreactions carried out on retinas derived from about a dozen species, confirmed the idea that these two complementary epitopes (COS-1 and OS-2 positive ones) are universally present in all mammalian retinas.

Comparing the immunocytochemical results obtained from the submammalian and mammalian animals, it can be established that mAb COS-1 is an antibody which is specific to a group of relatively homogeneous visual pigments in a phylogenetically wide range of animals. These photopigments are the middle-to-long wave length sensitive pigments of several species from turtle to human, covering a band of the visible spectrum between 535 and 620 nm. In spite of the relatively broad spectral range, this pigment family seems to be uniform, since these pigments occur exclusively in cones, albeit morphologically different ones, comprising the majority of colour-specific elements. Furthermore, all of them represent the longest wavelength sensitivities in the retina of each individual species.

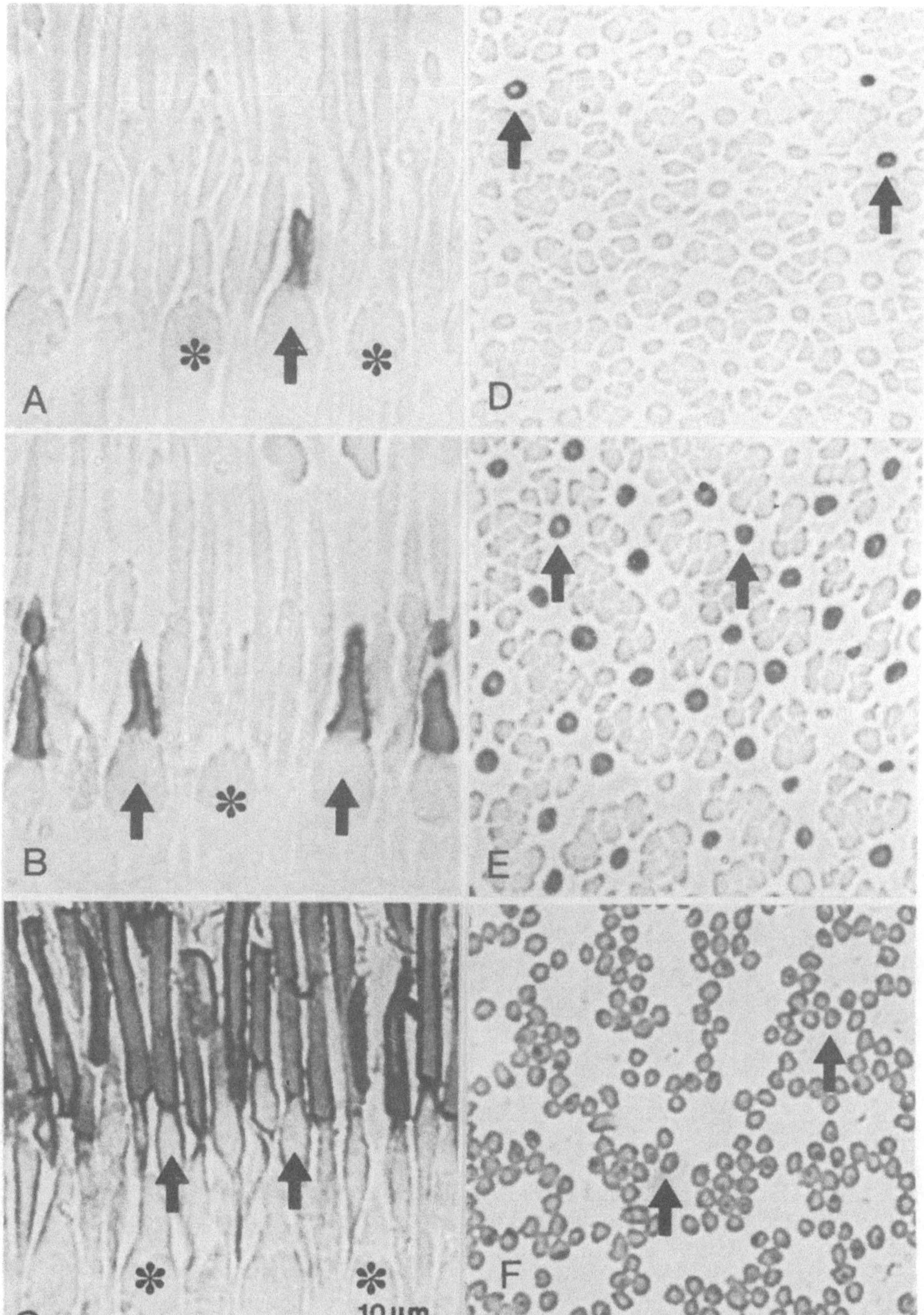

Fig.6 The antibodies, mAb OS-2 (A and D), mAb COS-1 (B and E), Ab RO-1 (C and F) were used for immunocytochemistry on sections of the porcine retina. Although the size and shape of the photoreceptors are different, the immunocytochemical staining pattern resembles greatly that of the bovine retina.

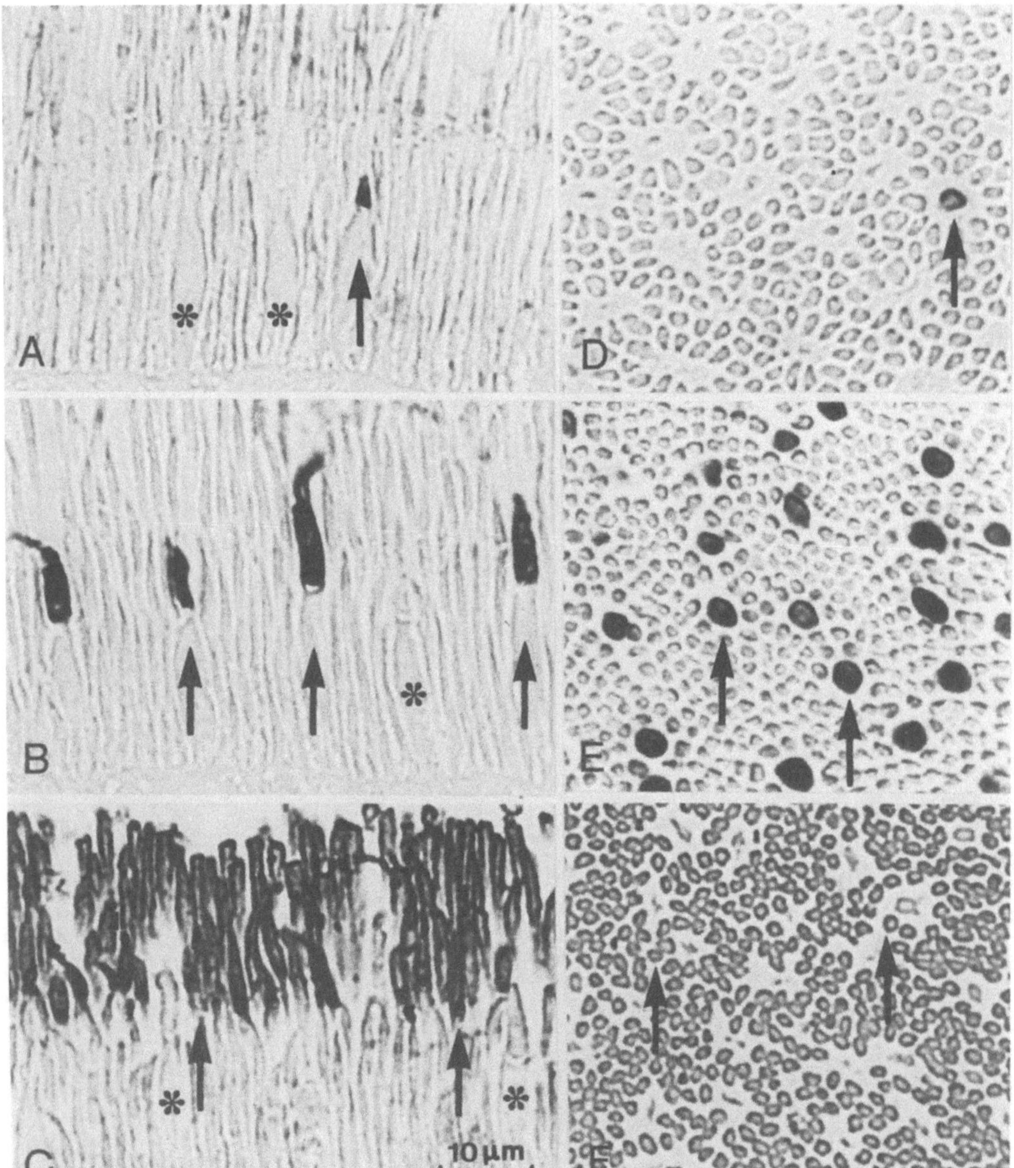

Fig.7. MAbs OS-2 (A and D) and COS-1 (B and E) together with Ab RO-1 (C and F) were used to label the photoreceptor subtypes of the dog retina. The same symbols are used as in Fig.5.

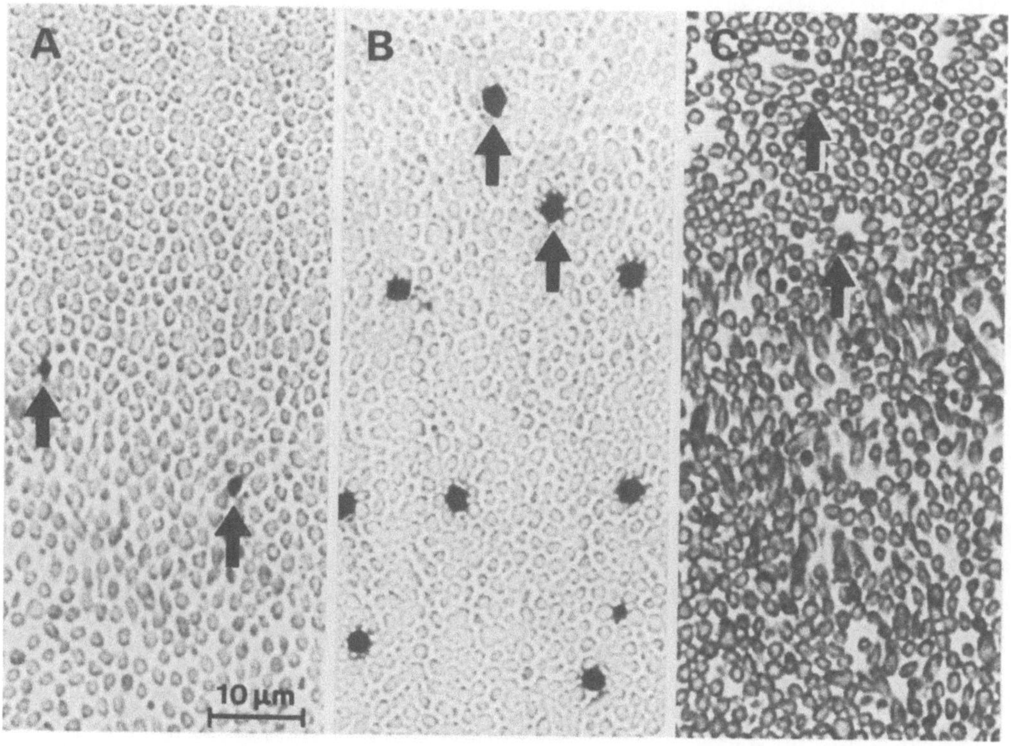

Fig.8. Tangential sections of the rat retina reacted with Ab OS-2 (A), mAb COS-1 (B) and Ab RO-1 (C). The number of cones related to that of the rods is the lowest among the mammalians included in this paper. The ratio of the two cone types, however, is similar to that measured in bovine, pig and dog retinas.

Besides this middle-to-long wavelength specific antibody, we possess another one (mAb OS-2) which specifically labels the mammalian blue cones, and a third one, the polyclonal anti-rhodopsin serum, recognizing the rod cells of every animal species studied, together with a fraction of reptilian and avian cone cells being sensitive to the shorter wavelengths of these retinas.

The advantages and perspectives of using the library of anti-visual pigment antibodies, are the following items: (1) The position occupied by each of the colour-specific elements in the neural network of the retina, together with the cell-to-cell connections among the visual cells can be detected. (2) Minute ultrastructural differences can be observed using electron microscopic immunocytochemistry. (3) The mosaic and the spatial organization of the photoreceptors can be mapped on large retinal areas. (4) The antibodies could also be used in the human ophthalmopathology for studying the colour vision disorders.

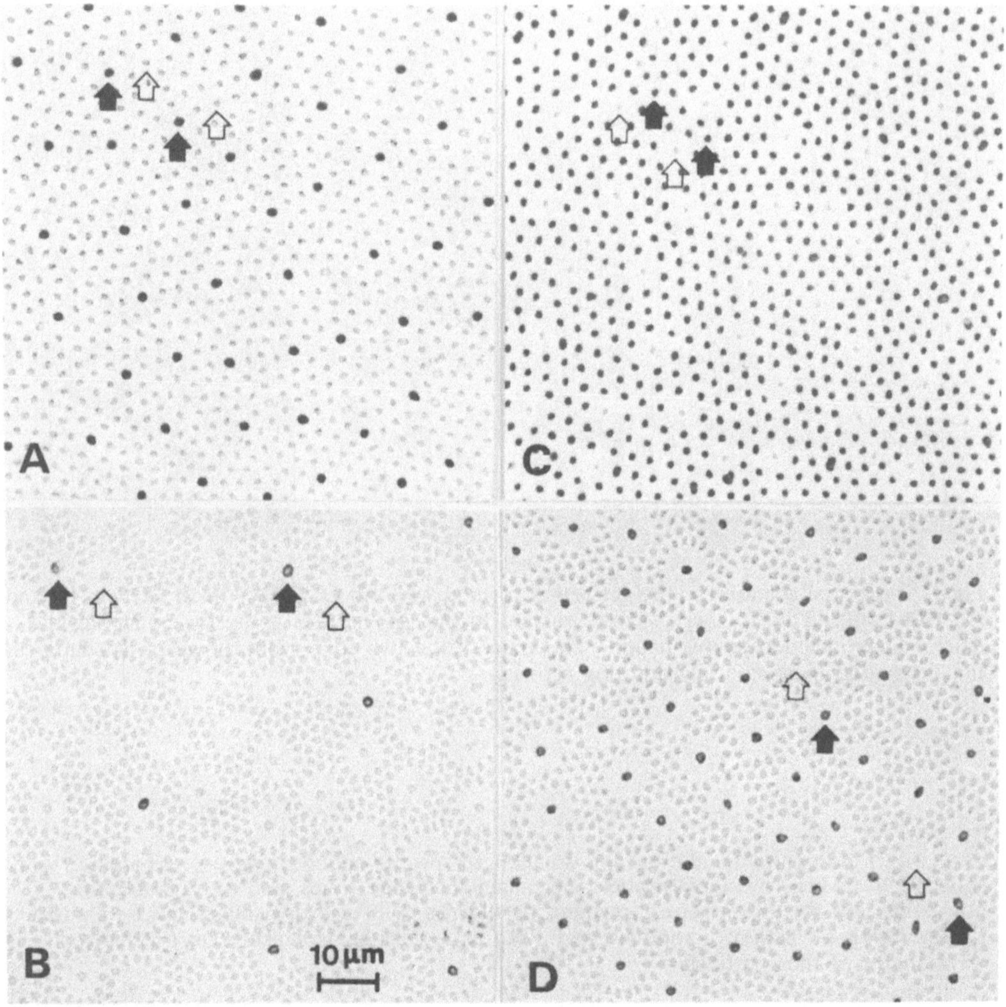

Fig.9. Tangential sections of the fovea centralis (A and C) and periphery (B and D) of the monkey retina, reacted with mAb OS-2 (A-B) and mAb COS-1 (C-D), respectively. Note that a small fraction of cones is stained by OS-2 in the rod-free foveal part of the retina and in the periphery as well. The majority of cone cells, in turn, are recognized by COS-1 in the whole retina. The ratio of the two cone types is relatively constant, independently from the presence (periphery) or absence (fovea) of rod cells. The positive elements are marked with filled arrows, while the open arrows point to the unstained ones.

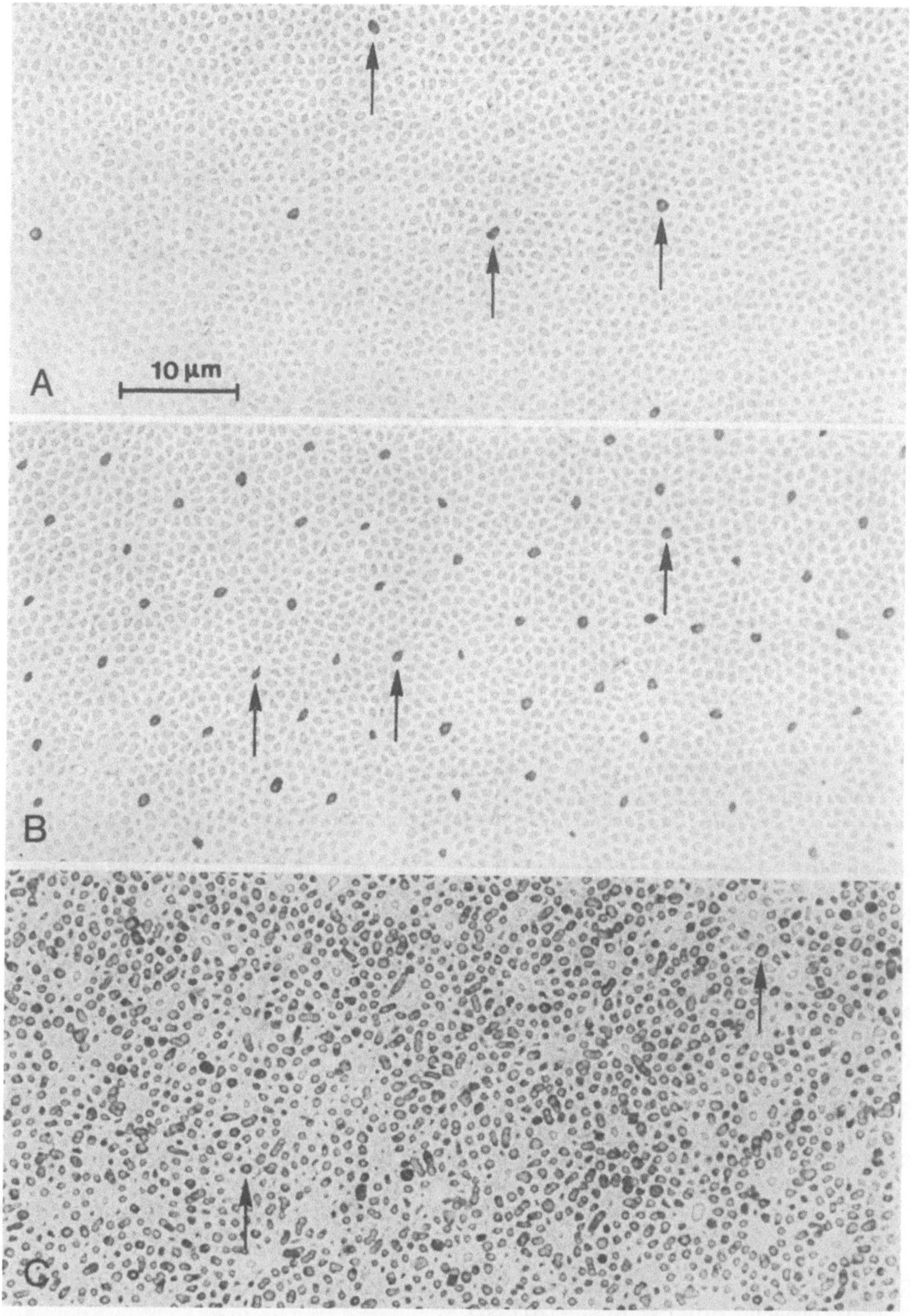

Fig.10. The peripheral part of the human retina. The tangential sections were reacted with mAb OS-2, COS-1 and RO-1, respectively. The frequency and the distribution of the three photoreceptor subpopulations strongly resemble those of the monkey retina. The cones are surrounded by several circles of rod cells. As the rod-cone distances are significantly greater, compared to those between the rods, the cone cells can be easily identified by the position occupied in the photoreceptor mosaic.

REFERENCES

Ahnelt, P. K., 1974, Characterization of the color related receptor mosaic in the ground squirrel retina, Vision Res., 25:1557-1567.

Bowmaker, J. K., 1977, The visual pigments, oil droplets and spectral sensitivity of the pigeon, Vision Res., 17:1129-1138.

Bowmaker, J. K., Knowles, A., 1977, The visual pigments and oil droplets of the chicken retina, Vision Res., 17:755-764.

Caldwell, J. H., Daw, N. W., 1978, New properties of rabbit retinal ganglion cells, J. Physiol. 276:257-276.

Cserhàti, P., Szél, À., Röhlich, P., 1988, Four cone types characterized by anti-visual pigment antibodies in the red area of the pigeon retina, Invest Ophthalmol. Vis. Sci., In Press

Daw, N. W., Pearlman, A. L., 1970, Cat colour vision: evidence for more than one cone process, J. Physiol., 211:125-137.

De Monasterio, F. M., Schein, S. J., McCrane, E. P., 1981, Staining of blue-sensitive cones of the macaque retina by a fluorescent dye, Science 213:1278-1281.

Gaur, V. P., Adamus, G., Arendt, A., Eldred, W., Possin, D. E., McDowell, J. H., Hargrave, P. A., Sarthy, P. V., 1988, A monoclonal antibody that binds to photoreceptors in the turtle retina, Vision Res., 28:765-776

Granda, A. M., Dvorak, C. A., 1977, Vision in turtles, Photoreceptors, in: "The Visual System in Vertebrates. Handbook of Sensory Physiology, Vol. VII/5," Crescitelli, F. ed., Springer Verlag, Berlin, Heidelberg, New York.

Kolb, H., Jones, J., 1987, The distinction by light and electron microscopy of two types of cone containing colorless oil droplets in the retina of the turtle, Vision Res., 27:1445-1458

Liebman, P., 1972, Microspectrophotometry of photoreceptors, in: "The Handbook of Sensory Physiology VII/1," Dartnal, H.J.A. ed., Springer, Berlin.

Liebman, P., Granda, A. M., 1971, Microspectrophotometric measurements of visual pigments in two species of turtle, Pseudemys scripta and Chelonia mydas, Vision Res., 11:105-114.

Mansfield, R. J. W., Levine, J. S., Lipetz, L. E., Collins, B. A., Raymond, G., MacNichol, E. F., 1984, Blue-sensitive cones in the primate retina: microspectrophotometry of the visual pigment, Exp. Brain. Res., 56:389-394.

Marc, R. E., Sperling, H. G., 1977, Chromatic organization of primate cones, Science, 196:454-456.

Mayor, H. D., Hampton, J. C., Rosario, B., 1961, A simple method for removing the resin from the epoxy-embedded tissue, J. Biophys. Biochem. Cytol., 9:909-910.

Nathans, J., Hogness, D., Hogness, D. S., 1986, Molecular genetics of human color vision: the genes encoding blue, green, and red pigments, Science, 232:193-202.

Nuboer, J. F. W., 1986, A comparative view on colour vision, Netherlands J. Zool., 36:344-380.

Nunn, B. J., Schnapf, J. L., Baylor, D. A., 1984, Spectral sensitivity of single cones in the retina of Macaca fascicularis, Nature, 309:264-266.

Ohtsuka, T., 1985, Relation of spectral types to oil droplets in cones of turtle retina, Science, 229:874-877.

Schnapf, J. L., Kraft, T. W., Baylor, D. A., 1987, Spectral sensitivity of human cone photoreceptors, Nature, 325:439-441.

Sperling, H. G., Johnson, C., Harwerth, R. S., 1980, Differential spectral photic damage to primate cones, Vision Res., 20:1117-1125.

Szél, À., Takàcs, L., Monostori, E., Vigh-Teichmann, I., Röhlich, P., 1985, Heterogeneity of chicken photoreceptors as defined by hybridoma supernatnats, Cell Tiss. Res., 240:755-741.

Szél, À., Röhlich, P., Govardovskii, V., 1986a, Moonoclonal antibody recognizing cone visual pigment, Exp. Eye Res., 43:871-883.

Szél, À., Takàcs, L., Monostori, É., Diamantstein, T., Vigh-Teichmann, I., Röhlich, P., 1986a, Monoclonal antibody recognizing cone visual pigment, Exp. Eye Res., 43:871-883.

Szél, À., Röhlich, P., Govardovskii, V., 1986b, Immunocytochemical discrimination of visual pigments in the retinal photoreceptors of the nocturnal gecko Teratoscincus scincus, Exp. Eye Res., 43:895-904.

Szél, À., Diamantsein, T., Röhlich, P., 1988, Identification of the blue-sensitive cones in the mammalian retina by anti-visual pigment antibody, J. Comp. Neurol., 273:593-602.

Szél, À., Röhlich, P., 1988a, Four photoreceptor types in the ground squirrel retina as evidenced by immunocytochemistry, Vision Res., In Press.
Szél, À., Röhlich, P., 1988b, The mosaic of the colour-specific photoreceptors in the mammalian retina as defined by immunocytochemistry, Acta. Morph., In press.
Van Veen, T., Vigh-Teichmann, I., Vigh, B., Hartwig, H. G., 1986, Light and electron microscopy of S--antigen- and opsin-immunoreactive photoreceptors in the retina of turtle, chicken, and hedgehog, Exp. Biol., 45:1-14.
Wald, G., Brown, P. K., Smith, P. H., 1953, Cyanopsin, a new pigment of cone vision, Science, 118:505-508.
Wald, G., Brown, P. K., Smith, P. H., 1955, Iodopsin, J. Gen. Physiol., 38:623-681.

CHEMICAL NEUROANATOMY OF THE INSECT VISUAL SYSTEM

Dick R. Nässel

Department of Zoology
University of Stockholm
S-106 91 Stockholm, Sweden

ABSTRACT

Biochemical, histochemical and immunocytochemical methods have demonstrated a number of neuroactive substances in the visual system of different insect species. Some of these are "classical neurotransmitters" such as acetylcholine, GABA, glutamate, taurine, octopamine, dopamine, noradrenalinee, serotonin and histamine; others are neuropeptides resembling FMRFamide, cholecystokinin (CCK), proctolin, vasoactive intestinal peptide (VIP) and adipokinetic hormone (AKH) The present paper gives an outline of some of the chemical circuits and pathways in the visual system of the blowfly *Culliphora erythrocephala* based on immunocytochemistry.

The neurotransmitter of the photoreceptors appears to be histamine. Many of the small field relay neurons at all levels of the visual system appear to contain GABA or glutamate, known to be inhibitory and excitatory transmitters, respectively. Large field projection neurons label with antisera against GABA, serotonin, dopamine, FMRFamide, CCK and VIP. Some of these projection neurons form extensive arborizations in various neuropils of the visual system as well as in other regions of the brain. Amacrine neurons are found at different levels of the visual system. In the medulla some large field amacrines were found to be FMRFamide- and serotonin-immunoreactive and other types of smaller field amacrines were found to be dopamine- and AKH-immunopositive. A number of neurons reacting with antisera to FMRFamide, proctolin, VIP, noradrenalinee, serotonin, histamine, GABA, glutamate, choline acetyltransferase, dopamine-β-hydroxylase and chromogranin A could not be resolved in complete detail.

Electron microscopic immunocytochemistry established that substances like FMRFamide, CCK, dopamine, serotonin and histamine are located in neural processes forming chemical synapses. The different substances may act either as neurotransmitters, neuromodulators or neurohormones, and a given substance probably has several functions at different levels of integration. Some of the chemically identified neurons appear to contain more than one neuroactive substance. It is clear that a large number of neurotransmitters and neuromodulators have not yet been detected in the blowfly visual system, but already it is apparent that chemical neurotransmission in the insect visual system is quite complex.

INTRODUCTION

Although many aspects of insect visual physiology are fairly well understood, we hardly know anything about the pharmacology of signal transfer and modulation in the insect visual system. Until recently most studies along this line have focused on demonstrating the presence of putative neuroactive substances or their synthetic enzymes. Several classical neurotransmitter substances such as acetylcholine, serotonin, dopamine, noradrenaline, GABA and glutamate have emerged as transmitter candidates in the insect visual system (reviewed by Klemm, 1976; Hardie, 1988; Nässel, 1988b). Immunocytochemistry has had a great impact on further studies of the cellular localization of putative neuroactive compounds. Several new substances have been indicated, notably neuropeptides which are known to be neurotransmitters, neuromodulators or neurohormones outside the insect visual system, and the resolution of detection could often be improved quite substantially. Nevertheless, we lack evidence that most of the detected substances are transmitters or modulators in the visual system. Some recent studies have approached the insect visual system pharmacologically. These studies have quite convincingly shown that histamine is the likely neurotransmitter of the retinal photoreceptors (Hardie, 1987) and have indicated that GABA may be an inhibitory substance in the lamina (Hardie, 1987) and part of the optomotor pathway (Bülthoff and Bülthoff, 1987a,b).

As will be shown in the present account it may not be easy to assess the nature and functions of neurotransmitters and neuromodulators in the complex insect visual system. At more accessible peripheral synapses and neurohormonal receptor sites pharmacological experiments have successfully established many neuroactive substances (see O'Shea and Schaffer, 1985; Evans and Myers, 1986). To understand chemical neurotransmission in the visual system it seems essential to analyze circuits that are well characterized physiologically and anatomically, as in the case of the photoreceptor/first order neuron synapses in the lamina or the motion-sensitive systems in the lobula plate. As a complement, immunocytochemistry of neuroactive substances can be employed to resolve the detailed morphology of pathways and circuits that are likely to use certain neuroactive substances in order to apply physiological and pharmacological analyses. Mappings of putative neuroactive substances have been made with immunocytochemistry in the visual system of a few insect species, but only a few systems have been resolved in any detail. These are neuronal systems reacting with antisera raised against serotonin (5-HT), dopamine (DA), histamine (HA), FMRFamide and GABA (Datum et al., 1986; Meyer et al., 1986; Schäfer and Bicker, 1986; Homberg et al., 1987; Nässel, 1988a; Nässel et al., 1988a,b,c).

The present account focuses on the chemical neuroanatomy of the visual system of flies such as the blowfly *Calliphora* and the fruit fly *Drosophila*, although some work has been published on other insects. The term chemical neuroanatomy used in the title includes localization of not only neuroactive substances, but also receptors, structural proteins, membrane surface glycoproteins, and metabolic enzymes. Emphasis will be on immunocytochemical detection of putative neurotransmitters and neuropeptides, but other findings will also be discussed when relevant. After a general presentation of neuroactive substances indicated in the optic lobes, I will discuss the detailed organization of neurons in the blowfly visual system reacting with antisera against serotonin, dopamine, FMRFamide, adipokinetic hormone (AKH) and vasoactive intestinal peptide (VIP). This is followed by a discussion of a few of the possible sites where pharmacology of visual processing can be studied or where immunocytochemical findings may tell us something about neuronal circuits. For more complete details as well as for a more comprehensive list of references some recent reviews will be referred to (Nässel, 1987a, 1988a,b; Hardie, 1988).

Overview of Transmitters in the Insect Visual System

Earlier accounts on the visual system have focused on the demonstration of so-called classical neurotransmitters. Small molecular weight substances, such as acetylcholine, serotonin, dopamine, noradrenaline, GABA, octopamine and histamine, were suggested as candidate neurotransmitters in the insect visual system since they could be detected biochemically or their synthesis could be demonstrated (see Klemm, 1976; Maxwell et al., 1978; Robertson, 1976; Dymond and Evans, 1979; Elias and Evans, 1983). The cellular localization of these compounds was initially investigated with histochemical methods and by autoradiography of radio-labeled substances taken up by the nervous system. As mentioned before, immunocytochemistry improved the resolution of the localization studies

Table 1. Putative neuroactive substances in the insect visual system

Putative neuroactive substance	Insect species	Methods
Acetylcholine	C,F,M	BC, EHC, EICC
GABA	F,B,M	BC, ICC, EICC
Glutamate	B,F	BC, ICC
Taurine	B	BC, ICC
Dopamine	F,C,L,M	BC, ICC, EICC (TH, DDC)
noradrenalinee	F,C,L	BC, ICC, EICC (DBH)
Serotonin	++	BC, ICC, EICC (DDC)
Octopamine	F,C,L	BC, ICC
Histamine	F,L,M	BC, ICC
Proctoline	F,C,Cr	ICC, BC*
AKH	F,L	ICC, BC*
VIP-like	F	ICC
FMRFamide-like	F,C,CB,L	ICC, BC*, mRNA
Corticotropin releasing factor-like	F	mRNA
Vasopressin/oxytocin-like	L	ICC, BC*
Somatostatin-like	Cr	ICC
Gastrin/cholecystokinin-like	F,Cr	ICC, BC**
Enkephalin-like	F	ICC, BC**

Abbreviations: B= honey bee; C= cockroach; CB= Colorado potato beetle; Cr= cricket; F= fly; L=locust; M= moth *Manduca sexta;* BC= biochemistry; ICC= immunocytochemistry; EICC= ICC of synthetic enzymes; EHC= enzyme histochemistry; mRNA= analysis of mRNA; TH= tyrosine hydroxylase; DDC= 5-HTP/dopamine decarboxylase; DBH= dopamine-beta-hydroxylase; *not in the visual system specifically, generally in the nervous system **amino acid composition known, not the sequence; ++= in more than 22 species. Data are derived from papers quoted in the text. None of the immunoreactive peptides has yet been isolated from the optic lobes.

and increased the number of substances that could be detected. Immunocytochemistry has some drawbacks, however, as to the specificity of the detection. With most polyclonal antisera or monoclonal antibodies, cross-reactivity with different cellular antigens may occur and critical specificity tests have to be pursued. Immunocytochemistry alone is not sufficient to establish the presence of a substance. In tissue such as the optic lobes, where individual cells cannot easily be microdissected for analysis, a combination of approaches has to be used. Throughout this paper it will therefore be made clear that the neurons detected with immunocytochemistry do not necessarily synthesize the transmitters or neuropeptides that the antisera were raised against. Nevertheless immunocytochemistry can be used as a first indicator of what substances to look for as neurotransmitters/modulators in different circuits of the visual system. The standard specificity tests employed in immunocytochemistry can give both false negatives and false positives, and further biochemical or immunoblot tests as well as *in situ* hybridization with cDNA probes should be performed (preferably using individual, identifiable micro-dissected neurons). In the meantime the suffix *-like immunoreactivity* will be used for neurons labeled with different antisera.

Apart from acetylcholine, serotonin, histamine, dopamine, noradrenaline, octopamine, GABA and glutamate no putative neuroactive substances that have been demonstrated conclusively in the insect visual system. Some neuropeptides have, however, been indicated by immunocytochemistry. These are listed in Table 1.

Immunocytochemically Identified Neurons in the Visual System of *Calliphora*

The blowfly visual system consists of a retina with about 4000 ommatidia whose receptor cells innervate the optic lobes. These consist of four columnar neuropils: the lamina (closest to the retina), the medulla, and the lobula complex with an anterior lobula and a posterior lobula plate (Fig. 1). The approximately 130,000 neurons of the optic lobes (Strausfeld, 1976) fall within a few major classes: (1) through-going relay neurons, which are columnar small-field units or wide-field elements, connecting the different neuropils; (2) different types of intrinsic or amacrine neurons connecting different layers of the same neuropil or different portions of the retinotopic array within a single neuropil; and (3) large projection neurons (ipsilateral, contralateral or bilateral), which often cover large portions of the visual field. From the medulla and the lobula complex there are large output neurons connecting to higher centers and to descending as well as motor pathways. Catalogues of cell types in these different groups and descriptions of their topographic relations to the projected mosaic and different pathways are to be found in several contributions (Strausfeld 1976, 1984; Strausfeld and Nässel 1980; Strausfeld and Bassemir, 1985; Strausfeld and Seyan, 1985; Hausen 1984).

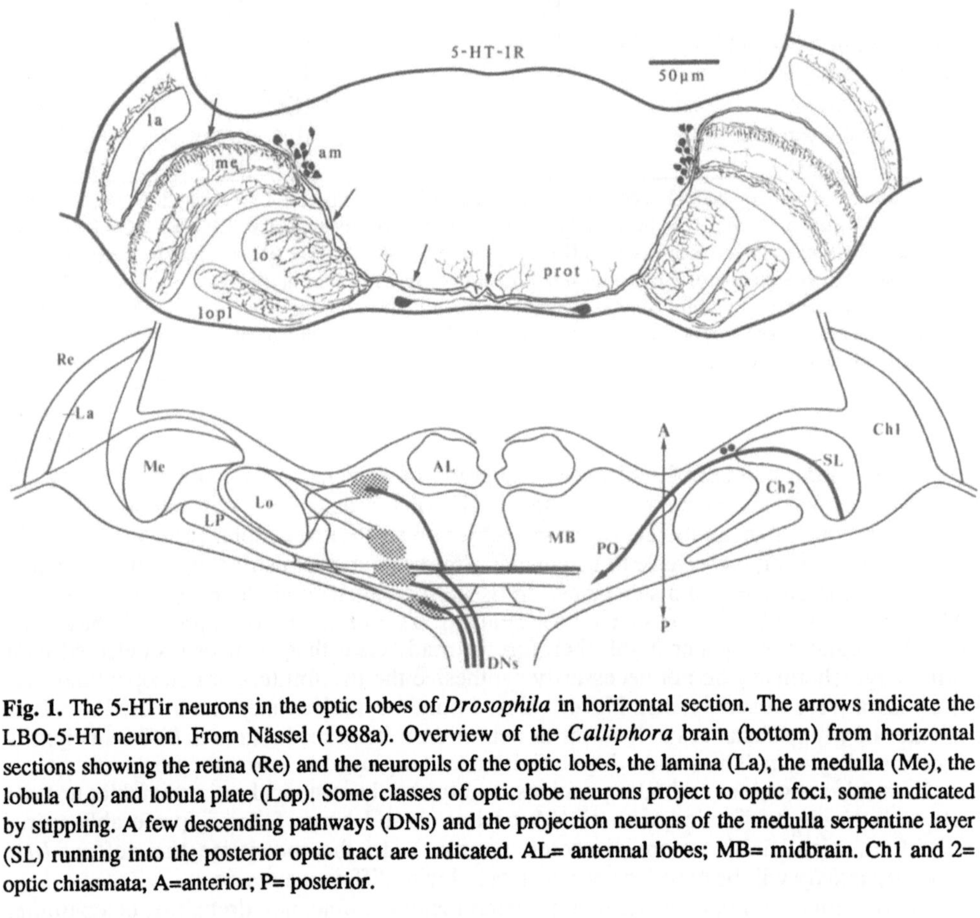

Fig. 1. The 5-HTir neurons in the optic lobes of *Drosophila* in horizontal section. The arrows indicate the LBO-5-HT neuron. From Nässel (1988a). Overview of the *Calliphora* brain (bottom) from horizontal sections showing the retina (Re) and the neuropils of the optic lobes, the lamina (La), the medulla (Me), the lobula (Lo) and lobula plate (Lop). Some classes of optic lobe neurons project to optic foci, some indicated by stippling. A few descending pathways (DNs) and the projection neurons of the medulla serpentine layer (SL) running into the posterior optic tract are indicated. AL= antennal lobes; MB= midbrain. Ch1 and 2= optic chiasmata; A=anterior; P= posterior.

Immunocytochemistry of putative neuroactive substances selectively labels different populations of neurons falling within the classes listed above. Only antisera against histamine, taurine and possibly GABA label photoreceptors (Datum et al., 1986; Schäfer et al., 1988; Nässel et al., 1988c), all others label relay- or interneurons. Commonly, each antiserum stains only a very small number of cell types ranging from one type (e.g. anti-AKH) to six types (e.g. anti-FMRFamide). Often the antisera also label a relatively small absolute number of neurons in the optic lobes. As an example there are two Leu-enkephalin immunoreactive and around 20 5-HT-immunoreactive neurons and a few thousand FMRFamide immunoreactive ones out of a total of about 130,000 neurons per optic lobe. For each immunoreactive substance a different distribution pattern of neurons and their processes can be distinguished (see Fig. 5). Hence, in some cases immunoreactive neurons can be seen only in one of the neuropil regions, e.g. in the lobula. In other cases all neuropils contain immunoreactive processes. Some antisera label several different types of neurons in one neuropil region, others just one type. Taken as a whole, each immunocytochemically identifiable population can be either homo- or heterogeneous with respect to morphological cell types (see Fig. 2). Here it should be emphasized that some antisera may recognize antigenic determinants in more than one substance (note that each antiserum, however, gives a reproducible and invariable staining). Thus the heterogeneity in identifiable neuron types may only be apparent. Furthermore, the distribution of the different immunoreactive substances outside the optic lobes varies substantially (Fig.6). Some putative neurotransmitters and neuropeptides may be distributed in large numbers of neurons outside the visual system even when they are rare in the visual system. No putative neurotransmitter or neuropeptide has yet been indicated only in the optic lobes. Interestingly, there are several substances that in addition to their location in interneurons of the optic lobes and elsewhere in the central nervous system seem to be present in motor neurons and neurosecretory neurons (see Nässel, 1987a, 1988a; Nässel and O'Shea, 1987; Duve et al., 1988). Still other substances, such as prothoracicotropic hormone (PTTH) and vasopressin-like peptide, are located in a very small number of neurons in the entire central nervous system (Rémy and Girardie, 1980; Mizoguchi et al., 1987; Nässel et al., 1988d). The different distribution patterns and their functional implications will be discussed in a later section.

MORPHOLOGY AND PROJECTIONS OF IMMUNOCYTOCHEMICALLY IDENTIFIABLE NEURONS

In this section I will outline the arrangements of some of the immunocytochemically identified neuron systems. The focus will be on systems identified with antisera against 5-HT, dopamine, FMRFamide and AKH, but also those that are proctolin- histamine-, and VIP-like immunoreactive will be discussed briefly. Only in a few exceptional cases it was possible to correlate immunolabeled neurons with neurons previously identified with neuroanatomical techniques. These cases will be mentioned in the text.

Serotonin-immunoreactive Neurons

The morphology of serotonin-immunoreactive (5-HTir) neurons in the visual system of insects has been dealt with in detail in a number of investigations (Nässel and Klemm, 1983; Schürmann and Klemm, 1984; Nässel et al., 1985, 1987; Nässel, 1987a, 1988a; Buchner et al., 1988). In flies the important findings can be summarized as follows (Figs. 1,2B,3A). There are very few 5-HTir neurons with processes in the visual system, but each neuron supplies large portions of the neuropil with varicose branches. These few neurons (in *Calliphora* and *Sarcophaga* about 23-26 on each side) are of three types: 2 large field projection neurons, about 20-22 amacrines in each medulla and 2 amacrines in each lobula. The two large projection neurons, called LBO-5HT, have cell bodies caudally in the protocerebrum whose processes bilaterally invade the lamina, medulla and lobula complex as well as portions of the protocerebrum with varicose processes. In the medulla these neurons supply three distinct layers with processes; in the lobula and lobula plate the arrangement of processes is not clearly patterned. The lamina 5-HTir processes reside mainly outside the synaptic neuropil in a layer containing the cell bodies of the lamina monopolar neurons and photoreceptor axon bundles (Nässel and Elekes, 1984; Nässel et al., 1985). Each of the LBO-5HT neurons, hence, can interact with neurons over the whole projected mosaic, at different integrational levels and on both sides of the brain. The morphology of the amacrines is not known in detail, but each neuron supplies a large portion, if not all, of the projected mosaic. The lamina portion of the LBO-5-HT neurons was earlier identified from Golgi impregnation and electron microscopy (Nässel et al., 1983).

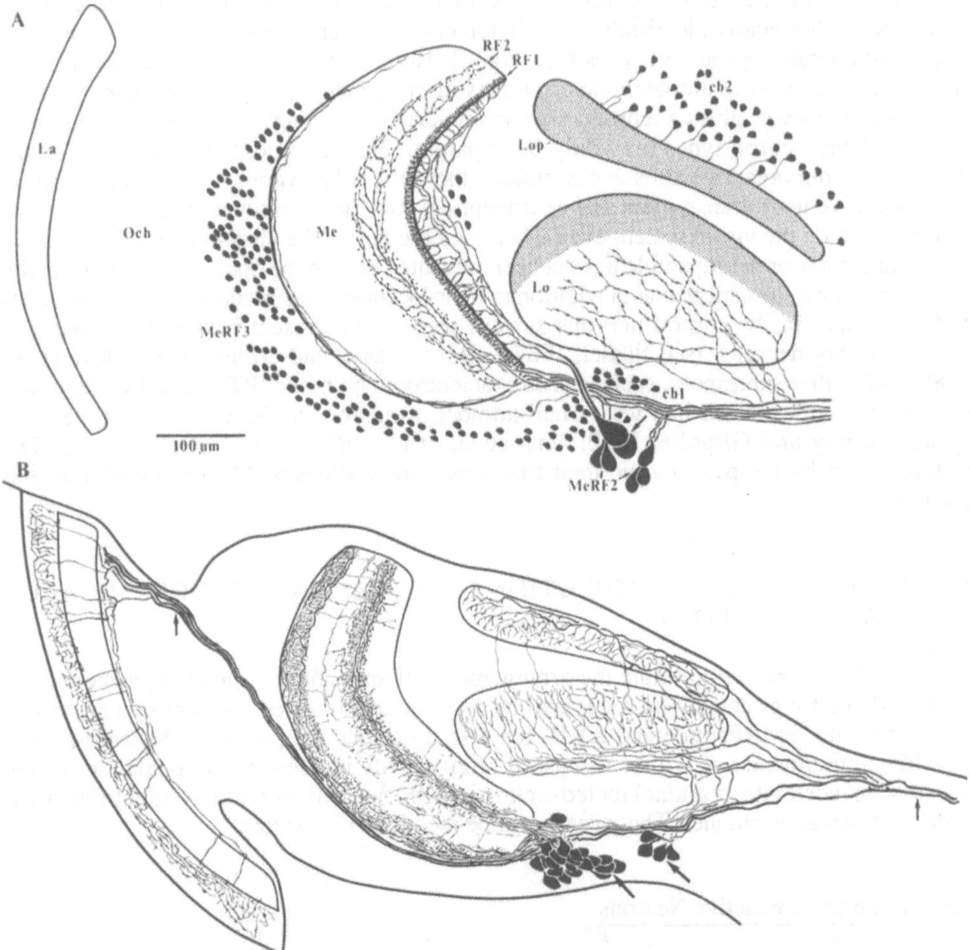

Fig. 2. Horizontal sections of optic lobes. **A.** FMRFir neurons in *Calliphora*. The cell types are described in the text.The cellbodies of the large amacrines MeRF1 are indicated by arrow (near MeRF2). The stippled areas in the lobula complex are derived from the processes of the cb2 neurons posterior to the lobula plate. (From Nässel et al., 1988b). **B.** The 5-HTir neurons in the fleshfly *Sarcophaga* The small arrows indicate the LBO-5-HT neurons with cell bodies in the protocerebrum (note the processes distal to the lamina neuropil); the large arrows indicate the two clusters of amacrines. (From Nässel et al., 1987).

Purely from the morphology it can be suggested that the serotonergic neurons are involved in signal processing or modulation over large volumes of neuropil. Among the 5-HTir neurons of the optic lobes no small-field relay neurons have been detected and no obvious polarity of the neurons can be seen morphologically. Hence, 5-HT possibly acts as a modulatory substance rather than as a neurotransmitter. It has previously been suggested that the 5-HTir neurons are involved in modulation of overall sensitivity or activity of visual interneurons, possibly related to different arousal states (Nässel et al., 1985; Nässel,

1988a). The two extensive LBO-5-HT neurons, which integrate all optic lobe neuropils on both sides with extensive areas of neuropil in the midbrain (protocerebrum), would be suitable candidates for such modulatory action. In the two systems of amacrines, 5-HT may be used as a transmitter or modulatory substance in more local circuits. To understand the role of the 5-HTir neurons in the visual system we need to identify their inputs and localize the 5-HT receptors in the visual system neuropils and to examine pharmacologically the action of serotonin in visual processing. Although serotonin has been shown to be a neuroactive compound in other parts of the insect nervous system (see Nässel 1988a), its action in the visual system remains to be studied.

Dopamine-immunoreactive Neurons

In the blowfly optic lobes dopamine-immunoreactive (DAir) neurons have been described in some detail (Nässel et al., 1988a). Two major types of DAir neurons can be seen in the optic lobes: (1) thousands of small field amacrine neurons in the medulla and (2) a small number of large projection neurons connecting the lobula and lobula plate on both sides (Fig. 4B).

The DAir amacrines form three layers of processes in the medulla. Each neuron seems to invade a few medulla columns only, but there is a more extensive lateral overlap between the processes of the amacrines in the deepest layer (the serpentine layer). It is not clear whether there is one DAir amacrine for each column (i.e. about 4,000 per medulla) or fewer. Neurons similar to the DAir amacrines have been described from Golgi preparations of *Musca domestica* (Strausfeld, 1976). Electron microscopy showed that the DAir amacrines may be centrifugal intrinsic neurons relaying signals from the serpentine layer of the medulla to the two more distal layers. In the serpentine layer the amacrines in addition are presynaptic to each other and hence may interact laterally. At the synaptic sites presynaptic specializations (Trujillo-Cenóz, 1965) and clear vesicles can be seen (Nässel et al., 1988a) in the DAir profiles. Only rarely were dense core vesicles encountered.

The DAir projection neurons bilaterally connect the lobula and lobula plate of the two hemispheres (Fig. 4). At least four such neurons can be seen with their cell bodies distributed two on each side of the lateral protocerebrum. These neurons form processes with different distributions in the brain proper so that a subdivision into two types can be made (Fig. 4C). Their processes in the lobula complex, however, seem to overlap totally and form the same type of branching patterns. In the lobula plate the DAir processes are mainly located within the posterior portion occupied by the dendrites of the large-field motion-sensitive neurons of the vertical system (Hausen, 1984). It is apparent that the dopaminergic projection neurons are connected to a much more selective portion of the optic lobe neurons than are the serotonergic ones. The lamina and medulla neuropils are excluded and the DAir lobula plate processes are in a certain layer. Furthermore central brain connections are of two distinct types rather than one for the 5-HTir projection neurons. It should be mentioned here that DAir neurons in other parts of the fly central nervous system form arborization and projection patterns quite distinct from the 5-HTir ones.

In the visual system dopamine may be a neurotransmitter in amacrines and projection neurons (Nässel et al., 1988a). The action of the the DAir neurons appear more localized than that of the 5-HTir neurons and the neurons seem from morphological features to be polarized in their signal transfer.

FMRFamide-immunoreactive Neurons, a Heterogeneous Group

FMRFamide (Phe-Met-Arg-Phe-NH_2) was originally isolated as a cardioexcitatory neuropeptide from molluscs (Price and Greenberg, 1977). Substances related to FMRFamide were found later in many insect species where they may play roles as neuroactive substances both in the central nervous system and at peripheral targets (Boer et al., 1980; Evans and Myers, 1986; Schneider and Taghert, 1988). As the title indicates,

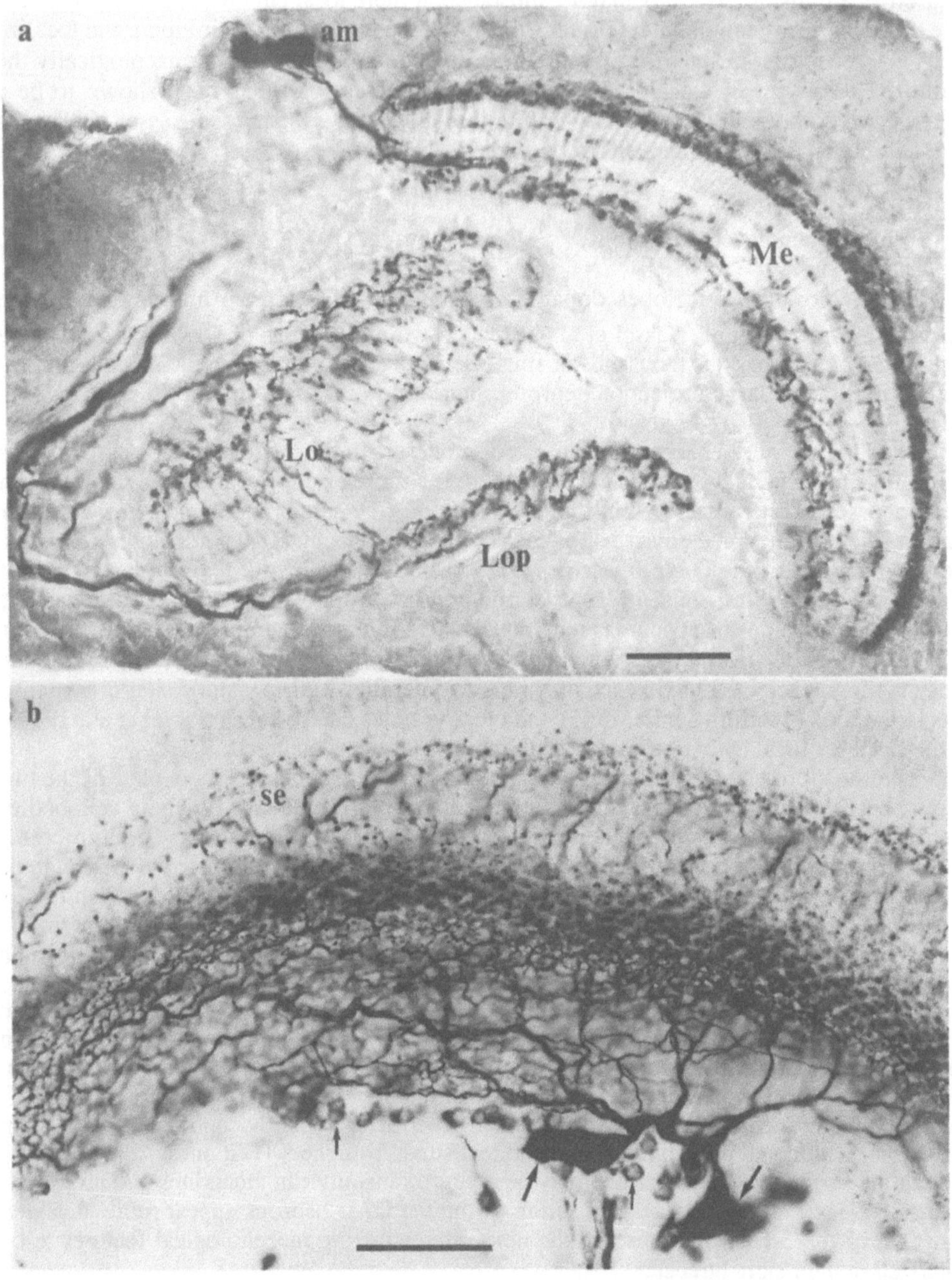

Fig. 3a. Serotonin-immunoreactive neurons in the *Calliphora* (70% pupal development) optic lobes seen in horizontal section. Some of the amacrine medulla neurons (am) can be seen. Scale: 50µm. From Ohlsson and Nässel (1987). **3b.** FMRFamide-immunoreactive neurons in the medulla seen in oblique frontal section. The two large amacrines (MeRF1) are indicated by large arrows. Note the dense distribution of fine processes of the amacrines in the innermost medulla layer. The serpentine layer (se) is invaded by processes of the MeRF2 neurons. Small arrows indicate small cell bodies of not classified neurons. Scale: 50 µm. From Nässel et al. (1988b).

different antisera to FMRFamide and its C-terminal fragment RFamide (Grimmelikhuijzen, 1985) selectively label a distinct population of fly optic lobe neurons of many morphological types (Fig. 2A). These neurons are, however, also immunocytochemically heterogeneous in the sense that a set of antisera against other neuropeptides recognize subpopulations of the FMRFamide-like immunoreactive (FMRFir) neurons (Nässel et al., 1988b; Ohlsson et al., 1988) similar to findings in other insects (Veenstra and Schooneveld, 1984; Verhaert et al., 1985; Myers and Evans, 1987). Antisera to bovine pancreatic polypeptide (BPP), gastrin/cholecystokinin (CCK), Leu-enkephalin and Met-enkephalin-Arg6-Phe7 (YGGFMRF) recognize subsets of the FMRFir neurons. In the following I will first treat the whole FMRFir population morphologically before distinguishing immunochemically identified subclasses. It is important to stress that we do not know whether any of the FMRFir neurons in the fly optic lobe contain authentic FMRFamide or one or several related compounds. Even if an FMRFamide-like peptide is present within some of the neurons, additional related or non-related neuropeptides may also be co-localized.

In the visual system of adult *Calliphora* eight types of FMRFir neurons have been distinguished (Nässel et al., 1988b). Of these, only two types could be resolved in more detail in adult insects. The remaining types were immunolabeled only in their cell bodies and in some of their varicose terminals and processes or were possible to trace in their entirety only in pupal stages (Ohlsson et al., 1988) (see Fig. 8). The two neuron types that could be resolved in detail are: (1) one type of large-field amacrine neuron (2 cells), termed MeRF1 (Fig. 3A), and (2) a type of large-field projection neuron (5 cells) called MeRF2 (Figs. 2A,4A). The amacrine neurons have their cell bodies at the base of the medulla and densely arborizing processes in the basalmost layer of the medulla neuropil. Each MeRF1 neuron seems to span with its processes the entire (or a very large portion of it) retinotopic mosaic. They are the most extensively arborizing optic lobe amacrines resolved with any technique so far. These neurons are both pre- and postsynaptic to the neurons in the basal medulla layer as demonstrated by electron microscopic immunocytochemistry (Nässel et al., 1988b). The FMRFir profiles in this layer contain clear vesicles and form typical presynaptic specializations.

The large FMRFir projection neurons, MeRF2, have their cell bodies anteriorly at the base of the medulla, near those of the amacrine neurons, and their processes connect the medulla and the lobula to a neuropil adjacent to the calyx of the mushroom body on the ipsilateral side (Fig. 4A). In the medulla the processes of the projection neurons are predominantly in and around the serpentine layer. In the lobula they form varicose processes diffusely in the basal half of the neuropil. Electron microscopic immunocytochemistry shows that the FMRFir processes in the serpentine layer contain two types of vesicles, clear ones and large granular ones. No clear-cut synaptic contacts could be resolved between the FMRFir profiles and other neurons in the serpentine layer, in contrast to the amacrine processes.

The FMRFamide-like peptide in the amacrines may be a neurotransmitter in local circuits in the basal layer of the medulla, involved in lateral interactions over extensive portions of the visual field. The projection neurons connect several neuropil areas. One of these the "lateral horn" (Strausfeld, 1976) adjacent to the mushroom body calyx, is innervated by neurons of the antenno-glomerular tract (chemosensory neurons), as well as interneurons of different kinds, some derived from the ocellar system others which connect to the central body complex (Strausfeld, 1976; Nässel, 1987b). Hence, the FMRFir projection neurons may be modulatory multimodal neurons.

There are some additional FMRFir neurons that could not be resolved in detail in the adult optic lobes. Distal to the medulla thousands of small FMRFir cell bodies (MeRF3) are located (Fig. 2A). These are probably amacrines of the medulla, but conclusive evidence is missing since the fibers from these cell bodies can be traced only into the distal medulla neuropil. Another set of small cell bodies (cb2) are located posterior to the lobula plate. These are most probably small field relay neurons connecting the lobula plate to the distal layer of the lobula (the stippled areas in Fig. 2A). Clusters of small cell bodies are located a the base of the medulla (MeRF4) and at the base of the lobula (cb1). Two additional neuron types could be seen in more detail during pupal development of the optic lobes when fiber patterns are less intricate (Ohlsson et al., 1988): (1) One neuron (PPOL) on each side of the posterior protocerebrum sends processes into ipsilateral protocerebral neuropil and into the lobula. These neurons were later recognized in the adult fly and their cell bodies are among two pairs of very prominent posterior FMRFir neurons (Fig.8) that also are CCKir and stain with the vital dye Neutral red. (2) The other neuron type has not been identified with certainty in the adult brain, but in pupae there are four neurons (PIOL) with cell bodies in the dorso-medial protocerebrum and processes running to the contralateral lobula neuropil (Fig. 8A).

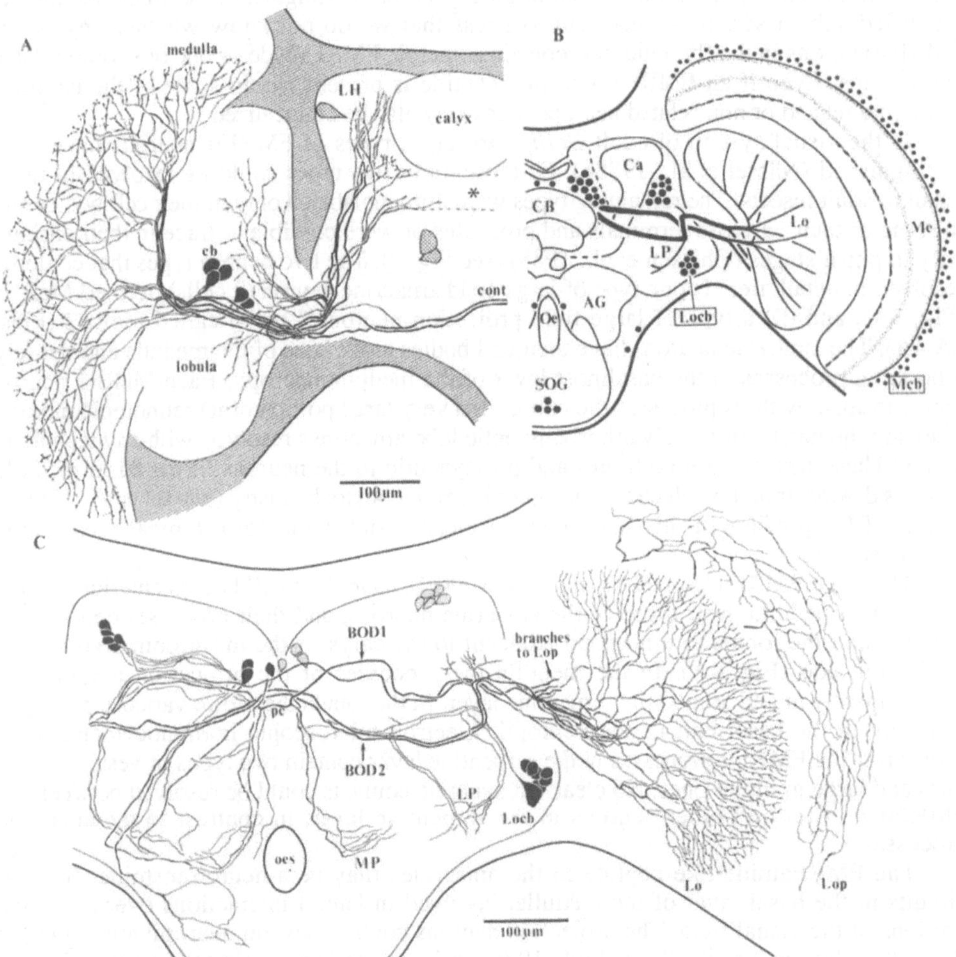

Fig. 4. Large field projection neurons in the optic lobe of *Calliphora* (frontal sections). **A.** FMRFir neurons of the MeRF2 type connect the medulla and lobula to the "lateral horn" neuropil (LH) adjacent to the mushroom body calyx in the dorsal protocerebrum. Only the basal portion of the medulla is shown. Contralateral fibers (cont) are probably derived from the PIOL neurons described in Fig.8. From Nässel et al. (1988b)**B.** Schematic drawing of DAir neurons in the optic lobes of the right hemisphere. The layering in the medulla is derived from the DAir amacrines (Mcb) and the lobula and lobula plate (not shown) processes are derived from bilateral neurons with cell bodies laterally in protocerebrum (Locb). Some other DAir cell bodies are shown as filled circles. LP= lateral protocerebrum; AG= antennal glomeruli; Ca= calyx; CB= central body. **C.** Detailed tracing of bilateral DAir neurons invading the lobula (Lo) and lobula plate (Lop). Two types can be distinguished (BOD1 and BOD2). In the left hemisphere some other DAir processes are shown that follow the sam trajectories as the optic lobe neurons. From Nässel et al. (1988a)

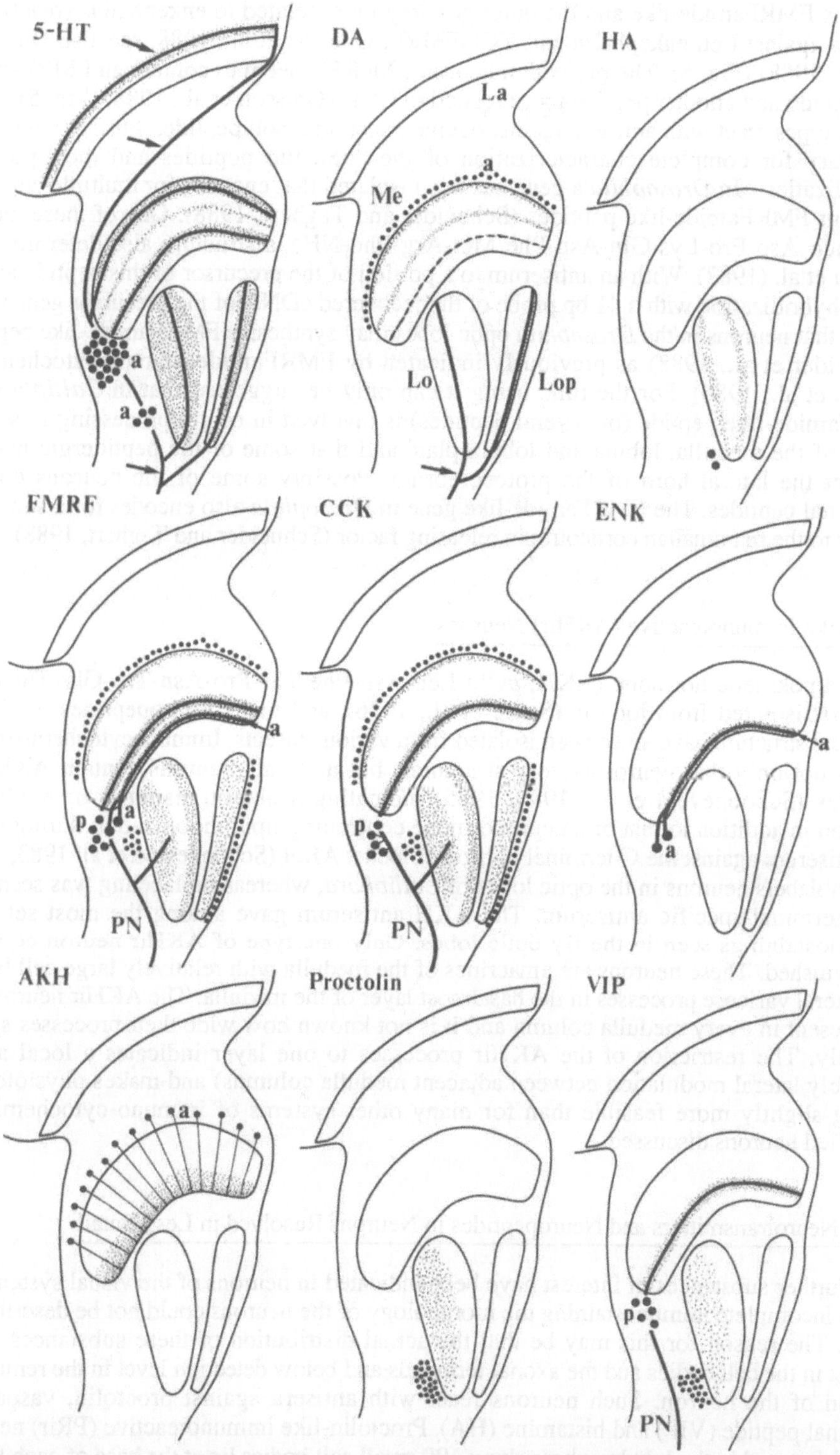

Fig. 5. Different patterns of immunoreactivity displayed with antisera against various neuroactive substances. Note that only 5-HTir neurons have processes in the lamina region. The arrows point at bilateral 5-HTir and DAir projection neurons with cell bodies in the midbrain (protocrebrum). HA= histamine, ENK= Leu-enkephalin and YGGFMRF; PN= projection neurons; p= cell bodies of projection neurons; a= amacrine neurons.

The MeRF1 amacrines seem to contain at least two different neuropeptides one of which is FMRFamide-like and the other possibly more related to enkephalins (reacts with antisera against Leu-enkephalin and YGGFMRF; Ohlsson et al., 1988; see also Duve and Thorpe, 1988) (Fig. 5). The projection neurons, MeRF2, seem to contain an FMRFamide-like peptide and another peptide more related to CCK (Ohlsson et al., 1988) (Fig. 5). Both neuron types react with antisera against bovine pancreatic polypeptide. Further studies are necessary for complete characterization of the authentic peptides and their possible colocalization. In *Drosophila* a gene has been isolated that encodes for multiple copies of different FMRFamide-like peptides (Schneider and Taghert, 1988). One of these has the sequence Asp-Pro-Lys-Gln-Asp-Phe-Met-Arg-Phe-NH_2, a structure also determined by Nambu et al. (1987). With an antiserum to a portion of the precursor of this peptide and by *in situ* hybridization with a 41 bp probe of the recovered cDNA of the precursor gene it was shown that neurons in the *Drosophila* optic lobes may synthesize FMRFamide-like peptides (Schneider et al., 1988) as previously indicated by FMRFamide-immunocytochemistry (White et al., 1986). For the time being it can only be suggested that in *Calliphora* an FMRFamide-like peptide (or several peptides) is involved in neural processing in several layers of the medulla, lobula and lobula plate and that some of the peptidergic neurons connect the lateral horn of the protocerebrum. Possibly some of the neurons contain additional peptides. The FMRFamide-like gene in *Drosophila* also encodes for a substance similar to the mammalian corticotropin releasing factor (Schneider and Taghert, 1988).

AKH-like Immunoreactive (AKHir) Neurons

Adipokinetic hormone (AKH; *p*Glu-Leu-Asn-Phe-Thr-Pro-Asn-Trp-Gly-Thr-NH_2) was first isolated from locusts (Stone et al., 1976), and many neuropeptides of similar chemical structure have since been isolated from various insects. Immunocytochemistry has shown not only that neurosecretory structures, but also interneurons contain AKH-like peptides (Schooneveld et al., 1983, 1985) indicating a neurotransmitter or modulator function in addition to that of a neurohormone controlling lipid mobilization during flight. An antiserum against the C-terminal portion of locust AKH (Schooneveld et al. 1983, 1985) immunolabels neurons in the optic lobes of *Calliphora*, whereas no labeling was seen with an N-terminal-specific antiserum. The AKH antiserum gave among the most selective immunostainings seen in the fly optic lobes. Only one type of AKHir neuron could be distinguished. These neurons are amacrines of the medulla with relatively large cell bodies and lateral varicose processes in the basalmost layer of the medulla. The AKHir neurons are not present in every medulla column and it is not known how wide their processes spread laterally. The restriction of the AKHir processes to one layer indicates a local action (possibly lateral modulation between adjacent medulla columns) and makes physiological testing slightly more feasible than for many other systems of immuno-cytochemically identified neurons discussed.

Other Neurotransmitters and Neuropeptides in Neurons Resolved in Less Detail

Further substances of interest have been indicated in neurons of the visual system, but due to incomplete immunostaining the morphology of the neurons could not be described in detail. The reason for this may be that the actual distribution of these substances is the highest in the cell bodies and the axonal terminals and below detection level in the remaining portion of the neuron. Such neurons react with antisera against proctolin, vasoactive intestinal peptide (VIP) and histamine (HA). Proctolin-like immunoreactive (PRir) neurons can be detected in the lobula, where about 100 small cell bodies lie at the base of each lobula neuropil (Nässel and O'Shea, 1987). PRir varicosities can be seen in the basal half of the neuropil (Fig. 5). It is hence not clear whether the PRir neurons are relay neurons or amacrines. The same is true for histamine-like immunoreactive (HAir) neurons (Fig. 5), which also have immunoreactive varicosities in two layers in the basal half of the lobula neuropil (Nässel et al., 1988c). The cell bodies of the HAir neurons are located at the base of the neuropil (probably only two cell bodies on each side). Vasoactive intestinal peptide-like immunoreactive (VIPir) neurons are seen both in the medulla and the lobula (Fig. 5). In the medulla they are large-field projection neurons with processes mainly in the serpentine layer. About five large neurons are seen on each side of the brain with cell bodies at the anterior base of the medulla (Holmqvist, Movérus and Nässel, in prep.). In the lobula the approximately 200 VIPir neurons strongly resemble the PRir neurons. We have not yet tested for co-localization of the two neuropeptides in these lobula neurons. Whereas proctolin and HA have been isolated from insects, the nature of the VIP-like substance indicated by immuncytochemistry remains unknown. We found that two other antisera against VIP do not label the optic lobe neurons in question (but do, however, label other neurons in the brain). Similarily antisera against peptide histidine-isoleucin (PHI), derived from the same precursor as VIP in mammals (see Sundler et al., 1988), as well as antisera against the related peptides glucagon, secretin and helodermine do not label the neurons in the optic lobes, although they label distinct cell populations in the remaining central nervous system (Nässel, Håkanson and Sundler, in prep.). These results may indicate that the immunoreactive substance in the visual system (in contrast to other parts of the brain) is not directly related to VIP, AKH or members of the glucagon family listed above, it may only share some antigenic determinants.

A number of antisera against mammalian type substances failed to reveal optic lobe neurons in *Calliphora* although they labeled neurons in other parts of the nervous system: melatonin, somatostatin, vasopressin, oxytocin, glucagon, secretin, peptide histidine-isoleucin, helodermin, insulin and substance P. Octopamine immunoreactive neurons have been described in the locust optic lobe (Konings et al., 1988), but so far no mapping of this kind has been performed for flies. Comparisons of immunolabelling with different antisera in the brain and optic lobes of *Calliphora* are shown in Figs. 5 and 6.

Other "Neurotransmitter-related" Substances

Antisera against synthetic enzymes or other components of the synaptic signal transfer system label specific neurons in the fly optic lobe. The heading may be misleading since it is not clear whether the authentic substances that are immunolabeled in the fly visual system are truly related to the mammalian substances against which the antibodies were raised. An antiserum against dopamine-β-hydroxylase, the synthetic enzyme of noradrenaline in mammals, labels neurons in the medulla and lobula of the fly (Klemm et al., 1985). Immunolabeling with antisera to noradrenaline was, however, not satisfactory in the visual system and a direct comparison could therefore not be made. Choline acetyltransferase the synthetic enzyme of acetylcholine could be demonstrated immunocytochemically in neurons of the *Drosphila* visual system (Buchner et al., 1986; Gorczyca and Hall, 1987). The presence of acetylcholine and cholin acetyltransferase in the brain can be determined by biochemistry (see Salvaterra and McCaman, 1985), α-bungarotoxin binding sites can be demonstrated in the *Drosophila* optic lobes (Dudai, 1980) and by immunocytochemistry putative acetylcholine receptors have been localized in the locust optic lobe neuropils (Vieillemaringe et al., 1987).

Chromogranins A and B are acidic proteins originally isolated from catecholamine storing chromaffin granules in cells of the adrenal medulla (Rieker et al., 1988). These substances have been identified in arthropods (Rieker et al., 1988). An antiserum raised against chromogranin A labels many neurons throughout the blowfly nervous system (Nässel, unpublished). In the optic lobes a distinct population of small neurons with their cell bodies at the base of the lobula invade the lobula neuropil. Their detailed morphology in the lobula cannot, however, be resolved. It is not clear what the antiserum recognizes in the neurons in question or what putative neurotransmitters may be involved. The labeled neurons, however, are distinct from any other population labeled so far with other antisera (including anti-dopamine).

In *Drosophila* several types of monoclonal antibodies produced against *Drosophila* nervous tissue specifically label neurons in the visual system (Fujita et al., 1982; Steller et al., 1987; Buchner et al., 1988). Also antisera against *Hydra* head activator and different

Ca2+-binding proteins label distinct sets of visual interneurons in *Drosophila* (Buchner et al., 1988). Immunocytochemical data of the kind presented in this section may be more informative for developmental studies and analysis of genetical mutants than for the understanding of neural transmission.

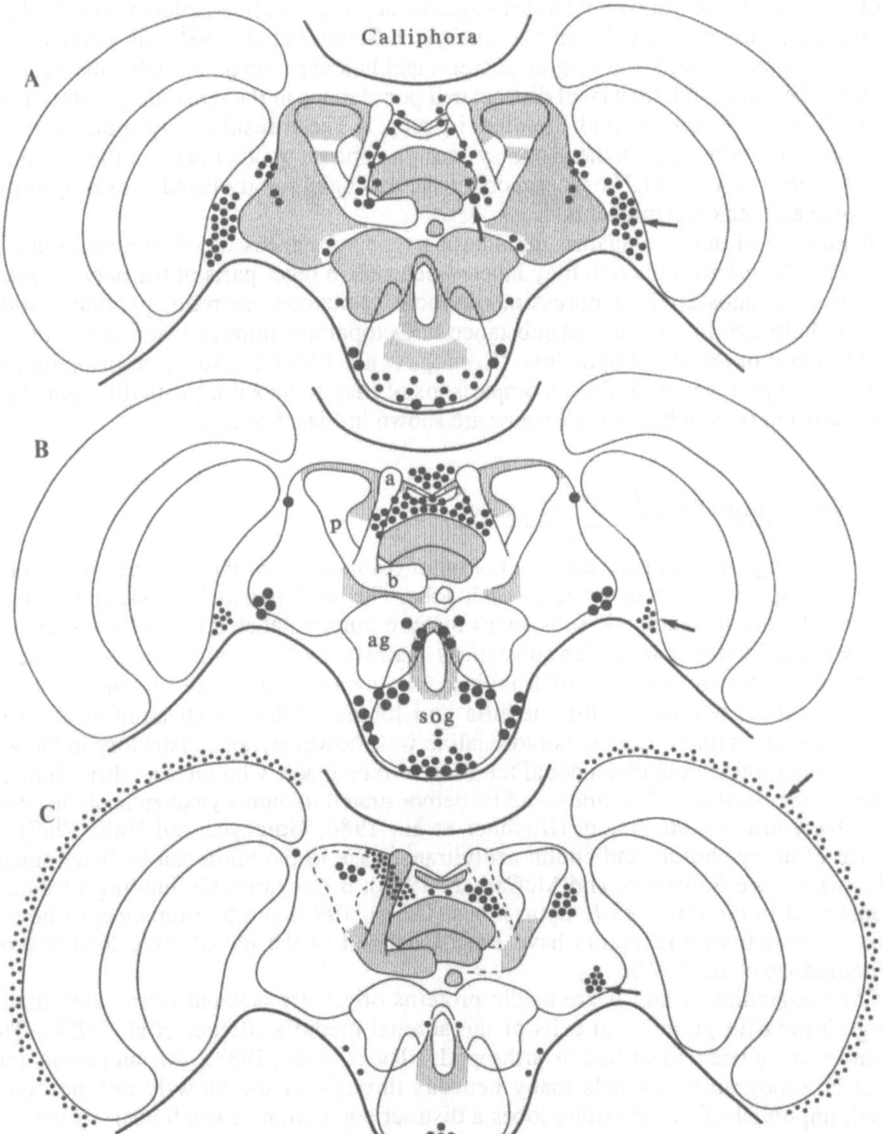

Fig. 6. The different distribution pattern of immunoreactive cell bodies seen with three different antisera to neuroactive substances.Differences are seen both in optic lobes and the remaining brain. Arrows indicate cell bodies of neurons innervating the optic lobes. In the midbrain the distribution of immunoreactive processes in neuropils is indicated by cross hatching (in optic lobes fibers are not shown). **A.** 5-HTir. **B.** Proctolin-like immunoreactivity. **C.** DAir (some midbrain regions in lateral protocerebrum not shown). Anterior cell bodies shown in left hemisphere, posterior ones in the right.

APPROACHES TO STUDYING THE ROLE OF NEUROTRANSMITTERS AND NEUROPEPTIDES

As indicated in the introduction the insect visual system is complex and compact and neuropharmacological studies may be more difficult than in the vertebrate retina. At a few sites in the insect visual system, however, analysis may be possible or the chemical anatomy itself may tell us something about neural processing. These are discussed below.

The Lamina Circuitry and Neurotransmitters

So far the lamina is the only part of the visual system where detailed synaptological data can be combined with physiology and pharmacology. This has made it possible to accumulate evidence that histamine is the transmitter of photoreceptors at their afferent synapses with large monopolar neurons in the lamina and that GABA may have an inhibitory role in this circuitry (Hardie, 1987). Since the synaptic wiring of the lamina has been studied in some detail and presumably all the major neuron types have now been classified (Strausfeld, 1976; Strausfeld and Nässel, 1980; Meinertzhagen and Fröhlich, 1983; Shaw, 1984) immunocytochemistry can help in understanding chemical transmission in the lamina. Unfortunately very few putative neurotransmitters and neuropeptides have been detected in the fly lamina. Morphologically only the 5-HTir, GABAir and HAir neurons have been described (Nässel et al.,1985, 1987, 1988c; Datum et al., 1986; Meyer et al., 1986) (Fig. 7). These neurons represent 4 of the 14 known types of neurons in the lamina (including the long and short photoreceptors): (1) the tangential LBO-5HT neurons with processes distal to the lamina synaptic neuropil; (2) the centrifugal small field columnar neurons, C2, which are GABAir; and (3-4) the histamine-immunoreactive photoreceptors R1-R6 and R8. Some of the monopolar neurons within the group L1-L5 may be glutamate-immunoreactive, in agreement with findings in the honey bee (Bicker et al., 1988). In spite of the large number of antisera tested against dopamine, noradrenaline and many neuropeptides (Table 1) no immunoreactivity was detected in the R7 photoreceptors, the lamina amacrine neurons, the two other types of tangential neurons, the centrifugal neuron C3 or the centripetal neuron T1. Possibly we have to search for other amino acid transmitters or novel neuropeptides. Improved glutamate and acetylcholine (choline acetyltransferase) immunocytochemistry has to be performed on *Calliphora.*

Identified Neuron Approach

Some of the large field neurons that can be labeled with antisera against serotonin, dopamine and FMRMamide have distinct and constant locations of their cell bodies and main axons. Such neurons possibly could be impaled for intracellular recordings combined with pharmacological approaches for a functional analysis. Localization of several of the FMRFir neurons for intracellular work is facilitated by the fact that they have large cell bodies that stain with the dye Neutral red *in vivo* (Fig. 8C). Similarily serotonergic neurons may be labeled *in vivo* with 6-hydroxydopamine (Rosza et al., 1986) or 5-7-dihydroxytryptamine (Vaney, 1986).

The Medulla and Lobula Layers with Diverse Neuroactive Substances

Some layers of the insect visual system appear richer in the different types of putative neurotransmitters and neuropeptides than others. These are the serpentine layer of the medulla (Fig. 9) and the basal half of the lobula. In other layers only a few or no substances have been indicated so far. There are probably several explanations for this. First, it must be assumed that until now we have accounted for only a fraction of the total number of different *neuroactive substances.* Hence, many of the layers that are apparently low in number of substances may contain compounds not yet screened for. Second, not all layers of the optic lobe neuropils are rich in synapses and synaptic processes and it is therefore not surprising if few neuroactive compounds are located in these layers. Third, there may be layers where substantial signal processing and modulation occur and where the number of neuroactive substances reflect this synaptic complexity. Here the number of local circuits involved is important; many classes of neurons and synapses may be reflected in numerous types of neuroactive substances.

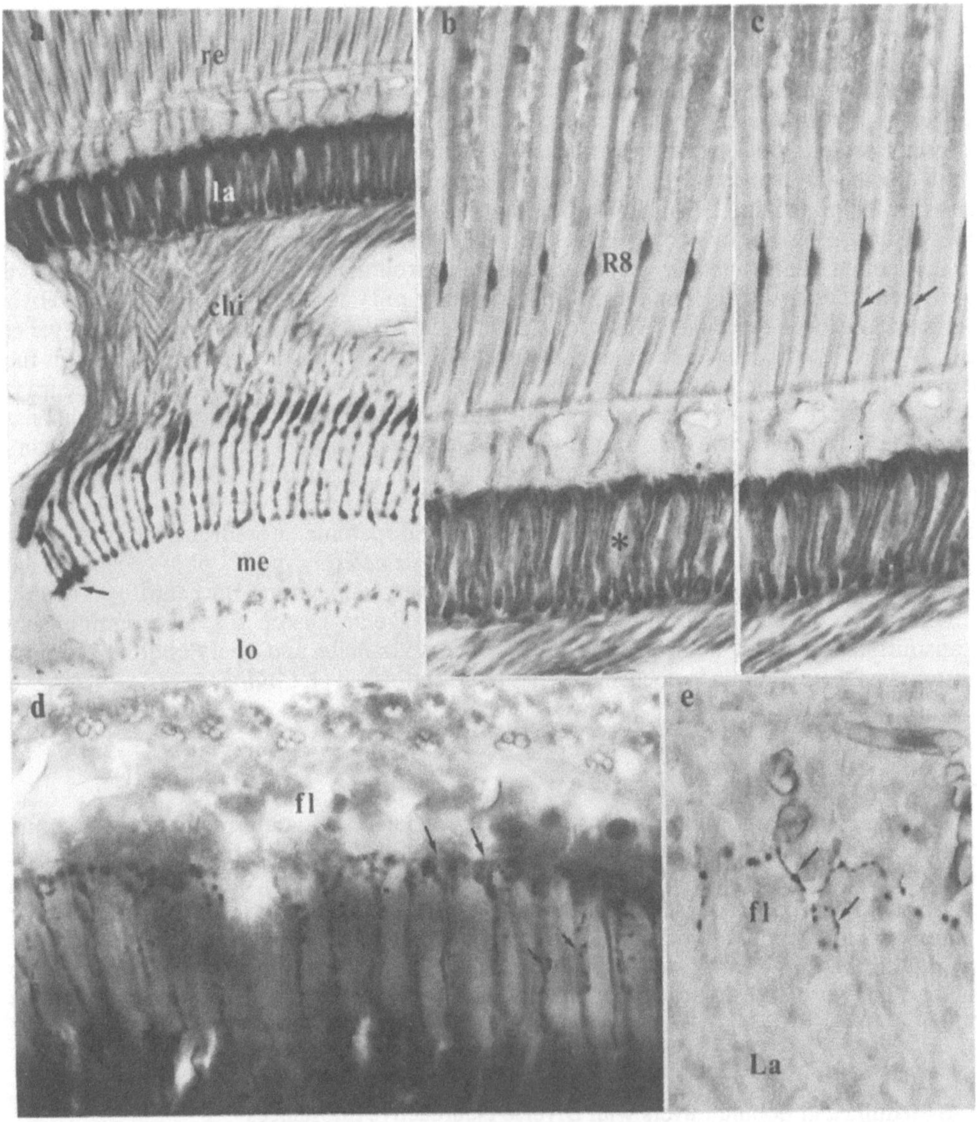

Fig. 7. Some of the identified neurons in the lamina of *Calliphora*. **a-c.** Histamine-immunoreactive photoreceptors R1-6 and R8. Micrographs from Nässel et al. (1988c). The R8 terminals in the medulla (me) are seen in **a.** In **b** and **c** the R8 are seen in the retina and the R1-6 terminals in the lamina (asterisk). **d.** The GABAir centrifugal neurons C2 are labeled in every cartridge of the lamina neuropil. Their distal swellings are indicated by long arrows. The shorter arrows point at varicosities inside the mid-portion of the synaptic neuropil. fl= fenestrated layer between lamina and retina. **e.** The 5-HTir processes in the fenestrated layer (arrows) distal to the lamina synaptic neuropil.

It has been shown earlier that some parts of the insect brain may use numerous neuroactive compounds in a dense synaptic neuropil. One such center is the so-called central body where electron microscopy and immunocytochemistry indicate a very complex chemical neurotransmission and modulation (Benedeczky 1981; Homberg, 1987; Nässel 1987b; Nässel and O'Shea, 1987). The function of the central body remains a mystery. Another such region is the serpentine layer of the medulla (Fig. 9), which as defined here, is a layer where many neuron types arborize. It is invaded by a large number of projection neurons of the Cuccati bundle, which connect the medulla to the brain and contralateral optic lobe, and it is the junction between the inner and outer medulla. Hence, it must be an active synaptic layer, possibly with numerous forms of local circuits, where many different types of signal transfer and modulation occur in a densely packed stratum. The basal portion of the lobula, which also contains many immunoreactive substances, is invaded by many types of projection neurons, including some ocellar interneurons (Nässel, 1987b). It can be assumed that the inner lobula contains multimodal neurons and that substantial modulation of visual and other signals occur there. Unfortunately the functional role of the different neuroactive substances in the medulla and lobula layers may be hard to test experimentally.

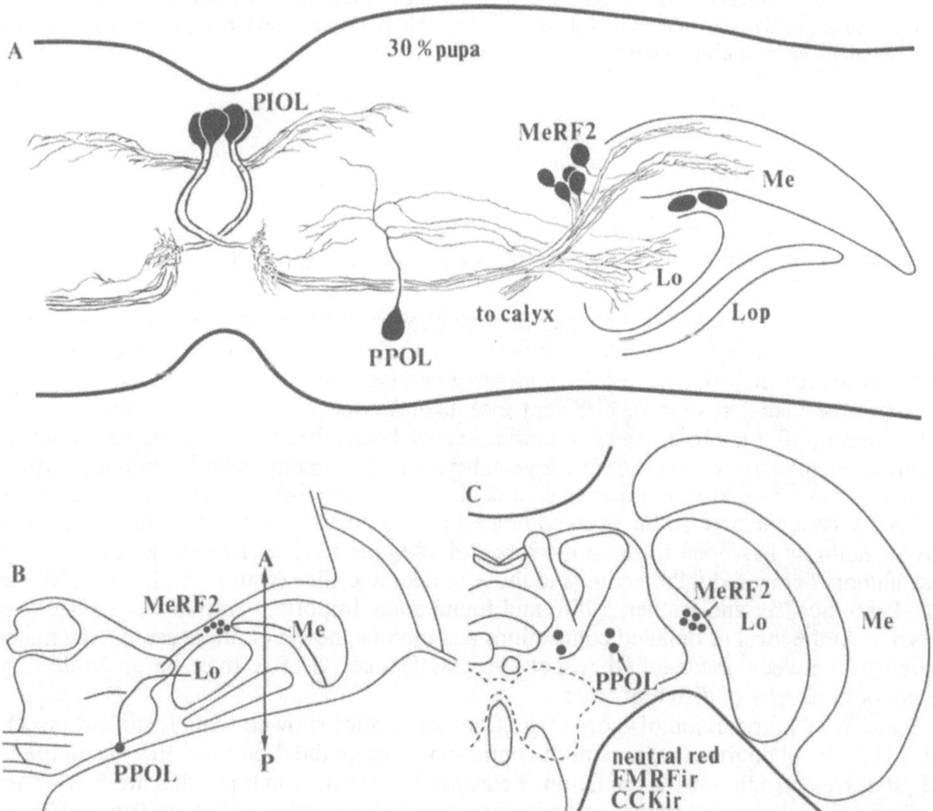

Fig. 8. Examples of identifiable neurons in the optic lobes. **A.** Horizontal section of developing *Calliphora* optic lobe. At 30% pupal development two neuron types reacting with anti-FMRFamide antisera could be detected that only thereafter could be identified in the adults.These are the PIOL and PPOL neurons innervating the lobula and midbrain. The posterior position of the PPOL neurons is also indicated in B and C. At this stage the processes of the MeRF2 neurons are not distributed over the entire medulla and the large medulla amacrines (MeRF1) have no immunoreactive processes. **B.** Schematic tracing of adult optic lobes in horizontal section indicating position of the PPOL and MeRF2 neurons. **C.** Interestingly, the PPOL and MeRF2 neurons are among a small number of neurons labeling with antisera against FMRFamide and CCK and that also stain with the vital dye Neutral red. This would simplify intracellular impalement of these particular neurons. The tracing shows a frontal view of the brain.

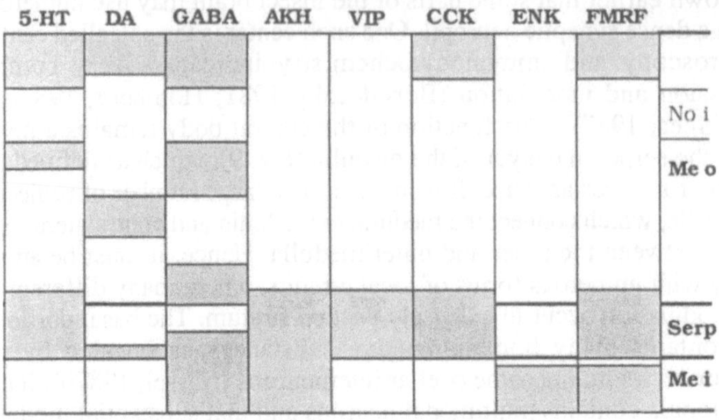

Fig. 9. Schematic illustration of the distribution of imunoreactive processes in different layers of the medulla seen with eight types of antisera. The main layers are the outer medulla (Me o), the serpentine layer (Serp) and the inner medulla (Me i). Further sublayers are formed by the immunoreactive processes. Note that some layers are devoid of any immunoreactivity (e. g. layer No i). Others, like the serpentine layer, contain many different immunoreactive substances.

COMPARATIVE ASPECTS OF NEUROTRANSMITTERS AND NEUROPEPTIDES

Unfortunately very few insect species have been studied in any detail with respect to localization of putative neuroactive substances. Older investigations using aldehyde-histofluorescence methods have provided some comparisons between catecholaminergic neurons in the visual system of different insects and crustaceans (Elofsson and Klemm, 1972; Klemm, 1976). Most insects studied, except honey bees, have catecholaminergic neurons in one pattern or the other in the visual system. Immunocytochemistry has provided the means for more detailed comparisons of neurotransmitter systems, but only the serotonin and GABA systems have so far been compared in any detail. The most thorough study of GABAir neurons has been in the optic lobes of *Manduca sexta* (Homberg et al., 1987). These authors compared their findings to those in bees and flies (Datum et al., 1986; Meyer et al., 1986; Schäfer and Bicker, 1986) and found some important similarities in GABAir cell types. At the level of detailed connections and circuits, however, there seem to be major differences between the studied insect species. Hence, GABA may be an inhibitory transmitter in circuits of different types.

Similarly a comparison of more than 20 insect species showed that all studied insects have 5-HTir in all neuropil regions of the optic lobe, except the dipterous flies where the 5-HTir processes branch outside the lamina neuropil. But the neuron types that are 5-HTir are of different morphological types, occur in varying numbers and are likely to form different types of connections in the different species (Nässel, 1988a). In spite of this serotonin may play some common role in at least some of the neurons in each insect species: wide-field serotonergic neurons connecting substantial portions of the visual system to the brain may be modulatory neurons controlling activity levels in other visual interneurons.

The occurrence of neuropeptides in the visual system appears to vary quite a bit between different insect species. This difference may in some cases only be apparent, since the studies have been undertaken in different laboratories, with different sources of antisera and with the use of different immunocytochemical protocols. In other cases studies of different insect species, using the same protocol, can show distinct differences between species in the neuropeptides present and in the types of neurons reacting to specific antisera.

An example of variation between species is seen when using an antiserum against gastrin/cholecystokinin: in *Calliphora* no immunoreactive neurons are seen in the lamina, whereas in the cricket, *Acheta*, extensive neurons connect the lamina and medulla. A similar type of neuron in the cricket lamina reacts with an antiserum against somatostatin, whereas no somatostatin-immunoreactive neurons are found in the *Calliphora* optic lobe (Johansson and Nässel, in prep.). One unique peptidergic system has been demonstrated in the locust optic lobe but could not be resolved in *Calliphora*. This consists of a pair of oxytocin/vasopressin-immunoreactive (VAir) neurons with their cell bodies in the subesophageal ganglion that appear to innervate substantial parts of the central nervous system including the optic lobes (Rémy and Girardie, 1980).

As seen in these examples there may be a species variation in the role played by a given neuroactive substance. Does this mean that certain functions are absent in those species lacking a given neuroactive substance or do other substances take over the role(s)? Here it should be born in mind that the neuronal architecture of the optic lobes is quite different between species. Some circuits may be added in more evolved visual systems. Gradual evolutionary changes in circuitry has, for instance, been demonstrated in the lamina of dipterous insects (Shaw and Meinertzhagen, 1986). It is important for the future to study the distribution of certain neurotransmitters and neuropeptides in the visual system of a variety of insects using the same antisera and the same protocol to establish true differences in chemical neuroanatomy and hence chemical neurotransmission.

DEVELOPMENTAL ASPECTS

One interesting aspect has emerged from studies of the postembryonic development of immunocytochemically identifiable neurons in the fly visual system. A small portion of the neurons found in the adult optic lobes are derived from metamorphosing larval neurons. This has been demonstrated for 5-HTir, FMRFir and DAir neurons in *Calliphora* (Ohlsson and Nässel 1987; Ohlsson et al., 1988; Ohlsson, 1988). These neurons change their arborization patterns and invade the primordia of the optic lobes early in pupal development. An example of metamorphosing neurons are the PIOL neurons shown in Fig. 8. The remaining neurons reacting with the above antisera seem to be derived from imaginal neuroblasts differentiating during postembryonic development. Interestingly, the redifferentiating larval neurons remain immunoreactive throughout metamorphosis, whereas the onset of immunoreactivity in the other neurons occurs later, possibly correlated with the onset of synaptogenesis (Ohlsson et al. 1988).

CONCLUSIONS AND OUTLOOK

As also indicated by Hardie (1988), it is clear that we are only at the beginning of the analysis of neurotransmitters and neuropeptides in the insect visual system. Immunocytochemistry may be one way to indicate which substances to study with pharmacological and physiological methods. With immunocytochemistry it is also possible to describe the morphology of neurons, neural circuits and pathways that possibly uses specific neuroactive substances; this anatomical framework may make possible a more precise physiological approach with employment of micro-iontophoresis for receptor characterization. Except for some small molecules, such as serotonin, GABA and dopamine, immunocytochemistry alone is, however, insufficient for the characterization of putative neuroactive compounds. Clearly the authentic substances need to be isolated and their structure determined. Analysis of neuronal mRNA, followed by *in situ* hybridization studies, or biochemical purification and sequencing of peptides should be attempted. Little is known about the receptors for neurotransmitters and neuropeptides within the CNS. Pharmacological characterization of receptors and localization studies by receptor autoradiography or use of anti-idiotypic antibodies (Venter et al. 1984; Vieillemaringe et al., 1987) should hence be performed. Since only a fraction of the total number of optic lobe neurons have so far been accounted for by immunocytochemistry of neuroactive substances, we need to search for the substances in the remaining neurons. Again, mRNA isolation and analysis may tell us what substances to look for with immunocytochemistry or *in situ* hybridization. Alternatively, one can search the vast number of catalogues on commercial immunochemicals or raid the freezers of neighboring laboratories for antisera not yet tested on the insect visual system. Another approach to the functional role of neuroactive substances is to isolate and analyze *Drosophila* mutants deficient in transmitter metabolism (Livingstone and Tempel, 1983; Tempel et al., 1984; Budnik et al., 1986; Chase and Kankel, 1987).

From the chemical neuroanatomy presented in this account a few points can be made. The lamina, which seems to be an excellent part of the visual system for analysis of neurotransmitters and neuropeptides, due to its well-described anatomy and physiology, has so far provided disappointingly few results with immunocytochemistry. Only histamine, serotonin and GABA have been indicated in this neuropil, whereas a large number of other substances have been indicated in the medulla and lobula where the anatomical and physiological framework is less good. Possibly the lobula plate with its large well-characterized neurons and a relatively large number of putative neuroactive substances may be another good site for pharmacological/physiological analysis.

Immunocytochemistry has also indicated that a given substance may have different functions in different neuron types and circuits of the visual system and in other parts of the CNS. Hence, gross interference with or application of a given substance may produce compound actions at many levels of integration. For instance FMRFamide-like peptide(s) may be used by different large projection neurons, by local amacrine neurons and by relay neurons, as well as possibly being a neurohormone. In some cases FMRFamide-like peptide(s) probably acts in conjunction with other substances.

Chemical neuroanatomy brings a new aspect to the classification of visual neurons. Morphologically similar neurons may use different neuroactive substances or combinations of substances, and a subdivision of neuron types may be necessary. Many neuron types were actually detected for the first time with immunocytochemistry (silver impregnation techniques may have resolved them only partially before or failed to stain them). The immunocytochemistry may also bring new aspects to the stratification of the synaptic regions of the optic lobe neuropils (receptor localization would possibly further emphasize this). Another interesting aspect of using immunocytochemistry is that whole populations of chemically defined neurons can be visualized, thus revealing distinct patterns of layers and projections and making developmental studies possible.

ACKNOWLEDGEMENTS

I thank Drs. Ian A. Meinertzhagen and Camilla Strausfeld for comments on the manuscript. Several people contributed to the research presented: K. Elekes, K. U. I. Johansson, N. Klemm, and L. G. Ohlsson. Gifts of antisera by the following people is gratefully acknowledged: C. J. P. Grimmelikhuijzen, M. O'Shea, H. Schooneveld, H. W. M. Steinbusch, and F. Sundler.

REFERENCES

Benedeczky, I., 1981, The ultrastructure and cytochemistry of aminergic and peptidergic terminals in Locusta migratoria migratorioides, *in:* "Neurotransmitters in invertebrates," K. S.-Rosza, ed., Pergamon Press, Budapest

Bicker, G., Schäfer, S., Ottersen, O. P., and Storm-Mathisen, J., 1988, Glutamate-like immunoreactivity in identified neuronal populations of insect nervous systems. *J. Neurosci.*, (in press).

Boer, H. H. , Schot, L. P. C., Veenstra, J. A., and Reichelt, D., 1980, Immunocytochemical identification of neuronal elements in the central nervous system of a snail, some insects, a fish, and a mammal with antiserum to the molluscan cardioexcitatory tetrapeptide FMRFamide. *Cell Tissue Res.*, 213: 21-27

Buchner, E., Buchner, S., Crawford, G., Mason, W. T., Salvaterra, P. M. and Sattelle, D. B., 1986, Choline acetyltransferase-like immunoreactivity in the brain of *Drosophila melanogaster*. *Cell Tissue Res.*, 246: 57-62.

Buchner, E., Bader, R. Buchner, S., Cox, J., Emson, P., Flory, E., Heizmann, C. W., Hemm, S, Hofbauer, A., and Oertel, W. H., 1988, Cell-specific immuno-probes for the brain of normal and mutant*Drosophila melanogaster*. *Cell Tissue Res.*,253: 357-370

Budnik, V., Martin-Morris, L., and White, K., 1986, Perturbed pattern of catecholamine-containing neurons in mutant *Drosophila* deficient in the enzyme dopa decarboxylase. *J. Neurosci.*, 6, 3682-3691

Bülthoff, H., and Bülthoff, I., 1987a, GABA-antagonist inverts movement and object detection in flies. *Brain Res.*, 407, 152-158

Bülthoff, H., and Bülthoff, I., 1987b, Combining neuropharmacology and behavior to study motion detection in flies. *Biol. Cybern.*, 55: 313-320.

Chase, B. A., and Kankel, D. R., 1987, A genetic analysis of glutamatergic function in *Drosophila. J. Neurobiol.*, 18: 15-41.

Datum, K.-H., Weiler, R., and Zettler, F., 1986, Immunocytochemical demonstration of γ-amino buturic acid and glutamic acid decarboxylase in R7 photoreceptors and C2 centrifugal fibers in the blowfly visual system. *J. Comp. Physiol.*, 159: 241-249.

Dudai, Y., 1980, Cholinergic receptors of Drosophila. In: "Receptors for neurotransmitters, hormones and pheromones in insects," D. B. Satelle, L. M. Hall, and J. G. Hildebrand, eds., pp. 92-110, Elsevier, Amsterdam.

Duve, H., and Thorpe, A., 1988, Mapping of enkephalin-related peptides in the nervous system of the blowfly, *Calliphora vomitoria*, and their colocalization with cholecystokinin (CCK)- and pancreatic polypeptide (PP)-like peptides. *Cell Tissue Res.*, 251: 399-415

Duve, H., Thorpe, A., and Nässel, D. R. , 1988, Light- and electron-microscopic immunocytochemistry of peptidergic neurons innervating thoracico-abdominal neurohaemal areas in the blowfly. *Cell Tissue Res.*, 253:583-595.

Dymond, G. R., and Evans, P. D., 1979, Biogenic amines in the nervous system of the cockroach *Periplaneta americana:* association with mushroom bodies and dorsal unpaired neurons. *Insect Biochem.*, 9:535-545.

Elofsson, R., and Klemm, N., 1972, Monoamine-containing neurons in the optic ganglia of crustaceans and insects. *Z. Zellforsch. mikrosk. Anat.*, 133: 475-499

Elias, M. S., and Evans, P. D., 1983, Histamine in the insect nervous system: distribution, synthesis and metabolism. *J. neurochem.*, 41: 562-568.

Evans, P., and Myers, C. M., 1986, Peptidergic and aminergic modulation of insect skeletal muscle. J. *Exp. Biol.*, 124: 143-176.

Fischbach, K. F., 1983, Neurogenetik am Beispiel des visuellen Systems von *Drosophila melanogaster*. Habilitation thesis. University of Würzburg.

Fischbach, K. F., and Heisenberg, M., 1984, Neurogenetics and behaviour in insects. *J. exp. Biol.*, 112: 65-93

Fujita, S. C., Zipurski, S. L., Benzer, S., Ferrus, A., and Shotwell, S. L., 1982, Monoclonal antibodies against the *Drosophila* nervous system. *Proc. Natl. Acad. Sci. USA*, 79: 7929-7933

Corczyca, M. G., and Hall, J. C., 1987, Immunohistochemical localization of choline acetyltransferase during development and in *Chat*^{ts} mutants of *Drosophila melanogaster*. J. Neurosci. 7: 1361-1369.

Grimmelikhuijzen, C. J. P., 1985, Antisera to the sequence Arg-Phe-amide visualize neuronal centralization in hydroid polyps. *Cell Tissue Res.*, 241: 171-182

Hardie, R., 1987, Is histamine a neurotransmitter in insect photoreceptors? *J. Comp. Physiol.*, 161: 201-213.

Hardie, R. C., 1988, Neurotransmitters in the compound eye, in: "Facets of Vision," Hardie, R. C. and Stavenga, D. G., eds. Springer, Berlin (in press)

Hausen, K., 1984, The lobula complex of the fly: Structure function and significance in visual behavior, in "Photoreceptions and vision in invertebrates," Ali, M. A., ed., pp. 523-560. Plenum Press, New York, London.

Holmqvist, B., Movérus, B., and Nässel, D. R., 1988, Metamorphosis of vasopressin and proctolin-like immunoreactive neurons supplying abdominal neurohaemal areas in the blowfly. *Europ. J. Neurosci. Suppl.* 1988:301

Homberg, U., 1987, Structure and functions of the central complex in insects, in: "Arthropod brain: its evolution, development, structure and function," A. P. Gupta, ed., pp. 347-368, John Wiley, New York.

Homberg, U., Kingan, T. G., and Hildebrand, J. G., 1987, Immunocytochemistry of GABA in the brain and suboesophageal ganglion of *Manduca sexta*. *Cell Tissue Res.*, 248: 1-24.

Klemm, N., 1976, Histochemistry of putative transmitter substances in the insect brain. *Progr. Neurobiol.* 7: 99-169.

Klemm, N., Nässel, D. R. and Osborne, N. N., 1985, Dopamine-β-hydroxylase-like immunoreactive neurons in two insect species, *Calliphora erythrocephala* and *Periplaneta americana*. *Histochemistry*, 85: 159-164

Konings, P. N. M., Vullings, H. G. B., Geffard, M., Nuijs, R. M., Diederen, J. H. B., and Jansen, W. F., 1988, Immunocytochemical demonstration of octopamine-immunoreactive cells in the nervous system of Locusta migratoria and Schstocerca gregaria. *Cell Tissue Res.*, 251: 371-379

Livingstone, M. S., and Tempel, B. L., 1983, Genetic dissection of monoamine neurotransmitter synthesis in *Drosophila. Nature*, 303: 67-70

Maxwell, G. D., Tait, J. F., and Hildebrand, J. G., 1978, Regional synthesis of neurotransmitter candidates in the CNS of the moth *Manduca sexta. Comp. Biochem. Physiol.*, 61C: 109-119

Meinertzhagen, I. A., and Fröhlich, A., 1983,The regulation of synapse formation in the fly's visual system. *TINS*, 6: 223-228

Meyer, E. P., Matute, C., Streit, P., and Nässel, D. R., 1986, Insect optic lobe neurons identifiable with monoclonal antibodies to GABA. *Histochemistry*, 84: 207-216.

Mizoguchi, A., Ishizaki, H., Nagasawa, H., Kataoka, H., Isogai, A., Tamura, S., Susuki, A., Fujino, M., Kitada, C., 1987, A monoclonal antibody against a synthetic fragment of bombyxin (4K-prothorccccacicotropic hormone) from a silkmoth, Bombyx mori: characterization and immunohistochemistry. *Molec. Cell. Endocrinol.*, 51: 227-235

Myers, C. M, and Evans, P. D., 1987, An FMRFamide antiserum differentiates between populations of antigens in the brain and retrocerebral complex of the locust, *Schistocerca gregaria*, *Cell Tissue Res.*, 250: 93-99.

Nambu, J. R., Andrews, P. C., Feistner, G. J., and Scheller, G. J., 1987, Purification and characterization of the FMRFamide-related peptide of *Drosophiola melanogaster*. *Soc. Neurosci. Abstr.*, 13: 1256

Nässel, D. R., 1987a, Neuroactive substances in the insect CNS, in: "Nervous Systems of Invertebrates," M. A. Ali, ed., pp. 171-212., Plenum Press, New York.

Nässel, D. R. , 1987b, Aspects of the functional and chemical anatomy of the insect brain, in: "Nervous systems of invertebrates," M. A. Ali, ed., pp. 353-392, Plenum Press, New York.

Nässel, D. R., 1988a, Serotonin and serotonin-immunoreactive neurons in the insect nervous system. *Progr. Neurobiol.*, 30: 1-85.

Nässel, D. R., 1988b, Immunocytochemistry of putative neuroactive substances in the insect brain, in: "Neurobiology of Invertebrates: Transmitters, modulators and receptors," J. Salanki and K. S.-Rosza, eds. pp. 147-160, Pergamon Press (Akadémiai Kiado), Budapest.

Nässel, D. R., and Elekes, K., 1984, Ultrastructural demonstration of serotonin-immunoreactivity in the nervous system of an insect *(Calliphora erythrocephala). Neurosci. Lett.*, 48: 203-210.

Nässel, D. R., and Klemm, N., 1983, Serotonin-like immunoreactivity in the optic lobes of three insects. *Cell Tissue Res.*, 232: 129-140

Nässel, D. R., and O'Shea, M., 1987, Proctolin-like immunoreactive neurons in the blowfly central nervous system. *J. Comp. Neurol.*, 265:437-454

Nässel, D. R., Elekes, K., and Johansson, K. U. I., 1988a Dopamine-immunoreactive neurons in the blowfly vidual system: light and electron microscopic immunocytochemistry. *J. Chem. Neuroanat.* (in press)

Nässel, D. R., Hagberg, M., and Seyan, H. S., 1983, A new, possibly serotonergic neuron of the blowfly optic lobe: an immunocytochemical and Golgi-EM study. Brain Res. 280: 361-367.

Nässel, D. R., Meyer, E. P., and Klemm, N., 1985, Mapping and ultrastructure of serotonin-immunoreactive neurons in the optic lobes of three insect species. *J. Comp. Neurol.*, 232: 190-204.

Nässel, D. R., Ohlsson, L., and Sivasubramanian, P., 1987, Postembryonic differentiation of serotonin-immunoreactive neurons in fleshfly optic lobes developing in situ or cultured in vitro without eye discs. *J. Comp. Neurol.*, 255: 327-340.

Nässel, D. R., Ohlsson, L. G., Johansson, K. U. I., and Grimmelikhuijzen, C. J. P., 1988b, Light and electron microscopic immunocytochemistry of neurons in the blowfly optic lobe reacting with antisera to RFamide and FMRFamide. *Neuroscience*, 27: 347-362

Nässel, D. R., Holmqvist, M. H., Hardie, R. C., Håkanson, R., and Sundler, F., 1988c, Histamine-like immunoreactivity in photoreceptors of the compound eyes and ocelli of the flies *Calliphora erythrocephala* and *Musca domestica. Cell Tissue Res.*, 253: 639-646

Ohlsson, L. G., 1988, Postembryonic development of identified neurons in the optic lobes of holometabolous insects, Thesis, Lund.

Ohlsson, L. G., and Nässel, D. R., 1987, Postembryonic development of serotonin-immunoreactive neurons in the central nervous system of the blowfly, *Calliphora erythrocephala*. I. The optic lobes. *Cell Tissue Res.*, 249: 669-679.

Ohlsson, L. G., Johansson, K. U. I., and Nässel, D. R., 1988, Postembryonic development of Arg-Phe-amide- and cholecystokinin-like immunoreactive neurons in the blowfly optic lobe. *Cell Tissue Res.*, in press.

O'Shea, M., and Schaffer, M., 1985, Neuropeptide function: the invertebrate contribution. *Ann. Rev. Neurosci.*, 8: 171-198.

Price, D. A., and Greenberg, M. J., 1977, Structure of a molluscan cardioexcitatory neuropeptide. *Science*, 197: 670-671

Rémy, C., and Girardie, J., 1980, Anatomical organization of two vasopressin-neurophysin-like neurosecretory cells throughout the central nervous system of the migratory locust. *Gen. Comp. Endocrinol.*, 40: 27-35.

Rieker, S., Fischer-Colbrie, R., Eiden, L., and Winkler, H., 1988, Phylogenetic distribution of peptides related to chromogranins A and B. *J. Neurochem.* 50: 1066-1073.

Robertson, H. A., 1976, Octopamine, dopamine and noradrenaline content of the brain of the locust, *Schistocerca gregaria. Experientia*, 32, 552-553.

Rosza, K. S., Hernadi, L., and Kemenes, G., 1986, Selective in vivo labelling of serotonergic neurones by 5,6-dihydroxytryptamine in the snail *Helix pomatia* L. *Comp. Biochem. Physiol.* 85C: 419-425.

Salvaterra, P. M., and McCaman, R. E., 1985, Choline acetyltransferase and acetyl choline levels in *Drosophila melanogaster:* a study using two temperature-sensitive mutants. *J. Neurosci.* 5: 903-910.

Schäfer, S., and Bicker, G., 1986, Distribution of GABA-like immunoreactivity in the brain of the honey bee. *J. Comp. Neurol.* 246: 287-300.

Schäfer, S., Bicker, G., Ottersen, O. P., and Storm-Mathisen, J., 1988, Taurine-like immunoreactivity in the brain of the honey bee. *J. Comp. Neurol.*, 268: 60-70.

Schneider, L. E., and Taghert, P. H., 1988, Isolation and characterization of a Drosophila gene that encodes multiple neuropeptides related to Phe-Met-Arg-Phe-NH2 (FMRFamide). *Proc. Natl. Acad. Sci.* 85: 1993-1997

Schneider, L. E., Sun, E., and Taghert, P. H., 1988, Cellular analysis of a neuropeptide gene expression in *Drosophila melanogaster. Soc. Neurosci. Abstr.* 14: 29

Schooneveld, H. , Tesser, G. I., Veenstra, J. A., and Romberg-Privee, H. M., 1983, Adepokinetic hormone and AKH-like peptide demonstrated in the corpora cardiaca and nervous system of *Locusta migratoria* by immunocytochemistry. *Cell Tissue Res.*, 230: 67-76.

Schooneveld, H., Romberg-Privee, H. M., and Veenstra, J. A., 1985, Adepokinetic hormone-immunoreactive peptide in the endocrine and central nervous system of several insect species: a comparative immunocytochemical approach. *Gen. Comp. Endocrinol.*, 57: 184-194.

Schürmann, F.-W., and Klemm, N., 1984, Serotonin-immunoreactive neurons in the brain of the honeybee. *J. Comp. Neurol.*, 225: 570-580.

Shaw, S. R., 1984, Early visual processing in insects. *J. exp. Biol.*, 112: 225-251

Shaw, S. R., and Meinertzhagen, I, A., 1986, *Proc. natn. Acad. Sci. U.S.A.*, 83: 7961-7965.

Stone, J. V., Mordue, W., Batley, K. E., and Morris, H. E., 1976, Structure of locust adipokinetic hormone, a neurohormone regulates lipid utilization during flight. *Nature*, 263: 201-211

Strausfeld, N. J., 1976, "Atlas of an Insect Brain," Springer, Berlin.

Strausfeld, N. J., 1984, Functional anatomy of the blowfly's visual system, in: "Photoreception and vision in invertebrates," M. A. Ali, ed., pp.483-522, Plenum Press, New York.

Strausfeld, N. J., and Bassemir, U. K., 1985, The organization of giant horizontal motion-sensitive neurons and their synaptic relations in the lateral deutocerebrum of *Calliphora* and *Musca*. *Cell Tissue Res.*, 242: 531-550.

Strausfeld, N. J., and Nässel, D. R., 1980, Neuroarchitectures serving compound eyes of crustacea and insects, in: "Handbook of sensory physiology VII/6B," Autrum H. ed., pp. 1-132, Springer, Berlin.

Strausfeld, N. J., and Seyan, H. S., 1985, Convergence of visual, haltere and prosternal inputs at neck motorneurons of *Calliphora. Cell Tissue Res.*, 240: 601-615.

Sundler, F., Ekblad, E., Grunditz, T., Håkanson, R., and Uddman, R., 1988, Vasoactive intestinal peptide in the peripheral nervous system, in: "Vasoactive intestinal peptide and related peptides," S. I. Said and V. Mutt, eds., *Ann. N. Y. Acad. Sci.*, Vol. 527.

Tempel, B. L., Livingstone, M. S., and Quinn, W. G., 1984, Mutations in the dopa decarboxylase gene effects learning in *Drosophila. Proc. Natl. Acad. Sci. USA*, 81, 3577-3581

Trujillo-Cenoz, O., 1965, Some aspects of the structural organization of the intermediate retina of dipterans. *J. Ultrastruct. Res.*, 13, 1-33

Vaney, D. I., 1986, Morphological identification of serotonin-accumulating neurons in the living retina, *Science*, 233: 444-446.

Veenstra, J. A., and Schooneveld, H., 1984, Immunocytochemical localization of neurons in the nervous system of The Colorado potato beetle with antisera against FMRFamide and bovine pancreatic polypeptide. *Cell Tissue Res.*, 235: 303-308.

Venter, J. C., Fraser, C. M., and Lindstrom, J. eds., 1984, "Monoclonal and anti-idiotypic antibodies:

probes for receptor structure and function," Alan R. Liss, New York.
Verhaert, P., Grimmelikhuijzen, C. J. P., De Loof, A., 1985, Distinct localization of FMRFamide- and pancreatic polypeptide-like material in the brain, retrocerebral complex and suboesophageal ganglion of the cockroach *Periplaneta americana* L. *Brain Res.*, 348: 331-338
Vieillemaringe, J., Souan, M. L., Grandier-Vazeilles, X., and Geffard, M., 1987, Immunocytochemical localization of acetylcholine receptors in locust brain using auto-anti-idiotypic acetylcholine antibodies. Neurosci. Lett. 79: 59-64.
White, K., Hurteau, T., and Punsal, P., 1986, Neuropeptide-FMRFamide-like immunoreactivity in *Drosophila*: development and distribution. *J. Comp. Neurol.*, 247: 430-438.

INSECT VISION AND OLFACTION:

COMMON DESIGN PRINCIPLES OF NEURONAL ORGANIZATION

Nicholas J. Strausfeld

Division of Neurobiology, Arizona Research Laboratories
611 Gould-Simpson Building, University of Arizona
Tucson, AZ 85721, USA

ABSTRACT. In insects, the organization of neurons in visual and olfactory neuropils is comparable to arrangements in analogous systems in the brains of higher vertebrates. However, although the two modalities of vision and olfaction are subjectively quite different from each other, in insects they are served by common neuroarchitectures, the glomeruli, which are here suggested to be paramount in the processing of qualitative information. Visual and olfactory systems show other specific similarities with respect to the parallel organization of large- and small-axoned neurons. In the visual system, two parallel channels comprise large color-insensitive and small color-sensitive relay neurons that are linked to two major descending pathways. Color insensitive pathways supply motor circuits mediating visually stabilized flight and optokinetic head movements. This pathway is distinct from the parallel subsystem comprising numerous smaller neurons and many synaptic stations that supply leg and direct flight muscle motor neuropils. These two subsystems provide a simple model of magno- and parvocellular organizations identified in the mammalian visual system. Surprisingly, there exists a similar parallel organization amongst large- and small-axoned neurons in the insect olfactory system. Magnocellular olfactory projection neurons provide a relatively direct route to descending pathways. Smaller parvocellular projection neurons provide the first step in a complex sequence of neurons in which higher brain centers play a cardinal role.

1. INTRODUCTION

Despite the preoccupation during the first decade of this century with applying the then revolutionary methods of Golgi and Ehrlich to reveal single neurons, only two scientists can be said to have had the benefit of comparative studies to draw analogies between the organization of nervous systems in disparate phyla. Cajal (with Sanchez 1915) emphasized the profound similarities in neuronal organization between the visual centers of vertebrates (Cajal 1911), insects, and cephalopods (on which he also published in 1917). His Russian contemporary, Alexander Zawarzin (1925), whose work is sadly neglected by many present day biologists, went much further in identifying the basic principle of design between the ventral segmental ganglia of insects and the vertebrate spinal cord.

Today, we are still much concerned by questions that address the functional significance of neuronal architectures, as expressed in different sensory systems of one genus and, in a broader context, between similar sensory systems that have arisen independently several times in evolution. Indeed, there is so much prima facie evidence that the idea of common design principles of neuronal connections has been much emphasized in a popular textbook of neurobiology (Shepherd 1988).

What, then, are the features that distinguish the organization of neurons serving different sensory systems and what are the common principles that unite them? Why do analogous sensory systems in different phyla have similar neuronal architectures?

In this chapter I shall duck the answers, but amplify the questions by reviewing two sensory systems in insects: vision and olfaction. As in their vertebrate counterparts, neuropils serving these modalities appear at first sight to be manifestly different in design. However, on closer scrutiny they share certain important features in common. Their differences can be related to the organization of their receptors and the importance of space in sensory processing. Their similarities relate to the enhanced representation in the brain of one or more specific features of the sensory milieu and their importance in adaptive behavior.

2. FUNCTIONAL ORGANIZATION OF SENSORY NEUROPILS

A. The Visual System

(1) Functional organization of receptors

In insects, the central organization of vision and olfaction provide simple "models" of their less accessible counterparts in vertebrates. From our anthropocentric point of view (and, be it noted, not point of smell), visual perception employs four dimensional cues: Cartesian coordinates and time. An image of the world is projected by our lens onto the retina where its spatial resolution depends on the number of photoreceptors that can be packed together at the focal plane of the lens (Land 1981). Temporal resolution depends on the speed of transduction, the reconversion of the photoisomerized visual pigment to its receptive state, and a variety of other factors that are common to all receptors, relating to opening and closing membrane channels, transmitter release, and so on (see Kirschfeld 1987; Pugh and Altman 1988). However, in the eye, only a limited number of properties distinguish between photoreceptor types: their working range, spectral sensitivity, and dichroism (Hardie 1984, 1986).

These aspects of the visual world -- intensity, color, and polarized light -- provide the raw data from which the underlying neuropils recompose those aspects of the environment that we can recognize, call by name, and employ to test the visual performance of other organisms.

The structures that perform this analysis share similar properties in different animal groups, irrespective of whether they are supplied by an optically sophisticated single lens, as in mammals, certain cephalopods, and arachnids, or by a multitude of lenses as in the compound eyes of crustaceans and insects (see Land 1981). In insects, a geometrically precise mosaic of photoreceptors is conferred retinotopically by successive interneurons into serially arranged strata of local circuits. At each level, columnar subunits composed of characteristic constellations of identified neurons represent specific areas of the visual field (for reviews see Strausfeld and Nässel 1980; Strausfeld 1989). In principle, this is a much simplified version of what occurs in mammals where areas of the visual field are represented by discrete parcellations of the visual cortex (Hubel and Wiesel 1977). It is now known that this modular design of the cortex is refined into a variety of discrete and repeating subunits attributable to specific functional roles -- ocular dominance, orientation, color perception, stereopsis, and form perception (Livingstone and Hubel 1984; Hubel and Livingstone 1987) -- all of which can be revealed by special silver methods (LeVay et al. 1975), cytochemical techniques (tritiated deoxyglucose: Hubel et al. 1977; lectins: Anderson et al. 1988; cytochrome oxidase: Wong-Riley 1979; Livingstone and Hubel 1984), and voltage-sensitive dyes (Blasdel and Salama 1986). As has been demonstrated in mammals (Hubel and Livingstone 1987), even when computational tasks are distributed in this fashion, positional information is retained deep in the system and, in some mammals, areas of high acuity, such as the area centralis and fovea, may be overrepresented centrally. This latter phenomenon has been realized since pioneering studies on patients with cortical lesions (Penfield and Rasmussen 1950) recognized that disproportionately large areas of the cortex were associated with the most sensitive surfaces of the body. More recently, experimental manipulation of somatosensory fields provided by the rows of vibrissae on the snouts of rodents has demonstrated some of the best known examples of "high acuity" enlargement amongst cortical subunits (Woolsey and van der Loos 1970; Steffen and van der Loos 1980). As described elsewhere in this volume (see the chapter by Pollak), in mustached bats the representation of a single frequency used for prey detection and location is enormously overrepresented in one strip of the inferior colliculus, which serves as an acoustic fovea (Pollak et al. 1986). Later, we shall encounter analogous areas in the visual system of male flies, and in the olfactory system of male moths.

The fine structure of retinotopic neuropils is characterized by discrete layers of synaptic microcircuits, the organization of which is reflected in the overall stratification of local and through-going neurons (e.g., primate retina: Dowling and Boycott (1966); mammalian cortex: Gilbert and Wiesel 1981; Katz 1987; Katz et al. 1989). In insects, as in mammals, discrete strata underlie successive synaptic interactions, sometimes of great complexity (for review see Strausfeld and Nässel 1980) and involving a staggering variety of local and retinotopic interneurons (Strausfeld 1976). Typically, in those insects for which vision plays a dominant role in survival, a great variety of complex visual cues are recognized and acted upon and it is thought that the number of layers of local circuits indicates the variety of computational tasks that are solved by the system. For example, bees, hoverflies, and dragonflies -- all of which perform sophisticated form detection, navigation, and aerobatics -- possess exquisitely stratified optic lobe neuropil, which are in crass contrast to the visual neuroarchitecture of certain nocturnal Lepidoptera and Coleoptera (Strausfeld and Blest 1970; Strausfeld and Nässel 1980).

The most intensely studied insect visual system is that of the blowfly *Calliphora erythrocephala*, and it is this species that will be focused on here. Its compound eyes, like those of many insect species, do not show obvious local specializations, except for a sex-specific dorsofrontal area of each eye that contains larger facets than elsewhere, and appears to be flatter (Beersma et al. 1977). In males this dorsal zone contributes to the 400 or so ommatidia that are absent from females. In certain hoverflies (syrphids), the eyes of males actually meet at the upper anterior margin (Collett and Land 1975).

But before discussing the significance of this male-specific area, and comparing it with male-specific regions of the olfactory system, it is first necessary to compare the organization of photoreceptors with that of olfactory receptors.

Blowfly retinae consist of between 4000-4400 ommatidia, each of which contains a group of receptors that, typical of apposition eyes, reach out to the inner surface of the crystalline cone and are optically isolated by screening pigments from the receptors of adjacent ommatidia. In most apposition eyes, such as those of the bee *Apis mellifera* (Ribi 1975), photoreceptors grouped beneath each lens are fused, share the same optical axis, and send their axons together to the same column in the underlying lamina neuropil. However, in flies, photoreceptors lie separately beneath each lens in an "open" configuration and each has a different optical alignment. Groups of eight receptors, distributed amongst different ommatidia, share an identical optical axis so that together they sample the same small area of the visual field (Kirschfeld 1967). The axons of this group, which has been called a visual sampling unit (Franceschini 1975), converge to the same retinotopic column in underlying neuropils (Braitenberg 1967). This adaptation for greater photon capture at low light intensities has been termed the neural superposition principle, and its unique optical properties have proven invaluable for single receptor studies of a variety of visual phenomena in behavior (Riehle and Franceschini 1984), physiological optics (for review see Franceschini 1985), and photochemistry (see Kirschfeld 1987).

R1-R6 photoreceptors have a dual-peak sensitivity to ultraviolet (UV) and green (Burkhardt 1962). R1-R6 of each visual sampling unit function over a wide working range (for review see Laughlin 1981) so that by analogy with the mammalian retina they correspond to a hybridization between scotopic rods and photopic cones. Typically, each optically coherent group of R1-R6 confers a precise map of sampling units into the lamina where their endings form open columns of receptor endings, termed optic cartridges (Braitenberg 1967; Kirschfeld 1967).

The two other members of each sampling unit (termed R7 and R8) have long axons that bypass the endings of R1-R6 to terminate in the second optic neuropil, the medulla (Campos-Ortega and Strausfeld 1972). R7 and R8 are photopic receptors that occur as color-sensitive pairs. In 30% of ommatidia R7 is sensitive to ultraviolet (UV) and R8 sensitive to blue; except for a special dorsal strip of the eye, where both R7 and R8 are UV-sensitive. In the rest of the eye, R7 is sensitive to visible purple and R8 to green/yellow (see Hardie 1986).

(2) Retinotopic organization of visual neurons and the segregation into color-sensitive and -insensitive pathways

In the lamina, six relay neurons (termed L1-L5 and T1) arise from each optic cartridge with their axons projecting alongside those of the optically relevant R7 and R8 photoreceptors (Strausfeld 1971; Campos-Ortega and Strausfeld 1972). Thus, in the medulla each visual sampling unit is represented by eight elements each of which penetrates to a characteristic depth in the neuropil where it gives rise to a stratified organization of terminal specializations.

Extensive studies employing Golgi and electron microscopy have shown that each lamina output neuron has a characteristic form and synaptic relationship with other cells (for review see Strausfeld and Nässel 1980). Rows of dendrites arising from three of the neurons (L1,L2, and T1) receive many hundred of synapses from the R1-R6 terminals. In contrast, the L3 neuron is distinguished by having a few dendrites contacting R1-R6 and an axon that projects outside the cartridge. Tracing axons into the medulla reveals that L3, R7, and R8 axons form a discrete pathway separate from other lamina elements so comprising a distinct trichromatic pathway: green, UV, and blue (30% of the medulla) or, green, purple, and green/yellow (70% of the medulla).

The remaining lamina neurons (L1, L2, L4, L5, and T1) have complicated synaptic interactions mediated laterally by processes of lamina amacrine cells. Indeed, L4 and L5 have no direct connections at all with receptor terminals but receive inputs indirectly via local circuits that involve groups of cartridges. L4 and L5 together with L1, L2, and T1 form a color-insensitive channel to the medulla that is associated only with R1-R6 receptors. They give rise to a tightly associated group of terminals in each medulla column.

The medulla comprises two neuropils: an outer division, which receives afferents from the lamina, and an inner division, which receives its inputs only from the outer part. The outer neuropil, which is as deep as R7 terminals are long, is highly stratified and separated from the inner division by a system of laterally directed axons, comprising what is known as the Cuccati bundle. These axons are derived from or give rise to large areas of tangential processes, which are situated at characteristic levels in the retinotopic mosaic (Cajal and Sanchez 1915). The axons of tangential cells provide the most peripheral output from the medulla into the mid-brain or contralateral lobes (Strausfeld and Blest 1970).

The terminals of lamina and retina afferents coincide in depth with discrete strata of processes that arise from two classes of nerve cells. These are (1) the through-going retinotopic interneurons (variously called transmedullary cells, T-cells, and Y-cells) and (2) neurons that provide local interactions within the stratum. The latter are termed local interneurons and comprise two types: axonal cells that link different groups of columns to each other laterally and anaxonal cells, reminiscent of amacrine cells of the vertebrate inner plexiform layer (Strausfeld 1976). In most diurnal flying insects, the medulla is typically divided into a great many discrete strata, far more than are typical of the inner plexiform layers of most vertebrates. However, an interesting multistratified vertebrate analogue is offered by the avian retina, originally described by Cajal in 1888. It is tempting to consider that insects and birds have to solve similar problems in a rapidly moving visual world where the ambient light intensities are continually changing, where motion plays a dominant role in stabilization, and where figure-ground and depth perception is vital both to avoid bumping into things and to land accurately.

The medulla surmounts two deeper neuropils, the lobula and the lobula plate. Each of these regions is composed of characteristic neural architectures from which originate two generic types of relay neurons. These are: large-diameter, wide-field tangential cells from the lobula plate, projecting to the brain's dorsal deutocerebrum and contralateral lobula plate, and palisades of small-field retinotopic neurons from the lobula terminating in lateral regions of the deutocerebrum (Strausfeld 1970). There are exceptions and variations on this basic structural theme, but they need not concern us here.

These two deepest regions of the optic lobes are supplied by neurons from the medulla. The most numerous inputs to the lobula plate are quartets of endings that originate distally from each retinotopic column. The four cells are called bushy T-cells due to the shape of their identical dendritic trees (Strausfeld 1970, 1976). One pair arises from each column deep in the inner division of the medulla, the other at a discrete superficial layer in the lobula. Like other retinotopic neurons, their dendrites have characteristic spatial relationships with other elements of the retinotopic mosaic. Bushy T-cells occupy small patches of the retinotopic mosaic consisting of six neighboring columns subtending vertically oriented areas of the visual field (Strausfeld 1984).

Each morphological species of retinotopic neuron has the following structural properties: (1) it is represented in each column; (2) it is associated with specific strata and thus a characteristic configuration of local interneurons; and (3) its shape is uniquely identifiable. Because the optic lobes consist of as many identical subunits (columns) as there are ommatidia, Golgi impregnations of each brain reveal many neurons and provide data that suggests precise depth relationships between different neurons, an essential guide for future electron microscopical studies. For example, in each column, each of the two pairs of bushy T-cell dendrites receives endings from a pair of narrow-field retinotopic neurons, the dendrites of which clasp the terminal complex comprising L1, L2, T1, L4, and L5. It should be noted here for later discussion, that the L1 and L2 axons and those of the pair of subsequent relays arising from them are larger than those of other retinotopic elements and their dendritic fields are smaller. Also, an important aspect to be considered later is that although the synaptic relationships between these neurons is not yet known, the same cellular relationships between these narrow-field but large-axon-diameter relay neurons are conserved in a variety of Diptera and have been identified in Lepidoptera (Strausfeld 1989). An analogous arrangement has been seen in Hymenoptera where a trivial variation in architecture is that neurons analogous to lobula plate tangentials are situated in a separate layer of neuropil beneath the lobula rather than over it (Strausfeld 1976).

In contrast to the large-axoned relays linking the color-insensitive R1-R6 system to the lobula plate, connections from the medulla to the lobula involve a great variety of small-axoned retinotopic relay neurons (the TM- [transmedullary], Y-, and T- cells). Studies on the housefly *Musca domestica* (Strausfeld 1976) demonstrated that medulla-to-lobula relay neurons branch mainly at the levels of L3 and R7,R8 terminals, suggesting that channels to the lobula are involved in color discrimination. Thus, at its most simple, the medulla can be viewed as a neuropil dominated by complex relays associated with trichromatic pathways, which is perforated by narrow-field columns that relay color-insensitive information to the lobula plate (Fig. 1). Outputs from the lobula plate segregate to descending pathways supplying motor circuits involved in visually stabilized flight and head movements (Strausfeld and Milde 1988; Milde et al. 1987), which are known to be color insensitive (see Egelhaaf et al. 1988). Outputs from the lobula are associated with quite a different set of descending neurons that project to leg motor circuits (Strausfeld and Milde 1988; see also Fig. 2).

A cardinal feature of the small-axoned relay neurons is that each morphological species terminates at a characteristic level in the lobula where it impinges on the dendrites of one or more layers (assemblies: Strausfeld and Hausen 1977) of lobula output neurons. Each assembly of lobula neurons thus represents a specific combination of medullary interneurons. In turn, each lobula assembly sends its axon to a discrete partition of neuropil in the lateral deutocerebrum of the brain, where they constitute the afferent components of a glomerular neuropil (Strausfeld and Bacon 1983). We shall return to the significance of this structure after considering the organization of olfactory glomeruli, in the next section.

Because the lobula plate lies superficially in the lobes and can be easily approached from the back of the head, electrophysiological studies of optic lobe neurons have focused almost entirely on this neuropil. Typically, each species of lobula plate neuron shows a characteristic color-independent (insensitive) response to motion, direction, and in some cases, to figure-ground cues (for review see Egelhaaf et al. 1988, Hausen and Egelhaaf 1989). Typically, the dendritic fields of lobula plate neurons are huge and visit very large domains of the projected retinotopic mosaic where they subtend characteristic areas of the visual field (Hausen 1982; Strausfeld and Bassemir 1985a). As demonstrated by intracellular (Hausen 1982) and muscle recordings (Milde et al. 1987) the shape and size of a dendritic field mapped into the pattern of retinotopic columns correspond to that cell's physiological receptive field.

Recordings from identified lobula neurons are rare by comparison and have been mainly obtained from the bee. However, because the shapes and organization of retinotopic elements in the bee lobula correspond to those in the blowfly we may assume, for the present, that this neuropil has similar properties in both species. Lobula neurons are characterized by their spectral selectivity, and restricted color-opponent receptive fields (Kien and Menzel 1977; Hertel 1980). Each assembly of lobula neurons receives a specific set of medulla relays originating from stratified networks of local circuits associated with retina/lamina afferents. This organization suggests that optic glomeruli in the lateral deutocerebrum, each of which represents a specific combination of circuits in the medulla, encode discrete qualitative aspects of the visual world.

B. The Olfactory System

(1) Functional organization of receptors

This section stresses one of the most significant differences between the functional organization of photoreceptors and olfactory receptors. Photoreceptors are the epitome of generalist receptors. Beneath them, optic lobe neuropils provide the filters for extracting feature- (or quality-) specific information from the visual world. In contrast, olfactory receptors are often highly tuned to specific chemicals (specialist or generalized-specialist receptors: Boeckh 1962; Schneider et al. 1964), and thus serve as feature detectors at the most peripheral level. It is suggested here, that in both systems, afferents carrying specific qualitative information about their respective environments project to similar neuroarchitectures -- the glomeruli.

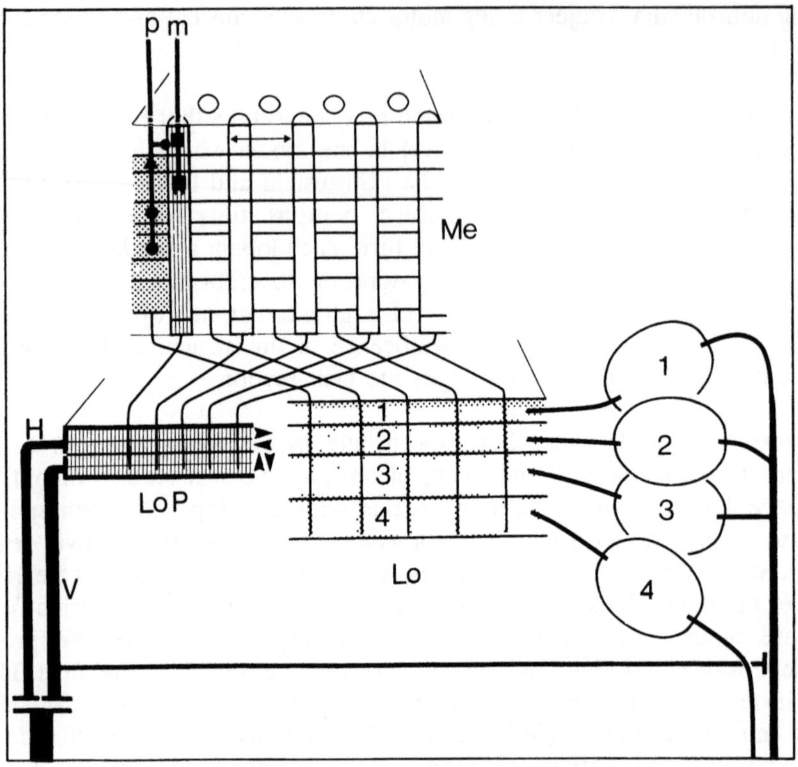

Fig. 1. Schematic showing the medulla (*Me*) perforated by magnocellular (*m*), color-insensitive pathways (*hatched*) to the lobula plate (*Lo P*) supplied by L1, L2 and T1 afferents. Relays from parvocellular (*p*) color-sensitive medulla neuropil (*stippled*), supplied by the R7,R8 and L3 lamina afferents, project to different levels in the lobula (*Lo*) from which specific layers of retinotopic neurons project into glomeruli (*1-4*), which are themselves invaded by the dendrites of descending projection neurons. Lobula plate outputs collate information about directionally selective motion and supply another group of descending neurons. *H, V* horizontal and vertical outputs from directionally selective layers (*arrowheads*) in the lobula plate, *doubleheaded arrow* width of one retinotopic column.

An easy way to compare the olfactory epithelium of a vertebrate with the antenna of an insect is to imagine that the nose has been turned inside out, and that the epithelium is now situated on, rather than in, an appendage. In insects the dendrites of the olfactory receptors are ensheathed in cuticle that is perforated by pores, which in electron micrographs sometimes appear to communicate with the lumen of the sensillum via electron-dense tubules (Steinbrecht 1973; Keil 1982). What role this tubule might have is uncertain, however (see Kaissling 1987).

Leaving aside the first two antennal segments, which are devoted to mechanoreception, the third segment (the flagellum) is, in many insects, richly

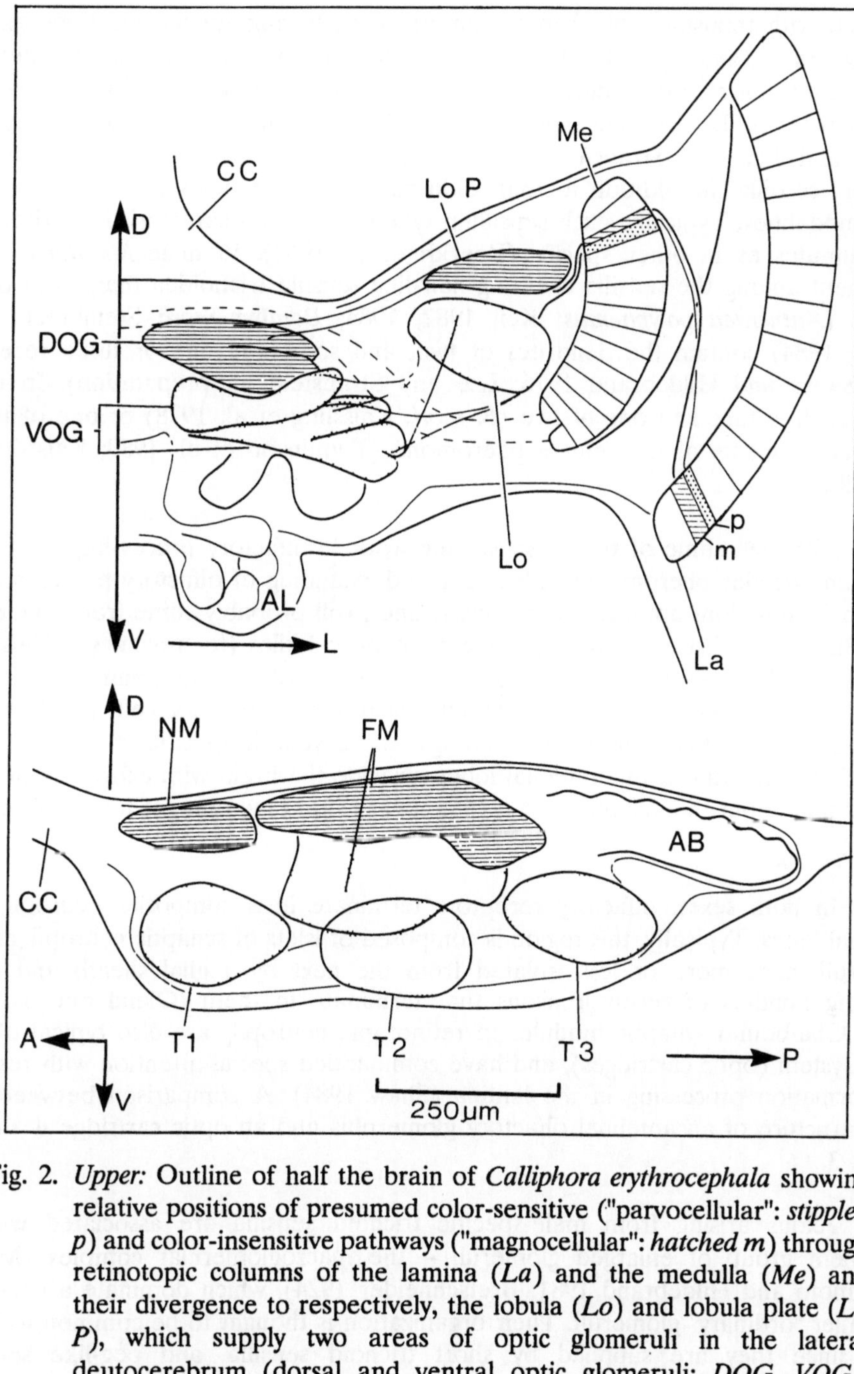

Fig. 2. *Upper:* Outline of half the brain of *Calliphora erythrocephala* showing relative positions of presumed color-sensitive ("parvocellular": *stippled p*) and color-insensitive pathways ("magnocellular": *hatched m*) through retinotopic columns of the lamina (*La*) and the medulla (*Me*) and their divergence to respectively, the lobula (*Lo*) and lobula plate (*Lo P*), which supply two areas of optic glomeruli in the lateral deutocerebrum (dorsal and ventral optic glomeruli: *DOG, VOG)*, situated above the antennal lobes (*AL*). *CC cervical connective.*
Lower: Longitudinal view of the fused thoracic-abdominal ganglia showing neck and flight motor centers (*NM, FM*), which are supplied by descending neurons arising from the DOG (*hatched*) and ventral neuropils of the leg motor (*T1, T2, T3*) and lateral flight motor areas, which are supplied by descending neurons arising from the VOG (*stippled*). *ABD* abdominal ganglia. *Arrows* give orientation: *D* dorsal, *V* ventral, *L* lateral, *A* anterior, *P* posterior.

endowed with thousands of chemoreceptors often having special locations on the antennae (see Kaissling 1987). In moths, the flagellum is annulate and considerable sexual dimorphism exists between male and female antennae (for review see Kaissling 1971). In the male sphinx moth *Manduca sexta* Sanes and Hildebrand (1976) and Lee and Strausfeld (in preparation) have identified five species of olfactory sensilla, in addition to four additional sensillum types that are typical of other modalities. Typically, each sensillum type occupies a characteristic territory on each annulus as in other species (Boeckh et al. 1960). In male *Manduca*, most prominent among the sensilla are long hair-like sensilla trichoidea that, as in other species (*Antheraea polyphemus*: Keil 1982, 1984; *Bombyx mori*: Steinbrecht and Gnatzy 1984) contain the dendrites of two, and sometime one or three receptor cells (Sanes and Hildebrand 1976; Lee and Strausfeld, in preparation). In male moths each is tuned to one of two (*B. mori*: Kaissling et al. 1978) or one of three active components of the female pheromone (Tumlinson et al. 1989; Kaissling et al. 1989).

Although some of the most exciting work in olfactory processing is on how the brain encodes pheromones, a broader understanding of olfactory perception, in terms of interactions amongst odor-specific lines, will probably come from studies of generalized-specialist receptors signaling common volatiles (for review see Kaissling 1971, 1987). For example, in the blow fly *Calliphora* 12 key compounds associated with putrescent meat or flower scents activate different receptor cells (Kaib 1974). We might expect that interactions amongst these volatile-specific channels at the level of interneurons in the antennal lobes, provide the brain with extremely precise information about the suitability of rotting meat as an environment for the female's progeny.

In both sexes, olfactory receptors terminate in a composite neuropil, the antennal lobes. Typically, this region is composed of islets of synaptic neuropil, called glomeruli, each more or less isolated from the next by a glial sheath and each receiving bundles of receptor axons that terminate in them (Oland and Tolbert 1989). Glia-bound synaptic modules in retinotopic neuropil are also typical of the visual system (optic cartridges), and have commanded special attention with respect to information processing in the lamina (Shaw 1984). A comparison between the basic structure of an antennal olfactory glomerulus and an optic cartridge is shown in Fig. 3.

Axons arising from male-specific trichoid sensilla are associated with a prominent group of enlarged glomeruli -- the macroglomerular complex (MGC: Matsumoto and Hildebrand 1981; Bretschneider 1924), which dominate a coalition of smaller "ordinary" glomeruli. Their organization is thought to be common to both sexes since they are supplied by short trichoid sensilla, and peg-like sensilla basiconica, present in both sexes in *Manduca*.

Later in this chapter, the macroglomerulus will be compared with a sex-specific region of neuropil associated with the male compound eye in order to emphasize similarities between the two sensory systems. It suffices to say at this point that in the antennal lobes sex-specific organization amongst interneurons occurs already at the first synaptic station, whereas in the visual system it is represented in deep retinotopic neuropil by fourth-order or even higher order interneurons.

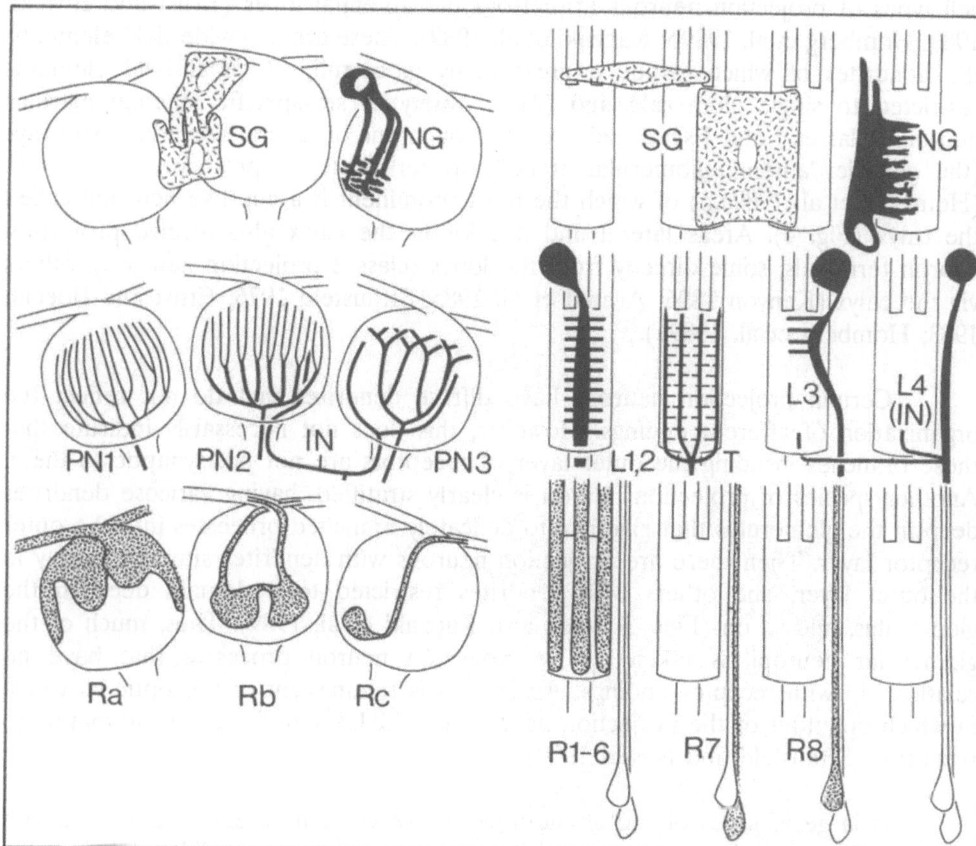

Fig. 3. First-order sensory neuropils serving olfaction and vision have analogous cellular constituents. *Left* Olfactory glomeruli of *Manduca sexta*. *Right* Lamina optic cartridges of *Calliphora erythrocephala*. *Top row*: Two types of glia: sheathing glia (*SG*), insulating synaptic modules, and neuropilar glia (*NG*) invading synaptic neuropil to give rise to lamellar processes between neurons. *Middle row*: Two basic classes: relay neurons (projection neurons *PN1-3*, or monopolar and T-cells *L1-3, T*), and local interneurons (*IN* and *L4*). *Bottom row*: Olfactory glomeruli and retinotopic columns receive different types of receptor endings occupying characteristic zones or levels (*Ra-c* in glomeruli, *R1-R6* in a cartridge with two deeper levels of *R7,R8* in the medulla).

(2) Cellular and functional organization of olfactory glomeruli

Cobalt fills into patches of sex-unspecific olfactory receptors demonstrate that their axons terminate within an outer petiolate cap of an ordinary glomerulus. However, cobalt fills also demonstrate local islets of receptor terminals deep within the glomerulus. Together, these results suggest that two, and possibly three, types of receptor axons can invade an ordinary glomerulus. Like optic cartridges in the lamina, each sex-unspecific (ordinary) olfactory glomerulus gives rise to a group of projection neurons, which exhibit a variety of dendritic tree geometries. Three major cell types of projection neurons arise from the antennal lobes (Ernst and Boeckh 1983; Homberg et al. 1989b; Kanzaki et al. 1989). These are: (1) wide-field elements, the dendrites of which invade a great many glomeruli; (2) small-field elements restricted to single glomeruli; and (3) a variety of sex-specific macroglomerular neurons. Classes 1 and 3 send their axons through one of several ascending pathways (the generic "antenno-glomerular tracts") to terminate in protocerebral centers (Homberg et al. 1989a,b) of which the most prominent is a cup-like neuropil called the calyx (Fig. 4). Areas lateral and inferior to the calyx also receive projection neuron terminals, some directly from the lobes (class 2 projection neurons), others via the calyx (Kenyon 1896; Arnold et al. 1985; Strausfeld 1976; Ernst and Boeckh 1983; Homberg et al. 1989a).

Certain projection neurons have diffuse dendrites that do not reflect the organization of afferent endings. However, this does not necessarily indicate that those branches invading the outer layer of receptors are not postsynaptic to them. Another species of projection neuron is clearly stratified, having varicose dendrites deep in the glomerulus that give rise to delicately branched processes into the outer receptor layer. Then there are projection neurons with dendrites situated mainly in the outer layer, and others with dendrites restricted to a domain deep in the glomerulus, and so on (Figs. 3-5; see also Kanzaki et al. 1989). Thus, much of the glomerular neuropil is taken up by projection neuron processes that have no relationship with receptor endings. Again, this is reminiscent of the optic cartridge in which only four of the projection neurons (L1,L2,L3 and T1) are postsynaptic to receptors (Strausfeld and Nässel 1980).

If large volumes of the glomerulus do not contain receptor terminals, with what do the projection neuron dendrites synapse? The antennal lobes are richly endowed with a variety of local elements, which can be classified into three basic types. (1) Small short-axoned local neurons (which are notoriously difficult to stain with the Golgi method) link a glomerulus with its immediate neighbors. (2) Larger neurons, showing a distinct polarity, give rise to dendrites within one glomerulus and an axon that branches into several distant glomeruli (see also Flanagan and Mercer 1989). (3) There is an amazing morphological variety of local interneurons the branches of which show no obvious polarity (e.g., Fig. 5). Non-polarized neurons imply that each type of neuron (characterized on the basis of its pattern of bifurcations, and detailed intraglomerular branches) occupies an equivalent part of each glomerulus it invades. Matsumoto and Hildebrand (1981) commented on the variety of such local neurons, and recent Golgi studies support the idea that several dozen distinct types of local "isomorphic" neurons supply the antennal lobes (Strausfeld 1988). Together, such neurons further subdivide the internal organization of an ordinary glomerulus into a number of discrete domains and laminae. For example, processes from some appear to climb exclusively amongst the main dendritic branches of projection neurons, whereas another type of local interneuron gives rise to dense arbors in neuropil near the margins of the glomerulus.

Surprisingly, despite this wealth of neuroanatomical detail it is not yet known if olfactory receptors are directly presynaptic to projection neurons leading into the brain. However, if they were not, this would be a radical and interesting departure from synaptic relationships seen in the lamina and be an obvious difference from connections seen in mammalian olfactory bulb (for review see Scott and Harrison 1987) where receptors are known to establish first-order connections with mitral

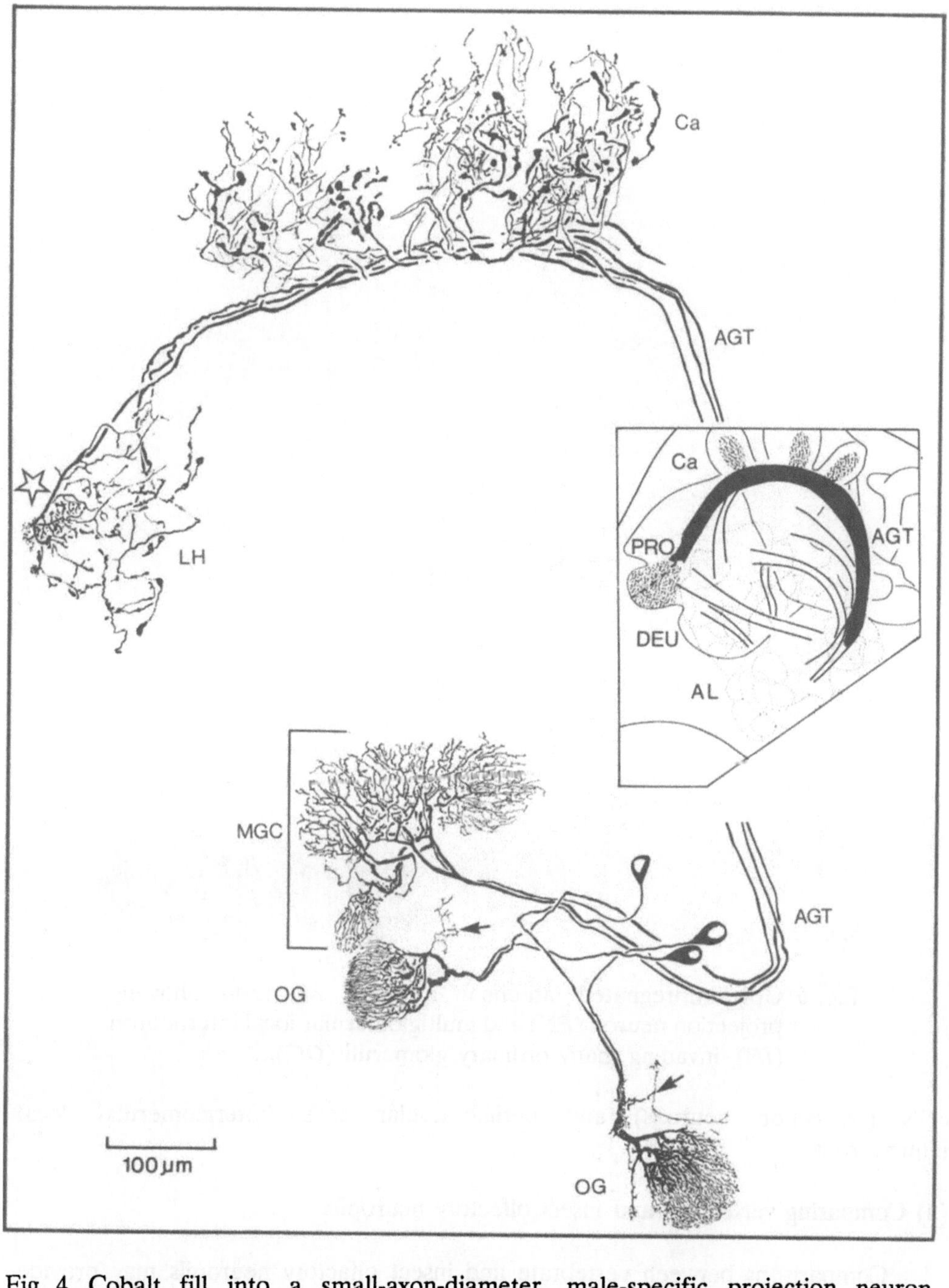

Fig. 4. Cobalt fill into a small-axon-diameter, male-specific projection neuron extending from the macroglomerular complex (*MGC*) of the sphinx moth *Manduca sexta* through the antennal glomerular tract (*AGT*) to the calyx (*Ca*) and male-specific zone of the lateral horn (*LH*) in the protocerebrum (*PRO*). Two types of projection neurons from ordinary glomeruli (*OG*), both with axon collaterals (*arrows*) project via the calyx into the more anterior sex-unspecific neuropil of the lateral horn (*asterisk*. Inset shows relative position of antennal lobes (*AL*, terminal areas (*shaded*) and the linking axon pathway (*black*) between them.

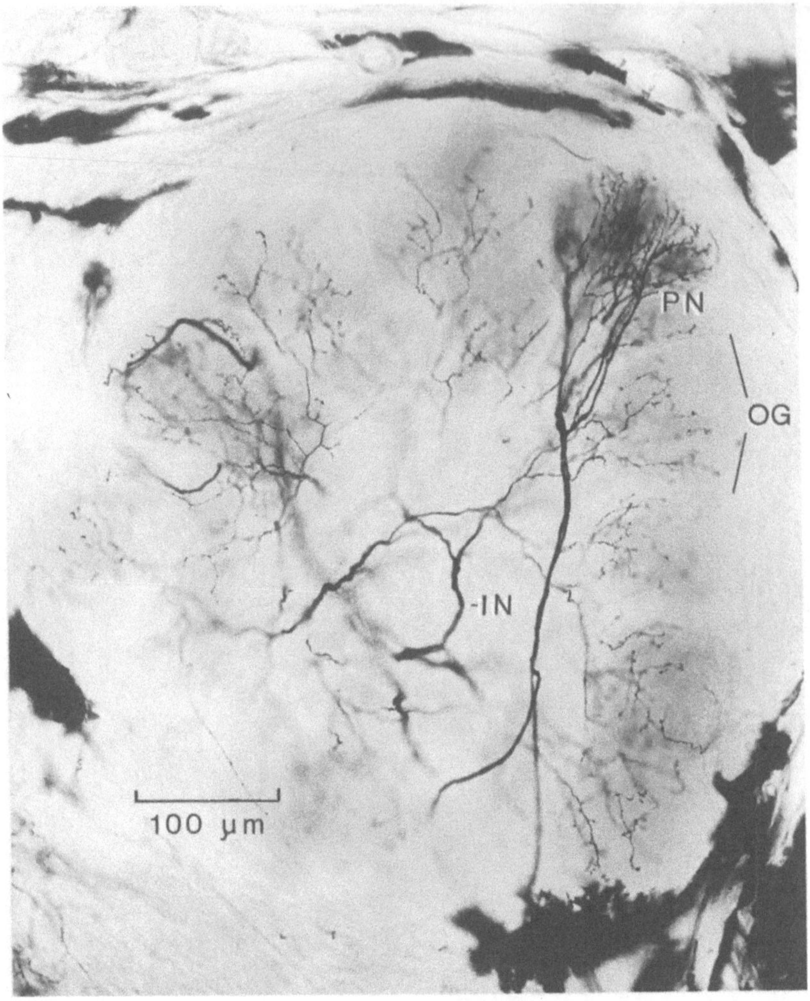

Fig. 5. Golgi-impregnated antennal lobe of *M. sexta* showing projection neuron (*PN*) and multiglomerular local interneuron (*IN*), invading many ordinary glomeruli (*OG*).

cells (projection neurons) and periglomerular cells (interglomerular local interneurons).

(3) Comparing vertebrate and insect olfactory neuropils

Comparisons between vertebrate and insect olfactory neuropils may provide useful search images in looking for synaptic interactions in the insect glomerulus. However, in doing so, it is important to note that whereas vertebrate olfactory bulb neuropil is usually composed of glomeruli surmounting plexiform layers (Macrides and Schneider 1982), the equivalent neuropil in an insect appears to be exclusively glomerular. Nevertheless, the two neuroarchitectures can be reconciled if we take into consideration the discrete layering within the insect glomerulus and attempt to equate this to synaptic strata beneath the glomeruli in the vertebrate olfactory bulb (Fig. 6). But before doing this, it is first necessary to contrast the different cell types.

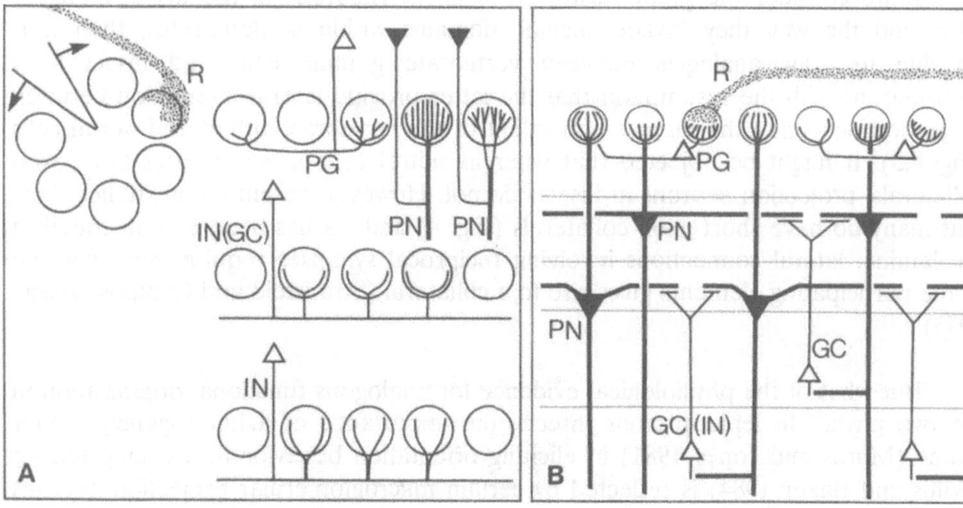

Fig. 6. Comparison between some basic cell types in insect olfactory (antennal) lobe (**A**) and reptilian/mammalian olfactory bulb (**B**). In **A**, the spherical lobe (*upper left*) has been opened up, and its glomeruli depicted three times. These layers are not equivalent to those shown in **B**. In **B**, the layer of olfactory glomeruli surmounts plexiform strata. Common elements are: the cap of receptor endings in each glomerulus (*shaded R*) and three major classes of neurons. (1) Local cells showing polarity link glomeruli (periglomerular neurons *PG*). (2) Local anaxonal interneurons (*IN*) in insects are suggested to be equivalent to granule cells (*GC*) in vertebrates and provide local circuits amongst (3) efferent projection neurons (*PNs*) which in vertebrates are tufted and mitral cells.

Different morphological species of mitral and tufted cells (e.g., see Mori et al. 1983) bear comparison with projection neurons from antennal lobe ordinary glomeruli (Burrows et al. 1982; Homberg 1984; Schildberger 1984; Kanzaki et al. 1989). Local neurons - the periglomerular and short-axoned elements in mammals (Price and Powell 1970; Pinching and Powell 1971a,b; Schneider and Macrides 1978)) - are comparable to certain types of insect olfactory lobe neurons that link one glomerulus with several others (Flanagan and Mercer 1989). In mammals, lateral inhibitory interaction between mitral cells of different glomeruli is mediated by local granule cells (Price and Powell 1970) originating from plexiform neuropil and sending their processes to mitral cell secondary dendrites in the plexiform layer where they form reciprocal "dendrodendritic" synapses (Reese and Shepherd 1972; for review see Mori 1987). In addition to their connections onto granule cells via their inner stratified dendrites, mitral cells also give rise to axon collaterals that are presynaptic onto granule cells in the deepest layer of the bulb. This situation is immediately reminiscent of reciprocal relationships between the L1, L2, and L4 cells in insect optic cartridges. L1 and L2 receive receptor inputs, and are pre- and postsynaptic to the L4 monopolar cells, which themselves receive no primary afferents (Strausfeld and Campos-Ortega 1973).

If we consider the great variety of local interneurons in the insect olfactory lobe, and the way they invade specific domains within a glomerulus, then it is possible to draw analogies between vertebrate granule cells and insect local interneurons with the assumption that the latter provide lateral interaction between projection neurons - the insect's equivalent of the vertebrate tufted and mitral cells (Figs.4,5). It might be objected that whereas mitral cells in vertebrates have axon collaterals, projection neurons in insects do not. However, recent Golgi studies show that many do have short axon collaterals (Fig. 4) and, as has been demonstrated in the lamina, lateral connections involving reciprocal synapses require that only one of the participating elements gives rise to a collateral (Strausfeld and Campos-Ortega 1973).

But what of the physiological evidence for analogous functional organization in the two phyla? In lepidopterous insects, the importance of a heterogeneous odor plume (Murlis and Jones 1981) in eliciting orientation behavior to a calling female (Willis and Baker 1984) is reflected by certain macroglomerular projection neurons registering temporal changes in the pheromone stimulus (Christensen and Hildebrand 1988; Christensen et al. 1989a). The activity of these cells is remarkably like that of certain mitral/tufted cells in salamander olfactory bulb glomeruli (Kauer and Hamilton 1987). In both insect macroglomerular projection neurons and salamander mitral/tufted cells, odor stimulation interrupts spontaneous activity with a brief inhibitory postsynaptic potential. This is succeeded by a phasic burst of excitation followed by a period of inhibition. Hamilton and Kauer suggest that the first component is derived, in salamanders, from inhibitory feed-forward by periglomerular cells, which are postsynaptic to receptors and presynaptic to mitral cells, which are themselves directly excited by the afferent input. Inhibition following the phasic excitation was suggested as being provided by granule cells. An almost identical response characterizes certain sex-specific macroglomerular projection neurons (Christensen and Hildebrand 1987) and sexually isomorphic ordinary glomerular neurons (Waldrop et al. 1987; Kanzaki et al 1989). Possibly, an analogous circuit may be operating in insects whereby projection neurons receive first-order afferent input as does a parallel element that provides feed-forward inhibition of the projection neuron. Models of signal integration by complex dendritic trees (Rall 1977) suggest that the dimensions and nodal points of a tree play an important role in temporal integration of postsynaptic potentials. Possibly, the time required to reach spike initiation within the finely branched and thin dendrites of the projection neuron is longer than that provided by the dual-synapse inhibitory pathway. It is significant that the initial ipsp seen in the projection neuron response is not due to inhibition by sensory inputs but instead may be mediated by local GABAergic neurons, for which morphological (Hoskins et al. 1986) and electrophysiological (Waldrop et al. 1987) evidence clearly exists. Furthermore, when the ipsp is abolished by bicuculine a depolarizing response of the projection neuron is uncovered (Waldrop et al.1987; Waldrop and Hildebrand 1989). However, the projection neuron is no longer able to follow the temporal properties of the signal (Waldrop et al. 1987; Christensen and Hildebrand 1988).

Glomerular synaptology is known to be extremely complex (Tolbert and Hildebrand 1981) and one of the most pressing needs in this area of research is to determine which cells are contacted by receptors. At present, the evidence for direct afferent input onto projection neurons in insects is equivocal, and recent reviews suggest models in which projection neurons receive no primary afferents at all

(Homberg et al. 1989a). The situation is made somewhat more complex in that mechanoreception obviously plays a central role in odor-induced behavior where wind is an important stimulus in eliciting and maintaining flight up an odor plume (Baker and Kuenen 1982; Baker et al. 1984). There is evidence that at least some glomeruli receive three types of receptor terminals, raising the interesting question of whether mechanoreceptors on the third antennal segment impart a second sensory modality to the glomerulus (Lee and Strausfeld, in preparation). Studies on Orthoptera (Schildberger 1984), Hymenoptera (Homburg 1984), and *Manduca sexta* (Kanzaki et al. 1989) describe both unimodal (mechanosensory or olfactory) and bimodal responses by identified projection neurons. In *Manduca*, bimodal responses are derived from projection neurons with bistratified dendritic trees whereas projection neurons responding exclusively to semiochemicals have diffuse non-stratified dendrites. Projection neurons having dendrites restricted to a subdomain in the middle of the glomerulus are activated by mechanosensory stimulation alone (Kanzaki et al. 1989). Although there are too few examples to make definitive statements, these results do suggest that the glomerulus may comprise a complex multimodal integration unit.

What is the functional significance of an olfactory glomerulus and why are specific numbers (Rospars 1983, see also this volume) and specific forms of glomeruli (Bretschneider 1924) found in specific locations in the lobes? Studies on the vertebrate olfactory system, using extracellular recordings (Kauer and Moulton 1974) and [^{3}H]2-deoxyglucose to record sustained neural activity (Lancet et al. 1982), suggest that discrete patches of olfactory receptors in the nasal epithelium are selectively sensitive to specific odorants and converge to discrete odor-specific glomeruli in the bulb. How many of these glomeruli show activity may depend on the concentration of the applied odor (Stewart et al. 1979). These results do not, however, imply that each olfactory glomerulus represents a discrete patch of the nasal epithelium. Although the early anatomists claimed that there exists a relatively simple geometrical relationship between receptor groups and glomeruli (Retzius 1892; Cajal 1911) recent horseradish peroxidase studies suggest that each glomerulus receives input from several patches of the epithelium, and that each patch is itself composed of a heterogeneous population of receptors diverging to several glomeruli (Jastreboff et al. 1984). Thus, in vertebrates it seems that although there exists a geometrical relationship between subsets of receptors and specific glomeruli this is not a simple somatotopic map. Again by analogy, studies on vertebrates raise the interesting question of whether there is a similar chemotopic organization in insects. In *Drosophila,* specific morphological types of sensilla have been shown to terminate in specific glomeruli. And although certain species of sensilla are located at characteristic patches on the antennae, the mode of projection into glomeruli is said to be type-specific rather than position-specific (Stocker and Singh 1983; Stocker et al. 1983; Borst and Fischbach 1987). [^{3}H]2-Deoxyglucose studies on *Drosophila* suggest that identifiable glomeruli situated at characteristic sites in the lobe are activated by certain odors and not others. In contrast to these "specialist" glomeruli others show "generalist" activity, responding to a variety of odors (Rodriguez 1988 and this volume). This result might suggest that any ordinary glomerulus is tuned to one or more specific odors, and the question arises how this tuning is accomplished. Evidence that specific types of projection neurons respond to or are inhibited by certain odorants typical of plant extracts comes from a number of studies (e.g., Homberg 1984; Schildberger 1984; Kanzaki et al. 1989). Specific tuning to blends of odorants is also known from recent studies on macroglomerular projection neurons

(Christensen and Hildebrand 1987) in *Heliothis* and *Manduca*, where a variety of specialist and generalist neurons originate from the same glomerular neuropil (Christensen et al. 1989a,b).

(4) Are olfactory receptors from an appendage topically mapped into the neuropil? Cerci versus antennae

Clearly, the discriminatory tasks performed by insect antennal lobes are varied and sophisticated: detection of conspecifics, discrimination of sympatric species, plant odors and the like. But is the receptor organization on the antennae in any way relevant to these functions?

Apart from studies on *Drosophila* and some knowledge about trichoid sensilla destined for the macroglomerular complex in *Manduca* (Camazine and Hildebrand 1979) we know very little about how receptors map into the antennal lobes from the antennae and whether such maps are function-specific or somatotopic. Possibly, some clues might be provided by comparing the organization and development of receptors from another (often annulate) appendage, the cercus, which in many pterygoid insects occurs in pairs, extending from the tip of the abdomen. Filiform wind-sensitive hairs are strategically organized into specific fields on the cercus, as are sensilla responding to gravity, touch, and chemical signals (Murphey 1981). Anatomical (Bacon and Murphey 1984; Walthall and Murphey 1986) and physiological studies (Jacobs et al. 1986) have shown that the wind-sensitive receptors project somatotopically into a special target neuropil (called by them a "glomerulus") where they are presynaptic onto specific dendrites of directionally selective ascending interneurons. Developmental studies (see Murphey et al. 1984) suggest that the precise mapping of receptors accomplishes a segregating out of different modalities to different target neuropils (olfactory receptors to one, mechanoreceptors to another) and the establishment of positional maps in which longitudinal rows of receptors send their axons to precise longitudinal zones of the glomerular neuropil (Bacon and Murphey 1984). A variety of inter- and interspecific transplantation experiments in developing crickets (Kamper and Murphey 1987) have demonstrated that each receptor has a specific positional address in the glomerular neuropil. Accurate sensory projection onto target neurons is largely responsible for proper computation of mechanosensory signals.

There is evidence from adult insects that mechanoreceptors arising from the first two segments of the "real" antennae project to discrete domains of mechanosensory neuropils lying dorsal to the antennal lobes (Sanchez 1937; Strausfeld and Bacon 1983) where different receptor types sort out into different neuropil partitions (Bacon and Strausfeld 1986). In addition, recent studies on the third segment of *Manduca* antennae (Lee and Strausfeld, in preparation) have resolved mechanosensory receptors situated at specific sites on the annuli, suggesting a receptor distribution very like that seen in the cerci. Although attempts to fill individual mechanoreceptors have so far failed, when annuli are treated with cobalt all their small-diameter axons project into some of the ordinary glomeruli. Assuming that mechanosensory axons are included amongst these terminals, the question arises of whether they confer a somatosensory map into the glomeruli that they invade. And, do the axons of the olfactory receptors on the same annulus follow the same trajectory as those of mechanosensory receptors? If they do, does this projection map longitudinal strips, as in the cerci? Developmental studies on *Manduca*, in which the number of annuli are reduced by about 80%, result in only a slight decrease in the number of glomeruli (Sirianni and Tolbert 1988) suggesting that if a somatotopic map exists at all, then it cannot relate to specific annuli being represented by specific glomeruli.

3. "PARVO-" AND "MAGNOCELLULAR" PATHWAYS IN INSECT VISUAL AND OLFACTORY NEUROPILS

A. Olfactory System

Central overrepresentation of part of sensory space is not restricted to the vertebrate brain but is found in insect olfactory and visual systems (moth olfactory systems: Bretschneider 1924; Boeckh and Boeckh 1979; Camazine and Hildebrand 1979; fly visual system: Strausfeld 1980; Hausen and Strausfeld 1980). In male *Manduca sexta* hugely enlarged glomeruli of the macroglomerular complex, devoted to male-specific receptors (Camazine and Hildebrand 1979), give rise to male-specific projection neurons possessing enormously enlarged dendritic fields (Christensen and Hildebrand 1987; Kanzaki et al. 1989). These male-specific neuropils, with their constituent interneurons, can also be induced to develop in females when male antennae are grafted onto them early in pupal development, leading to pheromone-induced male-like behavior in these chimeric animals (Schneiderman et al. 1982, 1986).

With the projection neurons from ordinary glomeruli, the majority MGC projection neurons project into the calyces of the mushroom bodies where they give rise to characteristic terminals before extending into sex-specific satellite neuropils of the protocerebrum (Kanzaki et al. 1989). However, certain large-axoned macroglomerular projection neurons have a different trajectory, sending their axons into the lateral protocerebrum and from there into areas of the deutocerebrum from which descending neurons arise (Strausfeld, in preparation and compare Figs. 4 and 7). A similar projection of large-diameter projection neurons has been noted in *Calliphora* (Strausfeld and Bacon 1983). This organization suggests that, as in the visual system, a subset of large sensory interneurons may provide a rapid transit of information to premotor descending pathways.

B. Visual System

In the visual system of male flies we find an analogous situation in which a special male-specific high-acuity area at the front of the eye (Land and Eckert 1985) contains sex-specific green-sensitive R8 receptors (Franceschini et al. 1981), which terminate not in the medulla, as they would from other areas of the eye, but in the lamina with R1-R6 (Hardie 1983). This results in a net gain by the color-insensitive system of object contrast (Hardie 1986). This is of particular advantage to male flies, which spend a large part of their lives chasing small contrasting profiles in the assumption that they are females (Land and Collett 1974). These rapine pursuits involve truly amazing aerobatics in which the male fly must compute the speed and direction of the pursued in order to intercept it (Land and Collett 1974; Wagner 1986).

The high-acuity zone of the male fly subtends an expanded male-specific area of the retinotopic mosaic in the lobula (Strausfeld 1979, 1980), which is associated with about 17 species of uniquely identifiable male-specific neurons (Strausfeld 1987). Interestingly, this area of the lobula does not seem to be supplied by anything other than the normal complement of medulla relay neurons providing identical inputs to the rest of the lobula. However, because male-specific lobula neurons occupy characteristic strata in the lobula they must interact with combinations of medulla inputs that are not found in other areas.

This male-specific subsystem gives rise to two classes of output neuron: columnar cells that project into optic glomeruli of the lateral deutocerebrum (Fig. 8A) and large-axoned tangential cells that project onto the outgoing axons of descending neurons (Strausfeld and Bacon 1983; Strausfeld 1987). This division into two diverging pathways, one carried by large-axoned tangentials, the other by small columnar units, is reminiscent of sex-unspecific differences between the projections from the lobula and the lobula plate. As intimated earlier, the lobula plate is associated with a color-insensitive pathway that supplies wide-field directionally motion-sensitive tangentials. Like the color-insensitive lobula plate tangentials (Strausfeld and Bassemir 1985b), large sex-specific tangential neurons from the lobula terminate on the trunks or axons of descending neurons that supply pterothoracic neuropil (Fig. 8B).

In summary, both the olfactory and visual systems contain large-axon-diameter pathways that provide direct routes, with few synaptic stations, to premotor descending neurons. In the visual system, and in the olfactory system, such pathways terminate on the axons of descending neurons, at a position below the dendritic tree and the optic glomeruli. Presumably, this subset of large neurons provides a rapid transit of information to premotor pathways.

4. GLOMERULI: A COMMON DESIGN SERVING COMMON INTEGRATIVE TASKS

A. Neuroarchitectural Similarities

The olfactory systems of vertebrates, crustacea (Arbas et al. 1988) and insects are characterized by glomerular first-order synaptic neuropil. In the insect olfactory system, the generic glomerulus can be thought of as a structurally complex module serving specific narrowly tuned olfactory receptors. The evidence so far suggests that a great variety of local interneurons provide substrates that refine this qualitative data, which is then carried by projection neurons centrally. Possibly, the cellular complexity of a glomerulus may be inversely related to the specificity of its input so that projection neurons serving labeled lines, as in the macroglomerular complex (Christensen et al. 1989a), would be expected to arise from less complex neuropil than projection neurons that may be involved in cross-fiber discrimination of odors from generalist or specialized-generalist receptors.

The suggestion is made here that, as in the olfactory system, glomerular neuropils receiving inputs from the lobula provide the substrate for discrimination between specific features of the visual environment.

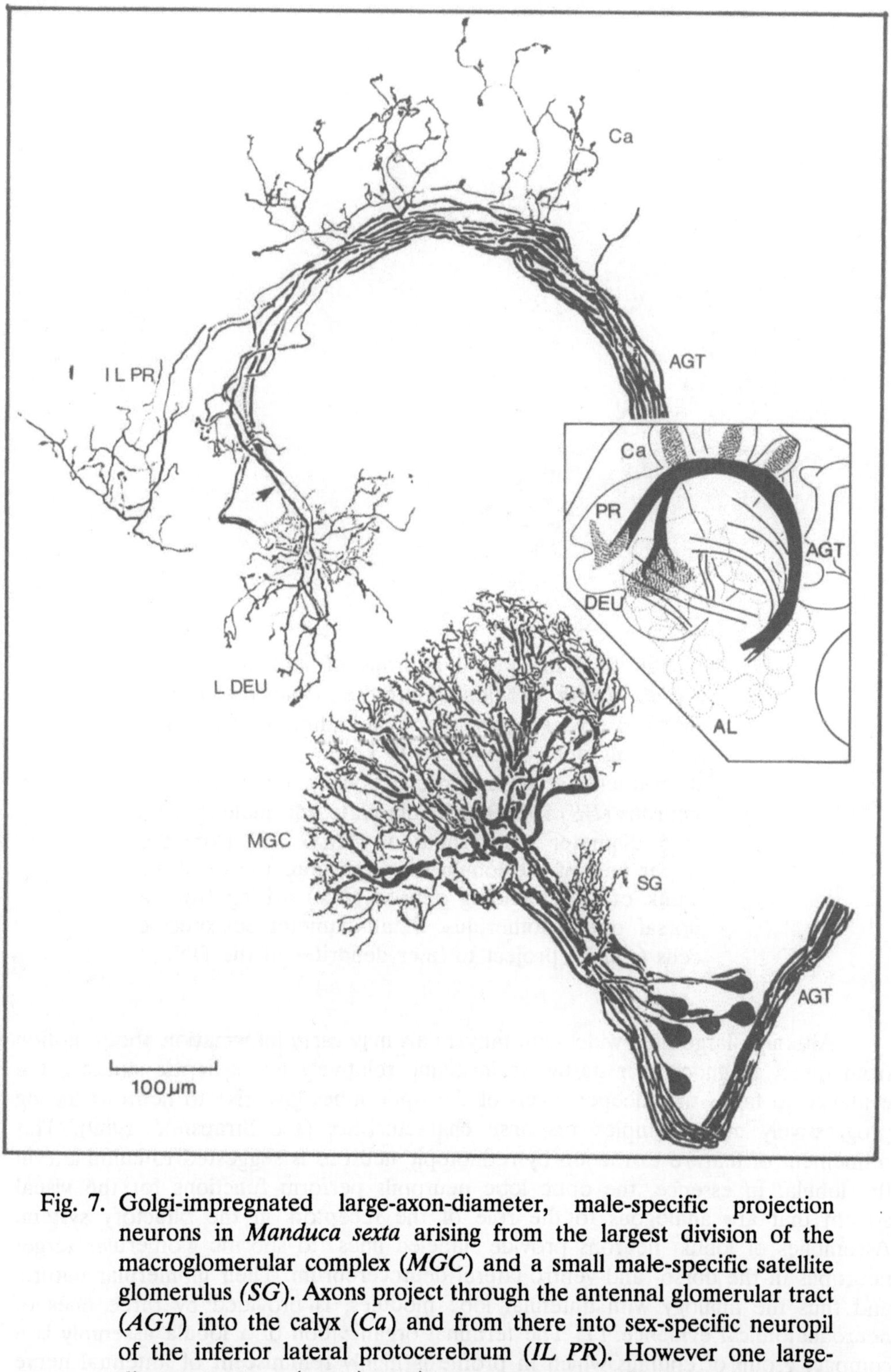

Fig. 7. Golgi-impregnated large-axon-diameter, male-specific projection neurons in *Manduca sexta* arising from the largest division of the macroglomerular complex (*MGC*) and a small male-specific satellite glomerulus (*SG*). Axons project through the antennal glomerular tract (*AGT*) into the calyx (*Ca*) and from there into sex-specific neuropil of the inferior lateral protocerebrum (*IL PR*). However one large-diameter axon (*arrow*) projects postero-ventrally into the lateral deutocerebrum (*L DEU*). The *inset* shows these trajectories against the outlines of the mentioned brain areas. *AL* antennal lobe.

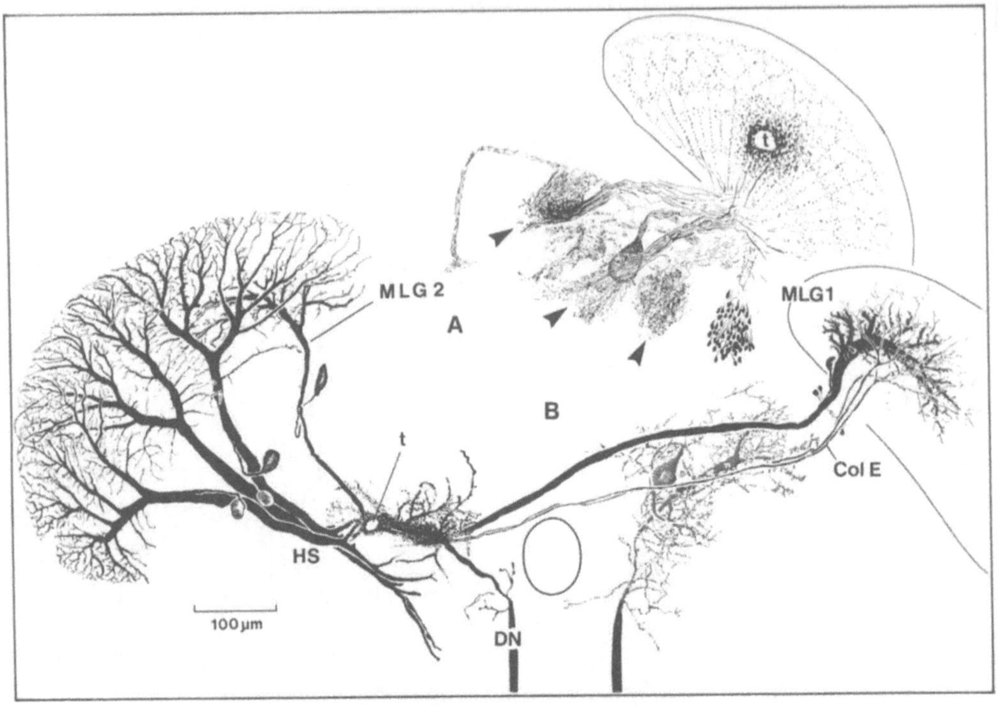

Fig. 8. Cobalt fills into visual neurons of male *Calliphora erythrocephala* **A** Mass fill into the lobula (*t*: electrode tract) to reveal groups of lobula endings (*arrowheads*) supplying optic glomeruli. **B** Large (magnocellular) color-insensitive wide-field directionally motion-sensitive neurons (*HS*) from the lobula plate and male-specific large-diameter lobula neurons (*MLG 1,2*) from the upper part of the lobula converge onto the dendritic trunk of a descending neuron (*DN*) arising from a dorsal optic glomerulus. Small-diameter sex-specific cells (*Col E*) project to finer dendrites of the DN.

Although large-field wide-axon tangentials may carry information about motion from direct magnocellular pathways involving relatively few synaptic stations, the evidence so far is that deeper levels of the optic lobes give rise to neurons having progressively more complex response characteristics (see Strausfeld 1989). This refinement of feature extraction by retinotopic neurons is suggested to culminate in the lobula. In essence, the optic lobe neuropils perform functions for the visual system that are analogous to the role of the receptors in the olfactory system. Assemblies of lobula neurons provide "labeled lines" to specific glomerular target neuropils in the dorso- and ventro-lateral deutocerebrum. Their glomerular nature, and thus the analogy with antennal lobe modules, is provided by three lines of neuroanatomical evidence. (1) The terminal organization of a lobula assembly is a compact group of endings which in profile is highly reminiscent of antennal nerve endings in the antennal lobes (Strausfeld and Bacon 1983). (2) Fluorescent stains differentiate between cell bodies of neurons and glia, and resolve lobula output neurons with projection neurons from these areas (Strausfeld and Seyan 1987). These stains have been used to reveal the locations of glia cells specifically associated with one or another group of lobula terminals. Corresponding Golgi impregnations and general stains have confirmed that islets of neuropils receiving lobula terminals are delineated from each other. The original name for these areas, "optic foci," which was based on the convergence of lobula endings into specific mid-brain regions, has been substituted by the term dorsal and ventral "optic glomeruli" (Strausfeld and Bacon 1983). (3) Recent intracellular and cobalt studies have identified a variety of local interneurons linking specific combinations of glomeruli. Projection neurons in optic glomeruli, like those originating from antennal lobe glomeruli have dendrites in one or several optic glomeruli, the latter arrangement suggesting convergence amongst several labeled lines onto the postsynaptic neuron. Recordings from descending neurons support the notion that, like antennal lobe projection neurons, they carry context-specific multimodal information from glomerular neuropil (Reichert and Rowell 1986; Milde and Strausfeld 1989 and in preparation; Gronenberg and Strausfeld 1989).

In conclusion, the segregation of different lobula assemblies into specific glomeruli suggests that different functional classes segregate out to specific targets in the brain in a manner analogous to the segregation of functional classes of olfactory receptors into different glomeruli of the antennal lobes (Camazine and Hildebrand 1978; Borst and Fischbach 1987; Rodriguez 1988; Christensen, personal communication). At its most simplistic, the compound eye and retinotopic optic lobes are to optic glomeruli what the olfactory sensilla on the antenna are to olfactory glomeruli.

B. Projections From Glomeruli: What Goes Up Must Come Down

1. Distributive outputs from the antennal lobes

In the olfactory and visual systems certain large-axon-diameter ("magnocellular") neurons provide direct routes to descending pathways. However, comparisons between systems of small-axon relay neurons suggest a fascinating structural difference between vision and olfaction. In vision, optic lobe outputs (and mechanoreceptor axons) impinge on projection neurons that descend down the cord to motor centers of the thorax, hence their generic name "descending neurons." In contrast, the vast majority of projection neurons from the antennal lobes ascend into the most anterior division of the brain, the protocerebrum (Fig. 9).

Outputs from the antennal lobes are distributed to a variety of processing centers: the lateral horn (Strausfeld 1976), and the inferior medial and the inferior lateral protocerebrum. Projection neuron axons reach these centers either directly through one of several antennal lobe tracts (Homberg et al. 1989b) or via an excursion into the cup-like neuropil of the "mushroom body's" calyx (Arnold et al. 1985; Homberg et al. 1989a). Because of this wide distribution, we do not have to assume that all these second tier antennal centers are necessarily involved in further discrimination amongst odors but that some may perform complex operations relating to odor concentration, memory, and location.

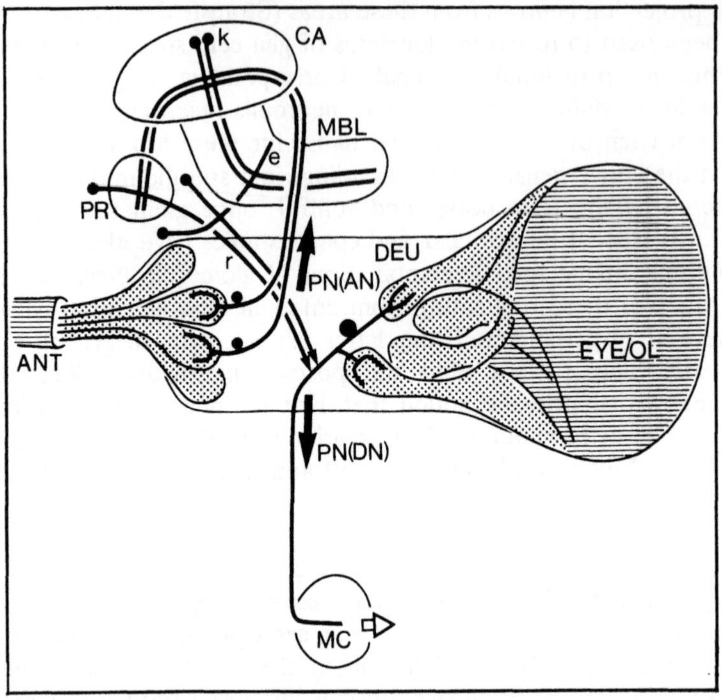

Fig. 9. Schematic showing olfactory ascending [*PN (AN)*] and visual descending projection neurons [*PN(DN)*] from olfactory and optic glomeruli (*stippled*) in the second brain segment, the deutocerebrum (*DEU*), supplied by the equivalent receptor organs, the antenna and eye/optic lobe (*ANT, EYE/OL*). The major difference between the two pathways is the complex series of protocerebral relay neurons serving antennal lobe projection neurons. These include: Kenyon cells (*k*) of the calyx (*Ca*) whose axons in the mushroom body lobes (*MBL*) intersect extrinsic neurons (*e*). These pass to recurrent neurons (*r*), which also receive axons from other protocerebral areas (*PR*) and which project to descending pathways leaving the brain for motor/local interneuron circuits (*MC*) of the thoracic ganglia.

A prominent protocerebral center has been implicated in olfactory memory. This is the (paired) mushroom body or corpora pedunculata from which neurons have been recorded that operate on an "if....then" logic (Gronenberg 1988; see review by Erber et al. 1987) and show long-term postsynaptic potentiation. Cryosurgical or genetic disruption of the mushroom bodies leads to deficiencies in olfactory learning and memory (Erber et al. 1980; Heisenberg et al. 1985). The structure of the mushroom body has also been compared recently (Erber et al. 1987) to associative centers in the mammalian brain, particularly the piriform cortex and entorhinal areas (Valverde 1965; Haberly 1985). Indeed, because of their prominence in social insects, the mushroom bodies attracted the attention of the earliest students of insect behavior, receiving from Dujardin (1850) the typically Gallic accolade of an "intelligence center." It is this structure that I shall review as the final element of this chapter.

2. The organization of the mushroom bodies: a substrate for multiple parallel processing

As is typical of any axonal neuron, different morphological types of projection neurons are classified according to their dendritic and their terminal morphologies. In the calyx, projection neurons give rise to an approximately radial arrangement of major axon collaterals from which arise clusters of side branches decorated with spines, blebs, varicosities and the like, according to the morphological cell type (Kenyon 1896; Schürmann 1973; Strausfeld 1976; Mobbs 1982). There is evidence that the terminal organization of projection neurons is independent of glomerular formation. In moths that have been deantennated early in pupal life, the antennal lobes do not contain glomeruli (Tolbert 1989), and small-field projection neuron dendrites fail to develop (Tolbert, personal communication). However, diffuse projection neurons in deafferented antennal lobes will still send axons into the calyx where they appear to form normal terminals (Strausfeld, in preparation).

The term "mushroom body" is derived from the general appearance of a pedunculus (stalk) surmounted by the calyx (for review see Schürmann 1987). Projection neuron terminals in the calyx interact with a dense population of relay neurons -- the Kenyon cells, named after their discoverer -- which send many thousands of parallel axons deeper into the brain. Studies on Orthoptera (Schürmann 1973), Diptera (Strausfeld 1976), and Hymenoptera (Mobbs 1982, 1985), have ascertained that Kenyon cell dendrites are organized into concentric laminae and clusters. In moths, Kenyon cell dendrites overlap each other and appear more diffuse (Pearson 1971), but this does not imply any lack of precision in the connections between projection neuron terminals and Kenyon cell dendrites to which they are presynaptic (Schürmann 1971). All the Kenyon cell axons project out of the base of the calyx, where they undergo rearrangement before projecting parallel to one another through the pedunculus. After a certain distance. each Kenyon cell axon divides into two tributaries, which form, collectively, the alpha and beta lobes. In certain species of Lepidoptera there is an additional appendage, the gamma lobe (Pearson 1971) which contains a small sub-population of unbranched Kenyon cell axons.

The trajectories of Kenyon cell axons from the calyx into the pedunculus apparently transform their radial coordinates into a rectilinear system (Mobbs 1982, 1985). Subsets of Kenyon cell axons give rise to patches and strata (output zones) of small collaterals and presynaptic spines at specific levels down their lengths in the pedunculus and lobes. Thus, although only about eight morphological types of Kenyon cell dendrites can be classified in the calyx (Pearson 1971; Schürmann 1971; Strausfeld 1976; Mobbs 1982), the organization of Kenyon cell terminals suggests a great many different connectivity patterns of these second-order relay neurons with third-order cells, the extrinsic neurons, that project from the calyx into the surrounding neuropil. It is this organization of many Kenyon cell subsets, all projecting in parallel with one another, that may hold the key to understanding the functional organization and behavioral role of the mushroom bodies.

If we trace back the Kenyon cell axons from any output zone, we see a corresponding set of Kenyon cell dendrites that interact with a certain portion of the antennal lobe input. Because this input is organized as a ring-like pattern of terminals distributed equally through the calyx neuropil, any subset of Kenyon cells is likely to contact representatives from all of the glomeruli. Thus although it would seem at first sight that there is a massive divergence from relatively few antennal lobe afferents onto an enormous number of Kenyon cells, if we consider the relationship between antennal lobe afferents and each discrete subset of Kenyon cells, then for any one subset there is convergence of antennal lobes afferents. There is a great variety of output zones and thus of the pattern of Kenyon cells contributing to any one of them. We may expect that each output zone reflects a specific quantitative relationship between the synaptic input from antennal lobe projection neurons onto a specific configuration of Kenyon cell dendrites.

Each Kenyon cell output zone, situated at a characteristic levels in the pedunculi and lobes, is intersected by dendrites originating from "extrinsic" cells (Kenyon 1896). Typically, extrinsic cell dendrites interweave amongst thousands of parallel fibers to reach their presynaptic target, where they then give rise to ornate filigrees of finer branches the shapes of which would send a topiarist into paroxysms of delight.

Extrinsic neurons are so named because they relay information from the lobes into the surrounding protocerebral neuropil (Schildberger 1984; Mobbs 1982). But from there on, our knowledge about this part of the olfactory pathway is very sketchy indeed. We know that olfactory information must reach the thoracic ganglia through descending pathways simply because insects react in a highly specific manner to olfactory cues received by their antennae (Schneiderman et al. 1982, 1986; for review see Kennedy 1983). And since extrinsic neurons from the mushroom bodies do not themselves reach descending neurons in the deutocerebrum, this implies that the neuropil surrounding the mushroom bodies must be involved in a further stage of higher order olfactory processing involving a further system of intermediate relays to descending neurons (Strausfeld et al. 1984).

4. CONCLUSION

This chapter demonstrates that the organization of the insect brain offers many accessible systems, comparable to those in vertebrates, for experimental analysis of sensory processing. However, it may seem ironic that the insect brain, once advertised as a "simple" system, has been shown to contain brain areas of great complexity, which may play roles similar to areas in vertebrates that are involved in memory and the internalization of sensory maps (see McNaughton and Nadel 1989). Possibly, Dujardin was not far off about the insect brain, if "intelligence" for an insect is to know where it is, where it was, and where it should be in the world about it.

Acknowledgements. I would like to thank Dr. Camilla Strausfeld for discussing versions of the manuscript and for editing it. Sections on the function of macroglomerular neurons have profited from discussions with Drs. Brian Waldrop and Tom Christensen. Research on which this chapter is based was supported by a grant from the National Institutes of Health National Eye Institute EY07151-03

5. REFERENCES

Anderson PA, Olavarria J, Van Sluyters RC (1988) The overall pattern of ocular dominance bands in cat visual cortex. J Neurosci 8:2183-2200

Arbas EA, Humphreys CJ, Ache BW (1988) Morphology and physiology properties of interneurons in the olfactory midbrain of the crayfish. J Comp Physiol A 164:231-241

Arnold G, Masson C, Budharugsa A (1985) Comparative study of the antennal lobes and their afferent pathway in the worker bee and the drone (*Apis mellifera*)

Bacon JP, Murphey RK (1984) receptive fields of cricket (*Acheta domesticus*) are determined by their dendritic structure. J Physiol (Lond) 352:601-616

Bacon JP, Strausfeld NJ (1986) The dipteran "Giant fibre" pathway: neurons and signals. J Comp Physiol A 158:529-548

Baker TC, Kuenen LPS (1982) Pheromone source location by flying moths: a supplementary non-anemotactic mechanism. Science 216:424-427

Baker TC, Willis MA and Phelan PL (1984) Optomotor anemotaxis polarizes self-steering zigzagging in flying moths. Physiol Entomol 9:365-376

Beersma DGM, Stavenga DG, Kuiper JW (1977) retinal lattice, visual field and binocularities in flies. J Comp Physiol A 119:207-220

Blasdel GG, Salama G (1986) Voltage-sensitive dyes reveal a modular organization in monkey striate cortex. Nature 321:579-585

Boeckh J, Kaissling K-E, Schneider D (1960) Sensillen und Bau der Antennegeissel von *Telea polyphemus*. Zool Anat 78:559-584

Boeckh J, Boeckh V (1979) Threshold and odor specificity of pheromone sensitive neurons in the deutocerebrum of *Antheraea pernyi* and *A. polyphemus*. J Comp Physiol A 132:235-242

Borst A, Fischbach K-F (1987) Golgi- and degeneration studies of the antennal lobes of *Drosophila melanogaster*. J Neurogenetics 4:115-127

Braitenberg V (1967) Patterns of projection in the visual system of the fly. I. Retina-lamina projections. Exp Brain Res 3:271-298

Bretschneider F (1924) Über die Gehirn des Eichenspinners und des Seidenspinners (*Lasiocampa quercus* und *Bombyx mori*). Z Wiss Zool A 60:563-578

Burkhardt D (1962) Spectral sensitivity and other response characteristics of single visual cells in the arthropod eye. Symp Soc Exp Biol 16:86-109

Burrows M, Boeckh J, Esslen J (1982) Physiological and morphological properties of interneurons in the deutocerebrum of male cockroaches which respond to female pheromone. J Comp Physiol 145:447-457

Cajal SR (1888) Estructura de los centros nerviosos de los aves. Rev Trim Histol Norm Patol 1:1-10

Cajal SR (1911) Histologie du Système Nerveux de l'Homme et des Vertébrés (1972 ed). Madrid CSIC

Cajal SR (1917) Contribucion al conocimiento de la retina y centros opticos de los cefalopodos. Trab Lab Invest Biol Univ Madrid 15:1-82

Cajal SR, Sanchez DS (1915) Contribucion al conocimiento de los centros nerviosos de los insectos. Parte I. Retina y centros opticos. Trab Lab Invest Biol Univ Madrid 13:1-168

Camazine SM, Hildebrand JG (1979) Central projections of antennal sensory neurons in mature and developing *Manduca sexta*. Soc Neurosci Abstr 5:155

Campos-Ortega JA, Strausfeld NJ (1972) The columnar organization of the second synaptic region of the visual system of *Musca domestica* L. I. Receptor terminals in the medulla. Z Zellforsch 124:561-585

Christensen TA, Hildebrand JG (1987) Male-specific, sex pheromone-selective projection neurons in the antennal lobes of the moth *Manduca sexta*. J Comp Physiol A 160:553-569

Christensen TA, Hildebrand JG (1988) Frequency coding by central olfactory neurons in the sphinx moth *Manduca sexta*. Chem Senses 13:123-130

Christensen TA, Mustaparta H, Hildebrand JG (1989a) Discrimination of sex pheromone blends in the olfactory system of the moth. Chemical Senses 14:122-136

Christensen TA, Hildebrand JG, Tumlinson JH, Doolittle RE (1989b) Sex pheromone blend of *Manduca sexta*: responses of central olfactory interneurons to antennal stimulation in male moths. Arch Insect Biochem Physiol 10: in press

Collett TS, Land MF (1975) Visual control of flight behavior in the hoverfly *Syritta pipiens*. J Comp Physiol A 99:1-66

Den Otter CJ, Scheil HA, Sander-Van Oosten A (1978) Reception of host-plant odours and female sex-pheromone in *Adoxophyes orana* (Lepidoptera: Tortricidae): Electrophysiology and morphology. Entomol Exp Appl 24:570-578

Dowling JE, Boycott BB (1966) Organization of the primate retina: electron microscopy. Proc Roy Soc B 166:80-111

Dujardin F (1850) Mémoire sur le système nerveux des Insects. Ann Sci Nat Zool 14:195-206

Egelhaaf M, Hausen K, Reichardt W, Wehrhahn C (1988) Visual course control in flies relies on neuronal computation of object and background motion. Trends Neurosci 11:351-358

Erber J, Masuhr T, Menzel R (1980) Localization of short-term memory in the brain of the bee *Apis mellifera*. Physiol Entomol 5:343-358

Erber J, Homberg U, Gronenberg W (1987) Functional roles of the mushroom bodies in insects. In: Gupta AP (ed) Arthropod brain. Its evolution, development, structure and function. Wiley, New York, pp 485-511

Ernst KD, Boeckh J (1983) A neuroanatomical study on the organization of the central antennal pathways in insects. III. Neuroanatomical characterization of physiologically defined response types of deutocerebral neurons in *Periplaneta americana*. Cell Tissue Res 229:1-22

Flanagan D, Mercer AR (1989) Morphology and response characteristics of neurones in the deutocerebrum of the brain in the honey bee *Apis mellifera*.

Franceschini N (1975) Sampling of the visual environment by the compound eye of the fly. Fundamentals and application. In: Snyder AW, Menzel R (eds) Photoreceptor optics. Springer, Berlin, pp 97-125

Franceschini N (1985) Early processing of color and motion in a mosaic visual system. Neurosci Res Suppl 2:17-49

Franceschini N, Hardie R, Ribi W, Kirschfeld K (1981) Sexual dimorphism in a photoreceptor. Nature 291:241-244

Gilbert CD, Wiesel TN (1981) Laminar specialization and intracortical projections in cat primary visual cortex. In: Schmitt FO, Worden FG, Adelman G, Dennis MG (eds). The organization of the cerebral cortex. MIT Press, Cambridge, pp 163-198

Grant AJ, O'Connell RJ, Hammond AM (1987) A comparative study of the neurophysiological response characteristics of olfactory receptor neurons in two species of noctuid moths. In: Rope SD, Atema J (eds) Olfaction and taste IX. Ann NY Acad Sci 510:311-314

Gronenberg W (1980) Anatomical and physiological properties of feedback neurons of the mushroom bodies in the bee brain. Exp Biol 46:115-125

Gronenberg W, Strausfeld NJ (1989) Descending neurons associated with wide-field motion-sensitive optic lobe afferents segregate to neck and flight motor neuropil and respond to panoramic cues. Soc Neurosci Abstr 15: in press

Haberly LOB (1985) Neural circuitry in olfactory cortex: anatomy and functional implications. Chem Senses 10:219-238

Hamilton KA, Kauer JS (1988) Responses of mitral/tufted cells to orthodromic and antidromic electrical stimulation in the olfactory bulb of the tiger salamander. J Neurophysiol 59:1736-1755

Hardie RC (1983) Projection and connectivity of sex-specific photoreceptors in the compound eye of the male housefly (*Musca domestica*). Cell Tissue Res 233:1-21

Hardie RC (1984) Properties of photoreceptors R7 and R8 in dorsal marginal ommatidia in the compound eyes of *Musca*and *Calliphora* J Comp Physiol A 154:157-165

Hardie RC (1986) The photoreceptor array of the dipteran retina. Trends Neurosci 9:419-423

Hausen K (1982) Motion sensitive interneurons in the optomotor system of the fly. II. The horizontal cells: Receptive field organization and response characteristics. Biol Cybern 46:67-79

Hausen K, Egelhaaf M (1989) Neural mechanisms in visual course control in insects. In: Stavenga DG, Hardie RC (eds) Facets of vision. Heidelberg, Springer, pp 391-424

Hausen K, Strausfeld NJ (1980) Sexually dimorphic interneuron arrangements in the fly visual system. Proc R Soc Lond B 208:57-71

Heisenberg M, Borst A, Wagner S, Byers D (1985) *Drosophila*mushroom body mutants are deficient in olfactory learning. Neurogenetics 2:1-30

Hertel H (1980) Chromatic properties of identified interneurons in the optic lobes of the bee. J Comp Physiol A 137:215-232

Homburg U (1984) Processing of antennal information in extrinsic mushroom body neurons of the bee brain. J Comp Physiol A 154:825-836

Homberg U, Christensen TA, Hildebrand JG (1989a) Structure and function of the deutocerebrum in insects. Annu Rev Entomol 34:477-501

Homburg U. Montague RA, Hildebrand JG (1989b) Anatomy of antenno-cerebral pathways in the brain of the sphinx moth *Manduca sexta* Cell Tissue Res 254:255-281

Hoskins SG, Homberg U, Kingan TG, Christensen TA, Hildebrand JG (1986) Immunocytochemistry of GABA in the antennal lobes of the sphinx moth *Manduca sexta* Cell Tissue Res 244:243-252

Hubel DH, Livingstone MS (1987) Separation of form, color, and stereopsis in primate area 18. J Neurosci 7:3378-3415

Hubel DH, Wiesel TN (1977) Functional architecture of macaque monkey visual cortex. Proc R Soc Lond B 198:1-59

Hubel DH, Wiesel TN, Stryker MP (1977) Orientation columns in macaque monkey visual cortex demonstrated by the 2-deoxyglucose autoradiographic technique. Nature 269:328-330

Jacobs GA, Miller JP, Murphey RK (1986) Integrative mechanisms controlling directional sensitivity of an identified sensory interneuron. J Neurosci 6:2298-2311

Jastreboff PJ, Pedersen PE, Greer CA, Stewart WB, Kauer JS, Benson TE, Shepherd GM (1984) Specific olfactory receptor populations projecting to identified glomeruli in the rat olfactory bulb. Proc Nat Acad Sci USA 81:5250-5254

Kaissling KE (1971) Insect olfaction. In: Beidler LM (ed) Handbook of sensory physiology IV. I. Springer, Berlin, pp 351-431

Kaissling KE (1987) R.H. Wright lectures on insect olfaction. Simon Fraser University Press, Burnaby

Kaissling KE, Kasang G, Bestmann HJ, Stransky W, Vostrowsky O (1978) A new pheromone of the silk worm moth Bombyx mori Sensory pathway and behavioral effect. Naturwissenschaften 65:382-384

Kaissling KE, Hildebrand JG, Tumlinson JH (1989) Pheromone receptor cells in the male moth Manduca sexta Arch Insect Biochem Physiol. in press

Kanzaki R, Arbas EA, Strausfeld NJ, Hildebrand JG (1989) Physiology and morphology of projection neurons in the antennal lobes of the male moth Manduca sexta J Comp Physiol A 165:427-453

Kamper G, Murphey RK (1987) Synapse formation by sensory neurons after cross-species transplantation in crickets: the role of positional information. Dev Biol 122:492-502

Katz LC (1987) Local circuitry of identified projection neurons in cat visual cortex brain slices. J Neurosci 7:1223-1249

Katz LC, Gilbert CD, Wiesel TN (1989) Local circuits and ocular dominance columns in monkey striate cortex. J Neurosci 9:1389-1399

Kauer JS (1987) Coding in the olfactory system. In: Finger TE, Silver WL (eds) Neurobiology of taste and smell. Wiley, New York, pp 205-231

Kauer JS, Hamilton KA (1987) Odor information processing in the olfactory bulb. Evidence from extracellular and intracellular electrodes and from 2-deoxyglucose mapping. In: Roper SD, Atema J (eds) Olfaction and taste IX. Ann NY Acad Sci 1987:400-402

Kauer JS, Moulton DG (1974) Responses of olfactory bulb neurones to odour stimulation of small nasal areas in the salamander. J Physiol 243:717-737

Kennedy JS (1983) Zigzagging and casting as a response to windborne odour: a review. Physiol Ent 8:109-120

Kenyon FC (1896) The brain of the bee. J Comp Neurol 6:133-210

Keil T (1982) Contacts of pore tubules and sensory dendrites in antennal chemosensilla of a silkmoth: demonstration of a possible pathway for olfactory molecules. Tissue and Cell 14:451-462

Keil T (1984) Reconstruction and morphometry of silkmoth olfactory hairs: a comparative study o0f sensilla trichoidea on the antennae of male Antheraea polyphemus and Antheraea pernyi (Insecta, Lepidoptera). Zoomorphology 104:147-156

Kien J, Menzel R (1977) Chromatic properties of interneurons in the optic lobes of the bee. I. Broad band neurons. J Comp Physiol A 113:17-34

Kirschfeld K (1967) Die Projektion der optischen Umwelt auf das Raster der Rhabdomere im Komplexauge von Musca. Exp Brain Res 3:248-270

Kirschfeld K (1987) Activation of visual pigment: Chromophore structure and function. In: Stieve H (ed) The molecular mechanism of photoreception. Dahlem Konferenzen Life Science Research Report 34. Springer, Berlin, pp. 29-49

Lancet D, Greer CA, Kauer JS, Shepherd GM (1982) Mapping of odor related activity in the olfactory bulb by high resolution 2-deoxyglucose autoradiography. Proc Natl Acad Sci USA 79:670-674

Land MF (1981) Optics and vision in invertebrates. In: Autrum H (ed) Handbook of sensory physiology VII 6 B. Invertebrate visual centers and behavior. Springer, New York, pp 471-594

Land MF, Collett TS (1974) Chasing behavior of houseflies (*Fannia canicularis*). A description and analysis. J Comp Physiol A 89:331-358

Land MF, Eckert H (1985) Maps of the acute zones of fly eyes. J Comp Physiol A 156:525-538

Laughlin SB (1981) Neural principles in the peripheral visual systems of invertebrates. In: Autrum H (ed) Handbook of sensory physiology VII 6 B. Invertebrate visual centers and behavior. Springer, New York, pp 130-280

LeVay S, Hubel DH, Wiesel TN (1975) The pattern of ocular dominance columns in macaque visual cortex revealed by a reduced silver stain. J Comp Neurol 159:559-576

Livingston MS, Hubel DH (1984) Anatomy and physiology of a color system in the primate visual cortex. J Neurosci 4:309-356

Macrides F, Schneider SP (1982) Laminar organization of mitral and tufted cells in the main olfactory bulb of the hamster. J Comp Neurol 208:419-430

McNaughton BL, Nadel L (1989) Hebb-Marr networks and the neurobiological representation of action in space. In: Gluck MA, Rumelhart DE (eds) Neuroscience and connectionist theory. Lawrence and Erlbaum, Hillsdale, in press

Matsumoto SG, Hildebrand JG (1981) Olfactory mechanisms in the moth *Manduca sexta* : response characteristics and morphology of central neurons in the antennal lobes. Proc Roy Soc Lond B 213:249-277

Milde JJ, Strausfeld NJ (1989) Cluster organization and response characteristics of the giant fiber pathway of the blowfly alliphora erythrocephala J Comp Neurol, submitted

Milde JJ, Seyan HS, Strausfeld NJ (1987) The neck motor system of the fly alliphora erythrocephala II. Sensory organization. J Comp Physiol A 160:225-238

Mobbs PG (1982) The brain of the honeybee, *Apis mellifera*. I. The connections and spatial organization of the mushroom bodies. Philos Trans R Soc Lond B 298:309-354

Mobbs PG (1985) Brain structure. In: Kerkut GA, Gilbert LI (eds) Comprehensive insect physiology, biochemistry and pharmacology, vol 5. Nervous systems, structure and motor function. Pergamon Oxford, pp 299-370

Mori K (1987) Membrane and synaptic properties of identified neurons in the olfactory bulb. Prog Neurobiol 29:275-320

Mori K, Kishi K, Ojima H (1983) Distribution of dendrites of mitral, displaced mitral, tufted and granule cells in the isolated turtle olfactory bulb. J Neurosci 2:497-502

Murlis J, Jones CD (1981) Fine-scale structure of odor plumes in relation to insect orientation to distant pheromone and other attractant sources. Physiol Entomol 6:71-86

Murphey RK (1981) The structure and development of a somatotopic map in crickets: the cercal afferent projection. Dev Biol 88:236-246

Murphey RK, Walthall WW, Jacobs GA (1984) Neurospecificity in the cricket cercal system. J Exp Biol 112:7-25

Oland LA, Tolbert LP (1989) Patterns of glial proliferation during formation of olfactory glomeruli in an insect. Glia 2:10-24

Pearson L (1971) The corpora pedunculata of *Sphinx ligustri* L. and other Lepidoptera: an anatomical study. Philos Trans R Soc B 259:477-516

Penfield W, Rasmussen T (1950) The cerebral cortex of man. Macmillan, new York,
Perret DI, Mistlin AJ, Chitty AJ (1987) Visual neurones responsive to faces. Trends Neurosci 9:358-364
Pollak GD, Weistrup JJ, Fuzessey Z (1986) Auditory processing in the mustache bat's inferior colliculus. Trends Neurosci 9:556-601
Pinching AJ, Powell TPS (1971a) The neuron types of the glomerular layer of the olfactory bulb. J cell Sci 9:347-377
Pinching AJ, Powell TPS (1971b) The neuropil of the periglomerular region of the olfactory bulb. J Cell Sci 9:379-409
Price JL, Powell TPS (1970) The synaptology of the granule cells in the olfactory bulb. J Cell Sci 7:125-155
Pugh E, Altman J (1988) A role for calcium in adaptation. Nature 334:16-17
Ribi WA (1975) The first optic ganglion of the bee. I. Correlation between visual cell types and their terminals in the lamina and medulla. Cell Tissue Res 165:103-111
Rall W (1977) Core conductor theory and the cable properties of neurons. In: Kandel E (ed) Handbook of physiology. The nervous system I. Amer Physiol Society, Bethesda, pp. 39-97
Reese TS, Shepherd GM (1972) Dendro-dendritic synapses in the central nervous system. In: Pappas GD, Pupura DP (eds) Structure and function of synapses. Raven, New York, pp 121-136
Reichert H, Rowell CHF (1986) Neuronal circuits controlling flight in the locust: how sensory information is processed for motor control. Trends Neurosci. 9:281-283
Retzius G (1892) Die Endigungsweise der Reichnerven. Biol Untersuch Neue Folge 3:25-28
Riehle A, Franceschini N (1984) Motion detection in flies: a parametric control over on-off pathways. Exp Brain Res 195:299-308
Rodriguez V (1988) Spatial coding of olfactory information in the antennal lobes of *Drosophila melanogaster*. Brain Res 543:299-307
Rospars P (1983) Invariance and sex specific variations of the glomerular organization in the antennal lobes of a moth *Mamestra brassicae*, and a butterfly *Pieris brassicae*. J Comp Neurol 220:80-96
Sanchez DS (1937) Sur le centre antenno-moteur ou antennaire postérieur de l'abeille. Trab Lab Invest Biol Univ Madrid 31:245-269
Sanes JR, Hildebrand JG (1976) Structure and development of antennae in a moth, *Manduca sexta*. Dev Biol 51:282-299
Schildberger K (1984) Multimodal interneurons in the cricket brain: properties of identified extrinsic mushroom body cells. J Comp Physiol A 154:71-79
Schneiderman AM, Matsumoto SG, Hildebrand JG (1982) Trans-sexually grafted antennae influence development of sexually dimorphic neurones in moth brain. Nature 298:844-846
Schneiderman AM, Hildebrand JG, Brennan MM, Tumlinson JH (1986) Trans-sexually grafted antennae alter pheromone-directed behavior in a moth. Nature 323:801-846
Schneider SP, Macrides F (1978) Laminar distribution of interneurons in the main olfactory bulb of the adult hamster. Brain Res Bull 3:73-82
Schürmann FW (1971) Synaptic contacts of association fibers in the brain of the bee. Brain Res 26:169-176
Schürmann FW (1967) The architecture of the mushroom bodies and related neuropils in the insect brain. In: Gupta AP(ed) Arthropod brain. Its evolution, development, structure, and functions.Wiley, New York, pp 231-264

Scott JW, Harrison TA (1987) The olfactory bulb: anatomy and physiology. In: Finger TE, Silver WL (eds) Neurobiology of taste and smell. Wiley, New York, pp 151-178

Shaw SR (1984) Early visual processing in insects. J Exp Biol 112:225-251

Shepherd GM (1988) Neurobiology, 2nd ed Oxford, New York

Sirianni PA, Tolbert LP (1988) Induction and stabilization of olfactory glomeruli in the developing insect brain. Soc Neurosci Abstr 14:423

Steffen H, van der Loos H (1980) Early lesions of mouse vibrissal follicles: their influence on dendrite orientation in the developing barrel field. Exp Brain Res 41:410-431

Steinbrecht RA (1973) Der Feinbau olfaktorischer Sensillen des Seidenspinners (Insecta, lepidoptera): Rezeptorfortsätze und reizleitender Apparat. Z Zellforsch mikrosk Anat 139:533-565

Steinbrecht RA, Gnatzy W (1984) Pheromone receptors of *Bombyx mori* and *Antheraea pernyi*. 1. Reconstruction of the cellular organization of the sensilla trichoidea. Cell Tissue Res 235:25-34

Stewart WB, Kauer JS, Shepherd GM (1979) Functional organization of the rat olfactory bulb, analyzed by the 2-deoxyglucose method. J Comp Neurol 185:715-734

Stocker RF, Singh RN (1983) Different types of antennal sensilla in *Drosophila* project into different glomeruli of the brain. Experientia 39:674

Stocker RF, Singh RN, Schorderet M, Siddiqi O (1983) Projection patterns of different types of antennal sensilla in the antennal glomeruli of *Drosophila melanogaster*. Cell Tissue Res 232:237-248

Strausfeld NJ (1970) Golgi studies on insects. II. The optic lobes of Diptera. Philos Trans R Soc Lond B 258:135-223

Strausfeld NJ (1971) The organization of the insect visual system (light microscopy). I. Projections and arrangements of neurons in the lamina ganglionaris of Diptera. Z Zellforsch 121:377-441

Strausfeld NJ (1976) Atlas of an insect brain. Springer, Heidelberg

Strausfeld NJ (1979) The representation of a receptor map within retinotopic neuropil of the fly. Verh Dtsch Zool Ges 73:167-179

Strausfeld NJ (1980) Male and female visual neurones in dipterous insects. Nature 283:381-383

Strausfeld NJ (1984) Functional neuroanatomy of the blowfly's visual system. In: Ali MA (ed) Photoreception and vision in invertebrates. Plenum, New York, pp 483-522

Strausfeld NJ (1987) Sex-specific neurons in the visual system of blowflies (*Calliphora erythrocephala*) represent concentrically organized retinotopic domains and are segregated out to end at specific regions of premotor descending neurons. Soc Neurosci Abstr 13:137

Strausfeld NJ (1988) Selective staining reveals complex microstructures within antennal lobe glomeruli of *Manduca sexta*. In: Elsner N, Barth FG (eds) Sense organs: interfaces between environment and behaviour. Thieme, Stuttgart, p 67

Strausfeld NJ (1989) Beneath the compound eye: Neuroanatomical analysis and physiological correlates in the study of insect vision. In: Hardie RC, Stavenga DG (eds) Facets of vision. Springer, Heidelberg, pp 317-359

Strausfeld NJ, Bacon JP (1983) Multimodal convergence in the central nervous system of insects. In: Horn E (ed) Multimodal convergence in sensory systems. Gustav Fischer, Stuttgart, pp 47-76

Strausfeld NJ, Bassemir UK (1985a) Lobula plate and ocellar interneurons converge onto a cluster of descending neurons leading to neck and motor neuropil in *Calliphora erythrocephala*. Cell Tissue Res 240:617-640

Strausfeld NJ, Bassemir UK (1985b) The organization of giant horizontal-motion-sensitive neurons and their synaptic relationships in the lateral deutocerebrum of *Calliphora* and *Musca*. Cell Tissue Res 242:531-550

Strausfeld NJ, Blest AD (1970) Golgi studies on insects.I. The optic lobes of Lepidoptera. Philos Trans R Soc Lond B 258:81-134

Strausfeld NJ, Campos-Ortega JA (1973) The L4 monopolar neurone: a substrate for lateral interaction in the visual system of the fly *Musca domestica* (L). Brain Res 59:97-117

Strausfeld NJ, Hausen K (1977) The resolution of neuronal assemblies after cobalt injection into neuropil. Proc R Soc Lond B 199:463-476

Strausfeld NJ, Milde JJ (1988) Descending neurons receiving common sensory inputs diverge from the insect brain to functionally distinct motor neuron pools in thoracic ganglia. Soc Neuroci Abstr 14:998

Strausfeld NJ, Nässel DR (1980) Neuroarchitectures serving compound eyes of crustacea and insects. In: Autrum H (ed) Handbook of sensory physiology VII/6B. Springer, Berlin, pp 1-32

Strausfeld NJ, Seyan HS (1987) Identification of complex neuronal arrangements in the visual system of *Calliphora erythrocephala* using triple fluorescence staining. Cell Tissue Res 247:5-10

Strausfeld NJ, Bassemir U, Singh RN, Bacon JP (1984) Organizational principles of outputs from dipteran brains. J Insect Physiol 30:73-93

Tolbert LP (1989) Afferent axons from the antenna influence the number and placement of intrinsic synapses in the antennal lobes of *Manduca sexta* Synapse 3:83-95

Tolbert LP, Hildebrand JG (1981) Organization and synaptic ultrastructure of glomeruli in the antennal lobes of the moth *Manduca sexta* a study using thin sections and freeze-fracture. Proc R Soc Lond B 213:279-301

Tumlinson JH, Brennan MM, Doolittle RE, Mitchell ER, Brabham A, Mazomenos BE, Baumhover AH, Jackson DM (1989) Identification of a pheromone blend attractive to *Manduca sexta*(L.) males in a wind tunnel. Arch Insect Biochem Physiol 10: in press

Valverde F (1965) Studies on the piriform lobe. Harvard University Press Cambridge

Wagner H (1986a) Fight-performance and visual control of the free-flying housefly (*Musca domestica*L.), II. Pursuit of targets. Phil Trans Roy Soc Lond B 312:553-579

Waldrop B, Hildebrand JG (1989) Physiology and pharmacology of acetylcholine responses of interneurons in the antennal lobes of the moth *Manduca sexta* J Comp Physiol A 164:433-441

Waldrop B, Christensen TA, Hildebrand JG (1987) GABA-mediated synaptic inhibition of projection neurons in the antennal lobes of the sphinx moth, *Manduca sexta* J Comp Physiol A 161:23-32

Walthall WW, Murphey RK (1986) Positional information, compartments and the cercal system of crickets. Dev Biol 113:182-200

Willis MA, Baker TC (1984) Effects of intermittent and continuous pheromone stimulation on the flight behaviour of the oriental fruit moth *Grapholita molesta* Physiol Entomol 9:341-358

Woolsey TA, van der Loos H (1970) The structural organization of layer IV in the somatosensory region (S1) of mouse cerebral cortex. The description of a cortical field composed of discrete cytoarchitectonic units. Brain Res 17:205-242

Wong-Riley MTT (1979) Changes in the visual system of monocularly sutured or enucleated cats demonstrable with cytochrome oxidase histochemistry. Brain Res 171:11-28

Zawarzin AA (1925) Der Parallelismus der Strukturen als Grundprinzip der Morphologie. Z wiss Zool 124:118-212

IDENTIFIED GLOMERULI IN THE ANTENNAL LOBES OF INSECTS: IN VARIANCE, SEXUAL VARIATION AND POSTEMBRYONIC DEVELOPMENT

Jean-Pierre Rospars* and Irène Chambille**

*Laboratoire de Biométrie, INRA, route de Saint Cyr
78000 Versailles, France
** Laboratoire de Physiologie Sensorielle, INRA
78350 Jouy-en-Josas, France

ABSTRACT

Sensory information, mainly olfactory, collected by each antenna is coded by numerous neuroreceptors. These neuroreceptors project to the ipsilateral antennal lobe (AL), a well-defined area of the insect brain, where they synapse with cerebral neurones within discrete, spheroidal knots of synaptic complexes, the glomeruli. We give evidence, based mainly on the cockroach *Blaberus craniifer* and the moth *Mamestra brassicae*, that the glomeruli are morphologically, morphometrically and ontogenetically identifiable units that are present in constant number and are arranged orderly in both species. However, the size of identified glomeruli varies significantly depending on species, sex and developmental stage. These size variations are related to variations in number of specific subsets of antennal neuroreceptors and give some insight into the functional role of the variant glomeruli. The most conspicuous sex-dimorphic glomeruli are the male macroglomerular complexes. They are present in species whose males detect the sex pheromone emitted by females and absent in species, such as the butterfly *Pieris brassicae*, which rely on visual cues for mating. During the postembryonic development of *Blaberus*, glomeruli increase exponentially in size at greatly varying growth rates. The glomeruli with the highest and lowest growth rates form 3 distinct subsets, which suggests that neuroreceptors of the same growth type (possibly of the same modality) project to specific areas. The significance of these results on the connections between antenna and antennal lobe, and the function of glomeruli is discussed.

INTRODUCTION

The deutocerebrum (Fig. 1) is the site of first-order synaptic processing of chemical, thermal, mechanical and proprioceptive information from the antennae (reviewed in Rospars, 1988a). The most prominent structures of the deutocerebrum are the paired antennal lobes (ALs). Synaptic interactions within the ALs are confined to spheroidal neuropilar structures partly enclosed in glial processes (e.g. Tolbert et al., 1983, Oland and Tolbert, 1986) called "glomeruli." The glomeruli contain the terminals of antennal sensory axons and the dendritic arborisations of the AL neurons. Terminals of non-antennal sensory axons are also found within them (e.g. Singh and Nayak, 1985) as are dendrites and terminals of central neurons (e.g. Ernst and Boeckh, 1983; Homberg, 1984).

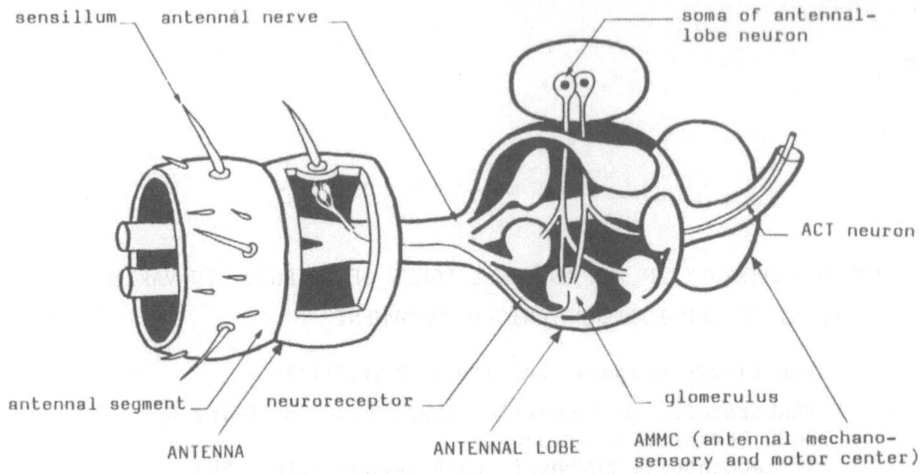

Fig. 1. Insect antennodeutocerebral system. Neuroreceptors from the 2 basal antennal segments (scape and pedicel, not shown) project to AMMC, whereas those from the other antennal segments (flagellum, 2 flagellar segments are shown) project to the ipsilateral antennal lobe (AL). The glomeruli contain synapses between flagellar and other sensory axons, AL local neurons and AL output neurons. Output neurons project via antennocerebral tracts (ACT) to higher-order centres in the protocerebrum such as mushroom bodies and lateral protocerebral lobe. Cerebral input neurons (not shown) are also present. (From Rospars, 1988a).

This review mainly emphasizes the glomerular organization in the cockroach *Blaberus craniifer*. The neonate nymph (L1) of this species is a small active animal, 8.5 mm long, which lives in the same environment as the adult. It becomes a 50-mm long adult within about 300 days, at the end of a continuous development with 9 moults for males and 10 for females. During the postembryonic development of the antennodeutocerebral system (Fig. 2) an increasing number of neuroreceptors (Urvoy, 1963, Lambin, 1973, Chambille, 1988) interact synaptically with a constant number of AL neurons within a constant number of glomeruli (Chambille and Rospars, 1985, Rospars and Chambille, 1986). Thus, the growth in size of the system results from an increase in number of antennal components and an increase in size of a constant number of AL components. Similar findings have been reported in the cockroach *Periplaneta americana* (Schaller, 1978; Prillinger, 1981).

These results introduce the 2 main problems we discuss:

(i) The constant number of glomeruli suggests that they might be individually identifiable units and neuropilar equivalents of identifiable neurons (e.g. Cohen and Jacklet, 1965; Rowell, 1976). This hypothesis (Chambille et al., 1978, Rospars et al., 1979), is examined in Section 1 using neuroanatomical and morphometric techniques. It gives support to the idea that glomeruli function as discrete, independent information-processing units.

(ii) The glomerular size clearly depends on the number of ingrowing neuroreceptors. Because variations in this number have been described depending on species, sex and stage of development, corresponding variations in size of some glomeruli may be expected. Section 2 illustrates how comparison of the volumes of identified glomeruli in these various categories gives clues to their functions and may prompt further physiological studies of information processing in ALs.

1. IDENTIFICATION OF GLOMERULI

We call the various copies of the same glomerulus identified in a series of ALs "homologs." In practice, ALs being compared two by two, we call any couple of identified glomeruli a "pair" and the identification process "matching."

Identification of a glomerulus requires that the following 2 conditions are met: (i) It must be a stable unit, i.e. present in all ipsilateral ALs of a given species, sex and stage of development. (ii) It must be recognized individually, i.e. similar to its homologs and distinct from all non-homologs according to one or more criteria.

The first condition is restrictive; the questions of knowing whether it applies also between right and left lobes, between sexes and between stages, and whether it applies to all glomeruli and to all species is left to experimental study. Conversely, the identification criteria mentioned in the second condition are not specified. In this work we have searched for morphological and morphometrical criteria, but other (biochemical, physiological, etc.) criteria should also be considered.

1.1. Evidence in Adult Male of Blaberus

Our results provide 5 morphological and morphometrical arguments as evidence of the identifiability of glomeruli in *Blaberus*:

Numerical constancy. The number of glomeruli per AL is nearly constant, close to 105-110 regardless of the brain side, the sex and the stage of development. One cannot achieve a better precision without identification because the boundaries of a few (2-5) glomeruli are not clear, e.g. when their surface of contact coincides with the sectioning plane. After complete identification we have found that the number of glomeruli is 107 in adult ALs (Rospars and Chambille, 1981) and 104-105 in L5 nymphs. It is 105-106 in L1 nymphs (without identification, Chambille and Rospars, 1985).

Morphological identification. Some glomeruli are directly identifiable on serial sections from their own morphological characteristics: shape, orientation, proximity to easily recognizable structures, e.g. tracts of fiber, and pattern of adjacent glomeruli (Fig. 3). Thus, one third of the glomeruli (37) was found to be repeatedly identifiable in frontally sectioned adult ALs. They are mainly located in the dorsal part of ALs, which is close to the protocerebrum and crossed by the main fiber tracts (Chambille and Rospars, 1981). But only 23 of these were followed through from L1 up adulthood. The 14 others could be identified gradually during development, indicating a progressive emergence of the adult morphology.

Visual identification from relative position. The most suitable identification criterion for the other featureless glomeruli is their relative position. Unfortunately, this criterion is not easy to use because the orientation of serial sections is not well reproducible, thus each AL is seen from a slightly different angle. This is one of the reasons why glomeruli first appear to be stacked at random. Similar glomerular patterns are still recognizable when unmistakable landmarks and features are present but are no longer so in their absence. We allowed for these orientational defects by computation (Fig. 4 left). Glomerular coordinates measured on sections were expressed in the same normalized coordinate system for all brains. Computer-aided reconstructions of ALs based on these coordinates can be obtained where glomeruli are schematized as circles (2D reconstructions, Figs. 4, 5 and 9), as spheres (3D reconstructions) or even with their original shapes (Rospars, 1988b).

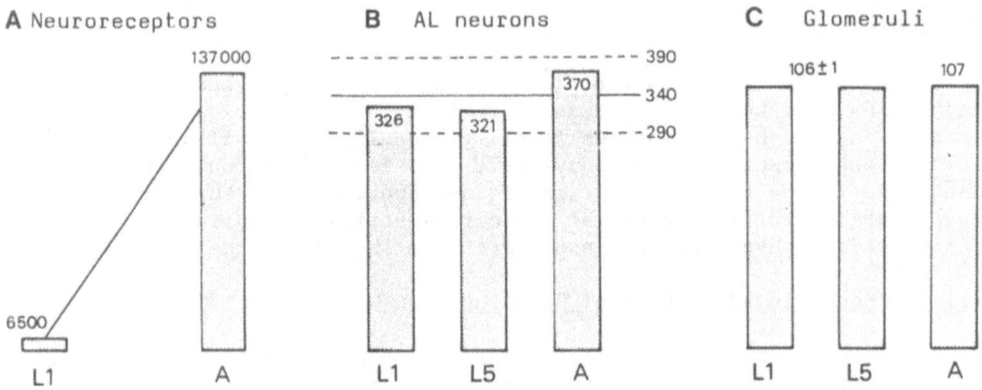

Fig. 2. Number of components and postembryonic development of antennodeutocerebral system. (A) The number of segments, sensilla and neuroreceptors of antenna increases greatly, e.g. neuroreceptors display a 20-fold increase in number from neonate nymph to adult. (B) Adult AL neurons gathered in 2 groups are present from hatching and increase in size afterwards. The number of nuclei remains within the limits of counting errors and give no evidence of a variation in neuron number. (C) Glomeruli are also present as soon as hatching. Their number is almost constant within the limits of delineation errors. Their growth in size is considerable: total glomerular volume undergoes a 22-fold increase from L1 to adulthood.

On these reconstructions step by step visual identification of all glomeruli can be achieved from their relative position (Fig. 4 right). The glomeruli with imprecise boundaries are also solved in this process. However, it can be objected that such matchings could be also achieved between sets including equal numbers of randomly stacked spheres and that even if the arrangement is orderly, there is no evidence that the matchings are right. These objections can be ruled out by a simple test. When a series of at least 3 ALs are matched to each other two by two (i.e. 1:2, 2:3 and 3:1) any matching, e.g. 3:1, can also be derived logically from the other two (e.g. 1:2 and 2:3 imply 1:3). If the stacking is random or the pairs mistaken, most of the visual pairs will be different from the logical ones. These inconsistent pairs were found to be quite rare and were easily corrected. Finally, homologous glomeruli were given a definitive identification number and the glomerular complement in adult was found to be 107.

Morphometrical identification. The variability in AL organization can be quantified from the normalized coordinates used in the computer reconstructions (Fig. 6) or from the interhomolog distances (Rank analysis, Fig. 7). In adult *Blaberus* 85% of the visual intrasexual pairs are unambiguous according to the latter stringent criterion (Rospars and Chambille, 1981; Table 1).

These results confirmed that visual matchings are valid and that the glomerular organization is built according to an invariant plan. This positional invariance has a practical significance because it opens the way to stereotactical applications and to identification of glomeruli from position alone. We have developed a first algorithm, derived from Rank analysis, which identifies 75% of adult glomeruli with about 5% erroneous identifications (Rospars and Chambille, 1981). A second algorithm finds among the 107 factorial (107! = 10^{172} possible matchings

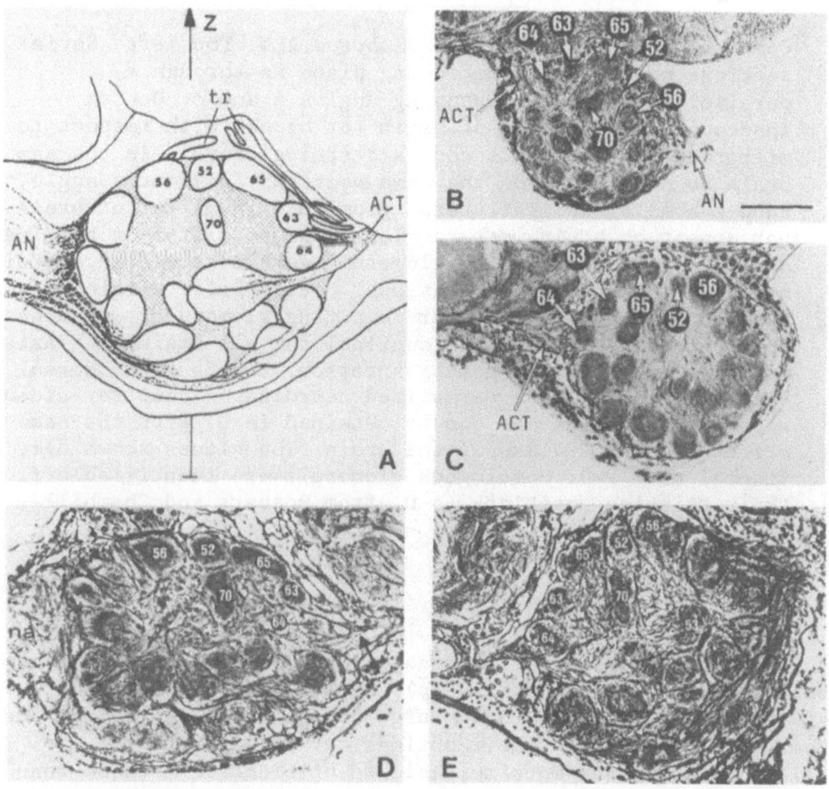

Fig. 3. Morphologically identified glomeruli in *Blaberus*. Nymphal and adult brains of both sexes have been serially sectioned in a frontal plane, impregnated with silver proteinate (Bodian), and photographed.Each AL is cut in about 50 sections. Each glomerulus is followed on an average 7 successive sections. Five morphologically identified glomeruli in L1 (B), L5 (C) and adults (D, E) are shown. AN, antennal nerve; ACT, tract of output fibers projecting to protocerebrum; tr, trachea. Scale bar: 100 lm. (After Chambille and Rospars, 1981, 1985).

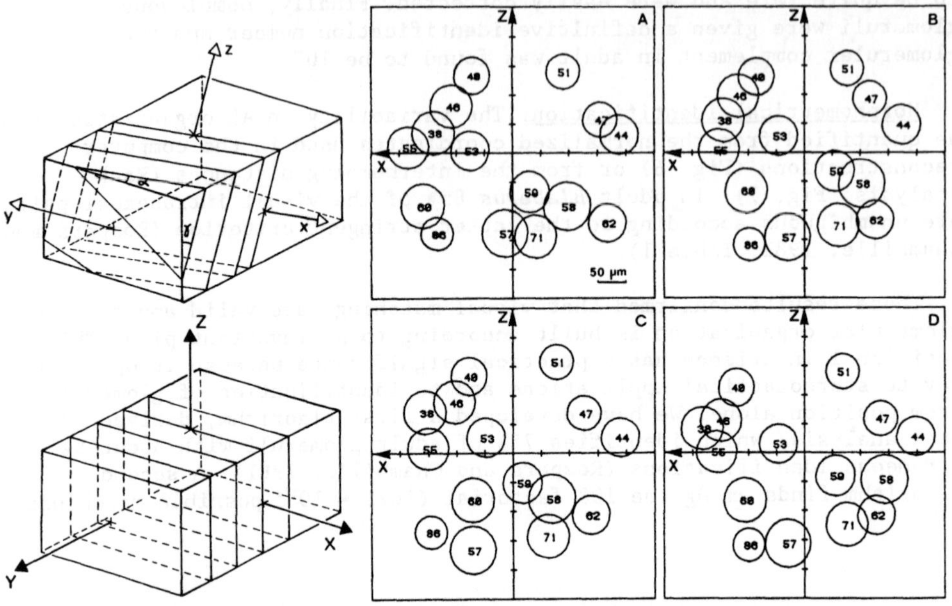

Fig. 4. Computer reconstructions of *Blaberus* ALs. *Top left*: Serial sections parallel to sectioning plane xz through a parallelepipedal block showing angles a and c. Due to inadequate orientation of brain (or block) with respect to microtome, 2 homologous contralateral glomeruli in the same brain do not appear on the same sections (asymmetry angle, a hand 2 homologous ipsilateral glomeruli in different brains do not appear on equivalent sections (due to different tilting angles, c. Coordinates of glomeruli were measured in coordinate system xyz bound to the sections. *Bottom Left:*: Angles a and c were determined and used for computing glomerular coordinates in a reference system XYZ identical for all brains. X-axis laterolateral, Y-axis posteroanterior, Z-axis ventrodorsal. Right: Based on these normalized coordinates computer-aided reconstructions of ALs can be obtained (A-D) with the same orientation regardless of the brain. The slices shown here are frontal with Y=0. Homologous glomeruli are identifiable from their relative positions (A-D, from Rospars and Chambille, 1981).

between 2 sets of 107 glomeruli the solution which minimizes the sum of interhomolog distances (Rospars,1988b). This algorithm is presently under study; it appears able to converge to the visual matchings and consequently to demonstrate their optimality. It suggests the possibility of faster and more reliable identification methods than are currently available. These are needed for applying routine anatomical and physiological techniques to identified glomeruli in such complex ALs.

However, rank analyses and automated identification show that the numbers of variant glomeruli in intersexual comparisons are higher than in intrasexual (male:male) ones (Rospars and Chambille, 1981; Table 1). This is the first indication for sexual dimorphism (see Section 2.2).

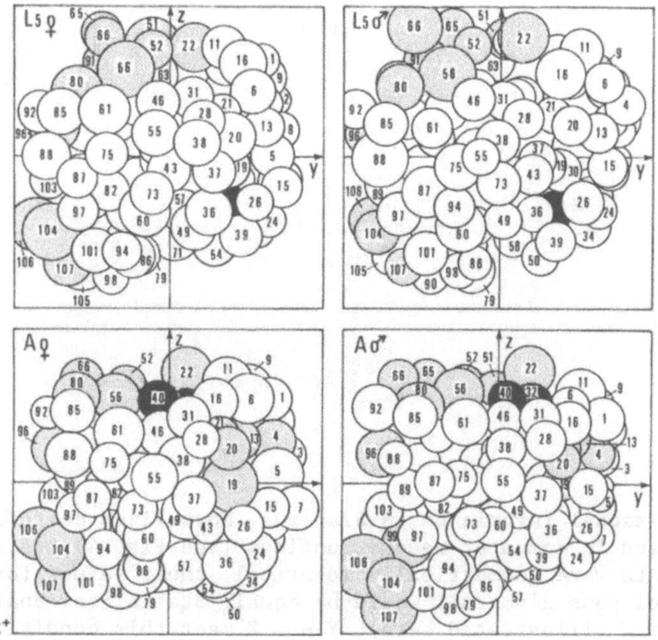

Fig. 5. Computer reconstructions of nymphal and adult *Blaberus* right ALs viewed from external side (Y-axis posteroanterior, Z-axis ventrodorsal). Unmatched nymphal glomeruli not found in adult (No. 108, in black) and adult glomeruli not found in L5 (Nos. 40, 32, in black, and 62 not shown) probably result from fusion and separation of glomeruli in the course of development. All other glomeruli (104) were matched. Glomeruli in grey are morphologically identifiable (23 in L5, 37 in adults). Size of square frame is 500 lm. (Modified from Chambille and Rospars, 1985).

The ALs of individuals at mid-development (L5) were also reconstructed and compared with the adult ALs (Chambille and Rospars, 1985). Some anomalous glomeruli were found and 104 glomeruli were matched (Fig. 5). The interstage positional variability is greater than the intrastage one (Table 1). It results from specific developmental displacements which maintain the invariance at each stage. This means that a non-random process is at work which moves the same or at least neighbouring glomeruli, in all developing ALs in the same way. It is very likely that the heterogeneous growth of glomeruli is the driving force behind these regular distortions (see Section 2.3).

Size constancy. The final argument (Fig. 8) is that the sizes of homologous glomeruli are statistically equal. This quantitative criterion gives a confirmation of matchings which is independent of position. However, in intersexual and interstage comparisons the correlations are less good. These apparent exceptions to invariance are discussed in Section 2.

In conclusion, glomeruli of *Blaberus* are identifiable units, constant in number, with stable position and size in ALs of the same sex and developmental stage. They can be uniquely identified by qualitative (morphological and from relative position) and quantitative (morphometrical) criteria. Variations in position and especially in size

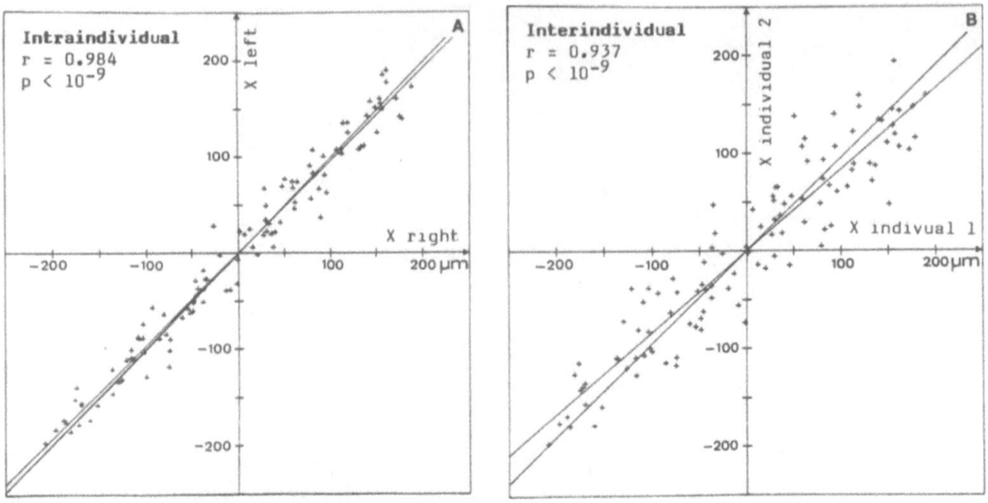

Fig. 6. Positional invariance in *Blaberus*. Normalized coordinates X, Y, Z used in computer reconstructions (see Fig. 4) are such that if ALs were geometrically identical, their values for homologous glomeruli would be equal. Statistical analysis shows that X (illustrated here), Y and Z meet this condition and fall close to the line of 45°-slope as expected. Deviation from this line in interindividual comparisons shows that the 95% confidence interval of the variation of a glomerulus around its mean position is about 27 lm. (Modified from Chambille et al., 1980).

occur between sexes and stages. As far as position is concerned differences between sexes are inconspicuous in this species and revealed only by quantitative analysis. Differences between stages are more conspicuous and shown by both qualitative and quantitative criteria.

1.2. Identified Glomeruli in Other Species

Recent studies made on various insect species indicate that glomerular identifiability is not exceptional in insects. Most species have between 50 and 200 glomeruli, some have none and others more than 1000 (Rospars, 1988a).

A more or less complete identification of glomeruli has been done in some species. In the cockroach *Periplaneta* ALs contain about 125 glomeruli in adults (Ernst et al., 1977), and 111-137 in L1 and L3 (Prillinger, 1981). Three glomeruli have been identified from their positions, shapes and sizes throughout development. In *Drosophila melanogaster* all 22 glomeruli have been identified (Stocker et al., 1983; Singh and Nayak, 1985; Rodrigues, 1988). In *Apis mellifica* 165-174 glomeruli have been counted in workers, 103 in males and 150-158 in queens (Arnold et al., 1984, 1988, see also Pareto, 1972), indicative of a great sexual dimorphism. Part of the glomeruli have been visually identified (57 in workers, 46 in males).

In the sphinx moth *Manduca sexta* ALs contain 57-65 glomeruli (macroglomerular complex not included; Schneiderman et al., 1983). Preliminary results point to intraindividual geometrical invariance (Rospars, unpublished results). The same glomerulus has been identified in 4 *Manduca* species (including *M. sexta*), *Antheraea*, Bombyx (Kent et al., 1986), *Rhodogastria* (Bogner et al., 1986) and *Pieris*

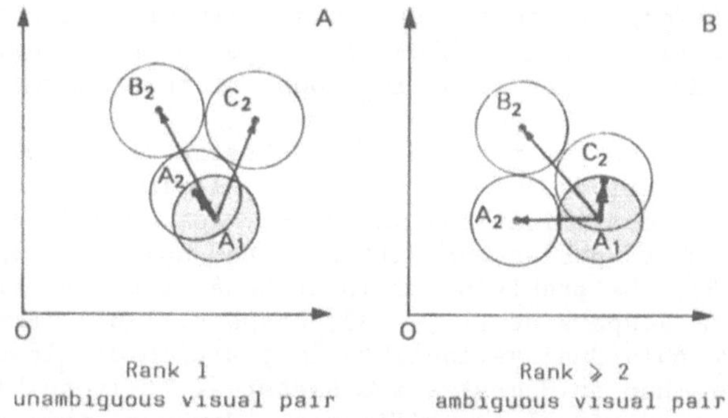

Fig. 7. Rank analysis. Glomerulus A_1 in lobe 1 can be regarded as invariant (A) when A_1 is nearer (rank 1) to its visual homolog A_2 than to any other glomerulus C_2 in lobe 2, and as variant (B) in the alternative cases (rank >= 2), i.e. $A_1A_2 > A_1C_2$. This criterion specifies quantitatively the condition of individual recognition used in the definition of identifiable glomeruli (Section 1). Small fluctuations in location that do not alter the glomerular pattern can be taken into account when superimposition of lobes 1 and 2 is not based on all visual pairs but only on those which are located inside a small sphere centered on A_1 (local rank analysis). (After Rospars and Chambille, 1985).

Table 1. Tests by local rank analysis of the invariance model[a]

Comparison	Blaberus	Mamestra	Pieris
Intraindividual	86 ± 9 (8) 107,100 lm	94 ± 4 (8) 66-68,65 lm	59 ± 6 (4) 59-66,55 lm
Interindividual & intrasexual	85 ± 5 (8) 107,110 lm	89 ± 2 (4) 66-67,75 lm	-
Intersexual	75 ± 7 (6) 107,110 lm	79 ± 4 (4) 63-65,75 lm	-
Interstage & intersexual	67 ± 1 (4) 104,130 lm	-	-

[a]See definition of Rank analysis in Fig. 7. First line of each cell: percentage of invariant glomeruli (rank 1) +^H_ standard deviation and number of comparisons (in parentheses). Second line: total number of glomeruli and radius of the local sphere (in italics). (From Rospars and Chambille, 1981, Chambille and Rospars, 1985, and Rospars, 1983).

staining of neuroreceptors coming from a sensory organ located on the labial palps (labial pit organ (LPO). It suggests that glomeruli can be interspecific invariants, at least in groups of related species.

Our investigation on the noctuid moth *Mamestra brassicae* (all 66-68 glomeruli were identified) and the butterfly *Pieris brassicae* (59-65 glomeruli, without identification) using the same methods as for *Blaberus*, showed that species can differ in the number of anomalous glomeruli (Fig. 9), the precision of invariance, and the extent of the sexual dimorphism (Rospars et al., 1983; Rospars, 1983). For example, in *Pieris*, the intraindividual variability in position of glomeruli (Table 1) is much higher than in *Mamestra* and *Blaberus*. It is so high that it prevents the interindividual identification. However, this does not prove that ALs in *Pieris* are randomly built, but simply indicates that identification is not feasible from position alone.

2. VOLUME AND PRESUMPTIVE FUNCTION OF GLOMERULI

Some identified glomeruli display significant variations in size between sexes and between developmental stages. These changes can be correlated with antennal or behavioural variations, and ultimately with variations in the number of neuroreceptors. These observations give a basis for screening those glomeruli which are the most suitable for physiological investigations.

2.1 Volume Comparison in ALs and Glomeruli

From a sensory point of view, notwithstanding phylogenetic relationships, the crepuscular and nocturnal moth *Mamestra* is closer to the cockroach *Blaberus*,which lives in caves and avoids light, than to the diurnal butterfly *Pieris*. The first two species mainly rely on olfactory cues to feed and mate while the dominant modality of the last one is vision. These sensory features are reflected in the brain structure and the relative size of ALs. The well-developed ALs and reduced optic lobes of *Mamestra* and *Blaberus* contrast with the small ALs and bulky optic lobes of *Pieris* (Fig. 10). The greater positional variability of glomeruli in *Pieris* may also be related to the lesser role of olfaction in this species.

These variations in gross size are also reflected in the relative size of glomeruli. The glomerular size has been measured as the radius (or diameter) of a sphere having the same volume as the glomerulus. The mean glomerular diameter varies with the species, but interspecific invariant features are present (Fig. 11) that probably bear witness to still unknown constant properties of the neuronal organization of glomeruli.

2.2. Sexual Dimorphism in Adults

Intersexual differences in ALs partly stem from sex-specific olfactory mechanisms (reviewed in Schneiderman and Hildebrand, 1985, Christensen and Hildebrand, 1987b and Rospars, 1985, 1988a). The most characteristic sex-specific behaviours are those bringing the sexes together for mating. One sex emits a calling signal which is received by the opposite sex, so that differences in the sensory capabilities of sexes are present. All types of propagated signals (sounds, colors, odors) are used in these long-range communications, but each species uses only one signal (e.g. Otte, 1977 for orthopteroid insects). In insects relying on olfactory signaling, attractant sex pheromones are generally released by females and captured by male-specific antennal

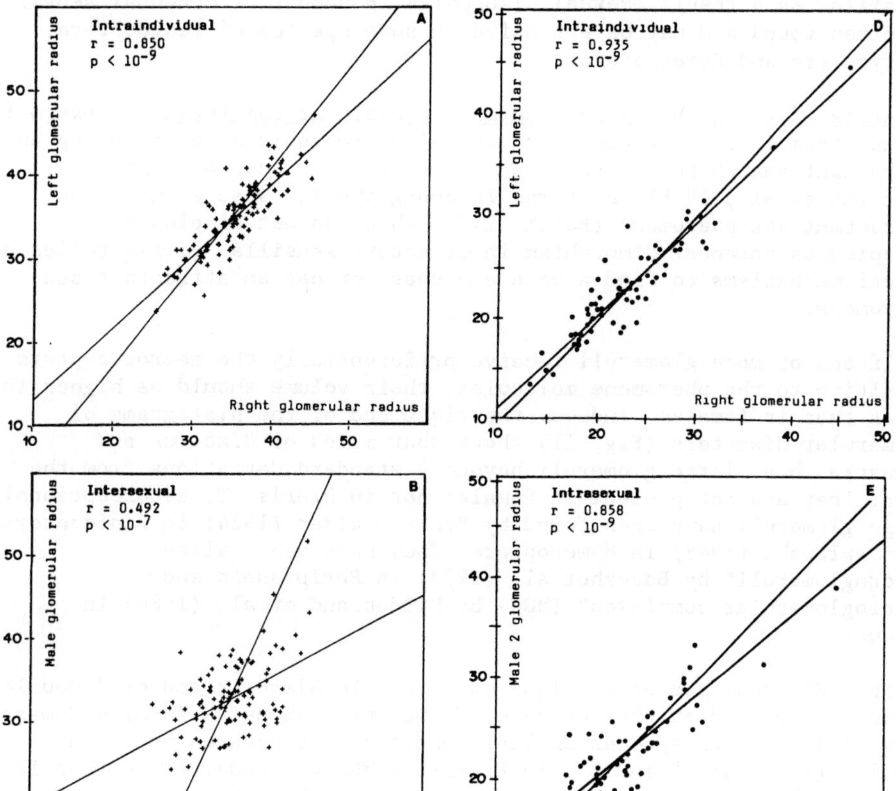

Fig. 8. Invariance and variability of glomerular radii in *Blaberus* (left column) and *Mamestra* (right column). Intraindividual (A, D) and intrasexual (E) variabilities are low and give evidence of size invariance. Intersexual (B, F) and interstage (C) variabilities are higher, showing that size invariance does not fit as well for ALs of different sexes or developmental stages. In F the coefficients of correlation and the regression lines do not take the MGC into account (squares). (After Chambille et al., 1980, Chambille and Rospars, 1985, Rospars, 1983, and unpublished data).

sensilla. As a result, sexual dimorphism in number of antennal sensilla is often found and may be extensive in some species of Lepidoptera, Dictyoptera and Hymenoptera.

Attractant sex pheromones and macroglomerular complexes. *Mamestra* is a good example of a species with sexually dimorphic antennae using an attractant sex pheromone (a blend of several hydrocarbon compounds, Descoins et al., 1978). *Blaberus* is among the few species using an attractant sex pheromone (Barth, 1970) which do not display a conspicuous antennal dimorphism in olfactory sensilla. *Pieris* relies on visual mechanisms to find a mate and does not use an attractant sex pheromone.

If one or more glomeruli receive preferentially the neuroreceptors sensitive to the pheromone molecules, their volume should be higher in males than in females. Indeed, the right end of the histograms of glomerular diameters (Fig. 11) shows that males of *Blaberus* and *Mamestra*, have large glomeruli beyond 3 standard deviations from the mean. They are not present in females nor in *Pieris*. These exceptionally large glomeruli have been found by Bretschneider (1924) in Lepidoptera and Jawlowski (1948) in Hymenoptera. They have been called "macroglomeruli" by Boeckhet al. (1977) in *Periplaneta* and "macroglomerular complexes" (MGC) by Hildebrand et al. (1980) in *Manduca*.

This MGC consists of a unique mass in male *Blaberus* and of 2 coupled glomeruli located at the entrance of the antennal nerve in male *Mamestra* (Fig. 12). In both species smaller homologous structures occur in females (see Figs. 4 and 8). In *Pieris* no MGC was observed, either in the area where the antennal nerve enters the lobe, or in any other location.

Three conclusions can be drawn from these data: (i) The presence of an MGC in species with sex pheromone and its absence in species like *Pieris* which produce no sex pheromone are in agreement with the hypothesis that the MGC plays a role in the processing of sex pheromone signals. Direct anatomical or physiological evidence has been given in *Manduca* (reviewed in Schneiderman and Hildebrand, 1985, and Christensen and Hildebrand, 1987b), Bombyx (Koontz and Schneider, 1987) and *Periplaneta* (reviewed in Boeck et al., 1984). (ii) As shown in *Blaberus* ALs are a more appropriate site than antennae to display sexual dimorphisms of small magnitude. (iii) A female homolog of the male MGC is presently known only in *Mamestra* and *Blaberus*. Its function is electrically responsive to stimulation by the sex pheromone (Priesner, 1979). Thus, the female homolog of the male MGC might also detect the pheromone. This would explain that pheromone emitting females tend to separate from each other. It might play a role in aggregative behaviour in female *Blaberus* (see Section 2.3).

Non-macroglomerular sexual dimorphisms. Besides the MGC other sexual dimorphisms are present.(i) Two glomeruli that are homologous in both sexes but larger in females have been described in Bombyx (large lateral glomeruli LLG, Koontz and Schneider, 1987). No such glomeruli have been clearly identified in *Mamestra* or *Blaberus*, although slight differences in size for a large part of the glomerular complement were shown statistically (Fig. 8). Some glomeruli without homologs in the opposite sex seem also present in *Mamestra* (Fig. 9).(ii) Glomeruli that differ in position have been found in the dorsoposteroexternal complex of *Mamestra* (Fig. 9). Statistical dimorphism in position, i.e. not allocated to any definite glomeruli, is also present, in *Mamestra* and *Blaberus*, because the intersexual variability in position is greater than the intrasexual one. It probably results from the above all-or-none and size dimorphisms, because a bulky glomerulus for example tends to push its neighbours.

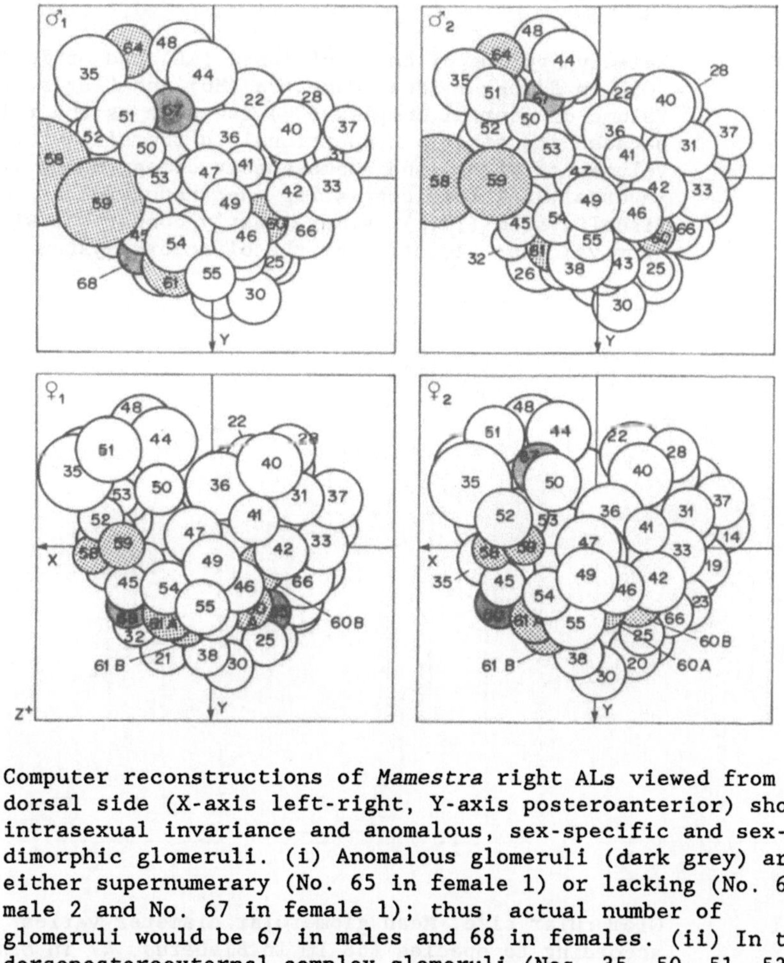

Fig. 9. Computer reconstructions of *Mamestra* right ALs viewed from dorsal side (X-axis left-right, Y-axis posteroanterior) showing intrasexual invariance and anomalous, sex-specific and sex-dimorphic glomeruli. (i) Anomalous glomeruli (dark grey) are either supernumerary (No. 65 in female 1) or lacking (No. 68 in male 2 and No. 67 in female 1); thus, actual number of glomeruli would be 67 in males and 68 in females. (ii) In the dorsoposteroexternal complex glomeruli (Nos. 35, 50, 51, 53) are present in both sexes but differences in position are such that intersexual homology is only a plausible hypothesis. (iii) Glomeruli without homologs are also present (light grey) in males (Nos. 63, not shown, and 64) and in females (Nos. 60B, 61B and 62B). (iv) Macroglomerular complex (light grey; Nos. 58 and 59). Size of square frame is 300 lm. (From Rospars, 1983).

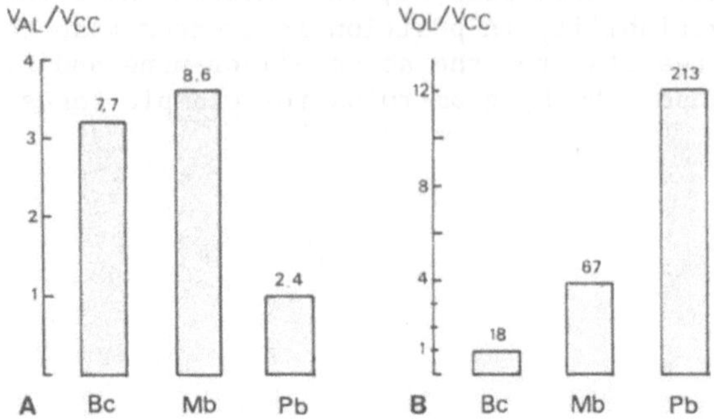

Fig. 10. Relative volume of antennal lobes (AL) and optic lobes (OL) in *Blaberus* (Bc), *Mamestra* (Mb) and *Pieris* (Pb). Volume of central complex (CC) is taken as the reference because its size is proportional to the size of the head. Volumes of visual and chemosensory compartments are inversely related, suggesting that reduction of the visual system (12:1) is more easily compatible with survival than reduction of the olfactory system (3.5:1).

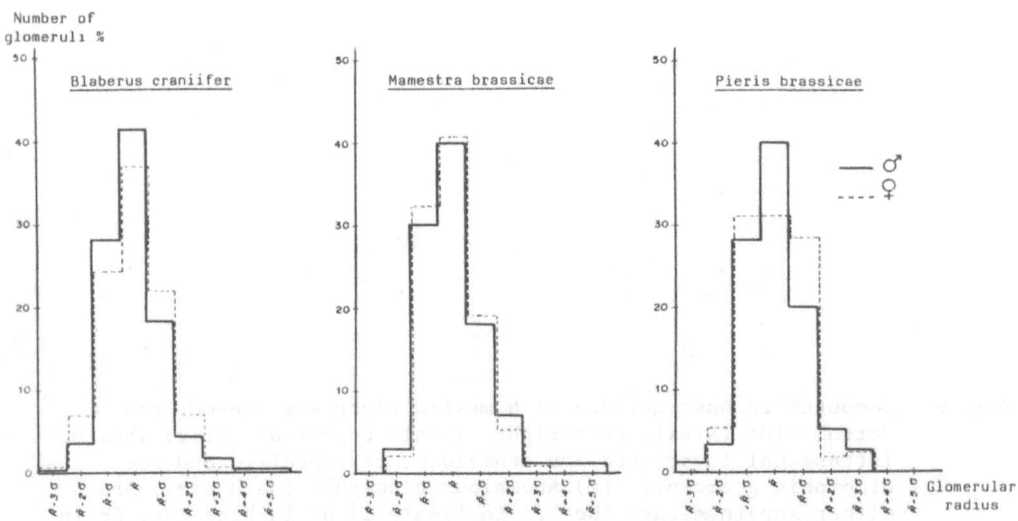

Fig. 11. Glomerular size. Mean glomerular diameter varies according to species (68 lm in *Blaberus*, 47 in *Mamestra* and 30 in *Pieris*). Relative to size of central complex (see Fig. 10) it is 30%, 40% and 25%, respectively. Within each of the 3 species the frequency distribution of glomerular radii is bell-shaped and slightly asymmetrical. Mean diameter is the most frequent, about 40% of the glomeruli are within half a standard deviation from the mean and nearly all glomeruli are within 3 standard deviations from the mean (exceptions are MGCs). (Adapted from Chambille et al., 1980 and Rospars, 1983).

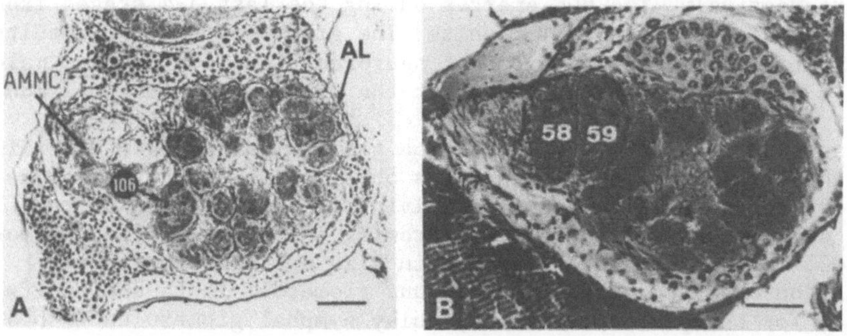

Fig. 12. Macroglomerular complexes (MGC). (A) In male *Blaberus*, the MGC is a unique mass (No. 106) located posteriorly in the AL. It is the most bulky glomerulus in males (diam. 110 lm); it possesses a female homolog (diam. 86 lm, 1/2 the male volume). (B) In male *Mamestra*, it consists of 2 coupled glomeruli (Nos. 58 and 59) located at the entrance of the antennal nerve. Females have 2 glomeruli in homologous positions, with a volume 12 times smaller (see Fig. 9). AL, antennal lobe, AMMC, antennal mechanosensory and motor centre. Scale bars: 100 lm. (After Chambille and Rospars, 1985, and Rospars, 1983).

These non-macroglomerular dimorphisms might be related to the known female-specific antennal sensilla and olfactory-controlled behaviours such as the detection by females of the "aphrodisiac" sex-pheromone emitted by males during courtship when partners are in close proximity or the selection of a host plant by ovipositing females.

2.3. Postembryonic Growth of Glomeruli

Unlike holometabolous insects, such as *Mamestra*, where ALs are reorganized during metamorphosis, in hemimetabolous insects such as *Blaberus* growth is the main change affecting ALs. However, the various types of antennal sensilla do not multiply at the same rate during postembryonic development in *Blaberus* (Rospars and Chambille, 1986; Chambille, 1988) and in the closely related species, *Leucophaea maderae* (Schafer, 1973). From hatching to adulthood the proprioceptive sensilla multiply less than 5-fold, the gustatory ones 11-fold and the olfactory ones more than 20-fold. These differences are likely to be reflected in the corresponding numbers of antennal neuroreceptors. We have shown that they are also reflected in the rate of glomerular growth in volume.

Glomeruli grow exponentially during the nymphal phase (Fig. 13). Each glomerulus is characterized by its specific growth rate. The dimorphism between the male MGC and its female homolog appears only in the preimaginal period between the last nymphal moult and the imaginal moult, due to an acceleration of the growth rate in the male, but not in the female (Rospars and Chambille, 1986). Remarkably, the ontogeny of the pheromone-dependent aggregative behaviour parallels the growth of this glomerulus, since in the male but not in the female, a strong increase in aggregation occurs during the preimaginal period (Brossut, 1970). Thus, in this species, the MGC might play a role in both interindividual and intersexual recognition mechanisms. In *Periplaneta* (Prillinger, 1981; Schaller-Selzer, 1984) the sexual dimorphism appears more progressively. The MGC emerges during the last 3-4 stages from 2-4 previously distinct glomerular units. The growth of the MGC results from that of AL neurons because axons of neuroreceptors coming from male-specific sensilla arrive in the AL at the time of the imaginal moult.

The growth rate of every identified glomerulus has been determined (Fig. 14A). The range of growth rates observed is wide; the volume of the slowest growing glomerulus is multiplied by less than 6 from L1 to adult, whereas that of the fastest growing one is multiplied by more than 100. This heterogeneous growth in size explains the poor correlation observed between homologous glomeruli in L5 and adults because, for example, a relatively bulky nymphal glomerulus with a low growth rate may give rise to a relatively small adult one.

The slowest growing glomeruli (S-glomeruli with rate below 8.3%, i.e. volume multiplied by less than 10 between L1 and adult) and fastest glomeruli (F-glomeruli with rate over 15.0%, i.e. volume multiplied by more than 50) are not located at random in the ALs. The S-glomeruli (except one) are gathered in a unique group comprising 6 glomeruli, the F-ones form two 4-glomerulus groups, one of which includes the remaining S-glomerulus and the male MGC (Fig. 14 B).

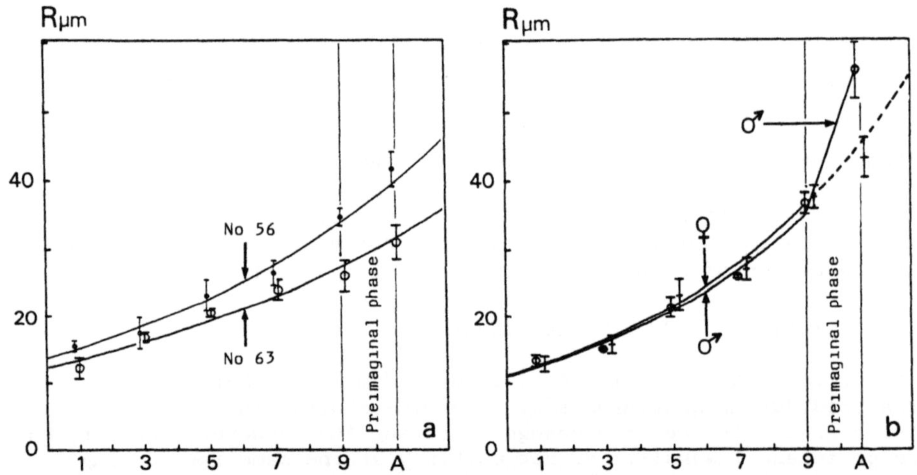

Fig. 13. Postembryonic glomerular growth in *Blaberus* illustrated by "ordinary" glomeruli Nos. 56 and 63 (a), and macroglomerulus No. 106 (b). Growth kinetics of glomeruli identified throughout development show 2 periods. During nymphal period (L1 to L9) radius R of each glomerulus follows a specific exponential curve characterized by initial size R0 and growth rate g; g is the relative size increase from one stage to the next: 9.3% for No. 63, 10.8% for No. 56. During preimaginal period (L9 to adult) g is the same or slightly higher. Male and female No. 106 follow the same exponential curves (no sexual dimorphism) with high g (14%). Dimorphism appears during preimaginal period from a differential growth in males and females. (Adapted from Rospars and Chambille, 1986).

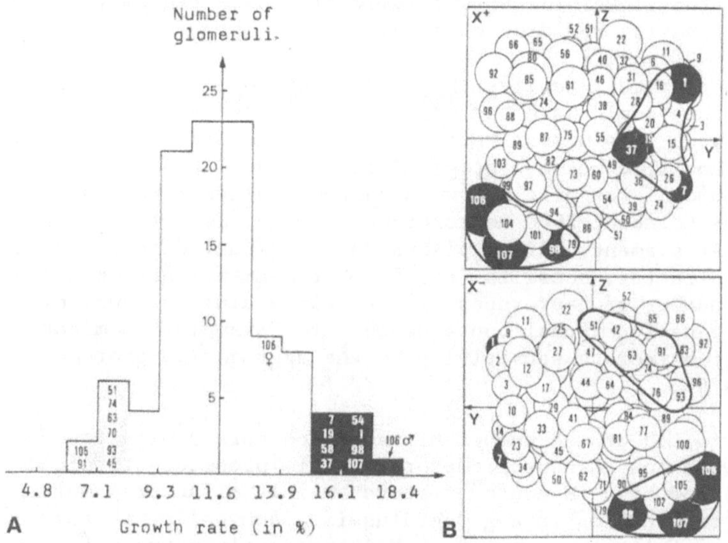

Fig. 14. Heterogeneous growth of glomeruli. (A) The growth rate g of every identified glomerulus was derived from its diameters at L5 and adult stages. For these 104 identified glomeruli, the distribution of g is approximately Gaussian. The range of g values observed, from 6.3% (glomerular volume multiplied 5.7-fold from L1 to adult) to 18.2% (x 117), is very wide. (B) The slow-growing S (light grey) and fast-growing F (black) glomeruli are gathered in 3 distinct subsets (After Rospars and Chambille, 1986).

3. DISCUSSION

3.1. Connections Between Antenna and AL

The projections to ALs of neuroreceptors innervating the pheromone-sensitive sensilla (Boeckh et al., 1977, Christensen and Hildebrand, 1987a, Koontz and Schneider, 1987) and other sensilla (Stocker et al., 1983) have been described in intracellular staining studies (reviewed in Rospars, 1988a). However, no complete description of the projections of neuroreceptor axons from the antenna to the AL has been given for any species. The results obtained for invariance and growth of identified glomeruli in *Blaberus* give information on the nature and specificity of connections between antenna and AL that cannot be easily obtained by cobalt and Golgi techniques:

The number of neuroreceptors and the glomerular volume are multiplied by almost equal factors between L1 and adults (see Fig. 2). This suggests that the volume of a glomerulus is almost proportional to the number of neuroreceptors which terminate in it. Such a relationship could not be predicted from the current knowledge of the neuronal structure of glomeruli.

Consequently, the number of neuroreceptors terminating in a given glomerulus is nearly constant at a given stage of development.

Otherwise, the glomerular volume would fluctuate randomly and no invariance in size would be observed.

Not all glomeruli receive the same number of neuroreceptors. Otherwise they would have had approximately the same size, which has not been observed. This number varies in the ratio 1:7 in adults (MGC not included) and 1:20 between stages (Rospars and Chambille, 1986). The excess volume of the male MGC with respect to its female homolog corresponds to about 2600 neuroreceptors, i.e. less than 10 sex-specific sensilla per segment. This explains why no sexual dimorphism in the number of olfactory sensilla (ca. 500 per segment) has been found. Also, the small number of neuroreceptors per glomerulus in young nymphs is likely to have functional consequences; for example, it might explain their lower olfactory sensitivity to the aggregative pheromone (Brossut et al., 1974).

The ratio L:H of the number of neuroreceptors coming from the sensilla with low (L-neuroreceptors, mainly proprioceptive and gustatory) and high (H-neuroreceptors, mainly olfactory) multiplication rates and terminating in a glomerulus is a convenient measure of the specificity of its innervation. L:H is presently unknown for any glomerulus but large variations in this ratio must be assumed to account for the wide range of glomerular growth rates which would otherwise be much restricted. Consequently, it is likely that slow growing S-glomeruli are preferentially innervated by L-neuroreceptors and fast growing F-glomeruli by H-neuroreceptors.

Subsets of neuroreceptors with similar growth rates (and possibly the same modality) terminate in neighbouring glomeruli. Otherwise, F- and S-glomeruli would not have been arranged in separate groups in the ALs. Whereas the various sensillar types are mixed on each antennal segment, the neuroreceptors that innervate them tend to project to separate areas in the ALs. This type of structure might merely result from an efficient use of positional information by ingrowing axons during embryogenesis.

3.2. Physiological Interpretation

The available physiological (e.g. Kaissling, 1987) and anatomical (e.g. Rospars, 1988a) data are compatible with 2 possible modes of functioning of an array of identified glomeruli (Rospars, 1983; see also Holley and Døving, 1977). These modes are an immediate extension of the well-known "labelled line" versus "across-fiber patterns" type of coding that occurs in many sensory systems. However, they markedly differ in that the basic coding units are no longer anonymous, sometimes unreliable, neuroreceptors but identified glomeruli that integrate the action of a number of neuroreceptors.

In the first mode each glomerulus functions as a separate unit and solves a special "problem." For instance, the MGC processes information sent by the male antennal neuroreceptors that code for the sex-pheromone components released by conspecific females (attractive) or by other related species (repulsive), and the LPO glomerulus (see Section 1.2) processes whatever information comes from the labial pit organ. Such a mode is rather costly in neurons because the number of problems solved cannot exceed the number of glomeruli.

A second more economical mode relies on the simultaneous functioning of a subset of interacting identified glomeruli. What is significant then is the pattern of activated and inhibited glomeruli ("across-glomeruli pattern," Rospars, 1983). For example, a small subset of 16 glomeruli that can each be in only 2 states, can generate 2^{16} (i.e. 65

536) different patterns, provided that every glomerulus be individually distinguished by the nervous system and not merely by the experimenter. Each of these patterns might code for a different olfactory or gustatory stimulus. Experimental evidence for such distinct, although overlapping, patterns of glomerular activity during olfactory stimulation has been given in *Drosophila* with labelled 2-deoxy-glucose (Rodrigues and Buchner, 1984, Rodrigues, 1988, and this volume). These patterns, when compared to reference patterns of genetic origin or memorized by experience, might finally contribute to trigger or modulate motor outputs.

ACKNOWLEDGEMENTS

We thank J.B. Denis, J.P. Vila and B. Dumortier for support. We are indebted to N. Hawlitzky for the use of the photomicrographic equipment, and J.G. Hildebrand for providing sections of *Manduca*. We thank M.F. Commeau and A. Ainsley for aid in the preparation of the English manuscript.

REFERENCES

Arnold, G., Masson, C. and Budharugsa, S., 1985, Comparative study of the antennal afferent pathway of the workerbee and the drone (*Apis* mellifera L.). Cell Tissue Res. 242: 593-605.

Arnold, G., Budharugsa, S. and Masson, C., 1988, Organization of the antennal lobe in the queen honey bee, Apis *mellifera* L. (Hymenoptera: Apidae). Int. J. Insect Morphol. Embryol. 17: 185-195.

Barth, R.H., 1970, The mating behavior of *Periplaneta americana* and *Blatta orientalis* with notes on 3 additional species of *Periplaneta* and interspecific action of female sex pheromone. Z. Tierpsychol. 27: 727-48.

Boeckh, J., Boeckh, V. and Kuhn, A., 1977, Further data on the topography and physiology of central olfactory neurons in insects, pp.315-21. In: Le Magnen, J. and Macleod, P. (eds.) Olfaction and Taste VI. Information Retrieval Ltd, London.

Bogner, F., Boppre, M., Ernst, K.D. and Boeckh, J., 1986, CO2 sensitive receptors on labial palps of *Rhodogastria* moths (Lepidoptera: Arctiidae): physiology, fine structure and central projection. J. Comp. Physiol. A 158: 741-49.

Bretschneider, F., 1924, Uber die Gehirne des Eichenspinners unddes Seidenspinners (*Lasiocampa quercus* L. und Bombyx mori). Jena Z. Naturw. 60: 563-78.

Brossut, R., 1970, L'interattraction chez Blabera *craniifer* Burm. (Insecta, Dictyoptera): sécrétion d'une phéromone par lesglandes mandibulaires. C.R. Acad. Sci. Paris D 270: 714-16.

Brossut, R., Dubois, P. and Rigaud, J., 1974, Le grégarisme chez *Blaberus craniifer*: isolement et identification de la phéromone. J. Insect Physiol. 20: 529-43.

Chambille, I., 1988, Les sensilles gustatives de l'antenne chez *Blaberus craniifer* Burm. (Insecta, Dictyoptera). Nombre, distribution et dimorphisme sexuel. C.R. Acad. Sci. Paris III 306:291-297.

Chambille, I., Rospars, J.P., and Masson, C., 1978, The organization of the sensory deutocerebral glomeruli of the cockroach *Blaberus craniifer*, p. 67. Abstracts of ECRO III. Pavia.

Chambille, I., Rospars, J.P., and Masson, C., 1980, The deutocerebrum of the cockroach *Blaberus craniifer* Burm. Spatial organization of the sensory glomeruli. J. Neurobiol. 11: 1-23.

Chambille, I., and Rospars, J.P., 1981, Le deutocérébron de la blatte *Blaberus craniifer* Burm. (Dictyoptera: Blaberidae). Etudequalitative et identification visuelle des glomérules. Int. J.Insect Morphol. Embryol. 10: 141-65.

Chambille, I. and Rospars, J.P., 1985, Neurons and identified glomeruli of antennal lobes during postembryonic development in the cockroach *Blaberus craniifer* Burm. (Dictyoptera: Blaberidae). Int. J. Insect Morphol. Embryol. 14: 203-26.
Cohen, M.J. and Jacklet, J.W., 1965, Neurones of insects: RNA changes during injury and regeneration. Science 148: 1237-40.
Christensen, T.A. and Hildebrand, J.G., 1987a, Male-specific, sex pheromone-selective projection neurons in the antennal lobes of the moth *Manduca sexta*. J. Comp. Physiol. A 160: 553-69.
Christensen, T.A. and Hildebrand, J.G., 1987b, Functions, organization,and Physiology of the Olfactory Pathways in the Lepidopteran Brain. pp. 457-484. In: Gupta, A.P. (ed.) Arthropod Brain: its Evolution, Development, Structure and Functions.
Descoins, C., Priesner, E., Gallois, M., Arn, H. and Martin, G., 1978, Sur la sécrétionphéromonale des femelles vierges de *Mamestra brassicae* L. et de *Mamestra oleracea* L. (Lépidoptères Noctuidae, Hadeninae). C.R. Acad. Sci. Paris D 286: 77-80.
Ernst, K.D., Boeckh, J. and Boeckh, V., 1977, A neuroanatomical study on the organization of the central antennal pathways in insects. II Deutocerebral connections in Locust amigratoria and *Periplaneta americana*. Cell Tissue Res. 176: 285-308.
Ernst, K.D. and Boeckh, J., 1983, A neuroanatomical study on the organization of the central antennal pathways in insects. III. Neuroanatomical characterization of physiologically defined response types of deutocerebral neurons in *Periplaneta americana*. Cell Tissue Res. 229: 1-22.
Hildebrand, J.G., Matsumoto, S.G., Camazine, S.M., Tolbert, L.P., Blank, S., Ferguson, H. and Ecker, V., 1980, Organization and physiology of antennal centres in the brain of the moth *Manduca sexta*, pp.375-82. In: Insect Neurobiology and Pesticide Action (Neurotox79). Society of Chemical Industry, London.
Holley A. and Døving K.V., 1977, Receptor sensitivity, acceptor distribution, convergence and neural coding in the olfactory system, pp. 113-24. In: Le Magnen, J. and MacLeod, P. (eds.) Olfaction and Taste VI. Information Retrieval Ltd, London.
Homberg, U., 1984, Processing of antennal information in extrinsic mushroom body neurons of the bee brain. J. Comp. Physiol. A 154:825-36.
Jawlowski, H., 1948, Studies on the insects brain. Ann. Univ. Mariae Curie Sklodowska C 3: 1-30.
Kaissling, K.E., 1987, R.H. Wright Lectures on Insect Olfaction. Colbow, K. (ed.) Simon Fraser University, Burnaby, B.C., Canada
Kent, K.S., Harrow, I.D., Quartaro, P. and Hildebrand, J.G., 1986, An accessory olfactory pathway in Lepidoptera: the labial pit organ and its central projections in *Manduca sexta* and certain other sphinx moths and silk moths. Cell Tissue Res. 245: 237-45
Koontz, M.A. and Schneider, D., 1987, Sexual dimorphism in neuronal projections from the antennae of silk moths (*Bombyx mori*, *Antheraea polyphemus*) and the gypsy moth (*Lymantria dispar*). Cell Tissue Res. 249: 39-50.
Lambin, M., 1973, Les sensilles de l'antenne chez quelques blattes et en particulier chez *Blaberus craniifer* Burm. Z. Zellforsch. 143: 183-206.
Lee, J.K. and Altner, H. 1986. Primary sensory projections of the labial palp-pit organ of *Pieris rapae* L. (Lepidoptera: Pieridae). Int. J. Insect Morphol. Embryol. 15: 439-48.
Oland, L.A. and Tolbert, L.P., 1986. Glial patterns during early development of antennal lobes of *Manduca sexta*: a comparison between normal lobes and lobes deprived of antennal axons. J. Comp. Neurol. 255: 196-207.
Otte, R.F., 1977, Communication in Orthoptera, pp. 334-61. In: Sebeok, T.A. (ed.) How Animals Communicate. Indiana University, Bloomington.
Pareto, A., 1972, Die zentrale Verteilung der Fuhlerafferenz bei Arbeiterinnen der Honigbiene, *Apis mellifera* L. Z. Zellforsch. 131: 109-40.

Priesner, E., 1979, Progress in the analysis of pheromone receptorsystems. Ann. Zool. Ecol. Anim. 11: 533-46.
Prillinger, L., 1981, Postembryonic development of the antennal lobes in *Periplaneta americana* L. Cell Tissue Res. 215: 563-75.
Rodrigues, V. and Buchner, E., 1984,(3H)2-Deoxyglucose mapping of odor-induced neuronal activity in the antennal lobes of *Drosophila melanogaster*. Brain Res. 324: 374-78.
Rodrigues, V., 1988, Spatial coding of olfactory information in the antennal lobe of *Drosophila melanogaster*. Brain Res. 453: 299-307.
Rospars, J.P., 1983, Invariance and sex-specific variations of the glomerular organization in the antennal lobes of a moth, *Mamestra brassicae* and a butterfly, *Pieris brassicae*. J. Comp. Neurol. 220: 80-96.
Rospars, J.P., 1985, Le Lobe Antennaire des Insectes: son Organization Glomérulaire Invariante, ses Variations Sexuelles et son Développement Postembryonnaire. Thèse Doct. Etat, Université Paris-Sud, Orsay.
Rospars, J.P., 1988a, Structure and development of the insect antennodeutocerebral system. Int. J. Insect Morphol. Embryol. 17:243-294.
Rospars, J.P., 1988b, Reconstruction, display and analysis of insect antennal lobes using a computerized digitizing system. Abstracts of ECRO VIII. Warwick.
Rospars, J.P., Chambille, I. and Masson, C.,1979, Invariance morphologique de l'organisation glomérulaire du deutocérébronchez la blatte *Blaberus craniifer* Burm. (Insecta, Dictyoptera) C.R. Acad. Sci. Paris D 288: 1043-46.
Rospars, J.P. and Chambille, I., 1981, The deutocerebrum of the cockroach *Blaberus craniifer* Burm. Quantitative study and automated identification of the glomeruli. J. Neurobiol. 12: 221-47.
Rospars, J.P., Jarry, C. and Chambille, I., 1983, Caractéristiques qualitatives et quantitatives de l'organisation glomérulairedes lobes antennaires de la Noctuelle *Mamestra brassicae* (Insecta, Lepidoptera). C.R. Acad. Sci. Paris D 296: 369-74.
Rospars, J.P. and Chambille, I., 1986, Postembryonic growth of antennal lobes and their identified glomeruli in the cockroach *Blaberus craniifer* Burm. (Dictyoptera: Blaberidae). Int. J. Insect Morphol. Embryol. 15: 393-415.
Rowell, H.F., 1976, The cells of the insect neurosecretory system: constancy, variability, and the concept of the unique identifiable neuron. Adv. Insect Physiol. 12: 63-123.
Schafer, R., 1973, Postembryonic development in the antenna of the cockroach, *Leucophaea maderae*: growth, regeneration, and the development of the adult pattern of sense organs. J. Exp. Zool. 183: 353-64.
Schaller, D., 1978, Antennal sensory system of *Periplaneta americana* L. Distribution and frequency of morphologic types of sensilla and their sex-specific changes during postembryonic development. Cell Tissue Res. 191: 121-39.
Schaller-Selzer, L., 1984, Physiology and morphology of the larval sexual pheromone-sensitive neurones in the olfactory lobe of the cockroach, *Periplaneta americana*. J. Insect Physiol. 30: 537-46.
Schneiderman, A.M., Hildebrand, J.G. and Jacobs, J.J., 1983, Computer-aided morphometry of developing and mature antennal lobes in the moth *Manduca sexta*. Soc. Neurosci. Abstr. 9: 834-35.
Schneiderman, A.M. and Hildebrand, J.G., 1985, Sexually dimorphic development of the insect olfactory pathway. Trends Neurosci. 8: 494-99.
Singh, R.N. and Nayak, S.V., 1985, Fine structure and primary sensory projections of sensilla on the maxillary palp of *Drosophila melanogaster* Meigen (Diptera: Drosophilidae). Int. J. Insect Morphol. Embryol. 14: 291-306.
Stocker, R.F., Singh, R.N., Schorderet, M. and Siddiqi, O., 1983, Projection patterns of different types of antennal sensilla in the antennal glomeruli of *Drosophila melanogaster*. Cell Tissue Res. 232: 237-48.
Tolbert, L.P., Matsumoto, S.G. and Hildebrand, J.G., 1983, Development of synapses in the antennal lobes of the moth *Manduca sexta* during metamorphosis. J. Neurosci. 3: 1158-75.
Urvoy, J., 1963, Etude anatomo-fonctionnelle de la patte et de l'antenne de la blatte *Blabera craniifer* Burmeister. Ann. Sci. Nat. Zool. 5: 287-413.

DROSOPHILA CHEMORECEPTORS

Renée Venard, Claude Antony and Jean-Marc Jallon

Laboratoire de Biologie et Génétique Evolutives
CNRS,
91198 Gif-sur-Yvette, France

INTRODUCTION

In recent years, chemoreceptor physiology has become a field of active research, on preparations ranging from ciliae to mammals. However, it is obvious that insects occupy a central place in the chemical world (Bell and Cardé, 1984). Olfactory or gustatory systems have to react to a considerable number of different molecules. One central question concerns the number of chemoreceptors in a given organ or organism, and the range of their specificity. If there are specific receptors for each molecule or even for each class of molecules, too many different receptors would have to exist, together with an extremely large number of genes coding for them. One possible solution is that there is some coordination of gene function involved.

Neurogenetics is an extremely powerful tool for investigating such problems. *Drosophila* is a particularly useful preparation: olfactometric and osmotropic studies have shown that the fruitfly can distinguish between attractants and repellents and is even able to recognize small differences in concentration (Begg and Hogben, 1946; Borst, 1983; Kikuchi, 1973; Fuyama, 1976). Siddiqi's group in Bombay has played a pioneering role in this field, by searching for, and finding, mutants anosmic for specific chemical functions (Rodrigues and Siddiqi, 1978).

Rodrigues has isolated several mutations on the X chromosome which lead to abnormal olfactory behavior of the adult fly in response to different odorants. The *olfA* and *olfB* mutants are anosmic for aldehydes, *olfC* mutants are anosmic for acetates and acetone, and *olfD* flies are anosmic for most odorants (Rodrigues, 1980).

Biochemical and molecular techniques now enable us to investigate the molecular components of chemoreception machinery. We report here a few initial steps in this field.

ANTENNAL RECEPTORS FOR SMALL MOLECULES

We have studied two simple chemical functions, acetates and aldehydes, attached to short carbon chains. Two mutant strains (*olfC* , *olfA*) exist which show anosmias for these substances. Our initial step was to develop a technique for recording the slow potentional response of the antenna (EAG) to olfactory stimulation, using a method originally developed for honey bees with a few modifications (Venard and Pichon, 1981).

Chemical stimulation leads to a tonic reponse, the amplitude of which was dependent on odor concentration. As there were no differences between the reponses of males and females to the tested odours,

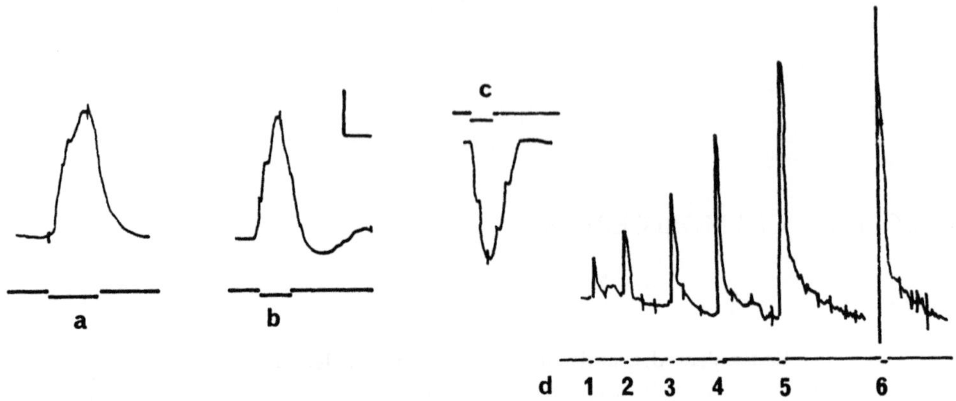

Fig. 1 Electroantennogram (EAG)-recording of *Drosophila melanogaster* responses to different compounds deposited on a piece of paper; a) butanol; b) ethyl acetate; c) ether; d) successive responses to increasing concentraions of butanol: 1) 2.3×10^{14}; 2) 8×10^{14}; 3) 2.7×10^{15}; 4) 8×10^{15}; 5) 2.2×10^{16}; 6) 5.7×10^{16} molecules under 2 seconds of stimulation. Length of horizontal bar: 12s; amplitude of vertical bar: 0.4 mV. An adult fly is immobilized in dental wax with its antennae lodged in their head concavity by fixing the aristae onto the eyes. Two glass microelectrodes filled with 0.17M sodium chloride were used, the reference microelectrode being inserted into the thorax and the recording electrode making a superficial contact near the base of the arista on the anterior face of the funiculus. the odor delivery system was a constant-flow olfactometer under a constant flux of nitrogen which is the odorant carrier gas.

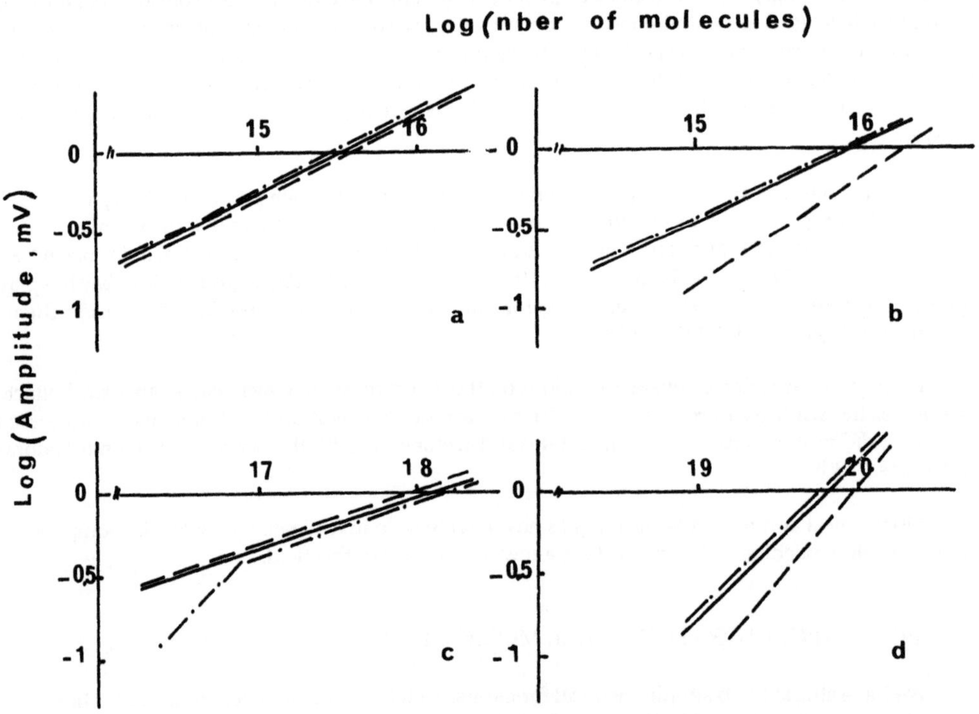

Fig 2 Dose-response curves to a) butanol; b) butyl acetate; c) benzaldehyde; d) 2-butanone, for wild-type (——), *olfC* (– – –) and *olfA* (—· —) strains of *Drosophila melanogaster*. Abscissae: logarithms of the number of molecules delivered during 2s stimulation. Ordinates: means of the logarithms of the EAG amplitudes.

we thereafter only tested male flies. The relationship between EAG amplitudes, recorded at the base of the funiculus as the sum of the responses of different receptor cells, and the concentrations of different odours, is shown in Figure 2.

For all chemicals, double logarithmic curves seem linear. We compared dose-response curves of wild flies with those of *olfC* mutants, which have a specific anosmia towards acetates, and with those of *olfA* mutants, which show a specific anosmia towards aldehydes (Venard and Pichon, 1984). Mutant EAGs appear to have an increased threshold for their specific odorant. *olfC* mutants show parallel responses for butyl acetate and 2-butanone, which suggests they have a defect in their response to the $CH_3\text{-}\overset{\overset{\displaystyle O}{\|}}{C}\text{-}$ group. The curve obtained with *olfA* mutants presented with benzaldehyde is biphasic, which might suggest the existence of two kinds of receptors, only one of them being affected by the *olfA* gene. Defects in both mutants seem to be at least peripheral.

Because of the small size of *Drosophila* antennal sensilla, single unit recordings are much more difficult than is the case in moths. Two methods have been used to get round this problem: competition experiments (Siddiqi, 1983) and binary odorant mixtures (Borst, 1984). By using such techniques, it is possible to determine whether two odors act on the same or different receptor types. It is also possible to interpret the results as suggesting that an odor pair acts on different receptor sites, or on the same receptor type but with different affinities. The results of such experiments on wild-type flies suggested that several types of acetate receptors exist on the antennae. This is in agreement with the fact that the EAG amplitudes we observed with *olfC* flies were decreased but not abolished.

To try to understand the defect present in *olfC* receptors, we first looked for possible differences in the morphology and topology of the hairs on the funiculus. Among a large number of uninnervated hairs (spinules) a characteristic distribution of three types of sensilla (basiconic, coeloconic and trichoid) was described by Mindek (1968) and confirmed by Venkatesh and Singh (1984). Using light and electron microscopy, Venkatesh and Singh showed that in wild type flies, basiconic and coeloconic sensilla are argyrophylic; moreover they distinguished two classes of basiconica, according to their size. The basiconica are single-walled and multiporous, with two or three branched dendrites. Coeliconica are double-walled and innervated by one to three branched dendrites. Link (1983) found numerous pores in the thick-walled trichoidea innervated by 1, 2 or 3 branched neurons. It seems probable that the three types of sensilla may play a rôle in olfaction. The three different types of sensilla project into 19-22 glomeruli in each antennal lobe. Only five glomeruli have an exclusively unilateral input (Stocker et al., 1983). Each glomerulus seems to correspond to one type of sensillum rather than to its location on the funiculus.

Using scanning electron microscopy, we re-examined the anterior faces of the funiculus from where we recorded the EAG ,for both wild-type and *olfC* flies. We did not detect any significant variation in the distribution of the three basic types of sensilla (Table 1).

Table 1 Analysis of the anterior side of the funiculus: A) diagram of the distribution of the different types of chemosensilla on a left antenna of a wild-type male. Basiconica: ◆ ; coeloconica ● ; trichoid ▲. B) After framing and subdivision into nine equal areas, the funiculus sensillum distribution was compared in wild-type and *olfC* flies. Ten antennae of each genotype were tested.

A

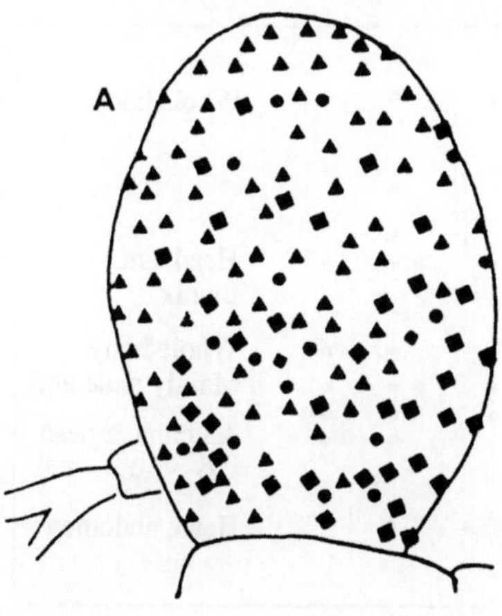

B

wild type				olf C		
9(2)	12(2)	7(2)	▲	9(2)	12(1)	5(2)
2(2)	4(2)	2(1)	◆	1(1)	4(2)	3(1)
0(1)	2(1)	0(1)	●	1(1)	1(1)	1(1)
11(4)	9(2)	7(2)	▲	12(5)	9(3)	4(2)
3(2)	5(2)	6(2)	◆	2(2)	7(2)	7(4)
2(1)	2(1)	1(1)	●	2(1)	2(1)	2(1)
4(1)	4(1)	0(1)	▲	11(3)	4(2)	0(1)
5(1)	5(1)	12(3)	◆	3(2)	7(2)	7(3)
2(1)	2(2)	2(1)	●	2(1)	1(1)	2(1)

Also using electron microscopy, Stocker and Gendre (1988) found a sexual dimorphism in the distribution of antennal sensilla in two wild-type strains, Sevelen and Oregon. Male and female antennae have about the same number of sensilla, but male funiculi tend to be smaller, and to have 30% more trichoid sensilla. This sex difference could explain our failure to observe a sensillar distribution similar to that reported by Mindek (1968).

Having found no difference in the antennal chemosensilla topography of the two genotypes, we then turned to the possible molecular differences which might account for the differences in the strains' responses. Electrophoresis—and in particular two-dimensional electrophoresis—provides a good way to compare the protein content of strains, and especially to detect proteins altered by a mutation. This technique does not, however, allow us to study the function of the protein. Because we are particularly interested in the functional characterization of receptor proteins, we used classical, one-dimensional electrophoresis in this study.

Known chemical transduction systems, such as the acetylcholine receptors of vertebrate neuromuscular junctions or the pheromone receptors of male moth antenna (Vogt et al., 1985), consist of not only a multiprotein complex involved in recognition and transduction mechanisms, but also catabolizing systems involved in recycling excess chemical messenger. We have hypothesized that esterase might play a role in the acetate olfactory system; we therefore studied antennal esterases.

Due to their extremely small size, we found it impossible to immediately isolate antennal chemosensilla, as is possible with other preparations (Klein, 1984, 1987). We therefore first cut antennae from frozen *Drosophila* heads, using microscissors. The antennae were then stocked in capillary vials in liquid nitrogen, each vial containing exactly 100 or 200 antennae, until enzyme assays were carried out. The antennae were homogenized in Ephrussi-Beadle Ringer solution containing 0.5% Triton X-100, 10% glycerol and 0.1% bromophenol blue. The consequent suspension was then clarified by centrifugation. The supernatant to be electrophoretically analyzed for esterase activity was loaded directly at the top of a 12% polyacrylamide gel. After electrophoresis, the gel was stained for esterase activities using artificial substrates such as α– or β- naphtyl acetates, following the method of Johnson et al. (1966). The hydrolysis of the acetates produces α– or β- napthol which may be trapped with diazonium salt to produce an insoluble stain. Two types of dye-couplers were used, fast blue and fast red, which yield different complexes with different colours.

Table 2 Comparison of the different esterasic bands in multiple-head or multiple-antennae homogenates characterized by various experimental conditions. Column 1: naphtyl acetate specificity in an α–/ β– acetate mixture;.Column 2: sensitivity to propanol, eserine, 60° C heating (activation: +; inactivation: -); Column 3: relative intensity of bands in head (He) and antenna (An) extracts of wild type (CS) and *olfC* (OC) flies. Column 4: localization. The equivalent of 0.5 abdomen, 1 head or thorax and 50 antennae was loaded onto the gel.

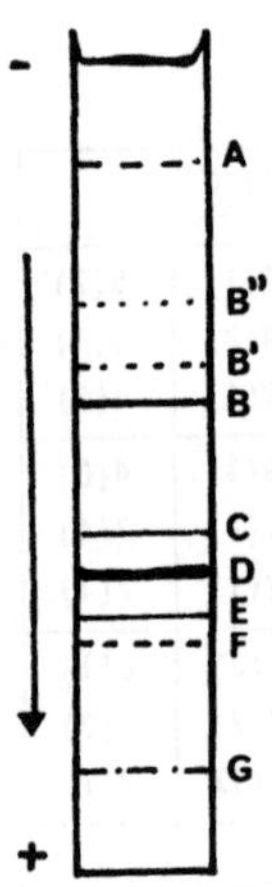

Band	SPECIFICITY	SENSITIVITY Propanol	Eserine	Heating	RELATIVE INTENSITY CS He	CS An	OC He	OC An	LOCALISATION
A	α	+			•	•	*	*	Whole body
B"	α				•		*		
B'	α				•		*		
B	α	+	-	-	••	•	**	*	Head, ant., thorax
C	α	+			•	•	*	*	Whole body
D	β	-			••	•	**	*	Mainly male abd.
E	α	+			•	•	*	*	
F	β	-					*	*	Antenna & head (OC only)
G	α				•		*		Head, abdomen

In a whole extract, we observed several zones of esterasic activity, characterized by their mobility and their specificity towards α– and β-naphtyl esters and their sensitivity to propanol and temperature, as shown in Table 2. The comparison of antennal extract of both genotypes revealed a few quantitative differences among bands A - D which we have not yet been able to satisfactorily quantify, together with one quantitative difference: there is an additional esterase, F, in antennal *olfC* extracts which is found in both sexes. This F band is close to band D, which corresponds to the polymorphic esterase 6 (Wright, 1963). However, unlike band D, band F is only observed after Triton treatment, suggesting that it may be membrane-linked. Band F was observed in heads and antennae of both *olfC* sexes, but never in the abdomens or thoraxes. Esterase 6 activity was particularly high in all abdomens of adult males, as expected.

The singular *olfC* protein with an esterasic activity (band F) might provide a simple explanation for the physiological defects previously reported; indeed it might catabolize in a different manner, for example more efficiently, the acetate molecules reaching the olfactory sensilla on the surface of the antenna, and thus decrease the number of stimulating molecules for the olfactory receptors, leading to a reduced transduction. It seems possible that we have found one element of the chemoreception machinery in *Drosophila*. In two other organisms, olfactory receptors seem to be associated with a specific nucleotide cascade, leading to depolarization, probably via phosphorylation of ion channel proteins (Chen, 1986; Lancet, 1987). These studies open up new frontiers in *Drosophila* research which will increasingly have to combine genetics, behavior, physiology, anatomy and biochemistry.

RECEPTORS FOR LONG—CHAIN MOLECULES

Flies not only react to short-chain acetates: a substantial amount of behavioral evidence suggests that they also respond to long-chain acetates. A male-specific lipid, *cis*-vaccenyl acetate (cVA), produced in the adult male ejaculatory duct and transferred to females during copulation, was characterized by Butterworth in 1969. Jallon et al. (1981) showed that in *in vitro* tests, cVA strongly inhibits the precopulatory behavior of males. Its "anti-aphrodisiac" role has been much studied, and is discussed in the accompanying paper by Ferveur et al.. Bartelt et al. (1985) used olfactometric studies to show that cVA also acts as a coattractant of food or acetone. In order to try and locate the receptors involved in the detection of such long-chain acetates, we undertook new electrophysiological experiments to test antennal electrical activity in response to stimulation with cVA. Given the low volatility of this substance, we could not use the same method of EAG described above. We therefore used a procedure developed for Lepidoptera, adapted to the smaller scale and particular problems presented by *Drosophila*.

The results of this study are presented in Figure 3. It is clear that cVA can stimulate receptors on the male antenna. Amplitudes in the range 0.2 - 1 mv were recorded. In contrast with moth EAGs in response to stimulation by their specific pheromone, our results do not suggest any extremely sensitive or specific mechanisms for the antennal detection of an aggregation pheromone.

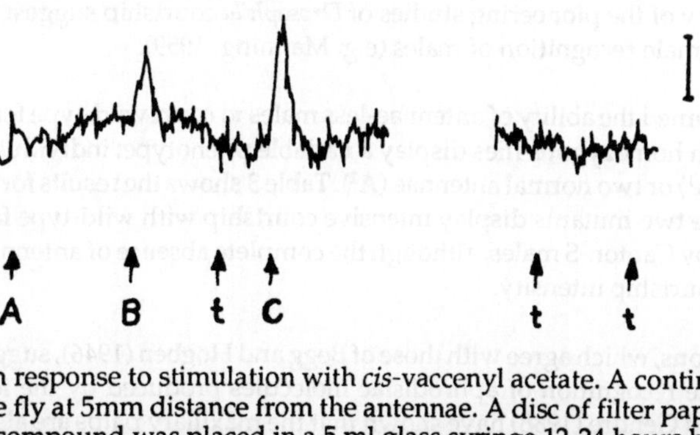

Fig. 3 EAG recorded in response to stimulation with *cis*-vaccenyl acetate. A continuous air-stream was directed onto the fly at 5mm distance from the antennae. A disc of filter paper impregnated with 100μg - 1 mg of compound was placed in a 5 ml glass syringe 12-24 hours before the experiment began, to allow for phase equilibration. By pushing the syringe piston, 0.1 ml (a), 0.2 ml (b) or 0.5 ml (c) of odor from the compound was mixed with the air stream and blown for 2 seconds onto the antennae. Tests were repeated at 30s intervals. (t): air stream without odor. Amplitude of vertical bar: 0.4 mv

Table 3 Compared vibration responses ($\overline{\Sigma T_N}$) of antennaless (A^0 or A^2) and wild-type (CS) males to mature virgin CS females. $\overline{T_L}$ = mean latency time; ΣT_N - mean cumulative wing vibration time of the tester male over N tests.

	N	$\overline{T_L}$ (sec)	$\overline{\Sigma T_N}$ (sec)	$\Sigma T/\Sigma T_{CS}$ (%)
CS	60	46	208 +/- 11.6	100
ant A^0	65	70	182 +/- 3.8	88
ant A^2	21		210	101

Stocker and Gendre (1988) have shown that wild-type males whose maxillary palps have been removed court virgin females in the usual fashion but cannot distinguish virgin females from females that had copulated 24 hours previously. Jallon et al. (1982) suggested that it is during this period that fertilised females may spray cVA on courting males whilst extruding their ovipositor (a behavior typically shown by fertilised females which tends to result in the male stopping courtship). Singh and Nayak (1985) report that basiconic-like sensilla exist on the maxilary palps, projecting sensory fibres to three of the antennal lobe glomeruli. These glomeruli also receive axons from antennal basiconic sensilla.

These results suggest that basiconic-like sensilla on the maxillary palps, associated with trichoid sensilla (which are pure mechanoreceptors), could be involved in the detection of cVA. This would not rule out the simultaneous, and perhaps linked, detection of cVA by antennal receptors.

One question which remains to be answered is whether the same receptors which detect short-chain acetates are also involved in the detection of long-chain acetates. Tompkins and Hall (1981) suggested that *olfC* males are unable to distinguish virgin females from mated females. However, this result will have to be checked in more controlled conditions, as the post-copulatory effect is complex.

HYDROCARBON PHEROMONE DETECTORS

Male sexual behavior in *Drosophila melanogaster* is mainly induced by female-specific chemical messages which consist of long-chain unsaturated hydrocarbons organized on the surface of the female's cuticle, as discussed in the accompanying article from our group (Ferveur et al.). Such heavy molecules have a low volatility and require an appropriate detection system on the male fly if they are to be detected during courtship. Many of the pioneering studies of *Drosophila* courtship suggest that the antennae play only a minor role in female recognition of males (e.g. Manning, 1959).

We first re-examined the ability of antennae-less males to court wild-type females, using the antennaless mutant in which homozygous flies display a variable phenotype: individuals may have no antenna (A^0), one antenna (A^1) or two normal antennae (A^2). Table 3 shows the results for A^0 and A^2 mutants and for wild-type flies. The two mutants display intensive courtship with wild-type females; both levels are similar to that shown by Canton-S males, although the complete absence of antennae in A^O flies does tend to slightly diminish courtship intensity.

These observations, which agree with those of Begg and Hogben (1946), suggest that male antennae are dispensable for the recognition of aphrodisiac molecules produced by the female. As pointed out previously, Stocker and Gendre (1988) have shown that the maxillary palps are also not necessary for the detection of virgin female aphrodisiac pheromones.

In an early description of *Drosophila melanogaster* sexual behavior, Manning (1959) mentions the male's "tapping" of the female as being an important element at the beginning of courtship. We examined

Table 4 Proportion of males showing wing vibration (% R) and the amount of wing vibration (ΣT_N) following coating of all 6 legs, or of only 5 legs. Leg left uncoated is indicated.

N coated legs	Uncoated leg	N	%R	ΣT_N (sec.)
6		50	27	2 +/- 0.7
5	prothoracic	48	60	33 +/- 2.8
	mesothoracic	27	48	20 +/- 4.1
	metathoracic	24	29	1.4 +/- 1.6

27 Canton-S courtship sequences using a Sony video recorder equipped with a frame-by-frame playback facility, which enabled us to detect rapid movements by the flies. We investigated whether the initiation of male wing vibration was necessarily preceded by a "tapping" behavior by the male, which we defined as the contact of one of the male's legs with any part of the female cuticle. In only two of the 27 courtships was no tapping detectable. In 16 courtships a "tapping" movement involving a prothoracic leg was clearly seen shortly before wing vibration.

These behavioral experiments suggest another candidate organ for the site of the male's pheromone receptors: the legs. In order to measure the role played by the legs in pheromone detection, we tried to alter the potential receptor properties by coating males' legs from the distal extremity to the tibia with a thin layer of dental wax. The courtship displayed by such treated males when faced with a virgin female are shown in Table 4.

It appears that i) male wing vibration is almost completely abolished when all six legs are coated; ii) if only five legs are coated the male's behavior is variable, depending upon which leg is left untreated. If the uncoated leg is one of the anterior appendages, relatively high amounts of wing vibration are observed. These results recall those of Schlein et al. (1981) on *Musca domestica* and *Glossina morsitans* which suggested that pheromone receptors are present on male tibia in these species. It seems probable that the legs of male *Drosophila*, especially the prothoracic legs, carry receptors which are sensitive to female-produced aphrodisiac hydrocarbons. A word of caution is necessary, however: it is difficult to avoid diffusion of the coating wax, especially given the small legs of *Drosophila.*

Hannah-Alava (1958) and Nayak and Singh (1983) have described numerous hairs on the legs of *Drosophila*, including two types of bristles on the basis of the presence/absence of bracts at the base of the bristle socket. These authors observed that on tarsal segments, only hairs without a bract had an apical core, which can be stained with silver nitrate. These bractless hairs of the trichoid type display a double lumen, one circle-shaped and the other crescent-shaped, which generally give rise to four dendrites. Nayak and Singh (1983) also observed marked differences between male and female prothoracic legs.

Using scanning electron microscopy and crystal violet diffusion, we checked the distribution of such chemoreceptors on each part of the legs - including the tibiae and the femurs - of flies of both sexes from our Canton-S strain. Quantitative results are shown in Figure 4. Although the femurs are devoid of chemosensitive hairs, such bristles are present on both tibial and tarsal segments in both sexes. However, as previously reported, there is also a striking sexual dimorphism as far as the forelegs are concerned: males present 30% more sensilla than females, especially as far as the basitarsal segment is concerned. This segment carries the male's sex-comb. Meso- and meta-thoracic legs do not display any difference. Similar results were found for the legs of the sibling species, *D. simulans*. On average 35.9 chemosensilla were found on male prothoracic legs, as compared to 26.1 on female prothoracic legs and 23.5 on male mesothoracic legs. We found fewer sensilla than Nayak and Singh (1983) on the proximal segments, perhaps because they are much smaller. Further, the extra chemoreceptors present on the male's foreleg are grouped, which would tend to favour inter-receptor interactions.

All these results are consistent with an aphrodisiac-specific gustatory function for these sensilla

A

			TARSAL SEGMENTS				
Leg	Sex	Tibia	1	2	3	4	5
PRO	Female	8.2 (0.5)	6.7 (0.5)	4.0 (0.2)	4.0 (0.0)	2.0 (0.0)	2
	Male	8.6 (0.7)	10.0 (0.7)	7.6 (0.6)	6.0 (0.2)	5.8 (1.4)	2
MESO	Female	8.2 (0.5)	5.0 (0.0)	4.0 (0.0)	1.8 (0.4)	2.0 (0.0)	2
	Male	9.0 (0.8)	5.2 (0.4)	4.0 (0.0)	1.7 (0.5)	2.0 (0.0)	2
META	Female	8.9 (1.0)	7.2 (0.6)	4.4 (0.5)	1.8 (0.4)	2.0 (0.0)	2
	Male	9.0 (0.5)	7.5 (0.5)	4.2 (0.5)	1.5 (0.5)	2.0 (0.0)	2

B

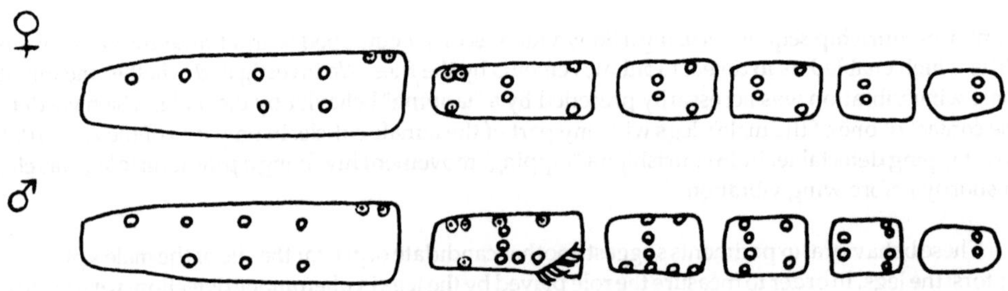

Figure 4 (A) Mean number of chemosensilla on the different parts of the prothoracic (PRO), mesothoracic (MESO) and metathoracic (META) legs in both male and female *Drosophila melanogaster* . Standard deviations are given in brackets. (B) Topographic distribution of trichoid sensilla on the left prothoracic leg of male and female *Drosophila melanogaster* (dorsal view). O - dorsally-oriented sensillum; ◉ - ventrally-oriented sensillum.

which are specific to the male's forelegs. This possibility raises a number of topics for future investigation:

1) How is such a lipophilic stimulus able to penetrate the sensilla and reach the membrane receptors?

2) Are these sensilla specialized in pheromone-detection and insensitive to salt and sugar, the target substances of most tarsal chemoreceptors (Dethier, 1971; Rodrigues and Siddiqi, 1978)? Or do some sensilla contain a receptor specific to aphrodisiac pheromones? (It is interesting to note that the stimulation of these sensilla, by sugar or by pheromones, triggers proboscis extension, in the form of the "tarsal reflex" or "licking".)

3) Are there similar receptors on the proboscis which could be stimulated during licking behavior? If the antennae carried such receptors, this might explain some of Robertson (1983)'s results.

4) Does the mechano-receptor which is normally found alongside the leg gustatory chemoreceptors play a role in inducing "licking" behavior?

5) Carson and Bryant (1979) observed significant changes in such sensilla during incipient speciation in *Drosophila silvestris*. Are there geographical variations in the distribution of such sensilla which correspond to the geographical hydrocarbon variants which exist in *D. melanogaster*? Do the closely-related species of the *D. melanogaster* species sub-group show variation for this character?

6) Where do the neurons involved in coding pheromone information project to in the thoracic ganglion? Are they organized differently in the two sexes, as in the case in the macroglomeruli found in the antennal lobe of the cockroach, where projections from a large number of pheromone sensilla converge?

The current research projects of our group are oriented towards finding answers to these questions.

REFERENCES

Bartelt, R. J., Schaner, A. M. and Jackson, L. L., 1985, *cis*-vaccenyl acetate as an aggregation pheromone in *Drosophila melanogaster*, *J. Chem. Ecol.*, 11:1747-1756.

Beckman, L. and Johnson, F., 1964, Esterase variations in *Drosophila melanogaster*, *Hereditas*, 51:212-220.

Begg, M. and Hogben, L., 1946, Chemoreceptivity of *Drosophila melanogaster*, *Proc. Roy. Soc. (London) B* ,133:1-19.

Bell, W. J. and Cardé, R. T.,1984, "Chemical Ecology of Insects", Chapman and Hall, London.

Borst, A, 1983, Computation of olfactory signals in *Drosophila melanogaster*, *J. Comp. Physiol*, 152:373-383

______, 1984, Identification of different chemoreceptors by electroantennogram-recording, *J. Insect Physiol.*, 30:507-510.

Butterworth, F. M., 1969, Lipids of *Drosophila*: a newly detected lipid in the male, *Science*, 163:1356-1357.

Carson, H.L. and Bryant, P.J., 1979, Change in a secondary sexual character as evidence of incipient speciation in *Drosophila silvestris*, *Proc. Natl. Acad. Sci USA*, 78:3940-3944.

Chen, Z., Pace, U., Ronen, D. and Lancet, D., 1986, Polypeptide gp95: a unique glycoprotein of olfactory cilia with transmembrane receptor properties, *J. Biol. Chem.* 261:1299-1305.

Dethier, V.G., 1971, "The Physiology of Insect Senses", Chapman and Hall, London.

Fuyama, Y., 1976, Behavior geentics of olfactory responses in *Drosophila*. I. Olfactory and strain differences in *Drosophila melanogaster*, *Behav. Genet.*, 6:407-420.

Hannah-Alava, A, 1958, Morphology and chaetotaxy of the legs of *Drosophila melanogaster*, *J. Morphol.*, 103:281-310.

Jallon, J-M., Antony, C. and Benamar, O., 1981, Un anti-aphrodisiaque produit par les mâles de *Drosophila* et transféré aux femelles lors de la copulation, *C. R. Acad. Sci. Paris.*, 292:1147-1149.

Johnson, F.M., Kanapi, C.G., Richardson, R.H., Wheeler, M.R. and Stone, W.S., 1966, An operational classification of *Drosophila* esterases for species comparisons, *Univ. Texas. Publ. Genet.* 6615:517-532.

Kikuchi, T., 1973, genetic alteration of olfactory function in *Drosophila melanogaster*, *Jap. J. Genet.*, 48:105-118.

Klein, U., 1987, Snesillum-lymph proteins from antennal olfactory hairs of the moth *Antheraea polyphemus* (Saturnidae), *Insect Biochem.*, 17:1194-1204.

______ and Keil, T.A., 1984, Dentritic membrane from insect olfactory hairs: isolation method and electron microscopic observations, *Cell. Mol. Neurobiol.*, 4:385-396.

Lancet, D. and Pace, U., 1987, The molecular basis of odor recognition, *Trends Biochem. Sci.*, 12:58-62.

Link, B., 1983, Raster und TEM Analyse der Sensillen des mesothorakalen Beines und des dritten Antennensegmentes von *Drosophila melanogaster*. Diploma thesis, University of Fribourg, Switzerland.

Manning, A., 1959, The sexual isolation between *Drosophila melanogaster* and *Drosophila simulans*, *Anim. Behav.* 7:60-65.

Mindek, G., 1968, Proliferationsund Transdeterminationsleistungen der weiblichen Genital-Imaginalscheiben von *Drosophila melanogaster* nach Kultur *in vivo*, *Wilhelm Roux Archiv.*, 161:249-280.

Nayak, S.V. and Singh, R.N., 1983, Sensilla on the tarsal segments and mouthparts of adult *Drosophila melanogaster* Meigen (Diptera: Drosophilidae), *Int. J. Insect. Morphol. and Embryol.*, 12:273-291.

Robertson, H.M., 1983, Chemical stimuli eliciting courtship by males in *Drosophila melanogaster*, *Experientia*, 39:333-335.

Rodrigues, V., 1980, Olfactory behaviour of *Drosophila melanogaster*, *in* "Development and Neurobiology of Drosophila", O. Siddiqi, Babu, P. and Hall, J., eds, Plenum, New York.

______ and Siddiqi, O., 1978, Genetic analysis of chemosensory pathway, *Proc. Indian Acad. Sci.*, 87b:147-160.

Schlein, Y., Galun, R. and Ben-Eliahu, M.N., 1981, Receptors of sex pheromone and abstinons in *Musca domestica* and *Glossina morsitans*, *J. Chem. Ecol.*, 7:291.

Siddiqi, O., 1983, Olfactory neurogenetics of *Drosophila*, pp. 242-261*in*, "Genetics: New frontiers", Chopra, V.L., Joshi, B.C., Sharma, R.P. and Bawal, H.C., eds, OUP and IBH, Oxford and New Delhi.

Singh, R.N. and Nayak, S.V., 1985, Fine structure and primary sensory projections of sensilla on the maxillary palps of *Drosophila melanogaster* Meigen, *Int. J. Insect Morphol. and Embryol.*, 14:291-306.

Stocker, R.F. and Gendre, N., 1988, Peripheral and central nervous system effects of Lozenge^3, a *Drosophila* mutant lacking basiconic antennal sensilla. *Dev. Biol.* 126.

______, Singh, R.N., Schorderet, M. and Siddiqi, O., 1983, Projection patterns of different types of antennal sensilla in the antennal glomeruli of *Drosophila melanogaster*, *Cell Tissue Res.*, 232:237-248.

Tompkins, L. and Hall, J. C., 1981, The different effects on courtship of volatile colpounds from mated and virgin *Drosophila* females, *J. Insect Physiol.* 27:17-21.

Venard, R. and Pichon, Y., 1981, Etude électroantennographique de la réponse périphérique de l'antenne de *Drosophila melanogaster* à des stimulations odorantes, *C.R. Acad. Sci. Paris* 232:839-849.

______, 1984, Electrophysiological analysis of the peripheral response to odours in wild-type and smell-deficient *olfC* mutants of *Drosophila melanogaster J. Insect Physiol.*, 30:1-5.

Venkatesh, S. and Singh, R.N., 1984, Sensilla on the third antennal segment of *Drosophila melanogaster* Meigen (Diptera: Drosophilidae), *Int. J. Insect Morphol. Embryol.*, 13:51-63.

Vogt, R.G., Riddiford, L.M. and Prestwich, G.D., 1985, Kinetic properties of a sex pheromone degrading enzyme: the sensillar esterase of *Antheraea polyphemus*, *Proc. Natl. Acad. Sci. USA*, 82:8827-8831.

Wright, T.R.F. 1963, The genetics of an esterase in *Drosophila melanogaster*, *Genetics*, 48:787-801.

THE ANTENNAL GLOMERULUS AS A FUNCTIONAL UNIT OF ODOR CODING IN DROSOPHILA MELANOGASTER

Veronica Rodrigues and Ludwin Pinto

Molecular Biology Unit
Tata Institute of Fundamental Research
Homi Bhabha Rd.
Bombay 5, India

ABSTRACT

Functional mapping experiments using 2-deoxyglucose autoradiography have led to the hypothesis that odor quality is represented as a spatial activity map among the neurons of the olfactory lobe. We are attempting to identify the neural basis of this sensory map in the antennal lobe of *Drosophila*. Our developmental and functional analysis suggests that the odor map is generated by the projection of broadly specific receptor neurons to the glomeruli. The activity generated among the glomeruli by these neurons is further tuned by the activities of the neurons local to the lobe.

INTRODUCTION

We are interested in how olfactory information is encoded and processed in the nervous system. In vertebrates as well as in insects, the first synaptic relay station of the olfactory pathway is the glomeruli of the olfactory lobes (Shepherd, 1972; Tolbert and Hildebrand, 1981; Chambille et al,1980). This organization of synaptic interactions between the receptor neurons and the lobe interneurons in discrete clusters makes the glomeruli a possible neural substrate for odor coding. In vertebrates the method of functional mapping using 2-deoxyglucose autoradiography has been used to localize the activity in the neurons of the olfactory pathway in response to olfactory stimulation. These results suggested that odor quality information is mapped spatially in the olfactory lobe (Jourdan et al, 1980; Lancet et al, 1982). In insects, electrophysiological recordings from the deutocerebral interneurons have revealed that the macroglomerulus of the male of several insect species is specifically involved in processing information about female sex pheromones (Boeckh and Boeckh, 1979; Matsumoto and Hidebrand, 1981). In *Drosophila melanogaster*, functional mapping experiments have shown that each odorant excites a specific subset of antennal glomeruli; the subsets are overlapping, but distinct for each odor class

(Rodrigues, 1988). This means that there exists in the brain of the fly a spatial map for each odor quality.

This proposal raises several interesting questions about infomation processing. For example, one can ask, if there are pre-determined sensory maps in the brain, to which incoming neuronal activity must be matched to give rise to perception. Are these maps genetically hard-wired or does plasticity during development play a significant role? These are some of the issues that we wish to address, and in this paper, we discuss our studies which begin to examine the basis of odor coding in the olfactory lobe. These questions are necessarily interdisciplinary, requiring details of structure and function as well as the development of the olfactory lobe. *Drosophila* offers the possibility of applying both classical as well as molecular genetic studies to study sensory coding.

Electrophysiogical recordings from the antennal sensilla of *Drosophila* have shown that most of the receptor neurons show a broad reaction spectra, responding to several odors, although at different thresholds (Siddiqi,1987). This implies that odor coding in these neurons occurs by an 'across-fibre' mechanism; that is the information is represented by activities in a population of neurons (Shepherd, 1981). How is the spatial map for each odorant generated? One possibility is that the sensory map reflects a precise connectivity of receptor neurons of determined specificity to distinct subsets of glomeruli. Postsynaptic neurons leaving each glomerulus would be expected to be generalized in their response patterns. From this it would appear that the glomeruli act merely as collection channels for olfactory information,without serving any integrative function. On the other hand, it is possible that the spatial activity domains arise, not only from the input information, but by complex excitatory and inhibitory influences of the interneurons in the antennal lobe.

RESULTS AND DISCUSSION

Each antennal lobe in *Drosophila* is 70μm in diameter and has been shown by cobalt backfilling studies to receive projections from the receptor neurons, which terminate in the glomeruli (Stocker et al, 1983). Each receptor neuron innervates a single glomerulus on the ipsilateral side and in most cases also sends a collateral branch to the corresponding glomerulus of the contralateral lobe. Two types of lobe interneurons have been described in the antennal lobe of *Drosophila* which are morphologically similar to those described in the moth (Fischbach and Borst, manuscript in preparation; Matsumoto and Hildebrand, 1981). The local lobe interneurons are anaxonic and arborize extensively in the lobe, interconnecting several glomeruli. Each output neuron originates from a single glomerulus and runs in the antennoglomerular tract to the calyx of the mushroom bodies and the lateral protocerebrum (Strausfeld,1976).

The glomeruli can be classified using a number of criteria as summarized in Fig. 1.

1. Toluidine blue-methylene blue staining of 3μm sections through the antennal lobe has revealed the presence of 22 glomeruli in each lobe. The glomeruli can be recognized by their shape and position in the lobe (Pinto et al,1988). The number of glomeruli is fairly invariant from individual to individual and between the sexes. The sizes of the glomeruli range between 9 and 25 μm in diameter. The presence of a small number of invariant glomeruli makes the antennal lobe in *Drosophila* an attractive system for the study of the genetic basis of odor coding.

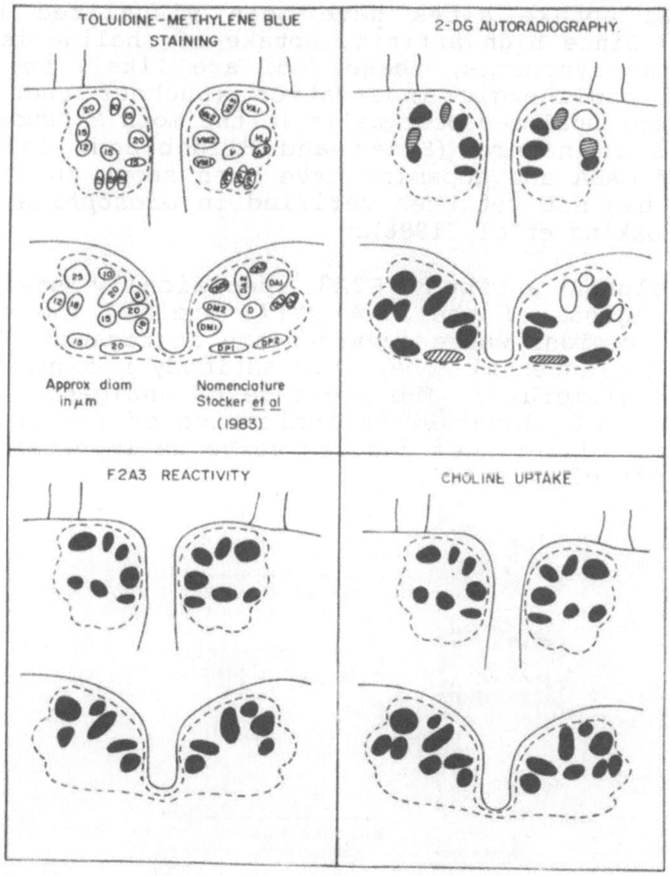

Figure 1. Identification of the antennal glomeruli in Drosophila using several different techniques. The total number of glomeruli visualized by toluidine blue-methylene blue staining is 22. Subsets of these are visualized by autoradiographic techniques and immunocytochemistry.

2. Stimulation of flies with six classes of odorants resulted in labelling of 16 of the 22 glomeruli in functional mapping experiments using 2-deoxyglucose autoradiography. The odors were classed according to the functional group of the chemical.

Behavioral, as well as electrophysiological experiments had previously established that the functional group of a chemical was an important parameter in determining its smell quality and each of these classes defined the sub- modality of olfaction (Borst, 1983; Siddiqi, 1987). The pattern elicited for each odorant class involves only a subset of the glomeruli and is overlapping between the different classes. The glomeruli that have not been labelled in these studies may be excited by stimuli other than those so far tested. Cobalt backfilling studies have shown that 3 glomeruli, VP1, VP2 and VP3 receive projections exclusively from the arista and are therefore unlikely to be excited by odors.

3. Specific uptake sites have been visualized in fifteen glomeruli. Since high affinity uptake of choline is linked to acetylcholine synthesis, these foci are likely to receive at least some cholinergic innervation (Buchner and Rodrigues, 1983). The antennal receptor cells in the moth *Manduca* sexta are cholinergic in nature (Sanes and Hildebrand, 1976). The presence of GABA and dopamine have been shown in other insect lobes, but has not yet been verified in *Drosophila* (Mercer et al, 1983; Hoskins et al, 1986).

4. A monoclonal antibody F2A3 specifically stains in the neuropile regions of the *Drosophila* brain. The staining is enhanced in regions where synapses are aggregated (Rane et al, 1987). In the antennal lobe, this antibody recognizes a subset of thirteen glomeruli. The presence of antigenic variability among subsets of glomeruli is indicative of functional and/or developmental differences and can serve as important molecular handles on the glomeruli.

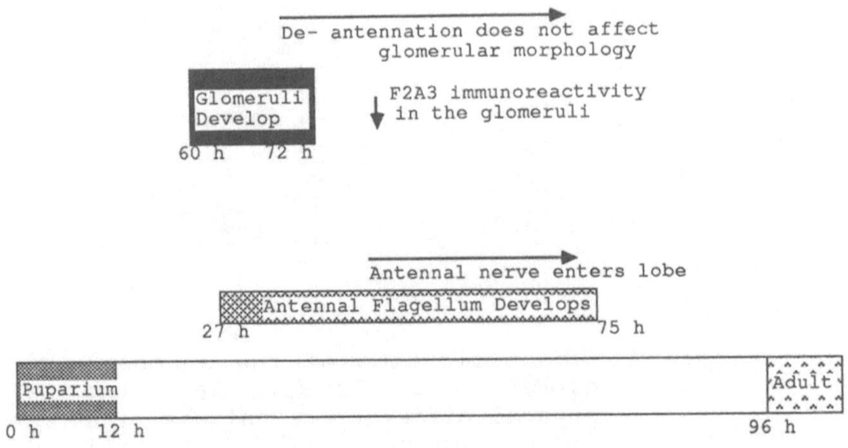

Figure 2 Development of the Drosophila antennal system. Staging of the pupae was done with respect to the time of formation of white pre-pupae.

We used histological methods (Pinto et al, 1988) to trace the development of the antennal glomeruli during pupation. Figure 2 summarizes these events, with reference to the formation of white pre-pupae. The formation of the glomeruli was observed to be complete by 72 hours after pre-pupa formation at 25°C. This time coincides fairly well with the arrival of the sensory axons into the lobe. F2A3 immunoreactivity develops after the glomeruli are morphologically comparable to the adult pattern. The cues involved in formation of the glomeruli in *Drosophila* are as yet unknown. In *Manduca sexta*, work by Hildebrand and his colleagues indicates that the lobe interneurons are pluripotent and the glomerular formation is influenced by the identity of the incoming neurons (Schneiderman et al,1982; Christensen and Hildebrand,1987a, b).

We tested the interactive influence of the afferent fibres by examining the glomerular morphology in flies in which the receptor neurons are either absent or reduced in number.Surgical removal of the antennal flagella was technically difficult in pupae less than 72 hours old. De-antennation in older pupae did not affect glomerular formation, even though the receptor terminals in the lobe were seen to have degenerated. The presence of sensory fibres in the lobe is therefore not required for the maintenance of the glomerular morphology. A similar finding was reported in the moth by Sanes and Hildebrand (1976). We have isolated an antennaless strain after screening of the progeny of males mutagenized with ethylmethanesulfonate. The phenotype is caused by mutation in two genes- a dominant autosomal and a recessive X-linked. Each of the mutant genes present alone does not result in any detectable phenotype in the antenna, but in combination lead to animals ranging from both antennae missing to relatively minor defects in a single antenna. Flies with either one or both antennae absent were processed for histology and 3µm sections through the antennal lobe were stained with methylene blue-toluidine blue. Reconstructions of camera lucidia drawings were examined and the results are summarized in Table 1. Flies in which both antennae are totally absent show an absence of all glomeruli in both lobes. Hence the presence of the incoming sensory neurons is essential for the formation of glomeruli. We have however not ruled out the possibility that the mutation directly affects the antennal lobe itself. In other insects, the formation of the glomerulus is believed to occur by the formation of an envelope of glial cells (Oland and Tolbert,1987).

Table 1. Effect of a lack of sensory neurons on the development of antennal glomeruli.

Antennal flagellae		Antennal nerve		Antennal glomeruli.	
right	left	right	left	right	left
present (6 preparations)	present	present	present	22	22
absent (5 preparations)	absent	absent	absent	0	0
present (18 preparations)	absent	present	absent	22	17*

* V,VP1,VP2,VP3, and D are not seen.

On the other hand, flies in which only one antenna was intact showed the presence of all glomeruli in the ipsilateral lobe. In the contralateral lobe, only the glomeruli receiving bilateral projections were detected (Stocker et al, 1983). The glomeruli designated V, D, VP1, VP2, VP3, (see fig.1) which have been found to receive mainly ipsilateral projections were not seen in any of our preparations. These results indicate that even a reduced amount of innervation is sufficient for the induction of glomeruli.

Previous work by Stocker, (1988) with the mutant *lozenge* had shown that the absence of subsets of sensilla on the antennal flagella leads to absence of a single glomerulus to which these sensilla project. These results are consistent with the assumption that the antennal lobe neurons act as collecting channels of information from sets of receptor neurons. The projection of the receptor neurons are convergent-- about 1200 sensory neurons from each antenna connect to a few hundred lobe output neurons in 22 glomeruli. The 2-deoxyglucose mapping data has allowed us to classify the receptor neurons into two classes on the basis of their excitation patterns (Rodrigues, 1988).

Class I neurons: Stimulation with most attractant odors e.g. acetate esters, leads to a labelling of the glomeruli in the ipsilateral lobe. On the contalateral side, the branches of the receptor neurons can be seen to be labelled on the peripheries of the glomeruli, but these evoke no excitation in the output neurons (Rodrigues and Buchner, 1984). The output neurons in the ipsilateral side are strongly labelled and these can be traced as they run in the antennoglomerular tract. Since most of these neurons innervate bilaterally, we speculate that they excite the ipsilateral lobe neurons but inhibit those on the contralateral side, presumably exerting their inhibitory influence through the local interneurons. This observation is akin to contralateral inhibition which is well known in sensory physiology, where it acts to amplify the difference in the signal received by two populations of receptor neurons.

Class II neurons: Stimulation with benzaldehyde leads to labelling of the corresponding glomeruli in both lobes. The output neurons are strongly labelled and these excite strong activity in both calyces. The receptor neurons excited by this repellent odor elicit excitation in the postsynaptic fibers on both sides.

A possible means to study neuronal integration is to examine the effect of odor mixtures on the sensory maps generated. If the glomeruli are merely collection stations for sensory input, then the map elicited by a mixture of odorants would be expected to be the sum of that elicited by each of the individual odors. Unilateral stimulation of ^{3}H-2-deoxyglucose flies with a mixture of ethyl acetate and benzaldehyde resulted in a complex pattern of labelling. On the ipsilateral side, all the foci observed with the individual odors could be recognized. On the contralateral side, labelling of discrete glomeruli could not be resolved. There was intense labelling of the lobe, but in none of the six preparations scrutinized could the output neurons be detected. In autoradiographs exposed for shorter times, it was possible to reconstruct some of the labelled profiles from camera lucida drawings of serial sections. These were comparable to the Golgi reconstructions of the local lobe interneurons. In Table 2a, the effects on three representative glomeruli are shown, although these are not the full pattern elicited by these odorants. 'Ester' neurons which activate ipsilateral glomeruli (denoted as + in the table) also induce strong inhibition in the corresponding glomerulus contralaterally. (-) DA2 is not excited by acetate esters (0) but is strongly excited bilaterally by aldehyde. (++) In combination, an inhibition of the contralateral glomerulus is observed. In DM1, labelling is observed by ester stimulation, but not by aldehyde. As expected, this glomerulus is excited only ipsilaterally. We propose that the class I receptor neurons lead to an inhibition of several output neurons. The inhibition is not restricted to the neurons with which they make contact. It is likely that the inhibition across glomeruli is mediated by the local lobe interneurons which have been described to be interconnecting several sets of glomeruli. A possible behavioral correlate for this observation is the finding that mixtures of chemicals are often more attractive than their components. In osmotropotactic experiments, Borst and Heisenberg (1982) found that the mixture of banana and benzaldehyde elicited stronger turning responses than banana alone. Benzaldehyde alone did not elicit any osmotropotactic response, but there was an inhibition of locomotion. In freely moving flies, in a Y maze olfactometer, benzaldehyde was found to elicit strongly negative chemotactic responses from flies (Siddiqi, 1987).

Table 2a. Effect of stimulation with odor mixtures on glomerular activity.

Stimulus	VA1		DA2		DM1	
	Ipsi.	Contra	Ipsi	Contra	Ipsi	Contra
Ester. (8 preps)	++	-	0	0	++	-
Aldehyde. (6 preps)	+-	+-	++	++	0	0
Ester+ Aldehyde (6 preps)	++	-	++	-	++	-

Table 2b. Evidence for inter-glomerular inhibition during stimulation with acetic acid.

Stimulus	VA1		VM1		DA3	
	Ipsi.	Contra	Ipsi.	Contra	Ipsi.	Contra
Air. (4 preps.)	+	-	+	-	0	0
Acetic acid (4 preps,)	0	0	++	-	++	-

Further evidence for information processing in the antennal lobe comes from the examination of the patterns generated by acetic acid and air (Table 2b). Stimulation of the fly with odor-free air resulted in activation of glomeruli VA1 and VMI. Since all the odorants are delivered in puffs of air, we expect these two glomeruli to be labelled as a background in all preparations. Stimulation with acetic acid even after several hours, did not result in labelling of VA1 . There is therefore evidence for inhibitory connections between glomeruli in the same lobe.

In summary, these preliminary studies lead us to postulate that while the receptor neurons dictate the bulk of the sensory map, this is tuned to a significant extent by excitatory and inhibitory interactions between the antennal lobe interneurons. The interpretation of these findings will remain speculative until these results can be analyzed in light of a detailed synaptology of the antennal lobe.

REFERENCES

Boeckh, J., and Boeckh, V., 1979, Threshold and odor specificity of pheromone-sensitive neurons in the deutocerebrum of *A. pernyi* and *A. polyphemus* (Saturnidae). J. Comp. Physiol.A. 132: 235-242.

Borst, A., 1983, Computation of olfactory signals in *Drosophila melanogaster*. J. Comp. Physiol.A., 152: 373-383.

Borst, A., and Heisenberg, M., 1982, Osmotropotaxis in *Drosophila melanogaster*. J. Comp. Physiol. A., 147: 479-484.

Buchner, E., and Rodrigues, V., 1983, Autoradiographic localization of ^{3}H-Choline uptake in the brain of *Drosophila melanogaster*.Neurosci. Letters 42: 25-31.

Chambille, I., Rospars, J. P., and Masson,C., 1980, The deutocerebrum of the cockroach *Blaberus craniifer* Burm. Spatial organisation of the sensory glomeruli. J. Neurobiol. 11: 135-157.

Christensen, T. A., and Hildebrand, J. G., 1987a, Male-specific, sex pheromone-selective projection neurons in the antennal lobes of the moth *Manduca sexta*. J. Comp. Physiol.A., 160:553-569.

Christensen, T. A., and Hildebrand, J.G., 1987b, Functions, organisation and physiology of the olfactory pathways in the lepidopteran brain. In: Arthropod brain: its evolution development, structure and function. Gupta, A.P. ed., Wiley and sons, New York.

Hoskins, S.G., Homberg, U., Kingan, T.A. Christensen, T.A., and Hildebrand, J.G., 1986, Immunocytochemistry of GABA in the antennal lobes of the sphinx moth *Manduca sexta*. Cell Tiss.Res. 244: 243-252.

Jourdan, F., Duveau, A., Astic, L., and Holley, A., 1980, Spatial distribution of ^{14}C-2-deoxyglucose uptake in the olfactory bulbs of rats stimulated with two different odors. Brain Res. 188 139-154.

Lancet, D., Greer, C.A., Kauer, J.S. and Shepherd, G., 1982, Mapping of odor-related neuronal activity in the olfactory bulb by high resolution 2-deoxyglucose autoradiography. Proc.Natl. Acad. Sci. U.S.A. 79: 670-674.

Matsumoto, S.G. and Hildebrand, J.G., 1981, Olfactory mechanisms in the moth *Manduca sexta;* response characteristics and morphology of central neurons in the antennal lobe. Proc. R.Soc. Lond. Ser. B. 213: 249-277.

Mercer, A.R., Mobbs, P.G, Davenport, A.P. and Evans, P.D., 1983 Biogenic amines in the brain of the honeybee, *Apis mellifera*.Cell Tiss. Res. 234: 655-677.

Oland, L. A., and Tolbert, L. P., 1987, Glial patterns during early development of antennal lobes of *Manduca sexta*: A comparison between normal lobes and lobes deprived of antennal axons. J. Comp. Neurol.255: 196-207.

Pinto, L., Stocker, R. F., and Rodrigues, V., 1988, Anatomical and neurochemical classification of the antennal glomeruli in *Drosophila melanogaster Meigen* (Diptera: Drosophilidae). Int. J. Ins. Morphol. and Embryol. in press.

Rane, N., Jithra, L., Pinto, L., Rodrigues, V., and Krishnan, K. S. , 1987, Monoclonal antibodies to synaptic macromolecules of *Drosophila melanogaster*. J. Neuroimmunol, 16: 331-344.

Rodrigues, V., 1988, Spatial coding of olfactory information in the antennal lobe of *Drosophila melanogaster*. Brain Res. 453: 299-307.

Rodrigues, V., and Buchner, E., 1984, 3-H-2-deoxyglucose mapping of odor-induced neuronal activity in the antennal lobes of *Drosophila melanogaster*. Brain Res. 324: 374-378.

Sanes, J.R., and Hildebrand, J.R., 1976, Acetylcholine and its metabolic enzymes in developing antennae of the moth *Manduca sexta*. Dev. Biol.52: 105-120.

Schneiderman, A.M. Matsumoto, S.G. and Hildebrand, J.G. 1982, Trans-sexually grafted antennae influence development of sexually dimorphic neurones in moth brain. Nature 298: 844-846.

Siddiqi, O., 1987,Neurogenetics of olfaction in *Drosophila melanogaster*. Trends Genet. 3: 137-142.

Shepherd, G.M.,1972, Synaptic organisation of the mammalian olfactory bulb. Physiol Rev. 864-917

Shepherd, G. M. 1981, The olfactory glomerulus: Its significance for sensory processing, In Brain mechanisms of sensation, Eds. Katzuki, Y., Nogren, K., and Sato. M., Pub. J. Wiley and sons pp 209-223.

Strausfeld, N.J., 1976, Atlas of an insect brain, Springer, New York.

Stocker, R. F., 1988, Peripheral and central nervous effects of *lozenge*[3] a *Drosophila* mutant lacking in basiconic sensilla. Dev. Biol. 127: 12-24

Stocker, R.F, Singh, R,N., Schorderet, M., and Siddiqi, O., 1983, Projection patterns of different types of antennal sensilla in the antennal glomeruli of *Drosophila melanogaster*. Cell Tiss. Res. 232: 237-248.

Tolbert L. P. and Hildebrand J. G. 1981, Organisation and synaptic ultrastructure of glomeruli in the antennal lobes of the moth *Manduca sexta*: A study using thin sections and freeze-fracture. Proc. R. Soc. Lond. B. 213: 23-32.

COMPLEX CHEMICAL MESSAGES IN *DROSOPHILA*

Jean-François Ferveur, Matthew Cobb and Jean-Marc Jallon

Laboratoire de Biologie et Génétique Evolutives
CNRS
91198 Gif-sur-Yvette, France

INTRODUCTION

Most animals use several types of sensory signals to communicate with homospecific and heterospecific individuals. The class of chemical messages known as *pheromones* was first established on the basis of a series of insect studies (Karlson and Butenandt, 1959). Pheromones were initially taken to be airborne substances; however, contact pheromones and pheromones transmitted in water have since been shown to exist throughout the animal kingdom. Pheromones may play various roles: for example, sexual attractants, or alarm or aggregation signals; further, they may be excitatory or inhibitory (Bell and Cardé, 1984).

Most studies of contact pheromones have concentrated on Dipteran species. In 1971, Carlson et al. identified the sex pheromone of the housefly *Musca domestica* as *cis*-9 tricosene (23 carbons), one of the main components of the female cuticle. Homologous substances have since been found in other species. They are all long-chain hydrocarbons with either double bonds or branchings. They are all necessary but not sufficient for the induction of male pre-copulatory behavior. Unlike butterfly or moth pheromones, these contact pheromones all have a relatively low activity, being active at a dose close to that present on a female. Finally, as contact pheromones, they are perceived only at very close distances or by direct contact.

Although the initial reasons for favouring *Drosophila* in studying the "genetic dissection of behavior" related to the mass of already-existing genetic data, and the relative ease with which mutations can be induced, several other advantages make this insect a preparation of choice for such an approach:.

1) courtship behavior generally lasts several minutes, allowing the discrete steps involved to be analysed;

2) low rearing-costs enable many behavioral comparisons to be made, and large-scale biochemical purification is possible;

3) there are a variety of closely-related species which show common qualitative traits in their courtship behavior, but with quantitative variations;

4) discrete mutations affecting sensory modalities (notably olfaction and gustation) have been isolated.

Drosophila pre-copulatory behavior has been particularly intensively investigated. Several kinds of sensory message are exchanged by sexual partners; chemical signals from females and acoustic signals from males seem to play major roles (Ewing, 1983). In this article we review aspects of the complex system of chemical messages found in *Drosophila melanogaster* and its sibling species *D. simulans*.

SEX PHEROMONES

The older literature is rich in indirect evidence for the exchange of chemical signals during *D. melanogaster* courtship. Direct evidence for a chemical message produced by females and directed towards males was provided by two articles published in 1980 (Venard and Jallon, 1980; Tompkins et al., 1980). These two groups used different bioassays for testing female extracts. Tompkins et al. (1980) used a "courtship index" (CI) which pooled all male behaviors, from the apparently passive Orientation through to Attempted Copulation. Our group chose a very specific response, male Wing Vibration, which is a key and discrete part of male courtship behavior, providing acoustic information for the female regarding at least the male's species, and perhaps his general "fitness". This behavior was used to calculate a "Sex Appeal Parameter" (SAP; Jallon and Hotta, 1979). Although the CI measure has been widely used, it confounds behaviors which can be induced solely by visual factors (e.g. Orientation; Jallon and Hotta, 1981) and other behaviors which probably require pheromonal stimulation before they are displayed (e.g. Licking).

Antony and Jallon (1982) clearly showed that the chemical signal involved is linked to the female's cuticle. The characterization of the chemical components was carried out by our own group, initially using the Canton-S (CS) strain of *D. melanogaster* (Antony and Jallon, 1982; Antony et al., 1985). Using a range of chromatographic methods, we were able to separate long-chain hydrocarbons, both saturated and unsaturated, with chain-lengths of between 23 and 29 carbons. The SAP bioassay showed that the most efficient "aphrodisiac" compound is *cis, cis*-7, 11 heptacosadiene (7,11 HD; 27 carbons), the most abundant hydrocarbon on the cuticle of females from this strain. Other substances, which are also potentially aphrodisiac, are also present on the female cuticle. In general they are between 25 and 29 carbons long, with at least one double-bond, in position 7 (Antony et al., 1985). These other substances do not reveal an *in vitro* stimulatory activity at a dose similar to that carried by a single female, but they may become stimulatory at higher doses. The most efficient substances, 7,11 dienes, are female-specific, but 7 monoenes such as 7-pentacosene (7-P; 25 carbons) are present on the cuticle of both sexes in similar amounts (see Table 1).

7-tricosene (7-T; 23 carbons), which does not induce male wing vibration in *D. melanogaster* CS males and which is relatively rare on *D. melanogaster* virgin females, is the most abundant hydrocarbon on the cuticle of *D. melanogaster* CS males (Antony and Jallon, 1982; Pechiné et al., 1985). Jallon (1984) raised the question as to whether this male-specific substance was a signal from males to females; no clear answer has yet been found. The possibility that 7-T serves an inhibitory function towards other males will be discussed in the next section.

These molecules, which have clear behavioral functions, are mixed in the cuticle with a number of other hydrocarbons whose physiological role has yet to be established. In *Musca domestica*, other hydrocarbons such as Z 14 tricosene-one, Z 9,10 epoxytricosene and various methyalkanes have recently been shown to play a behavioral role alongside 9 tricosene, especially in the control of male movement (Adams, 1986). This result suggests that other hydrocarbons may be found to play a role in *Drosophila* behavior, either singly or synergistically. We are probably only beginning to understand a complex pheromonal system.

Intraspecific Hydrocarbon Variation

All *D. melanogaster* females from African locations south of the Sahara (together with a few strains from equatorial islands in the Americas and Oceania) show high levels of the 5,9 form of heptacosadiene, and low levels of the 7,11 form which is predominant in all other strains, which, in this respect, are similar to CS (Jallon et al., 1986; Jallon, unpublished). Male flies show a latitudinal cline, with an equilibrium between two homologous 7 monoenes, 7-T and 7-P. As strain origin approaches the equator, 7-P becomes more abundant and 7-T less so, although intrastrain variation for this character is much greater than is the case for 5,9 HD / 7,11 HD variation in females (Luyten, 1982; van den Berg et al, 1983; Jallon, 1984; Jallon, unpublished) . The meaning of these variations is only partially understood. However, it seems clear from our experimental results that most males "prefer" females rich in 7,11 HD.

One problem is the accuracy of behavioral responses as a measure of stimulus efficacy.

Table 1 Main cuticular compounds for *D. melanogaster* and *D. simulans* males and females, and their behavioral consequences as assayed with SAP (see text for details).

	CUTICULAR PRODUCTION				STIMULATION OF MALE PRECOPULATORY BEHAVIOR	
	D. melanogaster		*D. simulans*		*D. melanogaster*	*D. simulans*
	MALE	FEMALE	MALE	FEMALE		
7-Tricosene 23:1 (7)	++	+	++	+++	-	++
7- Pentacosene 25:1 (7)	+	+	+	+	+	-
7,11 Heptacosadiene 27:2 (7,11)	-	++	-	-	++	-
7,11 Nonacosadiene 29:2 (7,11)	-	+	-	-	+	-

We and others describe mean responses of groups of males with a given genotype and history. This approach tends to undervalue inter-individual variability. It is well known that some pairs of flies copulate without any observable courtship. One possibility is that different males from a given strain may have different response thresholds to a given chemical. Further, synergies between cuticular compounds could induce different males to respond in different ways. This latter hypothesis has not been tested as all the substances have not yet been synthesized. Electrophysiological analysis of individual male responses should provide unambiguous answers. However, technical problems currently prevent this method from being used with this preparation. A final point that should be borne in mind is that although there is a clear correlation between induction of courtship and specific pheromonal molecules, chemical stimuli are not the only factors involved in the induction of courtship. Visual stimuli may induce at least early stages of courtship (Jallon and Hotta, 1981).

Interspecific Pheromone Variation

D. simulans is another cosmopolitan member of the *D. melanogaster* species sub-group which is often sympatric with *D. melanogaster*. 7,11 HD, the main component of *D. melanogaster* female cuticle, is completely absent from the cuticle of *D. simulans* females, which appear virtually unable to synthesize dienes. 7-T, the main hydrocarbon on *D. melanogaster* male cuticle, induces male courtship in *D. simulans* (Jallon, 1984). This substance is the major cuticular hydrocarbon in both female and male *D. simulans* flies. This phenomenon leads to more frequent male/male courtships than in *D. melanogaster*. However, the study of such male/male interactions also reveals the existence of inhibitory mechanisms tending to discourage courtship (Cobb and Jallon, submitted).

In both male and female *D. simulans* we have found a kind of hydrocarbon polymorphism similar to that observed in *D. melanogaster* males: there is an interstrain variation for the relative levels of 7-T and 7-P. Studies of interstrain and interspecific courtships suggest that 7-P may be able to induce courtship in *D. simulans* males (Cobb and Jallon, submitted).

Such studies have recently been extended to all eight related species in the *D. melanogaster* species sub-group. In three species there is a hydrocarbon sexual dimorphism, with female-specific signals which seem to play an aphrodisiac role, as in *D. melanogaster*. In the other five species, no qualitative differences are observed in the cuticular composition of the two sexes, 7-T being the main cuticular hydrocarbon(Jallon and David, 1987; Jallon et al., in preparation). This situation obviously does not allow for sex or species identification involving these substances. Nevertheless, 7-T appears to play a major role in inducing courtship in this group of species (Cobb and Jallon, submitted). Table 1 shows a summary of the situation in D. *melanogaster* and *D. simulans*.

Inhibitory Molecules

Two male-specific substances have been proposed as "anti-aphrodisiacs" able to influence male/male interactions and post-copulatory interactions in *D. melanogaster*. The first substance is a long-chain acetate, *cis*-vaccenyl actetate (cVA), which is synthesized in the ejaculatory bulb of mature males (Butterworth, 1969). Using the same bioassay involving a dead male covered with test substances as for sex pheromone characterization, Jallon et al (1981) showed that a male hexane wash contained strong inhibitory components, and that one of them, cVA, had an effect similar to that of the whole wash (around 60% reduction in wing vibration for a dose of 50 ng). This study also confirmed Butterworth's report that cVA was transferred to females during copulation. The authors suggested that the presence of this compound in the female might account for the widely-observed decrease in female attractiveness following copulation. However, three days later, cVA had disappeared, and the amount of sex pheromone had decreased, suggesting a new way for female "sex-appeal" to be controlled (Jallon et al., 1982).

The inhibitory effect of this substance was confirmed by two studies which used topical applications of an acetone solution containing cVA. Mane et al. (1983) observed a 30% decrease in the number of females mating in a group of wild-type females topically treated with 200 ng of cVA. Zawitowski and Richmond (1986) observed more marked effects, as measured by the CI (60% decrease) and the number of females copulating (50% decrease). Clear dose effects were also shown in the range 50 - 200 ng. These experiments clearly show that cVA has an inhibitory effect *in vitro* (Table 2).

Controversy has recently risen with regard to these results, initially due to experiments trying to describe the physiological consequences of this inhibitory property.

It had previously been shown that esterase 6, a polymorphic carboxylesterase found in *D. melanogaster* was also found in the ejaculate and was transferred to females. Mane et al. (1983) showed that 1) esterase 6 could hydrolyze cVA; 2) the reaction product, *cis*-vaccenol (cVOH), was anti-aphrodisiac, showing an efficiency equal to that of cVA in *in vitro* experiments similar to those described above. Zawitowski and Richmond (1986) confirmed this last result, and also showed that a possible synergy existed between cVOH and cVA, but found that at low doses cVOH showed a greater inhibitory effect than cVA. These data suggested that the transfer of esterase 6 was able to prolong the effect of cVA through the action of its hydrolysis product, cVOH. Vander Meer et al. (1986) found around 1.6 µg of cVA in Canton-S male ejaculatory bulbs, and33-55% of this level in females just after mating. However, they could not detect any cV0H in the reproductive tract of mated females at any time after mating, although cVA <u>was</u> present. Further, the decrease in cVA levels in females mated to wild-type males or males from a mutant strain bearing the esterase 6 nul allele were very similar, suggesting that esterase 6 does not play any role *in vivo* in the fate of cVA.

Scott and Richmond (1987) studied the timing of cVA transfer during copulation. Their results were somewhat variable, but they observed that at least four minutes copulation were required for 50 ng of cVA to be transferred to the female. Given that Tompkins and Hall (1981) reported that most of the reduction in female attractiveness occured in the first three minutes of copulation, Scott and Richmond (1987) claimed that they had found "evidence against an anti-aphrodisiac role for *cis*-vaccenyl acetate", even though this interpretation was only based on a correlation between two results obtained under different experimental conditions.

Despite the views of Scott and Richmond, cVA remains a good candidate for *in vivo* inhibition of male *D. melanogaster* courtship, given that it has a clear inhibitory effect at low doses *in vitro*. The means by which cVA might be transmitted from the mated female to the male has yet to be determined. We have previously suggested that the extrusion of the ovipositor, typical of fertilised females, might help release cVA in the first few hours after copulation. This suggestion was based on the fact that decaptitated females do not extrude and are courted similarly, whether they are virgin or have recently mated (Jallon, 1984). Supporting evidence can be found in Bartelt et al (1985)'s data; they found that females release cVA when they lay their eggs. Another question is whether cVA is involved in male/male courtship inhibition. If this is the case, it will require some kind of spraying mechanism.

Table 2 Comparative inhibitory effects of *cis*-vaccenyl acetate and 7-tricosene on *D.melanogaster* Canton-S male precopulatory behaviors, tested with mature females (see text for full explanation).
* 7-pentacosene rich Tai males were used. † - 7-pentacosene was the test substance.

BIOASSAY	QUANTITY (ng)	% INHIBITION
***cis*-VACCENYL ACETATE**		
SAP	55	60
(Jallon et al. 1981)	360	87
Mating level (Mane et al., 1983)	200	21
CI (normal female)	50	22
(Zawitowski and Richmond, 1986)	200	65
CI (headless female)	200	47
(Scott et al., 1988)	200*	29
7-TRICOSENE		
CI (normal female)	100	-
(Scott and Jackson, 1988)	400	44
CI (headless female)	50	-
(Scott and Jackson, 1988)	100	44
CI (headless females)	200	58
(Scott et al., 1988)	200*	10
	200†	19
	200*†	50

The second male-specific molecule which has been proposed to have an inhibitory role in *D. melanogaster* is 7-T, the most abundant cuticular hydrocarbon on mature males from non-equatorial *D. melanogaster* strains(400-500 ng per mature Canton-S male; Antony and Jallon, 1982; Jallon, 1984). As previously described, 7-T plays an aphrodisiac role in *D. simulans*.

Scott et al. (1988) showed that the topical application of 7-T in an acetone solution on *D. melanogaster* females tended to decrease their attractiveness for CS males. They also found dose-related effects. The smallest dose that significantly decreased the CI was 100 ng. With a dose of 400 ng the CI was reduced by 44%. The effect was more pronounced when the females were decapitated, and therefore immobile. The results for mating were somewhat different: over a 10-minute period, only females treated with 400 ng mated significantly less than the controls. When the observation period was extended to 20 or 30 minutes, no significant differences were found between control and experimental females. These effects were restricted to 7-T, as the position isomers 5-T and 9-T showed no such effects in similar doses and 7-P was much less inhibitory. Interestingly, whilst 7-T rich CS males appear to be inhibited by 7-T, this is not the case with males from the African Tai strain, which are rich in 7-P (Jallon, 1984). These males, which are not stimulated by 7-P at up to 1 µg (Jallon, unpublished) appear to be inhibited by 7-P and cVA, but not by 7-T (Scott and Jackson, 1988).

This apparently inhibitory role for 7-T in certain strains needs to be placed in the context of post-copulatory chemical communication between the sexes. Scott (1986) showed that 7-T is transferred from CS males to females during copulation. Whilst low levels of 7-T can be detected on virgin CS females (around 20 ng), this rises to 60-80 ng per female just after mating. Similar results were found by Camacho (1986), although absolute levels were somewhat lower (50 ng). Following intense courtship, but without copulation, the 7-T level may reach half this figure. However, 7-T levels decrease after copulation, especially during the first three hours.

Despite these results, the transfer of 7-T alone does not fully explain the marked decreased of female attractiveness which is observed after copulation. Indeed, in Scott's experiments the topical application of 70 ng of 7-T did not produce a large effect on the CI, even if the value is statistically significant from that observed in decapitated controls. Further, females tend to decrease their locomotor activity after mating.

One general methodological problem with such experiments is the difficulty of evaluating the real levels of a given molecule on the cuticle. Our understanding of the molecular dynamics of *Drosophila* cuticle is extremely incomplete: topical application may not lead to the molecule being present in the same amounts or with the same intermolecular interactions as in the intact fly. Further, the observed reduction in courtship may be due to a masking of the female's pheromone rather than to any genuine inhibitory effect.

Another problem for this interpretation of the role of 7-T - and indeed for our understanding of the pheromonal induction of *D. melanogaster* courtship in general - is that 7-T rich flies of both sexes may be courted by *D. melanogaster* CS males (Cobb and Jallon, submitted). The most important thing to remember is that in bioassays 7-T is not excitatory for CS males (Antony et al., 1985). Any observed courtships are probably due to the effect of other cuticular hydrocarbons, or to non-pheromonal factors.

AGGREGATION PHEROMONES

Using a Y-maze olfactometer, Venard (1980) observed that mature *D. melanogaster* Canton-S males - but not females - were able to attract mature females at a few cm distance, but that the reverse was not true. Moreover, males were also attracted by other males, although the effect was not very strong (attractivity index values: 0.25 for females, 0.08 for males) For females, a dose-response effect was observed (1-25 males). Maximum attraction was obtained 5-10 minutes after the beginning of the experiment.

Aggregation pheromones have been shown to exist in several species of *Drosophila*. In *D. simulans*, as in a number of other species, this substance is produced by males and attracts both male and female flies in an 80 cm wind-tunnel (Schaner et al., 1987). Male extracts are attractive to other flies, and also seem to increase activity levels, although quantitative responses are highly variable. Following fractionation, the most active component was shown to be cVA.

A clear dose-response relationship was established, with a threshold lower than 1 ng, and a plateau value close to a single-fly equivalent dose. cVA also acts as a powerful synergist of attractants such as fermented fruit and acetone (Table 3). Schaner et al. also showed that while cVA was absent from virgin females, half of the cVA content of a male was transferred to the female during copulation. The amount present in the female genital tract then decreased regularly, as in *D. melanogaster* females, and could subsequently be found in the medium upon which the females had been kept. Only 10% as much cVA was deposited by virgin males. This suggests an important ecological role in marking oviposition sites.

Bartelt et al. (1985) looked for a similar aggregation pheromone in *D. melanogaster* but could not find one in the true sense of the term. Mature male extracts were not attractive *per se*, but, as with *D. simulans*, they did increase the attractiveness of food and acetone (Table 3). This effect

Table 3 Mean numbers of flies caught in a windtunnel bioassay. Catches are out of 1000 flies tested. Data from Bartelt et al. (1985) and Schaner et al. (1987). n.s. - not significant; * around 1 fly equivalent; ** 150 µg

	D. simulans	*D. melanogaster*
Control	4.7	0.1 n.s.
Male extract	19.1 *	0.9 n.s.
cVA	24.8 *	-
Food	43.2	7.5
Acetone	-	3.0
Food + male extract	-	21.5 *
Food + low cVA	103.6*	4.2*
Acetone + high cVA	-	8.6**

was again clearly linked to cVA, and was absent when oleylacetate or cVOH were tested. The results for *D. melanogaster* are not as strong as for *D. simulans* . This may be because either the overall experimental conditions, or the age of the *D. melanogaster* flies used, was not optimal. It is probable that cVA was one of the substances which attracted *D. melanogaster* females in Venard (1980)'s experiments, which involved much shorter distances.

These data suggest a dual role for cVA, both as an attractant/synergist and as a close-range inhibitor of courtship. However, these functions are based on *in vitro* experiments, and the influence of cVA on *D. simulans* courtship has yet to be established.

ELEMENTS OF GENETIC CONTROL

Given that both *D. melanogaster* and *D. simulans* have strains showing different hydrocarbon phenotypes, it was a relatively simple matter to make intraspecific hybrids between the two hydrocarbon morphs, in order to get some idea of the nature of the genetic control of hydrocarbon production. It should be remembered that hydrocarbons are not proteins, and are therefore not primary gene products. The genes we are interested in therefore probably code for enzymes involved in the biosynthesis of these substances.

We initially studied the relationship of 7-T and 7-P levels, which vary in male *D. melanogaster* and in *D. simulans* flies of both sexes. Reciprocal F1 hybrids between strains often suggested that the production of 7-T was sex-linked, but important inter-strain variation exists, andthe quantitative importance of such sex-linked genes can be extremely variable. In *D. simulans* such genes appear to play an important role, whereas in *D. melanogaster* control appears to be largely autosomal (Jallon, 1984).

These results led us to use a slightly different approach, involving attached-X females, the male and female offspring of which bear the same X chromosome(s) as their same-sex parent. We studied the role of the X chromosome on 7-T and 7-P production in *D. simulans* flies, where both sexes show high 7-T or 7-P levels, using the 7-P rich Yaounde strain and the 7-T rich C(I)RM strain, in which the females have attached X chromosomes. This latter strain was a gift from Dr Watanabe.

Extracts of individual 4-day old flies were made according to the method of Jallon (1984) and injected into a gas chromatograph. Compounds were separated in a 25 m long capillar column (CP Sil 5), coupled to a hydrogen flame detector which was linked to a Spectraphysics analysis

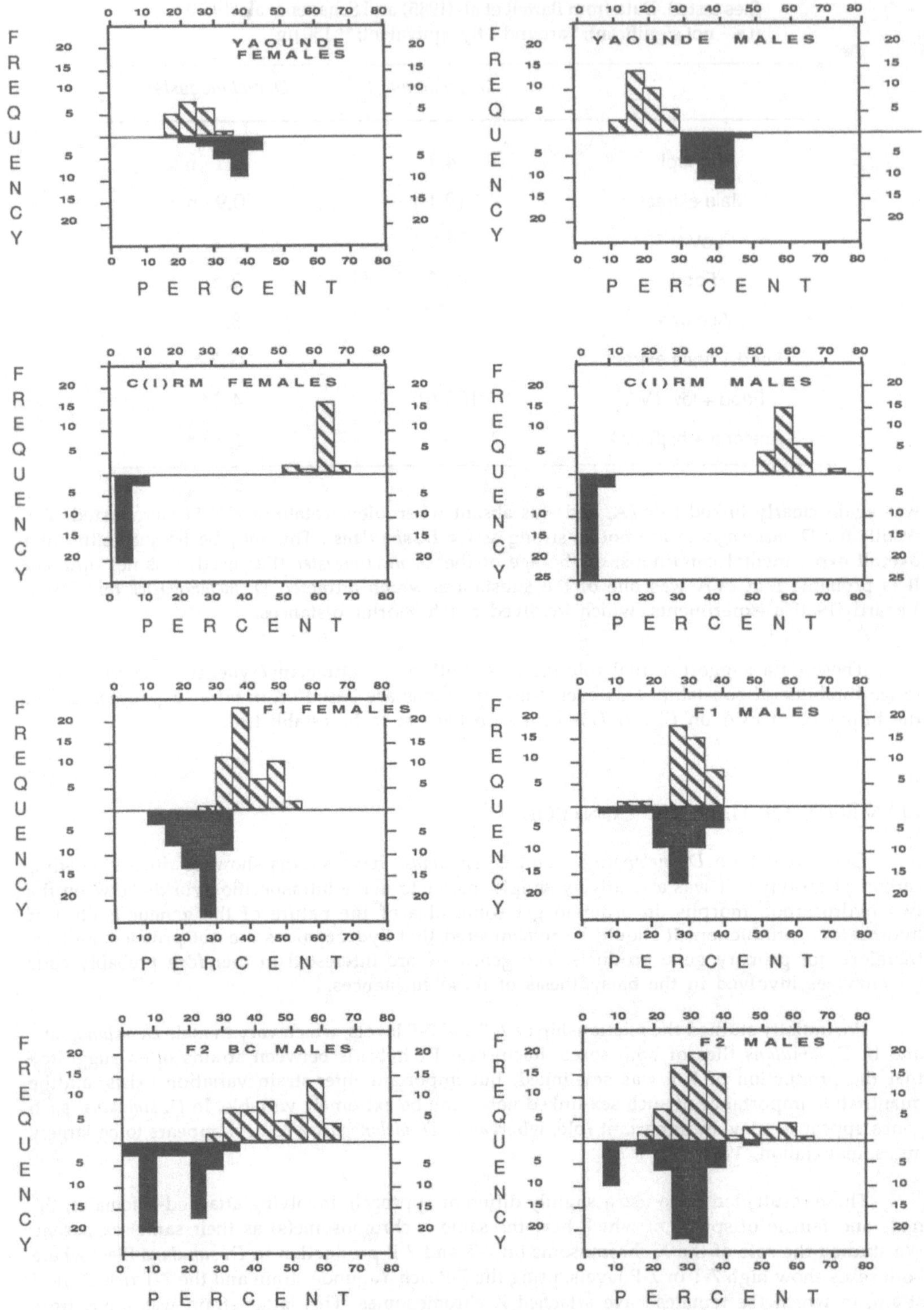

Fig. 1. Frequency histograms showing percentage of 7-tricosene (hatched) and 7-pentacosene (black). Percentages grouped in 5% classes. F1 flies are the product of C(I)RM females x Yaounde males. F2 males are the product of F1 females x F1 males. F2 BC males are the product of F1 males and C(I)RM females. Data for F2 and F2 BC females are not shown.

unit. Each compound was characterized by its retention time and by its quantity, in terms of the surface area of its peak. The proportions of 7-T and of 7-P were calculated on the basis of the total hydrocarbon mixture detected. Tests were carried out on both sexes of the parent strains and the F1 hybrids between C(I)RM females and Yaounde males, and on male F2 produced by F1 x F1 crosses and by F1 x parental strain (backcross - BC) crosses.

Histograms of the proportions of 7-T and 7-P were calculated for each group of flies and are shown in Figure 1. Female levels of 7-T tend always to be higher than those for male flies. In both sexes, parental levels are very different for both 7-T and 7-P: in C(I)RM flies 7-T is the main compound and 7-P levels are relatively low, whereas the reverse is true for Yaounde flies.

7-T Production

The histograms for both the parent strains and for the F1 flies are clearly unimodal. F1 males have 7-T levels which are different from those of their Yaounde fathers, but which nevertheless resemble Yaounde males more closely than they do C(I)RM males. F1 males are also different from their sisters, which show values more or less intermediate between the two parental values. Given that the two F1 sexes share their autosomes, the observed differences in 7-T must reflect the influence of the sex chromosomes, with the Yaounde chromosomes being dominant for this character. However the autosomes also play a role, as is shown by the fact that the F1 flies are not identical to their same-sex parents. The role of an autosome is also suggested by the fact that both F2 and F2 BC flies appear to show a bimodal distribution for 7-T values. A X^2 test confirmed that the observed distribution corresponds to that of a classic Mendelian distribution, with two equal modes for F2 BC males, and two modes, one of 75% (25+50), one of 25% for F2 males.

7-P Production

F1 males and females show 7-P values which are very similar. This indicates that the sex chromosomes are not directly implicated in determining the level of 7-P production. Both F2 and F2 BC flies show a bimodal distribution for 7-P values, with the same distribution as for 7-T. These results also seem to suggest a simple Mendelian model involving the segregation of one, or a very small number of genes, on a single pair of autosomes, with the gene(s) present on the Yaounde chromosomes being dominant. Figure 2 shows the ratios of 7-T : 7-P in the various crosses studied, plotted on a logarithmic scale; the existence of a bimodal distribution in the F2 is clearly supported by this presentation, supporting the argument that one, or a few, characters on one autosome are involved in determining the level of these two substances.

Although this result is statistically significant as far as qualitative effects are concerned, quantitative results are far more difficult to interpret, perhaps due to interactions between the autosomal factors controlling 7-P/7-T production and the sex-linked factors involved in 7-T production.

The autosome which plays a major role in determining 7-T and 7-P levels is probably the second chromosome, as suggested by recent experiments involving chromosome exchanges in *D. melanogaster* (Boukella, Ferveur and Jallon, in preparation). These experiments also suggest that loci on the third chromosome play a complementary role.

A complex genetic regulation of male-predominant pheromones in *D. melanogaster* has recently been suggested by Scott and Richmond (1988). They revealed the existence of X-linked loci and at least two different groups of autosomal loci.

BIOSYNTHETIC STUDIES

Most of the molecules involved in the *Drosophila* pheromone system described here have common structural features: a long linear chain with at least one double bond in position 7. They will therefore share a number of biosynthetic steps. Chan Yong and Jallon (1986) carried out the first study of *Drosophila* pheromone biosynthesis, using radioactive precursors. Using ^{14}C labelled

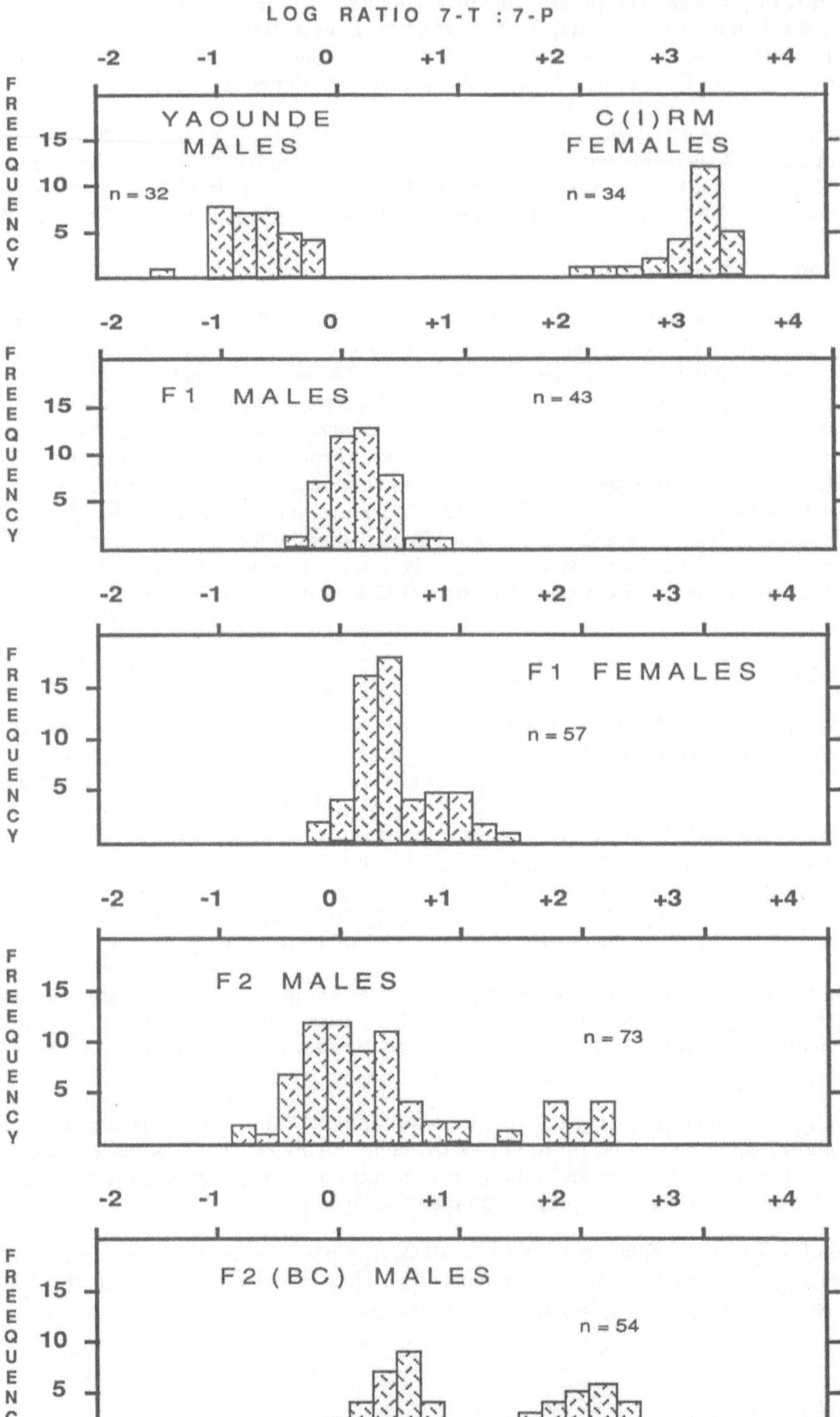

Fig. 2 Frequency histograms of the ratio of 7-T : 7-P, expressed on a logarithmic scale. For full details of the crosses involved, see text.

Table 4 Incorporation of radioactivity (in percent) into cuticular hydrocarbons of *D. melanogaster* virgin females.

	ALKANES	MONOENES + DIENES
Total hydrocarbons	34	66
Acetate ^{14}C	36	64
Palmitate ^{14}C 16:0	35	65
Stearate ^{14}C 18:0	95	5
Vaccenate ^{14}C 18.1 (7)	2	98

acetate, they revealed the *de novo* synthesis of cuticular hydrocarbons of virgin and mature CS males and females. Radioactivity can be found two hours after injection, with a 2-4% incorporation yield. This yield is maximal 2-3 days after imaginal eclosion, and does not vary much between males and females.

On the basis of previous studies, especially the pioneering work of Blomquist and Jackson (1979), we proposed a working hypothesis mainly involving fatty acids, the decarboxylation of which leads to hydrocarbons. Fatty acids are synthesized by an elongation mechanism involving the sequential addition of 2-carbon units to a given precursor (Jallon, 1984).

We tested this hypothesis by injecting ^{14}C-labelled palmitate, stearate and vaccenate into 2-day old virgin females. After two hours groups of flies were washed with hexane twice and the hexane extracts separated by two successive thin-layer chromatography procedures, as described in Antony and Jallon (1982), the first on silica, developed with 10% (w/v) diethyl ether in hexane. The faster-migrating spot, which contained hydrocarbons, was then microchromatographed on silver-nitrate impregnated silica and 8% (w/v) benzene in hexane. The radioactivity of both the saturated and the unsaturated fractions (monoenes and dienes were pooled) was then counted. Their proportions are shown in Table 4 After ^{14}C palmitate labelling, the situation is comparable to that after ^{14}C acetate injection: all hydrocarbons are labelled relative to their abundance, and levels are not different from those obtained from gas liquid chromatography (Antony and Jallon, 1982). However, after ^{14}C stearate injection, most of the radioactivity was recovered in saturated hydrocarbons; this tendency was reversed after ^{14}C vaccenate labelling, when almost all the radioactivity was found in monoenes and dienes.

Thus palmitate and not stearate is a common precursor for monoene and diene biosynthesis, as proposed in our 1984 working hypothesis. The simplest interpretation is that the first desaturation, leading to 7 unsatuated hydrocarbons, is introduced on palmitate by the activity of an enzyme with the same specificity as stearyl desaturase (Keith, 1967; Wang et al., 1982). This enzyme, which is being searched for in our laboratory, would be under the control of genes on the third chromosome, as clearly shown by interstrain chromosomal exchange (Boukella, Ferveur and Jallon, in preparation). Indeed, in both 5,9 HD and 7,11 HD the two double bonds remain at the same relative position, but the double bond closer to the methyl group is shifted by two carbon positions.

In the next step, palmitoleate would be elongated into vaccenate, which by sequential addition of 2-carbon units leads to fatty acids with a double bond in position 7 relative to their methyl end. These fatty acids are then decarboxylized to yield 7-monoenes. The biochemistry of elongation and decarboxylation is poorly understood. One particular problem is whether there are several specific enzymes or a single enzyme with a broad specificity for each catalytic function. All these steps are necessary in the production of 7-T and 7-P. It is therefore difficult to associate the genetic system described in the previous section with any particular step. However, as the 7-T/7-P ratio is explicitly genetically controlled, it may involve an elongation enzyme which would then have a critical role in fixing the proportions of two molecules which are behaviorally important.

Once the double bond in position 7 relative to the methyl end has been introduced, how is the second double bond positioned? Following biochemical principles established in studies on rat metabolism, we have suggested that vaccenate is a probable candidate to be desaturated by an enzyme different from the stearyl desaturase mentioned earlier (Jallon, 1984). This is a particularly exciting problem as such an enzyme and its structural gene would be sex- and species-specific. However, we are only at the beginning of this work; we have so far been able to show that the gene is regulated by the doublesex locus which, when mutated, turns it off completely, leading to a complete absence of the sort of dienes found in normal females (Jallon et al., 1988).

CONCLUSION

We have demonstrated that chemical communication between *Drosophila* is a complex phenomenon, which we do not yet fully understand. The role of some substances, such as cVA, may be multiple, or merely different from that initially proposed. Other substances, as yet unstudied, may play a role in inducing elements of behavior which have either already been investigated, or which have not yet been looked at. Although our understanding of the biosynthetic pathways involved appears relatively solid, the genetic picture, as our results indicate, appears both relatively straightforward (the influence of a relatively small number of autosomal genes in *D. simulans*), but also complex at the same time (role of sex chromosomes, difficulty of explaining quantitative effects). We can expect that future studies will cast light on what are currently somewhat obscure aspects of this subject.

REFERENCES

Adams, T. S., 1986, Effect of different female-produced pheromone components on male courtship behavior in the house fly, *Musca domestica*, pp. 297-304 *in* "Advances in Invertebrate Reproduction 4", M. Prochet, J. C. Andrews and A. Dhainaut (eds.), Elsevier, Amsterdam.

Antony, C., and Jallon, J-M., 1982, The chemical basis for sex recognition in *Drosophila melanogaster, J. Ins. Physiol.*, 28:873-880.

Antony, C., Davis, T. L., Carlson, D. A., Pechiné, J. M. and Jallon, J-M., 1985, Compared behavioral responses of males *Drosophila melanogaster* (CantonS) to natural and synthetic aphrodisiacs, *J. Chem. Ecol*, 11:1617-1629.

Bartelt, R. J., Schaner, A. M. and Jackson, L. L., 1985, *cis*-vaccenyl acetate as an aggregation pheromone in *Drosophila melanogaster, J. Chem. Ecol.*, 11:1747-1756.

Bell, W. J. and Cardé, R. T.,1984, "Chemical Ecology of Insects", Chapman and Hall, London.

Blomquist, G., and Jackson, L., 1979, Chemistry and biochemsitry insect waxes, *Prog. Lipid Res.*, 17:319-345.

Butterworth, F. M., 1969, Lipids of *Drosophila*: a newly detected lipid in the male, *Science*, 163:1356-1357.

Camacho, A., 1986, Effets post-copulatoires sur le comportement et les pheromones de *Drosophila melanogaster,* DEA report, Université Paris XIII.

Carlson, D., Mayer, M., Silhacek, D., James, J., Beroza, M. and Bierl, B., 1971, Sex attractant pheromone of the housefly: isolation, identification and synthesis, *Science*, 174:76-77.

Chan-Yong, T. P. and Jallon, J-M., 1986, Synthèse de novo d'hydrocarbures potentiellement aphrodisiaques chez les Drosophiles, *C. R. Acd. Sci. Paris*, 303:197-202.

Ewing, A., 1983, Functional aspects of *Drosophila* courtship, *Biol. Rev.*, 58:275-292.

Jallon, J-M., 1984, A few chemical words exchanged by *Drosophila* during courtship, *Behav. Genet.*, 14:441-478.

Jallon, J-M. and David, J. R., 1987, Variations in cuticular hydrocarbons along the eight species of the *Drosophila melanogaster* subgroup. *Evolution*, 41:294-302.

Jallon, J-M. and Hotta, Y., 1979, Genetic and behavioral studies of female sex-appeal in *Drosophila, Behav. Genet.*, 8:487-502.

Jallon, J-M. and Hotta, Y., 1981, Non-chemical messages of the female *Drosophila melanogaster, in*: "Taniguchi Symposia in Biophysics N° 7," Y. Hotta, ed., University of Tokyo, Tokyo.

Jallon, J-M., Antony, C. and Benamar, O., 1981, Un anti-aphrodisiaque produit par les mâles de *Drosophila* et transféré aux femelles lors de la copulation, *C. R. Acad. Sci. Paris.*, 292:1147-1149.

Jallon, J-M., Antony, C., Chan-Yong, T. P. and Maniar, S., 1986, Genetic factors controlling the production of aphrodisiac substances in *Drosophila. Adv. in Inv. Reprod.* 4:445-452.

Jallon, J-M., Benamar, O., Luyten, I. and Antony, C., 1982, Modulations de la production des hydrocarbures cuticulaires aphrodisiaques des Drosophiles resultant de perturbations génétiques et physiologiques , *in* "Les Mediateurs Chimiques", C. Descoins, (ed.), INRA, Paris.

Jallon, J-M., Lauge, G., Orssaud, L. and Antony, C., 1988, *Drosophila melanogaster* female pheromones controlled by the doublesex locus, *Genet. Res.*, 51:17-22.

Keith, A., 1967, Fatty acid metabolism in *D. melanogaster*: formation of palmitoleate, *Life Sciences*, 6:213-218.

Karlson, P. and Butenandt, A., 1959, Pheromones (ectohormones) in insects, *Ann. Rev. Entomol.*, 4:39-58.

Luyten, I., 1982, Variations intraspécifiques et interspécifiques des hydrocarbures cuticulaires chez *Drosophila simulans. C.R. Acad. Sci. Paris*, 295:723-736.

Mane, S. D., Tompkins, L. and Richmond, R. C., 1983, Male esterase 6 catalyzes the synthesis of a sex pheromone in *Drosophila melanogaster* females, *Science*, 222:418-421.

Pechiné, J. M., Perez, F., Antony, C. and Jallon, J-M., 1985, A further characterization of Drosophila cuticular monenes using a mass spectrometry method to localize double bonds in complex mixtures, *Anal.yt Biochem.*, 145:177-182.

Schaner, A. M., Bartelt, R. J. and Jackson, L. L., 1987, (Z)-11-Octadenyl acetate, an aggregation pheromone in *Drosophila simulans*, *J. Chem. Ecol.*, 13:1777-1786.

Scott, D., 1986, Sexual mimicry regulates the attractiveness of mated *Drosophila melanogaster* females, *Proc. Natl. Acad. Sci. USA*, 83:8429-8433.

Scott, D. and Jackson, L., 1988, Interstrain comparison of male-predominant antiaphrodisiacs in *Drosophila melanogaster.. J. Ins. Physiol.*, 34:863-871.

Scott, D. and Richmond, R. C., 1987, Evidence against an antiaphrodisiac role for *cis*-vaccenyl acetate in *Drosophila melanogaster*, *J. Insect Physiol.*, 33:363-369.

Scott, D. and Richmond, R. C., 1988, A genetic analysis of male-predominant pheromones of *Drosophila melanogaster*, *Genetics*, 119:639-646.

Scott, D., Richmond, R. C. and Carlson, D. A., 1988, Pheromones exchanged during mating: a mechanism for mate assessment in *Drosophila*, *Anim. Behav.*, 36:1164-1173.

Tompkins, L. and Hall, J. C., 1981, *Drosophila* males produce a pheromone which inhibits courtship, Z. *Naturf.*, 36c:694-695.

Tompkins, L., Hall, J. C. and Hall, L., 1980, Courtship stimulating volatile pheromones from formal and mutant *Drosophila*, *J. Ins. Physiol.*, 26:689-697

Vander Meer, R. K., Obin, M. S., Zawitowksi, S., Sheehan, K. B., Richmond, R.C., 1986, A reevaluation of the role of *cis*-vaccenyl acetate, *cis*-vaccenol and esterase 6 in the regulation of mated female sexual attractiveness in *Drosophila melanogaster*, *J. Insect. Physiol.*, 32:681-686.

van den Berg, M. J., Thomas, G., Hendriks, M., and Van Delden (1983), A reexamination of the negative assortative mating phenomenon and its underlying mechanism in *Drosophila melanogaster*, *Behav. Genet.*, 14:45-61.

Venard, R., 1980, Attractants in courtship mutants, *in* "Development and Neurobiology of *Drosophila*", O. Siddiqi, P. Babu, L.M. Hall and J.C. Hall (eds.), Plenum Press, New York.

Venard, R. and Jallon, J-M., 1980, Evidence for an aphrodisiac pheromone of female *Drosophila*, *Experientia*, 36:211-212.

Wang, D., Dillwith, J., Ryan, R., Blomquist, G. and Reitz, R., 1982, Characterization of the acyl-CoA desaturase in the housefly, *Musca domestica L.*, *Insect Biochem.*, 12:545-551.

Zawitowski, S. and Richmond, R. C., 1986, Inhibition of courtship and mating of *Drosophila melanogaster* by the male-produced lipid, *cis*-vaccenyl acetate, *J. Insect. Physiol.*,32:189-192.

THE ISOLATION OF ANTENNAL MUTANTS AND THEIR USE IN DROSOPHILA OLFACTORY GENETICS

Richard Ayer, Paula Monte, and John Carlson

Department of Biology
Yale University
New Haven, CT 06511 USA

ABSTRACT

A genetic approach to investigation of the *Drosophila* olfactory system, based on the use of mutations affecting antennal morphology, is discussed. We describe a convenient means of screening for mutants with antennal defects, and we describe the isolation of three such mutants. A new X-linked recessive mutant, *eab (extra antennal bristles)*, is found to exhibit characteristics suggestive of a homeotic transformation of antennae towards legs.

INTRODUCTION

This article concerns the use of genetics to study the olfactory system of *Drosophila melanogaster*. The principle organ of olfaction in *Drosophila* is the antenna, which consists of three main segments and a feather-like structure known as the arista. The first and second antennal segments bear bristles, unlike the third segment, which bears no bristles but is covered with several hundred fine sensilla. The distribution and morphology of these sensilla have been carefully described by Venkatesh and Singh (1984), and it is clear from their morphological and physiological properties (Siddiqi, 1984) that many of these sensilla are olfactory.

One genetic approach to the study of the *Drosophila* olfactory system uses behavior as a point of departure. Rodrigues and Siddiqi (1978) have illustrated the use of behavioral genetics to isolate mutants defective in olfactory response. Mutants defective at any level in the response pathway, including olfactory transduction or processing, may be isolated in behavioral screens.

A complementary genetic approach is based on the isolation of mutants bearing morphological defects affecting the olfactory system. Mutants with defects in antennal morphology may be useful both in studying the development and maintenance of the antenna and in studying the sensory functions which it supports.

Genetic analysis has already proven to be a powerful means of investigating pattern formation and determination in the antenna, as shown by studies of homoeotic mutations at the *Antennapedia* complex (Kaufman et al., 1980) and other loci. A detailed understanding of the molecular mechanisms underlying antennal development, however, will require a more complete description of the genes required in the process. The isolation and characterization of additional mutants bearing defects in antennal morphology may be a valuable step towards this end.

Mutants with abnormal antennal morphology may also be of great value in investigating the function of the olfactory system. Historically, the use of such mutants has played an important role in the study of *Drosophila* olfaction: localization of this sensory modality to the antenna was confirmed by early studies of the mutants *antennaless* and *aristapedia* (Begg and Hogben, 1943, 1946). The severe perturbations of antennae afforded by these mutations may still be useful as new means of measuring olfactory function become available. The fly contains a number of other sensory structures, some of which may also receive various forms of olfactory information (Harris, 1972; Singh and Nayak, 1985; Stocker and Gendre, in press; Monte and Carlson, unpublished). The roles of these other structures in response to various types of olfactory stimulation may in some cases be assessed more sensitively in mutants lacking antennal function.

Mutations disrupting antennal form at a finer level can be elegant tools for the study of the olfactory system, as shown by the work of Stocker and Gendre (1988) with *lozenge* mutants. These mutants lack one particular subset of antennal sensilla, the sensilla basiconica, and in addition are missing one subunit of the antennal lobes, glomerulus V, previously shown to be a target of neurons in the basiconica sensilla. Further analysis of these mutants has implied a role for sensory input in the development or maintenance of the antennal lobes. Moreover, a *lozenge* mutant has been studied behaviorally with the aim of investigating the functional role of the subset of the olfactory system which is affected by the mutation (Stocker and Gendre, in press).

Visual system mutants analogous to these two classes of mutants have been characterized in detail. Gross perturbations of the eye are effected by *sine oculis*, which drastically reduces the size of the eye, and *glass*, which both reduces the area of the eye and eliminates its light-stimulated electrical potential (Pak *et al.*, 1969). A finer perturbation is caused by the mutant *sevenless* (Harris *et al.*, 1977, Campos-Ortega *et al.*, 1979), in which one class of photoreceptor cells, the R7 cells, fails to develop, thereby eliminating one spectral component of the animal's visual response.

A third class of visual system mutants is exemplified by the mutant *rdgB* (Hotta and Benzer, 1970), which appears to develop normally but exhibits a morphological phenotype -- degeneration of photoreceptor cells -- following stimulation with light (Harris and Stark, 1977). Degeneration is postulated to result from the abnormal function of the light-stimulated transduction cascade. This class contains a number of visual system mutants which, like *rdgB*, show both defective visual system physiology and receptor cell degeneration. We do not know whether mutants defective in olfactory transduction may be identified by virtue of such a visible degenerative phenotype.

There is thus substantial precedent for the use of morphological mutants to study the olfactory system. Experiments described in this paper were designed to explore further the potential of this approach. We describe a simple method of screening for antennal defects, the isolation of three mutants with abnormal antennal morphology, and preliminary characterization of one mutant whose antennae show characteristics suggestive of a homoeotic transformation.

MATERIALS AND METHODS

Drosophila Stocks and Mutagenesis

Canton-S-5 (CS-5) is homozygous for a unique X chromosome, and was generated from a Canton-S stock obtained from O. Siddiqi, Tata Institute, Bombay, and from an FM7 balancer stock whose autosomes were of Canton-S origin, obtained from S.Benzer, California Institute of Technology. The CS-5 stock was mutagenized by means of EMS (Lewis and Bacher, 1968), X-irradiation (4300 rads), or hybrid dysgenesis. A behavioral enrichment for mutants defective in olfactory function, to be described elsewhere, was used following mutagenesis with X-irradiation and hybrid dysgenesis. For hybrid dysgenesis, CS-5 was used as an M strain, and was crossed to five P strains: Harwich, obtained from S. Artavanis-Tsakonas, Yale University, and π2, 78-61, 8-31-15, and Inbred Cage 3, from W. Engels, University of Wisconsin. In the case of EMS or X-rays, mutagenized males were crossed to *C(1)A y* virgin females, and male offspring were scored for abnormal antennal morphology. In the case of hybrid dysgenesis, F1 males were mated to *C(1)DX y f (P)* virgin females, and their male offspring were scored for antennal morphology. Putatively mutant males were crossed to *C(1)A y* or *C(1)DX y f (P)* females to establish lines. The mapping stock containing the *y cv v f* chromosome was obtained from L. Tompkins, Temple University.

Screening for Antennal Defects

A micropipette tip was severed so that the internal diameter of its small aperture -- approximately 1 mm -- was of a size such that the anterior surface of a fly head, including the antennae, could protrude slightly. The tip, fastened to a stand (Figure 1), was placed under a Zeiss light dissecting microscope with the tapered end upwards, in the beam of a light source positioned above. A fly, inserted into the base of the micropipette tip by means of a standard mouth aspirator, walked upwards, presumably driven by positive phototaxis and negative geotaxis. Air pressure was then applied with the mouth aspirator in order to lodge the fly at the top of the pipette tip, its head protruding slightly through the opening of the tip and its thorax resting beneath the aperture. In this position, antennae were examined at 128X magnification. The fly was removed from the apparatus without harm simply by applying gentle suction.

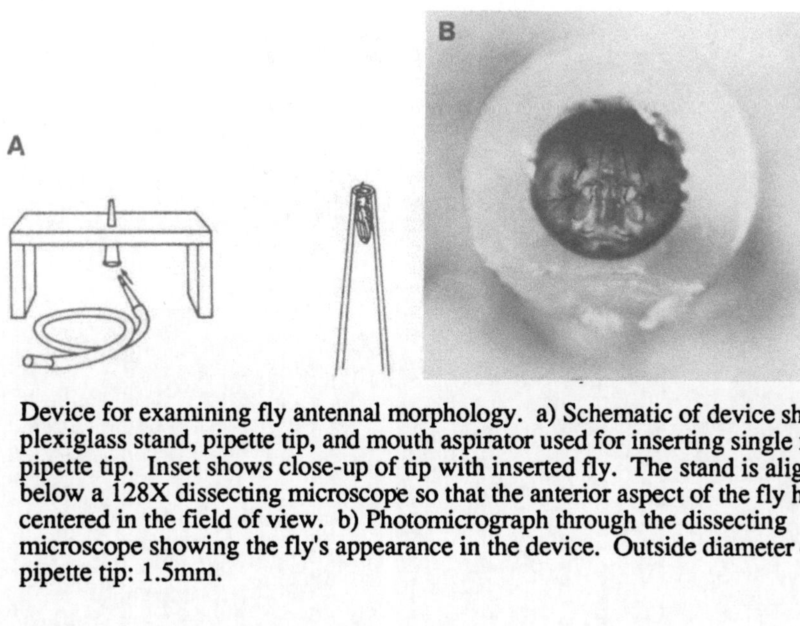

Fig. 1. Device for examining fly antennal morphology. a) Schematic of device showing plexiglass stand, pipette tip, and mouth aspirator used for inserting single flies into pipette tip. Inset shows close-up of tip with inserted fly. The stand is aligned below a 128X dissecting microscope so that the anterior aspect of the fly head is centered in the field of view. b) Photomicrograph through the dissecting microscope showing the fly's appearance in the device. Outside diameter of pipette tip: 1.5mm.

The following characteristics were scored: 1) presence, morphology, and relative positions of the second antennal segment, third antennal segment, and arista; 2) "hairy" texture of the third segment; 3) position, number, size, and shape of bristles on the second antennal segment; 4) movement of antennae in response to air currents; 5) antennal pigmentation. Flies were scored at a rate of approximately 100 per hour.

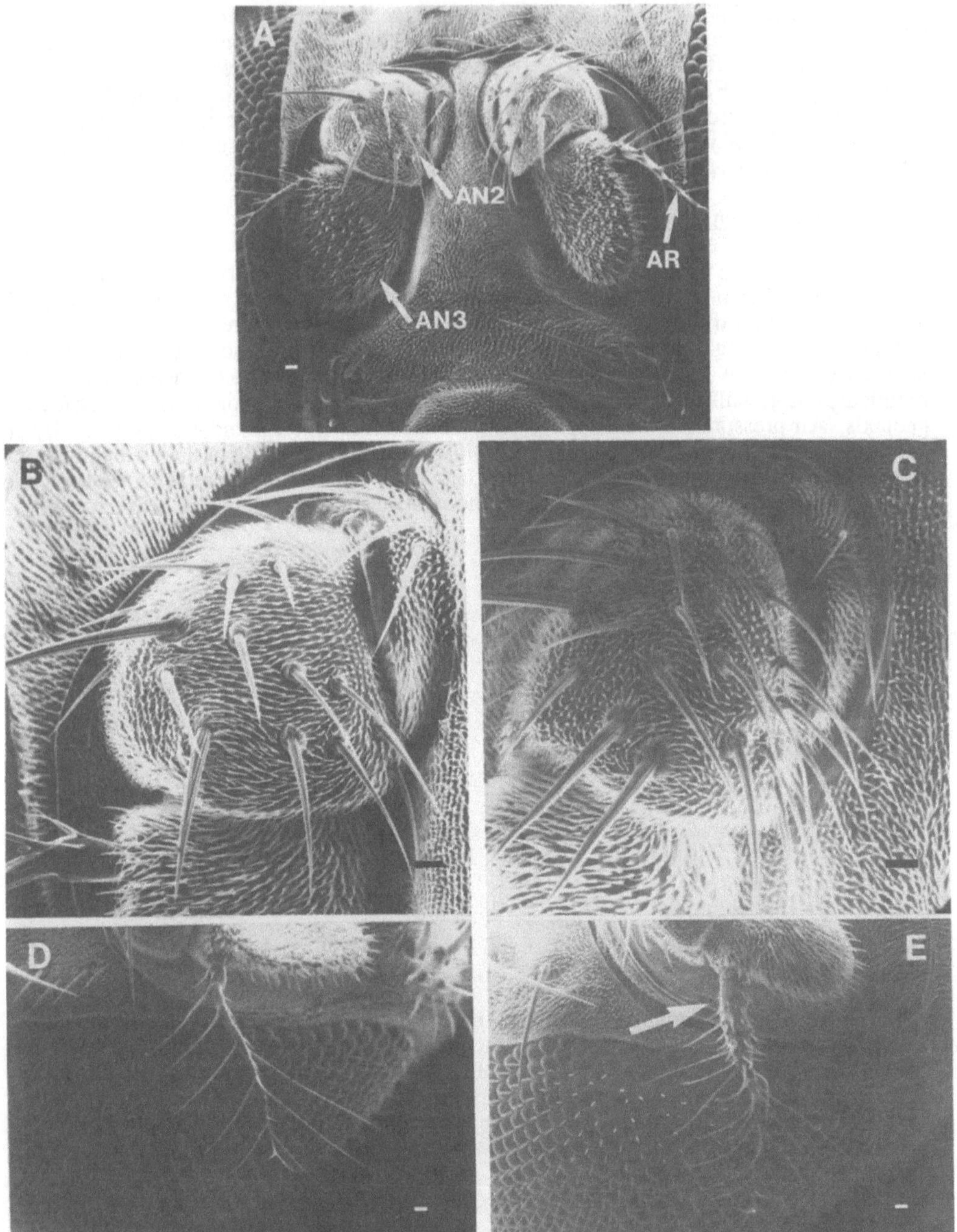

Fig. 2. Scanning electron micrographs of wild-type (CS-5) and mutant, (*eab*) antennae. a) Anterior aspect of a CS-5 head: second segment (AN2), bearing large bristles; third segment (AN3), covered with sensilla; feathery arista (AR). b) Second segment of wild-type. ~15 bristles are visible on the anterior aspect. c) Second segment of *eab*. ~27 bristles are visible on anterior aspect. d) Arista of wild-type. e) Arista of *eab*. Note thickening of base and alteration in appearance of its lateral hairs. Bar=10μ.

Fig. 2. cont'd. f) Third segment of wild-type. g) Third segment of *eab*. Note the alteration in shape and the indicated ectopic bristles. h) Third segment of *eab*, distended and bent. Bar=10μ.

RESULTS AND DISCUSSION

Screening of a total of 1048 mutagenized X chromosomes -- 335 mutagenized by EMS, 270 by X-irradiation, and 443 by hybrid dysgenesis -- has yielded three antennal mutants, one induced by X-irradiation, one by EMS, and one by hybrid dysgenesis.

The X-ray-induced mutant, X-M4, was isolated by virtue of its abnormally shaped third antennal segment, which was observed to bend laterally towards the eye. The EMS-induced mutant, EMS-M13, was isolated on account of lack of antennal movement upon stimulation with air currents. Closer examination revealed that antennal motions were hindered by the abnormal form of the head cuticle at the ptilinal suture: the prefrons was recessed with respect to the postfrons such that the antennae were locked in a fixed position. Although stable for many generations, neither of these mutants exhibited high penetrance, and neither has been maintained.

The dysgenic mutant exhibits several phenotypes, all recessive:

1) There is a $\geq$ 35% increase in the number of bristles on the anterior aspect of the second antennal segment, a phenotype whose penetrance is virtually 100% (Figure 2c).
2) The third segment is abnormally shaped (Figure 2, g and h), sometimes smaller than wild-type, and sometimes bent, distended, and shaped like a sock. Penetrance of this phenotype is approximately 15%.
3) The arista of mutant flies is thickened, especially at the base, and the long lateral hairs of the arista which lend a feathery appearance to wild-type are reduced in length and in regularity of pattern.
4) Ectopic bristles appear on the third antennal segment, although with low penetrance (Figure 2g). These bristles, up to nine in number, exist in a patch near the base of the arista; ectopic bristles may also be found on the basal cylinder from which the arista projects.
5) Irregular wing margins appear with complete penetrance (Figure 3), and the wings are divergent.
6) The metathoracic legs of a small fraction of flies are bent, such that the tarsal segments are at an angle to the tibia.
7) Viability is reduced among males carrying the mutant X chromosome.

The presence of the supernumerary bristles on the second antennal segment and the ectopic bristles on the third segment and basal cylinder of the arista have led us to name the mutant *extra antennal bristles* (*eab*).

In addition to these morphological phenotypes, *eab* has a behavioral phenotype. Unlike wild-type control males, which spend much of their time on the walls of the culture vials, *eab* males generally remain on the surface of the food. The mutants appear to be defective in the ability to climb up the walls. The penetrance of this behavioral phenotype exceeds that of the leg phenotype described above. Mutant males also exhibit a greater tendency to become mired in the culture medium.

As a preliminary means of determining whether the *eab* phenotypes result from mutation of a single locus, we attempted to map the antennal phenotype and the irregular wing margin phenotype by meiotic recombination mapping using the multiply marked chromosome *yellow crossveinless vermilion forked* (*y cv v f*). Limited data suggest a tentative localization of both phenotypes to the *cv* - *v* interval.

The distention of the third antennal segment, the thickening of the arista, and the appearance of ectopic bristles on the third segment and the basal cylinder of the arista, are all suggestive of a weak homoeotic transformation of the antenna towards a leg. A similar weak transformation has been extensively documented for *spineless-aristapedia of Bridges* (Villee, 1943). There are several homoeotic genes whose mutations produce transformations of antennae towards legs. Most, but not all of these mutations are dominant, and most of the genes are autosomal. *Antennapedex* maps to the X chromosome, but has been localized near the base of the chromosome, far from the *cv - v* interval (Ginter, 1969). Further genetic analysis, including determination of whether the various phenotypes of the *eab* mutant are effects of a single mutation, is required before definitive interpretations can be drawn as to the nature of the defect. We note, however, that at least one of the mutants exhibiting transformations of antennae to legs, *Antennapedia of Le Calvez*, also has divergent wings (Le Calvez, 1948), and *Brista* (from *Bristle on arista*) has effects on the distal portions of the legs (Sunkel and Whittle, 1987).

The *eab* chromosome was recovered from one of a limited number of dysgenic animals enriched for failure to respond to olfactory attractants. However, preliminary testing of *eab* has as yet revealed no olfactory abnormalities, either in a behavioral population assay designed to measure attraction to volatile chemicals, or in electrophysiological measurements of antennal response to ethyl acetate. We do not know whether a stock exhibiting greater penetrance and expressivity of the antennal defect would show an olfactory phenotype.

ACKNOWLEDGMENTS

We are grateful to Efrain Aceves-Pena for demonstrating the method of holding the fly during screening, and to Henry Sun, Stephen Helfand, and Craig Woodard for supplying mutagenized material. J. C. is an Alfred P. Sloan Research Fellow.

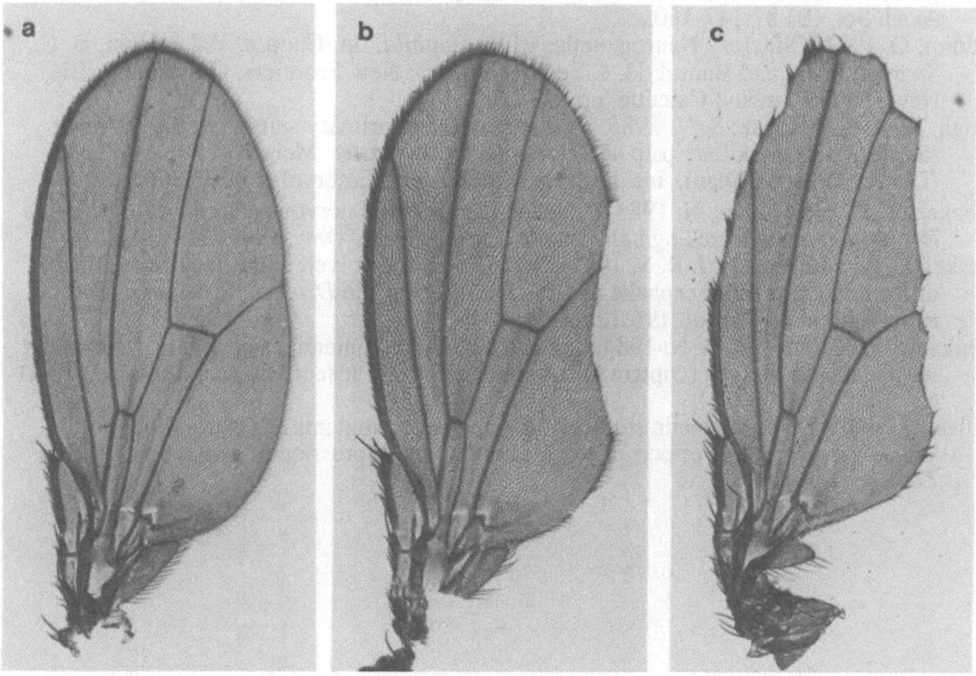

Fig. 3. Light micrographs of wild-type (CS-5) and mutant (*eab*) wings. a) wild-type. b) *eab*, low expressivity. c) *eab*, high expressivity.

REFERENCES

Begg, N., and Hogben, L. 1943. Localization of chemoreceptivity in *Drosophila*. Nature 152:535.

Begg, N., and Hogben, L. 1946. Chemoreceptivity of *Drosophila melanogaster*. Proc. Roy. Soc. (London) Ser. B 133:1-19.

Campos-Ortega, J. A., Jurgens, G., and Hofbauer, A. 1979. Cell clones and pattern formation: studies on *sevenless*, a mutant of *Drosophila melanogaster*. Wilhelm Roux's Arch. Dev. Biol. 186:27-50.

Ginter, E. K. 1969. Dros. Info. Service 44:50-51.

Harris, W. A. 1972. The maxillae of *Drosophila melanogaster* as revealed by scanning electron microscopy. J. Morph. 138:451-456.

Harris, W. A., and Stark, W. S. 1977. Hereditary retinal degeneration in *Drosophila melanogaster*. J. Gen. Physiol. 69:261-291.

Harris, W. A., Stark, W. S., and Walker, J. A. 1977. Genetic dissection of the photoreceptor system in the compound eye of *Drosophila melanogaster*. J. Physiol. 256:415-439.

Hotta, Y., and Benzer, S. 1970. Genetic Dissection of the *Drosophila* nervous system by means of mosaics. Proc. Natl. Acad. Sci. USA 67:1156-1163.

Kaufman, T. C., Lewis, R., and Wakimoto, B. 1980. Cytogenetic analysis of chromosome 3 in *Drosophila melanogaster* : the homeotic gene complex in polytene chromosome interval 84A-B. Genetics 94:115-133.

Le Calvez, J. 1948. In(3R) ss^{Ar}: mutation "*aristapedia* " heterozygote dominate, homozygote lethale chez *Drosophila melanogaster*. (inversion dans le bras droit du chromosome III). Bull. Biol. France Belg. 82:97-113.

Lewis, E. B., and Bacher, F. 1968. Dros. Info. Service 43:193.

Pak, W. L., Grossfield, J., and White, N. V. 1969. Nonphototactic mutants in a study of vision of *Drosophila*. Nature 222:351-354.

Rodrigues, V., and Siddiqi, O. 1978. Genetic analysis of chemosensory pathway. Proc. Ind. Acad. Sci. (B) 87:147-160..

Siddiqi, O. 1984. Olfactory Neurogenetics of *Drosophila*. In: Chopra, V. L., Joshi, B. C., Sharma, R. P., and Bansal, H. C. (eds) Genetics: New Frontiers. Oxford and IBH, New Delhi, Bombay, Calcutta, pp 243-261.

Singh, R. N., and Nayak, S. V. 1985. Fine structure and primary sensory progections of sensilla on the maxillary palp of *Drosophila melanogaster* Meigen (Diptera:Drosophilidae). Int. J. Insect Morphol. and Embryol. 14:291-306.

Stocker, R. F., and Gendre, N. 1988. Peripheral and Central nervous effects of *lozenge*^{3} : a *Drosophila* mutant lacking basiconic antennal sensilla. Dev. Biol. 127:12-24.

Sunkel, C. E., and Whittle, J. R. S. 1987. *Brista* : a gene involved in the specification and differentiation of distal cephalic and thoracic structures in *Drosophila melanogaster*. Roux's Arch. Dev. Biol. 196:124-132.

Venkatesh, S., and Singh, R. N. 1984. Sensilla on the third antennal segment of *Drosophila melanogaster* Meigen (Diptera:Drosophilidae). Int. J. Insect Morphol. Embryol. 13:51-63.

Villee, C. A. 1943. Phenogenetic studies of the homoeotic mutants of *Drosophila melanogaster* : I. The effects of temperature on the expression of aristapedia. J. Exp. Zool. 93:75-98.

ISOLATION OF AUTOSOMAL BEHAVIORAL MUTATIONS IN *DROSOPHILA*

Satpal Singh*, Maninder J. S. Chopra, Poonam Bhandari and Devasis Guha

School of Life Sciences, Guru Nanak Dev University, Amritsar 143 005, India

Drosophila can be used effectively for studying the nervous system by means of a wide variety of mutations that can be used to perturb and analyze individual components of the system. A number of behavioral mutations are available that affect membrane excitability and other neuromuscular phenomena in *Drosophila*. A striking example is provided by mutations at the *Shaker* (*Sh*) locus (Kaplan and Trout, 1969) that selectively disrupt the functioning of one type of potassium channel, I_A (Salkoff and Wyman, 1981; Wu et al., 1983; Wu and Haugland, 1985). *Sh* has been used to clone and analyze the properties of these channels (Kamb et al., 1987; Papazian et al., 1987; Baumann et al., 1987). There is now evidence for the existence of four potassium currents in the muscles of *Drosophila melanogaster* (Gho and Mallart, 1986; Wei and Salkoff 1986, Singh and Wu, 1987). These include two voltage activated K^+ currents, a fast I_A and a slow I_K, and two Ca^{++} activated K^+ currents, a fast I_{CF} and a slow I_{CS}. Mutations and pharmacological agents have been used to separate these currents from one another. By eliminating only I_A, *Sh* provides a distinction between the fast and the slow voltage activated K^+ currents I_A and I_K (Salkoff and Wyman, 1981; Wu et al., 1983; Wu and Haugland, 1985). The *slowpoke* (*slo*) mutation (Elkins et al., 1986) provides a similar separation between the fast and the slow Ca^{++} activated currents I_{CF} and I_{CS} (Singh and Wu, 1987). *Sh* and *slo* provide a distinction between the two fast currents I_A and I_{CF}; and quinidine at micromolar concentrations separates the two slow currents I_K and I_{CS} (Singh and Wu, 1987). A selective elimination of individual currents has been useful in identifying the single channels through which some of these currents pass and in analyzing the molecular properties of these channels (Komatsu and Wu, 1987; Solc et al., 1987; Komatsu. A., Singh. S., Rathe. P. and Wu. C.-F., Unpublished). It has also helped in understanding the role of individual currents and their interactions in membrane excitability (Elkins and Ganetzky, 1988; Singh et al., 1986).

A number of other *Drosophila* mutations have been used effectively in studies on development (Lewis, 1978), reproduction (Baker et al., 1976), neurogenesis (Campos-Ortega, 1985), neuromuscular excitability (Ganetzky and Wu, 1986), processing of visual information (Heisenberg and Wolf, 1984), olfactory perception (Siddiqi, 1987), learning (Dudai, 1985), and a number of other phenomena. However, this important strength of *Drosophila*, i.e. an availability of a variety of mutations, suffers from a serious limitation. It has been extremely difficult to obtain mutations on the autosomes, which constitute about 80% of the genome. The major difficulty arises from a need to make new recessive mutations homozygous before they express the mutant phenotype. This involves setting up of a large number of independent

* Present address: Department of Biology, University of Iowa, Iowa City, Iowa 52242, USA.

cultures for individual mutagenized chromosomes. On the other hand, mutations on the X-chromosome can express in males in the hemizygous condition, and can be identified by screening a large number of flies representing different mutagenized chromosomes without setting up independent lines. A similar method for screening autosomes without setting up lines can be extremely helpful in identifying new autosomal mutations in *Drosophila.*

The compound chromosomes, that have homologous copies of chromosomal arms attached at the centromere (Rasmussen, 1960; Holm, 1976), can provide a useful method for homozygosing newly induced mutations. A heterozygous mutation on one arm of such a chromosome can become homozygous during an event of meiotic recombination. Due to this property of spontaneous homozygosis, we do not have to set up independent lines for different mutagenized chromosomes (Singh, 1983). Mutagenized flies can instead be bred in mass cultures and screened for the phenotype of interest (Singh, 1983; Singh et al., 1987), as is done in case of the X-chromosome. We have used this method for obtaining a number of temperature-sensitive paralytic mutations on the second and the third chromosomes of *Drosophila.*

COMPOUND AUTOSOMES

The compound autosomes used in our study are the chromosomes which have the homologous copies of chromosomal arms attached at the centromere in a reverse metacentric configuration (Fig. 1A). Such chromosomes were first constructed by I. E. Rasmussen and E. Orias (Rasmussen, 1960; Lewis, 1967) for use in half-tetrad analysis in *Drosophila* (Chovnick, 1970). Their properties have been described by Holm (1976). They have a specific nomenclature. For example, *C(2L)RM, b;C(2R)RM, cn* denotes a compound (C) second (2) chromosome with the left arms (L) and the right arms (R) attached in a reverse metacentric (RM) configuration. *b* (*black* body) is a marker on the left arm and *cn* (*cinnabar* eyes) on the right arm. Description of markers used is given in Lindsley and Grell (1968).

Meiotic segregation pattern of the compound chromosomes is different from that of the normal ones. The two attached copies of the homologous arms in these chromosomes move together during segregation. A gamete receives either none or two copies of the genetic material associated with an arm and is aneuploid. A zygote is viable if it receives two copies of the left arm and two copies of the right arm and thus has a normal complement of the genome.

As the two attached copies of the homologous chromosomes move together during segregation, they carry together two meiotic products, i.e., a half tetrad, of a single event of

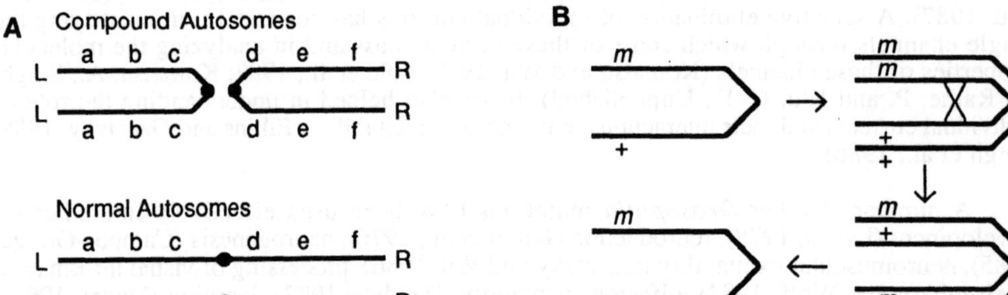

Fig. 1. **A:** Compound autosomes used in the experiment have homologous arms attached at the centromere. Both the copies of an arm go to the same gamete. The order of genetic markers a, b, c, d, e and f illustrates the reverse metacentric configuration. **B:** A mutation '*m*' present on one arm of a compound chromosome may become homozygous during meiotic recombination. L, Left arm; R, Right arm.

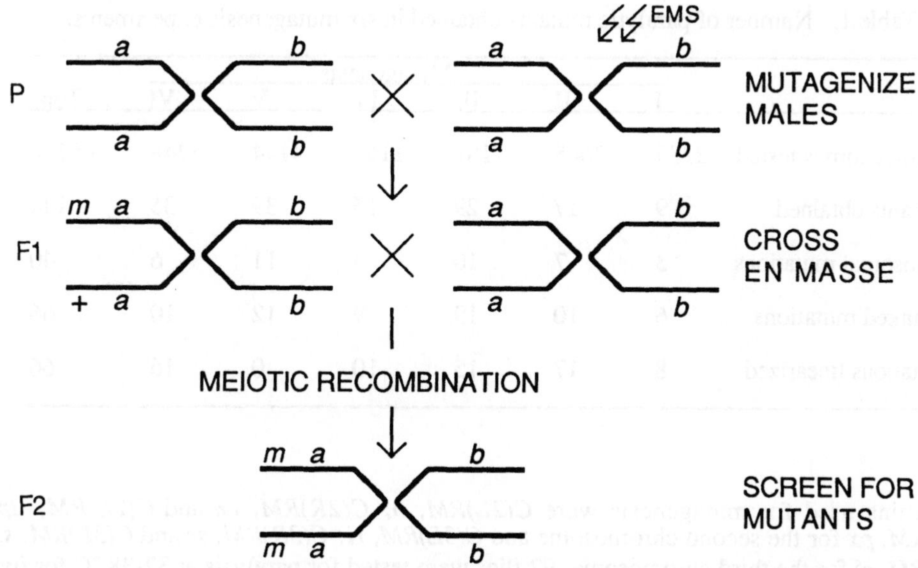

Fig. 2. Protocol for obtaining mutations. EMS treated males are crossed to females. F1 flies are crossed en masse. F2 flies are tested for the presence of mutants. *a* and *b*, visible genetic markers; *m*, a newly induced mutation; P, parents.

meiotic recombination. If a mutation is present in heterozygous form before recombination, this process can generate a half tetrad which is homozygous for the mutation (Fig. 1B).

IDENTIFICATION AND LOCALIZATION OF MUTATIONS

Homozygosis of a newly induced mutation in compound chromosomes, as discussed above, provides a simple method for identification of recessive mutations (Fig. 2). Males are fed EMS (Lewis and Bacher, 1968) and crossed to virgin females. F1 flies receive a copy of mutagen treated chromosomes from father and are likely to bear a mutation of interest. F1 flies are crossed en masse. F2 flies are tested for the phenotype of interest. If an event of meiotic recombination makes a particular mutation homozygous, it can be identified during this test.

We used the above method to obtain temperature-sensitive paralytic mutations on the second and the third chromosomes. Six independent rounds of mutagenesis were carried out.

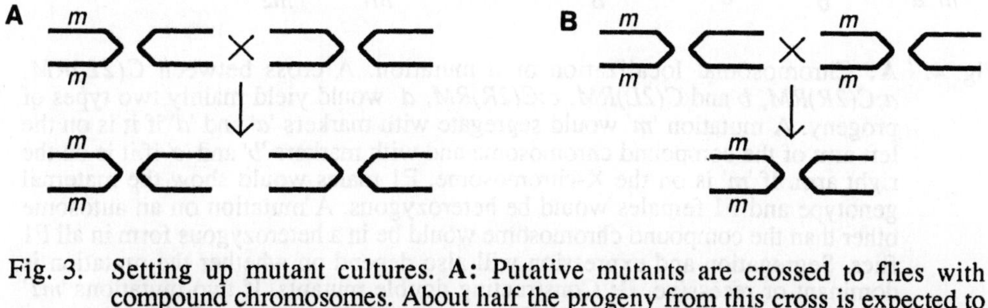

Fig. 3. Setting up mutant cultures. **A:** Putative mutants are crossed to flies with compound chromosomes. About half the progeny from this cross is expected to be mutant. **B:** Mutant cultures are set up by crossing mutant males and virgin mutant females.

Table 1. Number of paralytic mutants obtained in six mutagenesis experiments.

	Mutagenesis						
	I	II	III	IV	V	VI	Total
Chromosomes tested	2239	7648	7181	11166	4144	15748	48126
Mutants obtained	9	17	29	15	39	35	144
Autosomal mutations	3	7	10	3	11	6	40
X-linked mutations	6	10	19	9	12	10	66
Mutations linearized	8	17	15	10	0	16	66

The strains used for mutagenesis were *C(2L)RM, b; C(2R)RM, cn* and *C(2L)RM, dp; C(2R)RM, px* for the second chromosome and *C(3L)RM, ri; C(3R)RM, sr* and *C(3L)RM, st; C(3R)RM,* e^s for the third chromosome. F2 flies were tested for paralysis at 37-38 °C for five minutes. Progeny of about 48,000 F1 females were tested for mutants. These numbers represent the number of mutagenized chromosomes tested, because an F1 fly gets one independently mutagenized compound arm from its father and each such arm contains two copies of half a chromosome. Males do not contribute to homozygosis as *Drosophila* males do not show meiotic recombination. Number of mutants obtained is given in table 1.

Pure cultures for a recessive homozygous mutation on the compound chromosome obtained in the above screening can be set up by crossing this mutant to non mutant flies from the parent stock (Fig. 3). Half the progeny from this cross should be mutant. A cross between virgin mutant females and mutant males from this progeny would yield a pure culture. Our mutagenesis protocol also yields recessive mutations on the X-chromosome as well as dominant mutations on any of the chromosomes. Procedure for establishing pure lines for such mutations is different and is mentioned later.

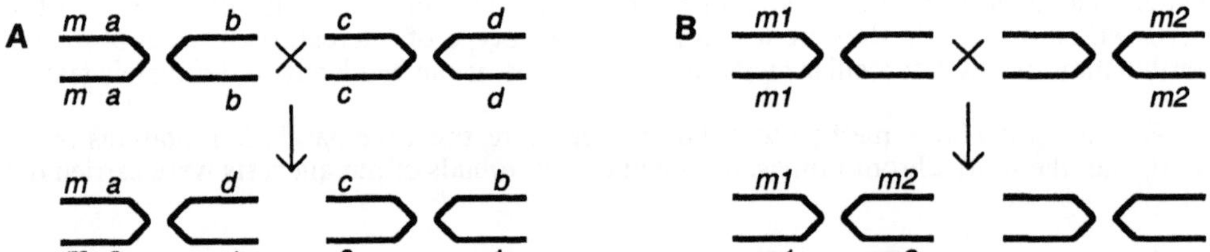

Fig. 4. **A:** Chromosomal localization of a mutation. A cross between *C(2L)RM, a;C(2R)RM, b* and *C(2L)RM, c;C(2R)RM, d* would yield mainly two types of progeny. A mutation '*m*' would segregate with markers '*a*' and '*d*' if it is on the left arm of the compound chromosome and with markers '*b*' and '*c*' if it is on the right arm. If '*m*' is on the X-chromosome, F1 males would show the maternal genotype and F1 females would be heterozygous. A mutation on an autosome other than the compound chromosome would be in a heterozygous form in all F1 flies. Segregation and expression will also depend on whether the mutation is dominant or recessive. **B:** Constructing double mutants. If two mutations '*m1*' and '*m2*' are located on two different arms of the compound chromosomes, double mutants can be constructed by a single cross.

It is easy to identify the chromosome which bears a new mutation obtained by the above method (Fig. 4A). Flies bearing the new mutation, and a marker on each compound arm, are crossed to flies bearing a different set of markers on the compound arms. Nature of segregation of the new mutation with respect to markers reveals the chromosome on which the mutation is present; whether it is present on the left or the right arm in case of the compound chromosome; and whether it is recessive or dominant (Fig. 4A). Table 1 gives the chromosomal localization of different mutations obtained in our experiments. These numbers refer to the number of independently identified mutations. F1 females were grouped in batches of about one hundred each. Mutations obtained from the progeny of different batches were taken to be of independent origin. The total number of mutations given in the table is more than the number of mutations noted for the X-chromosome and the autosomes because some of the mutations are yet to be localized.

The meiotic segregation pattern of compound chromosomes provides an easy method for constructing double mutants in certain cases. If the two mutations lie on two different arms of the compound chromosome, half the progeny of a simple cross between the two would carry both the mutations (Fig. 4B). It is likewise simple to combine two mutations if one of them lies on, say, the compound second chromosome and the other one lies on another chromosome with the strain bearing a compound second chromosome. Mutations obtained by using the compound chromosomes would generally meet these criteria, as many of our mutations do.

A number of genetic experiments require the mutations to be present on normal linear chromosomes. It is simple to linearize the compound chromosomes. Virgin females bearing compound chromosomes are X-irradiated and crossed to males bearing linear chromosomes (Chovnick et al., 1970). This cross primarily produces aneuploid zygotes that are inviable. However, an egg containing a chromosome linearized by X-irradiation can give rise to a viable zygote. Male and female flies arising out of this cross can be crossed between themselves to give rise to flies with linear chromosomes and carrying the new mutation in the homozygous form (Fig. 5). Number of mutations for which the chromosomes have been linearized are given in table 1.

DISCUSSION

Compound chromosomes provide an efficient way for obtaining recessive autosomal mutations. They remove the major problem in existing protocols, that of having to establish independent lines from individual mutagenized chromosomes. Compound autosomes make it easy to screen tens of thousands of mutagenized chromosomes within a short time. We have used this method mainly for obtaining temperature-sensitive paralytic mutations. It has also

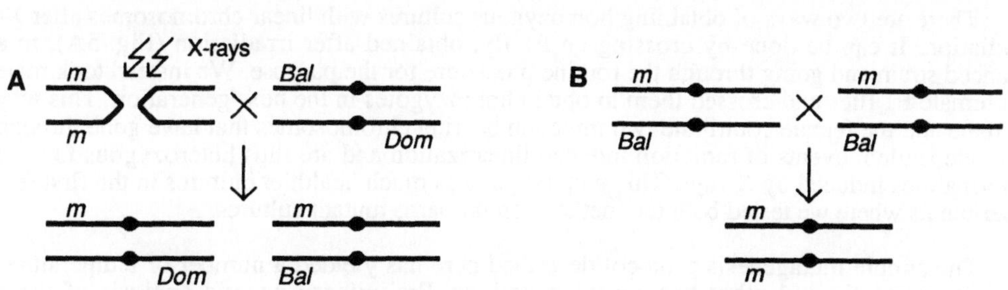

Fig. 5. Linearizing compound chromosomes. A cross between flies with compound chromosomes and those with linear chromosomes normally produces inviable aneuploids. **A:** X-irradiation can linearize the compounds yielding viable zygotes. **B:** Males and females obtained after X-irradiation can be crossed to yield mutant flies with linear chromosomes. *Bal*, a balancer chromosome; *Dom*, a dominant marker.

given us other behavioral and morphological mutations. It can be easily used for obtaining a variety of mutations, for example those affecting development, visual system, flight or a number of other functions that can be tested in single flies.

The method can also be used for obtaining mutations, for example those affecting olfactory response, that can be identified only by testing populations of flies. One way to do so is to mutagenize flies with a heterozygous marker near the proximal end of each of the compound arms. Homozygosis of a marker will identify F2 flies in which the compound arms have undergone meiotic recombination proximal to the marker. These F2 flies can be used to set up independent cultures by crossing them to flies from the original stock. Half the progeny from this cross should be homozygous for a newly induced mutation. Depending on the nature of the required mutation, the mutant phenotype can be tested with the whole progeny as such or with only the homozygous flies selected on the basis of visible markers. Although this procedure involves setting up independent cultures, it is much simpler than the existing procedures for isolating autosomal mutations which can be recognized only on the basis of population behavior. Compound chromosomes can also be used to identify sterile mutations. This can be done by starting in the same way as in the case of population behavior and then using F2 males for female sterile mutations and F2 females for male sterility.

Probability of homozygosis in compound arms increases with increase in distance from the centromere. Chances of recovering proximal mutations can be increased by testing a higher number of F2 flies for each F1 female. As suggested by A. Chovnick, the probability of recombination can be increased by X-irradiating F1 flies. One way to increase the overall efficiency of the screening procedure would be to use proximal mutations that impart resistance to a certain treatment, say a drug. Use of a heterozygous resistance mutation would allow picking up only those chromosomes in F2 that have undergone homozygosis proximal to this marker. A dominant temperature-sensitive mutation can serve a similar purpose.

An interesting possibility would be to combine the features of the compound chromosomes and the P-elements for mutagenesis. This would allow us to exploit the advantages offered by P-element mutagenesis.

Procedure described in figure 3 fails to give pure mutant cultures in certain cases. In such a case it is likely that we are not dealing with a recessive mutation on the compound chromosome, but either with a recessive mutation on the X-chromosome or a dominant mutation on any of the chromosomes including the compound autosome. If a dominant autosomal mutation is to be maintained and propagated as a pure culture, it can be done after linearizing the compound chromosome. Cultures for mutations on the X-chromosome can be set up by crossing appropriate flies without linearizing the compounds.

There are two ways of obtaining homozygous cultures with linear chromosomes after X-irradiation. It can be done by crossing an F1 fly, obtained after irradiation (Fig. 5A), to a balanced strain and going through the routine procedure for the purpose. We instead took male and female F1 flies and crossed them to obtain homozygotes in the next generation. This way the male and the female contribute two mutation bearing chromosomes that have gone through two independent events of radiation induced linearization and are thus heterozygous for any abnormalities induced by X-rays. This method gave us much healthier cultures in the first few experiments where we tested both the methods on the same mutant cultures.

The simple mutagenesis protocol described here has yielded a number of temperature-sensitive paralytic and other behavioral mutations. Preliminary genetic analysis of these mutations in underway. Some of the existing behavioral mutations that have been studied till now show interesting effects on membrane excitability and other neuromuscular phenomena (Hall, 1982). It is clear from the spectrum of defects in these mutants, and from the number and allelic distribution of the mutations, that we have hardly exploited the immense potential of *Drosophila* for answering some key questions related to the nervous system. The large number of mutations that our method has yielded, and is continuing to yield, will provide a useful material for exploring some of these questions.

ACKNOWLEDGEMENTS

This work was supported by grant number 4/13/84 G from the Department of Atomic Energy, Government of India, to S. S. and research fellowships from the Council of Scientific and Industrial Research, India, to M. J. S. C., P. B. and D. G.

REFERENCES

Baker, B. S., Carpenter, A. T. C., Esposito, M. S., Esposito, R. E., Sandler, L. (1976) The genetic control of meiosis. Ann. Rev. Genet. **10:** 53-134

Baumann, A., Krah-Jentgens, I., Muller, R., Muller-Holtkamp, F., Seidel, R., Kecskemethy, N., Casal, J., Ferrus, A., Pongs, O. (1987) Molecular organization of the maternal effect region of the *Shaker* complex of *Drosophila* : characterization of an I_A channel transcript with homology to vertebrate Na^+ channel. EMBO J. **6**, 3419-3429.

Campos-Ortega, J. A. (1985) Genetics of early neurogenesis in *Drosophila melanogaster*. Trends in Neurosci. **8:** 245-250

Chovnick, A., Ballantyne, G. H., Baillie, D. L., Holm, D. G. (1970) Gene conversion in higher organisms : Half tetrad analysis of recombination within the *rosy* cistron of *Drosophila melanogaster*. Genetics **66:** 315-329

Dudai, Y (1985) Genes, enzymes and learning in *Drosophila* . Trends in Neurosci. **8:**18-21

Elkins, T., Ganetzky, B. (1988) The roles of potassium currents in *Drosophila* flight muscles. J. Neurosci. **8**, 428-434.

Elkins, T., Ganetzky, B., Wu, C.-F. (1986) A *Drosophila* mutation that eliminates a calcium-dependent potassium current. Proc. Natl. Acad. Sci. USA **83**, 8415-8419.

Ganetzky, B., Wu, C.-F. (1986) Neurogenetics of membrane excitability in *Drosophila.* Ann. Rev. Genetics **20**: 13-44

Gho, M., Mallart, A. (1986) Two distinct calcium-activated potassium currents in larval muscle fibers of *Drosophila melanogaster.* Pfluegers Arch. **407**, 526-533.

Hall, J. C. (1982) Genetics of the nervous system in *Drosophila* . Quart. Rev. Biophysics **15:**223-479

Heisenberg, M., Wolf, R. (1984) Vision in *Drosophila*.: Genetics of microbehavior. Springer-Verlag

Holm, D. G. (1976) Compound Autosomes. In: The Genetics and Biology of *Drosophila* . Vol. **1b:** pp 529-561. Edited by M. Ashburner, E. Novitsky. Academic Press

Kamb, A., Iverson, L.E., Tanouye, M.A. (1987) Molecular characterization of *Shaker*, a *Drosophila* gene that encodes a potassium channel. Cell **50**, 405-413.

Kaplan, W. D., Trout III, W. E. (1969) The behavior of four neurological mutants of *Drosophila.* Genetics **61**, 399-409.

Komatsu, A., Wu, C.-F. (1987) Single-channel potassium currents in membrane vesicles derived from normal and mutant *Drosophila* larval muscles. Soc. Neurosci. Abstr. **13**, 530.

Lewis, E. B. (1967) Genes and gene complexes. In: Heritage from Mendel. pp 17-47. Edited by R. A. Brink. Univ. of Wisconsin press, Madison, Wisconsin.

Lewis, E. B. (1978) A gene complex controlling segmentation in *Drosophila* . Nature **276:** 565-570

Lewis, E. B., Bacher, F. (1968) Method of feeding ethyl methanesulfonate (EMS) to *Drosophila* males. *Drosophila* Inform. Serv. **43:** 193

Lindsley, D. L., Grell, E. H. (1968) Genetic variations of *Drosophila melanogaster* . Carnegie Inst Wash publ 627

Papazian, D.M., Schwarz, T.L., Tempel, B.L., Jan, Y.N., Jan, L.Y. (1987) Cloning of genomic and complementary DNA from *Shaker*, a putative potassium channel gene from *Drosophila.* Science **237**, 749-753.

Rasmussen, I. E. (1960) *Attached - 2R;2L* . *Drosophila* Inform. Serv. **34:** 53

Salkoff, L., Wyman, R. (1981) Genetic modification of potassium channels in *Drosophila Shaker* mutants. Nature **293**, 228-230.

Siddiqi, O. (1987) Neurogenetics of olfaction in *Drosophila melanogaster.* Trends in Genetics **3**: 137-142

Singh, S. (1983) A mutagenesis scheme for obtaining autosomal mutations in *Drosophila* . Indian J. Exp. Biol. **21:** 635-636

Singh, S., Wu, C.-F. (1987) Genetic and pharmacological separation of four potassium currents in *Drosophila* larvae. Soc. Neurosci. Abstr. **13**, 579.

Singh, S., Wu, C.-F., Ganetzky, B. (1986) Interactions among different K^+ and Ca^{++} currents in normal and mutant *Drosophila* larval muscles. Soc. Neurosci. Abstr. **12**, 559.

Singh, S., Bhandari, P., Chopra, M. J. S., Guha D (1987) Isolation of autosomal mutations in *Drosophila melanogaster* without setting up lines. Mol. Gen. Genet. **208**: 226-229

Solc, C.K., Zagotta, W.N., Aldrich, R.W. (1987) Single-channel and genetic analyses reveal two distinct A-type potassium channels in *Drosophila*. Science **236**, 1094-1098.

Wei, A., Salkoff, L. (1986) Occult *Drosophila* calcium channels and twinning of calcium and voltage-activated potassium channels. Science **233**, 780-782.

Wu, C.-F., Ganetzky, B., Haugland, F.N., Liu, A.-X. (1983) Potassium currents in *Drosophila*: Different components affected by mutations of two genes. Science **220**, 1076-1078.

Wu, C.-F., Haugland, F. N. (1985) Voltage clamp analysis of membrane currents in larval muscle fibers of *Drosophila* : Alteration of potassium currents in *Shaker* mutants. J. Neurosci. **5**, 2626-2640.

PROJECTIONS AND FUNCTIONAL IMPLICATIONS OF LABELLAR NEURONS FROM INDIVIDUAL SENSILLA OF DROSOPHILA MELANOGASTER

Shubha R. Shanbhag and R. Naresh Singh

Molecular Biology Unit, Tata Institute of Fundamental Research
Homi Bhabha Road, Navy Nagar, Colaba, Bombay 400 005, India

ABSTRACT

Our previous studies using Golgi silver impregnations from labellar sensilla of adult Drosophila melanogaster had revealed seven distinct neuronal types projecting in the suboesophageal ganglion of the brain. These are : coiled fibre (type I); shrubby fibre (type II); ipsilateral ventral fibre (type III); ipsilateral dorsal fibre (type IV); contralateral ventral fibre (type V); contralateral dorsal fibre (type VI) and central fibre (type VII) (Nayak and Singh, 1985).

We have attempted to identify the neurons present in a single taste sensillum using the neuronal marker horse radish peroxidase (HRP). The success rate of single sensillum fillings with HRP is about 20%, but the fibre projections obtained could be identified without ambiguity. Although a single sensillum in question has five neurons, yet at a time, only one or at the most two neurons are labelled in any given experiment. The type of neuron labelled was found to be specific to the stimulus solute present in the HRP solution. (i) The presence of 0.1 M potassium chloride in HRP stains type II and type VI fibres separately or both together with an equal probability in any preparation. (ii) HRP dissolved in 0.1 M sucrose solution mainly showed type IV (60%) and less frequently type II (20%) fibres, while in the remaining 20% both types II and IV are labelled. (iii) When 0.1 M sodium chloride solution containing HRP is used, the type of fibres stained are very similar to those of HRP dissolved in 0.1 M sucrose, however, the frequency of fibres stained is different - type IV (35%), type II (35%) and types II and IV together (30%). (iv) HRP dissolved in distilled water alone, revealed type I (30%) and type II (20%) fibres separately or together (35%). Occasionally type V or type VII (<6% each) also get stained.

It is known from behavioural and electrophysiological studies in our laboratory that at lower concentrations flies are attracted by sodium chloride ($\leq$ 0.1 M NaCl) but are not by potassium chloride (Arora 1985; Swati Joshi, personal communication).

In the present study type II fibres get stained irrespective of the stimulant present in HRP solution. However, when attractants are used as stimulant solute in HRP solution, only type IV fibres are labelled. Type VI fibres are stained when the stimulant is a repellant.

INTRODUCTION

During the last few years we have been interested in the anatomical aspects of the chemosensory systems of Drosophila melanogaster (Nayak & Singh, 1983, 1985; Singh & Nayak, 1985; Singh & Singh, 1984; Venkatesh & Singh, 1984; Stocker & Singh, 1983; Stocker et al., 1983).

The labial palp of the wild type Canton S strain of Drosophila consists of taste hairs and taste pegs (Falk et al., 1976; Nayak & Singh, 1983). On the labellum the taste hairs are arranged in a rather constant pattern of two ill defined medial rows and a peripheral row. The hairs in two medial rows, in addition to one mechanosensory neuron have predominantly four chemosensory neurons each. The peripheral hairs usually contain one mechanosensory and two chemosensory neurons (Nayak & Singh, 1983). The axons from these sensilla project mainly within the anterior-medial region of the suboesophageal ganglion (SOG) (Stocker & Schorderet, 1981; Nayak & Singh, 1985). The analysis of primary labellar sensory projections using Golgi silver impregnations revealed seven distinct neuronal types I-VII projecting in the SOG of the brain (Nayak & Singh, 1985). Incidently, this study correlates well with the number of chemosensory neuron-types present in different types of sensilla of the labellum (Nayak & Singh, 1983, 1985).

We undertook the present study in order to understand, whether the four functionally distinct neurons of a single labellar sensillum are projected in the brain according to their function or are represented spatially depending upon the location of the individual sensilla.

We used one of the refined techniques of horse radish peroxidase (HRP) injection described by Nässel (1983). Since it is known that, HRP as a neuronal marker provides valuable information concerning shape, function and chemical identy of the neuron. In addition it also gives enough contrast both at light- and electron microscope levels (Seyan et al., 1983). In the present work, we also provide evidence to show that when different stimulus-conditions in HRP solution are used, different types of neurons are stained. It was also found that in the labellar sensory neuropile of SOG a specific glomerular organization exists.

MATERIALS AND METHODS

Three-to-five day old D. melanogaster Canton S females were used. Sensory projections of selected taste hair on the labellum (Fig. 1a,b) was examined after the uptake of HRP (Sigma, type VI), dissolved in the following solutions: (i) 0.1 M potassium chloride (A.R.), (ii) 0.1 M sucrose (A.R.), (iii) 0.1 M sodium chloride (A.R.), (iv) distilled water. Solutions (i)-(iii) were prepared in glass distilled water.

Immobilization of the fly and HRP injections were according to the schedules described by Nässel (1983) and the concentration of HRP in solutions was 2-3%. The hair-shaft of the sensillum was cut with a pair of iris scissors and the stub was inserted in a capillary filled with HRP solution (Fig.1) of the appropriate composition. The flies were kept with the capillary stuck on them, overnight at 5° C in the dark. Subsequent dissection, fixation and enzyme reactions were carried out according to Nässel (1983), followed by dehydration and embedding in Durcupan ACM (Fluka). Fifteen µm thick serial sections were cut either in frontal or sagittal planes on a sliding microtome with steel knife. Sections were mounted with 'Permount' under coverslip.

For making methylene blue whole-mount preparations, the entire fly was immersed for two days in 1% methylene blue dissolved in 70% ethanol. The labellum was removed, dehydrated, cleared and mounted in 'Permount'.

Camera lucida tracings were done on a Leitz Vario-Orthomat 2 photomicroscope at a final magnification of 1250x. The same microscope was used for photomicrography.

RESULTS

A single sensillum which was used in different flies for labelling with HRP (Fig. 1b,c), has 5 neurons but at a time only one or at the most 2 neurons are labelled in any given experiment. If results of many such experiments are put together, we find mainly 4 types of fibres projecting in the different regions of the SOG are labelled (Table 1). Rarely, in two out of 94 preparations, we also find either type V or type VII fibre getting stained (Table 1). All these fibres fall morphologically into one of the seven types I-VII, characterized earlier by us (Figs. 3, 4) (Nayak & Singh, 1985). More than 200 preparations were made for each stimulant, out of which about 20% preparations were successful for each solute. In all experimental conditions the same sensillum (Figs. 1b, 2a marked arrow) was used for labelling the neurons with HRP.

Results of Neuron Labellings in a Sensillum by HRP, when Different Stimulant Solutes are Present in Solution with HRP

(i) HRP dissolved in 0.1 M potassium chloride solution. Type II and type VI fibres are stained either separately or together with equal probability in any individual preparations (Table 1, Figs. 5b, 6a, d). In

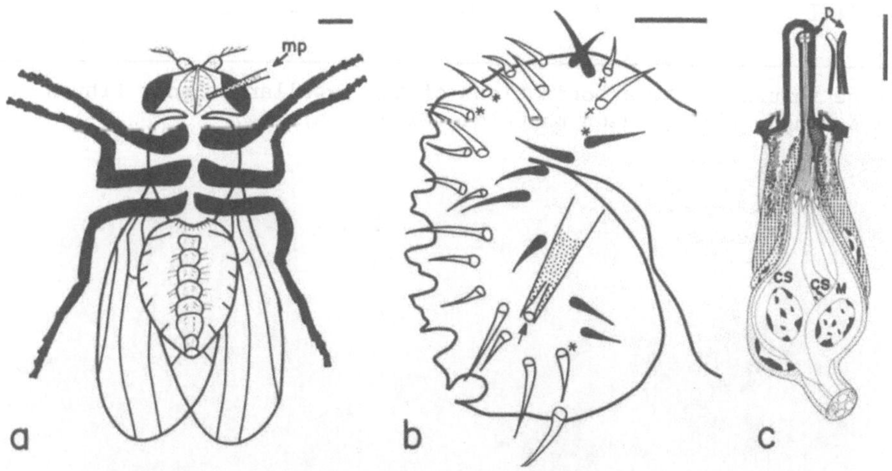

Fig. 1. (a) Diagram of a Drosophila melanogaster fly immobilized with plasticine, with dorsal side down for filling HRP solution in a single sensillum, on the left labellum.
mc = microcapillary filled with HRP solution.
Scale = 200 μm.
(b) Magnified view of the labellum, showing position and neuronal composition of individual sensilla.
Unshaded = 4CS + 1M; solid filled = 2CS + 1M; marked * = variable composition, either 2-4CS + 1M; marked † = composition not determined; marked arrow = HRP filled sensillum.
CS = chemosensory neuron/s; M = mechanosensory neuron.
Scale = 25 μm.
(c) Detailed diagram of the labellar sensillum containing 4CS + 1M neurons, chosen to be filled with HRP solution.
CS = chemosensory neuron/s; M = mechanosensory neuron.
Scale = 5 μm.

Table 1. Types of fibres from a single sensillum labelled with HRP in different experiments.

Type of labellar sensory fibre	Volume of terminal arborizations μm^3 *	Stimulant solute in H R P solution			
		0.1 M KCl	0.1 M Sucrose	0.1 M NaCl	H_2O
I-Coiled	20,500	-	-	-	4
II-Shrubby	12,100	11	6	6	3
III-Ipsilateral ventral	3,200	-	-	-	-
IV-Ipsilateral dorsal	3,000	-	18	6	-
V-Contralateral ventral	3,400	-	-	-	1
VI-Contralateral dorsal	3,500	9	-	1	-
VII-Central	3,500	-	-	-	1
Types I+II	-	-	-	-	5
Types II+IV	-	-	6	5	-
Types II+VI	-	10	-	1	-

* Average volume of the arborizations of the labellar sensory fibres (Nayak & Singh, 1985), based on Golgi silver staining.

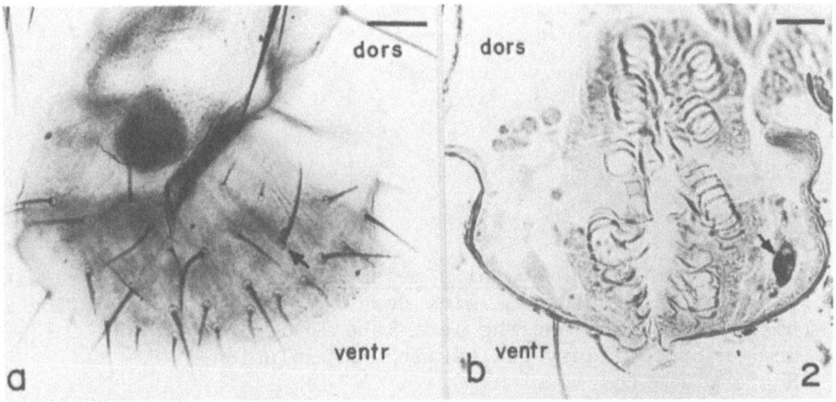

Fig. 2. (a) Lateral view of a whole-mount preparation of methylene blue-stained labellum of Drosophila with taste hairs. Sensillum marked with arrow was chosen for labelling with HRP. dors = dorsal; ventr = ventral.
Scale = 20 µm.
(b) Photomicrograph of a 15 µm thick frontal section of the labellum after HRP staining, showing the stain in the sensillar sac(arrow).
Scale = 20 µm.

almost all the preparations, the type II fibres stained intensely (dark brown to black) (Fig. 6a), whereas the type VI fibre usually get labelled weakly (Fig. 6d).

<u>(ii) HRP dissolved in 0.1 M sucrose solution</u>: Filling the same hair with this solution labelled mainly type IV fibre (Table 1, Figs. 5c, 6e,h). The terminal arborizations of the lateral branches of type IV fibre have many boutons (Figs. 5c marked arrow; 6h marked *) and these endings are located in the peripheral region of the ventrolateral glomerulus (VL) of the labellar sensory neuropile (Fig. 6e,h). HRP dissolved in sucrose solution also stains type II fibre either separately (Table 1) or together with type IV fibre (Table 1, Figs. 5c, 6b marked *).

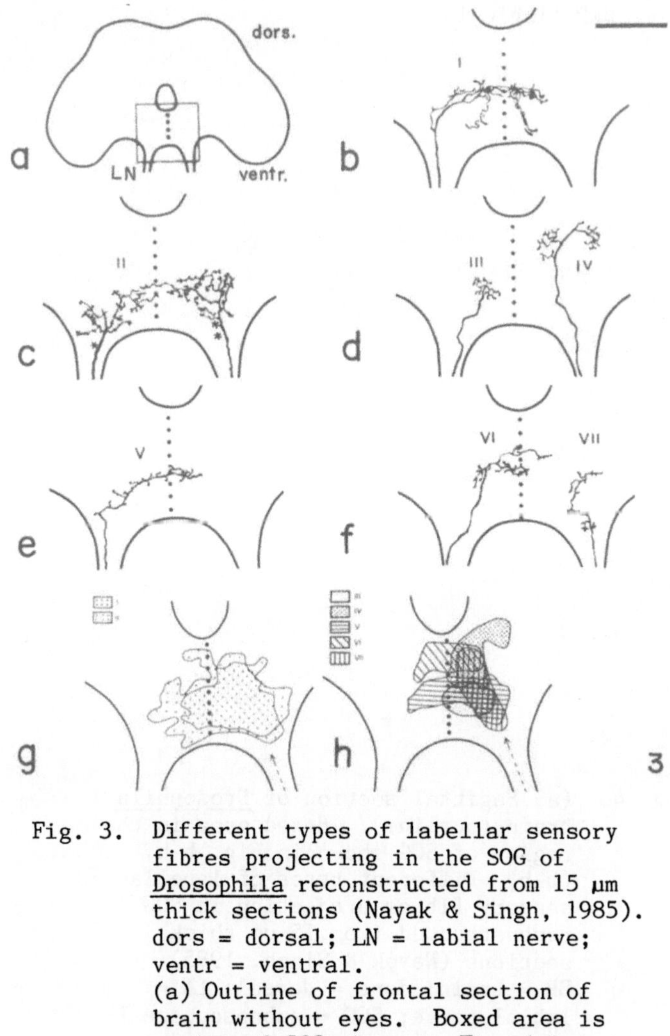

Fig. 3. Different types of labellar sensory fibres projecting in the SOG of <u>Drosophila</u> reconstructed from 15 µm thick sections (Nayak & Singh, 1985). dors = dorsal; LN = labial nerve; ventr = ventral.
(a) Outline of frontal section of brain without eyes. Boxed area is region of SOG shown in Fig. b-h.
(b) Type I fibre. (c) Type II fibre (* marked is ipsilateral and ** marked extend to contralateral side).
(d) Type III and type IV fibres.
(e) Type V fibre. (f) Type VI and type VII fibres. (g) and (h) Diagrams showing regions occupied by arborizations of the labellar sensory fibres types I-VII in SOG.
Scale = 50 µm.

(iii) HRP dissolved in 0.1 M sodium chloride solution. This solution labels type II (Figs. 5d, 6c) and type IV (Figs. 5d, 6g marked *) fibres, either separately (Fig. 6c) or together (Table 1). However, the terminal arborizations of the lateral branch ending at the ventrolateral glomerulus in type IV have a few boutons (Figs. 5d, 6g marked *) as compared to the endings of type IV fibres stained with HRP dissolved in sucrose solution (Fig. 6h marked *).

(iv) HRP dissolved in distilled water. Fillings with HRP dissolved in distilled water alone revealed type I coiled fibre, type II shrubby fibre either separately or together (Table 1, Figs. 5e, 6f,i). However, occasionally (<6%) we do find type V or type VII getting stained with aqueous solution of HRP (Table 1).

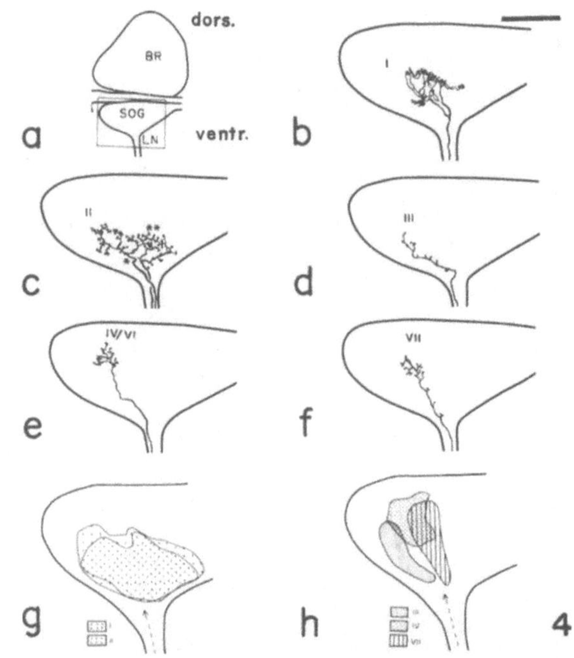

Fig. 4. (a) Sagittal section of Drosophila brain (outline). Boxed area is the region of SOG shown in Fig. b-h. (b-h) Different types of labellar sensory fibres projecting in SOG, reconstructed from 15 µm thick sections (Nayak & Singh, 1985). BR = brain; dors = dorsal; LN = labial nerve; SOG = suboesophageal ganglion; ventr = ventral. (b) Type I fibre. (c) Type II fibre. (d) Type III fibre. (e) Type IV/VI fibres. (f) Type VII fibre. (g-h) Regions occupied by the arborizations of different types of fibres in SOG. Scale = 50 µm.

present HRP staining also revealed the glomerular organization of the labellar sensory neuropile in the SOG (Fig. 7). According to the location, size and shape at least 7 glomeruli like structures are visible in the anteromedial region of the SOG (Figs. 7,8). Altogether there are two dorsoanterior glomeruli (DA), 30-40 µm diameter; two large ventrolateral (VL) glomeruli about 50 µm diameter; two small ventromedial (VM) glomeruli 20x50 µm cross-section and a single large medial (M) glomerulus about 50 µm diameter. Except for the medial glomerulus (M) which is common to both ipsi- and contralateral sides, rest of the glomeruli are present one each on either side of the labellar sensory neuropile in the SOG (Figs. 7,8).

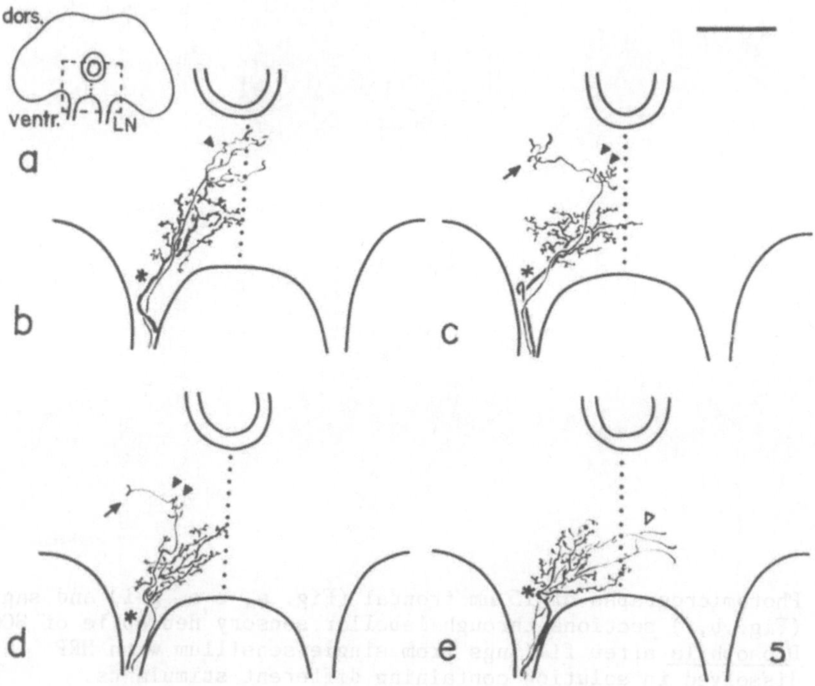

Fig. 5. (a) Outline of a frontal section of Drosophila brain minus eyes. Boxed area is the region of SOG shown in Fig. b–e, where types of fibres stained from a single labellar sensillum with HRP dissolved in solution containing the following stimulant-solutes are shown. (b) 0.1 M potassium chloride. (c) 0.1 M sucrose. (d) 0.1 M sodium chloride. (e) Distilled water.
dors = dorsal; LN = labial nerve; ventr = ventral. Δ = type I fibre; ✱ = type II fibre; ▲▲ = type IV fibre; ▲ = type VI fibre.
Scale = 50 µm.

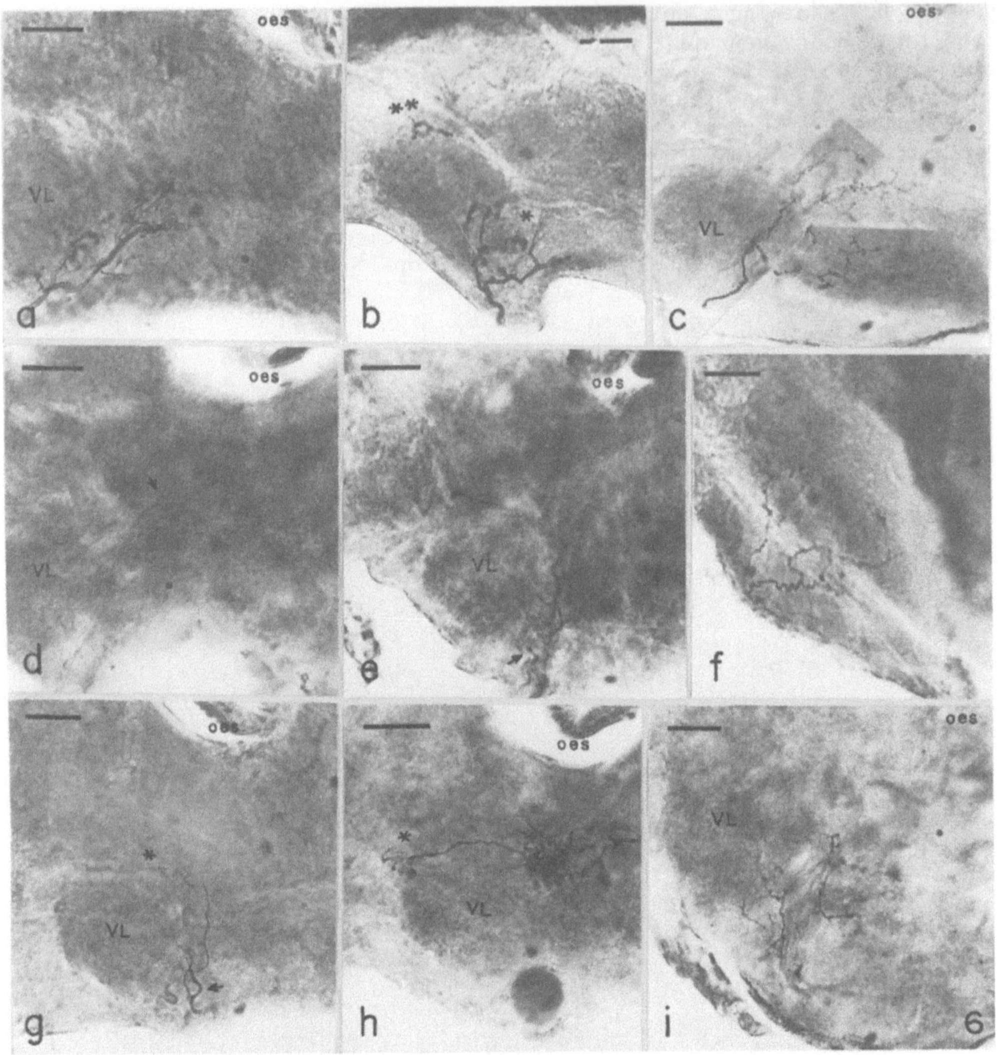

Fig. 6. Photomicrographs of 15 µm frontal (Fig. a, c–e, g–i) and sagittal (Fig. b,f) sections through labellar sensory neuropile of SOG of Drosophila after fillings from single sensillum with HRP dissolved in solution containing different stimulants.
a and d, 0.1 M potassium chloride; b,e and h, 0.1 M sucrose; c and g, 0.1 M sodium chloride; f and i, distilled water. Filled sensillum on the left side. Dorsal on top; oes = oesophagus, VL = ventrolateral glomerulus.
Scale = 20 µm.
(a) Type II fibre. (b) Type II fibre marked * and type IV fibre marked **. (c) Type II fibre. (d) Type VI fibre stained very weakly (arrow). (e) Type IV fibre with stained terminals of the lateral branch projecting near dorsal region of VL glomerulus. Type II fibre is also stained (arrow) but the terminals are present in the adjacent sections. (f) Type I fibre. (g) Type IV marked *, and part of type II fibre (marked arrow). Note the terminals of type IV projecting in the same region in both e and g. Most of the terminals of type II fibre are contained within the adjacent sections. (h) Type IV fibre showing many terminals of the lateral branch in the periphery of VL glomerulus. (i) Type II fibre, many terminals are present in adjacent sections. A few project in the medioventral region of VL glomerulus.

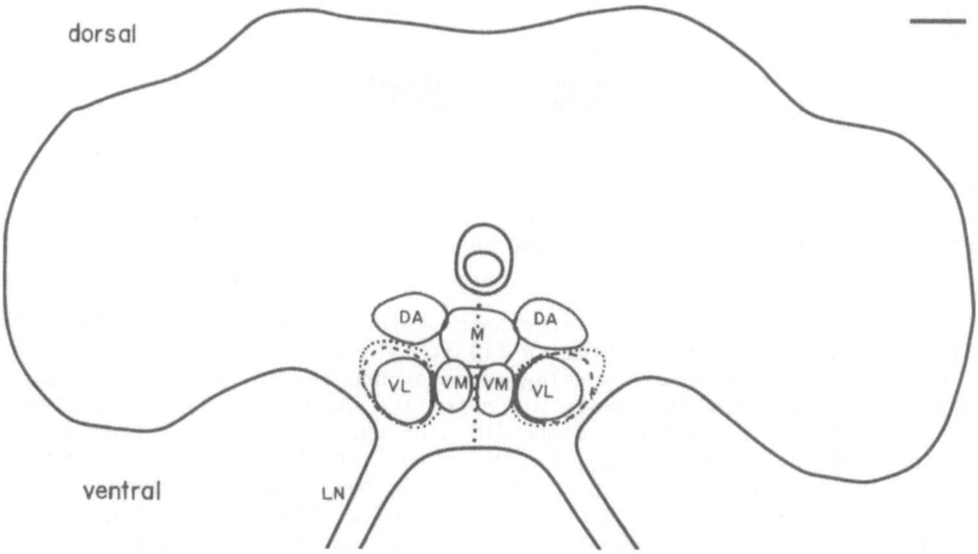

Fig. 7. Diagram of Drosophila brain through the labellar sensory region of the SOG from anterior to posterior levels, in frontal plane showing relative position, shape and size of 7 glomeruli identified by HRP staining. Continuous line —— = more anterior section; broken line --- = middle section; dotted line ...... = posterior section. The terminology roughly indicates relative positions of glomeruli within the labellar sensory neuropile.
(A = anterior; D = dorsal; L = lateral; LN = labial nerve; M = medial; Oes = oesophagus; V = ventral).
Scale = 25 µm.

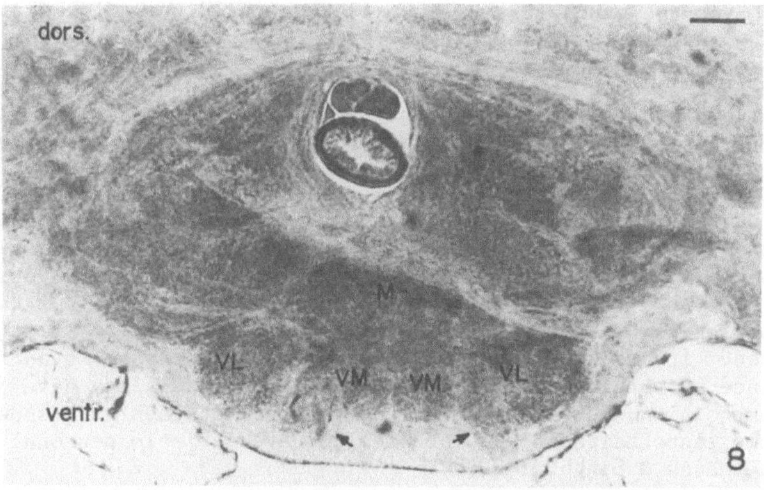

Fig. 8. Photomicrograph of 15 µm thick frontal section through labellar sensory neuropile in SOG after HRP staining showing 5 out of 7 glomeruli.
dors = dorsal; M = medial glomerulus; ventr = ventral; VL = ventrolateral glomeruli; VM = ventromedial glomeruli; arrow indicates the labial nerve tract in the SOG.
Scale = 25 µm.

DISCUSSION

The main findings of this study are the followings : (1) HRP dissolved in solutions containing various stimulants, labels only one or at the most two neurons from a single sensillum containing 4 chemosensory and one mechanosensory neurons. (2) The type of neuron labelled is specific to the stimulant solute which could be either an attractant or a repellant present in HRP solution. (3) We have identified 7 glomeruli in the suboesophageal ganglion of the brain which form the main projection areas of the labellar sensory neurons.

We selected a particular large hair of the medial row on the labellum (Figs. 1b, 2a marked arrow) because (i) it is quite isolated from the rest of the hairs, giving enough working space around it, so that a glass-microcapillary filled with HRP can be placed on its stub with greater confidence and ease. (ii) The neuronal composition of the hair is known from our earlier studies (Nayak & Singh, 1983), as well as the functional aspects of each of the 4 chemosensory neurons present in this sensillum are also characterised (Arora, 1985; Arora et al., 1987). Electrophysiologically a neuron designated as S, present in this sensillum is sensitive to sugars. Whereas two neurons respond to salt - one of them designated as L1 responds to lower concentrations of salt and the other designated as L2 to higher salt concentrations. A fourth neuron called W is inhibited by sugars and salts and is believed to be a water-cell (Rodrigues & Siddiqi, 1978, 1981; Arora 1985; Arora et al., 1987).

Irrespective of the stimulant present in HRP solution, most of the times we find type II fibre gets labelled (Table 1). Since the terminals of the fibre show profuse branching and occupy a major region of the labellar sensory neuropile (Figs. 3,4 and 5), we suspect this fibre could be functioning as a 'vigilant' neuron. Perhaps the vigilant neuron alerts the fly for any type of incoming sensory stimulus. However, the attraction or repulsion response of the fly is determined mainly by either type IV or type VI neurons respectively.

The behavioural studies have shown that the wild type fly is attracted by sugars and repelled by salts (Rodrigues & Siddiqi, 1978, 1981; Siddiqi & Rodrigues, 1980). Though both sodium chloride and potassium chloride are repellant but at lower than 0.1 M concentrations *Drosophila* flies show attraction response for sodium chloride in contrast to potassium chloride, which is repellant at that concentration (Arora 1985, Swati Joshi personal communication). The most striking observation in the present study is the staining of type IV fibre when HRP solution contains attractants (0.1 M sucrose or 0.1 M sodium chloride), and type VI fibre is stained when the stimulant is a repellant (0.1 M potassium chloride). With HRP dissolved in distilled water, apart from type II fibre, usually type I fibre gets labelled (Table 1). Probably type I fibre correlates to the water cell of the earlier electrophysiological studies. At times we also find type V or VII (<6% each) fibre stained. It could be because the electrode might have moved from its original place of capping or rarely this sensillum itself may have a variable neuronal composition. Our earlier ultrastructural studies have shown that at times some taste hairs on the labellum do have variable neuronal composition (Nayak & Singh, 1983)

The type IV fibre stained with HRP solution containing attractants (either 0.1 M sucrose or 0.1 M sodium chloride) terminate in the periphery of the ventrolateral glomerulus (VL) (Fig. 6e,g,h). The morphology of the terminal branches show certain subtle differences. In the former the lateral branch of this fibre (Fig. 5c marked arrow; 6h marked *) has many arborizations, whereas in the latter only one or two arborizations are observed (Figs 5d marked arrow; 6g marked *). If this difference is of significance, the type IV fibres comprise of two sub-populations of neurons — one class responding to sugars only, the so called S neuron (Arora et al., 1987), and the other class responding to low concentrations of sodium chloride, designated as L_1 neuron also mediating attraction response (Arora, 1985; Swati Joshi personal communication).

We infer from the present data that this technique of labelling neurons with HRP dissolved in different solutions of stimulant is demonstrating functional neuroanatomy in *Drosophila* brain. The present study also supports that the neurons of a sensillum are represented functionally in the brain at distinct locations.

ACKNOWLEDGEMENTS

We thank Dr. Veronica Rodrigues, Mrs. Kusum Singh, Miss Swati Joshi and Miss Seema Deshpande for helpful suggestions.

REFERENCES

Arora, K. 1985. Neurogenetic studies on taste mechanism of *Drosophila melanogaster*. Ph.D. Thesis, The University of Bombay, Bombay.

Arora, K., Rodrigues, V., Joshi, S., Shanbhag, S. and Siddiqi, O. 1987. A gene affecting the specificity of chemosensory neurons of *Drosophila*. *Nature*, 330: 62-63.

Falk, R., Bleiser-Avivi, N. and Atidia, J. 1976. Labellar taste organs of *Drosophila melanogaster*. *J. Morphol*. 150: 327-342.

Nässel, D.R. 1983. Horse radish peroxidase and other heme proteins as neuronal markers, pp. 44-91. In N.J.Strausfeld (ed.) *Functional Neuroanatomy*, Springer-Verlag, Berlin, Heidelberg, New York, Tokyo.

Nayak, S.V. and Singh, R.N. 1983. Sensilla on the tarsal segments and mouthparts of adult *Drosophila melanogaster* Meigen (Diptera: Drosophilidae). *Int. J. Insect Morphol. Embryol*. 12: 273-279.

Nayak, S.V. and Singh, R.N. 1985. Primary sensory projections from the labella to the brain of *Drosophila melanogaster* Meigen (Diptera: Drosophilidae). *Int. J. Insect Morphol. Embryol*. 14: 115-129.

Rodrigues , V. and Siddiqi, O. 1978. Genetic analysis of chemosensory pathway. *Proc. Ind. Acad. Sci*. (B) 87: 147-160.

Rodrigues, V. and Siddiqi, O. 1981. A gustatory mutant of *Drosophila* defective in pyranose receptors. *Mol. Gen. Genet*. 181: 406-408.

Seyan, H.S., Bassemir, U.K. and Strausfeld, N.J. 1983. Double marking for light and electron microscopy, pp. 112-131. In N.J.Strausfeld (ed.) *Functional Neuroanatomy*, Springer- Verlag, Berlin, Heidelberg, New York, Tokyo.

Siddiqi, O. and Rodrigues, V. 1980. Genetic analysis of a complex chemoreceptor, pp. 347-359. In O. Siddiqi, P.Babu, L.M. Hall and J.C. Hall (eds) *Development and Neurobiology of Drosophila*, Plenum, New York, London.

Singh, R.N. and Nayak, S.V. 1985. Fine structure and primary sensory projections of sensilla on the maxillary palp of *Drosophila melanogaster* Meigen (Diptera: Drosophilidae). *Int. J. Insect Morphol. Embryol*. 14: 291-306.

Singh, R.N. and Singh, K. 1984. Fine structure of the sensory organs of *Drosophila melanogaster* Meigen larva (Diptera: Drosophilidae). *Int. J. Insect Morphol. Embryol*. 13: 255-273.

Stocker, R.F. and Schorderet, M. 1981. Cobalt filling of sensory projections from internal and external mouthparts in *Drosophila*. *Cell Tissue Res*. 216: 513-523.

Stocker, R.F. and Singh, R.N. 1983. Different types of antennal sensilla in *Drosophila* project into different glomeruli of the brain. Experientia, 39: 674.

Stocker, R.F., Singh, R.N., Schorderet, M., Siddiqi, O. 1983. Projection patterns of different types of antennal sensilla in the antennal glomeruli of *Drosophila melanogaster*. Cell Tissue Res. 232: 237-248.

Venkatesh, S. and Singh, R.N. 1984. Sensilla on the third antennal segment of *Drosophila melanogaster* Meigen (Diptera: Drosophilidae). Int. J. *Insect Morphol.Embryol*. 13: 51-63.

CATIONIC ACCEPTOR SITES ON THE LABELLAR CHEMOSENSORY NEURONS OF DROSOPHILA MELANOGASTER

Swati Joshi, Kavita Arora and Obaid Siddiqi

Molecular Biology Unit
Tata Institute of Fundamental Research
Homi Bhabha Road, Bombay 400 005, India

ABSTRACT

We are studying the molecular basis of salt reception in *Drosophila melanogaster*. The gustatory sensilla of *Drosophila* contain two different neurons designated L1 and L2 which respond to salts. Previous work has shown that the L1 neuron responds principally to cations and carries at least two acceptor sites that distinguish different cations (Arora,1985). We have further obtained biochemical and genetic evidence that at least two types of cationic sites are involved. One of these responds selectively to Na^+ while the other is relatively non-specific and responds to a variety of cations. The response of L1 to salts can be inactivated by trypsin. Following inactivation, the response to Na^+ and K^+ recover with a different time course.

In feeding preference test normal flies show a slight preference for concentration of Na^+ below 0.1M. At concentrations above 0.1M both Na^+ and K^+ are repellents. We have obtained several gustatory mutants which are strongly attracted by low concentrations of Na^+ but not by K^+.

The electrophysiological response of L1 to K^+ saturates at a frequency which is about 30% of the maximal response evoked by Na^+. The mutation *gustE^V86^* eliminates the extra Na^+ response leaving the K^+ response unchanged. In feeding experiments *gustE^V86^* does not show the normal attraction to Na^+. *GustR* a dominant mutation on the III-chromosome shows strong attraction to Na^+ together with enhanced firing of the L1 neuron to NaCl. This mutant is also insensitive to attraction by sugars-sucrose, fructose and trehalose.

K^+ fails to evoke an attraction response either in the wildtype or in the mutants. In this study we try to explain the basis of different electrophysiological and behavioural responses the fly exhibits towards different cations, with the help of gustatory mutants.

INTRODUCTION

It was evident as early as 1926, that insects possess chemoreceptors on their mouthparts and legs when Minnich (1926,1929) showed that stimulation of tarsal hairs of lepidopterans results in an extension of proboscis. Since then a wealth of information has been added to our understanding by V. G. Dethier (Dethier, 1955, 1976). Dethier measured gustatory response in the blowfly *Phormia* using the simple proboscis extension reflex. Behavioural studies were coupled with the study of electrical responses of the chemoreceptors using the tip recording method of Hodgson, Lettvin and Roeder (1955). This method was used by several workers (Wolbarsht,1957; Evans and Mellon,1962; Dethier,1976; Rodrigues and Siddiqi,1978; Crnjar et al.,1983; Fujishiro et al.,1984) to classify specificities of the labellar chemosensory neurons.

Over the past ten years we have been investigating the molecular basis of taste perception in *Drosophila*. We have studied the behavioral and electrophysiological responses of the wildtype flies to salts,sugars and other chemicals. We have obtained several gustatory mutants of *D. melanogaster* with altered taste responses to sugars and salts (Rodrigues and Siddiqi, 1978). By comparing the behavioural and electrophysiological responses of wildtype and mutant flies we hope to identify the genes that specify and regulate the acceptor sites for salts and sugars.

In the wildtype Canton S strain of *Drosophila* each half labellum carries 34 gustatory sensilla (Nayak and Singh, 1983). The neuronal organisation of the chemoreceptors is very similar to that of the larger dipterans. Most of the labellar sensilla contain five bipolar neurons- four chemosensory and one mechanosensory. The sugar sensitive neuron designated as S carries independent acceptor sites for pyranose, furanose and trehalose sugars (Tanimura and Shimada, 1981; Rodrigues and Siddiqi, 1981; Tanimura et al., 1982; Tanimura et al., 1988). Two other neurons designated as L1 and L2 respond to salts (Rodrigues and Siddiqi, 1978; Crnjar et al, 1983; Fujishiro et al, 1984). The ionic specificities of these neurons and their possible role in discriminating anions and cations is not known. The W cell which is inhibited by higher concentrations of salt and sugar is believed to mediate the detection of water.

In *Drosophila melanogaster* the L1 cell responds principally to cations. The electrophysiological response of L1 to K^+ saturates at a frequency which is about 30% of the maximal response evoked by Na^+. Behaviourally, *Drosophila* exhibits a bimodal response to Na^+. At lower concentrations flies show a preference for Na^+ but not for K^+. At higher concentrations, both Na^+ and K^+ are repellents. We have obtained biochemical and genetic evidence which indicates that the L1 neuron carries two types of acceptor sites that distinguish different cations. With the help of gustatory mutants we try to explain the basis of different electrophysiological and behavioural responses the fly exhibits towards different cations.

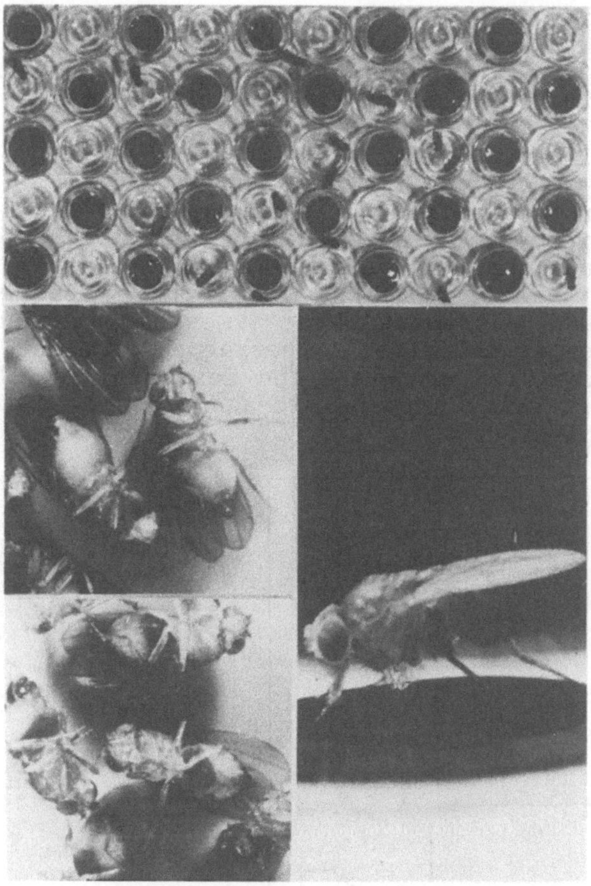

Fig. 1. Feeding behaviour test:
Top:Flies feeding on test plate
Left: Flies with white and coloured abdomens.Right: Fly feeding on coloured well.

Measurement of Gustatory Behaviour

A simple yet sensitive behavioural assay was developed for the measurement of taste responses of *Drosophila* by Tanimura et al. (1982). Tanimura's method has been adapted by us to measure concentration dependent responses of adult *Drosophila* to chemicals which elicit acceptance responses as well those which are rejected.

Flies are given a choice between two agar based solutions alternately filled in a 6 X 10 well micro-test plate. One of the choices is marked with a non-tasting food dye- Carmoisine red. Prior to the test flies are starved for 18-20 hours in a bottle containing moist filter-paper. Starved flies are allowed to feed on these plates in dark for 1 hour. After feeding they are immobilised and scored on the basis of the colour in their abdomen (Fig. 1). In the absence of any cue flies eat randomly from both the wells and appear red. To measure attraction response the stimulus is added to the uncoloured well and flies with uncoloured or white abdomens give an index of attraction response. In order to measure rejection response the stimulus is added to the coloured well. Here, the proportion of uncoloured flies indicate the rejection response.

Response of L1 to Ionic Stimuli

Kavita Arora examined the electrophysiological responses of the L1 neuron to various monovalent and divalent salts. The results indicate that the response of L1 is determined primarily by the cation (Arora,1985). The impulse frequency of L1 in response to NaCl was about three times higher than the impulse frequency to other chlorides. The response to NaCl saturates at about 45 spikes/0.5 sec while the response to KCl, LiCl and a number of other salts of metallic and organic cations saturates at about 15 spikes/0.5 sec.

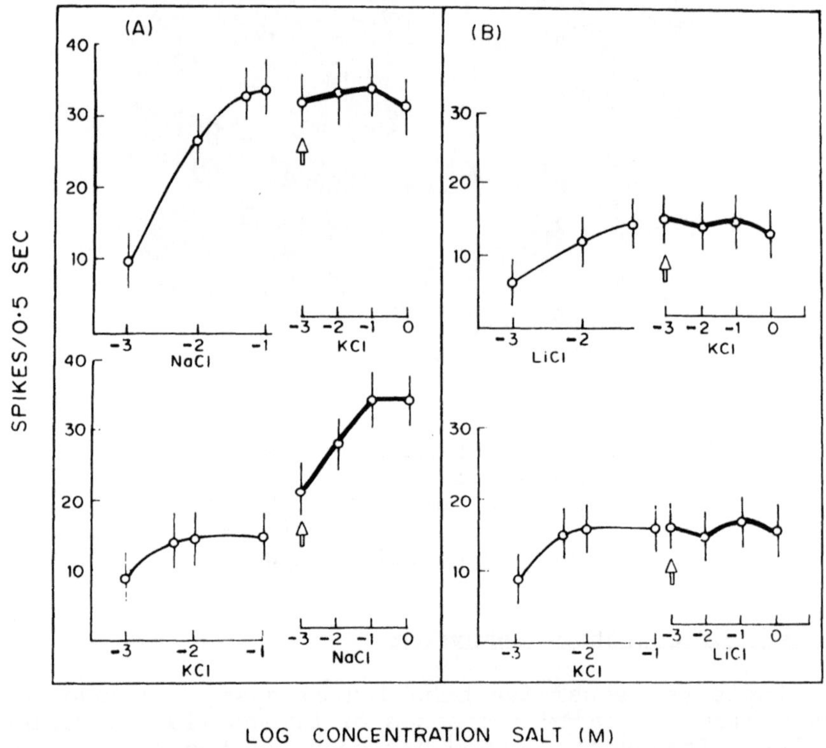

Fig. 2. Saturation of acceptor sites for cations: (A) Top: L1 saturated by NaCl gives no additional response to KCl. Bottom: NaCl elicits additional firing after saturat- -ion by KCl. (B) KCl and LiCl mutually saturate the acceptor sites for each other.

Interaction between pairs of ionic stimuli viz. Na^+ and K^+, K^+ and Li^+ was studied by measuring the response at increasing concentration of one ion in the presence of saturating concentration of the other (Fig. 2). The analysis was based on the premise that stimulation by a pair of ions acting at independent acceptor sites would elicit totally additive response while those acting at a common site would elicit intermediate responses depending on the affinity of the site for the ions.

Competition studies between Na^+ and K^+ & K^+ and Li^+ show that there are at least two types of acceptor sites on the L1 neuron. We have designated these as A and B. The B site can be saturated by Na^+, K^+ and Li^+. We attribute the additional firing of the L1 to NaCl in a KCl saturated neuron to be due to activity of Na^+ at the A site.

Further evidence that A and B are two separate sites has been obtained from the trypsin digestion experiment. Treatment of labellar hairs with trypsin abolishes the L1 response to Na^+ and K^+. The differential recovery of the response to Na^+ and K^+ following trypsin treatment indicates that the L1 cell carries separate sites for these ions. We infer that the recovery is due to the site common to K^+ and Na^+ i.e. the B site.

Gustatory Mutants of Drosophila melanogaster

Several groups have reported gustatory mutants of *Drosophila* (Isono and Kikuchi,1974; Falk and Atidia,1975; Rodrigues and Siddiqi, 1978; Tompkins et al., 1982; Morea,1985). A set of four X-linked genes that alter the fly's response to attractants and repellents were isolated in our laboratory (Rodrigues and Siddiqi, 1978; Siddiqi and Rodrigues,1981). The two salt mutants *gustB* and *gustC* were found to be insensitive to repulsion by NaCl. Wildtype flies show a slight preference for NaCl when tested for their attraction behaviour by the feeding preference assay. This attraction response is enhanced in the mutants *gustB* and *gustC*. The attraction to NaCl in *gustB* is due to NaCl induced firing of the S cell (Arora et al., 1987).

We have isolated several new mutants of *Drosophila* with altered response to salts and sugars. The mutant *GustR* was isolated using the feeding preference assay. *GustR* is autosomal dominant, located on chromosome III between the markers roughoid and hairy.

The mutation in the gene *GustR* affects the taste perception of the fly to several classes of chemicals. *GustR* flies show an increased threshold for repulsion by NaCl. In fact they are strongly attracted by NaCl solutions (Fig. 3). Flies carrying this mutation are less sensitive to attraction by sucrose, glucose, fructose and trehalose.

The electrophysiological response of *GustR* to NaCl and sucrose was measured. As indicated in Figure 4, there is an enhanced firing of L1 in the mutant. The electrophysiological response of the S cell is indistinguishable from that of the wildtype flies.

Electrophysiological screening of X chromosome mutagenised lines for specific defect in the L1 firing yielded the mutant $gustE^{V86}$. Figure 5 describes the electrophysiological response of $gustE^{V86}$ to NaCl and KCl. The response of L1 in the wildtype flies saturates at 0.1M concentration of NaCl with a firing frequency of 45 spikes/0.5 sec. $gustE^{V86}$ flies show a

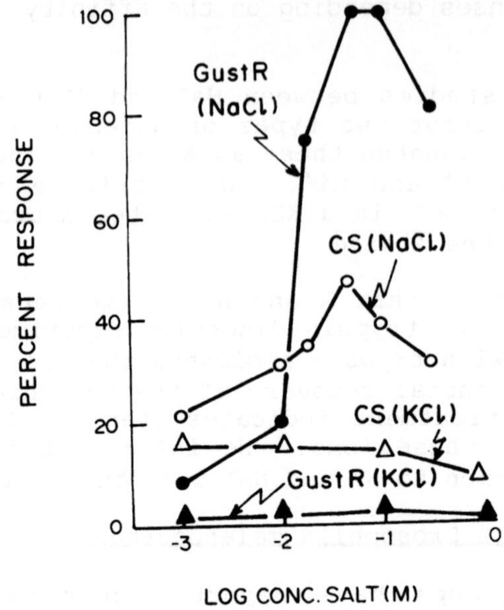

Fig.3. Wildtype flies (CS) exhibit slight preference for NaCl. *GustR* flies are strongly attracted by NaCl. KCl fails to evoke attraction in CS or *GustR*.

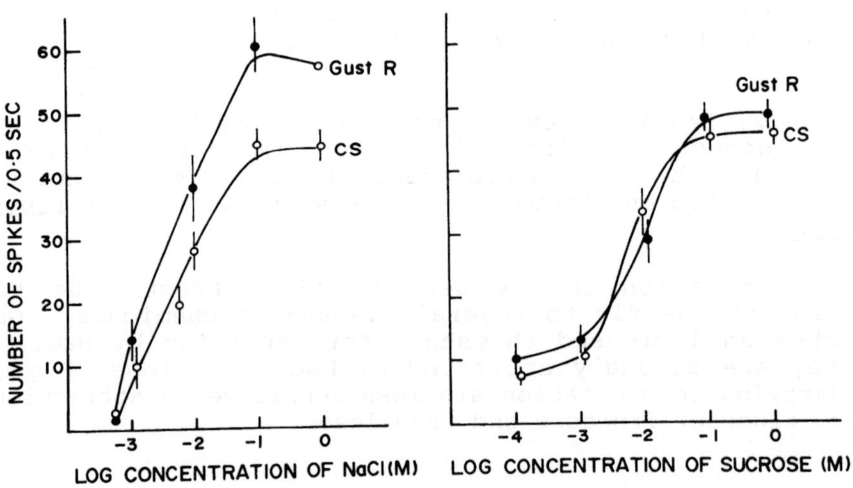

Fig. 4. Electrophysiological responses of wildtype (CS) and *GustR* to NaCl and Sucrose.

strong reduction in Na^+ induced firing at all concentrations of NaCl. Electrical responses to K^+ remain unchanged.

Paradoxically with reduction in L1 firing *gustE*^{V86} flies become more sensitive to rejection by Na^+. In fact they do not exhibit the normal attraction response to Na^+. The behavioural response to K^+ remains unchanged.

It is evident from the electrophysiological response of *gustE*^{V86} that the mutation eliminates the Na^+ induced firing of L1 corresponding to the A site. This has resulted in the elimination of attraction response to Na^+. The strong attraction to Na^+ in *GustR* is accompanied by an enhanced activity of the L1 neuron. On the other hand the pleiotropy exhibited by the mutant suggests that the effects are central rather than peripheral. The enhanced attraction to Na^+ in *GustR* has a peripheral neural correlate but this by itself cannot account for the insensitivity of the mutant to sugars inspite of normal S cell activity.

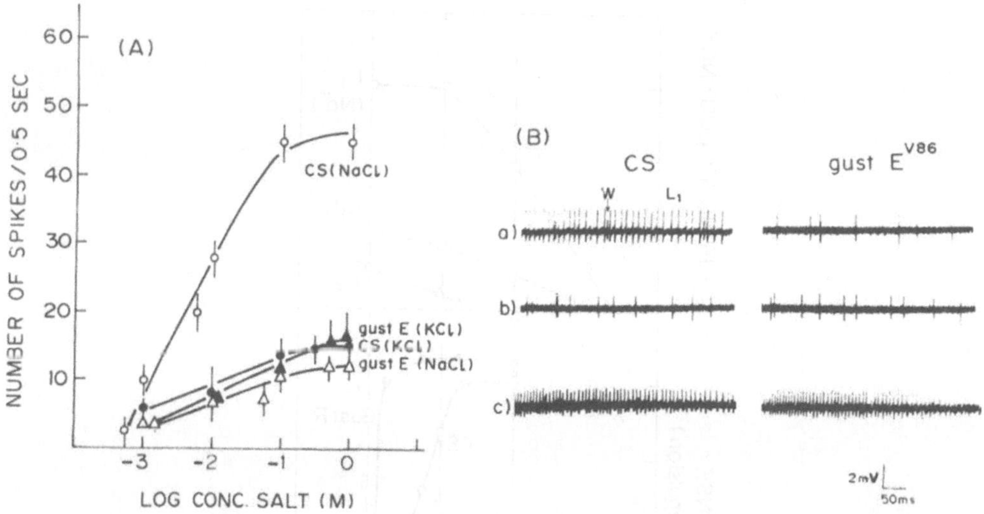

Fig. 5 (A) Extracellular sensory responses of wildtype and *gustE*^{V86} to NaCl and KCl. (B) Impulses from wildtype (CS) and *gustE*^{V86} flies: a) 0.1M NaCl b) 0.1M KCl c) 0.1M Sucrose in 5mM Choline Chloride.

The kinetics of behavioural response of wildtype and *GustR* are summarised in Figure 6 along with the electrophysiological response of the two L neurons. At concentrations below 0.1M the increasing activity of the L1 neuron is paralleled by the attractive phase of the behavioural curve. In the concentration range above 0.1M where most of the repulsion occurs the activity of L1 has reached saturation. The inhibitory phase of the response curve coincides with increasing activity of L2.

Although aversive behaviour to salts was attributed to the activity of salt cells in *Phormia* (Dethier,1976) the evidence assigning a functional role to each of these is lacking. It was believed that behavioural rejection could be due to the stimulation of salt cells or inhibition of water or sugar cell.

In *Drosophila* our observations indicate that the principal role of L1 is to mediate attraction to Na^+. In mutants such as *GustR* the enhanced attraction to Na^+ is accompanied by increased firing of L1. The loss of site A mediated response in *gustE^{V86}* blocks attraction to Na^+. The impulses from L1 do not contribute to the repulsion response because in *gustE^{V86}* the sensitivity to Na^+ is increased. Although at present we do not have a direct evidence supporting the role of L2 in mediating behavioural rejection this hypothesis can be tested by mutational analysis.

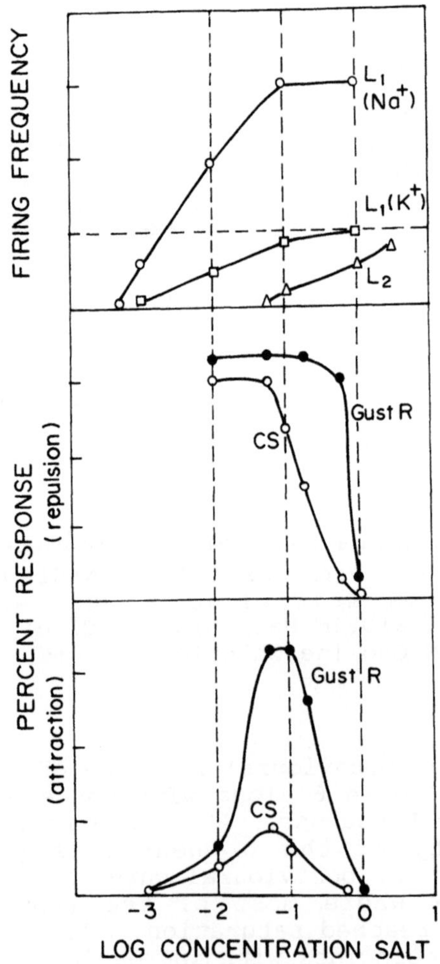

Fig.6. Kinetics of behavioural and electrophysiological responses of *Drosophila* to NaCl.

ACKNOWLEDGEMENTS

The authors are grateful to Dr. Veronica Rodrigues for useful suggestions during the course of this work.

REFERENCES

Arora, K., 1985, Neurogenetic studies on taste mechanisms of *Drosophila melanogaster*, Ph.D. Thesis, Bombay University.

Arora, K., Rodrigues, V., Joshi, S., Shanbag, S., and Siddiqi, O., 1987, A gene affecting the specificity of chemosensory neurons of *Drosophila*, Nature, 330: 62-63.

Crnjar, R., Cancedda, A., Angioy, A. M., Liscia, A., and Pietra, P.,1983, Electrophysiological responses of labellar chemosensilla of a wildtype and a salt tolerant mutant (lot-94) of *D. melanogaster* Meigen, Monitore Zool. Ital., (N.S.) 17: 387-394.

Dethier, V. G.,1955, The physiology and histology of the contact chemoreceptors of the blowfly, Quart. Rev. Biol. 30: 348-371

Dethier, V. G., 1976, The Hungry Fly, Harvard University Press, Cambridge, Massachusetts.

Evans, D. R., and Mellon, Jr., D.,1962, Stimulation of the primary taste receptors by salts, J. Gen. Physiol., 45:651-661.

Falk, R., and Atidia, J., 1975, Mutation affecting taste perception in *Drosophila melanogaster*, Nature, 254: 325-326

Fujishiro, N., Kijima, H., and Morita, H., 1984, Impulse frequency and action potential amplitude in labellar chemosensory neurons of *D. melanogaster*, J. Insect Physiol., 30: 317-325.

Hodgson, E. S., Lettvin, J. Y., and Roeder, K. D.,1955, Physiology of a primary chemoreceptor unit, Science, 122: 417-418.

Isono, K., and Kikuchi, T., 1974, Autosomal recessive mutation in sugar response of *Drosophila*, Nature, 248: 243-244.

Minnich, D. E., 1926, The organs of taste on the proboscis of the blowfly *Phormia regina*, Anat. Rec., 34: 126.

Minnich, D. E., 1929, The chemical sensitivity of the legs of the blowfly *Calliphora vomitoria* to various sugars, Zeit. Vergl. Physiol., 11: 1-55.

Morea, M., 1985, Deletion mapping of a new gustatory mutant in *Drosophila melanogaster*, Experientia, 41: 1381-1384 .

Nayak, S. V., and Singh, R. N., 1983, Sensilla on the tarsal segments and mouthparts of adult *D. melanogaster*, Int. J. Insect Morphology and Embryology, 12: 273-291.

Rodrigues, V., and Siddiqi, O.,1978, Genetic analysis of chemosensory pathway, Proc. Ind. Acad. Sci. (B), 87: 147-160.

Rodrigues, V,. and Siddiqi, O., 1981, A gustatory mutant of *Drosophila* defective in pyranose receptors, Mol. Gen. Genet., 181: 406-408.

Siddiqi, O,. and Rodrigues, V., 1980, Genetic analysis of a complex chemoreceptor. In: Development and Neurobiology of *Drosophila*. Siddiqi, O., Babu, P., Hall, L., and Hall, J. (eds.) pp 347-359, Plenum, N. Y. and London.

Tanimura, T,. and Shimada, I., 1981, Multiple receptor proteins for sweet taste in *Drosophila* discriminated by papain treatment, J. Comp. Physiol., 141: 265-269.
Tanimura, T., Isono, K., Takamura, T., and Shimada, I., 1982, Genetic dimorphism in taste sensitivity to trehalose in *D. melanogaster*. J. Comp. Physiol., 147: 433-437.
Tanimura, T., Isono, K., and Yamamoto, M., 1988, Taste sensitivity to Trehalose and its alteration by gene dosage in *Drosophila melanogaster*, Genetics, 119: 399-406.
Tompkins, L., Cardosa, J., White, F., and Sanders, T. G., 1979, Isolation and analysis of chemosensory behaviour mutants in *D. melanogaster*, Proc. Nat. Acad. Sci. U. S. A., 76: 884-887.
Wolbarsht, M. L., 1957, Water taste in *Phormia*, Science, 125: 1248.

DROSOPHILA HOMOLOGS OF VERTEBRATE SODIUM CHANNEL GENES

Mani Ramaswami, Ali Lashgari and
Mark A.Tanouye

Division of Biology
California Institute of Technology, 216-76
Pasadena, CA 91125, USA

SUMMARY

We have screened D. melanogaster genomic libraries for sequences similar to vertebrate sodium channel genes. We describe two transcription units that are strikingly homologous to a rat sodium channel cDNA and are conserved in a phylogenetically distant species of Drosophila, D. virilis. They appear to encode the major subunits of two distinct sodium channel proteins. A partial sequence for one of these has been previously reported (Salkoff et al., 1987a). The other transcription unit maps to position 14C/D, close to the *paralyzed* (*para*) gene. Mutations in *para* have previously been shown to affect excitability in the central and peripheral nervous systems of Drosophila (Ganetzky and Wu, 1986; Burg and Wu, 1986). Sequence comparisons suggest that both genes might have arisen before the divergence of vertebrate and invertebrate species. This finding has implications for the diversity of sodium channels in invertebrate and vertebrate nervous systems.

INTRODUCTION

Ion channels are a diverse group of integral membrane proteins that regulate the passage of ions through cell membranes. Electrophysiological, pharmacological and biochemical methods have distinguished several types of sodium and calcium channels, generally involved in the depolarization of excitable membranes (Barchi, 1987; Noda et al., 1986a; Catterall,1986). An even larger number of distinct potassium channel types involved in membrane repolarization has also been detected (Rudy, 1988). The differential distribution of these channel types underlies the range of electrical responses of neurons (Hille, 1984).

The cloning of ion channel genes allows the detailed biochemical and biophysical characterization of channel proteins. Recent results have shown that voltage-gated sodium channels, potassium channels, and possibly calcium channels, share several sequence motifs (Noda et al., 1984; Noda et al., 1986a; Aulds et al., 1988; Salkoff et al., 1987a,b; Kamb et al., 1987; Kamb et al.,1988; Papazian et al., 1987; Tempel et al., 1987; Baumann et al., 1987; Schwarz et al., 1988; Tanabe et al., 1987). Sodium channels contain four internal repeats about 250 residues in size that are called homology domains (Fig 1). A putative calcium channel has an identical structural organization. All potassium channels cloned so far are similar in structure to a single homology domain of a sodium channel. These data suggest that different channel types may have evolved from a single ancestral voltage-gated channel that arose early in phylogeny.

The isolation of ion channel genes in an organism like Drosophila, where genetic manipulations are feasible, facilitates the identification of other genes involved in ion channel function (see Discussion). Several mutations affecting neural excitability have been identified in Drosophila (Tanouye et al., 1986; Ganetzky and Wu, 1986). Mutations that affect specific classes of potassium channels *(Shaker, ether a go go, slowpoke* and *hyperkinetic)* cause abnormalities associated with increased membrane excitability (Ganetzky and Wu, 1986). Conversely, mutations that affect sodium currents *(para ts, nap ts* and *tip-E ts)* cause an overall decrease in membrane excitability (Tanouye et al., 1986; Ganetzky and Wu, 1986; Ganetzky, 1984; Ganetzky, 1986). These mutations might identify structural genes for ion channels, or genes for proteins involved in the synthesis, membrane distribution or modulation of ion channels.

We have searched D. melanogaster genomic libraries, for sequences similar to an mRNA for the large subunit of a voltage-gated sodium channel from rat brain, with goals to: a) identify all sequences in the relatively small genome of Drosophila that are homologous to a rat sodium channel gene; b) analyze the diversity of sodium channels in Drosophila; c) further understand the structure and evolution of ion channels; and d) complement the genetic approaches that have been used to identify genes involved in membrane excitability. In this paper, we report results from the screen, and the partial characterization of two transcription units that appear to encode the major subunits of distinct sodium channel proteins; one of these genes probably corresponds to the *para* locus.

RESULTS

Isolation of Drosophila Sequences Similar to a Rat Sodium Channel Gene

Three cDNA clones encoding parts of the major subunit of the rat brain sodium channel, RatIIA (Fig. 1; Aulds et al., 1988), were used as hybridization probes. The clones AG141, EAF1, and EAF8 contained coding sequences for approximately 75% of RatIIA, and included all of homology domains A, C, and D (Fig. 1). Blots of Drosophila genomic DNA were probed under conditions of reduced stringency designed to detect sequences approximately 55% homologous to the cDNA probes (Experimental Procedures). However, no reproducible signal was detected above the background hybridization. Forty two recombinant phage clones were isolated when Drosophila genomic libraries were screened under similar hybridization conditions. These were placed into 21 groups of non-overlapping clones. The initial group assignments were based on a comparison of their restriction maps, and on hybridization experiments in which DNA from all 42 clones was probed with restriction fragments purified from particular clones. In some cases the assignments were verified by in-situ hybridizations of clone DNA to polytene chromosomes in the salivary glands of <u>Drosophila</u>. The groups have been given the name of a representative clone in the group (Table 1).

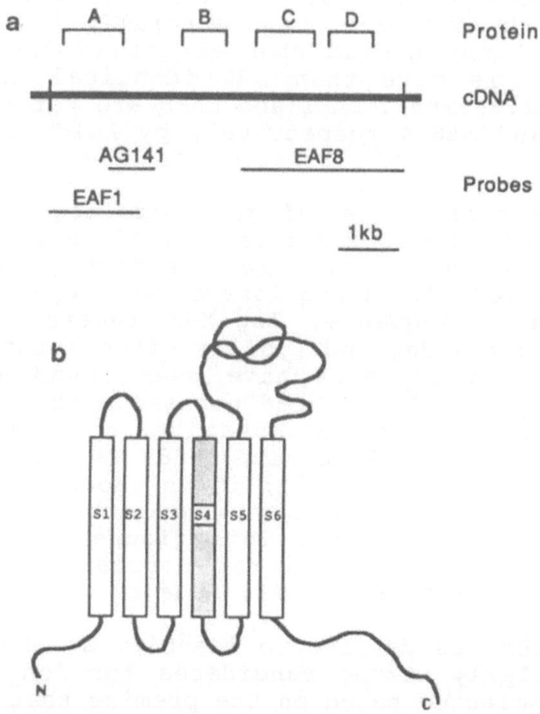

Fig.1 Representations of the rat sodium channel RatIIA, and the cDNA clones used as hybridization probes.

a. The major subunit of the rat sodium channel is a 2000 amino acid polypeptide with four internal repeats called homology domains. Each repeat unit is 250-300 residues in size and approximately 50% identical to the others. In the figure the homology domains are labelled as A,B,C, and D from amino to carboxyl terminus. The repeat units are presumed to assemble in the membrane as "pseudo subunits" around an ion selective pore.

The cDNA probes used in the experiments reported here are shown at their apposite positions on the channel gene. EAF1 is a cDNA extending from nucleotide -10 to 1511 (Aulds et al., 1988). It encodes amino acid residues 0 to 500 and includes all of homology domain A. EAF8 is a cDNA extending from nucleotide 3361 to 5868. It encodes amino acid residues 1122 to 1956 including all of homology units C and D. AG141 is a 558 b.p. cDNA beginning at nucleotide number 980, and

(Fig.1 Cont.)

encodes 186 residues starting at the end of segment S5 of homology domain A. The sequences most highly conserved between different sodium channels lie largely within the homology domains (Noda et al., 1986a). The major subunit of sodium channels is sufficient to form a voltage-gated, sodium selective pore. In some tissues (including rat brain), two additional small subunits have been found whose structures and functions are unknown (Noda et al., 1986b; Aulds et al., 1988; earlier work is reviewed in Catterall, 1986). The sequence of the sodium channel RatIIA (Aulds et al., 1988) is more than 99% identical to RSC2 (Noda et al., 1986). EAF1 and EAF8 are referred to as NA2.2 and NA8.4 respectively by Aulds et al., (1988).

b. A cartoon of a single homology domain. Multiple hydrophobic segments (S1, S2, S3, S5, and S6) flank a characteristic positively charged segment called S4. S4 consists of 4-8 repeats of a 3 amino acid sequence, Arg-X-X (where X is a hydrophobic residue and Lys. is often substituted for Arg.). S4 segments have been found in the sequence of a voltage-gated potassium channel and in the sequence of a putative voltage-gated calcium channel (Kamb et al., 1988; Tempel et al., 1987; Tanabe et al., 1987); it is thought to form the voltage sensing unit of these channels (Guy and Seetharamulu, 1986). This figure is adapted from data taken from several sources (Aulds et al., 1988; Noda et al., 1984; Noda et al., 1986a).

A second screen was devised to identify a subset of these 21 groups as particularly strong candidates for ion channel genes. Five groups were selected based on the premise that D.melanogaster genes with conserved homologs in rat are likely to be very highly conserved in D.virilis, a distant species of Drosophila. It has been shown that about 70% of single copy DNA from D.melanogaster does not form stable hybrids with D.virilis DNA under conditions corresponding to roughly 60% homology (Zwiebel et al., 1982).

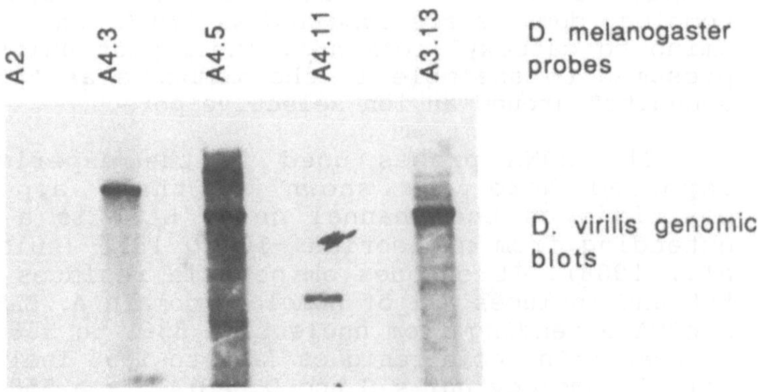

Fig.2. Southern blots at reduced stringencies to genomic DNA from Drosophila virilis.

Our preliminary results showed that virtually all D.melanogaster genomic clones that we isolated (about 16 kb on an average) had some sequences that were conserved at this level in D.virilis. However, only 5 of the 21 groups had specific restriction fragments (1.5 kb to 3.5 kb in size) that crosshybridized with both rat sodium channel cDNAs and D.virilis genomic DNA (Fig.2; Table 1; Experimental Procedures). Hence onlythese five groups (A20, A3.13, A4.3, A4.5 and A4.11) contained phylogenetically conserved segments of DNA that were similar in sequence to the rat sodium channel gene (Table 1). Two groups that crosshybridized most strongly with the rat sodium channel sequences (A3.13 and A4.3) were chosen for further characterization.

Table 1.

Clone	Probe	Overlapping clones	D.virilis conservation
A1	AG141	6	-
A2	AG141	1	-
A4	AG141	1	-
A8	AG141	1	-
A9	AG141	3	-
A10	AG141	1	-
A11	AG141	1	-
A20	AG141	1	+
A29	AG141	2	-
A46	AG141	1	-
L1	AG141	1	-
L2	AG141	1	-
L3	AG141	2	-
A2.30	EAF8	1	-
A2.35	EAF8	2	-
A3.1	EAF8	2	-
A3.13	EAF8	8	+
A4.1	EAF8	2	-
A4.3	EAF8	2	+
A4.5	EAF8	2	+
A4.11	EAF8	1	+

The probes used are purified restriction fragments from various D.melanogaster genomic clones: a 4 kb SalI-EcoRI fragment from clone A2 (lane 1), a 1.8 kb SalI-HindIII fragment from clone A4.3 (lane 2), a 1.7 kb EcoRI fragment from clone A4.5 (lane 3), a 3.8 kb SalI fragment from clone A4.11 (lane 4) and a 2.5 kb SalI-HindIII fragment from A3.13 (lane 5). Each of the restriction fragments used as probe contains sequences homologous to the rat cDNA, EAF8. The restriction fragments from A4.3, A4.5, A4.11 and A3.13, in lanes 2 to 5 respectively, are strongly conserved in D. virilis. The fragment from A2 in lane 1 is not conserved. The autoradiogram shows a 24 hour exposure. For conditions of hybridization and washing, see Experimental Procedures. (Continued)

Results from screen for Drosophila ion channel genes. The first column shows the name of a representative clone from a group of overlapping genomic clones isolated. The third column shows the total size of the group. The group size may not indicate the representation of clones in the library as number of isolates from any particular locus roughly corresponds with the strength of the signal obtained on hybridization to heterologous rat cDNA probes (see Discussion). Thus, the largest number of isolates are obtained with the most strongly crosshybridizing sequences and vice versa. The last column indicates clones in which the restriction fragments that crosshybridized with rat sodium channel cDNAs were strongly conserved in D.virilis. Clones that gave weak hybridization signals on genomic blots of D.virilis DNA are not distinguished from those that gave no signal at all. For more details see text and Experimental Procedures.

A3.13 and A4.3 are Transcribed in Adult Drosophila Heads

Genomic restriction maps representing 42 kb of A3.13 DNA and 17 kb of A4.3 DNA are shown in Figure 3. In A3.13 three non-contiguous restriction fragments, 3.5 kb, 1.7 kb and 1.3 kb in size were identified which crosshybridized with EAF8 and D.virilis genomic DNA (Fig. 3a). In A4.3, a single 1.8 kb SalI-HindIII fragment crosshybridized strongly with EAF8 and with D.virilis genomic DNA (Fig. 3b). Each of these restriction fragments was used as a probe to screen Drosophila cDNA libraries. No cDNA clones were found among about 500,000 recombinant phage in a 9-12 hour embryonal cDNA library. However, 30 cDNA clones from A3.13 and 14 from A4.3 were isolated from an adult head cDNA library. Thus sequences in A3.13 and A4.3 appear to be transcribed in adult Drosophila heads. The sizes of the cDNA clones from A3.13 are small (0.5 kb to 2.5 kb) in comparison to those from A4.3 (0.8 kb to 7.0 kb). The restriction maps of the cDNA clones from A4.3 indicate that a transcription product from the locus is at least 9kb in length (data not shown). One cDNA clone from each locus which crosshybridized strongly with rat sodium channel sequences was identified (P15 from A3.13 and B1 from A4.3), and used in further studies.

A3.13 Defines a Drosophila Sodium Channel Gene

The nucleotide sequence of a 660 b.p. cDNA from A3.13 (P15) was determined. The cDNA is apparently incomplete as a single open reading frame extends through its entire length. The deduced amino acid sequence shows an S4-like segment (4-8 repeats of a 3 amino acid motif, X-X-Arg, in which X is a hydrophobic amino acid and Lys often substitues for Arg) with 8 positively charged residues, that is flanked by hydrophobic, potential membrane-spanning domains (Fig. 4a). A comparison of nucleotide sequences shows that the P15 sequence is roughly 68% identical to a region of the rat sodium channel gene RSC2 (Noda et al., 1986) which encodes homology domain D (data not shown). A majority of nucleotide differences between P15 and the rat sodium channel gene are in "wobble base" positions and do not alter the predicted amino acid sequence significantly. The amino acid sequence deduced from P15 is shown aligned with sequences from RSC2 (Noda et al., 1986) and DSC, a putative Drosophila sodium channel (Salkoff et al., 1987a; Figure 4a). A total of 141 residues are identical between rat and P15 sequences over the 212 residues shown, while only 108 residues are identical between P15 and DSC in the same region. An S4- like segment with eight positively charged residues found in P15, is present in domain D of all rat sodium channels, the eel sodium channel and in DSC.

The genomic clone A3.13 was mapped in situ onto polytene chromosomes from larval salivary glands (Pardue and Gall, 1975). It hybridized to a single site between bands 14C and 14D on the X chromosome (Fig. 5a). Earlier genetic work on neural excitability mutants in Drosophila has identified a gene named *paralyzed (para)*

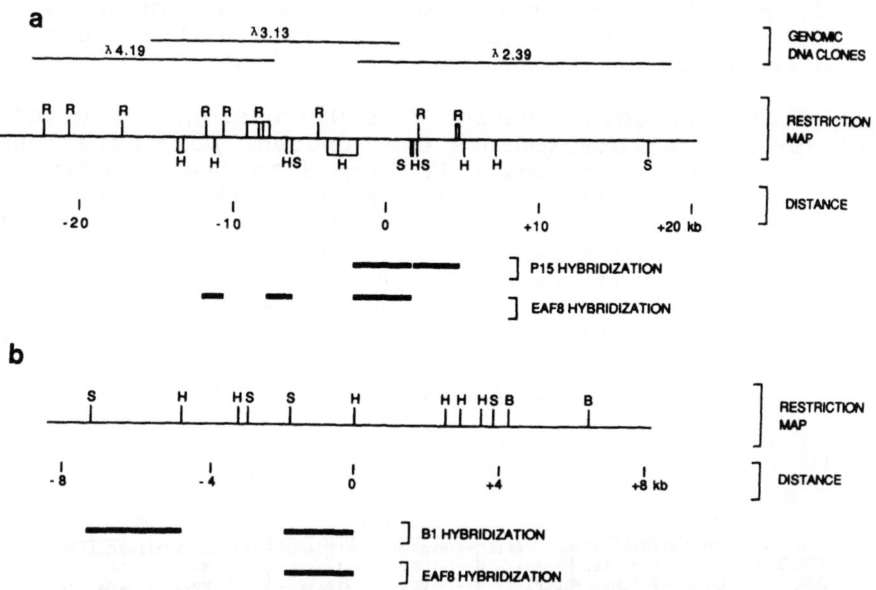

Fig.3 Restriction maps of two genomic loci that appear to encode different Drosophila sodium channels.

a. Map of restriction enzyme cleavage sites derived from eight genomic clones of the A3.13 group; 3 representative clones (A4.19, A3.13 and A2.39) are shown. Indicated are 3 non-contiguous restriction fragments that crosshybridize with EAF8 and with D.virilis genomic DNA: Each of these restriction fragments was used as a hybridization probe to screen D.melanogaster cDNA libraries. The cDNA (P15) from A3.13 described in the text hybridizes to 4 genomic fragments, a 3.5 kb SalI-HindIII fragment at map position 0, a 2.2kb EcoRI fragment at map position 3.0 and two small restriction fragments (EcoRI and HindIII-EcoRI) centered at about map position 4.3. Distance is shown in kilobases of DNA, R is an EcoRI restriction site, S is a SalI restriction site and B is a BamHI site.

(Fig.3. Cont.)

b. Map of restriction enzyme cleavage sites in the genomic clone A4.3. A solid black bar indicates a 1.8 kb SalI HindIII fragment centered at about map position -1, that crosshybridizes with the rat cDNA, EAF8 and blots of D.virilis genomic DNA. The fragment was used as a hybridization probe to screen D.melanogaster cDNA libraries. The A4.3 cDNA described in the text hybridizes to 2 genomic restriction fragments: the 1.8 kb SalI-HindIII fragment at position -1, and a 2.4kb SalI-HindIII fragment centered at approximately map position -6. The restriction map depicted here is similar to a portion of the DSC map (Salkoff et al., 1987a) but there are some differences.It is possible that some restriction site polymorphisms exist between the two clones (Salkoff, personal communication), and also that the genomic sequences we have cloned do not overlap enough with those depicted by Salkoff et al., for us to rationalize the differences by visual examination alone.

that maps close to this location (see Discussion). More recently, several chromosomal breakpoints that uncover *para* mutations have been generated and cytologically mapped to the interval 14C6-8 (Ganetzky and Wu, 1986). A chromosomal walk through *para* has included many of these translocation breakpoints (Ganetzky et al., 1986; Loughney and Ganetzky, 1988). DNA from A3.13 hybridizes to cloned DNA from this walk (Loughney and Ganetzky, personal communication). When taken together with the genetic and electrophysiological phenotypes of *para* (see Discussion), this indicates quite strongly that A3.13 derives from the *para* locus and that *para* encodes a Drosophila sodium channel.

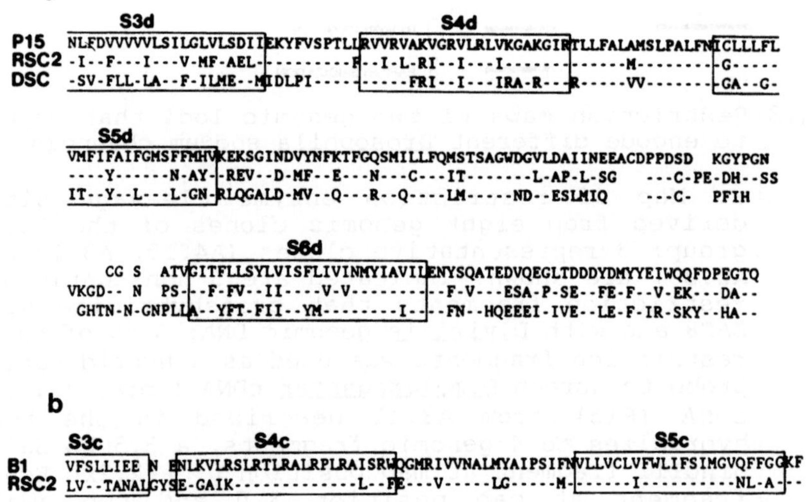

Fig.4 Amino acid sequences of putative Drosophila sodium channels deduced from cDNAs.

a. The protein sequence deduced from P15, a 660 base pair cDNA from A3.13, is shown in the figure.
(Fig.4. Cont.)

For comparison, it is aligned with homologous sequences encoded in RSC2, a rat sodium channel gene, and in DSC, a putative _Drosophila_ sodium channel gene. Amino acid identities with P15 are indicated by dashes. Gaps are inserted in the sequence for optimal alignment. Segments S3 to S6 of homology domain D of the rat sodium channel (see text and Fig. 1) are shown in boxes. For reference, the RSC2 and DSC sequences shown in the figure are from amino acid 1596 to 1812 of RSC2 (Noda et al., 1986a) and 1383 to 1595 of DSC (Salkoff et al., 1987a). The rat sodium channel RatIIA (Aulds et al., 1988) is more than 99% identical to RSC2 (Noda et al., 1986).

b. The protein sequence derived from B1, a cDNA from A4.3. For comparison it is aligned with homologous sequences in RSC2. Amino acid identities are shown by dashes. The S3, S4 and S5 segments from homology domain C of the rat sodium channel are shown in boxes. 52 out of 76 amino acids are identical between the two sequences. The rat sequence shown is from amino acid 1287 to 1364 of Noda et al., (1986a). The B1 sequence is identical to a sequence in DSC from amino acid 1080 to 1155 (Salkoff et al., 1987a).The genomic clone A4.3 was mapped in situ on polytene chromosomes from the salivary glands of third instar _Drosophila_ larvae. It hybridized to a single site at 60E (Fig. 5b), which is virtually identical to the reported cytogenetic map position of the DSC gene (Salkoff et al., 1987a). It appears that A4.3 and DSC define the same _Drosophila_ gene. This conclusion is based on (a) the sequence identity we find between the two genes; (b) the identical cytological map positions of A4.3 and DSC; and (c) the unique hybridization signals obtained with probes from A4.3, both on blots of genomic DNA (data not shown), and on polytene chromosomes from larval salivary glands.

A4.3 Defines a Second Drosophila Sodium Channel

The partial nucleotide sequence of a cDNA from A4.3 (B1) was determined. A comparison of nucleotide sequences shows that B1 is about 59% identical to a sequence in the RSC2 gene that encodes a part of homology domain C of the rat sodium channel (data not shown).The deduced amino acid sequence (Fig. 4b) contains an S4 like segment with 6 positively charged residues that is followed by a hydrophobic, putative membrane-spanning domain. This amino acid sequence is identically contained in homology domain C of DSC, a putative _Drosophila_ sodium channel whose partial sequence has been reported recently (Fig. 4b; Salkoff et al., 1987a). A comparison of the nucleotide sequence shows that the B1 cDNA and DSC sequences are identical over approximately 240 nucleotides (Salkoff et al., 1987b; Experimental Procedures). Limited single stranded sequence 3' to this region of the cDNA shows some interesting deviations. A 78 nucleotide insertion is found in B1, nucleotide insertion is found in B1 that is not included in DSC. An essentially identical 78 base sequence has been designated an intron (Salkoff et al., 1987a; 1987b), based on the presence of consensus splice sites flanking the segment, the presence of stop codons in frame with the predicted protein sequence, and a sharp divergence from the rat and eel sodium channel sequences. This insertion in the B1 cDNA, might be a result of incomplete processing of RNA. Some cDNA clones from apparently incompletely processed transcripts have also been detected from the Drosophila *Shaker* locus that encodes components of a voltage-gated potassium channel (Kamb et al., 1988).

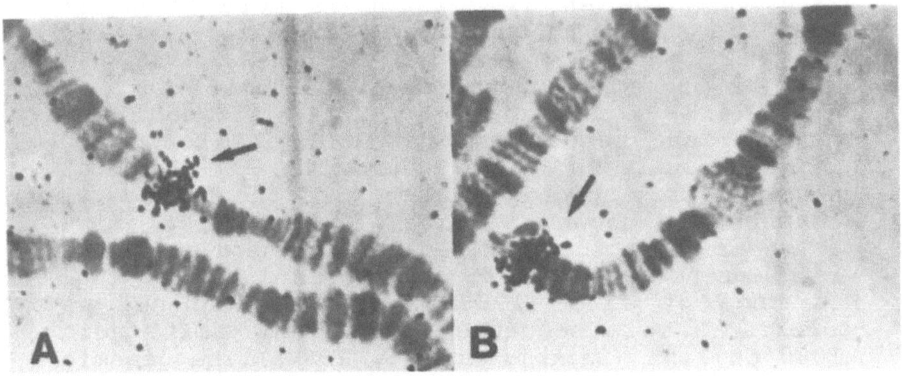

Fig.5. Cytological mapping of Drosophila genomic clones by in situ hybridizations to polytene chromosomes

a. The genomic clone A3.13 maps to the region between 14C and 14D on the X chromosome. Chromosome rearrangements that uncover mutations in the gene *para* have been mapped to 14C6-8 (Ganetzky and Wu, 1986; see Discussion).

b. Clone A4.3 maps to 60E at the tip of the right arm of chromosome 2. We are unaware of any neural excitability mutant in Drosophila that maps to this location. However, DSC (a putative Drosophila sodium channel gene) has been mapped to 60D-E (Salkoff et al., 1987a).

DISCUSSION

Drosophila offers several advantages for the study of genes involved in nervous system function and development. First, it has a nervous system with diverse ion channel and neurotransmitter populations, and complex patterns of neuronal development and connectivity. Second, Drosophila allows genetic manipulations to identify the biological functions of any cloned gene. Similar genetic experiments also permit the isolation of mutations in genes involved in any given function. Third, the small size of the fly genome (about 160,000kb) makes it relatively easy to construct and screen genomic libraries with confidence that the screen includes every segment of the fly genome. This bypasses the problems of low levels of transcription, or tightly regulated transcription that are inherent in screening cDNA libraries.

Drosophila genomic libraries have been screened at low stringency to identify families of genes of related function (Frigerio et al., 1986; McGuiness et al., 1984; Zuker et al., 1985; O'Tousa et al., 1985). We have similarly attempted to identify a family of ion channel genes in the organism. We screened the equivalent of twelve independent Drosophila genomes with cDNA clones encoding portions of a rat sodium channel (Fig. 1; Experimental procedures.). Table 1 shows the frequencies at which the different sequences were isolated in our screen. Due to weak and variable signals usually obtained in low stringency hybridizations (Experimental Procedures), the figures in Table 1 do not indicate the relative representations of the sequences in the genomic libraries. If we assume (as has been our experience) that the different sequences are present at roughly equal frequencies in this library, the figures imply that we have probably identified all sequences in the Drosophila genome as similar to the rat sodium channel gene as A3.13 (eight independent isolates were obtained). However, other Drosophila genes as homologous to the rat cDNAs as A4.11 and A20 (one clone was isolated in each case) might not have been recovered in our screen (Table 1).

para Encodes a Drosophila Sodium Channel

Many lines of evidence suggested previously that *para* could be a structural gene for a Drosophila sodium channel:
1. Several *para* alleles show rapid and reversible temperature sensitive paralysis above their characteristic restrictive temperatures (Suzuki et al., 1971; Ganetzky and Wu, 1986).
2. The paralysis of *para ts* mutants is associated with a temperature dependent block in action potential propagation in some Drosophila neurons (Siddiqi and Benzer, 1976; Benshalom and Dagan, 1981, Wu and Ganetzky, 1980; Ganetzky et al., 1986).
3. In vitro cultures of dissociated neurons from *para ts* larvae show a temperature dependent resistance to veratridine, a neurotoxin that binds and stabilizes sodium channels in an "open" conformation (Suzuki and Wu, 1984).
4. Electrophysiological recordings from dissociated neurons of wild type and *para* embryos show that the number of neurons expressing a sodium current is significantly reduced by *para* mutations (O'Dowd et al., 1987).
5. Null mutations in *para* are lethal (Ganetzky and Wu, 1986).
All of these phenotypes are consistent with the proposal that *para* encodes a sodium channel protein essential for the normal development and function of the organism. However, the data do not rule out alternate functions for the *para* gene product.

In this report we provide molecular genetic evidence to support the proposal that *para* encodes a sodium channel. We have shown that A3.13 encodes a protein highly homologous to vertebrate sodium channels (Fig 5a). By in-situ hybridizations, this clone maps very close to the location of *para* (14C6-8). These results are in agreement with those of Loughney and Ganetzky (1988, personal communication) who have recently cloned the *para* gene using genetic criteria, and have shown that the locus encodes a protein homologous to vertebrate sodium channels. Results from the two laboratories indicate that A3.13 and *para* represent identical genes and that *para* encodes the major subunit of a Drosophila sodium channel.

Genes for two neural excitability mutants in Drosophila (*para* and *Shaker*), that were implicated in ion channel function, have now been cloned (Kamb et al., 1987; Papazian et al., 1987; Baumann et al., 1987). Both appear to encode ion channel proteins. This increases the likelihood that other similar behavioral mutants, particularly those that interact genetically with *Shaker* and *para* (see below), identify additional ion channel genes.

The Diversity of Sodium Channels in Drosophila

We have provided molecular evidence that at least two independent sodium channel genes, *para* and DSC (Salkoff et al., 1987a), are present in Drosophila. Previous genetic evidence supports our suggestion of Drosophila sodium channel diversity. In particular, it has been shown that some neural circuits in Drosophila are relatively unaffected by *para* mutations (Williamson et al., 1974; Nelson and Baird, 1985; Elkins and Ganetzky unpublished results). DSC, in which no mutations have been isolated, may affect responses in these circuits. Other candidates for Drosophila sodium channel genes are defined by mutations that show behavioral phenotypes similar to *para ts*, such as *nap ts* and *tipE ts* (Tanouye et al., 1986; Ganetzky, 1984; Ganetzky, 1986; Ganetzky and Wu, 1986). Double mutant combinations of *para, nap* and *tipE* show enhanced phenotypes ranging from lowered restrictive temperatures and reduced viability, to unconditional lethality. The synergistic interactions between these mutations suggests that the wild type genes are involved in the same physiological function. As sodium channels and potassium channels have antagonistic roles in neural excitability, one might predict that, in double mutant combinations, sodium channel defects could partially suppress physiological abnormalities caused by defective potassium channels. Indeed, some alleles of *nap, tip-E* and *para* suppress behavioral phenotypes of *Sh,* a mutation that has been shown to affect a voltage-gated potassium channel in Drosophila (Ganetzky and Wu, 1982; Jackson et al., 1986; Ganetzky and Wu, 1986; Kamb et al., 1987; Kamb et al., 1988; Iverson et al., 1988; Baumann et al., 1987; Papazian et al., 1987; Tempel et al., 1987; Schwarz et al., 1988; Timpe et al., 1988). Other less well characterized loci that suppress the behavioral defect of *Shaker* flies have also been identified (Ganetzky and Wu, 1986). The discovery that *para* is probably a sodium channel gene makes it likely that *tip-E, nap* and the other suppressors of potassium channel mutations, are closely involved in sodium channel function. The genes defined by these mutations remain candidates for other sodium channel genes in Drosophila.

Implications for Sodium Channel Evolution and Diversity

There is increasing evidence that vertebrate sodium channels constitute a diverse family of proteins (Barchi, 1987). Electrophysiological and pharmacological methods have distinguished two classes of sodium current in rat skeletal muscle membranes, and a slightly different current in rat brain (Barchi, 1987). Distinctive sodium currents have been seen in the Purkinje cells of guinea pigs (Llinas and Sugimori, 1980) and in type I astrocytes in the rat optic nerve (Barres et al., 1987). Sodium channels in the rat muscle membrane are different from those in T-tubular membranes (Jaimovich et al., 1982; Haimovich et al.,

1987). In part this diversity may be due to the existence of multiple sodium channel genes, as messages for three distinct sodium channel genes are simultaneously expressed in rat brain (Noda et al., 1986a; Aulds et al., 1988). The sequences of the three proteins encoded by these transcripts are about 80% identical to each other (Noda et al., 1986).

Table 2 shows sequence homologies between four putative sodium channel proteins. These figures have been computed over about 210 residues for which sequence is available for all the channel proteins. The predicted protein sequence of para is 67% identical to the rat sodium channel (RSC2) and 62% identical to the eel channel. It is, however, only 51% identical to DSC over the same sequence. The DSC sequence is 54% and 56% identical to the rat and eel channels respectively. All four sodium channel proteins are

<u>Table 2.</u>

	para	DSC	RSC2	EELSCH	DHPR
para	-	51	67	62	30
DSC		-	54	56	29
RSC2	-	-	-	75	31
EELSCH				-	29

Percent homologies between ion channels. The homologies have been computed over about 210 amino acids for which sequence information is available for *para*. RSC2 represents a sodium channel gene from rat brain reported by Noda et al., 1986a. EELSCH represents the sodium channel gene from electric eel reported in Noda et al., 1984. The sequence of the dihydropyridine receptor, DHPR, was taken from Tanabe et al., 1987, and the DSC sequence used in these calculations was obtained from Salkoff et al., 1987a. Identical amino acids are scored as 1 and all substitutions are scored as 0. An insertion or deletion is scored as a single substitution independent of its size. We believe it is unlikely that the results of our limited sequence comparisons will be drastically changed when more sequence from *para* becomes available. This is based on the observation that the relative homologies among RSC2, EELSCH and DSC remain more or less constant in all four homology domains of the channel proteins (Salkoff et al., 1987). For reference, the sequences are from amino acid 1386 to 1602 for EELSCH (Noda et al., 1984) and 1183 to 1415 for DHPR (Tanabe et al., 1987). The sequences for DSC and RSC2 are as in Fig.4.

about 30% identical to the dihydropyridine receptor that has been proposed to function as a voltage-gated calcium channel (Tanabe et al., 1987). The striking sequence similarity between these genes, suggests that they share a common evolutionary origin. Voltage-gated calcium channels have been detected in protists, while voltage-gated sodium channels appear to have evolved more recently

(Hille, 1984). Thus, the sequence data is consistent with an evolutionary scenario in which the first ancestral sodium channel arose by duplication of, and divergence from, a voltage-gated calcium channel. The strong homology between the sodium channel genes in vertebrates and in Drosophila, suggests that voltage-gated sodium channels evolved before the divergence of vertebrates and invertebrates about 600 million years ago (Salkoff et al., 1987a). The fact that *para* is more closely related to the known vertebrate channels than it is to DSC is easily explained if DSC and *para* diverged from each other before *para* diverged from the vertebrate sodium channels. Though our arguments are based on limited sequence information, it appears that two distinct sodium channel genes existed in the early Cambrian period, even before the divergence of vertebrates from invertebrates. On this basis one would predict that DSC might define a subfamily of vertebrate sodium channels that are yet to be molecularly identified.

EXPERIMENTAL PROCEDURES

Standard Techniques

The following standard methods are described by protocols and combined references in Davis et al.,(1980), and Maniatis et al., (1982): screening of recombinant DNA libraries at high stringency, preparation of plasmid and phage DNA, restriction enzyme analysis, agarose gel electrophoresis, Southern blot transfers to nitrocellulose filters, DNA fragment isolation, hybridizations to DNA on filters, and subcloning of DNA fragments into plasmid vectors. Radiolabelled DNA probes were synthesised using the random hexamer primer method (Feinberg and Vogelstein,1983). DNA restriction endonucleases and polymerases were obtained from Boeringer Mannheim. Labelled nucleotide triphosphates were purchased from Amersham and ICN. Random hexanucleotide primer was obtained from Pharmacia. Nitrocellulose filters used in screening phage plaques were BA85 filters from Schleicher and Schuell. Nitrocellulose membranes for Southern blots were obtained from Amersham.

Screening Recombinant DNA Libraries at Reduced Stringencies

Lambda phage libraries were constructed in the EMBL3 vector with wild type Canton-S (CS) and Oregon-R (OR) genomic DNA by Alexander Kamb. (Kamb et al., 1987) They were screened with radiolabelled rat sodium channel cDNA probes. The hybridizations were carried out at 25°C in 5 x SSPE, 0.1% SDS, 50mM Tris (pH7.5), 100 ug per ml denatured salmon sperm DNA, and 10% dextran sulfate. The filters were first rinsed at room temperature in 2 x SSC and 0.1% SDS and then washed for half an hour at 55°C in the same solution. No crosshybridization was observed when the wash temperature was raised to 65°C. Weak and variable signals were usually obtained in these screens. The hybridization signals are weak due to the reduced stability of the DNA heteroduplexes. They are variable for many reasons; variations in plaque size which result when phage libraries are plated at relatively high densities, is possibly the most important factor .

Southern Blots to D.virilis Genomic DNA

D. virilis genomic DNA was prepared by the following procedure. About 100 flies were homogenized in 2 mls of 0.1M Nacl, 0.1M Tris (pH7.5), 0.05M EDTA and 0.5% SDS at 0^oC. The homogenate was incubated at 65^0C for 30 minutes. To the suspension, 0.3 mls of 8M CH_3COOK was added and the mixture was kept on ice for 10 minutes. After centrifugation, the supernatent was mixed with an equal volume of 100% ethanol. The precipitated DNA was pelleted in a microcentrifuge, suspended in 100 µl of water and treated with RNAse. Fifteen µl of this suspension was cut with two resriction endonucleases, Eco R1 and Hind III. The digest was run out on a 0.8% agarose gel and blotted onto Amersham Hybond membranes in the manner described by Maniatis et al. (1982). The hybridization to D. melanogaster probes was done under reduced stringency conditions described earlier. The washes were done at 60^0C in 2x SSC and 0.1% SDS.

In-situ Hybridizations to Polytene Chromosomes

Salivary glands were dissected out from wild type (Oregon R) flies and squashed in a solution of lactic acid, acetic acid and water (1:2:3). Hybridization probes were prepared using the random hexamer method (Feinberg and Vogelstein, 1983) with ^3H dCTP and dTTP, and detected by autoradiography as described in Pardue and Gall (1975).

Screening cDNA Libraries

The cDNA library from which cDNAs to DSC, *para,* A4.11, A4.5 were obtained was a gift from Dr. Paul Salvaterra. It was constructed from poly A+ RNA from adult fly heads, in lambda gt11. Fourteen clones representing DSC, 10 representing *para*, 45 representing A4.11 and 7 representing A4.5 were identified from about 500,000 clones screened. A 9-12 hour embryonal cDNA library (a gift from Dr. K. Zinn) was screened with DSC and *para* but no positive clones were identified from about 500,000 recombinant phage clones.

DNA Sequence Analysis

The sequences of both strands of the cDNA clones were deduced by the dideoxy chain termination method (Sanger et al., 1977) using templates prepared from recombinant M13 virions and synthetic primers prepared in the Caltech microchemical facility. A modified bacteriophage T7 DNA polymerase from the sequenase kit from U.S. Biochemicals was used in the reactions instead of Klenow enzyme. Only single stranded sequence from B1 was obtained, and the figure shows sequence that is identical to that reported for DSC (Salkoff et al., 1987). Computer analysis of the DNA sequences was done on an IBM XT using software written by Dr. Al Goldin at Caltech, and on a Macintosh 512 using the DNA Inspector software package.

ACKNOWLEDGEMENTS

We thank A. Goldin, V. Aulds, N. Davidson and R. Dunn for the rat sodium channel clones that were used as hybridization probes in our experiments. We are grateful to K. Loughney and B. Ganetzky for making their unpublished results available to us. We thank U. Banerjee, M. Gautam, L. Iverson, A. Kamb, M. Mathew, K. McCormack, J. Robinson, B. Rudy and W.W.Trevarrow for helpful discussions throughout the course of the work. We also thank R. McMahon for excellent technical assistance and Chi Bin Chien for expert assistance with word-processing programs. This research was supported by the Pfeiffer Research Foundation, and by USPHS grant NS21327-01 to M.T. M.R. was supported by fellowships from the Evelyn Sharp Foundation and the Markey Charitable Trust. M.T. is a McKnight Foundation Scholar and a Sloan Foundation Fellow.

REFERENCES

Aulds, V.J., Goldin, A., Krafte, D., Marshall, J., Dunn, J.M., Catterall, W.A., Lester, H.A., Davidson, N. and Dunn, R.J., 1988. A rat brain sodium channel alpha subunit with altered voltage responses. Neuron, 1:449-461.

Barchi, R.L., (1987). Sodium channel diversity: subtle variations on a complex theme. TINS.10:221-223.

Barres, B.A., Chun, L.L.Y. and Corey, D.P., 1987. Are glial and neuronal sodium channels the same? Soc Neurosci. Abs., 1:577.

Baumann, A., Krah-Jentgens, I., Mueller, R., Mueller-Holtkamp, F., Seidel, R., Kecskemethy,N., Ferrus, A. and Pongs, O., 1987. Molecular organization of the maternal effect region of the Shaker complex of Drosophila: characterization of an IA channel transcript with homology to vertebrate sodium channels. EMBO J., 6:3419-3429.

Benshalom, G. and Dagan, D., 1981. Electrophysiological analysis of the temperature sensitive paralytic Drosophila mutant, para ts. J. Comp. Phys., 144:409-417.

Burg, M.G. and Wu, C.F., 1986. Differentiation and Central Projections of Peripheral Sensory Cells with Action-Potential Block in Drosophila Mosaics. J. Neurosci., 6(10):2968-2976.

Catterall, W.A., 1986. Molecular properties of voltage-gated sodium channels. Ann. Rev. Biochem. 55:953-985.

Davis, R.W., Botstein, D. and Roth, J.R., 1980. Advanced bacterial genetics. A manual for genetic engineering. (Cold Spring Harbor, New York: Cold Spring Harbor Laboratory).

Feinberg, A.P. and Vogelstein, B., 1983. A technique for radiolabelling DNA restriction endonuclease fragments to high specific activity. Anal. Biochem., 132:6-13.

Frigerio, G., Burri, M., Bopp, D., Baumgartner, S. and Noll, M., 1986. Structure of the segmentation gene paired and the Drosophila *prd* gene as a part of a gene network. Cell, 47:735-746.

Ganetzky, B., 1984. Genetics of membrane excitability in Drosophila: lethal interactions between two temperature sensitive paralytic mutations. Genetics, 108:897-911.

Ganetzky, B., 1986. Neurogenetic analysis of Drosophila mutations affecting sodium channels: synergistic effects on viability and nerve conduction in double mutants involving tip-E. J. Neurogenet., 3:19-31.

Ganetzky, B., Loughney, K. and Wu, C.F., 1986. Analysis of mutations affecting sodium channels in Drosophila. In Tetrodotoxin, Saxitoxin and the molecular biology of the sodium channel, ed. C.Y.Kao, S.R.Levinson. New York: NY. Acad. Science, pp 325-327.

Ganetzky, B. and Wu, C.F., 1982. Indirect suppression involving behavioral mutants with altered nerve excitability in Drosophila melanogaster. Genetics, 100:597-614.

Ganetzky, B. and Wu, C.F., 1986. Neurogenetics of membrane excitability in Drosophila. Ann. Rev. Genet., 20:3-44.

Guy, H. and Seetharamulu, P., 1986. Molecular model of the action potential sodium channel. Proc. Natl. Acad. Sci. U.S.A., 83:508-512.

Haimovich, B., Schotland, D.L., Fieles, W.E. and Barchi, R.L., 1987. Localization of sodium channel subtypes in adult rat skeletal muscle using channel specific monoclonal antibodies. J. Neurosci., 7:2957-2966.

Hille, B. (1984). Ionic channels of excitable membranes. (Sunderland, Ma: Sinauer).

Iverson, L.E., Tanouye, M.A., Davidson, N.A., Lester, H.A. and Rudy, B.R., 1988. Expression of A-type potassium channels from Shaker cDNAs. Proc. Natl. Acad. Sci. U.S.A., 85:5723-5727.

Jackson, F.R., Wilson, S.D. and Hall, L.M., 1986. The tip-E mutation of Drosophila decreases saxitoxin binding and interacts with other mutations affecting nerve membrane excitability. J. Neurogenet., 3:1-17.

Jaimovich, E., Ildefonse, M., Barnahim, J., Rougier, O. and Lazdunsky, M., 1982. Centruriodes toxin, a selective blocker of surface sodium channels in skeletal muscle: voltage clamp analysis and biochemical characterization of the receptor. Proc. Natl. Acad. Sci. U.S.A., 79:3986-3900.

Kamb, A., Iverson, L.E. and Tanouye, M.A., 1987. Molecular characterization of Shaker, a Drosophila gene that encodes a potassium channel. Cell, 50:405-413.

Kamb, A., Tseng-Crank, J. and Tanouye, M.A., 1988. Multiple products of the Drosophila Shaker gene contribute to potassium channel diversity. Neuron, 1:421-430.

Llinas, R. and Sugumori, M., 1980. Electrophysiological properties of in vitro purkinje cell somata in mammalian cerebellar slices. J. Phys., 305:171-195.

Loughney, K. and Ganetzky, B., 1988. The Drosophila *para* locus is a sodium channel structural gene. Soc. Neurosci. Abs. 14:577.

Maniatis, T., Fritsch, E.F., and Sambrook, J., 1982. Molecular cloning: A Laboratory Manual (Cold Spring Harbor, New York: Cold Spring Harbor Laboratory).

McGuiness, W., Levine, M.S., Hafen, E., Kuroiwa, A. and Gehring, W.J., 1984. A conserved DNA sequence in homeotic genes of the Drosophila Antennapoedia and Bithorax complexes. Nature, 308:428-433.

Nelson, J. and Baird, D.H., 1985. Action potentials persist at restrictive temperatures in temperature sensitive paralytic mutants of adult Drosophila. Soc. Neurosci. Abs., 11:313.

Noda, M., Ikeda, T., Kayano, T., Suzuki, H., Takeshima, H., Kurasaki, M., Takahashi, H. and Numa, S., 1986a. Existence of distinct sodium channel messenger RNAs in rat brain. Nature, 320:188-192.

Noda, M., Ikeda, T., Suzuki, H., Takeshima, H., Takahashi, T., Kuno, M. and Numa, S., 1986b. Expression of functional sodium channels from cloned cDNA. Nature, 322:826-828.

Noda, M., Shimizu, S., Tanabe, T., Takai, T., Kayano, T., Ikeda, T., Takahashi, H., Nakayama, Y., Kanaoka, Y., Minamino, N., Kangawa, K., Matsuo, H., Raftery, M.A., Hirose, T., Inayama,

S., Hayashida, H., Miyata, T. and Numa, S., 1984. Primary structure of the Electrophorus electricus sodium channel deduced from cDNA sequence. Nature, 312:121-127.

O'Dowd, D., Germeraad, S. and Aldrich, R., 1987. Expression of sodium currents in embryonic Drosophila neurons: differential reduction by alleles of the para locus. Soc. Neurosci. Abs., 13:577.

O'Tousa, J.E., Baehr, W., Martin, R.L., Hirsch, J., Pak, W.L. and Applebury, M.L., 1985. The Drosophila *ninaE* gene encodes an opsin. Cell, 40:839-850.

Papazian, D.M., Schwarz, T.L., Tempel., Jan, Y.N. and Jan, L.Y., 1987. Cloning of genomic and complementary DNA sequences from Shaker, a putative potassium channel gene. Science, 237:749-753.

Pardue, M., and Gall, J., 1975. Nucleic Acid Hybridizations to the DNA of Cytological Preparations. In Methods in Cell Biology, Vol 10, D. Prescott, ed. (New York; Academic Press), pp 1-17.

Rudy, B., 1988. Diversity and ubiquity of potassium channels. Neuroscience, 25:729-749.

Salkoff, L., Butler, A., Wei, A., Scavarda, N., Giffen, K., Ifune, K., Goodman, R. and Mandel, G., 1987a. Genomic organization and deduced amino acid sequence of a putative sodium channel gene in Drosophila. Science, 237:744-749.

Salkoff, L., Butler, A., Scavarda. N. and Wei, A., 1987b. Nucleotide sequence of the putative sodium channel gene from Drosophila: the four homologous domains. Nucleic Acids Res., 15:8569-8573.

Schwarz, T.L., Tempel,B.L., Papazian, D.M., Jan, Y.N. and Jan L.Y., 1988. Multiple potassium-channel components are produced by alternate splicing at the Shaker locus in Drosophila. Nature, 331:137-142.

Siddiqi,O. and Benzer,S., 1976. Neurophysiological defects in temperature sensitive paralytic mutants of Drosophila melanogaster. Proc. Natl. Acad. Sci. U.S.A., 73:3253-3257.

Suzuki, N. and Wu, C.F., 1984. Altered sensitivity to sodium channel specific neurotoxins in cultured neurons from temperature sensitive paralytic mutants of Drosophila. J. Neurogenetics, 1:225-238.

Suzuki, D.T., Grigliatti,T. and Williamson, R., 1971. Temperature sensitve mutations in Drosophila melanogaster, VII. A mutation (*para ts*) causing reversible adult paralysis. Proc. Natl. Acad. Sci. U.S.A., 68:890-893.

Tanabe., T., Takeshima,H., Mikami, A., Flockerzi, V., Takahashi, H., Kangawa, K., Konjima, M., Matsuo, M., Hirose, T. and Numa, S., 1987. Primary structure of the receptor for calcium channel blockers from skeletal muscle. Nature, 328:313-318.

Tanouye, M.A., Kamb, C.A., Iverson, L.E., and Salkoff, L., 1986. Genetics and molecular biology of ionic channels in Drosophila. Ann. Rev. Neurosci., 9:225-276.

Tempel, B.L., Papazian, D.M., Schwarz, T.M., Jan, Y.N. and Jan, L.Y., 1987. Sequence of a probable potassium channel component encoded at the Shaker locus of Drosophila. Science, 237:770-775.

Timpe, L.C., Schwarz, T.L., Tempel, B.L., Papazian, D.M., Jan, Y.N. and Jan, L.Y., 1988. Expression of functional potassium channels from Shaker cDNA in Xenopus oocytes. Nature, 331:143-145.

Williamson, R., Kaplan, D.W and Dagan, D., 1974. A fly's leap from paralysis. Nature, 252:224-226.
Wu, C.F. and Ganetzky, 1980. Genetic alteration of nerve membrane excitability in temperature-sensitive paralytic mutants of *Drosophila melanogaster*. Nature, 286:814-816.
Zuker, C.S., Cowman, A.F. and Rubin, G.M., 1985. Isolation and structure of a rhodopsin gene from D. melanogaster. Cell, 40:851-858.
Zwiebel, L.J., Cohn, V.H., Wright, D.R. and Moore, G.P., 1982. Evolution of single-copy DNA and the ADH gene in seven Drosophilids. J. Mol. Evol. 19:62-71.

THE FUNCTIONAL ORGANIZATION OF THE AUDITORY BRAINSTEM IN THE MUSTACHE BAT AND MECHANISMS FOR SOUND LOCALIZATION

George D. Pollak

Department of Zoology
The University of Texas at Austin
Austin, Texas 78712 USA

INTRODUCTION

The ability of bats to orient and successfully avoid obstacles in total darkness has been of interest to scientists for more than two centuries. Although audition has always been strongly associated with this ability, general acceptance of orientation by sound came only around 1940 with the elegant studies of Griffin and his colleagues (Griffin and Galambos 1941; Galambos and Griffin 1942; an excellent summary of this work is provided in Griffin 1958). They showed that bats are not only able to navigate through complex environments but they also can detect, identify and locate prey in the night sky by emitting loud ultrasonic calls and listening to the echoes that are reflected from nearby insects. Griffin (1944) coined the term echolocation to describe this form of biological sonar. However, it was not until the early 1960s that Alan Grinnell (Grinnell 1963 a,b,c,d) and Nobuo Suga (1964a, b) published the first reports of neural processing of ultrasonic signals by echolocating bats. A few years later, in 1967, Grinnell (1967) reported that the evoked potentials from the inferior colliculus of the mustache bat, a species that had not previously been studied, differed from comparable neural potentials seen in any other animal. The unique feature was that the thresholds of the neural potentials were very sharply tuned to about 60 kHz, the dominant frequency of the mustache bat's orientation calls.

The discovery of the sharply tuned neural thresholds was important because it showed that the mustache bat's auditory system has some extraordinary specializations, and in subsequent studies those specializations have proven to be of major experimental and conceptual importance. In this chapter I consider the mustache bat's inferior colliculus, and how my colleagues and I have exploited the special adaptations at 60 kHz to reveal its functional organization, with particular attention given to physiological mechanisms for sound localization. Before discussing the organizational features, I turn first to the mustache bat's biosonar system and explain why 60 kHz plays such an important role in the life of this animal.

The Doppler-Based Biosonar System of the Mustache Bat

Mustache bats belong to a non-taxonomic group of bats known as the long constant frequency bats. The name derives from the type of echolocation calls they emit. The calls are characterized by an initial long constant frequency (CF) component, effectively a tone burst, that has a duration as long as 30 msec, and each call is terminated by a 2-4 msec, downward sweeping frequency modulated (FM) component (Novick and Vaisnys 1964; Henson et al. 1980; Suga 1984). The CF component of the mustache bat's orientation call is emitted with four harmonics, at 30, 60, 90 and 120 kHz (Fig. 1). Most energy is in the 60 kHz second harmonic. The FM of the second harmonic. The FM of the second harmonic sweeps from about 60 kHz down to about 45 kHz, and is used for target ranging, and probably for detecting other target attributes (Simmons et al., 1975). The key feature of this biosonar system, however, is the long 60 kHz CF component because it is through this frequency that this animal "sees" much of its world.

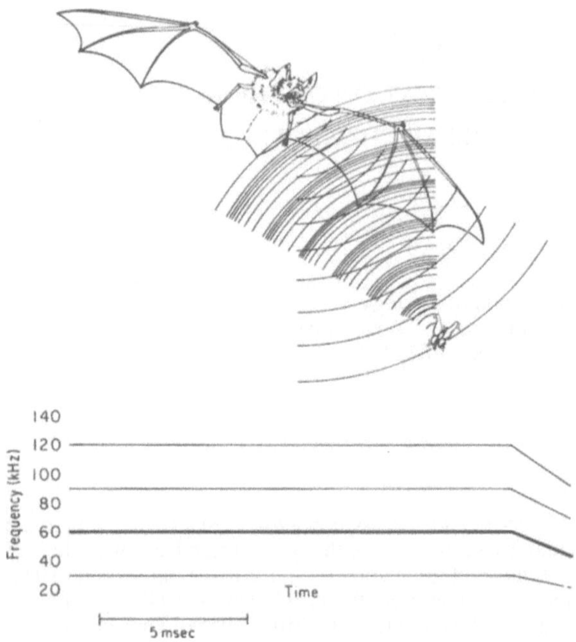

Fig. 1. At the top is a drawing of an echolocating mustache bat and the frequency modulations in the echo due to the wing movements of a nearby moth. The emitted constant frequency component is depicted as a series of regularly spaced waves whereas the echo reflected from the moth is shown as a series of irregularly spaces waves (i.e., the frequency modulations) that are repeated periodically. Below is a time-frequency analysis (sonogram) of one echolocation call. Each call has a relatively long constant frequency component that can be as long as 30 msec and is terminated with a brief frequency modulated portion. Four harmonics are usually present but the 60 kHz second harmonic is always dominant, as indicated by the thicker line.

As bats pursue their targets, they emit a steady stream of orientation calls that the animal modifies in various ways at different stages of the pursuit. The most remarkable modification is that mustache bats, and other long CF bats such as horseshoe bats, adjust the frequency of the CF component of their pulses to compensate for Doppler shifts in the echoes that reach their ears (Schnitzler 1967, 1970; Schuller et al. 1974; Simmons 1974; Henson et al. 1980; Henson et al. 1982) (Fig. 2). This behavior, called Doppler-shift compensation, was discovered by Hans-Ulrich Schnitzler (1967) and is the expression of an extreme sensitivity for motion.

The advantage of Doppler shift compensation is that it enhances the ability to hunt in acoustically cluttered environments by maximizing the bat's ability to distinguish a fluttering insect from background objects and by providing an effective means for recognizing particular insect species (Neuweiler 1983, 1984a,b). All long CF/FM bats that have been studied hunt for flying insects in dense foliage beneath the forest canopy (Bateman and Vaughan 1974; Neuweiler 1983, 1984b). As they fly in this environment, the echo CF component from stationary background objects, such as trees, leaves, and bushes, is Doppler shifted upward due to the relative motion of the bat towards stationary background (Trappe and Schnitzler 1982). Long CF bats compensate for the Doppler shifts in the echoes by lowering the frequency of subsequent emitted pulses by an amount nearly equal to the upward frequency shift in the echo. Consequently, these bats clamp the echo CF component and hold it within a narrow frequency band that varies only slightly from pulse to pulse. However, flight speed differences between the bat and stationary objects do

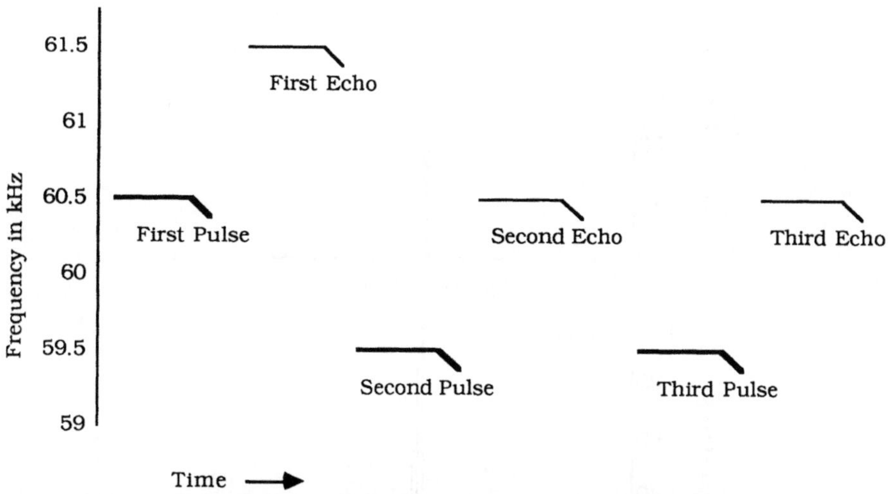

Fig. 2. Schematic illustration of Doppler-shift compensation in a flying mustache bat. The first pulse is emitted at about 60.5 kHz, and the Doppler shifted echo returns with a higher frequency, at about 61.5 kHz. The mustache bat detects the difference between the pulse and echo and lowers the frequency of its subsequent emitted pulses by an amount almost equal to the Doppler-shift. Thus the frequencies of the subsequent echoes are held constant and return at a frequency very close to that of the first emitted pulse.

not represent the total extent of movements a hunting bat would encounter. A small fluttering insect that happens to cross the path of a mustache bat will generate echoes substantially different from those of trees and other background objects. As the emitted CF component strikes and is reflected from the insect, the motion of the wings creates Doppler shifts which vary with the velocity, and hence the movements of the wings (Schnitzler et al. 1983; Schuller 1984). These Doppler shifts create periodic frequency modulations that are superimposed on the CF component of the echoes (Figs. 1 and 4). In addition, the wing motion, or flutter, presents a reflective surface area of alternating size, thereby also creating periodic amplitude modulations, or "glints", in the echo CF component. Thus the echoes from fluttering insects return with a center frequency close to the frequency of echoes from stationary background objects, but are distinguished by the rich pattern of frequency and amplitude modulations. There is now substantial evidence that the periodic frequency and amplitude changes allow the bat to distinguish fluttering insects from background and provide the information for recognizing different insect types (Goldman and Henson 1977; Schnitzler et al. 1983; Schnitzler and Flieger 1983; Link et al. 1986).

Adaptations of the Peripheral Auditory System for Processing 60 kHz

Because the long CF/FM bats "see" much of their world through the narrow window of a small band of frequencies, around 60 kHz for mustache bats, many of the pronounced specializations in their auditory systems are regions designed for processing the CF component of the echo (e.g., Suga 1978; Schnitzler and Ostwald 1983; Neuweiler 1984a; Pollak et al. 1986). The special adaptations are evident in the cochlea, which is greatly hypertrophied and densely innervated by the auditory nerve in the basal region representing 60 kHz (Henson 1978; Kossl and Vater 1985; Vater 1987; Zook and Leake 1988). The cochlea has a prominent resonance at 60 kHz, which can be seen when monitoring cochlear microphonic potentials (Suga and Jen 1977; Pollak et al. 1979; Henson et al. 1985), i.e., the summed responses of the outer hair cells in the organ of Corti. The resonance is manifest in the sharp tuning at 60 kHz of the mustache bat's cochlear microphonic audiogram (Pollak et al. 1972) (Fig. 3 upper panel), and is of particular interest because the resonant frequency and the frequency at which mustache bats clamp their echo CF component when compensating for Doppler shifts are nearly identical (Henson et al. 1980, 1982). The densely

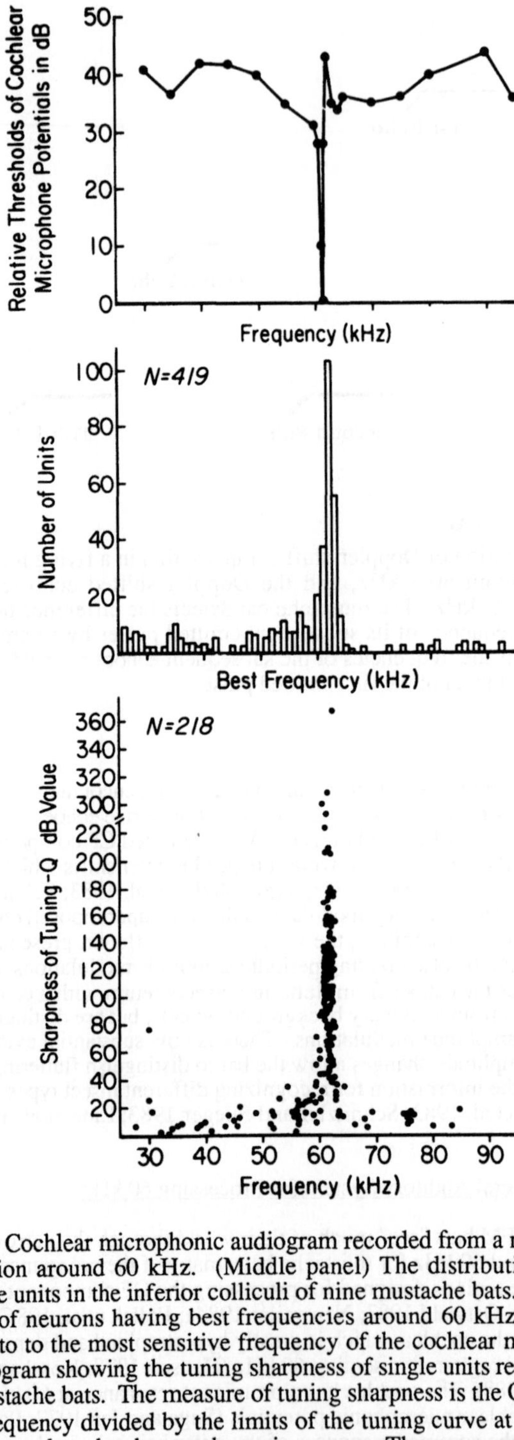

Fig. 3. (Upper panel) Cochlear microphonic audiogram recorded from a mustache bat. Note the sharp sensitive region around 60 kHz. (Middle panel) The distribution of best frequencies recorded from single units in the inferior colliculi of nine mustache bats. There is a pronounced overrepresentation of neurons having best frequencies around 60 kHz, and these frequencies correspond closely to to the most sensitive frequency of the cochlear microphonic audiogram. (Lower panel) Histogram showing the tuning sharpness of single units recorded from the inferior colliculi of nine mustache bats. The measure of tuning sharpness is the Q_{10dB} value, defined as the neuron's best frequency divided by the limits of the tuning curve at 10 dB above threshold. The larger the Q_{10dB} value, the sharper the tuning curve. The overrepresented 60 kHz neurons have much sharper tuning curves than do neurons having higher or lower best frequencies.

innervated 60 kHz region of the cochlea and the resonance at that frequency are expressed as an overrepresentation of very sharply tuned primary auditory nerve fibers having their lowest thresholds, i.e., best frequencies, at the resonant frequency of the bat's cochlea (Suga et al. 1975; Suga and Jen 1977).

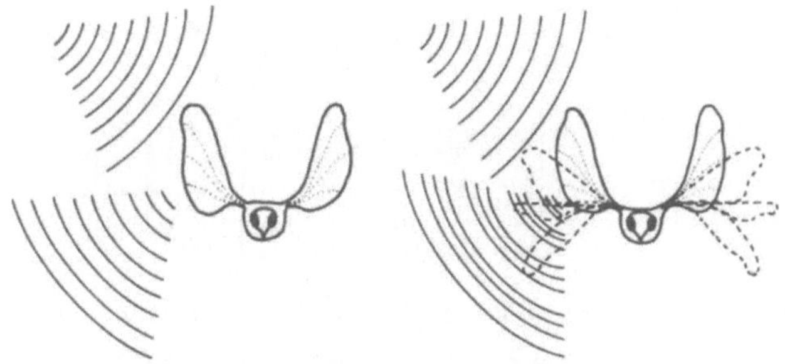

Fig. 4 Drawings to illustrate the echoes reflected from an insect in which the wings are stationary (left) and from an insect whose wings are in motion (right). Note that the echo returning from the nonflying moth is simply an unmodulated tone, whereas the echo reflected from a fluttering moth has pronounced frequency modulations due to the Doppler shifts induced by the velocity of the moving wings.

The sharp tuning of the 60 kHz neurons appears to be a specialization for the fine analysis of the frequency modulation patterns imposed on the echo CF component from a fluttering insect (Suga et al. 1975; Suga and Jen 1977; Schuller and Pollak 1979; Schuller 1979a,b, 1984; Pollak and Schuller 1981; Bodenhamer and Pollak 1983). Suga and Jen (1977) were the first to describe the coding of such modulation patterns in the auditory nerve of the mustache bat. They mimicked insect echoes by modulating the frequency or amplitude of a 60 kHz carrier tone with low frequency sinusoids of 50-100 Hz (Fig. 5). The low frequency sinusoid is the modulating waveform, and is distinguished from the 60 kHz carrier frequency. This arrangement creates either sinusoidally amplitude modulated (SAM) or sinusoidally frequency modulated (SFM) signals (Fig.5). With SFM signals the depth of modulation, the amount by which the frequency varies around the carrier, mimics the degree of Doppler shift created by the motion of the wings. The other modulating parameter, the modulation rate, simulates the insect's wingbeat frequency.

Suga and Jen (1977) observed that 60 kHz filter neurons routinely discharge in register with the phase of the modulating waveform of an SFM signal, and some do so even when the modulation depth varies by as little as ± 10 Hz around the 60 kHz carrier frequency. Such sensitivity for frequency modulation is truly remarkable. Although primary auditory neurons in other animals also display phase locking to SFM signals, such high sensitivity occurs only in filter neurons.

The sharply tuned neurons at all levels of the mustache bat's auditory system are extremely sensitive to the periodic frequency modulations created by insect wingbeats. When presented with sinusoidal frequency modulations that sweep as little as ± 50 Hz around a 60 kHz carrier frequency, sharply tuned neurons in the mustache bat's inferior colliculus also respond with discharges that are phase-locked to the periodicity of the modulation waveform (Bodenhamer and Pollak 1983) (Fig. 6). Neurons having best frequencies above or below the resonant frequency of the cochlea are one or two orders of magnitude less sensitive to sinusoidal modulations than are the 60 kHz neurons, and their tuning curves are correspondingly wider (Fig. 3 lower panel). These results provide strong support for the idea that the sharp tuning is a specialization for the fine analysis of frequency modulation patterns, and thus is the neural code for target recognition.

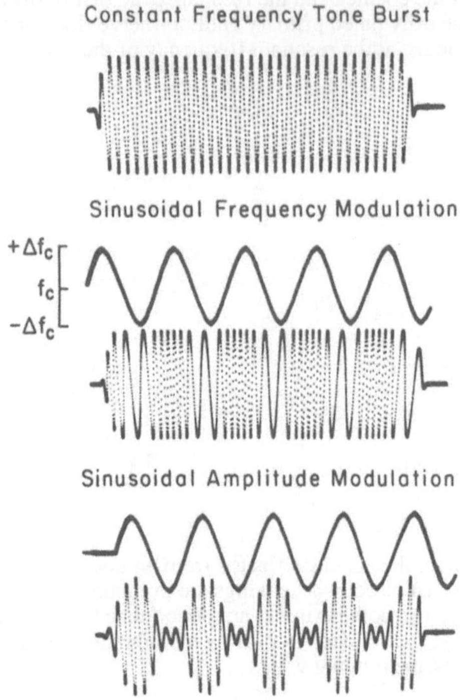

Fig. 5. Illustration of fine structure of a tone burst (upper record), a sinusoidally frequency modulated (SFM) burst (middle record) and a sinusoidally amplitude modulated (SAM) burst (lower record). The tone burst in upper record is a shaped sine wave, where the frequency, or fine structure, of the signal remains constant. The fine structure of the SFM burst changes sinusoidally in frequency. The modulation waveform is shown above the record of the signal's fine structure. In the SAM burst, the fine structure of the signal is a constant frequency, but the amplitude varies with the sinusoidal modulating waveform.

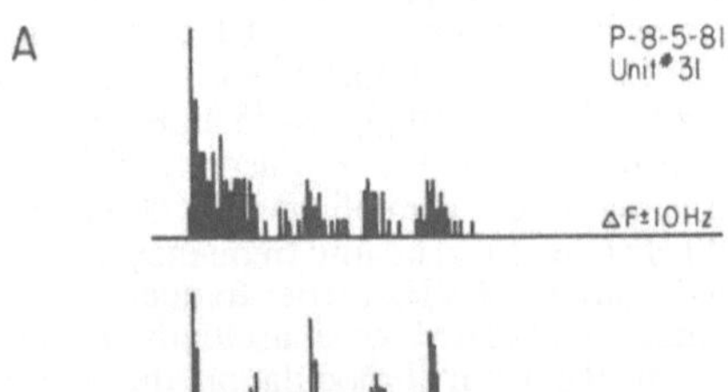

Fig. 6. Peri-stimulus time histograms of a sharply tuned 60 kHz neuron in the mustache bat's inferior colliculus that phase-locked to SFM signals. Signal envelope and SFM waveform are shown below. This was an unusually sensitive neuron and phase-locked when the frequency swings (Δf) were as small as ± 10 Hz around the 62.87 kHz center frequency. Scale bar is 20 msec.

Peripheral Adaptations are Conserved in the Auditory Pathway

The primary auditory pathway consists of a series of ascending parallel pathways in which the cochlear surface is remapped upon each succeeding auditory region. The parallel pathways originate as each auditory nerve fiber enters the brain, where it divides into an ascending branch that innervates the anteroventral cochlear nucleus and a descending branch that innervates the posteroventral and dorsal cochlear nucleus. The orderly representation of frequency established in the cochlea is preserved in each division of the cochlear nucleus as three separate tonotopic maps. The principal auditory pathways originate from these three major divisions of the cochlear nucleus (Fig. 7). Each pathway ascends in parallel with the others, and has a unique pattern of connectivity with the subdivisions of the superior olivary complex and the nuclei of the lateral lemniscus. Many of these centers, such as the superior olivary nuclei, are the initial sites of convergence of information from the two ears, and thus play an important role in sound localization, while others, such as the cochlear nuclei, process information from only one ear (e.g., Goldberg 1975; Roth et al. 1978; Brunso-Bechtold et al. 1981; Aitkin 1985; Irvine 1986; Zook and Casseday 1982, 1985, 1987; Ross et al. 1988). All of these pathways ultimately terminate in an orderly fashion in the central nucleus of the inferior colliculus (ICc), where the multiple tonotopic maps of the lower nuclei are reconstituted into a single tonotopic arrangement.

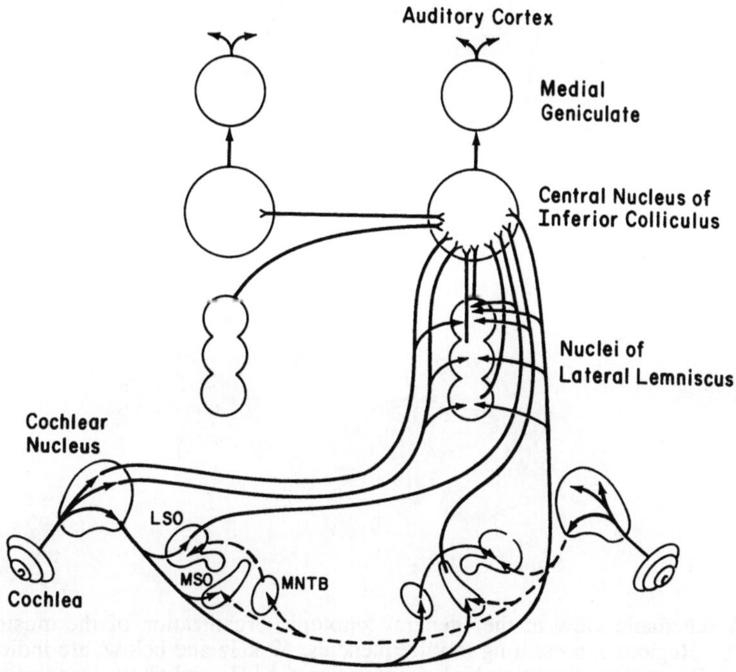

Fig. 7. Schematic diagram of the principal connections of the brainstem auditory system. Note the parallel pathways that originate from the cochlear nucleus. Some pathways converge on nuclei that receive innervation from the two ears, such as the MSO and LSO , while others project directly to the inferior colliculus. MNTB, medial nucleus of the trapezoid body; MSO, medial superior olivary nucleus; LSO lateral superior olivary nucleus.

One of the distinguishing features of the mustache bat's auditory system is that the overrepresentation and sharp tuning of 60 kHz neurons described above are conserved throughout the auditory system, including the ICc (Suga et al. 1975, 1976; Suga 1984; Kossl and Vater 1985; Feng and Vater 1985; Neuweiler 1980; Pollak 1980; Pollak and Bodenhamer 1981; Pollak and Casseday 1988; Zook et al. 1985). The overrepresentation is reflected in the tonotopy. Figure 8 shows the general tonotopic arrangement in the mustache bat's brainstem, and illustrates the overrepresentation of 60 kHz in each auditory nucleus.

The 60 kHz overrepresentation is most dramatically expressed in the central nucleus of the inferior colliculus. The tonotopy of the ICc in less specialized mammals is manifest as an orderly stacking of sheets of neurons (Rockel and Jones 1973; FitzPatrick 1975; Oliver and Morest 1984; Aitkin 1985), which imparts a laminated appearance to the ICc in Golgi impregnated material. The neuronal population of any one sheet, or lamina, is most sensitive to a particularfrequency, and each lamina, therefore, represents a segment of the cochlear surface (Merzenich and Reid 1974; FitzPatrick 1975; Semple and Aitkin 1979; Serviere et al. 1984). The orderly arrangement of isofrequency laminae recreates the cochlear frequency-place map along one axis of the ICc. In mustache bats, however, the orderly tonotopic sequence of isofrequency laminae is distorted due to the overrepresentation of 60 kHz (Pollak et al. 1983; Zook et al. 1985) (Fig. 9).

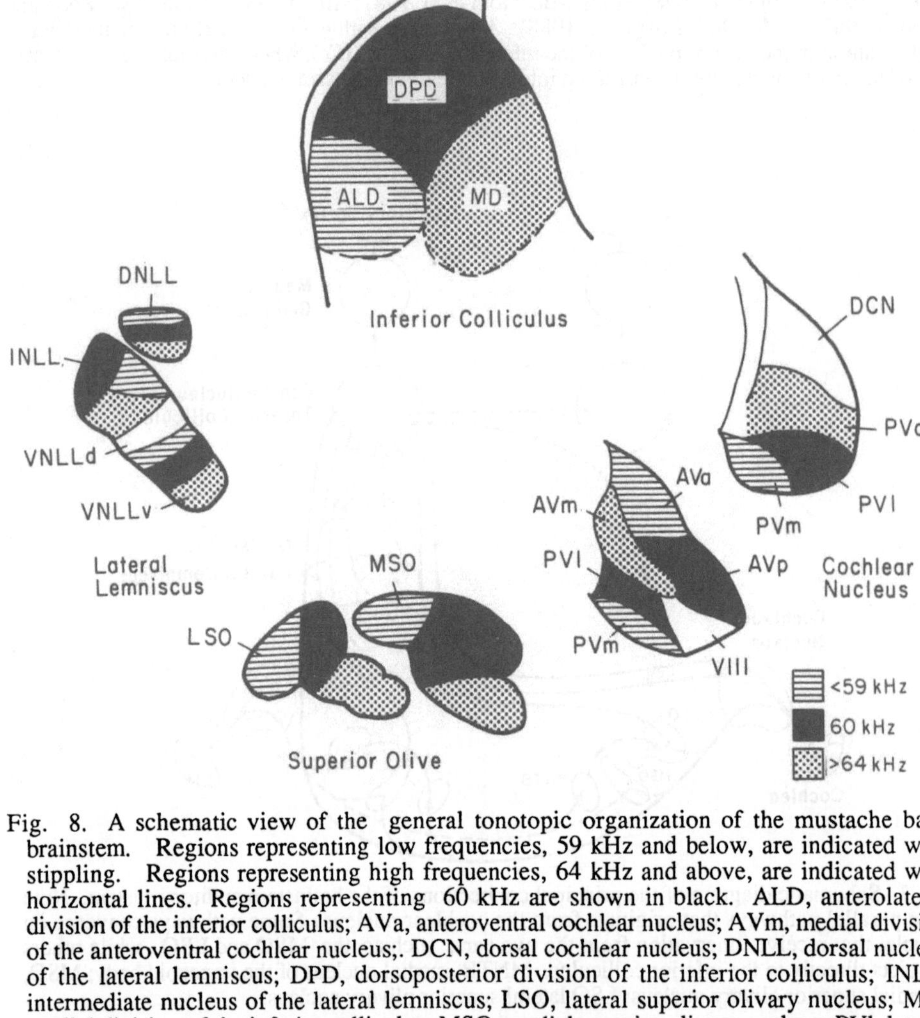

Fig. 8. A schematic view of the general tonotopic organization of the mustache bat's brainstem. Regions representing low frequencies, 59 kHz and below, are indicated with stippling. Regions representing high frequencies, 64 kHz and above, are indicated with horizontal lines. Regions representing 60 kHz are shown in black. ALD, anterolateral division of the inferior colliculus; AVa, anteroventral cochlear nucleus; AVm, medial division of the anteroventral cochlear nucleus;. DCN, dorsal cochlear nucleus; DNLL, dorsal nucleus of the lateral lemniscus; DPD, dorsoposterior division of the inferior colliculus; INLL, intermediate nucleus of the lateral lemniscus; LSO, lateral superior olivary nucleus; MD, medial division of the inferior colliculus; MSO, medial superior olivary nucleus; PVl, lateral division of posteroventral cochlear nucleus; PVm, medial division of posteroventral cochlear nucleus: PVc, caudal division of posteroventral cochlear nucleus; VIII, eighth cranial nerve. (From Ross et al. 1988)

The representation of frequency in the mustache bat's ICc is split into three divisions: the anterolateral division, the medial division and the dorsoposterior division. Each division represents a different region of the cochlear surface. The sheet-like isofrequency laminae in the anterolateral division provide an orderly representation of only the lower frequencies, from about 10 kHz to about 59 kHz; the laminae of the medial division contain an orderly representation only of the high frequencies, ranging from about 64 kHz to over 120 kHz; and the dorsoposterior division (DPD), which occupies about a third of the ICc volume, contains exclusively the 60 kHz representation. The isofrequency composition of the dorsoposterior division is seen in the similarity of the best frequencies of its sharply tuned neurons, which differ by only $\pm$ 300 Hz among the population (Pollak and Bodenhamer 1981; Zook et al. 1985; Wenstrup et al. 1986a). Additionally, the best frequencies coincide closely with the resonant frequency of the individual bat's cochlea (Fig. 4, middle record). The dorsoposterior division, then, is the midbrain representation of the "resonant" segment of the bat's cochlear partition, and corresponds to the frequency at which the bat clamps the echo CF component when Doppler compensating.

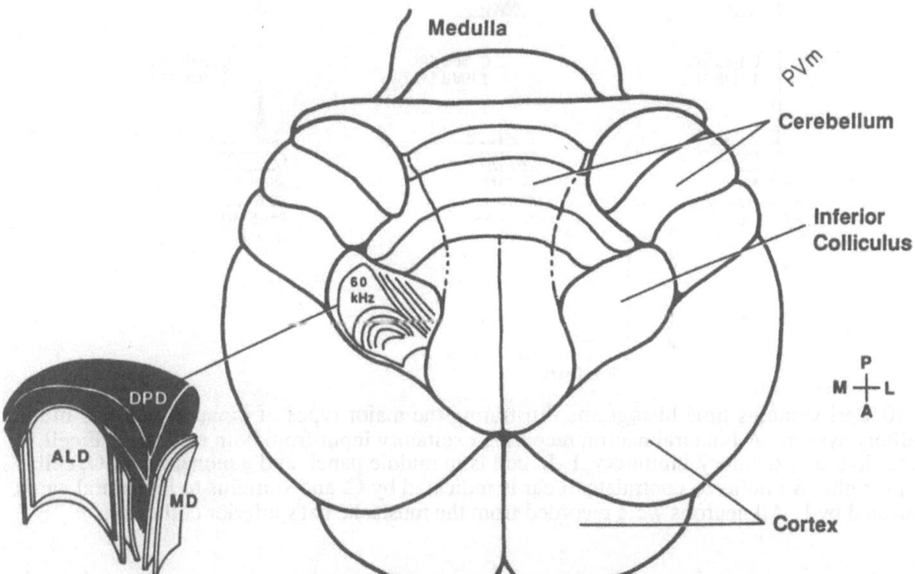

Fig. 9. Schematic drawing of a dorsal view of the mustache bat's brain to show the location of the 60 kHz region in the inferior colliculus. The hypertrophied inferior colliculi protrude between the cerebral cortex and cerebellum. In the left colliculus are shown the isofrequency contours, as determined in anatomical and physiological studies. A three dimensional representation of the laminar arrangement in the inferior colliculus is shown on far left. Low frequency contours, representing an orderly progression of frequencies from about 59 kHz to about 10 kHz, fill the anterolateral division (ALD). High frequencies, from about 64 kHz to over 120 kHz, occupy the medial division (MD). The 60 kHz isofrequency contour is the dorsoposterior division (DPD), and is the sole representation of the sharply tuned neurons in the bat's colliculus. Data from Zook et al. (1985). Drawing from Pollak et al. (1986).

Collicular Features Derived from Processing in the Central Auditory Pathway: Monaural and Binaural Representation

A number of important features of the ICc are derived from processing that occurs within lower central auditory structures. Among these derived characteristics are binaural response types. Unlike the visual system, which first combines information from the two eyes in the cortex, binaural processing occurs early in the auditory pathway, initially in the superior olivary nuclei in the brainstem (Boudreau and Tsuchitani 1968; Goldberg 1975; Irvine 1986). Since these, as well as monaural regions such as the cochlear nucleus, send projections to the ICc,

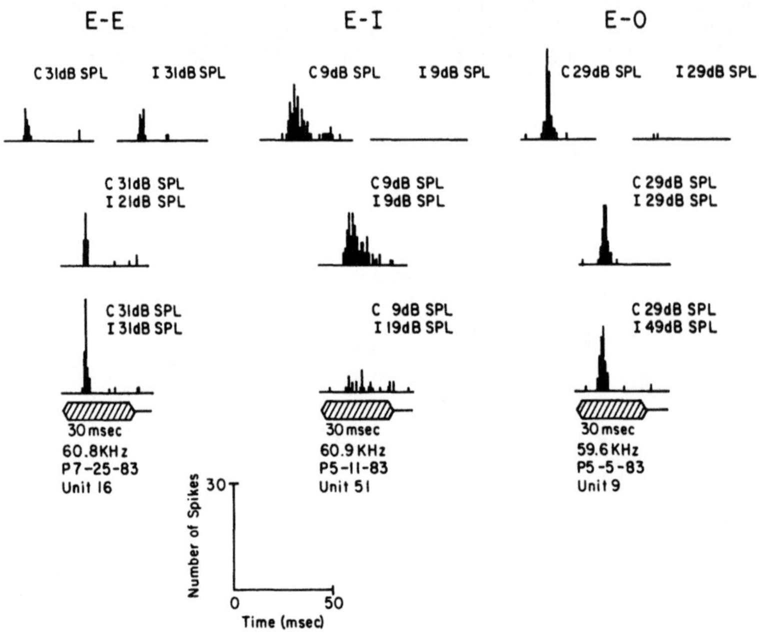

Fig. 10. Peri-stimulus time histograms illustrating the major types of binaural neurons in the auditory system. A binaural neuron receiving excitatory input from both ears, an E-E cell, is on far left; an excitatory-inhibitory, E-I, cell is in middle panel; and a monaural, E-O, cell is on far right. Stimulus to contralateral ear is indicated by C, and stimulus to ipsilateral ear is indicated by I. All neurons were recorded from the mustache bat's inferior colliculus.

neurons having various monaural or binaural response properties are common in the inferior colliculus (Roth et al. 1978; Semple and Aitkin 1979; Irvine 1986; Aitkin 1985). Monaural ICc neurons almost always receive excitatory input from the contralateral ear, and are designated as E-O neurons, the E referring to contralateral excitation and the O referring to the absence of ipsilateral influence (Fig. 10). There are two major types of binaural cells: 1) E-E neurons receive excitatory inputs from both ears; and 2) E-I neurons receive excitation from the contralateral ear and inhibition from the ipsilateral ear (Fig. 10).

The enlarged 60 kHz region of the mustache bat's inferior colliculus is an excellent model for the analysis of convergence and separation of inputs within one isofrequency contour of the inferior colliculus. Injections of HRP confined to this isofrequency contour (Fig. 11) reveal that the dorsoposterior division receives projections from the same set of lower auditory nuclei that project to the entire central nucleus of the inferior colliculus (Ross et al. 1988). These include: 1) contralateral projections from the cochlear nucleus and inferior colliculus; 2) ipsilateral projections from the medial superior olive, and from the ventral and intermediate nuclei of the lateral lemniscus; and 3) bilateral projections from the lateral superior olive and dorsal nucleus of the lateral lemniscus. The projections arise from discrete segments of the various projecting nuclei, each of which presumably represents 60 kHz.

Since the dorsoposterior division receives projections from several lower nuclei that are monaural (e.g., cochlear nucleus and ventral nucleus of the lateral lemniscus) as well as several

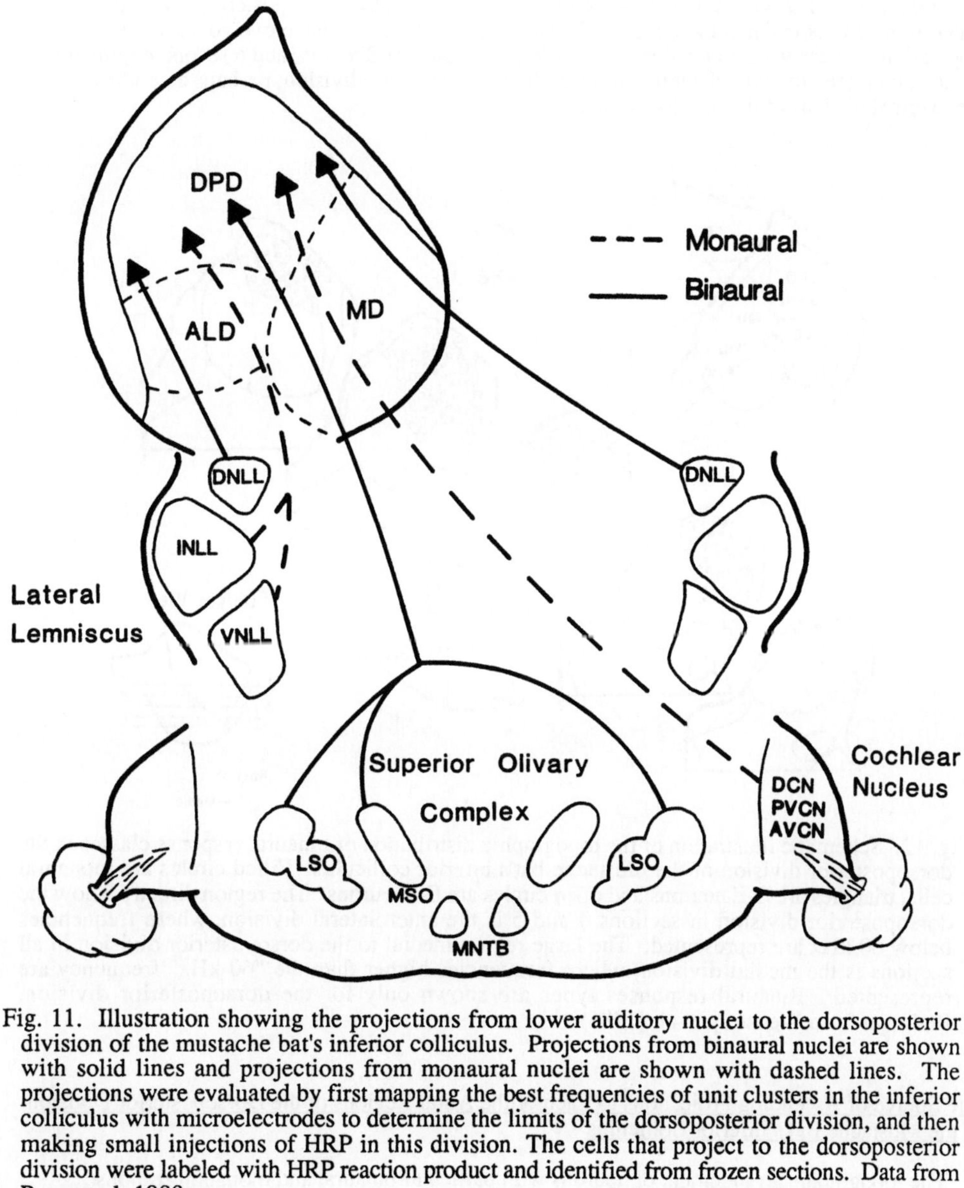

Fig. 11. Illustration showing the projections from lower auditory nuclei to the dorsoposterior division of the mustache bat's inferior colliculus. Projections from binaural nuclei are shown with solid lines and projections from monaural nuclei are shown with dashed lines. The projections were evaluated by first mapping the best frequencies of unit clusters in the inferior colliculus with microelectrodes to determine the limits of the dorsoposterior division, and then making small injections of HRP in this division. The cells that project to the dorsoposterior division were labeled with HRP reaction product and identified from frozen sections. Data from Ross et al. 1988.

that are binaural (e.g., medial and lateral superior olives), it is not surprising that monaural neurons and the two major binaural types are common in the dorsoposterior division, as they are in other isofrequency contours of the mustache bat's inferior colliculus (Fuzessery and Pollak 1985; Wenstrup et al. 1986a). However, neurons having common aural properties are topographically organized within the dorsoposterior division (Wenstrup et al 1986a), and probably in other isofrequency contours of the mustache bat's inferior colliculus as well (Wenstrup et al. 1986b). There are several zones in the dorsoposterior division each having a predominant aural type (Fig. 12). Monaural (E-O) neurons are located along the dorsal and lateral parts of the dorsoposterior division. E-E cells occur in two regions, one in the ventrolateral dorsoposterior division and the other in the dorsomedial dorsoposterior division. E-I neurons also have two zones: the main population is in the ventromedial region of the dorsoposterior division, and a second population occurs along the very dorsolateral margin of the dorsoposterior division, perhaps extending into the external nucleus of the inferior colliculus.

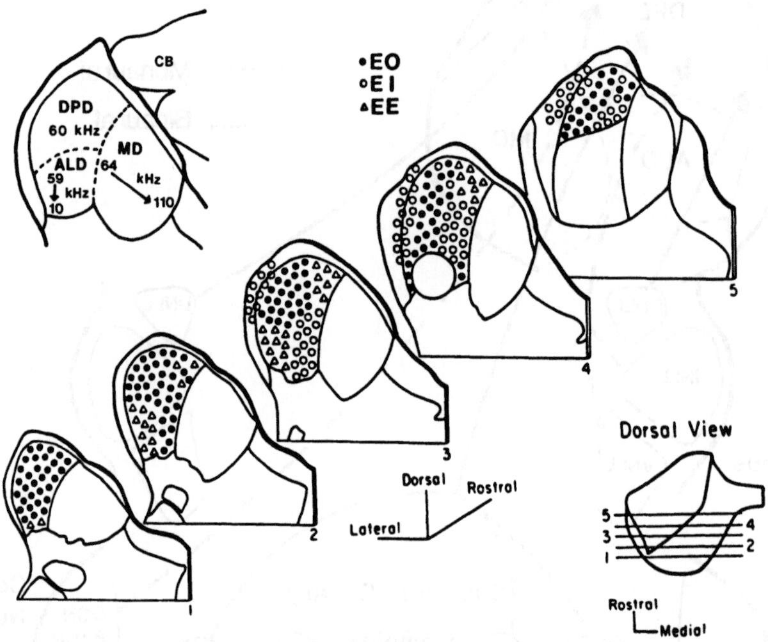

Fig. 12. Schematic illustration of the topographic distribution of binaural response classes in the dorsoposterior division of the mustache bat's inferior colliculus. Filled circles are monaural cells, triangles are E-E neurons and open circles are E-I neurons. The region directly below the dorsoposterior division in sections 4 and 5 is the anterolateral division, where frequencies below 60 kHz are represented. The large region medial to the dorsoposterior division in all sections is the medial division, where frequencies higher than the "60 kHz" frequency are represented. Binaural responses types are shown only for the dorsoposterior division. Redrawn from Wenstrup et al. (1986).

<u>The Source of Ascending Projections to Each of the Aural Regions of the Dorsoposterior Division of the Mustache Bat's Inferior Colliculus</u>

The topographic arrangement of neurons with particular binaural and monaural response properties in the dorsoposterior division suggests that the aural properties in each region are a consequence of the projections that terminate in each region. Linda Ross and I exploited the greatly enlarged 60 kHz region to determine the source of ascending projections to the four different aural regions of the dorsoposterior division (Ross and Pollak 1988). Small iontophoretic deposits of HRP were made within each of the physiologically defined aural regions, and the locations and numbers of retrogradely labeled cells in the auditory brainstem nuclei were determined. With this method, we demonstrated that each monaural and binaural region within the dorsoposterior division receives its chief inputs from a different subset of nuclei in the lower brainstem. For most regions, the response properties reflect the subset of inputs.

Below I describe the nuclei that project to three binaural areas of the dorsoposterior division: the projections to the monaural regions are not considered here. Each binaural area has a distinctive pattern of inputs from monaural and binaural nuclei. Considered first are the projections to the ventromedial part of the 60 kHz contour, an area in which the neurons are excited by sound to the contralateral ear and inhibited by sound to the ipsilateral ear (E-I cells). The lower panel in Fig. 13 shows that the input to this area comes largely from binaural nuclei, especially the dorsal nucleus of the lateral lemniscus and less so from the lateral superior olive. A major input also originates in the intermediate nucleus of the lateral lemniscus, which is a monaural nucleus. The robust inputs from the dorsal nucleus of the lateral lemniscus and lateral superior olive distinguish the E-I region from the other aural regions of the 60 kHz contour.

The other two binaural areas contain neurons that are excited by sound at either ear (E-E cells). One E-E area is situated ventrolaterally and the other dorsomedially in the dorsoposterior division as shown in Fig. 12. Figure 13 (top panel) shows that the two major inputs to the ventrolateral E-E area are from the medial superior olive and the ventral nucleus of the lateral lemniscus. There is little or no input from lateral superior olive. This pattern is in marked contrast to projections to the E-I area, which arise largely from the dorsal and intermediate nuclei of the lateral lemniscus.

Finally, the projections to the dorsomedial E-E area (Fig. 13, middle panel) differ substantially from both the E-I region and ventrolateral E-E region. The dominant input arises from the contralateral inferior colliculus. Substantial input also arises from both the ventral and intermediate nuclei of the lateral lemniscus, but there is almost no input from the binaural centers below the tectum (LSO, MSO and DNLL).

These projection patterns suggest that the binaural properties of the medial E-E area arise from interaction via the two colliculi, whereas the binaural properties in the dorsolateral E-E area arise from the binaural centers in the lower brainstem, especially MSO (Fig. 14). In contrast, the binaural properties of the E-I area are shaped largely by the binaural nucleus of the lateral lemniscus, the DNLL, and to a lesser extent by the LSO (Fig. 14).

These studies show that each aural region of the dorsoposterior division is distinguished both by its neural response properties and by the unique pattern of ascending projections it receives. The connectional differences among the various aural regions are also reflected in functional differences. Some functional distinctions are apparent, such as between monaural compared to binaural neurons and between E-I and E-E neurons. Below, the different ways in which E-E and E-I cells code for sound location are considered in detail. However, our understanding of how the response properties of individual neurons are shaped by the particular set of connections they receive does not reach much beyond the more obvious differences. For example, it is clear just from the connectional patterns that there must be at least two major subtypes of E-E neurons, but we have little insight into exactly how the connectional differences are expressed physiologically. Moreover, we presently can only speculate about how the convergence of inputs from several lower nuclei could shape the response characteristics of a collicular neuron expressing a particular aural type. Clarifying these issues represents a major challenge for the future.

The Population of E-I neurons Have Different Sensitivities for Interaural Intensity Disparities

The population of E-I neurons is of particular interest because they differ in their sensitivities to interaural intensity disparities (IIDs) (Wenstrup et al. 1986a; 1988a). These E-I neurons compare the sound intensity at one ear with the intensity at the other by subtracting the activity generated in one ear from that in the other. Supra-threshold sounds delivered to the excitatory (contralateral) ear evoke a certain discharge rate that is unaffected by low intensity sounds presented simultaneously to the inhibitory (ipsilateral) ear. However, when the ipsilateral intensity reaches a certain level,

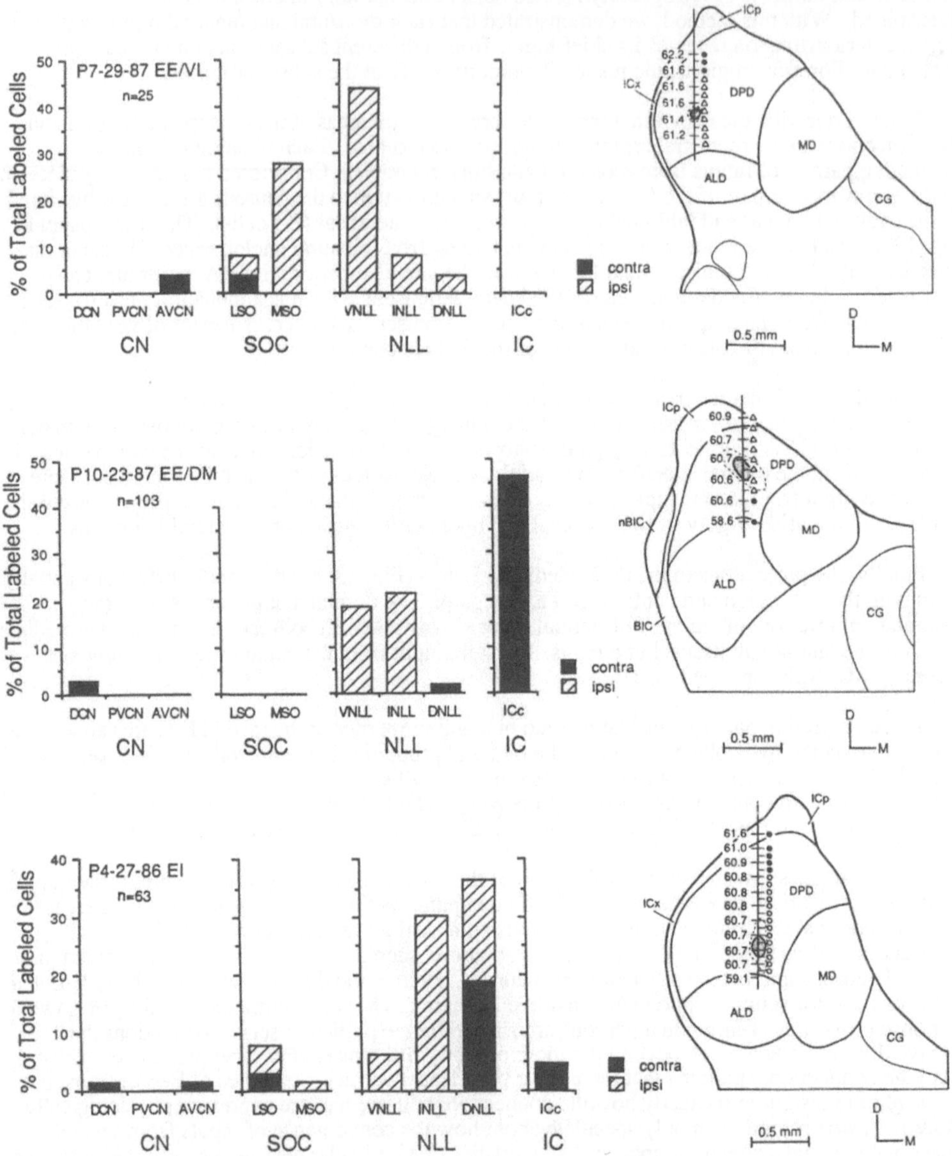
P7-29-87 EE/VL
n=25
% of Total Labeled Cells
DCN PVCN AVCN
LSO MSO
VNLL INLL DNLL
ICc
CN
SOC
NLL
IC
contra
ipsi
ICp
ICx
62.2
61.6
61.6
51.6
61.4
61.2
DPD
MD
ALD
CG
0.5 mm
D
M
P10-23-87 EE/DM
n=103
ICp
60.9
60.7
60.7
60.6
60.6
58.6
nBIC
BIC
P4-27-86 EI
n=63
61.6
61.0
60.9
60.8
60.8
60.8
60.7
60.7
60.7
60.7
59.1

Fig. 13

Fig. 13. Percentage of labeled cells in different auditory nuclei in the lower brainstem following a small injection of HRP in three of the binaural regions in the dorsoposterior division of the mustache bat's inferior colliculus. The reconstructions of the electrode tracts are shown on the right. The best frequencies of the neurons at each location along the electrode tract are noted to the left of each tick mark along the tract. The symbol to the right of each tic mark indicates the aural type: monaural sites are shown as filled circles, E-I sites as open circles and E-E sites as triangles. The size of the deposits are indicated by the small stippled circles. The percentage of the total number of labeled cells that was found in each nucleus is shown in the bar graphs on the left. Striped bars refer to labeled cells in ipsilateral nuclei and black bars indicate labeled cells in contralateral nucleus. (Top Panel): Percentage of labeled cells in different lower auditory nuclei following a small deposit of HRP in ventrolateral E-E region. Note large percentage of labeled cells in VNLL and MSO. This case was unusual in that the LSO has a small percentage of labeled cells. No labeled cells were found in the LSO in four other ventrolateral E-E cases. (Middle Panel): Percentage of labeled cells in different lower auditory nuclei following a small injection of HRP in dorsomedial E-E region. Note the high proportion of labeled cells in the opposite inferior colliculus and the absence of labeled cells in the LSO and MSO. (Bottom Panel); HRP injection in ventromedial E-I region. Note the high proportion of labeled cells in the DNLL and INLL. Smaller percentages of labeled cells were seen in the VNLL, opposite inferior colliculus and LSO. Abbreviations: ALD; anterolateral division of inferior colliculus; AVCN, anteroventral cochlear nucleus; BIC, brachium of the inferior colliculus; CG, central grey; CN, cochlear nucleus; DCN, dorsal cochlear nucleus; DNLL, dorsal nucleus of the lateral lemniscus; DPD, dorsoposterior division of inferior colliculus; ICc, central nucleus of the inferior colliculus; ICp, pericentral region of inferior colliculus; ICx, external nucleus of inferior colliculus; INLL, intermediate nucleus of the lateral lemniscus; LSO, lateral superior olivary nucleus: MSO, medial superior olivary nucleus; MD, medial division of inferior colliculus; NLL, nuclei of the lateral lemniscus; PVCN, posteroventral cochlear nucleus; SOC, superior olivary complex; VNLL, ventral nucleus of the lateral lemniscus. (Data from Ross and Pollak 1988).

Fig. 14. Chief projections that determine the binaural properties of each of the three aural regions of the dorsoposterior division in the mustache bat's inferior colliculus. Ipsilateral projections are shown in white, contralateral projections are black and projections from nuclei that receive input from the two ears are shown with stripes. (Top Panel): The binaural properties of neurons in the ventrolateral E-E region are largely determined by inputs from the medial superior olivary nucleus. (Lower Left Panel): Binaural neurons in the dorsomedial E-E area are constructed differently. Input from the contralateral ear is conveyed from two monaural nuclei, the intermediate and ventral nuclei of the lateral lemniscus. Input from the ipsilateral ear arises via projections from the dorsoposterior division on the opposite side. (Lower Right Panel): E-I properties are constructed via bilateral projections from two binaural nuclei; the dorsal nucleus of the lateral lemniscus (DNLL) and the lateral superior olivary nucleus. Abbreviations are given in caption of Fig. 13. (From Ross and Pollak 1988).

(Continued)

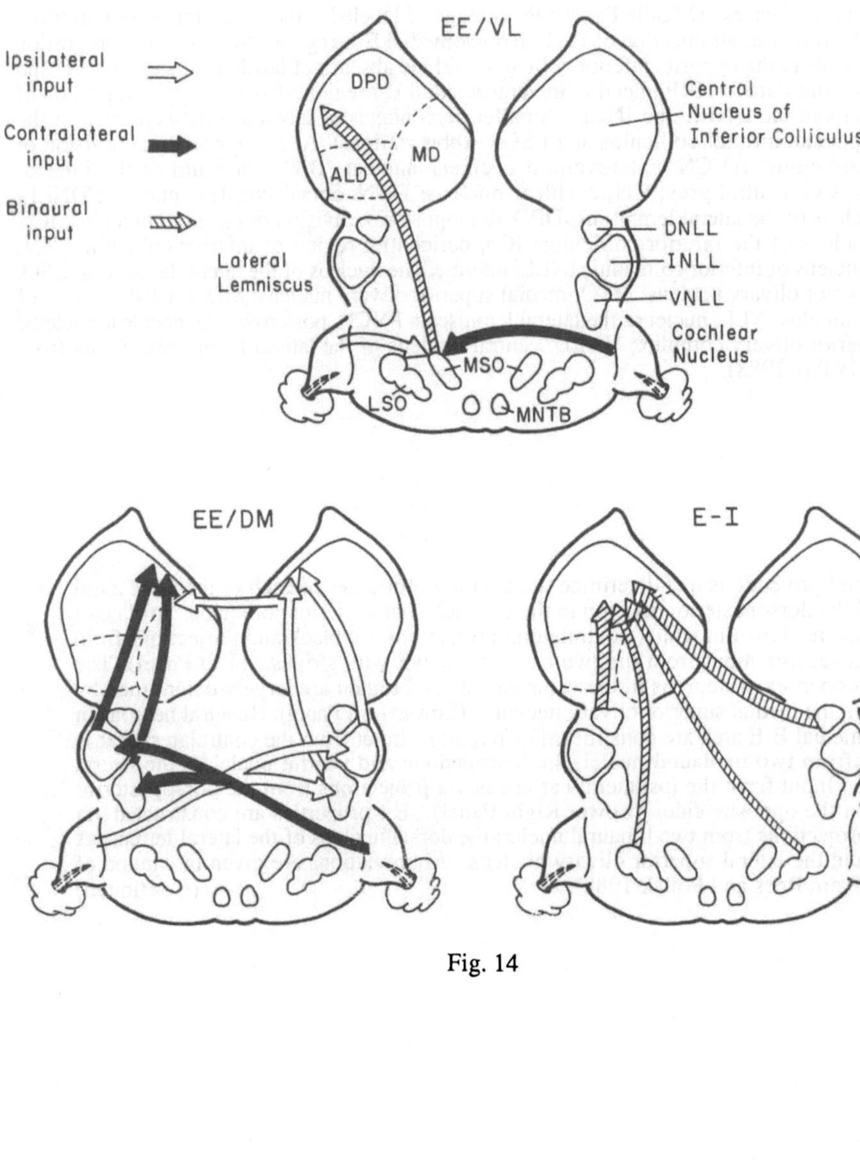
EE/VL
Ipsilateral input
Contralateral input
Binaural input
DPD
MD
ALD
Central Nucleus of Inferior Colliculus
DNLL
INLL
VNLL
Lateral Lemniscus
Cochlear Nucleus
MSO
LSO
MNTB
EE/DM
E-I

Fig. 14

and thus generates a particular IID, the discharge rate declines sharply, and even small increases in ipsilateral intensity will, in most cases, completely inhibit the cell (Fig. 15). Thus each E-I neuron has a steep IID function and reaches a criterion inhibition at a specified IID that remains relatively constant over a wide range of intensities (Wenstrup et al. 1986a, 1988a). The criterion we adopted is the IID that produces a 50% reduction in the discharge rate evoked by the excitatory stimulus presented alone. This IID is called the neuron's inhibitory threshold. An inhibitory threshold has a value of 0 dB if equally intense signals in the two ears elicit the criterion inhibition. An inhibitory threshold is assigned a positive value if the inhibition occurs when the ipsilateral signal is louder than the contralateral signal, and is assigned a negative value if the intensity at the ipsilateral ear is lower than the contralateral ear when the discharge rate is reduced by 50%. The inhibitory thresholds of E-I neurons in the dorsoposterior division vary from +30 dB to -20 dB (Fuzessery and Pollak 1984, 1985), encompassing much of the range of IIDs that the bat would experience.

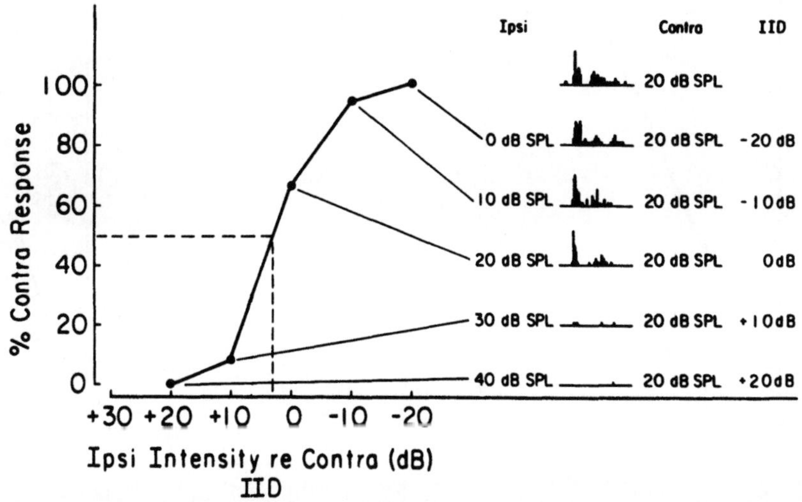

Fig. 15. Responses of a single E-I neuron as a function of interaural intensity disparity (IID). On the right are shown the peri-stimulus time histograms of discharges evoked by the contralateral stimulus alone (top histogram), and the inhibition produced as the intensity of the ipsilateral (inhibitory) signal was increased. The best frequency of this neuron was 62 kHz and its threshold was 10 dB SPL. The inhibitory threshold, defined as the IID value at which the discharge rate declined by 50%, was +4 dB (at an ipsilateral intensity 4 db higher then the contralateral intensity).

Ultrasonic Frequencies Generate Large Interaural Intensity Disparities

Frequencies having wavelengths shorter than the animal's head generate substantial IIDs due to acoustic shadowing and the directional properties of the ear. Moreover, the value of an IID varies with the location of the sound along the azimuth (Fuzessery and Pollak 1984, 1985; Wenstrup et al. 1988b), as shown in Fig. 16 for 60 kHz in the mustache bat. When the sound emanates from directly in front of the bat, at 0° elevation and 0° azimuth, equal sound intensities reach both ears, and an interaural intensity disparity of 0 dB is generated. The largest interaural intensity disparities originate at about 40° azimuth and 0° elevation, where the sounds are about 30 dB louder in one ear than in the other ear. Within the azimuthal sound field from roughly 40° on either side of the midline, the range of interaural intensity disparities created by the head and ears at 60 kHz is about 60 dB (+30 dB to -30 dB). Thus the interaural intensity disparities change on the average by about 0.75 dB/degree. If recordings were made on the left side of the brain and the sound source was located in the left hemifield at 40° azimuth and 0° elevation, the terminology would refer to this as a +30 dB interaural intensity disparity. Figure 16 (right panel) shows a schematic representation of the interaural intensity disparities generated by 60 kHz sounds in both azimuth and elevation. Notice that an interaural intensity disparity is not uniquely associated with one position in space, but rather a given intensity disparity can be generated by 60 kHz from several spatial locations, a feature that we shall address in a later section. Since the intensity disparity at the ears varies with the position of a sound source in space, a neuron's sensitivity to interaural intensity disparities is suggestive of how it will respond to sounds emanating from various spatial locations.

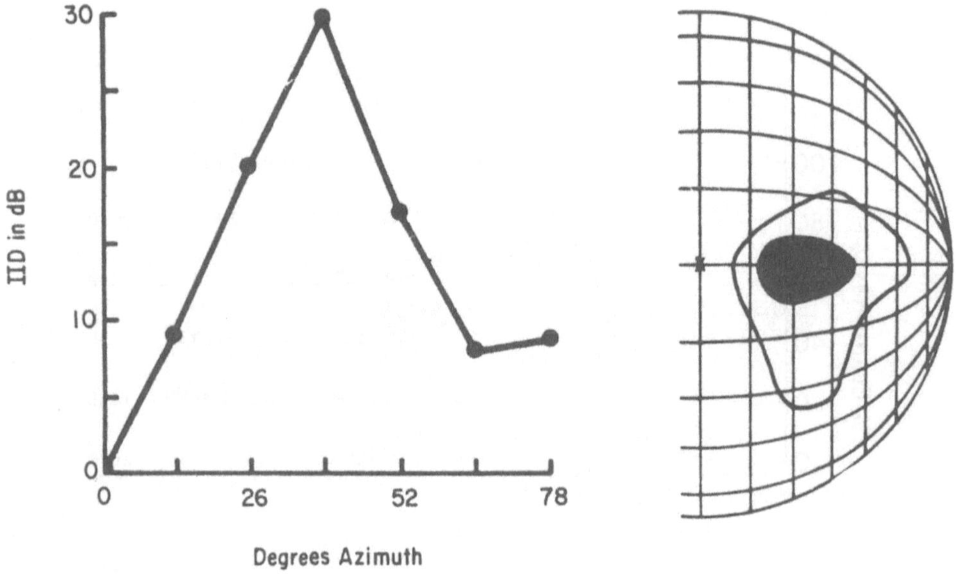

Fig. 16. Left Panel: Interaural intensity disparities (IID) of 60 kHz sounds for azimuthal locations at 0 degrees elevation. Right Panel: Schematic illustration of IIDs of 60 kHz sounds at different azimuths and elevations. Blackened area is region of space where all IIDs are 20 dB or greater. The area enclosed by the solid line is region of space from which all IIDs are 10 dB or greater. Data from Fuzessery and Pollak (1985).

The Spatial Selectivities of 60 kHz E-I Neurons are Determined by the Disparities Generated by the Ears and the Neuron's Inhibitory Threshold

To determine more precisely how binaural properties shape a neuron's receptive field, Fuzessery and his colleagues (Fuzessery and Pollak 1984, 1985; Fuzessery et al. 1985; Fuzessery 1986; Wenstrup et al. 1988b) first determined how interaural intensity disparities of 60 kHz sounds vary around the bat's hemifield, as shown in Fig. 16. They then evaluated the binaural properties of collicular neurons with loudspeakers inserted into the ear canals, and subsequently determined the spatial properties of the same neurons with free-field stimulation delivered from loudspeakers around the bat's hemifield. With this battery of information, the quantitative aspects

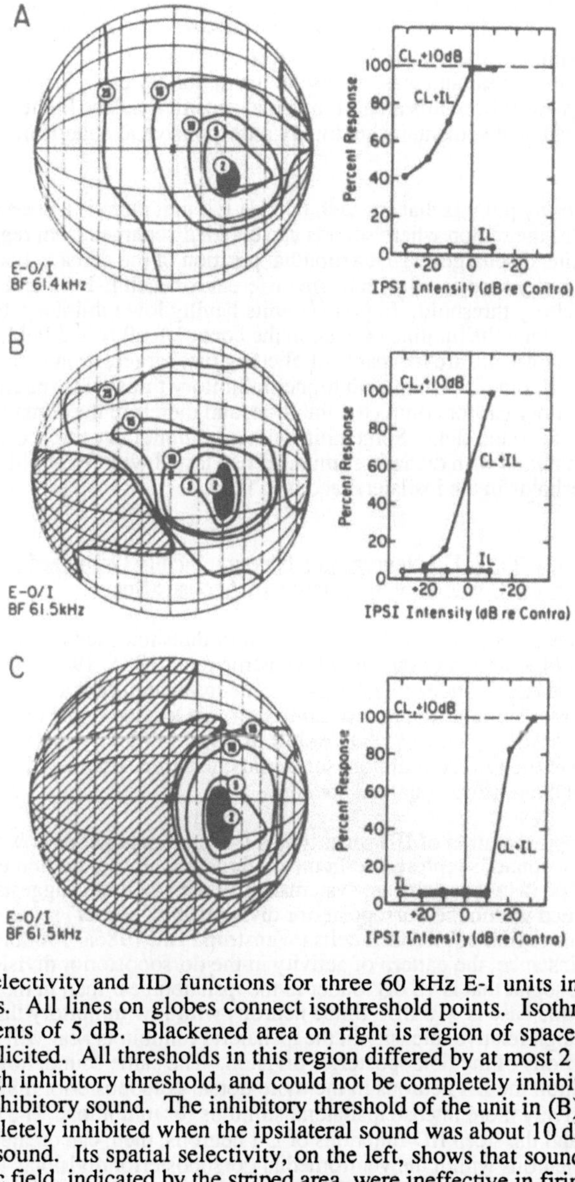

Fig. 17. Spatial selectivity and IID functions for three 60 kHz E-I units in the mustache bat's inferior colliculus. All lines on globes connect isothreshold points. Isothreshold contours are drawn in increments of 5 dB. Blackened area on right is region of space from which lowest thresholds were elicited. All thresholds in this region differed by at most 2 dB. The uppermost unit (A) had a high inhibitory threshold, and could not be completely inhibited regardless of the intensity of the inhibitory sound. The inhibitory threshold of the unit in (B) was lower, and the neuron was completely inhibited when the ipsilateral sound was about 10 dB more intense than the contralateral sound. Its spatial selectivity, on the left, shows that sounds in portions of the ipsilateral acoustic field, indicated by the striped area, were ineffective in firing the unit even with intensities of over 100 dB SPL. Similar arguments apply to the unit in (C). The inhibitory threshold of this unit was even lower than for the unit in B. Correspondingly, the region of space from which sounds could not evoke discharges in this unit was also expanded in both the ipsilateral and into the contralateral acoustic field. From Fuzessery and Pollak (1985).

of binaural properties could be directly associated with the neuron's spatially selective properties.

The binaural properties of three 60 kHz E-I units and their receptive fields are shown in Fig. 17. The first noteworthy point is that the spatial position at which each unit had its lowest threshold was the same among all 60 kHz E-I units, at about -40° azimuth and 0° elevation. This location corresponds to the position in space where the largest interaural intensity disparities are generated, i.e., the position in space at which the sound is always most intense in the excitatory ear and least intense in the inhibitory ear. The thresholds increase almost as circular rings away from this area of maximal sensitivity. The highest thresholds are in the ipsilateral sound field, and some units are totally unresponsive to sound emanating from these regions of space.

The second noteworthy point is that for each 60 kHz E-I unit there is a position along the azimuth that demarcates the region where sounds can evoke discharges from regions where sounds are ineffective in evoking discharges. That azimuthal position of the demarcation, and the interaural intensity disparity associated with it, are different for each E-I cell and correlates closely with the neuron's inhibitory threshold. In 60 kHz units having low inhibitory thresholds, the demarcating loci occur along the midline or even in the contralateral sound field, and sounds presented ipsilateral to those loci are incapable of eliciting discharges, even with intensities as high as 110 dB SPL (Fig. 17B and C). Units with higher inhibitory thresholds require a more intense stimulation of the inhibitory ear for complete inhibition, and therefore the demarcating loci of these units are in the ipsilateral sound field. Some units with high inhibitory thresholds (Fig. 17A) could never be completely inhibited with dichotic stimuli. When tested with free-field stimulation these units display high thresholds in the ipsilateral acoustic field.

Inhibitory Thresholds of 60 kHz E-I Neurons are Topographically Organized Within the Dorsoposterior Division and Create a Representation of Acoustic Space

A finding of particular importance is that the inhibitory thresholds are topographically arranged within the ventromedial E-I region of the DPD (Wenstrup et al. 1985; 1986a) (Fig. 18). E-I neurons with high, positive inhibitory thresholds (i.e., neurons requiring a louder ipsilateral stimulus than contralateral stimulus to produce inhibition) are located in the dorsal E-I region. Subsequent E-I responses display a progressive shift to lower inhibitory thresholds. The most ventral E-I neurons have the lowest inhibitory thresholds; they are suppressed by ipsilateral sounds equal to or less intense than the contralateral sounds.

The topographic representation of IID sensitivities has implications for how the azimuthal, i.e., horizontal, position of a sound is represented in the mammalian inferior colliculus. The fact that intensity disparities of 60 kHz tones change systematically with azimuth suggests that the value of an IID can be represented within the dorsoposterior division as a "border" separating a region of discharging cells from a region of inhibited cells (Wenstrup et al. 1986a, 1988b; Pollak et al. 1986). Consider, for instance, the pattern of activity in the dorsoposterior division on one side generated by a 60 kHz sound that is 15 dB louder in the ipsilateral ear than in the contralateral ear (Fig. 19). The IID in this case is +15 dB. Since neurons with low inhibitory thresholds are situated ventrally, the high relative intensity in the ipsilateral (inhibitory) ear will inhibit all the E-I neurons in ventral portions of the dorsoposterior division. The same sound, however, will not be sufficiently loud to inhibit the E-I neurons in the more dorsal dorsoposterior division, where neurons require a relatively more intense ipsilateral stimulus for inhibition. The topology of inhibitory thresholds and the steep IID functions of E-I neurons, then, can create a border between excited and inhibited neurons within the ventromedial dorsoposterior division. The locus of the border, in turn, should shift with changing IID, and therefore should shift correspondingly with changing sound location, as shown in Fig. 19.

One Group of E-E Neurons Code for Elevation Along the Midline

A population of E-E units has been found that are most sensitive to sounds presented close to or along the vertical midline. However, the elevation at which these cells are most sensitive is

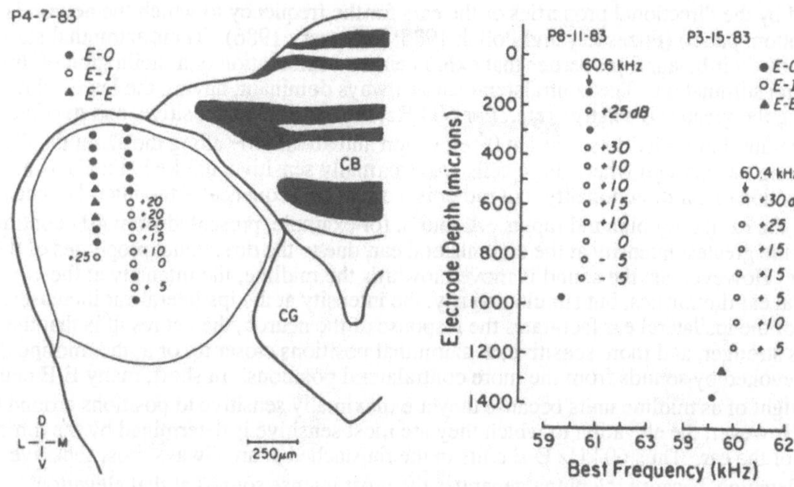

Fig. 18. Left Panel: Systematic shift in the inhibitory thresholds of 60 kHz E-I unit clusters, shown in a transverse section through the inferior colliculus of the mustache bat. The value of the inhibitory threshold at each locus is indicated at right. Positive values of the inhibitory threshold indicate that the sound at the inhibitory ear had to be louder than the sound at the excitatory ear to produce a 50% suppression of the response evoked by the stimulus at the excitatory ear. All unit clusters were sharply tuned to 63.1-64.0 kHz. Right Panel: Systematic decrease in inhibitory thresholds of unit clusters with depth in two dorsoventral electrode penetrations from different mustache bats. Inhibitory thresholds are shown at left of each recording locus. (From Wenstrup et al. 1985)

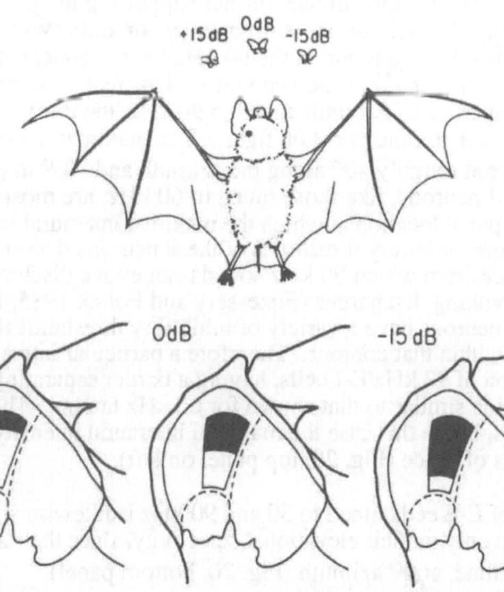

Fig. 19. Schematic illustration of the relationship between the value of the interaural intensity disparity produced by a sound source at a given location (moths at top of figure) and the pattern of activity in the ventromedial E-I region of the left dorsoposterior division, where IID sensitivities are topographically organized. The activity in this region, indicated by the blackened area, spreads ventrally as a sound source moves from the ipsilateral to the contralateral sound field. From Wenstrup et al. (1986).

determined by the directional properties of the ears for the frequency to which the neuron is tuned (Fig. 20 bottom panel) (Fuzessery and Pollak 1985; Fuzessery 1986). Their azimuthal selectivities are shaped by their binaural properties that exhibit either a summation or a facilitation of discharges with binaural stimulation. The contralateral ear is always dominant, having the lowest threshold and evoking the greatest discharge rate. For 60 kHz, the ear is most sensitive, and generates the greatest interaural intensity disparities at 0° elevation and about -40° along the azimuth. The position along the azimuth where these cells are maximally sensitive, unlike E-I cells, is not so much a function of ear directionality, but rather is a direct consequence of the interplay between the potency of the excitatory binaural inputs. A sound, for example, presented from 40° contralateral will create the greatest intensity at the contralateral ear, due to the directional properties of the ear for 60 kHz. However, as the sound is moved towards the midline, the intensity at the contralateral ear diminishes, but simultaneously, the intensity at the ipsilateral ear increases. Since excitation of the ipsilateral ear facilitates the response of the neuron, the net result is that the response is stronger, and more sensitive, at azimuthal positions closer to, or at the midline then are responses evoked by sounds from the more contralateral positions. In short, many E-E neurons can be thought of as midline units because they are maximally sensitive to positions around 0° azimuth. However, the elevation to which they are most sensitive is determined by the directional properties of the ear. Thus 60 kHz E-E units in the mustache bat are always most sensitive at about 0° elevation, because the pinna generates the most intense sounds at that elevation.

Spatial Properties of Neurons Tuned to Other Frequencies

The chief difference among neurons tuned to other harmonics of the mustache bat's echolocation calls, at 30 and 90 kHz, is that their spatial properties are expressed in regions of space that differ from 60 kHz neurons, a consequence of the directional properties of the ear for those frequencies (Fig. 20, top, panel) (Fuzessery and Pollak 1985; Fuzessery 1986). The binaural processing of neurons tuned to those frequencies, however, is essentially the same as those described for 60 kHz neurons. Moreover, there is even evidence for an orderly representation of interaural intensity disparities in the 90 kHz isofrequency laminae, further supporting the generality of a topology of inhibitory thresholds among E-I cells across isofrequency contours (Wenstrup et al. 1986b). In short, the sort of binaural processing found in the 60 kHz lamina appears to be representative of binaural processing within other isofrequency contours. This feature is most readily appreciated by considering the spatial properties of E-I units tuned to 90 kHz, the third harmonic of the mustache bat's orientation calls (Fig. 20, middle panel on right). The maximal interaural intensity disparities generated by 90 kHz occur at roughly 40° along the azimuth and -40° in elevation (Fig. 20 top panel on right). The 90 kHz E-I neurons, like those tuned to 60 kHz, are most sensitive to sounds presented from the same spatial location at which the maximal interaural intensity disparities are generated. Additionally, the inhibitory thresholds of these neurons determine the azimuthal border defining the region in space from which 90 kHz sounds can evoke discharges from the region where sounds are incapable of evoking discharges (Fuzessery and Pollak 1985; Fuzessery et al. 1985). The population of 90 kHz E-I neurons have a variety of inhibitory thresholds that appear to be topographically arranged within that contour. Therefore a particular interaural intensity disparity will be encoded by a population of 90 kHz E-I cells, having a border separating the inhibited from the excited neurons, in a fashion similar to that shown for 60 kHz lamina. The same argument can be applied to the 30 kHz cells, but in this case the maximal interaural intensity disparity is generated from the very far lateral regions of space (Fig. 20, top panel on left).

The spatial behavior of E-E cells tuned to 30 and 90 kHz is likewise similar to those tuned to 60 kHz. The distinction is only in their elevational selectivity, since their azimuthal sensitivities are for sounds around the midline, at 0° azimuth (Fig. 20, bottom panel).

The Representation of Auditory Space in the Mustache bat's Inferior Colliculus

We can now begin to see how the cues for azimuth and elevation are derived. The directional properties of the ears generate different interaural intensity disparities among frequencies at a particular location. The neural consequence is a specific pattern of activity across each isofrequency contour in the bat's midbrain. Figure 21 shows a stylized illustration of the interaural intensity disparities generated by 30, 60 and 90 kHz within the bat's hemifield, and below is shown the loci of borders that would be generated by a biosonar signal containing the three harmonics emanating from different regions of space. Consider first a sound emanating from 40° along the azimuth and 0° elevation. This position creates a maximal interaural intensity disparity at 60 kHz, a lesser interaural intensity disparity at 30 kHz and an interaural intensity disparity close to

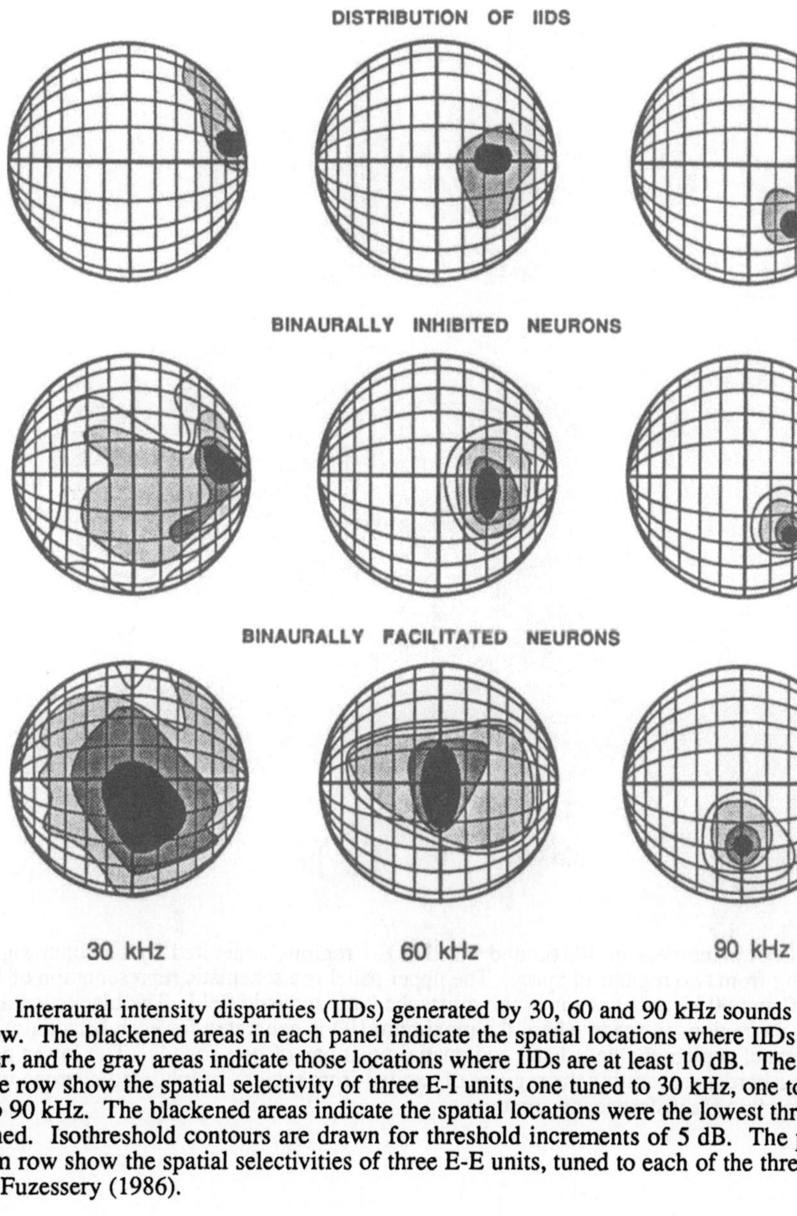

Fig. 20. Interaural intensity disparities (IIDs) generated by 30, 60 and 90 kHz sounds are shown in top row. The blackened areas in each panel indicate the spatial locations where IIDs are 20 dB or greater, and the gray areas indicate those locations where IIDs are at least 10 dB. The panels in the middle row show the spatial selectivity of three E-I units, one tuned to 30 kHz, one to 60 kHz and one to 90 kHz. The blackened areas indicate the spatial locations were the lowest thresholds were obtained. Isothreshold contours are drawn for threshold increments of 5 dB. The panels in the bottom row show the spatial selectivities of three E-E units, tuned to each of the three harmonics. From Fuzessery (1986).

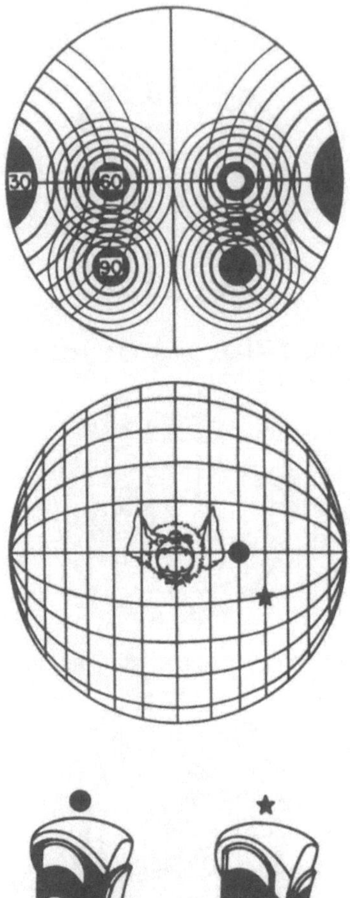

Fig. 21. Loci of borders in 30, 60 and 90 kHz E-I regions generated by biosonar signals emanating from two regions of space. The upper panel is a schematic representation of IIDs of 30, 60 and 90 kHz that occur in the mustache bat's acoustic field. The blackened areas indicate the regions in space where the maximum IID is generated for each harmonic. The middle panel depicts the bat's head and spatial positions of two sounds. The lower panel shows the borders separating the regions of excited from regions of inhibited neurons in the 30, 60 and 90 kHz isofrequency contours.

0 dB at 90 kHz. The borders created within each of the isofrequency contours by these interaural intensity disparities are shown in Fig. 21 (bottom panel). Next consider the interaural intensity disparities created by the same sound but from a slightly different position in space, at about 45° azimuth and -20° elevation. In this case there is a decline in the 60 kHz interaural intensity disparity, an increase in the 90 kHz interaural intensity disparity, but the 30 kHz interaural intensity disparity will be the same as it was when the sound emanated form the previous position. The constant interaural intensity disparity at 30 kHz is a crucial point, and it occurs because for a given frequency an interaural intensity disparity is not uniquely associated with one position in space, but rather can be generated from a variety of positions. It is for this reason that the accuracy with which a sound can be localized with one frequency is ambiguous (Blauert 1969/70; Butler 1974; Musicant and Butler 1985; Fuzessery 1986). However, spatial location, in both azimuth and elevation, is rendered unambiguous by the simultaneous comparison of three interaural intensity disparities, because their values in combination are uniquely associated with a spatial location.

This representation becomes ineffective along the vertical midline, at 0° azimuth, where the interaural intensity disparities will be 0 dB at all frequencies and all elevations. The borders among the E-I populations will thus not change with elevation because the interaural intensity disparities remain constant. The population of E-E neurons tuned to different frequencies may be important in this regard (Fuzessery and Pollak 1984, 1985; Fuzessery 1986). E-E units tuned to different frequencies exhibit selectivities for different elevations, and thus as the elevation along the vertical meridian shifts, the strength of responding also shifts for E-E units as a function of their frequency tuning. In combination then, the two major binaural types provide a neural representation of sound located anywhere within the bat's acoustic hemifield.

A striking feature of the above scenario is its similarity, in principle, to the ideas about sound localization proposed previously by Pumphery (1948) and Grinnell and Grinnell (1965). What these recent studies provide are the details of how interaural disparities are encoded and how they are topologically represented in the acoustic midbrain.

CONCLUSIONS

The overrepresentation and sharp tuning of 60 kHz neurons are the most prominent adaptations of the mustache bat's auditory system. It is noteworthy that these adaptations are functionally similar to aspects of the visual and somatosensory systems. In both of these sensory systems the portions of the sensory surface subserving fine discrimination magnified in the central nervous system (Hubel and Wiesel 1977; Sur et al. 1978). The analogy with vision is even clearer when the integrated actions of both sensory and motor systems are considered. In vision, changing the direction of gaze keeps images of interest fixated on the fovea. Similarly, Doppler compensation ensures that acoustic signals of interest are processed by a region of the sensory surface innervated by a large number of primary fibers having exceptionally narrow tuning curves. The resonant segment of the bat's cochlea, then, is essentially an "acoustic fovea", and the central representations of that region, such as the dorsoposterior division, can be thought of as the foveal regions of the auditory system (Schuller and Pollak 1979; Pollak and Casseday 1988).

The acoustic features requiring fine resolution are for prey recognition and identification. With a constant frequency sonar system, a fluttering insect's "acoustic signature" is conveyed by the particular modulation patterns in the echo CF component, and is displayed prominently in the inferior colliculus by a unique spatiotemporal pattern of activity across the dorsoposterior division. Although the chief function of the expanded frequency representation in long CF/FM bats is to represent an insect's acoustic profile, it is not the only attribute represented in the dorsoposterior division. The reception of the echo CF component at the two ears creates an interaural intensity disparity whose value is a function of the spatial position of the target. The coding of sound location is represented, at least in part, by the locus of the border separating populations of E-I neurons discharging in register with the modulation waveforms from inhibited populations of E-I neurons in the ventromedial dorsoposterior division.

These experiments have shown how features of biologically important acoustic stimuli are encoded by the nervous system. The dorsoposterior division has been of value in these studies because its specialized features provide an enlarged "picture window" through which we obtain a magnified view of how multiple attributes of the external world are represented in the mammalian auditory midbrain.

Acknowledgment

I thank Carl Resler for technical assistance and Janet Young for much of the artwork. Supported by grant NS 21826 from the Public Health Service.

REFERENCES

Aitkin, L.M. (1985) The Auditory Midbrain: Structure and Function in the Central Auditory Pathway. Humana Press, Clifton, New Jersey.
Bateman, G.C. and Vaughan, T.A. (1974) Nightly activities of mormoopid bats. J. Mammal. 55: 45-65.
Blauert, J. (1969/1970) Sound localization in the median plane. Acoustica 22: 205-213.
Bodenhamer, R.D. and Pollak, G.D. (1983) Response characteristics of single units in the inferior colliculus of mustache bats to sinusoidally frequency modulated signals. J. Comp. Physiol. 153: 67-79.
Boudreau, J.C. and Tsuchitani, C. (1968) Binaural interaction in the cat superior olive S segment. J. Neurophysiol. 31: 445-454
Brunso-Bechtold, J.K., Thompson, G.C. and Masterson,R.B. (1981) HRP study of the organization of auditory afferents ascending to the central nucleus of the inferior colliculus in the cat. J. Comp. Neurol. *97*: 705-722.
Butler, R.A. (1974) Does tonotopy subserve the perceived elevation of a sound? Federation Proc. 33: 1920-1923.
Feng, A.S. and Vater, M. (1985) Functional organization of the cochlear nucleus of rufous horseshoe bats (*Rhinolophus rouxi*): frequencies and internal connections are arranged in slabs. J. Comp. Neurol. 235:529-553.
FitzPatrick, K.A. (1975) Cellular architecture and topographic organization of the inferior colliculus of the squirrel monkey. J. Comp. Neurol. 164: 185-208.
Fuzessery, Z.M. and Pollak, G.D. (1984) Neural mechanisms of sound localization in an echolocating bat. Science 225: 725-728.
Fuzessery, Z.M. and Pollak, G.D. (1985) Determinants of sound location selectivity in bat inferior colliculus: A combined dichotic and free-field stimulation study. J. Neurophysiol. 54: 757-781.
Fuzessery, Z.M., Wenstrup, J.J. and Pollak, G.D. (1985) A representation of horizontal sound location in the inferior colliculus of the mustache bat (*Pteronotus p. parnellii*). Hearing Res. 20: 85-89.
Fuzessery, Z.M. (1986) Speculations on the role of frequency in sound localization. Brain Behav. Evol. 28: 95-108.
Galambos, R. and Griffin, D.R. (1942) Obstacle avoidance by flying bats. J. Exp. Zool. 89: 475-490.
Goldberg, J.M. (1975) Physiological studies of the auditory nuclei of the pons. In: Keidel, W.D. and Neff, W.D. (eds.), Handbook of Sensory Physiology, Vol. V. Auditory System, Part 2. Springer-Verlag, New York, pp. 109-144.
Goldman, L.J. and Henson, O.W., Jr. (1977) Prey recognition and selection by the constant frequency bat, *Pteronotus p. parnellii*. Behav. Ecol. Sociobiol. 2:411-419.
Griffin, D.R. (1944) Echolocation by blind men and bats. Science 100: 589-590.
Griffin, D.R. (1958) Listening in the Dark. Yale University Press, New Haven, Conn.
Griffin, D.R. and Galambos, R. (1941) The sensory basis of obstacle avoidance by flying bats. J. Exp. Zool. 86: 481-506.
Grinnell, A.D. (1963a) The neurophysiology of audition in bats: Intensity and frequency parameters. J. Physiol. 167: 38-66.
Grinnell, A.D. (1963b) The neurophysiology of audition in bats: Temporal parameters. J. Physiol. 167: 67-96.
Grinnell, A.D. (1963c) The neurophysiology of audition in bats: Directional localization and binaural interactions. J. Physiol. 167: 97-113.
Grinnell, A.D. (1963d) The neurophysiology of audition in bats: Resistance to interference. J. Physiol. 167: 114-127.
Grinnell, A.D. (1967) Mechanisms of overcoming interference in echolocating animals. In: Busnel R-G (ed) Animal Sonar Systems, Vol I. Laboratoire de Physiologie Acoustique, Jouy-en-Josas 78, France, p. 451-481.

Grinnell, A.D. and Grinnell V.S.(1965) Neural correlates of vertical localization by echolocating bats. J. Physiol. 181: 830-851.
Harnischfeger, G., Neuweiler, G. and Schlegel, P. (1985) Interaural time and intensity coding in the superior olivary complex and inferior colliculus of the echolocating bat, *Molossus ater*. J. Neurophysiol. 53: 89-109.
Henson, M.M. (1978) The basilar membrane of the bat, *Pteronotus parnellii*. Anat. Rec. 153: 143-158.
Henson, O.W., Jr., Henson, M.M., Kobler, J.B. and Pollak, G.D. (1980) The constant frequency component of the biosonar signals of the bat, *Pteronotus p. parnellii*. In: Busnel, R.-G. and Fish, J.F. (eds), Animal Sonar Systems, Plenum Press, New York, p. 913-916.
Henson, O.W., Jr., Pollak, G.D., Kobler, J.B., Henson, M.M. and Goldman, L.J. (1982) Cochlear microphonics elicited by biosonar signals in flying bats, *Pteronotus p. parnellii*. Hearing Res. 7: 127-147.
Henson, O.W., Jr., Schuller, G. and Vater, M. (1985) A comparative study of the physiological properties of the inner ear in Doppler shift compensating bats (*Rhinolophus rouxi* and *Pteronotus parnellii*). J. Comp. Physiol. 157: 587-607.
Hubel, D.H. and Wiesel, T.N. (1977) Functional architecture of macaque monkey visual cortex. Ferrier Lecture, Proc. R. Soc. Lond. 198: 1-59.
Irvine, D.R.F. (1986) The Auditory Brainstem. Progress in Sensory Physiology 7, Autrum, H. and Ottoson, D. (eds.), Springer-Verlag, Berlin-Heidelberg.
Kossl, M. and Vater, M. (1985) The frequency place map of the bat, *Pteronotus parnellii*. J. Comp. Physiol. 157: 687-697.
Link, A., Marimuthu, G. and Neuweiler, G. (1986) Movement as a specific stimulus for prey catching behavior in rhinolophid and hipposiderid bats. J. Comp. Physiol. 159: 403-413.
Merzenich, M.M. and Reid, M.D. (1974) Representation of the cochlea within the inferior colliculus of the cat. Brain Res. 77: 397-415.
Musicant, A.D. and Butler, R.A. (1984) The psychophysical basis of monaural localization. Hearing Res. 14: 185-190
Neuweiler, G. (1980) Auditory processing of echoes: Peripheral processing. In: Busnel, R.-G. and Fish, J.F. (eds) Animal Sonar Systems. Plenum Press, New York, p. 519-548.
Neuweiler, G. (1983) Echolocation and adaptivity to ecological constraints. In: Huber, F., Markl, H. (eds) Neuroethology and Behavioral Physiology: Roots and Growing Pains. Springer-Verlag, Berlin Heidelberg New York Tokyo, p. 280-295.
Neuweiler, G. (1984a) Auditory basis of echolocation in bats. In: Bolis, L., Keynes, R.D., Maddrell S.H.P. (eds) Comparative Physiology of Sensory Systems. Cambridge University Press, Cambridge, p. 115-141.
Neuweiler, G. (1984b) Foraging, echolocation and audition in bats. Naturwissenschaften 71: 46-455.
Novick, A. and Vaisnys, J.R. (1964) Echolocation of flying insects by the bat, *Chilonycteris parnellii*. Biol. Bull. 127: 478-488.
Oliver, D.L. and Morest, D.K. (1984) The central nucleus of the inferior colliculus in the cat. J. Comp. Neurol. 222: 237-264.
Pollak, G.D., Henson, O.W., Jr. and Johnson, R. (1979) Multiple specializations in the peripheral auditory system of the CF-FM bat, *Pteronotus parnellii*. J. Comp. Physiol. 131: 255-266
Pollak, G.D. (1980) Organizational and encoding features of single neurons in the inferior colliculus of bats. In: Busnel, R.-G. and Fish, J.F. (eds) Animal Sonar Systems. Plenum Press, New York, p. 549-587.
Pollak, G.D. and Bodenhamer, R.D. (1981) Specialized characteristics of single units in inferior colliculus of mustache bat: frequency representation, tuning, and discharge patterns. J. Neurophysiol. *46*: 605-619.
Pollak, G.D. and Schuller, G. (1981) Tonotopic organization and encoding features of single units in the inferior colliculus of horseshoe bats: Functional implications for prey identification. J. Neurophysiol. 45: 208-226.
Pollak, G.D. and Casseday, J.H. (1989) The Neural Basis of Echolocation in Bats. Springer-Verlag, New York, (in press).
Pollak, G.D., Henson, O.W., Jr. and Novick, A. (1972) Cochlear microphonic audiograms in the pure tone bat, *Chilonycteris parnellii parnellii*. Science 176: 66-68.
Pollak, G.D., Bodenhamer, R.D. and Zook, J.M. (1983) Cochleotopic organization of the mustache bat's inferior colliculus. In J.-P. Ewert, R.R. Capranica and D.J. Ingle (eds): Advances in Vertebrate Neuroethology. New York, Plenum Press, p. 925-935.
Pollak, G.D., Wenstrup, J.J. and Fuzessery, Z.M. (1986) Auditory processing in the mustache bat's inferior colliculus. Trends in Neurosci. 9: 556-561.
Pumphery, R.J. (1947) The sense organs of birds. Ibis 90: 171-199.
Rockel, A.S. and Jones, E.G. (1973) The neuronal organization of the inferior

colliculus of the adult cat. I. The central nucleus . J. Comp. Neurol. 147: 11-60.
Ross, L.S., Pollak, G.D. and Zook, J.M. (1988) Origin of ascending projections to an isofrequency region of the mustache bat's inferior colliculus. J. Comp. Neurol. 270: 488-505.
Ross, L.S. and Pollak, G.D. (1988) Differential projections to aural regions in the 60 kHz isofrequency contour of the mustache bat's inferior colliculus. J. Neurosci. (in press).
Roth, G.L., L.M. Aitkin, R.A. Andersen, and M.M. Merzenich (1978) Some features of the spatial organization of the central nucleus of the inferior colliculus of the cat. J. Comp. Neurol. *182*: 661-680.
Semple, M.N. and Aitkin, L.M. (1979) Representation of sound frequency and laterality by units in the central nucleus of the cat's inferior colliculus. J. Neurophysiol. *42*: 1626-1639.
Serviere, J., Webster, W.R. and Calford, M.B. (1984) Iso-frequency labelling revealed by a combined [^{14}C]-2-deoxyglucose, electrophysiological and horseradish peroxidase study of the inferior colliclus of the cat. J. Comp. Neurol. 228: 463-477.
Schnitzler, H.-U. (1967) Discrimination of thin wires by flying horseshoe bats(Rhinolophidae). In: Busnel, R.-G. (ed) Animal Sonar Systems, Vol I. Laboratoire de Physiologie Acoustique, Jouy-en-Josas 78, France, p. 68-87.
Schnitzler, H.-U. (1970) Comparison of echolocation behavior in *Rhinolophus ferrumequinum* and *Chilonycteris rubiginosa*.. Bijdr Dierkd 40: 77-80.
Schnitzler, H.-U. and Flieger, E. (1983) Detection of oscillating target movements by echolocation in the greater horseshoe bat. J. Comp. Physiol. 153: 385-391.
Schnitzler, H.-U. and Ostwald, J. (1983) Adaptations for the detection of fluttering insects by echolocation in horseshoe bats. In: Ewert, J.-P., Capranica, R.R. and Ingle, D.J. (eds) Advances in Vertebrate Neuroethology. Plenum Press, New York, p. 801-827.
Schnitzler, H.-U., Menne, D., Kober, R. and Heblich, K. (1983) The acoustical image of fluttering insects in echolocating bats. In: Huber, F., Markl, H. (eds) Neuroethology and Behavioral Physiology: Roots and Growing Pains. Springer-Verlag, Berlin Heidelberg New York Tokyo, p. 235-250.
Schuller, G. (1979a) Coding of small sinusoidal frequency and amplitude modulations in the inferior colliculus of the CF-FM bat, *Rhinolophus ferrumequinum* . Exp. Brain Res. 34: 117-132.
Schuller, G. (1979b) Vocalization influences auditory processing in collicular neurons of the CF-FM bat, *Rhinolophus ferrumequinum* . J. Comp. Physiol. 132: 39-46.
Schuller, G. (1984) Natural ultrasonic echoes form wing beating insects are coded by collicular neurons in the CF-FM bat, *Rhinolophus ferrumequinum*. J. Comp. Physiol. 155:121-128.
Schuller, G. and Pollak, G.D. (1979) Disproportionate frequency representation in the inferior colliculus of horseshoe bats: Evidence for an "acoustic fovea". J. Comp. Physiol. 132: 47-54.
Schuller, G., Beuter, K. and Schnitzler, H.-U. (1974) Response to frequency shifted artificial echoes in the bat, *Rhinolophus ferrumequinum*. J. Comp. Physiol. 89: 275-286.
Simmons, J.A. (1971) The sonar receiver of the bat. Ann. NY. Acad. Sci. 188: 161-184.
Simmons, J.A. (1973) The resolution of target range by echolocating bats. J. Acoust. Soc. Amer. 54: 157-173.
Simmons, J.A. (1974) Response of the Doppler echolocation system in the bat, *Rhinolophus ferrumequinum*. J. Acoust. Soc. Amer. 56: 672-682.
Simmons, J.A., Howell, D.J. and Suga, N. (1975) Information content of bat sonar echoes. Amer. Sci. 63: 16-21.
Suga, N. (1964a) Recovery cycles and responses to frequency modulated tone pulses in auditory neurons of echolocating bats. J. Physiol. 175: 50-80.
Suga, N. (1964b) Single unit activity in the cochlear nucleus and inferior colliculus of echolocating bats. J. Physiol. 172: 449-474.
Suga, N. (1978) Specialization of the auditory system for reception and processing of species-specific sounds. Fed. Proc. 37: 2342-2354.
Suga, N. (1984) The extent to which biosonar information is represented in the bat auditory cortex. In: Edelman, G.M., Gall, W.E. and Cowan, W.M. (eds) Dynamic Aspects of Neocortical Function. John Wiley & Sons, New York, pp 315-374.
Suga, N. and Jen, P.H.-S. (1977) Further studies on the peripheral auditory system of "CF-FM" bats specialized for the fine frequency analysis of Doppler-shifted echoes. J. Exp. Biol. 69: 207-232.
Suga, N., Simmons, J.A. and Jen, P.H.-S. (1975) Peripheral specializations for fine frequency analysis of Doppler-shifted echoes in the CF-FM bat, *Pteronotus parnellii*. J. Exp. Biol. 63: 161-192.

Suga, N., Neuweiler, G. and Moller, J. (1976) Peripheral auditory tuning for fine frequency analysis by the CF-FM bat, *Rhinolophus ferrumequinum*. IV Properties of peripheral auditory neurons. J. Comp. Physiol. 106: 111-125.

Sur, M., Merzenich, M.M. and Kass, J.H. (1980) Magnification, receptivefield area and "hypercolumn" size in areas 3b and 1 of somatosensory cortex in owl monkeys. J. Neurophysiol. 44: 295-311.

Trappe, M. and Schnitzler, H.-U. (1982) Doppler-shift compensation in insect-catching horseshoe bats. Naturwissenschaften 69: 193-194.

Vater, M. (1987) Narrow-band frequency analysis in bats. In: Fenton MB, Racey P and Rayner JMV (eds) Recent Advances in the Study of Bats. Cambridge University Press, Cambridge, p.. 200-210.

Wenstrup, J.J., Fuzessery, Z.M. and Pollak, G.D. (1986a) Binaural response organization within a frequency-band representation of the inferior colliculus: Implications for sound localization. J. Neurosci. 6: 692-973.

Wenstrup, J.J., Ross, L.S. and Pollak, G.D. (1986b) Organization of IID sensitivity in isofrequency representations of the mustache bat's inferior colliculus. In: IUPS Symposium on Hearing, University of California, San Francisco, CA. Abstract 415.

Wenstrup, J.J., Fuzessery, Z.M. and Pollak, G.D. (1988a) Binaural neurons in the mustache bat's inferior colliculus: I. Responses of 60 kHz EI units to dichotic sound stimulation. J. Neurophysiol. 60: 1369-1383.

Wenstrup, J.J., Fuzessery, Z.M. and Pollak, G.D. (1988b) Binaural neurons in the mustache bat's inferior colliculus: II. Determinants of spatial responses among 60 kHz EI units. J. Neurophysiol. 60: 1384-1404.

Zook, J.M. and Casseday, J.H. (1982) Origin of ascending projections to inferior colliculus in the mustache bat, *Pteronotus parnellii*. J. Comp. Neurol. 207: 14-28.

Zook, J.M. and Casseday, J.H. (1985) Projections from the cochlear nuclei in the mustache bat, *Pteronotus parnellii* . J. Comp. Neurol. 237: 307-324.

Zook, J.M. and Casseday, J.H. (1987) Convergence of ascending pathways at the inferior colliculus of the mustache bat, *Pteronotus parnellii*. J. Comp. Neurol. 261: 347-361.

Zook, J.M. and Leake, P.A. (1988) Correlation of cochlear morphology specializations with frequency representation in the cochlar nucleus and superior olive of the mustache bat, *Pteronotus parnellii*. J. Comp. Neurol. (in press).

Zook, J.M., Winer, J.A., Pollak, G.D. and Bodenhamer, R.D. (1985) Topology of the central nucleus of the mustache bat's inferior colliculus: Correlation of single unit properties and neuronal architecture. J. Comp. Neurol. 231: 530-546.

AUDITORY HABITUATION AND EVOKED POTENTIALS IN A LEARNING RESPONSE

Ayako Shikata[1,3], Yoshihiro Shikata[2],
Takashi Matsuzaki[1], and Satoru Watanabe[3]

[1]Research Section, Matsuzaki Hospital, Osaka, Japan
[2]Department of Mathematics, Nagoya University, Nagoya, Japan
[3]RIEM, Nagoya University, Nagoya, Japan

INTRODUCTION

In humans event-related potentials (ERP) are correlated with psychological processes, underlying language and learning. The slow wave components of the ERP with latencies greater than 50 ms appear to be most closely correlated with these processes. Stuss et al. (1983) have shown that a slow negative wave with a peak latency of 250-500 ms appears when subjects require semantic processing of visual stimuli. Our earlier study (Linke et al. 1980) showed that after semantic auditory stimuli, which are the auditory counterpart of the above visual stimuli, there is a slow negative wave with a peak latency of 350-700 ms. In this paper we term this slow response, the semantic negative (SN). We have observed that at the first presentation of a stimulus (in our experiments the presentation of a foreign word) there is no obvious SN, but upon repetition of the stimulus during the course of an experiment the SN becomes more distinct. Psychological habituation associates a semantic image in the memory with the stimulus, resulting in the SN. This may also be the case for learning a foreign word.

In this paper we calculate the habituation index as a measure of the habituation of the ERP, which is correlated with the psychological habituation that occurs during the learning of a foreign word. We use this index instead of the amplitude of the ERP because the ERP is not very stable. ERP amplitude decreases after repeated stimuli due to a phenomenon known as desensitization, whereas the psychological habituation increases. To obtain the habituation index we first used a norm obtained by integration of the ERP over a time domain corresponding to the SN, the semantic domain, so as to reduce the effect of fluctuations. Then we chose another time domain as a reference and normalized the integrated norm for the semantic domain over that of the reference domain.

Harris (1943) defines habituation as the response decrement resulting from repeated stimulation. Dishabituation is defined as the recovery of the response decrement with the presentation of another stimulus and is one of the characterizing properties of habituation (Humphrey 1930, 1933). Sokolov and Voronin (1960) reported that varying the intensity of an auditory stimulus results in dishabituation of the habituated response. If the variation in the stimulus is infrequent, it generally results in a slow positive wave, termed P300, with a peak latency of about 300 ms associated with the orienting response. Using the "missing paradigm" method in which the less frequent stimulus (rare stimulus) is obtained by omitting part of the regularly delivered stimulus (regular stimulus), Ritter et al. (1968) and Simson et al. (1976) reported the appearance of a slow positive wave P300 in the evoked potential. Presentation of the rare stimulus resulted in both the recovery of the ERP as dishabituation and the P300 of the orienting response. Both effects may be considered equivalent in terms of characterizing habituation.

We have chosen the time domain of the orienting response as the reference domain. We used a foreign word as the regular stimulus and defined the habituation index as the logarithm of the ratio of the norm for the semantic domain to the norm for the P300 domain. Our study attempts to determine whether this index of habituation eliminates possible effects from causes other than habituation and reflects the actual psychological habituation by comparing the index with the accuracy of pronunciation the word that the subjects are learning.

METHODS

Experimental Protocol

The subjects were nine Japanese student nurses, aged 15 to 22. Before the experiment we made sure that they knew no German. The regular stimulus was a German imperative phrase "Licht an" meaning turn on the light; the rare stimulus was "Licht" obtained by leaving off the "an". Before stimulus presentation we presented a conditioning signal, which was a 1000 Hz beep of 240 ms duration. Figure 1 shows the timing of the conditioning and the test signals. These signals were delivered through a headphone at the normal listening level of about 10 dB to subjects lying quietly on a bed. One experiment consisted of five to eight runs; each run contained 20 stimuli, either regular or rare, with at least 3 and at most 6 rare stimuli presented at random. The task of the subject was to count the number of rare stimuli in the run, to report it, and to reproduce the full "Licht an" sentence after each run during the 2-min pause between runs, which lasted about 5 min. We recorded the subject's reproduction of the phrase, and two persons who had lived in Germany for 1 to 3 years judged the subject's pronunciation on a scale of 0 to 5. We used the sum of these scores for each subject as the measure for psychological habituation.

We used Ag-AgCl disk electrodes and picked up the responses between the vertex (Cz) and the frontal (Fz) with the right ear as the indifferent electrode using an NEC San-Ei portable EEG amplifier with a time constant of 10 s and a high cut-off frequency of 25 Hz, according to the results on vertex dominance of slow waves (Goff GD et al. 1977; Goff WR et al. 1977). The amplifier output was connected to an NEC San-Ei 7T17 signal processor computer system with a built-in analog-digital converter, which made it possible to record the responses on floppy disk. We made chart recordings of the EEG, EMG, and EOG to check muscle or eye movement, and if one of these recordings was unusual, we excluded the corresponding ERP from our averaging. Figure 2 shows a block diagram of the experimental set-up.

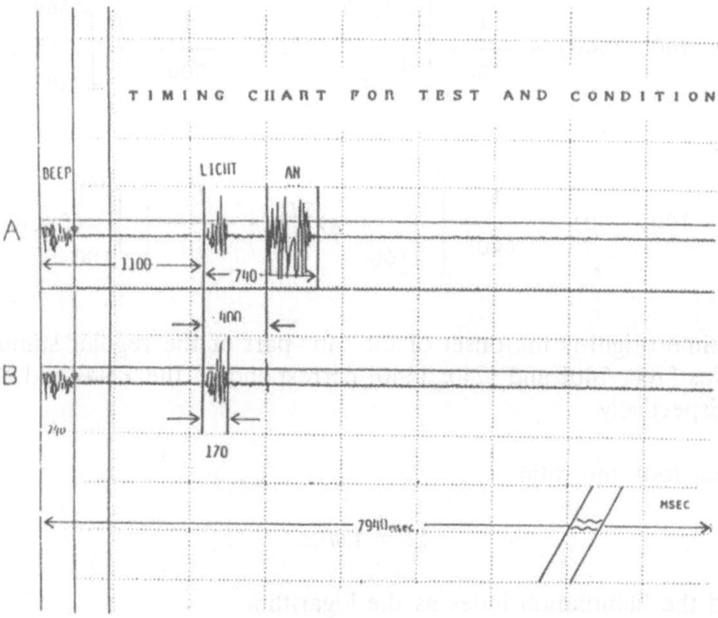

Fig. 1

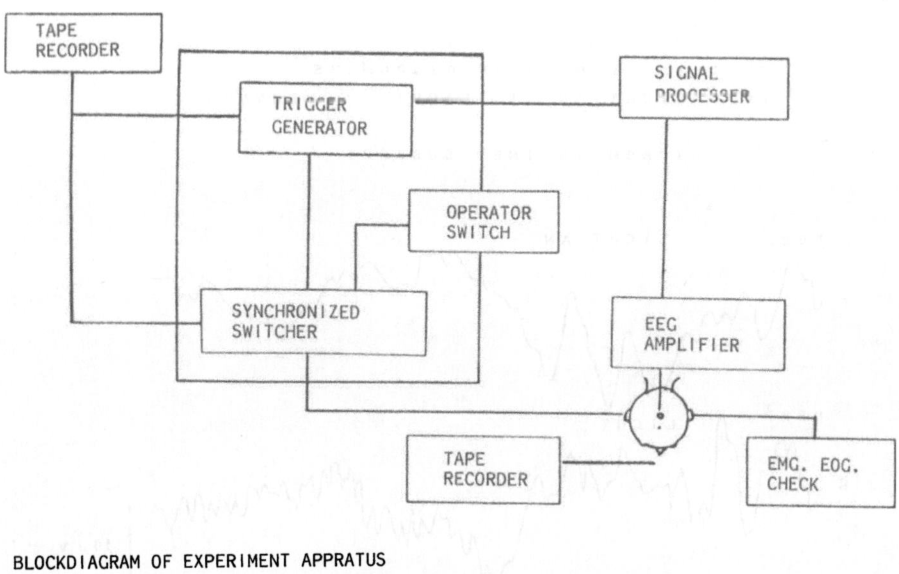

Fig. 2

Data Processing

After reducing the alpha component with a digital filter, we calculated for each run the average evoked potential for the regular stimuli (f) and for the rare stimuli (g). We then computed the norms, Ls and Lo, for the run by the following integration of the difference between the regular and rare stimuli, f-g:

$$\mathrm{Ls{=}L}(f-g:\ 740,\ 1540) = \frac{1}{800}\left[\int_{740}^{1540}(f-g)^2dt - \frac{1}{800}\left\{\int_{740}^{1540}(f-g)d\right\}^2\right]$$

$$\mathrm{Lo{=}L}(f-g:\ 100,\ 740) = \frac{1}{640}\left[\int_{100}^{740}(f-g)^2dt - \frac{1}{640}\left\{\int_{100}^{740}(f-g)dt\right\}^2\right]$$

where the time origin is the onset of the "an" part of the regular stimulus and the time domains [100, 740] and [740, 1540] correspond to the P300 and the semantic domains, respectively.

We then took the ratio:

$$H = Ls/Lo$$

and defined the habituation index as the logarithm

$$h = 20 \log H$$

The processor completed all these computations within 10 s.

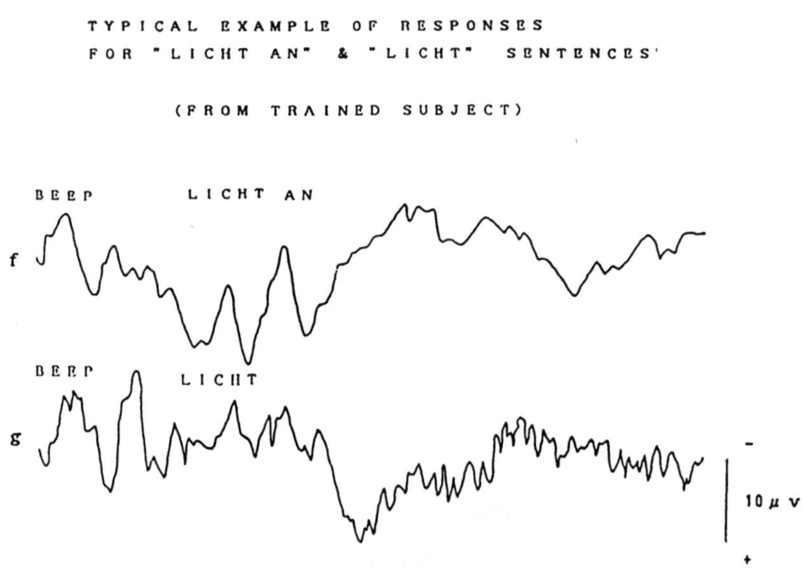

Fig. 3

RESULTS

Figure 3 shows the ERP recorded from one of the judges during a single run. The upper trace is the ERP for the regular stimulus (f) and the lower trace for the rare stimulus (g). The upper trace clearly shows the SN with a peak latency of 700 ms after the offset of the regular stimulus. The habituation index was high.

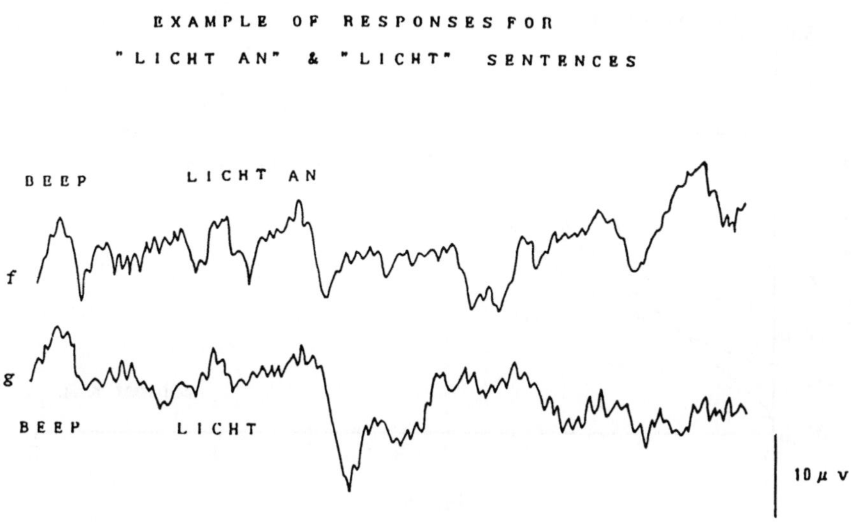

Fig. 4

Figure 4 shows an ERP from an early run from a naive subject. The SN was not clear; the habituation index was negative.

Figures 5 and 6 are plots of the habituation index (solid line marked by squares) and of the pronunciation grades (broken line marked by white dots). Figure 5 is typical for naive subjects, while Fig. 6 is from a subject who had a good command of English. There was good correlation between the index and the pronunciation grade for each subject. The correlation coefficients for the subjects in Figs. 5 and 6 were 0.61 and 0.93, respectively. The correlation coefficient was over 0.5 for all except one subject, but the correlation coefficient across all subjects was only 0.3. Excluding the lowest subject gave a correlation coefficient of 0.34.

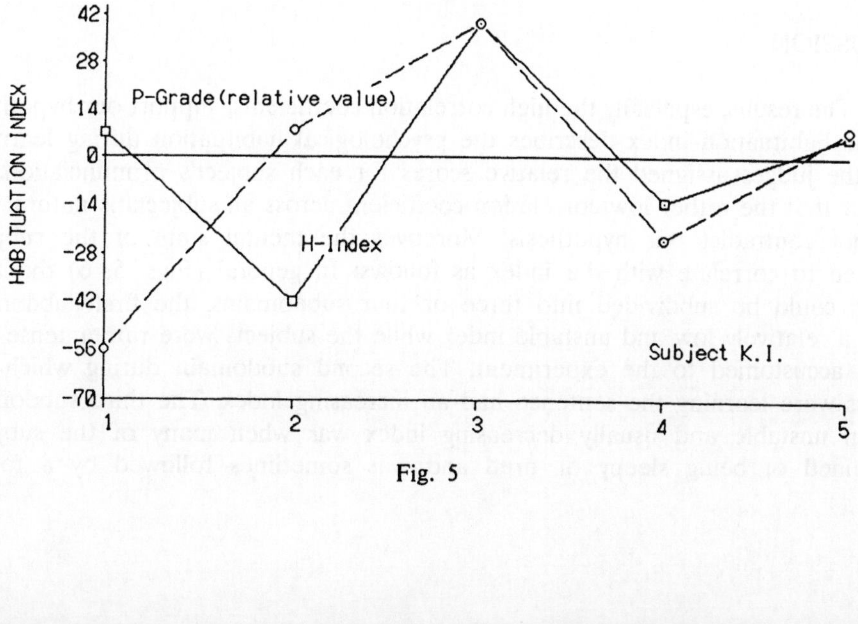

Fig. 5

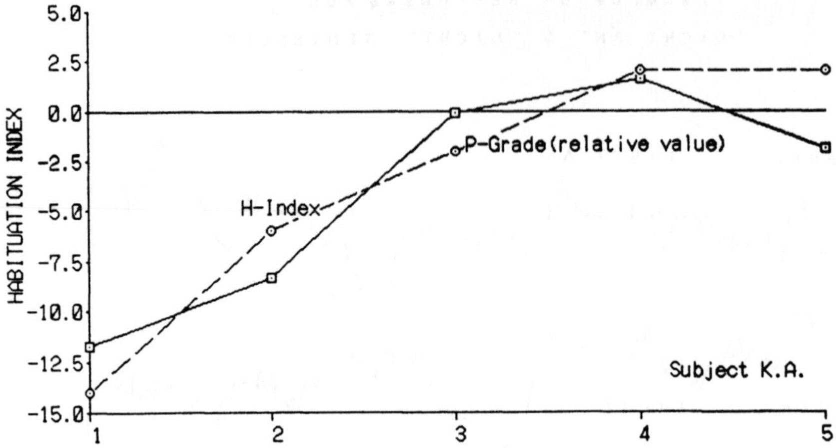

Fig. 6

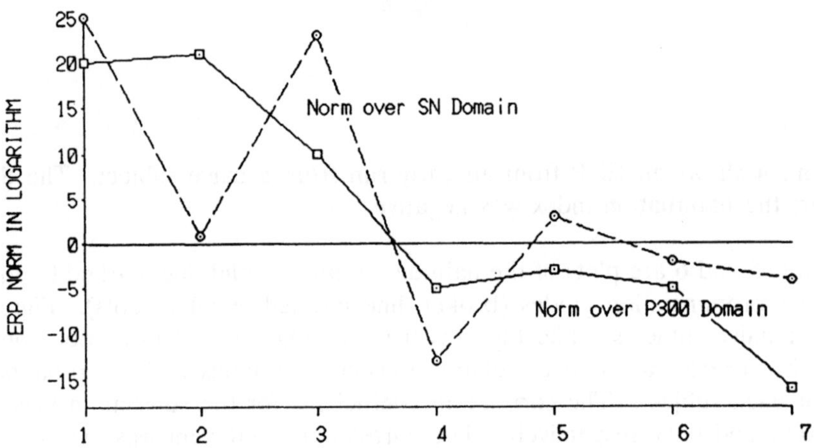

Fig. 7

Figure 7 shows a plot of the norm Ls over the SN domain (broken line marked by white dots) and the norm Lo over the P300 domain (solid line marked by squares) in a logarithmic scale.

DISCUSSION

1. The results, especially the high correlation coefficients, support our hypothesis that the habituation index describes the psychological habituation during learning. Since the judges assigned the relative scores for each subject's pronunciation, we consider that the rather low correlation coefficient across all subjects is natural and does not contradict our hypothesis. Moreover the mental state of the subjects appeared to correlate with the index as follows: In general (Figs. 5, 6) the time domain could be subdivided into three or four subdomains, the first subdomain having a relatively low and unstable index while the subjects were rather tense and getting accustomed to the experiment. The second subdomain during which the subjects were learning the sentence had an increasing index. The third subdomain with an unstable and usually decreasing index was when many of the subjects complained of being sleepy or tired and was sometimes followed by a fourth

subdomain during which the index increased as did accuracy of pronunciation, perhaps corresponding to renewed effort by the subject.

2. In connection with the study of habituation Fruhstofer (1971) reported that a decrease in the amplitude of the primary component of the evoked potential follows a negative exponential curve over time under constant experimental conditions. As Fig. 7 shows, the norms over the SN and the P300 domains also followed negative exponential curves, which may imply that the norm is natural with respect to habituation.

3. The norm used here has the following general expression:

$$L(f: [s, t]) = \frac{1}{s - t}\left[\int_t^s f^2\,dt - \frac{1}{s - t}\left\{\int_t^s f\,dt\right\}^2\right]$$

for time interval [s, t] and response f. It has the following general properties:

L(af,[s, t]) = L(f,[s,t])..................(1)

L(f + a, [s, t]) = L(f,[s, t])..........(2)

for any constant a > 0, response f, and interval [s, t].

Therefore the ratio

L(f,[s, t])/L(f,[s', t'])

is independent of multiplication by or addition of a constant f for any pair of intervals [s, t], [s', t'], indicating that the ratio is independent of any changes in amplifier gain and DC drift that may occur. The ratio can serve as an intrinsic measure for the phenomenon itself. The latter property (2) indicates that the norm can act as a low-pass filter with a suitable choice of the time interval for the integration. Since integration in general cuts down fast responses, the ratio may pick up the desired phenomenon as a stabilized band-pass filter, making possible the application of this method to other problems.

REFERENCES

Fernandez-Peon R (1967) Neurophysiological correlates of habituation and other manifestations of plastic inhibition. Electroencephalogr Clin Neurophysiol (Suppl) 13:101-114

Fruhstofer H (1971) Habituation and dishabituation of the human vertex response. Electroencephalogr Clin Neurophysiol 30:306-313

Goff GD, Matsumiya, Allison T, Goff WR (1977) The scalp topography of human somatosensory and auditory evoked potentials. Electroencephalogr Clin Neurophysiol 42:57-76

Goff WR, Allison T, Lyons W, Fisher TC, Conte R (1977) Origins of short latency auditory evoked potentials in man. In: Desmedt JE (ed) Progress in clinical neurophysiology, vol 2. Karger, Basel, pp 30-40

Harris JD (1943) Habituation response decrement in the intact organism. Psychol Bull 40:385-422

Humphrey G (1930) Extinction and negative adaptation. Psychol Rev 37:361-363

Humphrey G (1933) The nature of learning. Harcourt Brace, New York

Linke D, Shikata Y, Shikata A (1980) Semantik korrelierte cerebrale Potentiale bei der Sprechwahrnehmung. Vortrag auf dem Nervenarzt, Bonn, 1-3 Februar

Prosser S, Arslan E, Michelini S (1981) Habituation and rate effect in the auditory cortical potential evoked trains of stimuli. Arch Otorhinolaryngol 233:179-187

Ritter W, Vaughan HG Jr, Costa LD (1968) Orienting and habituation to auditory stimuli: a study of short term changes in average evoked responses. Electroencephalogr Clin Neurophysiol 25:550-556

Simson R, Vaughan HG Jr, Ritter W (1976) The scalp topography of potentials associated with missing visual or auditory stimuli. Electroencephalogr Clin Neurophysiol 40:3-42

Sokolov EN, Voronin V (1960) Neuronal models and the orienting influence. In: Brazier MAB (ed) The central nervous system and behavior, vol III. Macy Foundation, New York

Stuss DT, Sarazin FF, Leech EE, Picton TW (1983) Event related potentials during naming and mental rotation. Electroencephalogr Clin Neurophysiol 56:133-146

WATER WAVE ANALYSIS WITH THE LATERAL-LINE SYSTEM

Andreas Elepfandt

Faculty of Biology
University of Konstanz
D-7750 Konstanz, W-Germany

ABSTRACT

The organization of water surface wave analysis with the lateral-line system has been investigated by electrophysiological recording and behavioral testing in the clawed frog, Xenopus laevis. In the afferent fibers, wave frequency is encoded by phase coupling of the discharges, while stimulus intensity is encoded over a range of 80 dB by phase coupling and mean firing rate. Additionally, cardioid directional sensitivity is found. Behavioral tests after partial lesions show that small groups of the animal's lateral-line organs are sufficient for localizing waves from any direction; this suggests that the high number of lateral-line organs in Xenopus may serve more complex wave analyses. The capability for complex analysis is demonstrated in the animal's ability to detect the direction and the frequency of component waves in wave superpositions. In the midbrain, a topological organization with regard to wave direction is found that does not reflect the topology of the organs on the body. Wave frequency discrimination is found: its accuracy is comparable to that in hearing, and wave frequency memory is of absolute pitch quality. The results correspond to stimulus processing in the auditory system and give new support to the octavolateralis hypothesis of a common evolutionary origin of the lateral-line and eighth-nerve sensory systems.

INTRODUCTION

The lateral-line system is a mechanoreceptive sensory system of fish, tadpoles, and aquatic adult amphibians - the primary aquatic vertebrates - that is specialized for the detection and analysis of water movements along the animal's body. In some species, parts of the system are modified for electroreception (Bullock and Heiligenberg, 1986) which, however, will not be considered here. The lateral-line organs are distributed as small patches (neuromast organs) over the animal's head and trunk, usually - in particular along the flanks - in some linear arrangement, which gave the system its name. The number of organs comes typically to some 50 to 1000, dependent on the species. Their receptive elements are sensory hair cells capped by a gelatinous flag-shaped matrix, the cupula, that protrudes for fractions of a millimeter into the water. Occasionally, as in the clawed frog Xenopus, *several cupulae are built* per organ (Fig.1). Deflection of the cupula leads to deflection of the sensory hairs which in turn affects the sensory cells'

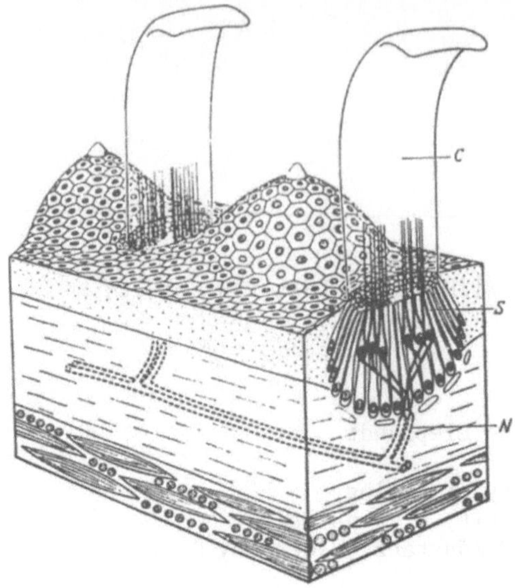

Fig. 1. Diagrammatic section of a lateral-line organ in Xenopus laevis. An organ consists of 4 - 12 groups of sensory hair cells (S) in the epidermis, each group being capped by a flag-shaped cupula (C) that is deflected by impinging waves. The organ's hair cells are divided into two populations of opposite directional sensitivity. Information from each population is conveyed to the medulla by a separate afferent neuron (N) (after Görner, 1963).

receptor potential. The morphology and physiology of the sensory hair cells are the same as in the vertebrate auditory and vestibular systems: on the cell's apical surface one kinocilium stretches lateral to a group of stereocilia, bending the hairs toward the kinocilium causes depolarization of the potential, reverse bending produces hyperpolarization, and hair deflection into other directions affects the potential in correspondence to the cosine between the cell's axis of maximal sensitivity and the direction of the deflection (Russell, 1981; Hudspeth, 1983). Within each organ, the 100 to 600 sensory hair cells are either all depolarized maximally by deflection in the same direction or, in most species, form two populations with mutually opposite polarity. Information about cupular deflection at an organ is conveyed to the medulla by usually two to ten afferent neurons. Since each of these neurons contacts only hair cells of identical polarity, the sinusoidal directional sensitivity of the hair cells is reflected in the afferent response (Görner, 1963).

Whereas the characteristics of the lateral-line organs - which show numerous species-specific variations - have been investigated in some detail (see Russell, 1976, 1981; Hudspeth, 1983; Sand, 1984), knowledge of the system as a whole lags behind. By means of the lateral-line system, water flow along the body is detected as well as the direction of impinging waves, the system is used for spatial analysis of obstacles, in intraspecific communication, for maintaining position in schooling, and several other purposes (see Disler, 1960; Dijkgraaf, 1962; Bleckmann, 1986), but the underlying analytical capabilities of the system and its functional organization have been relatively little explored (Bleckmann, 1985, 1986).

This chapter reports recent insights into the functional properties of the system that have been gained by electrophysiological and behavioral investigations of wave analysis in the clawed frog, Xenopus laevis. This purely aquatic frog retains its lateral-line system after metamorphosis, and it can by means of the system localize and identify impinging water waves (Kramer, 1933; Elepfandt et al., 1985). Four aspects will be considered here: stimulus parameter encoding in the afferent neuron, peripheral and central organization of wave localization, wave frequency analysis, and analysis of wave superpositions. In a final section, parallels to stimulus processing in the auditory system will be discussed that give support to the octavolateralis hypothesis, which claims a common evolutionary origin of the vertebrate auditory and lateral-line systems.

STIMULUS PARAMETER ENCODING IN THE AFFERENT NEURON

Stimulus parameter encoding in the afferent fibers is the basis for central nervous processing: only encoded parameters can be analysed further, and the code's characteristics form constraints for the analysis. The response properties of the lateral-line organs in Xenopus have been studied by several authors (e.g., Görner, 1963; Pabst, 1977; Strelioff and Honrubia, 1978; Kroese et al., 1978, 1980), but these studies were made with isolated skin preparations and the responses were analysed with regard to the local water flow at the organ. The biological function of the system is, however, to provide the animal with information about stimuli originating at some distance.

Therefore, Elepfandt and Wiedemer (1987) investigated the response characteristics of the lateral-line organ under more naturalistic conditions. The immobilised frog was suspended on its head in the center of a circular basin filled with water 7 cm deep so that the head was just protruding and the body floating perpendicular in the water, a posture typical for freely moving Xenopus. At a radial distance of 12.5 cm, wave stimuli were produced by means of an electromagnetic vibrator whose probe extended into the water; the waves' peak-to-peak amplitudes were measured 7.5 cm distant from the source. The responses of caudal lateral-line organs were investigated by single-cell recordings from the afferent nerve. The afferent fibers fire spontaneously 5 - 25 spikes/s at irregular intervals. In response to wave stimuli the discharges become phase coupled to the wave, thus enabling frequency analysis (Fig.2a). The degree of phase coupling increases linearly with the logarithm of the wave amplitude, and the slope of the increase is independent of stimulus frequency (Fig.2b). Phase coupling saturates at approximately 25 dB above threshold, but these and larger wave amplitudes are encoded by the fiber's mean firing rate. Low stimulus intensities hardly affect the mean firing rate, but at about 20 dB above the threshold for phase coupling, the firing rate begins to increase. As in phase coupling, this increase is linear with the logarithm of wave amplitude and its slope independent of stimulus frequency. Together, phase coupling and mean firing rate allow for stimulus intensity encoding over at least 80 dB, which spans the range of naturally occurring wave amplitudes. By considering only the maximum discharge rate during the response cycles, phase coupling and mean firing rate can be combined into one uniform code for wave amplitude.

Directional response properties were determined by presenting identical stimuli from various directions around the frog. In contrast to the isolated-skin preparations, in which a sinusoidal directional sensitivity has been found (corresponding to an 8-shaped directional sensitivity for waves in polar plots), naturalistic stimulation reveals cardioid directional response characteristics with a sensitivity minimum for waves that impinge from the opposite side of the organ's location with respect to the head (Fig.2c). That is, the directional response characteristic is essentially determined

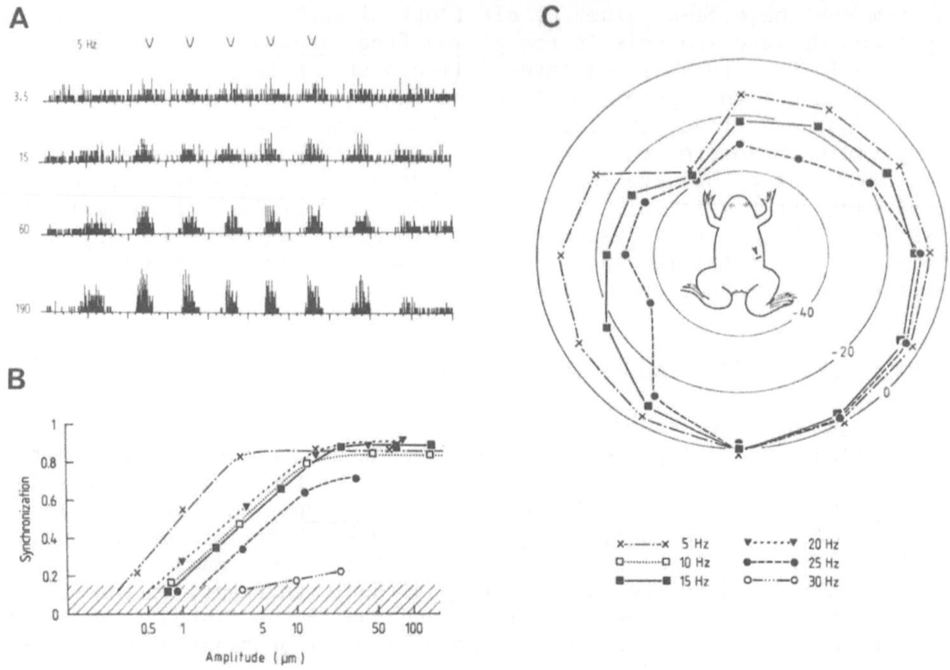

Fig. 2. Response characteristics to water waves of lateral-line afferent neurons in Xenopus laevis. (a) Post-stimulus histograms of responses to 5-Hz waves of different amplitudes. The indentations on the top indicate the wave cycle, the numbers on the left side indicate p-p wave amplitudes in µm. (b) Degree of phase coupling, or synchronization, as a function of wave amplitude. Synchronization increases linearly with the logarithm of wave amplitude, the slope is independent of stimulus frequency. (c) Polar plot of directional sensitivity relative to maximal sensitivity at the respective frequency, the numbers at the circles indicate dB difference. The location of the organ and the fiber's axis of maximal sensitivity for local water flow are indicated by the mark and the arrow head on the frog's right side (from Elepfandt and Wiedemer, 1987).

by the wave shadow of the animal, which can amount to 30 - 40 dB. Compared to that shadow, the effect of the organ's sinusoidal sensitivity for local water flow turns out to be relatively insignificant.

WAVE LOCALIZATION

Xenopus can, by means of its lateral-line system, determine the direction of impinging water waves. Such waves may be produced by insects dropped into the water and struggling to get out, or in the laboratory by dipping lightly a small rod into the water, and Xenopus responds to them by turning toward the wave origin, or by swimming forward if the wave arrives from the front (Kramer, 1933). By measuring the turn angles as a function of stimulus direction it was found that Xenopus can localize waves at an accuracy of 5°. Often, turns toward posterior waves undershoot somewhat (see, e.g., the responses to 120° - 180° in fig.3b), but this can be attributed to the turn behavior and is not due to sensory inaccuracy.

Peripheral Organization of Localization

Because each individual lateral-line organ can be stimulated by waves from all directions, accurate wave localization can be achieved only by comparing the inputs from several organs, just as interaural comparison is required for sound localization. Xenopus possesses approximately 180 lateral-line organs, and the organization of the comparison between these organs was investigated by testing animals in which lateral-line organs had been destroyed (Elepfandt, 1982, 1984; Görner et al., 1984). It turns out that no particular group of lateral-line organs is essential for localization. Instead, various local groups of the organs are each sufficient to detect all wave directions with undiminished accuracy. The groups may be small - in some cases less than 10% of the organs - and they need not be symmetric with regard to the animal: even completely unilateral groups suffice for accurate localization of ipsilateral and contralateral waves. Only after destruction of all or all but one lateral-line organ does localization accuracy deteriorate and responsiveness drop drastically.

What might be the functional meaning of this peripheral 'redundancy' of wave localization? The accurate turn responses demonstrate that small numbers of lateral-line organs are sufficient for the analysis of local water movements. If this can be done simultaneously by groups of organs on different areas of the body, this might enable the frog to analyze more complex waves than the simple circularly expanding waves presented in the tests.

Central Organization of Localization

The central nervous organization of wave localization was investigated by testing animals that had experienced various brain lesions (Elepfandt, 1988a,b). The midbrain is essential for localization: after midbrain removal, the animals respond to waves but cannot localize them, but when only di- and telencephalon are removed, localization is unimpaired for all directions. Tests after partial midbrain lesions reveal a topological organization that is strikingly different from the peripheral organization: unilateral destruction of the sensory midbrain - that is, tectum and torus semicircularis - selectively abolishes localization of all contralateral waves, and local midbrain lesions produce localization failures within angular sectors of the contralateral hemifield (Fig.3). Thus, in contrast to the periphery, where local and unilateral groups of organs are sufficient to localize all waves, the midbrain has a topology that is organized according to stimulus direction, that is, to an external parameter rather than to organ topology.

The details of this midbrain topology are still being explored. By means of electrophysiological recording, a neural map for wave directions has been found in the tectum that is in register with the tectal visual map (Zittlau et al., 1986), but the contribution of this map to wave localization is as yet undecided. From differences in the types of localization defects after the lesions, it can be concluded that several areas with different functions are involved in wave localization, all of which seem to be organized topologically with respect to stimulus direction (Elepfandt, 1988b).

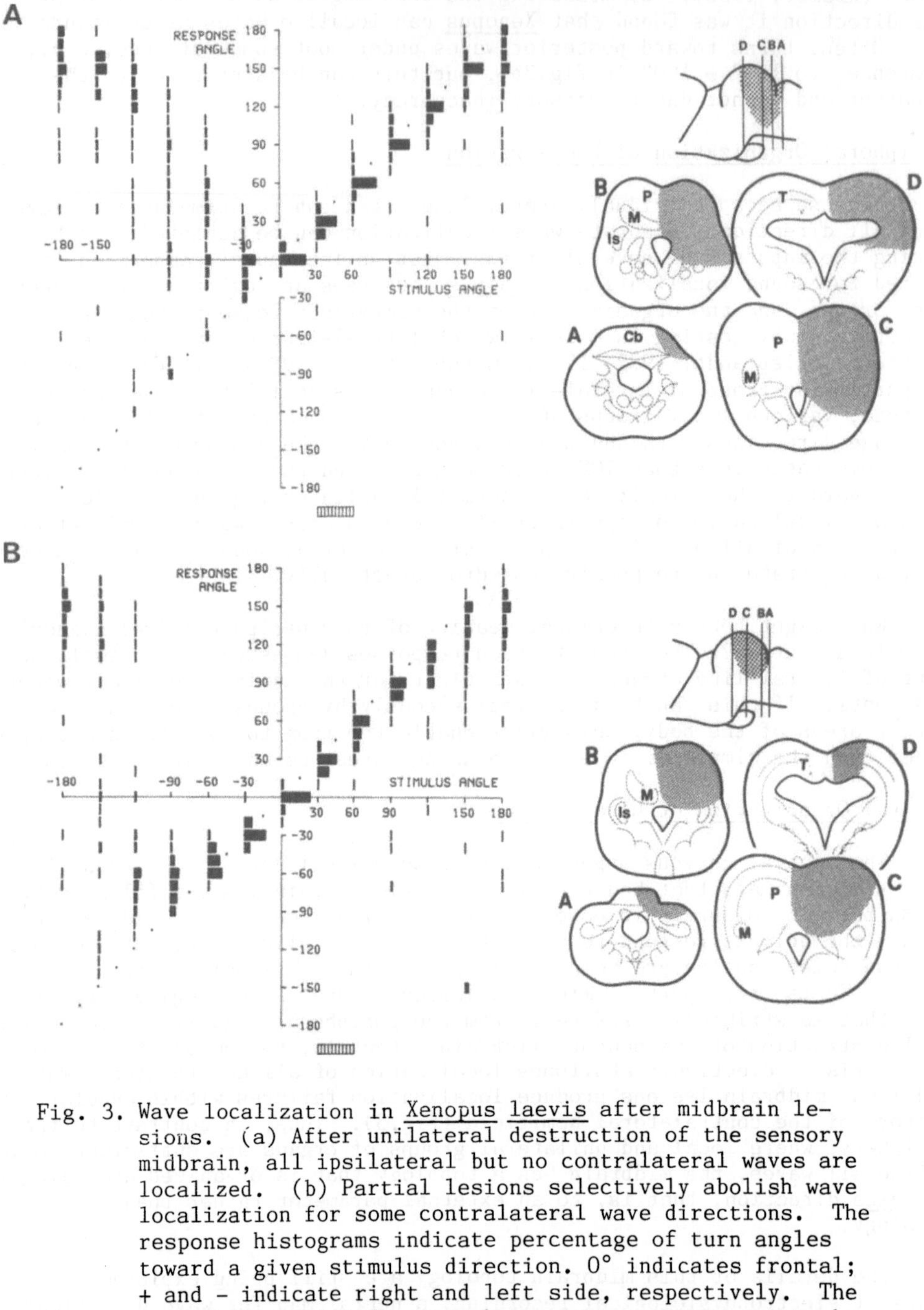

Fig. 3. Wave localization in Xenopus laevis after midbrain lesions. (a) After unilateral destruction of the sensory midbrain, all ipsilateral but no contralateral waves are localized. (b) Partial lesions selectively abolish wave localization for some contralateral wave directions. The response histograms indicate percentage of turn angles toward a given stimulus direction. 0° indicates frontal; + and - indicate right and left side, respectively. The dotted diagonal indicates where correct turns were to be expected. The length of the hatched bar on the bottom indicates 100%. On the right side, the lesion site is indicated by the stippled area; the top figure is a left side view of the frog's midbrain indicating the section levels of the figures shown underneath. Cb - Cerebellum, Is - nucleus isthmi, M - magnocellular nucleus, P - principal nucleus, T - tectum (from Elepfandt, 1988a,b; S. Karger AG, Basel).

WAVE FREQUENCY DISCRIMINATION

Previous recordings from the afferent nerve had shown that all lateral-line organs of *Xenopus* have identical frequency tuning: sensitivity is maximal at 20 - 25 Hz, at lower frequencies it decreases with about 6 dB/oct, whereas at higher frequencies sensitivity decreases somewhat slower and the phase coupling of the discharges becomes blurred (Kroese et al., 1978).

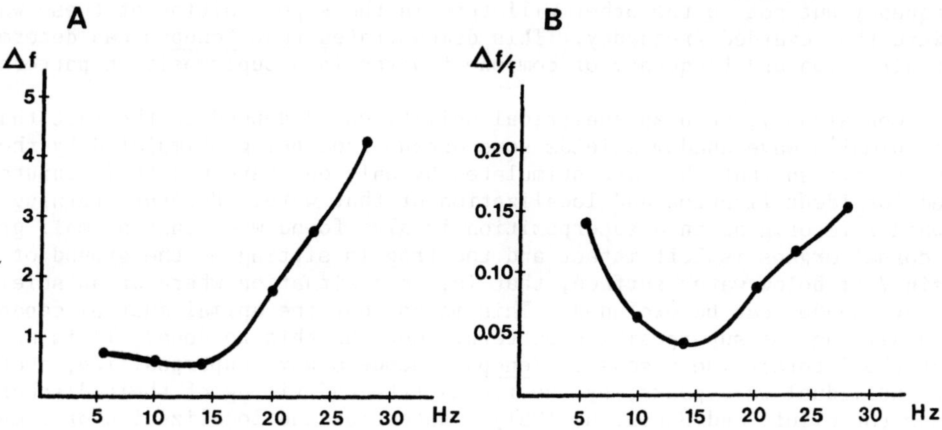

Fig. 4. Absolute (a) and relative (b) wave frequency discrimination limens in *Xenopus laevis*. Relative discrimination is optimal at 14 Hz (from Elepfandt et al., 1985).

From this identical tuning of the organs, it had been speculated for some time that *Xenopus* might be incapable of frequency discrimination. However, conditioning tests revealed good wave frequency discrimination in *Xenopus* (Elepfandt et al., 1985). In these tests, *Xenopus* was presented monofrequent waves of 3 s duration at 1 min interstimulus intervals. Stimulus frequency was alternated in a pseudorandom fashion, turns toward one frequency were rewarded with food, responses to the other punished by a stroke with a rod on the frog's head. Stimulus discrimination on the basis of intensity was prevented by continuously varying wave amplitudes. Frequency discrimination is found in the range 5 - 30 Hz (Fig.4). At 13 - 15 Hz discrimination is optimal having a relative discrimination limen of 4%, which means that 14 Hz can be discriminated from 14.6 Hz. The tests also showed 'absolute pitch' for wave frequencies: the absolute frequency of a rewarded wave stimulus was recognized after three or more weeks with the same accuracy as during the discrimination test (Elepfandt, 1986a). In addition, *Xenopus* discriminates easily between monofrequent and frequency modulated waves (Elepfandt, unpublished observation).

Because all lateral-line organs in *Xenopus* have identical frequency tuning, this discrimination cannot be based on differential frequency sensitivity of the organs, but must be based on the analysis of the discharge volleys that result from afferent phase coupling.

ANALYSIS OF WAVE SUPERPOSITIONS

The tests described so far have been made with single waves expanding circularly on an otherwise undisturbed smooth water surface. In nature, the demands for stimulus analysis will often be more complicated: there is background noise from water ripples produced by wind or other sources, and there are wave reflexions and superpositions of waves from different sources.

To test for the ability to analyze more complex stimuli, Xenopus was presented two waves simultaneously from different directions so that the frog was stimulated by the superposition of the waves. In such a situation, Xenopus responds by turning toward one of the two wave sources (Elepfandt, 1986b). A Xenopus that had been trained with single waves to respond to one frequency but not to the other will turn in the superposition of these waves toward the rewarded frequency. This demonstrated that Xenopus can determine the direction and frequency of component waves in a superposition pattern.

Conceivably, such an analytical ability could depend on the fact that the animal's wave shadow shields some organs from being stimulated by the second wave so that they are stimulated by only one wave and their inputs used for identification and localization of that wave. However, turning toward wave origins in a superposition is also found when only a small group of dorsal organs is left intact and the frog is sitting on the ground of the basin 7 cm below water surface, that is, in a situation where organ shielding by wave shadow can be excluded. This means that the animal must be capable of analyzing the superposition pattern. How can this be done? It is known that the lateral-line organs of Xenopus encode a wave superposition, that is, its individual wave peaks, by the firing rate and timing of their discharge peaks (Strelioff and Sokolich, 1981). Detection and localization of component waves in the superposition is thus possible by comparing the superposition patterns at the individual organs, as reflected in the afferent discharges. The combination of these discharge pattern analyses, taking into account each organ's location on the body, amounts to a combined spatiotemporal input analysis.

PARALLELS TO AUDITORY PROCESSING

At the end of the last century, it was hypothesized on the basis of anatomical comparisons that the auditory and lateral-line systems have a common evolutionary origin. Since the auditory organs are innervated by the eight nerve, this hypothesis was called the octavolateralis hypothesis (octavus = eighth). It remains controversial to the present day (Northcutt, 1980; Boord and McCormick, 1984). So far, the comparison has been limited essentially to anatomy and embryology, due to a lack of knowledge of lateral-line physiology. Only with regard to the receptive elements has a physiological comparison become possible, revealing striking correspondence in fine structure and mechanoelectric transduction of the systems' sensory hair cells (Russell, 1981; Hudspeth, 1983). In this section, parallels of the results on wave stimulus processing to auditory stimulus processing will be pointed out.

Comparison of Organ Morphology

For a better understanding of the physiological comparison, a look at some gross morphological features of the organs is useful. Lateral-line organs exist in two basic forms: superficial organs (as in Xenopus), which are inserted in the epidermis and whose cupulae protrude into the water surrounding the animal, and canal organs, which are located in fluid-filled canals that extend submerged in the dermis and connect to the aquatic environment through small pores. During ontogenesis, canal organs originate as superficial organs and the formation of the canal occurs later. If we continue in our mind this canal development by closing the pores and bending the canal into a circle, we obtain the semicircular canals with their ampullary organs. Indeed ampullary organs show the morphological and functional characteristics of lateral-line organs. Similarly, one can imagine the cochlea as a modified canal, with the receptors - now covered by a basilar membrane instead of many cupulae - forming a continuum along the canal. Sound analysis is done by analyzing fluid waves traveling in the cochlea, mediated by the mechanics of the middle and inner ear whose function is to transform sound frequency differences into spatial differences of stimulation along the basilar membrane.

Physiological Comparison

Stimulus parameter encoding. As indicated in the preceding paragraph, the mechanisms by which a stimulus leads to a deflection of the receptor's sensory hairs are quite different in the lateral-line and auditory systems: in the case of superficial lateral-line organs, an impinging wave directly deflects the cupula and the underlying sensory hairs, whereas in hearing, sound is transformed into traveling fluid waves that differentially stimulate the receptors depending on frequency. Despite this considerable difference, encoding of stimulus parameters in the lateral-line afferent fibers is very similar to that in low-frequency auditory afferent neurons below their best frequency. In both, frequency is reflected in the sequence of discharge volleys. The above description of wave amplitude encoding by phase coupling and mean firing rate could be taken, word for word, as a description of auditory intensity encoding. The similarity encompasses further details such as, e.g., the intensity above threshold at which synchronization saturates and the degree of phase coupling at the point of saturation (Elepfandt and Wiedemer 1987). In addition, the directional sensitivity of the afferent neurons is cardioid in both systems. Consequently, central nervous processing in the lateral-line and auditory systems is based on the same afferent code of stimulus parameters. Such a degree of correspondence despite the differences in the organs' mechanics seems to be more than just accidental and difficult to explain as the result of mere convergence.

Wave localization. In the periphery, various local groups of lateral-line organs suffice for accurate localization of waves from any direction. An auditory parallel to this exists in our ability to localize high-frequency as well as low-frequency tones, that is, tones that stimulate different populations of auditory receptors. In contrast to input comparison in the lateral-line system, however, comparison of auditory inputs is always bilateral because on each side there is only one bottle-neck, the middle ear, through which the sound can enter to the auditory receptive elements. This may be considered the trade-off for using the original space analysis as an additional mechanism of frequency analysis.

A midbrain topology with regard to stimulus direction, as described here with regard to wave direction, is also known in the auditory system. Famous representatives are the space map in the owl (Knudsen and Konishi, 1978) or the encoding of sound space in the bat (Pollak, this volume). It should be emphasized that in both systems - in contrast to other sensory systems of vertebrates - the space topology does not reflect the topology of the receptors, but has to be established by neural computation.

Frequency discrimination. In hearing, it is still discussed whether frequency discrimination is based on a time principle (i.e., analysis of discharge timing) or a space principle (i.e., differential tuning of the receptors along the basilar membrane). Protagonists of either principle often argue that the other principle is insufficient for explaining the accuracy of frequency discrimination that is found (Pickles, 1982). The accuracy of wave frequency discrimination in Xenopus equals at its optimal frequency that of auditory frequency discrimination. Due to the identical frequency tuning of all lateral-line organs in Xenopus, wave frequency discrimination in this case must be based on a temporal analysis of discharges. This demonstrates that frequency discrimination of auditory accuracy can be based on a time principle.

Analysis of wave superpositions. An auditory parallel to the analysis of wave superpositions exists in the analysis of some complex tones. If two tones of similar frequency - e.g., a second or a third - are presented simultaneously, frequency distribution along the cochlea is insufficient to let the receptors be stimulated only by their best frequency, respectively (except at very low intensities close to threshold). Instead the stimulation results from a superposition of the two waves that correspond to the two stimulus frequencies. The form of the superposition changes along the cochlea, and it has been shown that the time course of local superposition is reflected in the discharge pattern of the pertaining auditory afferent fiber (Brugge et al., 1969). Our ability to recognize the two component frequencies thus implies the ability to compare the superposition patterns at various receptor sites, after taking into account the location of those sites along the cochlea. This is a spatio-temporal analysis just as described above for the analysis of wave superposition with the lateral-line system.

Analysis of complex tones certainly also involves other mechanisms of analysis. It would be interesting to see to what degree analogues can be found in the mechanisms of wave analysis with the lateral-line system.

The octavolateralis hypothesis. The comparison shows that at all levels auditory parallels exist to the lateral-line organization of wave analysis in Xenopus. This gives strong support to the octavolateralis hypothesis of a common evolutionary origin of the lateral-line and eighth-nerve sensory systems. As yet, however, the evidence is not conclusive, and more data are needed before the hypothesis can be confirmed or disproved.

ACKNOWLEDGEMENTS

The experiments of A.E. were supported by DFG grant El 75/2, C. Wynne revised the English text.

REFERENCES

Bleckmann, H., 1985, Perception of water surface waves: how surface waves are used for prey identification, prey localization, and intraspecific communication, Progr. Sensory Physiol., 5:147.

Bleckmann, H., 1986, Role of the lateral line in fish behaviour, in: "The Behaviour of Teleost Fishes," T. J. Pitcher, ed., Croom Helm, London.

Boord, R. A., and McCormick, C. A, 1984, Central lateral line and auditory pathways: a phylogenetic perspective, Amer. Zool., 24:765.

Brugge, J. F., Anderson, D. J., Hind, J. E., and Rose, J. E., 1969, Time structure of discharges in single auditory nerve fibers of the squirrel monkey in response to complex periodic sounds, J. Neurophysiol., 32:386.

Bullock, T. H., and Heiligenberg, W., 1986, "Electroreception," Wiley, New York.

Dijkgraaf, S., 1962, The functioning and significance of the lateral-line organs, Biol. Rev., 38:51.

Disler, N. N., 1960, "Lateral Line Sense Organs and their Importance in Fish Behavior," Acad. Sci. USSR, Leningrad (translated from Russian: Israel Program for Scientific Translations, Jerusalem 1971).

Elepfandt, A., 1982, Accuracy of taxis response to water waves in the clawed toad (Xenopus laevis Daudin) with intact or with lesioned lateral line system, J. Comp. Physiol., 148:535.

Elepfandt, A., 1984, The role of ventral lateral line organs in water wave localization in the clawed toad (Xenopus laevis Daudin), J. Comp. Physiol. A, 154:773.

Elepfandt, A., 1986a, Wave frequency recognition and absolute pitch for water waves in the clawed frog, Xenopus laevis, J. Comp. Physiol. A, 158:235.

Elepfandt, A., 1986b, Detection of individual waves in an interference pattern by the clawed frog, Xenopus laevis Daudin, Neurosci. Lett. (Suppl.), 26:S380.

Elepfandt, A., 1988a, Central organization of water wave localization in the clawed frog, Xenopus. I. Involvement and bilateral organization of the midbrain, Brain Behav. Evol., 31:349.

Elepfandt, A., 1988b, Central organization of water wave localization in the clawed frog, Xenopus. II. Midbrain topology for wave directions, Brain Behav. Evol., 31:358.

Elepfandt, A., and Wiedemer, L., 1987, Lateral-line responses to water surface waves in the clawed frog, Xenopus laevis, J. Comp. Physiol. A, 160:667.

Elepfandt, A., Seiler, B., and Aicher, B., 1985, Water wave frequency discrimination in the clawed frog, Xenopus laevis, J. Comp. Physiol. A, 157:255.

Görner, P., 1963, Untersuchungen zur Morphologie und Elektrophysiologie des Seitenlinienorgans vom Krallenfrosch (Xenopus laevis Daudin), Z. Vergl. Physiol., 47:316.

Görner, P., Moller, P., and Weber, W., 1984, Lateral-line input and stimulus localization in the african clawed toad Xenopus sp, J. Exp. Biol., 108:315.

Hudspeth, A. J., 1983, Mechanoelectrical transduction by hair cells in the acousticolateralis sensory system, Annu. Rev. Neurosci., 6:187.

Knudsen, E. I., and Konishi, M., 1978, A neural map of auditory space in the owl, Science, 200:795.

Kramer, G., 1933, Untersuchungen über die Sinnesleistungen und das Orientierungsverhalten von Xenopus laevis Daud, Zool. Jb. Physiol., 52:629.

Kroese, A. B. A., van der Zalm, J. M., and van den Bercken, J., 1978, Frequency response of the lateral-line organ of Xenopus laevis, Pflügers Arch., 375:167.

Kroese, A. B. A., van der Zalm, J. M., and van den Bercken, J., 1980, Extracellular receptor potentials from the lateral-line organ of Xenopus laevis, J. Exp. Biol., 86:63.

Northcutt, R. G., 1980, Central auditory pathways in anamniotic vertebrates, in: "Comparative Studies of Hearing in Vertebrates," A. N. Popper, R. R. Fay, eds., Springer, Berlin.

Pabst, A., 1977, Number and location of the sites of impulse generation in the lateral-line afferents of Xenopus laevis, J. Comp. Physiol., 114:51.

Pickles, J. O., 1982, "An Introduction to the Physiology of Hearing," Academic Press, London.

Russell, I. J., 1976, Amphibian lateral line receptors, in: "Frog Neurobiology," R. Llinas, W. Precht, eds., Springer, Berlin.

Russell, I. J., 1981, The responses of vertebrates hair cells to mechanical stimulation, in: "Neurons without Impulses," A. Roberts, B. M. H. Bush, eds., Cambridge Univ. Press, Cambridge.

Sand, O., 1984, Lateral-line systems, in: "Comparative Physiology of Sensory Systems," L. Bolis, R. D. Keynes, S. H. P. Maddrell, eds., Cambridge Univ. Press, Cambridge.

Strelioff, D., and Honrubia, V., 1978, Neural transduction in *Xenopus laevis* lateral line system, *J. Neurophysiol.*, 41:432.
Strelioff, D., and Sokolich, W. G., 1981, Stimulation of lateral-line sensory cells, *in*: "Hearing and Sound Communication in Fishes," W. N. Tavolga, A. N. Popper, R. R. Fay, eds., Springer, Berlin.
Zittlau, K. E., Claas, B., and Münz, H., 1986, Directional sensitivity of lateral line units in the clawed toad *Xenopus laevis* Daudin, *J. Comp. Physiol. A*, 158:469.

FUNCTIONAL ROLES OF MECHANOSENSORY AFFERENTS IN SEQUENTIAL MOTOR ACTS DURING COPULATION IN MALE CRICKETS

Masaki Sakai and Takahiro Ootsubo

Department of Biology, Okayama University

Tsushima-Naka-3-1-1, Okayama, Japan 700

INTRODUCTION

Copulation is a consummatory behavior consisting of relatively simple motor acts. In arthropods, the male usually plays a major role in copulation. For example, the male performs grasping the female, inserting the penis and transferring a spermatophore during copulation. On the other hand, the female is normally quiet and even somewhat hypnotic during copulation. We assumed that the male certainly required sensory feedback to carry our each motor act sequentially. But how does the male get information to start or stop a motor act while the female is inactivated? To answer this, we studied functional roles of tactile input in copulation behavior in the male cricket.

MATERIALS AND METHODS

Subjects: Crickets *Gryllus bimaculatus* reared under a constant light and dark condition (L:D = 12:12) at 27±2°C.

Operations

Lesion

Figure 1 shows the posterior abdominal segment and sensory organs that were lesioned. 1) The cerci were cut to different lengths with fine scissors. 2) Severance of the cercal sensory nerve was made by inserting a tungsten needle through the cut end of the cercus to scratch all the axon bundles inside the cercus. 3) The cercus was rotated 180 after incising the surrounding cuticle and pushed back into the opening. Control was the male whose cerci were first rotated 180° and then returned to their original positions. 4) The primary afferent nerves from the periproct (10th abdominal tergite) were removed by scratching off the nerves attached to the periproct. 5) The pouch (9th abdominal sternite) was cut off on its top margin where a number of bristle hairs were located. 6) The nerve bundles form the epiphallus were severed by the same technique as in (2).

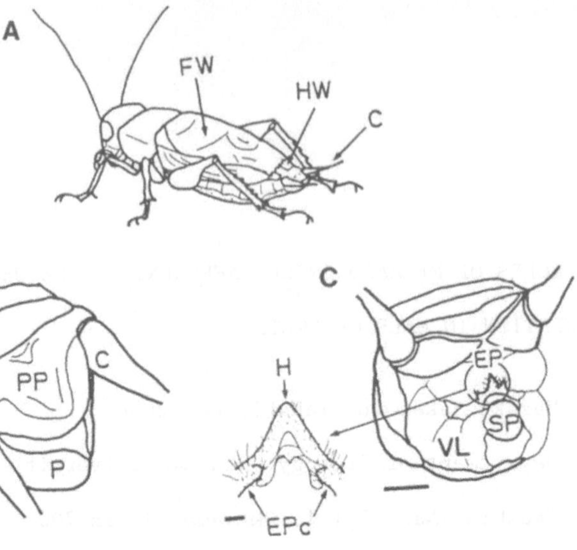

Fig.1A-C. Body parts of the male cricket. A Resting posture. The forewings and hindwings cut off on the rear end. B A dorsal view of the posterior abdomen in the resting male. C The same as B but during the final stage of copulation. Inset show the detail of the epiphallus. Scale bar, 1 mm for C and 100 μm for the inset. AT abdominal tergite; C cercus; VL epiphallic ventral lobe; EP epiphallus; EPc epiphallic convexity; FW forewing; H hook; HW hindwing; P pouch; PP periproct; SP spermatophore. From Sakai and Ootsubo (1988).

Hair removal. Four types of sensilla are present on the cercus (Fig.2B); wind receptive filiform hairs (Edwards and Palka 1974), gravity receptive clavate hairs (Murphey 1981; Sakaguchi and Murphey 1983), contact and possibly chemo-sensitive trichoid hairs (Schmidt and Gnatzy 1972) and hair socket-deflection sensitive campaniform hairs (Bischof 1975; Dumpert and Gnatzy 1977; Heusslein and Gnatzy 1987). 1) Filiform hairs were plucked out as many as possible with forceps, but hairs (50%) less than 500 μm were left. 2) Trichoid hairs were cut off together with the clavate and filiform hairs since they were interspersed with numerous filiform hairs. Most of the campaniform sensilla around the cut sockets of the filiform hairs were left intact. 3) Sensilla on the convexities (lateral process, Fig.2A) of the sclerotized epiphallus were cut off with a razor.

Behavioral test. The standard test procedure was as follows: The male was paired with a female in a 100 ml beaker on the day following the operation and a trial started when the male showed backward slipping (BWS). If the male succeeded in hanging his hook onto the female's subgenital plate located beneath the base of the ovipositor and remained there for 1-2s, the trial was regarded as a successful one. The male was immediately separated from the female to begin another trial soon. If the male came out under the female without successful hooking, the trial was regarded as a failure. The hanging ability (success in hooking) was represented as an average of 10 trials from 10 males in each experiment and the statistical significance was calculated by the

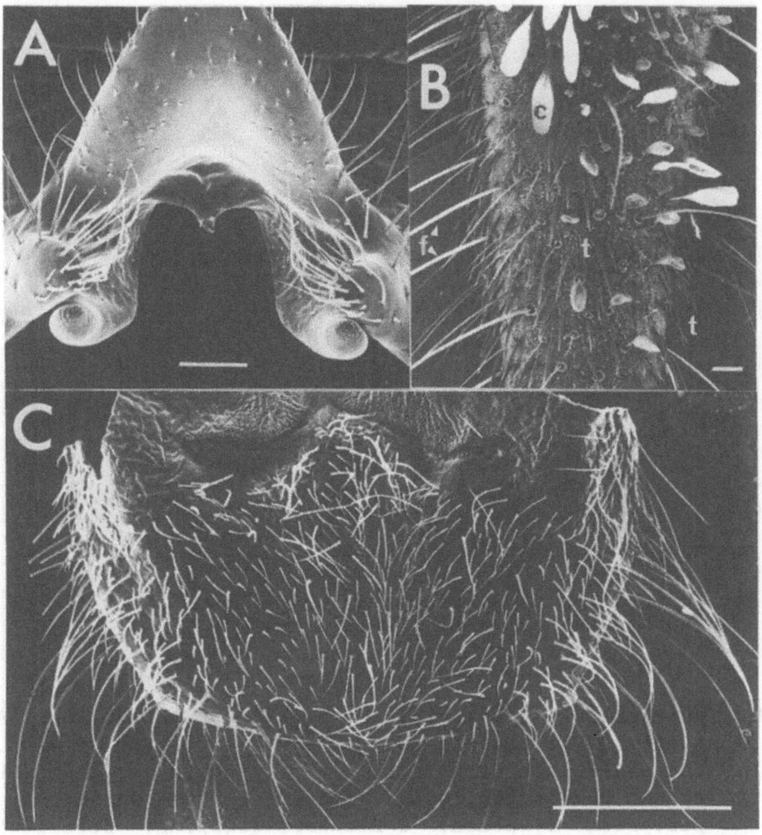

Fig.2. Scanning electron micrograph of the sensilla related to copulation acts. A epiphallus; B cercus; C periproct; f filiform hair; c clavate hair; t trichoid hair. Scale bar 100 μm for A and B, and 500 μm for C.

t-test. The males which received hair ablation of the epiphallus had different behavioral criteria since they had no deficiency in hanging the hook: A trial started after the male hung his hook onto the subgenital plate. If spermatophore extrusion (SPE) was not made within five consecutive trials or required longer than a normal SPE, the male was regarded as having deficits in SPE.

The male was also tested to examine what response could be elicited with the contact stimulus in a restrained condition on a polystyrene ball with the pronotum fixed with a flexible iron wire. Different regions of the body surfaces were touched with bars in different sizes. To stimulate only the filiform hairs, audio sounds in different frequencies (10-1000Hz) were delivered from a loud speaker at levels up to 100dB. For activation of the campaniform sensilla, wind puffs were given manually to the cerci through a nozzle of a blower. The sockets of the filiform hairs on whose bases the campaniform sensilla were located were observed to move.

Hair morphology and distribution. To determine the size of the trichoid hairs and their distribution, a soft cercus from a freshly moulted adult male was placed between two cover glasses and gently pressed to make it flat. Both sides of the cercal surface were examined

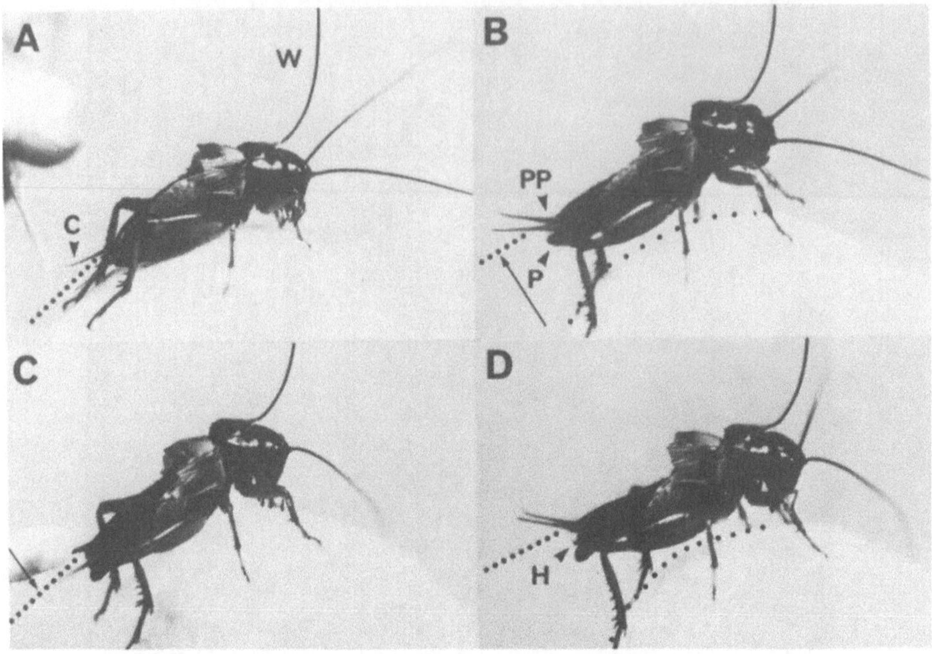

Fig.3A-D. Photographs of artificially evoked copulation acts. A resting. Stimulus is not given yet. W, iron wire; C, cercus. B Intense posture. The picture was taken after the cotton ball stimulus was removed from the dorsum. PP, periproct; P, pouch. C Backward slipping. Stimulation was given on the middle and distal regions of the cerci just after B. D Hooking. The picture was taken just after a cotton ball stimulus was removed which had been applied on the periproct. H, hook. From Sakai and Ootsubo (1988).

under a light microscope (Fig.4 left). The total number of the trichoid hairs on a single cercus was 1465±156 (n=10) and the average density was 341±43 per 1 mm segment. No difference was found in the density among the 8 segments.

RESULTS

Copulation acts

Several terms describing copulation acts are defined. First, intense posture (IP) is a posture with the body raised and the cerci oriented slightly upward (Fig.3B). IP sometimes accompanies the protrusion of the genitalia, opening the periproct and pouch. Second, backward slipping (BWS) is a movement to slip underneath the female with the abdomen extended downward (Fig.3C) Third, the term hooking means coordinated movements to hang the epiphallic hook onto the female's subgenital plate. Hooking consists of body thrusting movements,

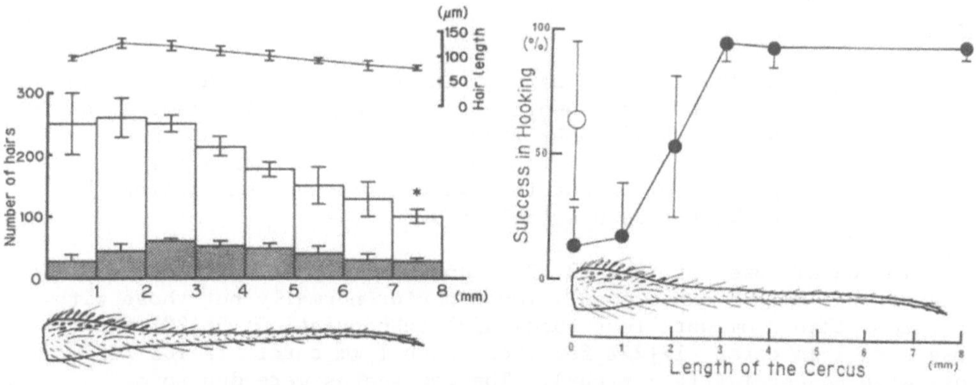

Fig.4. Left: Number of trichoid hairs and their length on a cercus. Bar graph in the white is the average number of trichoid hairs in each cylindrical segment of 1 mm (except*) and bar graph in grey is the average number in the dorsomedial quadrant. Line graph is the average length of hairs in each segment (scale, right). Note that the average number in each segment increases gradually as the cercal diameter increases from peripheral to proximal but decreases in the most proximal 1 mm region where the clavate hairs occupy its medial surface. Right: Effect of successful hooking expressed as the percentage of 10 trials from 10 males. This convention will be the same in the following figures. Abscissa points to where cerci were cut off. From Sakai and Oosubo (1988).

vigorous cercal oscillation, opening the periproct and pouch, protruding the epiphallic complex and everting the hook (Fig.3D). Fourth, spermatophore extrusion (SPE) is the act of pushing out the spermatophore by contracting the spermatophore sac and by expanding the epiphallic ventral lobe and contiguous membrane structure like a water pillow in the pouch.

Wings and dorsum

The previous ablation studies indicated that neither the elytra nor antennae played a crucial role in copulation (Huber 1955; Beck 1974; Loher and Rence 1978). The removal of the two pairs of wings did not affect hooking behavior: success in hanging the hook was 96±5% before and 95±7% after the operation. The time spent for hooking did not change significantly from 3.2±2.1s before to 2.9±1.0s after the operation. The contact with the anterior parts of the elytra or the hindwings elicited only escape behavior while the stimulation to the posterior parts of the elytra induced IP.

Then, the dorsal regions of the abdomen were examined for their ability to elicit IP with a small piece of clay (2x2 mm tip) attached to the bar. The palpation on the notum or the 1st-3rd abdominal tergites did not elicit IP while the palpation of the 4th-9th abdominal tergites did. If the male was highly excited, he opened the pouch and protruded the epiphallic complex. The larger stimulus (5x10mm) induced hooking or body thrusting. these results indicated that a apart of the elytra and most of the dorsum participated in initiated the 1st step of the copulation behavior.

Cerci

Transection. The cerci, a pair of antenna-like appendages on the rear end of the abdomen were known to be necessary for the male to carry out copulation (Sihler 1924; Huber 1955; Beck 1974; Loher and Rence 1978). The cerci obviously seem to function as a searching device for the female's subgenital plate but its exact manner of controlling the motor performance has not been established.

The cerci were cut to 4, 3, 2, 1 and 0 mm. Males with cerci longer than 3 mm performed copulation behavior normally but those with cerci less than 2 mm were less successful in copulation: 54±28% for those with 2 mm cerci, 17±22% for those with 1 mm cerci, 13±16% for those with no cerci (Fig.4 right). The low scores were due to a difficulty in slipping underneath the female after the mad had a contact with the female. The male, instead, rushed into hooking with the abdomen highly raised and was unable to align in parallel with the female even if he happened the get under her by chance (Fig.5 left). The previous study indicated that the cerciectomized male often lifted the female into awkward positions by pushing his abdomen (Huber 1955; Loher and Rence 1978). This abnormal posturing was primarily caused by the failure in BWS.

Stimulation experiments with a cotton bar (4x11 mm) indicated that a gentle touch on the dorsal regions of the middle and distal cerci elicited BWS. It occurred, however, when the stimulus was given during IP: otherwise evasive responses were induced.

The extirpation of one cercus produced a moderate deficiency in hanging the hook with a success rate of 63±32% (Fig.4 right). An analysis of the time spent from the start of BWS to the success in hanging indicated that the operated males spent significantly more time (4.9±2.6s) than the intact males (3.2±1.2s).

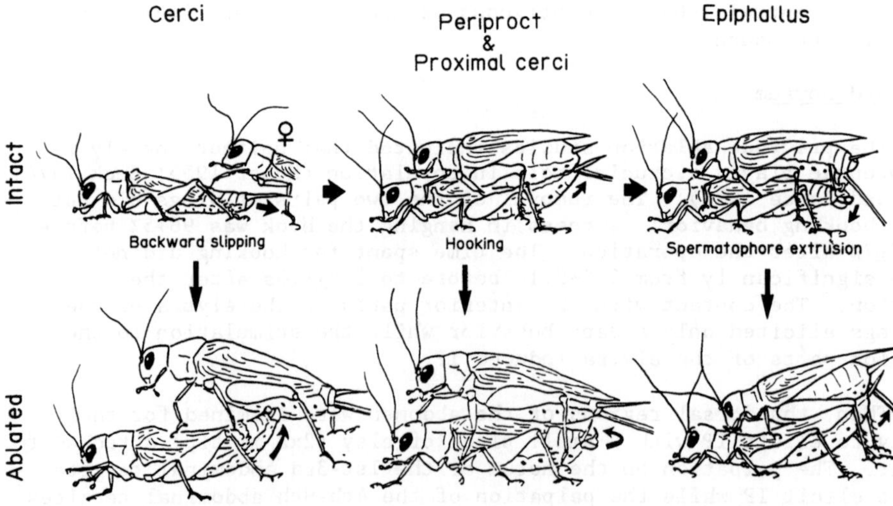

Fig.5. Three types of deficit in copulation produced by deafferentation on the cerci, periproct and epiphallus. See text for details.

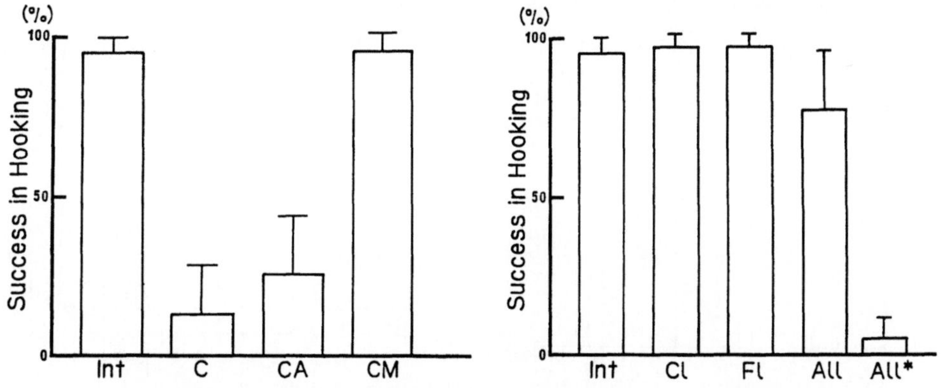

Fig.6. Left: Effect of different treatments of the cerci on copulation behavior. Int intact; C cercal cut; CA cercal sensory nerve cut; CM cercal immobilization. Right: Effects of cercal hair removal on copulation behavioral. Int intact; Cl clavate hair removal; Fl filiform hair removal; All clavate, filiform and trichoid hair removal. All* the same as All except the surfaces from which the hairs were removed were coated with wax. From Sakai and Ootsubo (1988).

Stimulation of either one of the cerci in the middle and distal regions induced a lowering of the abdomen on the touched side and its turning movements toward the stimulus.

Additionally, the possible afferent contribution of the cercal muscles or pericecal regions were tested by lesioning the cercal sensory nerve or immobilizing the cerci with wax by fixing the cercal base to the periproct and paraproct (Fig.6 left). Males with cerci 4 mm long behaved normally (Fig.4 right) but the scores of the sensory nerve-lesioned group were as low (12±1%) as the non-cercal males (13±16%). On the other hand, males whose cercal oscillatory movements were restrained were not handicapped in their hooking ability (97±6%). These results indicated that the input through the cercus itself was crucial for the male to carry out BWS.

<u>Sensilla contributing to BWS</u>. Both the clavate hair removal (98±4%) and the filiform hair removal (98±4%) groups showed nearly normal BWS (Fig.6 right). The all hair removal group showed only a slight decrease in successful rate (78±19%) but a large decrease (5±7%) occurred after wax treatment. However, the contribution of the filiform hairs remained uncertain because most of the small hairs of this type were left intact. All of the filiform hairs were vibrated by sounds from a loud speaker during IP. The male never showed BWS to the sound intensity even up to 100dB. Air puffs, used for stimulation of the campaniform sensilla, were applied during IP. The male again showed no sign of BWS. Together with all of the above results, trichoid hairs are the most plausible origins of afferent input for BWS.

<u>Regional differences</u>. The non-selective hair removal was performed on the circumferential regions of the cerci; dorsal, ventral, lateral and medial (Fig.7 left). The ventral removal group (V, 96±19%)

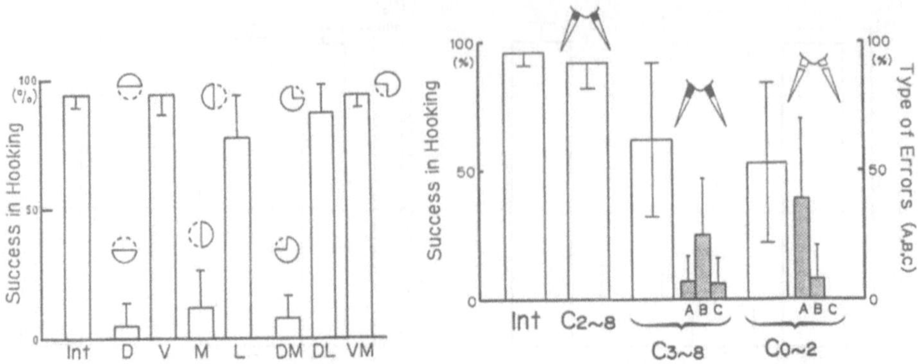

Fig.7. Left: Effect of hair removal from the circumferential regions of cerci on hooking. Int intact males. Capitals on abscissa represent ablated regions. D dorsal; V ventral; M medial; L lateral; DM dorsomedial; DL dorsolateral; VM ventromedial. Inset: Circles represent cross sections of the cercus viewed from behind in which the parts drawn by broken lines represent ablated regions. Right: Effect of sensory hair ablation of the proximal regions of the dorsal part of cerci on hanging the hook and comparison of the deficiencies by ablation of the proximal and distal cerci. Int intacts; C2-8 males with cerci whose proximal 2 mm regions are hair-ablated. C3-8 males with cerci whose proximal 3 mm are hair-ablated. C0-2 males with cerci of 2 mm length. Black parts of cerci in the inset are hair-ablated regions. A Failure in hook hanging due to difficulty in BWS. B Failure in hook hanging due to initiation of improperly positioned hooking under the female after BWS. C Failure in hook hanging for unknown reasons. Right ordinate represents number of failed trials as a percentage for A, B and C. From Saka and Ootsubo (1988).

was not affected by the treatment and removal group showed a minor decrease (L, 79±17%) while the dorsal removal (D, 5±9%) and medial removal (M, 12±17%) groups had extremely low scores. Then, the cercus was subdivided into three quadrants; dorsomedial (DM), dorsolateral (DL) and ventromedial (VM). The DL and VL showed no difference from the intact group while the VM scored (8±9%), as low as the group of males without cerci.

These results raised a question of whether the dorsomedial quadrant of the cercus was a specific region for initiating BWS or whether it was the region that was predominantly stimulated by the female's belly. Thus, the cerci were rotated 180 to an upside down position. The results indicated that most of the males with rotated cerci failed in hanging the hook (7±19%) while the sham operated group (82±9%) was not significantly different from the intact group (96±5%).

Bilateral contact stimulation of the dorsolateral regions of the middle and distal cerci with two wooden bars 1 mm wide induced BWS while stimulation of the ventral regions always elicited an evasion by forward locomotion or kicking. This result was in good agreement with the cercal rotation experiment.

To mark the contact areas of the cerci with the female, the female's belly was powdered with crystals of procion yellow. Just before the male hung his hook after the start of BWS, he was picked up to examine the distribution of procion yellow spots attaching to the

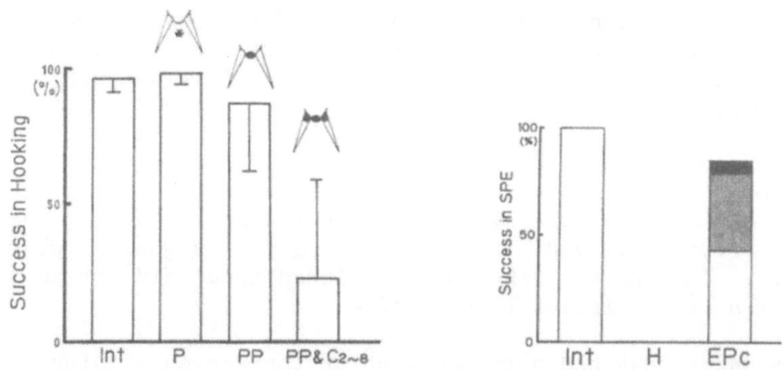

Fig.8. Left: Effects of ablation of the periproct and pouch on hooking behavior. Int intact; P pouch-ablated (asterisk in inset): PP periproct-ablated (black part); PP and C2-8 periproct (black part) and hairs from the proximal 2 mm of cerci ablated. From Sakai and Ootsubo (1988). Right: Effects of epiphallic nerve transection (H) and hair removal of the convexities (EPc). In EPc, blank means males succeeded in SPE in the 1st trial; grey, 2nd trial; black, 7th trial. All EPc males had shown SPE in the 1st trial before the operation.

periproct and cerci. The results revealed that the dorsolateral region scarcely touched the female during BWS while the dorsomedial regions touched her closely. From this, it can be concluded that the dorsal half of the cercus is involved in BWS but contact is mostly from its medial side during copulation.

Then, the regional difference was examined in the longitudinal aspect. One group had hairs of the proximal 2 mm removed so that the tactile input came through the intact regions of 2-8 mm (C2-8) and the other had the proximal 3 mm regions removed (C3-8). Success in hanging the hook for the former group (92$\pm$1%) did not differ from that of the intact males while that of the latter group (62$\pm$30%) was nearly that of the males (54$\pm$28%) with cerci 2 mm long (CO-2, A). The males in the C3-8 group accomplished BWS with ease but could not begin hooking at the proper position under the females.

Stimulation experiments indicated that the male could elicit BWS when a pair of thin bars 0.5 mm wide touched the dorsal regions both cerci between 2 and 8 mm from the base. On the other hand, stimulation of the proximal 2 mm regions elicited hooking with vigorous oscillation of the cerci. This indicates that the proximal regions of the cercus are functionally different from the rest and rather close to the periproct.

<u>Periproct, pouch and genitalia</u>

Numerous bristle hairs were present on the periproct (Fig.2C) and pouch. The males deafferented on this region did not indicate a difficulty in accomplishing BWS (87$\pm$30%) (Fig.8 left). However, the male spent significantly more time (8.6$\pm$5s) for hanging the hook onto the subgenital plate than the intact (3.2$\pm$2s). When additional ablation of the hairs on the proximal 2 mm of the cerci was made, hooking was severely aborted. Such a male could adjust his body axis to the female and began hooking at nearly the correct position but could not aim at the subgenital plate. The hook came off even if it was hung by chance (Fig. 5 middle).

Periproct stimulation during IP elicited hooking but it induced kicking for forward locomotion if stimulated during resting. The abdominal end moved vigorously toward the touched direction indicating that the periproct played a role in finding the subgenital plate accurately. In contrast, the pouch lesioned group did not show any deficiencies in copulation (Fig.8 left).

The males that had a transection of the nerves arising from both the epiphallic hook (central process) and from its convexities (lateral process) were unable to push out a spermatophore (Fig.5 right). They could not make SPE even when the abdomen was pulled down to the horizontal position after the successful hanging of the hook while the males with only the convexities ablated could succeed in SPE though being less efficient (Fig.8 right).

DISCUSSION

Instinct behavior involves a sequence of events. In courtship, each mating act is triggered by particular sensory input coming from the partner and proceeds like a chain reaction between male and female (Tinberen 1951). On the other hand, copulation is normally carried out by a male. Our study demonstrated that each motor act is controlled by sensory input coming through the contact sensitive hairs on the body surfaces, cerci and genitalia and these receptors were successively activated in combination through the contact brought about by the male's own action.

Copulation sequence

Three basic motor acts are present in the process of unifying the genitalia. One is an IP that can be elicited by touching the posterior parts of the elytra or 4th to 9th abdominal tergites. Another motor act is BWS that is a movement to get underneath the female. BWS is initiated by touching the dorsal regions of the cercus between 2 and 8 mm. The other motor act, hooking, can be initiated during IP by contact stimulation of either the dorsum, periproct or proximal cerci. The combined stimulation of the above three regions can lead to hooking accurately. It should be emphasized that BWS and hooking occur only when the stimulus is given during IP: otherwise, evasion responses take place.

These results allowed us to explain the mechanisms of copulation under natural conditions (Fig.9). First, the courting male makes IP when the elytra and dorsum are slightly pressed by the foreleg of mounting female (Fig.9-1). Second, the male subsequently initiates BWS since the middle and distal regions of the dorsal surface of the cerci make contact with he female's belly (Fig.9-2). Third, the BWS stops when the cerci protrude backward by 5-6 mm from under the female's abdomen; at the moment the input from the trichoid hairs on the middle and distal regions of the cerci's reduced to almost nothing (Fig.9-3). The male immediately makes an enhanced IP by gaining a large amount of input from the back. Fourth, the male starts hooking because of a tight contact of the periproct and proximal cerci with female's belly (Fig.9-3). The lowering of the ovipositor of the female helps for a tight contact between the periproct and subgenital plate. Finally, after hanging, the abdomen of the female is pulled down to the horizontal

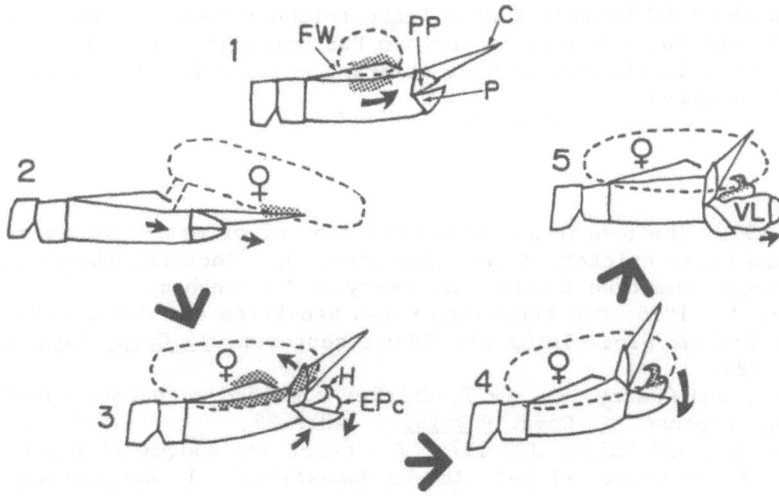

Fig.9. Schematic illustration of the input and output relations on copulation acts in the male cricket. Dotted regions (1-5) indicate those relevant to the initiation of particular movements shown by small arrows. See text for details. From Sakai and Ootsubo (1988).

position (Fig.9-4) and the spermatophore is extruded (Fig.9-5). SPE is not triggered by the input from the cerci or periproct because the cerciectomized or tergite ablated male can still carry out SPE as long as the hook hangs on the subgenital plate. On the other hand, males with the epiphallic nerve deafferented never succeeded in normal SPE.

Cercal function and trichoid hairs

Cerci are known to be indispensable for copulation behavior by male crickets (Sihler 1924; Huber 1955; Beck 1974; Loher and Rence, 1978) and locusts (Mika 1959; Gregory 1965) but not cockroaches (Roth and Barth 1967). Our study demonstrated that cerci control of the vertical movements through antagonistic actions of IP and BWS that are initiated by inputs from the proximal and distal regions of cerci. The vertical movements are also controlled by the input from the periproct when the male nearly reaches the position where it starts hooking. On the other hand, the sideways movements are controlled by inputs from the left and right cerci.

The present study demonstrated that there were regional differences in cercal function between the dorsal ventral halves and between the proximal and distal regions of the dorsal half. According to a recent study (Murphey 1985), the cercal projection of the trichoid hairs on the dorsal and ventral halves terminated differently in the anterior part of the terminal abdominal ganglion. Our results suggest that some modulating mechanisms are present in the dorsal terminal field since the reflex of BWS can be changed by the posture (IP) in the sexually excited state.

Finally, it would be worthwhile to consider the number of trichoid hairs needed for the male to execute BWS. Since the males with cerci of 3 mm showed normal BWS while the males with cerci of 2 mm had a deficit we assumed about 60 hairs (4% of all the trichoid hairs in one cercus) were sufficient for the male to perform BWS normally. This is the number of hairs in the dorsomedial quadrant between 2 and 3 mm (see Fig. 4, 3rd bar in grey).

REFERENCES

Beck, R., 1974. The neural and endocrine control of mating behavior on the male house cricket, *Acheta domesticus* L, Doctoral theses of physiology submitted to the University of Nottingham.

Bischof, H.-J., 1975, Die keulenformingen Sensillen auf den Cerci der Grille *Gryllus bimaculatus* als Schwerezeptoren, J. Comp. Physiol., 98:277-288.

Dumpert, K., and Gnatzy, W., 1977, Cricket combined mechanoreceptors and kicking response, J. Comp. Physiol., 122:9-25.

Edwards, J. S., and Palka, J., 1974, The cerci and abdominal giant fibres of the house cricket, *Acheta domesticus*. I. Anatomy and physiology of normal adults, Proc. R. Soc. Lond. B., 185:83-103.

Gregory, G. E., 1965, On the initiation of spermatophore formation in the African migratory locust, *Locusta migratioria migratoriodes*, Reiche and Fairmire, J. Exp. Biol., 42:423-435.

Heusslein R., and Gnatzy W., 1987, Central projections of campaniform sensilla on the cerci of crickets and coackroaches, Cell Tissue Res., 247:591-598.

Huber, F., 1955, Sitz and Bedeutung nervoser Zentren fur Instinkthandulungen beim Mannchen von *Gryllus campestris* L. Z. Tierpsychol., 12:12-48.

Loher, W., and Rence, B., 1978, The mating behavior of *Teleogryllus commodus* (Walker) and its central and peripheral control, Z Tierpsychol., 46:225-259.

Mika, G., 1959, Uber das Paarungsverhalten der Wanderheuschreck *Locusta migratioria* R. and F. und deren Abhangigkeit vom Zustand der inneren Geschlechtsorgane, Zool. Beitr. Berl., 4:153-203.

Murphey, R. K., 1981, The structure and development of a somatotopic map in crickets: the cerca afferent projection, Dev. Biol., 88:236-246.

Murphey, R. K., 1985, A second cricket cercal sensory system: bristle hairs and the interneurons they activate, J. Comp. Physiol. A., 156:357-367.

Roth, L. M. and Barth, R. H., 1967, The sense organs employed by cockroaches in mating behavior, Behavior, 28:58-94.

Sakaguchi, D. S., and Murphey R. K., 1983, the equilibrium system of the cricket: Physiology and morphology of an identified interneuron, J. Comp. Physiol., 150:141-152.

Sakai and Ootsubo, 1988,

Schmidt, K., Gnatzy, W., 1972, die Feinstruktur der Sinneshaare auf den Cerci von *Gryllus bimaculatus* Deg. (*Salatoria, Gryllidae*). III. Die kurzen Borstenhaare, Z. Zellforsch, 126:206-222.

Sihler, H., 1924, Die Sinnesorgane an den Cerci der Insekten, Z. Jb. Anat., 45:519-580.

Tinbergen, N., "The study of instinct," Oxford University Press, New York, London. (1951)

CORRECTIVE FLIGHT STEERING IN LOCUSTS: CONVERGENCE OF EXTERO- AND PROPRIOCEPTIVE INPUTS IN DESCENDING DEVIATION DETECTORS

Klaus Hensler

Zoologisches Institut
Rheinsprung 9
4051 Basel, Switzerland

INTRODUCTION

Our knowledge about the neural basis of locust flight has improved since the first publication of intracellular recordings from flight neurons (Robertson and Pearson 1982), by the characterization of some 100 central neurons which are somehow involved in flight, and by unraveling many of their connections. It is now clear that the basic motor pattern underlying the wingbeat is generated by a network of central neurons (Robertson and Pearson 1985; Stevenson and Kutsch 1987). This pattern is entrained, and stabilized by proprioceptive feedback monitoring the wingbeat (Wendler 1983; Möhl and Neumann 1983; Horsmann et al. 1983; Reye and Pearson 1987; Elson 1987; Wolf and Pearson 1988), and is steadily controlled, and if necessary modulated, by exteroceptive inputs reporting movements and orientation in space (reviewed by Rowell 1988). Thus unintended course deviations are rapidly counteracted by modulating the amplitude, the angle of attack, and the timing of the wingbeat (Möhl and Zarnack 1977; Thüring 1986), by rudder-like movements of abdomen and legs (Camhi 1970; Taylor 1981a; Arbas 1986), and possibly by changing the center of mass through deflection, retraction, or extension of the abdomen (cf. Zanker 1988).

Course deviations may result from gusty winds and from faulty motor performance, but also from corrective steering itself. This is indicated by the observation that locusts, like other insects (Heisenberg and Wolf 1984; Wagner 1986), do not follow a perfectly straight path while flying under closed-loop conditions but oscillate slightly along a straight course (Baker 1979, Robert, in press; Hensler and Robert, in prep.). Thus straight flight is a regulated process, and the ideal

---

Abbreviations: AP action potential
DN, descending movement detector neuron
PSP, postsynaptic potential
TIN, thoracic interneuron

orientation in space is not restored perfectly by a single steering manoeuvre. This is typically either overshooting or insufficient, so that subsequent manoeuvres are necessary.

Deviations can be translatory, along the three orthogonal body axes (lift, thrust, slip), or rotatory, around these axes (roll, pitch, yaw); the present paper concentrates on the latter. Especially roll and pitch deviations must be carefully prevented as they change the orientation with respect to gravity, and can lead to a crash. Yaw deviations carry the danger of disorientation but they do not affect flight stability.

Course deviations are detected visually, as movements of the panorama (Rowell and Reichert 1986; Robert, in press), and by mechanoreception, by measuring the direction of wind using the antennae (Gewecke 1974; Arbas 1986), together with wind-sensitive hairs on the head (Camhi 1969a), the sternum (Pflüger 1984), the cerci (Altman 1983), and probably other parts of the body. In addition wing strain is perceived by campaniform sensilla in the wing veins (Elson 1987). Course control can not, however, be based on movement detection alone, as even minute mistakes in successive correction manoeuvres could accumulate and lead to disorientation or crash. Orientation in space needs absolute references. These are objects of interest for yaw control (Robert, in press), and ultimately the reference must be gravity for roll and pitch control. Gravity itself is not a suitable measure, as flight manoeuvres produce interfering accelerations, and accordingly, gravity orientation has never been found in flying insects, although this was looked for in various species (reviewed in Hengstenberg 1988). However, the gravity vector is indicated reliably by the gradient of light between sky and ground which is determined by the orientation of the horizon, at least in open country and during high-level flight. Flying locusts, like many other species (Wehner 1981), show in fact a 'dorsal light reaction' which is independent of gravity, and they stabilize themselves with respect to an artificial horizon (Goodman 1965; Robert, in press).

Orientation with respect to wind is far less reliable than visual orientation, although locusts can measure the direction of wind (Camhi 1969a, b; Bacon 1983), and stabilize an upwind course when mounted in front of a windtunnel (Camhi 1970; Möhl, in press). However, an untethered flying animal will rapidly adopt the movement of the surrounding air after a deviation from course. Hence, wind receptors perceive course deviations only transiently, and tonic wind input reports the animals's own locomotion but not a false course or wind drift.

Consequently, eyes and ocelli are essential for course control (cf. Preiss and Gewecke 1988; Riley et al. 1988), but these have no fixed spatial relation to the aerodynamic organs (wings, abdomen, legs). Instead, course deviations regularly elicit compensatory head movements which stabilize the retinal image (e.g. Taylor 1981a, b), and flying locusts occasionally make head movements in various directions without obvious external trigger (Hensler and Robert, in prep.). The actual course, or a deviation from course, can thus only be determined by the convergent evaluation of exteroceptive inputs signalling the orientation of the head with respect to the environment, and proprioceptive inputs signalling the relative orientation of head and body.

The main topic of this paper is to describe this convergence at the level of single neurons, which were recorded intracellularly in the central nervous system of the desert locust *Locusta migratoria* by conventional techniques (Fig. 1). It turned out that this convergence takes place at the level of descending deviation detectors, which are known to act as interfaces between the exteroceptors of the head and the

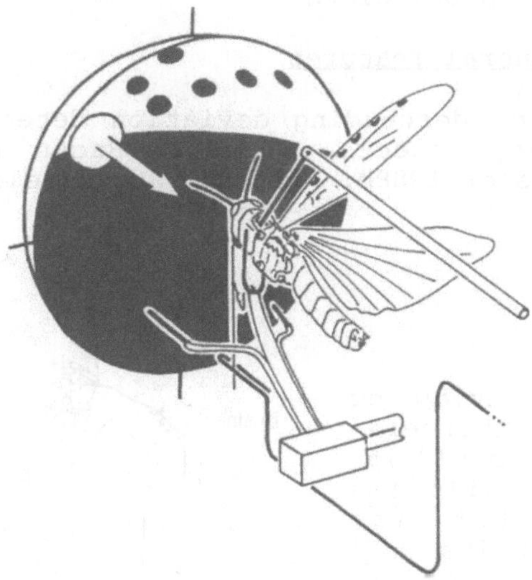

Fig. 1. (A) Experimental setup. Locusts with the rear halves of the eyes covered with black enamel faced an artificial semi-horizon. This could be rotated by means of servos around the roll, yaw, and pitch axes to simulate course deviations. Frontal wind could be blown onto the animal through a central opening in the horizon (arrow). Head rolling was either prevented, or imposed using a vertical lever glued onto the head, engaged with the axis of a servo. Lever and axis could be disengaged during experiments to allow for active head movements. Then the lever functioned as part of a capacitive transducer (Sandeman 1968) measuring head position around the roll axis. Intracellular recordings were made from axons in the neck connectives, through a small hole cut into the pronotum of otherwise minimally dissected locusts. The gut was removed, and the connectives were stabilized on a metal spoon. The thoracic box and the neck joint were left intact, thus allowing normal wing and head movements. The joint between pro- and mesothorax was immobilized. The left wings are not drawn to allow a better view. Flight activity was recorded from the ventral insertion points of the first basalar muscles (M97). The same setup was also used with more heavily dissected animals, the pterothorax of which was opened to expose the thoracic ganglia (cf. Robertson and Pearson, 1982).

thoracic motor centers controlling steering movements of wings, abdomen and legs (Rowell and Reichert 1986; Reichert and Rowell 1985). Further on, the article describes the neural control of steering reactions, and compensatory head movements, and finally discusses the neural events responsible for corrective course control, and recent ideas about the function of compensatory head movements. The presented data deal chiefly with deviations around the longitudinal body axis (roll). However, I assume that deviations in other directions are controlled by similar neural mechanisms.

DESCENDING DEVIATION DETECTORS

Morphology and general features

Populations of descending deviation detectors (DNs) were identified in many insects (locusts: Bacon 1983; Griss and Rowell 1986; Hensler 1988b; flies: Strausfeld et al. 1984;

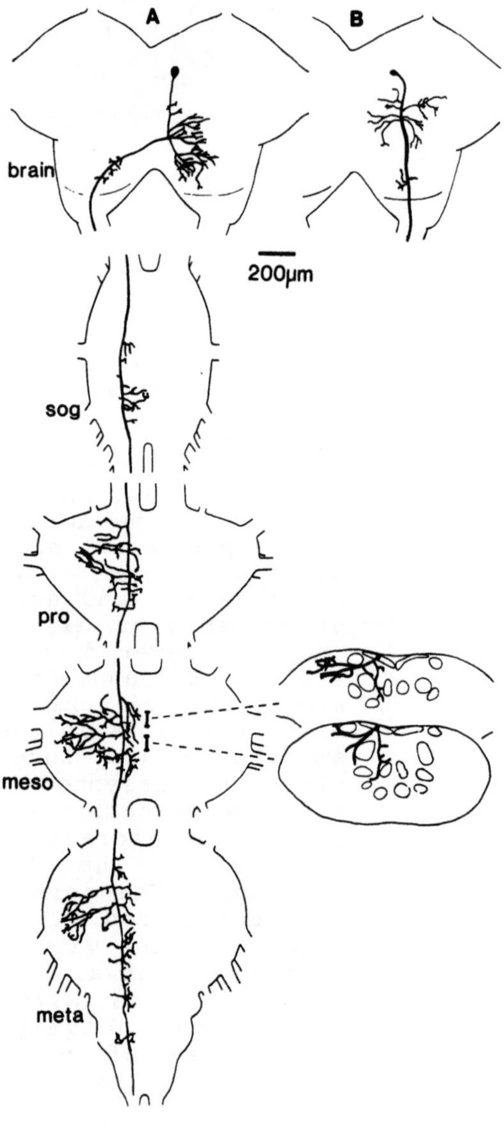

Fig.2. (A) Dorsal view of the pars intercerebralis neuron PI(2)5, named by Williams (1975) after the position of its cell body in the median pars intercerebralis. According to the conventions adopted here this is defined as a left neuron. The morphology shown in vertical sections of the mesothoracic ganglion is representative for all ganglia posterior to the brain. (B) Brain structure of a descending deviation detector neuron with similar response properties to those of PI(2)5, except for the inverted direction specificity for roll and yaw deviations. Posterior to the brain its anatomy is similar to that shown in (A).

dragonflies: Olberg 1986; bees: Goodman et al. 1987). Typical locust DNs are shown in Fig. 2. Input regions appeared to be confined to the brain, as no subthreshold activity is seen in recordings posterior to the brain, and as current injection (up to 15 nA) at these locations has no effect on the discharge pattern. According to their different input sources the brain anatomies of different DNs differ (Fig. 2A, B; Bacon 1983; Griss and Rowell 1986). The structure in the brain is the main criteria for morphological identification, because DNs look quite uniform in the suboesophageal and thoracic ganglia. The branching patterns are not obviously specific, and the only differences are that axons may run in different tracts, or that projections are uni- or bilateral. Dendrites of all DNs project predominantly dorsally where they overlap branches of neck motoneurons (Honegger et al. 1984), wing motoneurons (Tyrer and Altman 1974), abdominal motoneurons (Baader 1988a), and premotor interneurons (Reichert and Rowell 1985; Pearson and Robertson 1987).

The morphology and physiology of five locust DNs is known in detail. These are the pars intercerebralis neuron PI(2)5 (Fig. 2A; Hensler 1988b), the tritocerebral commissure giant neuron (TCG; Bacon 1983), and DNM, DNI, and DNC, which are characterized after strong input either from the <u>m</u>edian ocellus, or one of the ocelli <u>i</u>psi- or <u>c</u>ontralateral to the axon (Rowell and Reichert 1986). Other DNs with so far unknown brain structure have been identified physiologically (K. Hensler in prep.; A. Baader in prep.). Hence, we know 10-15 pairs of DNs involved in the control of corrective flight steering. The actual number is probably much higher, because mass stainings revealed some 70 pairs of long descending fibers (Altman and Kien 1985), and many of these may play a role in course control.

DNs may be unimodal, like TCG[1] which codes tonically for the direction of wind (Bacon 1983), but most are multimodal. Some are exclusively exteroceptive, others receive also input from proprioceptors of the neck, coding for movements and position of the head. Purely proprioceptive DNs were not found. Here I describe the response properties of several proprioceptive DNs, with emphasis put on PI(2)5 which is best known concerning this aspect. PI(2)5 is described in more detail elsewhere (Hensler 1988b), and the description of DNI, DNC, and DNM responses to visual and wind inputs are mainly reviewing the results of Rowell and Reichert (1986), whereas proprioceptive inputs to these neurons were tested in my own recordings. Data from other DNs, with as yet unknown brain anatomy, are not yet published elsewhere. In what follows the origin of data is not further mentioned. Unless otherwise stated the terms ipsi- and contralateral are used with respect to the axon, and the terms exteroceptive and proprioceptive deal with visual and wind receptors of the head on the one hand, and proprioceptors of the neck on the other.

---

[1]TCG is sometimes described as multimodal because of its response to light-on/off stimuli. My own recordings failed to prove any influence of movements and/or position of horizon and head. TCG is thus unimodal with respect to deviation detection.

Visual deviation detection

The DNs described here respond to movements of large portions of the panorama (which indicate self-movement) but not to small objects moving in an arbitrary direction, as for example DMD neurons do (Rowell 1971). However, visual neurons of higher order may be highly specific in their preferences of form and/or movements (Olberg 1986; see also Hubel and Wiesel 1979), and sophisticated tests might reveal further specific preferences of these DNs.

Visual DNs are phaso-tonically excited either by light-on or light-off stimuli. The former is typical for PI(2)5 and other DNs receiving input from compound eyes alone (e.g. Fig. 2B). Light-off responses are typical for units like DNI, DNC, and DNM receiving input from compound eyes and ocelli. The present data are, however, not sufficient to state this as a general rule.

Coding of horizon movements. Course deviations were simulated by rotatory movements of an artificial horizon (Fig. 1), and it must be noted that horizon movements to the left simulate deviations to the right and vice versa. All directions given in the following refer to the direction of simulated deviations, and not to the direction of horizon movements.

Most visual DNs respond with bursts of action potentials to course deviations in specific directions (direction specificity; Figs. 3, 4, 6A, B, 8A). Deviations in the 'antipreferred' direction are often inhibitory (or they suppress tonic excitation) as can be seen in DNs with tonic background activity (Fig. 8C). Threshold, strength, and directionality of the response may vary in different DNs. They are excited either by ipsi- or contralateral roll/yaw (generally the preferred roll and yaw directions are identical), and either by pitch up or pitch down. This makes four possible combinations and DNs of each category have been recorded. For example DNI prefers ipsilateral roll/yaw, and DNC prefers contralateral roll/yaw, but both DNs prefer pitch down. PI(2)5 and the DN of Fig. 2B also prefer ipsi- and contralateral roll respectively, but both are excited by pitch up. The response patterns to roll and pitch are simply explained by the facts that DNI and DNC are excited by light-off stimuli at the ipsi- and contralateral ocellus respectively, whereas PI(2)5 and the DN of Fig. 2B are excited by light-on stimuli at the contra- and ipsilateral eye respectively. Roll deviations darken the side to which the animal rolls, and illuminate the other, whereas pitch deviations illuminate or darkens both sides at the same time. Thus simple light-on or light-off responses in a pair of contralateral homologues is sufficient to distinguish between roll and pitch deviations. However, deviations are also recognized through the detection of moving cues in the panorama. This is indeed the only way to detect yaw deviations.

One further PI(2)5-like DN is known (prefering pitch-up/ipsilateral roll) which fires tonically, also in the dark. The brain morphology is unknown but the thin axon winds closely along that of PI(2)5, and it would be interesting to know whether this DN is identical to PI(2)1, the finest of the three

PI(2) neurons crossing the midline in the brain (Williams 1975). In addition, at least two fibers of the pitch up/contralateral roll category are known physiologically (one is shown in Fig. 2B). Unfortunately the recordings from many other DNs are not sufficient to distinguish them clearly as individuals, but they demonstrate the existence of further DNs with similar response properties.

Occasionally, DNs were found with much more restricted preferred inputs. For example the DN of Fig. 4 ignores pitch deviations, and its weak modulation by roll can be neglected. However, contralateral yaw elicits strong activity. Such

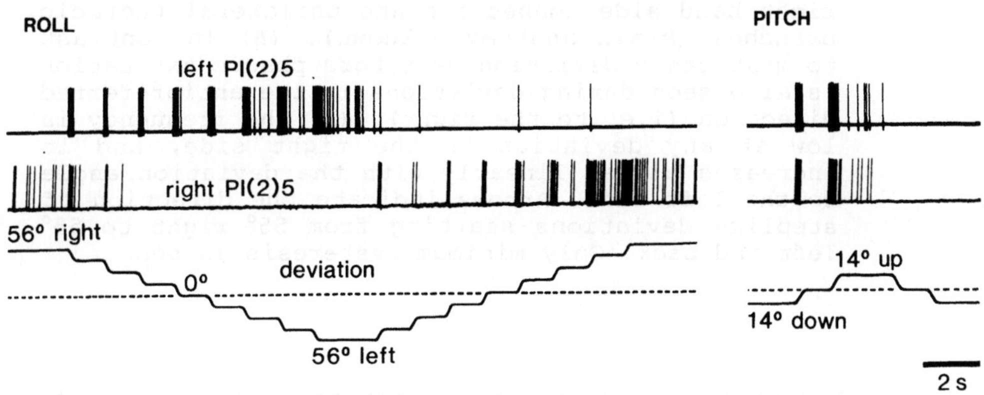

Fig. 3. Response of the left and right PI(2)5 in a quiet locust to simulated roll and pitch deviations, the artificial horizon being moved in successive steps of 14°. The response is direction specific, i.e., PI(2)5 is excited by roll deviations to the side of the axon, and by pitch-up deviations. The response to identical steps increases the more the initial deviation shifts towards the preferred direction (sector specific response; cf. Fig. 9 A).

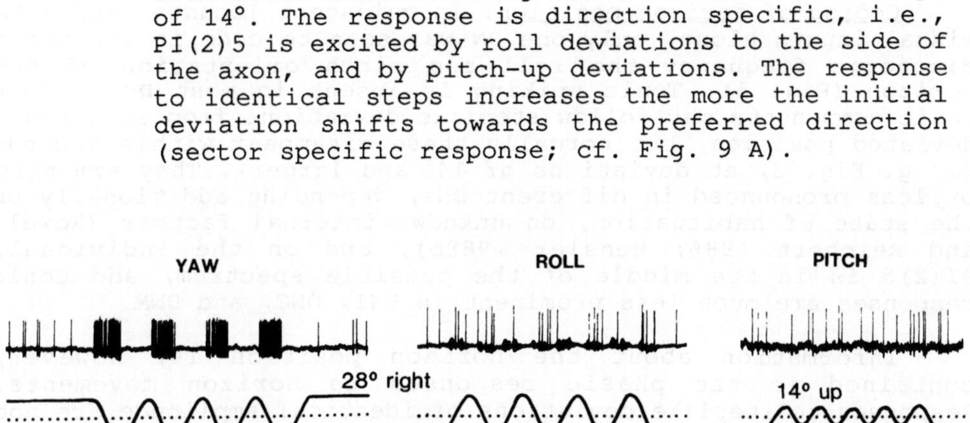

Fig. 4. Responses to simulated course deviations of a yaw-selective, descending deviation detector with its axon in the right hand side connective and with unilateral thoracic branches (brain anatomy unknown). Pitch deviations are ignored, and roll weakly modulates the basic tonic activity. The reduction of tonic spiking during yaw in the 'antipreferred' direction indicates inhibition.

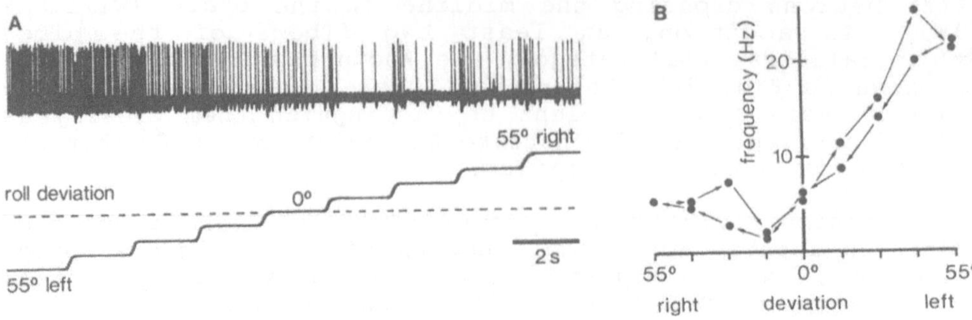

Fig. 5. Frequency coding of the horizon position in a descending deviation detector with its axon in the right hand side connective and unilateral thoracic branches (brain anatomy unknown). (A) In contrast to most other deviation detectors phasic excitation is also seen during deviations in the antipreferred direction (i.e. to the right). (B) The frequency is low at any deviation to the right side, and it increases almost linearly with the deviation angle to the left side. Arrows indicate the direction of steplike deviations starting from 55° right to 55° left and back. Only minimum hysteresis is seen.

specific DNs were only rarely seen, but their true proportion within the entire pool of DNs may be underestimated, because not all possible (combinations of) stimuli were presented.

Coding of horizon position. In quiescent locusts, and with visual inputs alone, only one DN was seen to code in its tonic discharge frequency the roll and pitch orientation of the horizon (Fig. 5). Tonic spiking is absent in most DNs. Short tonic components may follow steplike deviations from an already deviated position, but normally these disappear within seconds (e. g. Fig. 3, at deviations of 14° and larger). They are more or less pronounced in different DNs, depending additionally on the state of habituation, on unknown internal factors (Rowell and Reichert 1986; Hensler 1988b), and on the individual. PI(2)5 is in the middle of the possible spectrum, and tonic responses are much less prominent in DNI, DNC, and DNM.

Information about the horizon position is, however, contained in the phasic responses to horizon movements. Responses to steplike deviations of identical amplitude are not constant, but increase the more the horizon is initially oriented into the preferred direction (Fig. 3, 9A). This feature is called sector specificity, and is characteristic for almost all visual DNs. Sector specificity of PI(2)5 comes about by subthreshold input coding for the overall illumination of the essential eye (ipsilateral to the main arborizations in the brain, cf. Fig. 2A). This input determines the excitability, i.e. the number of action potentials during phasic responses. However, the effect of changed illumination on the excitability does not remain constant over indefinite time but vanishes slowly within about ten minutes, most probably because of visual adaptation. Sector specificity in DNC, DNI, and DNM depends at least partially on the illumination of the ocellus providing the main input (in contrast to PI(2)5 these DNs are excited when dimming the light). Presently we do not know whether the horizon position is also measured by neurons which are sensitive to specific orientations of edges or longitudinal structures in the visual field. To my knowledge no such neurons are yet described for insects.

A flying locust adopts that orientation in space at which activity in all DNs with opposite direction specificity is balanced. Because of sector specificity this is the case when the horizon is oriented normally in the visual field. The phasic nature of many DNs suggests that information about the horizon position is expressed only in the responses to horizon movements and not as a constant flow of information. However, some DNs, which in quiescent locusts respond phasically, receive additional input during flight so that they become more tonic (Figs. 10, 11B, C, and section 'Wind input...'). Further, straight flight actually means slight oscillation of the locust along the ideal path. I postulate that many DNs respond to this small course deviations, and are thus apparently tonically active.

Input from neck proprioceptors

Many, though not all DNs are influenced by proprioceptive input reporting head position and movements. Important proprioceptors are hair plates on the cervical sclerites (Goodman 1959; Haskell 1959), and the chordotonal organ associated with neck muscle 54 (Shepheard 1974). The receptor axons enter the prothoracic ganglion, but a small effect of head movements/position remains even after complete denervation of this ganglion. Hence, other receptors, most probably those on the posterior part of the head, play a minor role as well. None of the known or presumed receptors projects into the brain (Bräunig et al. 1983), and as inputs to DNs appeared to be restricted to the brain, ascending interneurons must be interposed between these two elements.

Coding of head movements. Proprioceptive inputs to DNs can only rarely be demonstrated by moving the head in the dark, but frequently by their modulatory effect on the response to another stimulus. For example, PI(2)5 responds weakly to head movements in the dark in only some 50 % of trials, but it responds strongly and reliably in the presence of light (visual stimulation was excluded by either using an unstructured panorama, or by moving head and horizon simultaneously into the same direction). The presence of input from neck proprioceptors is demonstrated when the response to horizon movement relative to the fixed head is smaller than the response to head movement causing the same visual input (Fig. 6A-C; it has to be noted that this neuron did not respond to head roll in the dark, Fig. 6D). Mostly, direction specificity is weak or missing in the responses to head movement. Such responses to head movements were seen in at least six DNs including PI(2)5, DNC, and the DNs of Figs. 2B and 6. DNI and DNM are not yet tested sufficiently, and proprioceptive inputs were not seen in TCG.

Compensatory head movements reduce the visual information about the original deviation (cf. Fig. 7C). In the six DNs mentioned above, this deficit is balanced to a various degree

Fig. 6. (A, B) Response of a descending deviation detector with its axon on the left hand side of the thoracic ganglia, and with unilateral thoracic branches (brain anatomy unknown) to simulated roll deviations of 28° to the left while the head is immobilized. (C) The response is larger when rolling the head instead of the horizon thus causing the same visual input as in A. (D) In the dark, the same head roll as in C elicits no response. (B-D) Histograms show mean responses to 15 stimulus presentations. The lower trace in (A) indicates the course of the simulated deviations and head movements.

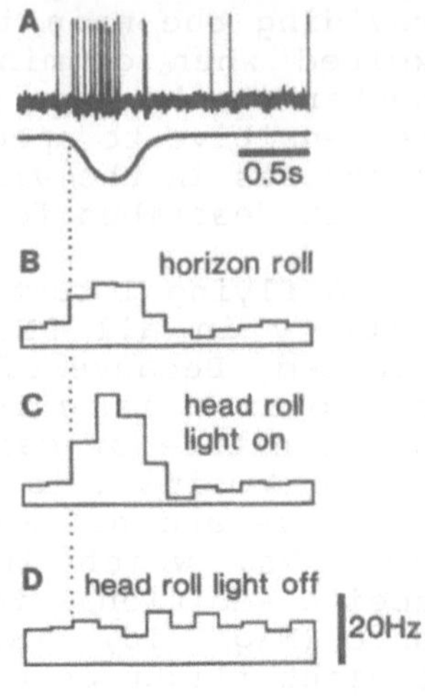

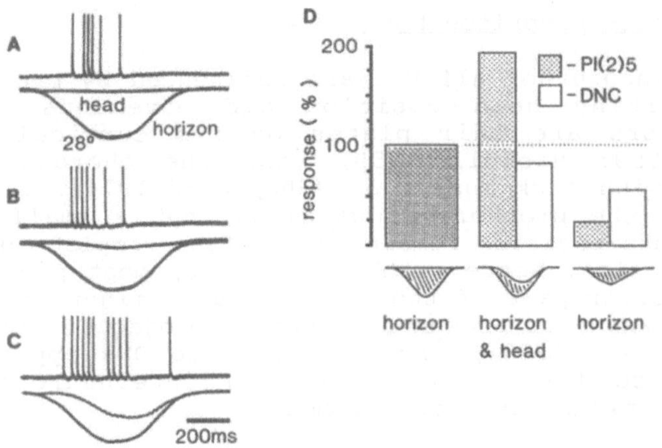

Fig. 7. Response of PI(2)5 to horizon rolling (A) while the head is immobilized, (B) during compensatory head rolling performed by the animal (due to dissection the amplitude is only about 5°), and (C) while compensatory head rolling of 14° is imposed. Note that head movements increase the response. (D) *Middle columns*: quantitative presentation of data such as C, normalized with respect to responses such as A (N=15). Compensatory head movements increase the response of PI(2)5, and they decrease the response of DNC. *Right columns*: same net visual input as in the middle column but head movements prevented. The differences between the middle and the right columns represent the input from neck proprioceptors, thus revealing that the response of DNC, although reduced by the visual effect of the head movement, also contains proprioceptive excitation. *Crosshatched areas* below the histogram indicate visual input.

by proprioceptive input reporting the head movement itself. In PI(2)5 the loss of visual input is obviously more than compensated no matter whether the head is moved by the animal itself, or by the experimenter (compare Fig. 7A with 7B, C). Compensatory head rolling of more than 8° may double the response to a simulated deviation of 40° (Fig. 7D). In the case of DNC, compensatory head rolling reduces the response (Fig. 7D, middle column). Nevertheless, the response decreases further when the same nett visual stimulus is presented while the head is immobilized (Fig. 7D, right column).

Some DNs process proprioceptive inputs in a different way. E.g. the DN of Fig. 8 responds to simulated deviations with excitation and inhibition respectively (Fig. 8A, C), just as seen in other DNs. However, the response is almost negligible when moving the head instead of the horizon (Fig. 8B, D).

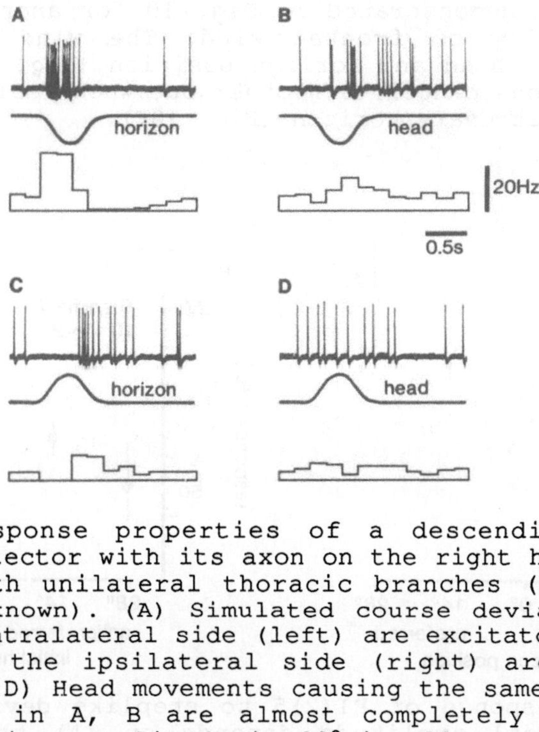

Fig. 8. Response properties of a descending deviation detector with its axon on the right hand side, and with unilateral thoracic branches (brain anatomy unknown). (A) Simulated course deviations to the contralateral side (left) are excitatory, (B) those to the ipsilateral side (right) are inhibitory. (C,D) Head movements causing the same visual input as in A, B are almost completely ignored. The maximum amplitude is 28° in all cases. *Histograms* show mean responses to 15 stimulus presentations.

<u>Coding of head position</u>. Only two visual DNs were found to code for head position in their tonic discharge frequency when other inputs are absent (one is the DN of Fig. 5). In both cases the discharge frequency changes linearly over a range of ± 30°, with minimal hysteresis. The response persists in the dark, and the slope of the characteristic curve is somewhat steeper when the visual field is illuminated. Inputs from horizon and head position summate, so that the visual response is either amplified, or suppressed when the head is deflected

in one or the other direction. Suppression is complete at head roll of 30° to the left.

In most other proprioceptive DNs the head position, like the horizon position, is coded by subthreshold, tonic inputs. The excitability of PI(2)5 changes linearly between head positions of ± 30°, from about 50% to 150 % of the response at the normal head position (Fig. 9B; the visual situation was the same at every head positions, cf. legend). The slope of the corresponding curve is almost identical to that of sector specificity (Fig. 9A). This has interesting consequences: when the head rolls with respect to the thorax it rolls automatically with respect to the horizon but with the opposite sign. Hence, the modulatory effect of the head position is balanced by the modulatory effect of the horizon position. In other words, sector specificity relates to the body (which must be kept in line with the horizon) and not to the head (a deflection of which has no aerodynamic effect). The same effect is more directly demonstrated in Fig. 10 for another DN which responds tonically to frontal wind. The wind response is modulated by both head and horizon position (Fig. 10C, D), but the two modulations cancel each other out when rolling the head in front of a stationary horizon (Fig. 10E).

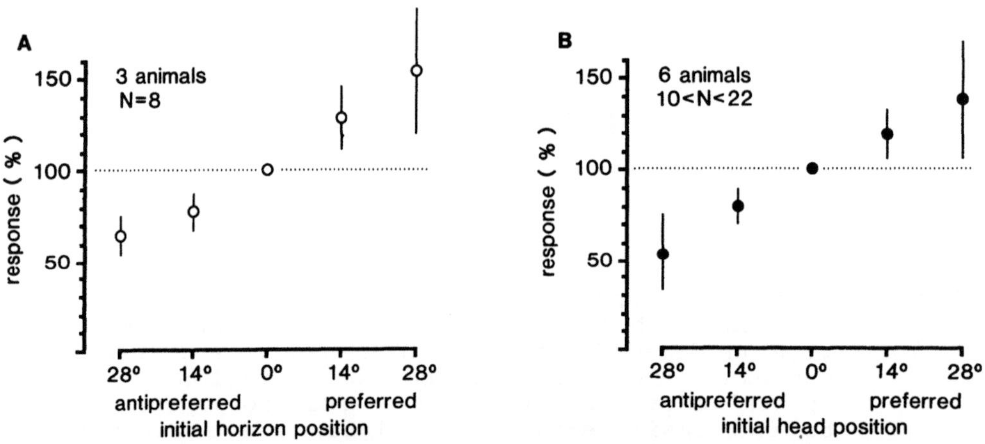

Fig. 9. The response of PI(2)5 to steplike deviations of identical amplitude depends on (A) the initial position of the horizon in the visual field (sector specificity), and (B) the relative roll position of the head with respect to the prothorax (the visual situation was constant at any head angle because horizon and head were aligned to each other before simulating deviations; cf. sketches in Fig. 14). The response increases the more horizon or head are oriented in the preferred direction (cf. Fig. 3). Note that the position information about horizon and head is manifested only in responses to deviations, and not as tonic spiking. N is the number of single experiments, each one representing 10 successive stimuli.

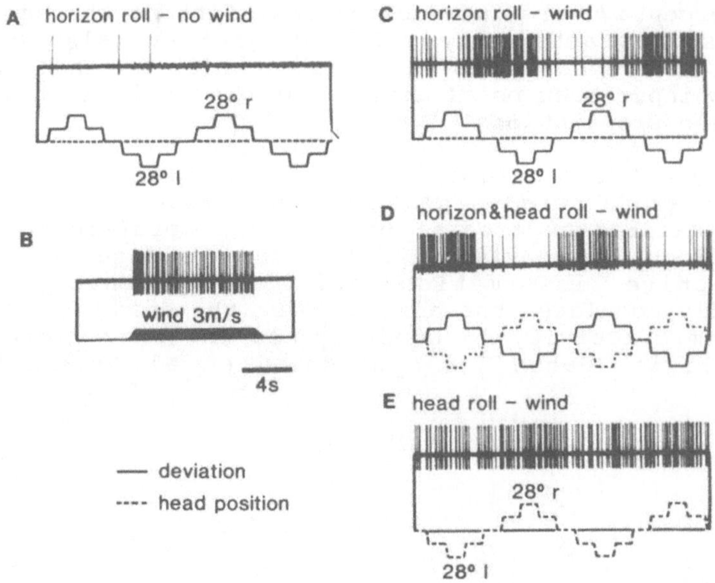

Fig.10. Response properties of a descending deviation detector with its axon on the right hand side, and with bilateral thoracic branches (brain anatomy unknown). (A) Simulated deviations alone elicit no response (and nor do mere head movements). (B) Frontal wind causes tonic spiking activity. (C, D) The wind response is modulated by the roll position of horizon and head respectively. In D head and horizon were moved in parallel keeping the visual situation constant. (E) Both modulatory effect balance each other when moving the head with respect to a stationary horizon (thus also shifting the retinal image in the opposite direction). The wind velocity was 3 m/s in all cases.

Wind responses, central inputs, and combination with visual stimuli

Free flight is associated with wind from an anterior direction, which influences many though not all DNs. Extracellular recordings from connectives revealed 60 % of 124 neurons (some might have been recorded more than once) to be excited, 13 % to be inhibited and the rest to be unaffected by stimulation of the wind-sensitive head hairs (Varanka and Svidersky 1974a, b; see also Williamson and Burns 1982). Excitatory wind responses of locust neurons were classified by Camhi (1969b) into slowly and rapidly adapting types. The former monitor tonically the presence of wind, either without any obvious directionality, or showing a preference for wind from a specific direction. The phasic neurons indicate changes of direction and/or acceleration of wind. None of these neurons is known morphologically but many of them may belong to the population of DNs described here. My own recordings revealed a similar percentage of wind-excited DNs, fitting the same scheme of response characteristics. Flight activity also causes in some 50 % of the DNs tonic excitation which is independent of wind. This input must be of central origin as it also occurs in deafferented preparations.

TCG, the tritocerebral commissure giant, is the best described among wind-sensitive, non-visual DNs. TCG is tonically excited by frontal wind, and it prefers ipsilateral yaw, and pitch down deviations. The wind response also contains proprioceptive information about the wingbeat, which rhythmically modulates the airflow over the head. During flight this rhythmic activity is superimposed on tonic excitation of central origin (Bacon 1983; Horsmann et al. 1983; Horsmann 1985).

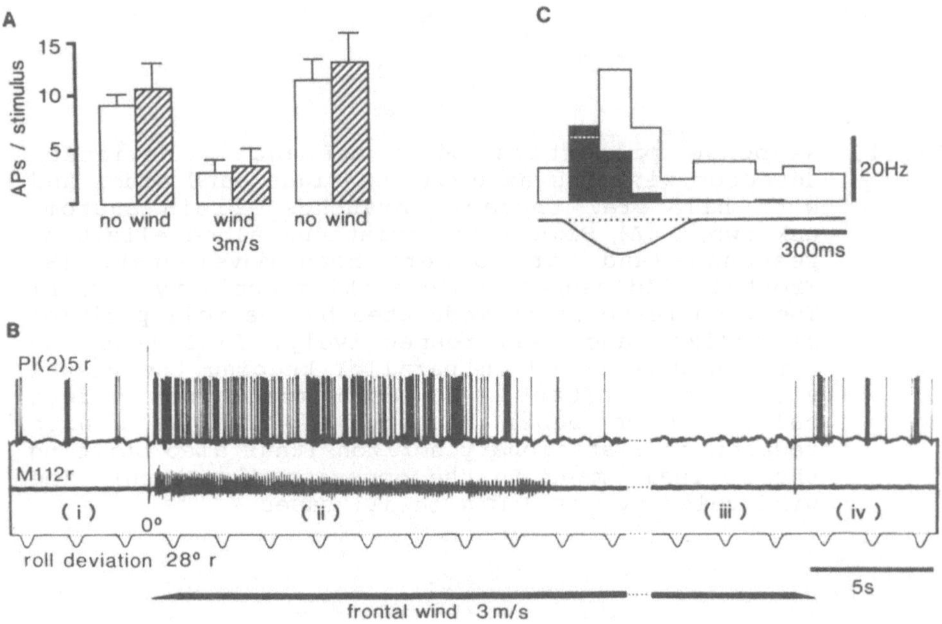

Fig. 11. (A) In quiescent locusts the response of PI(2)5 to roll movements of horizon (clear columns) or head (hatched columns) is inhibited by frontal wind (N=20, error bars = 1 s.d.). (B, i): Response to simulated deviations at rest. (B, ii) Response to simulated deviations during flight activity, started and maintained by frontal wind. All flight muscles are denervated except for M 112 (dorsal longitudinal wing depressor) which serves as a monitor of flight. (B, iii) The response to deviations is completely suppressed by wind inhibition when flight activity ceases. (B, iv) The response recovers immediately after switching off the wind. (C) Mean response to the simulated deviations as in B, i; (black histogram; N=6), and B, ii (clear histogram; N=8). Note that both the amplitude and phase of the response have been changed by the wind.

Most wind-sensitive DNs respond to visual stimuli as well. DNC, DNI, and DNM are phasically excited at the onset of wind. The response disappears rapidly in DNI, but tonic firing is regularly seen in DNM, and occasionally in DNC. The directionality of the wind response is congruent with that of the visual response. The wind response increases with decreasing illumination of the ocellus, then providing the major input to the DN. It is modulated by the same inputs as cause sector specificity of the visual responses. Fig. 10 demonstrates such an effect for a different DN which codes tonically for the velocity of frontal wind (Fig. 10B) but does not respond to visual and proprioceptive inputs alone (Fig. 10A). Inputs due to horizon position and head position only become obvious in the modulation of the wind response (Fig. 10C, D). The wind response remains unchanged when moving the head against a stationary horizon (thus moving the retinal image in the opposite direction). Under these conditions both modulatory inputs cancel each other out (Fig. 10E; cf. section 'Coding of head position').

Inhibition by frontal wind is only clearly demonstrated for PI(2)5 (Fig. 11A, B-iii), although I have preliminary evidence for other DNs as well. The amount of inhibition depends on the wind velocity, and is independent of wind direction, at least within the range of about ± 30°. Wind inhibition is overridden during flight activity by tonic excitation from central sources (Fig. 11B). This excitation increases the number of spikes following simulated deviations, and makes PI(2)5 a more tonic neuron, but at the cost of a reduced signal-to-noise ratio for phasic responses (Fig. 11B, C).

<u>Bulk activity of all DNs</u>

The above data suggest that the course of flying locusts is controlled by a number of more or less specialized DNs (probably exceeding 20 pairs), coding phasically for movements of horizon and head, and tonically for the actual course and for head position with respect to the thorax. The relative strength of these inputs may vary considerably in different DNs. Some respond to exteroceptors alone, but purely proprioceptive DNs were never found. Some 50% of all recorded DNs have proprioceptive inputs. The responses of most of them are qualitatively similar to these shown in Figs. 6, 7, 9, 10, but this does not tell us much about the importance of proprioceptive inputs for flight steering. This question was approached by recording the bulk activity of all steering neurons as expressed in the steering reactions of wing muscles. Corrective torque comes about principally by shifts of the relative timing of wing muscle activity (Möhl and Zarnack 1977; Taylor 1981b). It turned out that shifts of the 'relative latency' between left and right muscles 97 (first basalar, direct depressors; Fig. 12A, B) are proportional to corrective torque following a simulated course deviation (Hensler and Robert, in prep.). Such shifts are larger when the head is fixed to the pronotum than in the presence of compensatory head movements (Fig. 12B, C). Nevertheless, some 25% of the latency shift (thus of the torque) results from proprioceptive input (Hensler and Robert, in prep.), similar to what was shown for the response of DNC (Fig. 7).

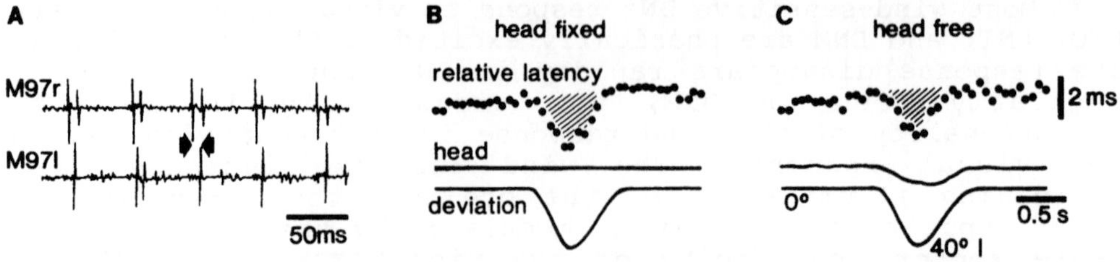

Fig. 12.(A) Myograms from the left and right muscles 97 (first basalar: direct depressor muscles) during flight. Arrows indicate the relative latency used as measure for steering responses. Large and and small spikes within one recording are both from the same motoneuron (data not shown). (B, C) The relative latency shifts in response to simulated roll deviations. This shift is proportional to corrective torque (Hensler and Robert in prep.). Note that the response (indicated by the hatched area) is larger when the head is fixed when compensatory head movements are permitted. Relative latencies are averaged responses (N=22).

Motor effects

Although the response properties and morphology of the DNs strongly suggest a crucial role in course control, the ultimate proof requires either the demonstration of a continuous chain of synaptic connections from DNs to motoneurons (as done for ocellar DNs: Reichert and Rowell 1985), or the evocation of the expected motor effect by stimulation of single DNs (i.e. shift of the relative latency into the same direction as during simulated deviation in the preferred direction of the respective DN). These latter effects may be expected to be small and blurred by the normal fluctuation of relative latencies, so that a clear demonstration is only to be achieved by averaging several responses. This was first done for TCG, revealing an effect on the relative timing of a number of wing musles (Bacon 1983). TCG can easily be recorded and stimulated extracellularly from the tritocerebral commissure, whereas selective stimulation of other DNs requires penetration with intracellular electrodes. Unfortunately averaging is almost impossible in the deafferented preparations normally used for intracellular recordings, because flight sequences are very short, and scatter of the relative latency is large. For these reasons I developed the preparation shown in Fig. 1; this allows one to record or stimulate a neuron while still permitting the locust to move its wings and head freely.

Stimulation experiments are here illustrated through PI(2)5. Short, depolarizing pulses, causing spike frequencies of some 200 Hz, do indeed shift the relative latency in the expected direction (Fig. 13A), by the same amount as during a simulated deviation of some 40° (cf. Fig. 12B, C). Some DNs (mostly wind-sensitive ones, and including TCG) elicit steering movements of the abdomen as well (Baader 1988b; and pers. comm.), but this was not observed for PI(2)5 and DNC. However,

PI(2)5 and many other visual neurons also elicit a compensatory head movement (Fig. 13B). Thus, they participate in the control of head movements which subsequently modulate their own excitability (see also section 'Conclusions ...'). Although we have no comprehensive picture yet about the motor effects of the DNs, different DNs in different combinations, and at different relative strengths seem to elicit shifts of the wings, and movements of the abdomen, legs, and head .

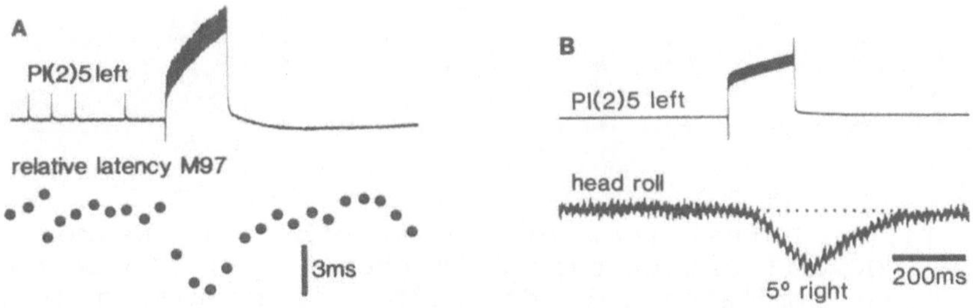

Fig. 13. Electrical stimulation of PI(2)5 causes (A) shifts the relative latency in M97 (mean values from 9 stimuli), and (B) compensatory head rolling. PI(2)5 was penetrated in the neck connective, and stimulation was achieved by taking advantage of the 'post inhibitory rebound effect', i.e. a depolarizing DC-pulse was applied after 10-30 minutes of hyperpolarization with some 10 nA.

THORACIC INTERNEURONS

Some DNs synapse directly upon wing motorneurons (Tyrer 1981; Simmons 1980; Rowell and Pearson 1983), but these connections are weak (PSPs mostly less than 0.5 mV), and not reliably established (Tyrer pers. comm.; Hensler unpubl.). The main pathway from DNs to motoneurons involves at least one level of intercalated thoracic interneurons (TINs; Reichert and Rowell 1985). TINs are a population of morphologically heterogeneous neurons projecting into one or both halves of one or several ganglia. There they synapse onto a various number of motoneurons which are thus modulated in a coordinated way. Typically this pathway is not effective in non-flying animals, as DN-input alone is not sufficient to excite TINs beyond their threshold. This pathway is only gated during the excitatory phases of the rhythmic input from the flight oscillator occuring during flight activity (Reichert and Rowell 1985; cf. Fig. 14C). More data are necessary about the connections between DNs, TINs, and motoneurons, to establish whether this is in fact the main neuronal pathway for corrective course control. However, the present data strongly support this view.

If TINs are really combining the activity of different DNs they should display a similar spectrum of response properties. This has in fact been found, and although not all TINs respond to head movements/position, six are known to do so (one was identified by Elson 1987). One of them, the mesothoracic neuron 761, is briefly introduced in Fig. 14. It receives monosynaptic input from an ocellar DN, most probably from DNC (Rowell,

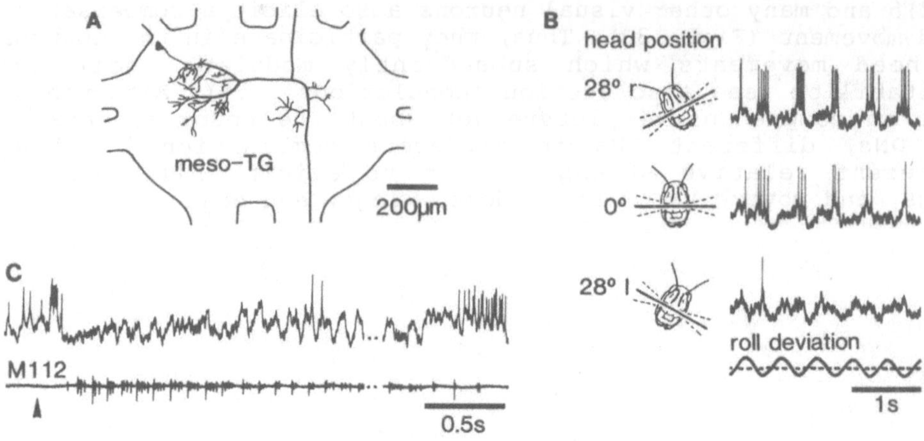

Fig. 14.(A) Dorsal view of the mesothoracic neuron 761 (Rowell and Reichert, in prep.). (B) The response to simulated roll deviations is modulated by the head position relative to the prothorax. In all cases the visual situation is identical, as indicated in the *insets*, showing the head position, the orientation of the horizon (solid line), and the maximum amplitudes of simulated deviations (dashed lines). (C) 761 is rhythmically inhibited during flight activity. Flight activity is monitored as rhythmic activity in wing muscle 112.

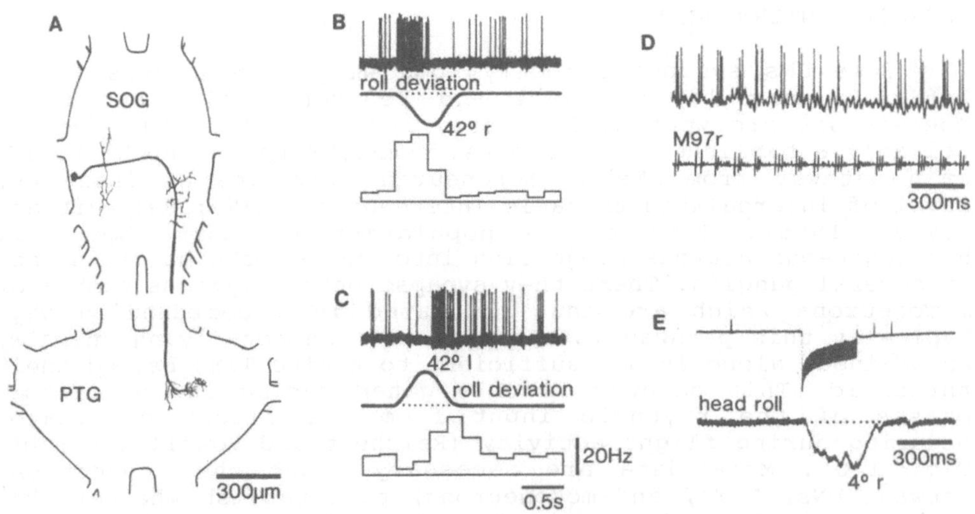

Fig. 15.(A) Dorsal view of a thoracic neuron connecting the suboesophageal- and the prothoracic ganglion. (B, C) Responses to simulated roll deviations. (D) The neuron is rhythmically excited during flight activity. Flight activity is indicated by the rhythmic activity in wing muscle 97. (E) electrical stimulation of the neuron causes head rolling to the right.

personal communication), and responds sector specifically to simulated course deviations (not shown). Its excitability is modulated by the relative head position (Fig. 14B), in the same way as shown for DNs (cf. Figs. 9B, 10D).

For reasons of segmental homology one would assume that neck motoneurons are coordinated by a similar population of premotor interneurons (here also called TINs, although neck motoneurons are located in the suboesophageal ganglion as well). Such neurons are already described for another orthopteran species, the cricket *Gryllus campestris* (Hensler 1988a), and one such neuron is known in locusts. This is rhythmically modulated by flight activity (Fig. 15D), and causes head rolling when stimulated electrically (Fig. 15E). It responds to simulated course deviations (Fig. 15B, C), but proprioceptive inputs were not observed.

CONCLUSIONS AND COMMENTS

The neural connections underlying flight steering are summarized in the simplified block diagram of Fig. 16: course deviations are counteracted by a negative feedback loop involving DNs, and TINs (Fig. 16, point 1; scheme modified after Rowell 1988). The same DNs also elicit compensatory head movements which reduce the perception of course deviations *via* a negative feedback loop (Fig. 16, point 2), and increase the DN response by proprioceptive input *via* a positive feedback loop (Fig. 16, point 3). This system functions only when the

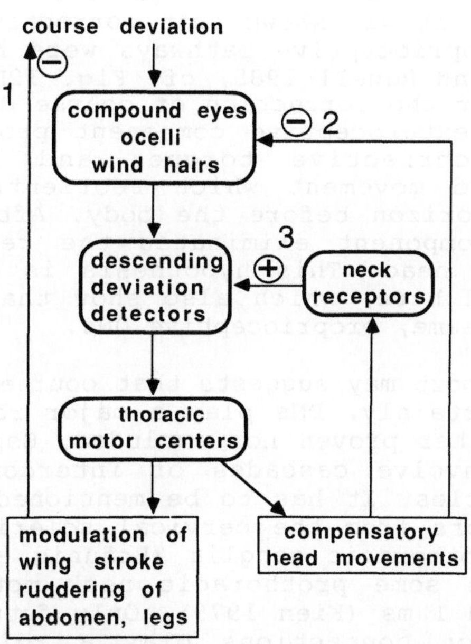

Fig. 16 Schematic summary of the neuronal connections known to underly corrective course control. For details see text.

overall effect of the positive feedback does not exceed that of the negative feedback. Indeed, compensatory head movements reduce the phasic steering responses to deviations (Fig. 12C), as would be expected if the negative feedback component is the stronger.

A locust can fly straight even when its head is immobilized, and the question remains as to why they reduce their steering responses by compensatory head movements. The common belief is that head movements stabilize the retinal image, thus improving visual exploration of the world. A further explanation comes from observations on locusts flying under closed-loop conditions. These oscillate along the straight path while stabilizing themselves with respect to their visual surround (Baker 1979; Robert, in press; Hensler and Robert in prep.). The oscillations are more pronounced when the head is immobilized than when it is free to move (Hensler and Robert, in prep.). Thus head movements seem to reduce overshooting steering reactions by influencing phasic DN responses to deviations. Hence, head movements stabilize the retinal image not only directly, but also by smoothing out the flight path.

The tonic coding of horizon and head position acts more slowly than the phasic responses to movements, which rapidly provide the animal with a corrective impulse. Tonic activity keeps the locust's body close to the normal position, either directly, *via* tonically spiking DNs, or indirectly, by adjusting the strength of phasic DN responses to horizon and head. An old model of course control assumed that deviations first realign the head with respect to the world, and that subsequently the body is reoriented by means of neck proprioceptors (Mittelstaedt 1950; Goodman 1965). This model was rejected when it was shown that corrective steering still occured after proprioceptive pathways were by-passed (Taylor 1981b; Reichert and Rowell 1985; cf. Fig. 12B). This led to a two-step model for the correction of course deviations (Taylor 1981b): a fast, exteroceptive component provides the animal with initial, corrective torque, and also elicits a compensatory head movement which reorients the head with respect to the horizon before the body. Afterwards, a slow, proprioceptive component eliminates the resulting mismatch between body and head. This hypothesis is supported by the results presented here, which also show that both steps are performed by the same, proprioceptive DNs.

The above report may suggests that course control is based on DNs alone. Certainly, DNs play a major role, but parallel pathways are neither proven nor excluded. Especially the slow pathway might involve cascades of interconnections before reaching the muscles. It has to be mentioned in this context that proprioceptors from the cervical sclerites project into the pro- and mesothoracic ganglia (Bräunig et al. 1983), and that they excite some prothoracic neck motoneurons with a latency of around 1 ms (Kien 1979). Only future research will show whether such connections play a role during flight steering, but our present knowledge provides a broad basis which will ensure that this is a very interesting task.

I like to thank Hugh Rowell for critically reading the manuscript. Supported by the Swiss Nationalfonds

References

Altman J S (1983) Sensory inputs and the generation of the locust flight motor pattern: from the past to the future. In: Nachtigall W (ed) Biona report 2 - Insect flight II, pp 127-136, Fischer Verlag, Stuttgart New York.

Altman J S, Kien J (1985) The anatomical basis for intersegmental and bilateral coordination in locusts. In: Bush B M H, Clarac F (eds) Coordination of motor behaviour. Cambridge University Press, Cambridge London New York, pp 91-119.

Arbas E A (1986) Control of hindlimb posture by wind-sensitive hairs and antennae during locust flight. J comp Physiol A 159: 849-857.

Baader A (1988a) Some motor neurones of the abdominal longitudinal muscles of grasshoppers and their role in steering behaviour. J exp Biol 134: 455-462.

Baader A (1988b) Activity of neck motor neurons and flow-field modulated interneurons in flying locusts. In: Elsner N, Barth F G (eds) Sense organs. Proceedings of the 16th Göttingen Neurobiology Conference, p 121. Georg Thieme Verlag, Stuttgart New York.

Bacon J (1983) The identified neurone approach to the study of insect flight. In: Nachtigall W (ed) Biona report 2 - Insect flight II, pp 1-9, Fischer Verlag, Stuttgart New York.

Baker P S (1979) Flying locust visual responses in a radial wind tunnel. J comp Physiol 131: 39-47.

Bräunig P, Pflüger H-J, Hustert R (1983) The specificity of central nervous projections of locust mechanoreceptors. J comp Neurol 218: 197-207.

Camhi J M (1969a) Locust wind receptors. I. Transducer mechanics and sensory response. J exp Biol 50: 335-348.

Camhi J M (1969b) Locust wind receptors. II. Interneurons in the cervical connective. J exp Biol 50: 349-362.

Camhi J M (1970) Yaw-correcting postural changes in locusts. J exp Biol 52: 519-531.

Elson R C (1987). Integration of wing proprioceptive and descending exteroceptive sensory inputs by thoracic interneurones of the locust. J exp Biol 128: 193-217.

Gewecke M (1974) The antennae of insects as air-current sense organs and their relationship to the control of flight. In: Barton Browne L (ed) Experimental analysis of insect behaviour, pp 100-113. New York, Heidelberg: Springer Verlag.

Goodman L J (1959) Hair receptors in locusts. Nature Lond 183: 1108-1109.

Goodman L J (1965) The role of certain optomotor reactions in regulating stability in the rolling plane during flight in the desert locust, *Schistocerca gregaria*. J exp Biol 42: 385-407.

Goodman L J, Fletcher W A, Guy R G, Mobbs P G, Pomfrett D J (1987) Motion sensitive descending interneurons, ocellar L_D Neurons and neck motoneurons in the bee: A neural substrate for visual course control in *Apis mellifera*. In: Menzel R, Mercer A (eds) Neurobiology and behavior in honeybees, pp 159-171. Springer Verlag, Berlin Heidelberg New York.

Griss C, Rowell C H F (1986) Three descending interneurons reporting deviation from course in the locust. I. Anatomy. J comp Physiol A 158: 765-774.

Haskell P T (1959) Function of certain prothoracic hair receptors in the desert locust. Nature Lond 183: 1107.

Heisenberg M, Wolf R (1984) Vision in *Drosophila*. Genetics of micro-behavior. In: Braitenberg V (ed) Studies of brain function, vol XII. Springer Verlag, Berlin, Heidelberg, New York.

Hengstenberg R (1988) Mechanosensory control of compensatory head roll during flight in the blowfly *Calliphoraerythrocephala*. J comp Physiol 163: 151-165.

Hensler K (1988a) Intersegmental interneurons involved in the control of head movements in crickets. J comp Physiol A 162: 111-126.

Hensler K (1988b) The pars intercerebralis neurone PI(2)5 of locusts: convergent processing of inputs reporting head movements and deviations from straight flight. J exp Biol (in press).

Honegger H-W, Altman J S, Kien J, Müller-Tautz R, Pollerberg E (1984) A comparative study of neck muscle motor neurons in a cricket and a locust. J comp Neurol 230: 517-535.

Horsmann U (1985) Der Einfluß propriozeptiver Windmessung auf den Flug der Wanderheuschrecke und die Bedeutung descendierender Neuronen der Tritocerebralkommissur. Dissertation, Universität Köln, Germany.

Horsmann U, Heinzel H-G, Wendler G (1983) The phasic influence of self-generated air current modulations on the locust flight motor. J comp Physiol 150: 427-438.

Hubel D H, Wiesel T N (1979) Brain mechanisms of vision. Sci Amer 241: 150-162.

Kien J (1979) Variability of locust motoneuron responses to sensory stimulation: A possible substrate for motor flexibility. J comp Physiol 134: 55-68.

Mittelstaedt H (1950) Physiologie des Gleichgewichtssinnes bei fliegenden Libellen. Z vergl Physiol 32: 422-463.

Möhl B. Short term learning during flight control in *Locusta migratoria*. J comp Physiol (in press).

Möhl B, Neumann L (1983) Peripheral feedback-mechanisms in the locust flight system. In: Nachtigall W (ed) Biona report 2 - Insect flight, pp 81-87. Fischer Verlag, Stuttgart New York.

Möhl B, Zarnack W (1977) Flight steering by means of time shifts in the activity of the direct downstroke muscles in the locust. In: Nachtigall W (ed) Physiology of movement - Biomechanics, pp 333-339, Gustav Fischer Verlag, Stuttgart New York.

Olberg R M (1986) Identified target-selective visual interneurons descending from the dragonfly brain. J comp Physiol 159: 827-840.

Pearson K G, Robertson R M (1987) Structure predicts synaptic function of two classes of interneurons in the thoracic ganglia of *Locusta migratoria*. Cell Tissue Res 250: 105-114.

Pflüger H J (1984) The large fourth abdominal intersegmental interneuron: A new type of wind-sensitive ventral cord interneuron in locusts. J comp Neurol 222: 343-357.

Preiss R, Gewecke M (1988) Visually induced wind compensation in the migratory flight of the desert locust,

Schistocerca gregaria. In: Elsner N, Barth F G (eds) Sense organs. Proceedings of the 16th Göttingen Neurobiology Conference, p 38. Georg Thieme Verlag, Stuttgart New York.
Reichert H, Rowell C H F (1985) Integration of nonphaselocked exteroceptice information in the control of rhythmic flight in the locust. J Neurophysiol 53: 1216-1233.
Reichert H, Rowell C H F (1988) Neuronal circuits controlling flight in the locust: How sensory information is processed for motor control. Trends Neurosci 9: 281-283.
Reye D N, Pearson K G (1987) Projections of the wing stretch receptors to central flight neurons in the locust. J Neurosci 7: 2476-2487.
Riley J R, Krueger U, Addison C M, Gewecke M (1988) Visual detection of wind-drift by high flying insects at noon: a laboratory study. J comp Physiol A 162: 793-798
Robert D. Visual steering under closed-loop conditions by flying locusts: flexibility of optomotor response and mechanisms of correctional steering. J comp Physiol (in press).
Robertson R M, Pearson K G (1982) A preparation for the intracellular analysis of neural activity during flight in the locust. J comp Physiol 156: 311-320.
Robertson R M, Pearson K G (1985) Neural networks controlling locomotion in locusts. In: Selverston A I (ed) Model neural networks and behavior, pp 21-35, Plenum, New York London.
Rowell C H F (1971) The orthopteran descending movement detectors (DMD) neurones: A characterization and review. Z vergl Physiol 73: 167-194:
Rowell C H F (1988) Mechanisms of flight steering in locusts. Experientia 88: 389-395.
Rowell C H F, Pearson K G (1983) Ocellar input to the flight motor system of the locust: structure and function. J exp Biol 103: 265-288.
Rowell C H F, Reichert H (1986). Three descending interneurons reporting deviation from course in the locust. II. Physiology. J comp Physiol 158: 775-794.
Sandeman, D.C. (1968). A sensitive position measuring device for biological systems. Comp. Biochem. Physiol. 24, 635-638
Shepheard P (1974) Control of head movement in the locust, *Schistocerca gregaria*. J exp Biol 60: 735-767.
Simmons P (1980) A locust wind and ocellar brain neurone. J exp Biol 85: 281-294.
Stevenson P A, Kutsch W (1987) A reconsideration of the central pattern generator concept for locust flight. J comp Physiol A 161: 115-129
Strausfeld N J, Bassemir U, Singh R N, Bacon J P (1984) Organizational principles of outputs from dipteran brains. J Insect Physiol 30: 73-93.
Taylor C P (1981a) Contribution of compound eyes and ocelli to steering of locusts in flight. I. Behavioural analysis. J exp Biol 93: 1-18.
Taylor C P (1981a) Contribution of compound eyes and ocelli to steering of locusts in flight. II. Timing changes in flight motor units. J exp Biol 93: 19-31.
Thüring D A (1986) Variability of motor output during flight steering in locusts. J comp Physiol 158: 653-664.
Tyrer N M (1981) Transmission of wind information on the head of the locust to flight motor neurons. In: J. Salánki

(ed) Adv. Physiol. Sci., Vol. 23, Neurobiology of invertebrates, pp 557-570.

Tyrer N M, Altman J S (1974) Motor and sensory flight neurones in a locust demonstrated using cobalt chloride. J comp Neurol 157: 117-138.

Varanka I, Svidersky V L (1974a) Functional characteristics of the interneurons of wind-sensitive hair-receptors on the head in *Locusts migratoria* L. _ I. Interneurons with excitatory responses. Comp Biochem Physiol 48A: 411-426.

Varanka I, Svidersky V L (1974b) Functional characteristics of the interneurons of wind-sensitive hair-receptors on the head in *Locusts migratoria* L. _ II. Interneurons with inhibitory responses. Comp Biochem Physiol 48A: 411-426.

Wagner H (1986) Flight performance and visual control of flight of the free-flying housefly (*Musca domestica* L.). I. Organization of the flight motor. Phil Trans R Soc Lond B 312: 527-551.

Wehner R (1981) Spatial vision in arthropods. In: Autrum H (ed) Handbook of sensory physiology, Volume VII/6C, Comparative Physiology and Evolution of Vision in Invertebrates. Springer Verlag, Berlin Heidelberg New York, pp 287-616.

Wendler G (1983) The interaction of peripheral and central components in insect locomotion. In: Huber F, Markl H (eds) Neuroethology and behavioral Physiology. pp.42-53. New York, Heidelberg: Springer Verlag.

Williams J L D (1975) Anatomical studies of the insect central nervous system: A ground-plan of the midbrain and an introduction to the central complex in the locust, *Schistocerca gregaria* (Orthoptera). J Zool Lond 176: 67-86.

Williamson R, Burns M D (1982) Large neurones in locust neck connectives. I. Responses to sensory inputs. J comp Physiol 147: 379-388.

Wolf H, Pearson K G (1988) Proprioceptive input patterns elevator activity in the locust flight system. J Neurophysiol 59: 1831-1853.

Zanker J M (1988) How lateral abdomen deflection contributes to flight control of Drosophila melanogaster. J comp Physiol A 162: 581-588.

SENSORY CONTROL OF LOCAL REFLEX - MOVEMENTS IN LOCUSTS

Hans-Joachim Pflüger

Freie Universität Berlin, FB Biologie
Neurobiologie, Königin-Luise-Str. 28-30
D - 1000 Berlin 33, FRG

ABSTRACT

This article describes the central nervous connectivity of single identified receptor cells such as the tibial spur receptor and tibial campaniform sensilla of a locust hind leg. An identified tibial spur receptor makes direct, monosynaptic connections with a local spiking interneurone, an identified tibial campaniform sensillum makes direct connections with motor neurones and local nonspiking interneurones. Other motor neurones are most likely excited by a disinhibitory pathway. The behavioural significance of these reflexes is discussed.

INTRODUCTION

The study of the neural control of locomotion in locusts has been a focal point of research in insect neurobiology. For a long time all approaches and discussions in this field were somewhat overshadowed by the central versus peripheral controversy which in the light of new results has lost much of its sharpness. Most researchers in this field would now agree that sensory receptors play an important, if not key role in the proper execution of locomotory movements in insects. Therefore, the study of how sensory activity can influence motor commands is of great interest.

Insects, for example in the locust, offer several experimental advantages, of which I would like to mention the following: i) they possess identifiable neurones which can be recorded intracellulularly with microelectrodes, stained individually by the injection of dyes and, in many cases, have been shown to have unique function in behaviour, ii) they have a remarkably rich behavioural repertoire, and in particular many different locomotor patterns which they will express under restrained and partially dissected conditions, iii) they are easily bred and therefore make excellent laboratory animals.

For our studies we have selected local reflexes which are reflex arcs running exclusively within one body segment. This has the advantage that such reflex movements can easily be elicited in a locust which is dissected for intracellular recordings and fixed except for one or several appendages (Fig. 1A). One of the first aims is the identification of pathways between sensory receptors, interneurones and motor neurones. In the following article I shall describe some of the pathways that have been identified. Much of the emphasis is laid on the connectivity of identified receptors.

RESULTS AND DISCUSSION

An enormous wealth of sensory receptors on the legs provide the locust with information on external and internal proprioceptive stimuli. For animals with an exoskeleton it is important to have receptors that are stimulated when any obstacle touches the surface of the body.

This information is provided by long, stiff tactile hairs scattered over the surface. Each tactile hair possesses one receptor cell that responds to touch in a phasic, non directional way. The axons of these cells project into the ventral neuropiles of the thoracic ganglion (Pflüger et al. 1988) and there is evidence that the projections are somatotopically arranged (Pflüger 1980a, Pflüger et al. 1981, Johnson and Murphey 1985) Reflexes elicited by tactile hairs on a leg are avoidance responses whereby the leg is moved away from the stimulus and muscles of several joints of a leg are co-activated. Even stimulating a single hair can be sufficient to elicit the avoidance movement. All avoidance responses are subject to an enormous flexibility, apparently depending on the behavioural state, or general activity of the animal. They are most reliably expressed in a resting animal, or in an animal whose corresponding segment has been completely isolated by cutting all connectives to other segments (Pflüger 1980b).

Avoidance movements of the locust hind leg have been studied in great detail (Siegler and Burrows 1983). The receptor cells in tactile hairs make direct, excitatory monosynaptic connections with a population of local spiking interneurones. Each of the latter receives inputs from many hairs of a relatively well defined receptive field. The surface of the leg is thus mapped on this group of interneurones. These physiological results correspond to anatomical findings. There is for example a considerable overlap between interneuronal branches and the sensory terminals within the ventral neuropiles of a ganglion. That the spiking interneurones are indeed the primary integrators of mechanosensory information is underlined by the fact that they also receive inputs from many other receptors. This will be described in the next section.

At the distal end of the tibia there are two pairs of movable spurs (Burrows and Pflüger 1986, Fig. 1B). A specialized campaniform sensillum (the so called "spur receptor") is situated at the base of each spur and excited phasically when the latter is passively displaced (Figs. 1C, 2A,B). Afferent spikes from the receptors of the two anterior spurs directly excite local spiking interneurones which are unaffected by those from the two posterior spurs. Other local spiking interneurones receive an opposing pattern of inputs (Fig. 2B,C). The afferent spikes of the two spur receptors of one side evoke epsps of different amplitude in a particular spiking interneurone. This could result from an uneven synapse distribution of the two sensory axons onto the spiking interneurone, i.e. the larger epsp being closer to the recording site in the soma of the interneurone. However, it could also be a special feature of the synapses between the interneurone and one of the two spur receptors, showing that the two spurs are not equally important for eliciting a local reflex. The spiking interneurones are also directly excited by tactile hairs on the tibia.

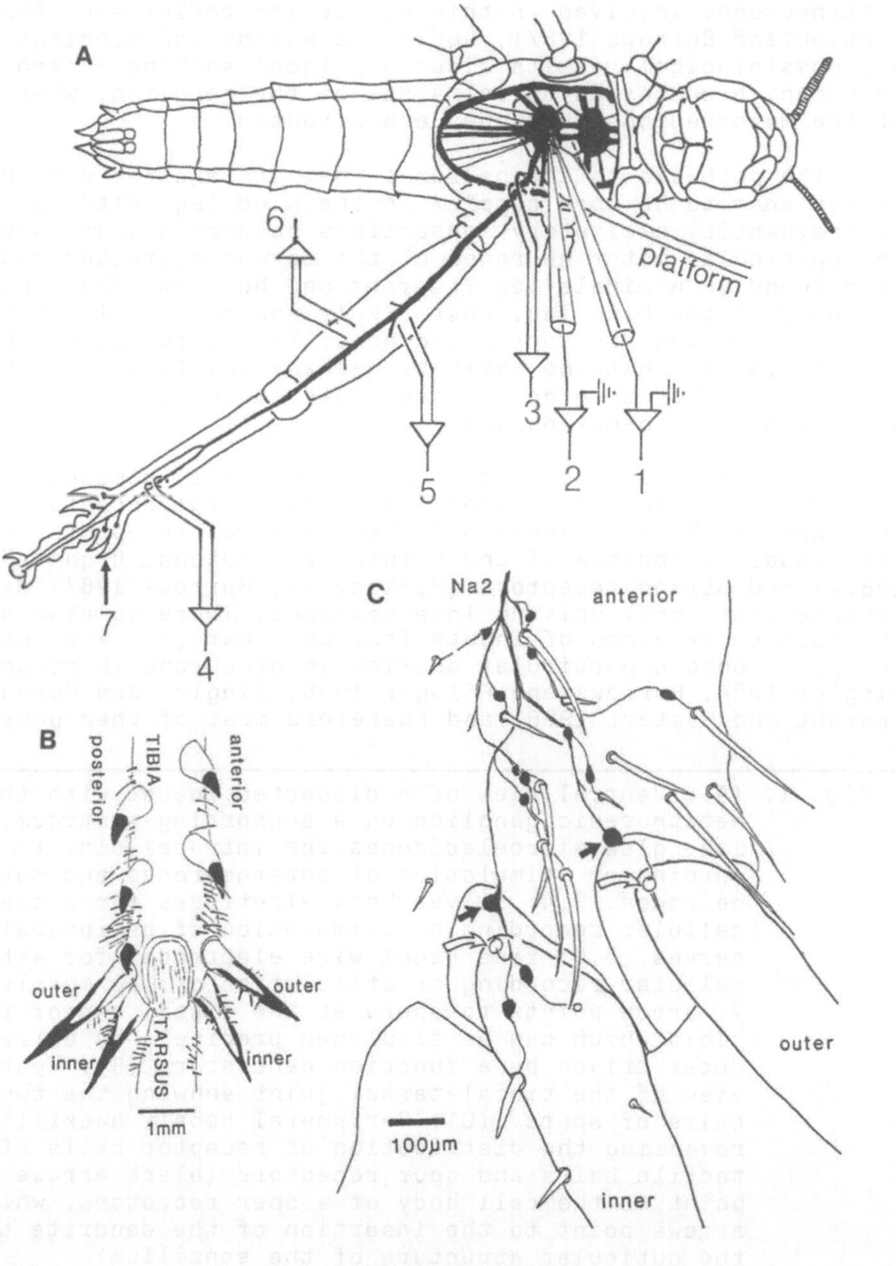

Fig. 1 Continued Next page

The reflex movement itself consists of an inhibition of the tonically active single tarsus levator motor neurone, as a result of which the tarsus is lowered (Fig. 2A). It turns out that the local spiking interneurones mediate this inhibition. Each spike within the interneurone evokes an inhibitory post-synaptic potential (ipsp) in the tarsus levator motor neurone which prevents it from spiking (Burrows and Siegler 1982, 1984). Therefore, this reflex is an example of a three-neurone-arc (Fig. 2D). Immunocytochemical studies by Watson and Burrows (1987) show that at least some of the local spiking interneurones show GABA-immunoreactivity and that the application of picrotoxin, which blocks the effects of GABA, abolishes this inhibition. Therefore, it can be concluded that the local spiking interneurones involved in this particular reflex are GABAergic (Watson and Burrows 1987). Again the anatomical findings support the physiological results since all local spiking interneurones have many branches in dorsal areas of the ganglion, where most of the motorneuronal branches are situated.

The pathways to the neurones that are excited during this reflex seem to be more complex in the hind leg. Although direct (monosynaptic) excitatory connections between a spur receptor and particular motor neurones of the tarsus depressor muscle were found in a middle leg (Laurent and Hustert 1988) none could be found in the hind leg. Most likely another local interneurone, a nonspiking one, is interposed here. Therefore, in the hind leg it appears that the excitory pathway involves at least four neurones, two of them being local interneurones, one spiking and the other nonspiking one.

It soon became clear that not only hairs, campaniform sensilla (Burrows and Siegler 1985) and tarsal receptors (Laurent and Hustert 1988) but also internal receptors such as individual scolopidia of the femoral chordotonal organ (Burrows 1987a) and strand receptors (Pflüger and Burrows 1987) directly synapse onto local spiking interneurones. There is also an enormous convergence of inputs from both external and internal receptors onto a particular spiking interneurone (Burrows and Siegler 1984, Burrows and Pflüger 1986, Siegler and Burrows 1984, Laurent and Hustert 1988) and therefore most of them possess

---

Fig. 1. (A): Ventral view of a dissected locust with the metathoracic ganglion on a supporting platform. 1,2: glassmicroelectrodes for intracellular recording or stimulation of interneurones and motor neurones. 3,4: silver hook electrodes for extracellular recording or stimulation of peripheral nerves. 5,6: fine steel wire electrodes for extracellular recording or stimulation of leg muscles. 7: arrow points to spurs at the distal end of the tibia which can be displaced precisely by a transducer driven by a function generator. (B): Ventral view of the tibial-tarsal joint showing the two pairs of spurs. (C): Peripheral cobalt backfill revealing the distribution of receptor cells of tactile hairs and spur receptors (black arrows point to the cell body of a spur receptors, white arrows point to the insertion of the dendrite to the cuticular structure of the sensillum).

elaborate receptive fields (Burrows 1985). All output connections which have been identified so far are inhibitory and are either made directly to motor neurones (Burrows and Siegler 1982, 1984), to local nonspiking interneurones (Burrows 1987b) or to intersegmental interneurones (Laurent 1987).

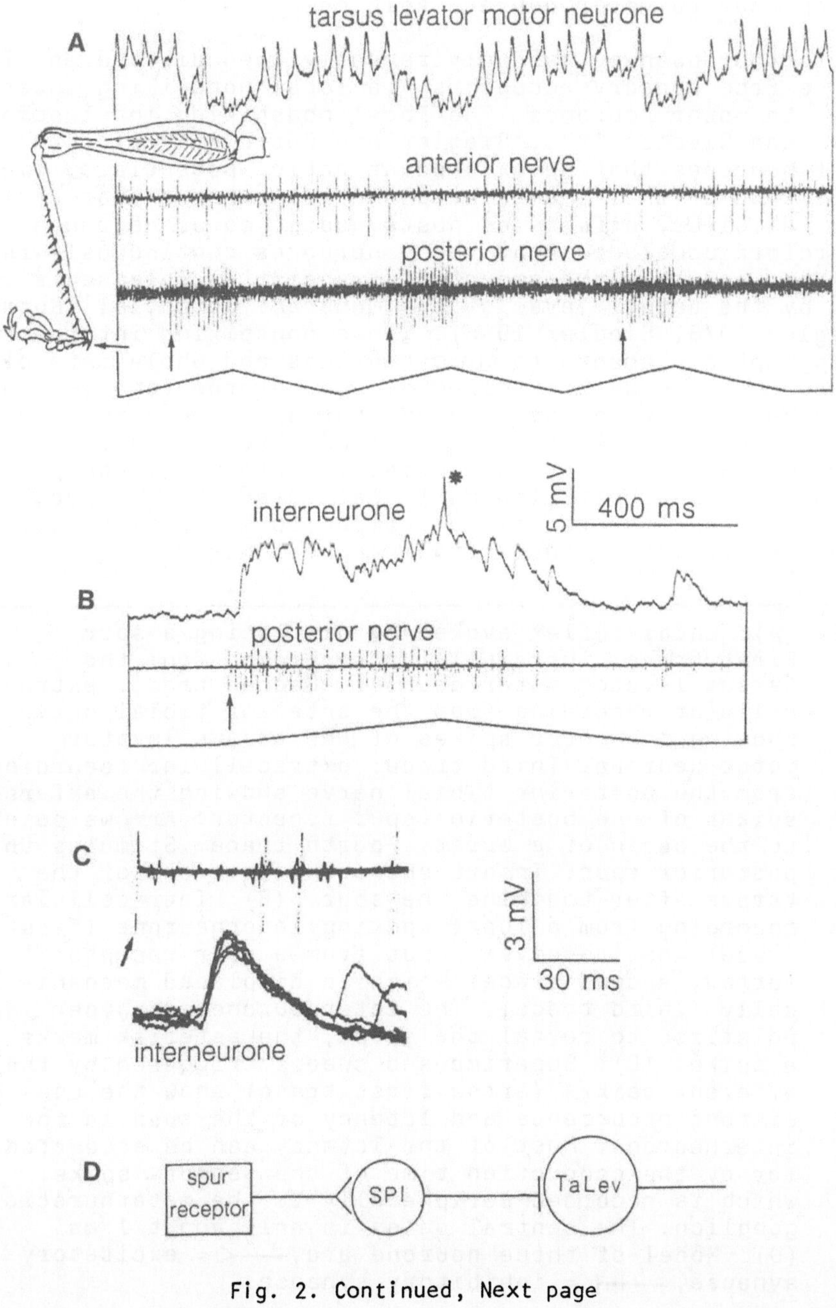

Fig. 2. Continued, Next page

The pathway via the local spiking interneurones is not the only one for sensory receptors. The simplest kind of reflex is a direct connection between a sensory receptor and a motor neurone (monosynaptic reflex). In the locust an increasing number of such direct, monosynaptic connections have been described (wing stretch receptor, Burrows 1975; individual scolopidia of the femoral chordotonal organ, Burrows 1987a; tibial campaniform sensilla, Burrows and Pflüger 1988; tarsal receptors, Laurent and Hustert 1988 and hair-plate sensilla in cockroaches, Pearson et al. 1976). Each afferent spike evokes an epsp in the motor neurone. Under favourable conditions summation of such epsps can lead to motor neurone spiking.

A further pathway has only recently been discovered. This being one from sensory receptors via local nonspiking interneurones to motor neurones. The local nonspiking interneurones (Burrows and Siegler 1976, Siegler and Burrows 1979) form a class of neurones that never produce action potentials. Small displacements of their membrane potential exert either excitatory or inhibitory effects on postsynaptic motor neurones. It was therefore concluded that these neruones continuously release transmitter and that the amount of transmitter release is controlled by the actual level of the membrane potential (Burrows and Siegler 1978, Siegler 1985). Known nonspiking interneurones are presynaptic elements to motor neurones and whole sets of motor neurones can be controlled by one interneurone (Burrows 1980). Recently, it was discovered that afferences from chordotonal organs (Burrows et al. 1988) and tactile hairs (Laurent and Burrows 1988) also synapse directly onto local nonspiking interneurones. In the following I shall describe a direct synaptic pathway from two tibial campaniform sensilla (CS) of the hind leg to local nonspiking interneurones.

Fig. 2. (A): Local reflex evoked by displacing a spur. First trace: intracellular recording from the tarsus levator motor neurone. Second trace: extracellular recording from the anterior tibial nerve showing the motor spikes of the tarsus levator motor neurone. Third trace: extracellular recording from the posterior tibial nerve showing the afferent spikes of one posterior spur receptor. Arrows point to the begin of a burst. Fourth trace: Stimulus to posterior spur. Insert shows the movement of the tarsus after touching the spur. (B): Intracellular recording from a local spiking interneurone (first trace) that receives input from a spur receptor (arrow, second trace) which is displaced mechanically (third trace). The interneurone was hyperpolarized to reveal the epsps, the asterisk marks a spike. (C): Superimposed sweeps triggered by the afferent spikes (arrow first trace) show the consistent occurrence and latency of the epsp in the interneurone. Most of the latency can be accounted for by the conduction time of the sensory spike, which is recorded peripherally to the metathoracic ganglion. The central delay is only about 1 ms. (D): Model of three neurone arc, ——< = excitatory synapse, ——┤ = inhibitory synapse.

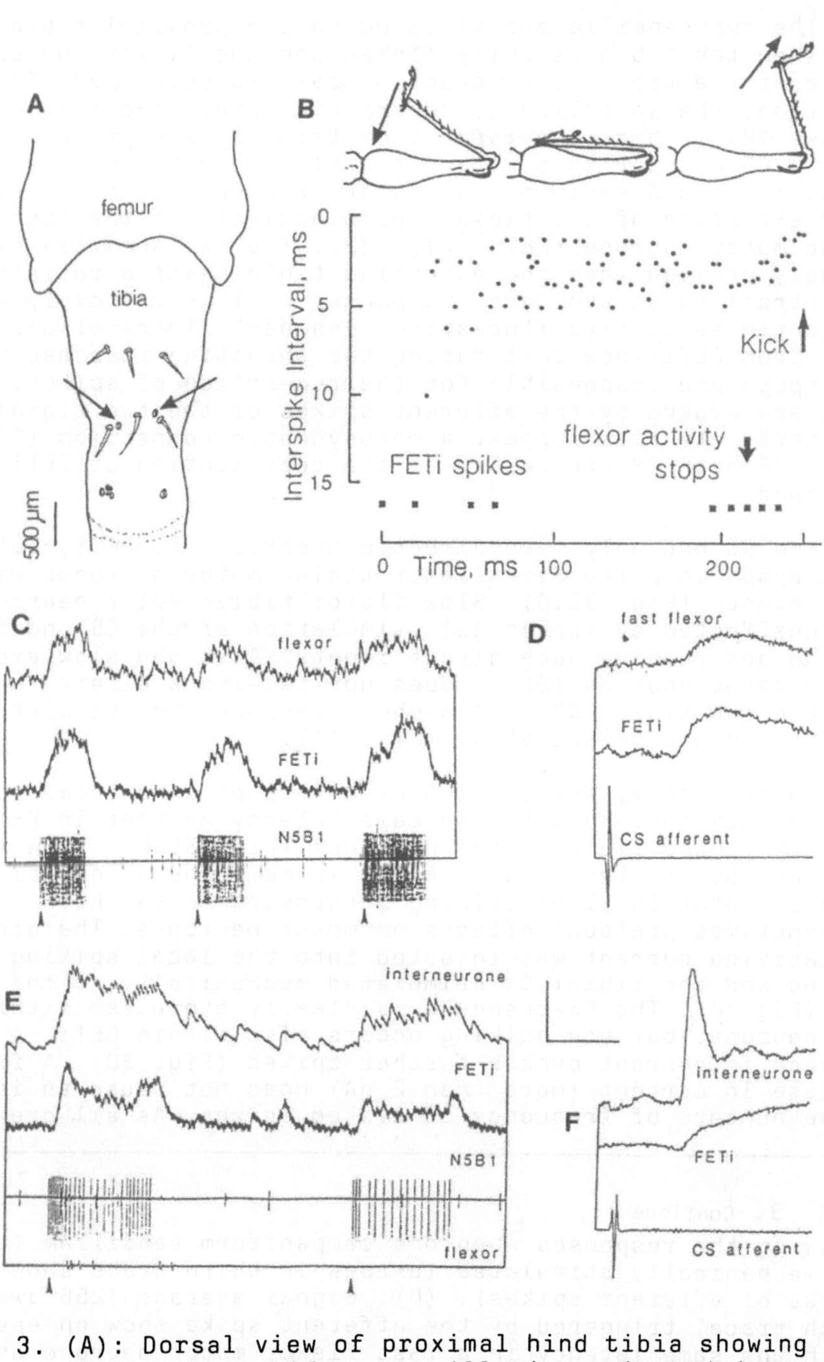

Fig. 3. (A): Dorsal view of proximal hind tibia showing the two campaniform sensilla investigated (arrows). (B): Interspike interval plot of afferent burst during the co-contraction phase. The motor pattern which precedes a jump is shown in the insert, it consists of an initial flexion phase, the co-contraction phase with the tibia fully flexed and the kick (Heitler and Burrows 1977). (C): Intracellular recordings from a flexor tibiae motor neurone (first trace) and FETi (second trace)

(Continued)

The two sensilla are situated on the proximal tibia (Fig. 3A). When the tibia is fully flexed and the flexor and extensor muscle of the tibia co-contract, strain is developed. This situation, the so called co-contraction phase (Heitler and Burrows 1977), produces force isometrically and precedes a jump. The two CS monitor this strain exactly, whereby their spike frequency reaches a maximum just before the occurrence of the rapid extension of the tibia due to activity of the fast extensor tibiae motor neurone (FETi, Fig. 3B): The two sensilla are also strongly excited when the extending tibia meets a resistance. In contrast to an avoidance response, FETi is reflexely excited and continues to fire (thrusting response). Intracellular recordings from FETi show that during the thrusting response summating epsps are responsible for the reneration of spikes. These epsps are evoked by the afferent spikes of the two tibial CS and the short latencies suggest a monosynaptic connection (Figs. 3C-F). IF both CS are destroyed the reexcitation of FETi is abolished.

The CS not only make direct connections to FETi, but also evoke epsps in a few fast flexor tibiae motor neurones with the same latency (Fig. 3C,D). Slow flexor tibiae motor neurones remain unaffected by mechanical stimulation of the CS and therefore do not receive such direct inputs. Also the slow excitor tibiae motor neurone (SETi) does not receive a direct input from the two tibial CS, although it responds to stimulation of the CS with a depolarization (Fig. 4A).

Surprisingly, the CS also evokes epsps in a local nonspiking interneurone with the same latency as that in FETi or the fast flexor tibiae motor neurones (Fig. 3E-F). It was mentioned before that small shifts in membrane potential of these premotor local nonspiking interneurones can have clear and sometimes profound effects on motor neurones. Therefore, depolarizing current was injected into the local spiking interneurone and the tibial CS stimulated mechanically at the same time (Fig. 4). The CS-response is clearly expressed within the interneurone, but now spiking occurs also within SETi. A gradual increase in current evokes further spikes (Fig. 3C). A further increase in current (more than 2 nA) does not cause an increase in the numbers of frequency of evoked spikes. As all previous

---

Fig. 3. Continued) showing the responses when one campaniform sensillum (CS) is mechanically stimulated (arrows in third trace show burst of afferent spikes). (D): Signal average (256 sweeps each trace) triggered by the afferent spike show an epsp with the same latency in a fast flexor motor neurone and FETi. (E): Afferent CS-spikes (third trace and arrows) cause a depolarization within a local spiking interneurone (first trace, neurone hyperpolarized for better visualization of epsps) and FETi (second trace). A flexor tibiae motor neurone also responds (see electromyogram, fourth trace). (F): Signal average again shows epsps with the same latency in the spiking interneurone and FETi (see also D). Calibration: horizontal (C) 500 ms, (E) 250 ms, (D) 18 ms, (F) 17 ms, vertical (C) flexor 4 mV, FETi 2 mV, (E) 10 mV.

tests to show a connection between the two tibial CS and SETi failed, the suggestion in that these excitatory effects result from a disinhibitory pathway (Fig. 3D). The impaled local nonspiking interneurone (interneurone 1 in Fig. 4D) inhibits a second local nonspiking interneurone (interneurone 2 in Fig. 4D), that does not itself receive direct CS-inputs but exerts a tonic inhibitory effect on SETi. If nonspiking interneurone 1 is depolarized, for example by epsps from the afferent CS-spikes or by current injection, nonspiking interneurone 2 is inhibited and therefore its inhibitory influence on SETi is removed. This will cause a depolarization within SETi and thus a "response" occurs as if it were the result of CS-stimulation.

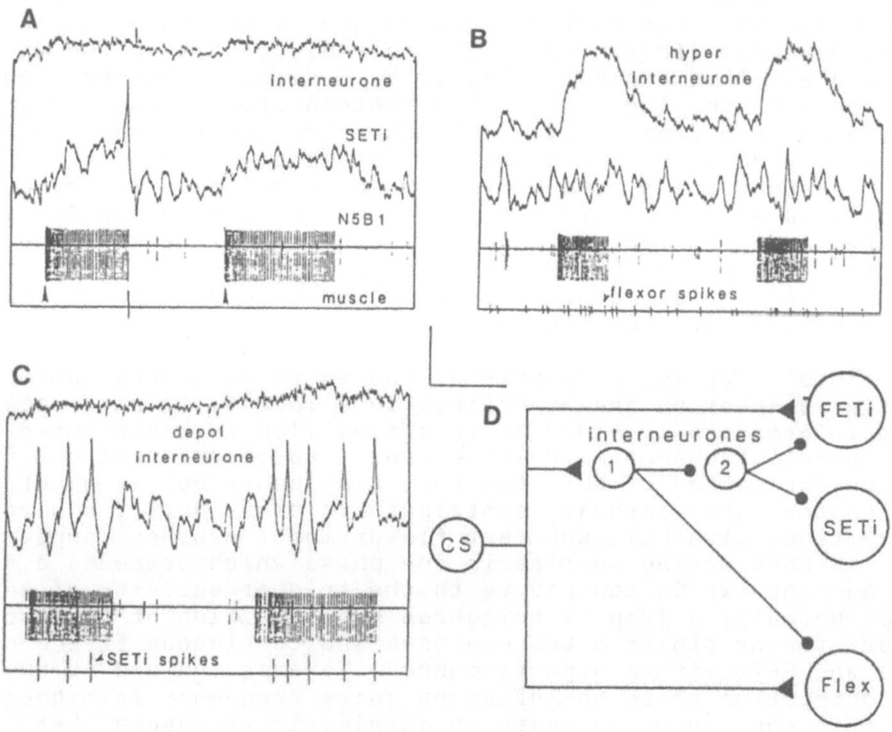

Fig. 4. (A-C): Manipulating the membrane potential of the nonspiking interneurone (first traces) alters the response of SETi (second traces) to stimulation of one CS (third traces, arrows in A show afferent burst). Electromyogram recordings from extensor and flexor tibiae muscle (fourth trace). (D): Diagram showing the connections that could explain these observations. Intracellular recordings are from interneurone 1 ——◀ = excitatory synapses, ——● = inhibitory synapses. Calibration: horizontal (A,C) 250 ms, (B) 500 ms, vertical interneurone 10 mV, SETi 4 mV.

This disinhibitory pathway hypothesis can be tested experimentally. By injecting hyperpolarizing current into nonspiking interneurone 1, the "response" within SETi should not occur, since interneurone 2 can now fully exert its inhibitory action. Although stimulation of the tibial CS will depolarize nonspiking interneurone 1, this is insufficient to override the hyperpolarization due to current injection. The result of the experiment is exactly as predicted (Fig. 4B). Under these conditions SETi should not "respond". This experiment also has effects on other motor neurones which do receive direct CS-inputs. Injecting hyperpolarizing current into nonspiking interneurone 1 reduces the amplitudes of epsp within FETi, which are evoked by the afferent CS-spikes, by as much as 40%.

Effects also occur within flexor tibiae motor neurones. When nonspiking interneurone 1 is at its "resting" membrane potential or depolarized, stimulation of the CS does not evoke spikes within these flexor motor neurones but causes only subthreshold depolarizations in those which receive a direct input from CS. Hyperpolarizing interneurone 1, however, causes spikes within flexor motor neurones, which can be recorded in the flexor electromyogram (fourth trace in Fig. 4B), each time the CS are stimulated. This can be explained by a direct inhibitory connection between the local nonspiking interneurone 1 and flexor tibiae motor neurones (Fig. 4D). When the interneurone is hyperpolarized and unable to exert its inhibitory effect, the direct afferent connection between the CS and the flexor motor neurones is able to make some of them spike. As the flexor muscle is innervated by at least 9 motor neurones we do not know whether the motor spikes revealed in the electromyogram are from those fast flexor tibiae motor neurones which have been previously shown to receive a direct CS-input.

This study is an example of the enormous divergence of a single receptor on the hind tibia in a locust. It also shows that information provided by this sensillum is distributed over many parallel channels. We have mentioned before that due to the re-excitation of FETi the thrusting behaviour is generated. How do the other pathways contribute to behaviour? The direct connections with FETi and fast flexor motor neurones support both co-contracting muscles in the phase which precedes a jump. How can the two CS contribute to the trigger activity of the jump? Normally a jump is triggered by inhibition of the excitatory flexor tibiae motor neurones and continuous firing of FETi and SETi with a high frequency. Towards the end of the co-contraction phase the afferent spike frequency is highest, due to a continuous increase in strain. If we assume that the summating epsps within the local nonspiking interneurone 1 have to reach a threshold before the inhibitory action of this neurone is exerted, then all excitory flexor motor neurones will be inhibited, due to a direct connection between interneurone 1 and them, but FETi and SETi will continue their firing as the disinhibitory effects will be greatest.

This shows how, by using different paralles channels, one receptor can contribute to a number of behavioural events. It also reveals how important local nonspiking interneurones are in gating or modulating such influences.

ACKNOWLEDGEMENTS

I thank Paul A Stevenson, Berlin, for correcting my English version. The support of my research by the DFG (Deutsche Forschungsgemeinschaft) and by an EEC laboratory twinning grant to Malcolm Burrows and myself is gratefully acknowledged.

LITERATURE

Burrows, M., 1975. Monosynaptic connexions between wing stretch receptors and flight motorneurones of the locust. J exp Biol 62: 189-219.

Burrows, M., 1980. The control of sets of motoneurones by local interneurones in the locust. J Physiol (Lond) 298: 213-233.

Burrows, M., 1985. The processing of mechanosensory information by spiking local interneurones in the locust. J Neurophysiol 54: 463-478.

Burrows, M., 1987a. Parallel processing of proprioceptive signals spiking local interneurones and motor neurones in the locust. J Neurosci 7: 1064-1080.

Burrows, M., 1987b. Inhibitory interactions between spiking and nonspiking local interneurones in the locust. J Neurosci 7: 3282-3292.

Burrows, M., Pflüger, H.J., 1986. Processing by Local Interneurons of Mechanosensory Signals Involved in a Leg Reflex of the Locust. J Neurosci 6: 2764-2777.

Burrows, M., Pflüger, H.J., 1988. Positive feedback loops from proprioceptors involved in leg movements of the locust. J comp Physiol A 163: 425-440.

Burrows, M., Siegler, M.V.A., 1976. Transmission without spikes between locust interneurones and motoneurones. Nature Lond 262: 222-224.

Burrows, M., Siegler, M.V.S., 1978. Graded synaptic transmission between local interneurones and motoneurones in the metathoracic ganglion of the locust. J Physiol Lond 285: 231-255.

Burrows, M., Siegler, M.V.S., 1982. Spiking local interneurones mediate local reflexes. Science NY 217: 650-652.

Burrows, M., Siegler, M.V.S., 1984. The morphological diversity and receptive fields of spiking local interneurones in the metathoracic ganglion of the locust. J comp Neurol 224: 483-508.

Burrows, M., Siegler, M.V.S., 1985. Organization of receptive fields of spiking local interneurones in the locust with inputs from hair afferent. J Neurophysiol 53: 1147-1157.

Burrows, M., Laurent, G. and Field, L.H., 1988. Proprioceptive inputs to nonspiking local interneurones contribute to local reflexes of a locust hindleg. J Neurosci (in press).

Heitler, W.J., Burrows, M., 1977. The locust jump. I. The motor programme. J Exp Biol 66: 203-219.

Johnson, S.E. and Murphey, R.K., 1985. The afferent projection of mesothoracic bristle hairs in the cricket. Acheta domesticus. J comp Physiol A 156: 369-379.

Laurent, G., 1987. Parallel effects of joint receptors on motor neurones and intersegmental interneurones in the locust. J comp Physiol A 160: 341-353.

Laurent, G. and Hustert, R., 1988. Motor neuronal field delimit patterns of motor activity during locomotion of the locust. J Neurosci (in press).
Pearson, K.G., Wong, R.K.S., Fourtner, C.R., 1976. Connexions between hair plate afferents and motoneurones in the cockroach leg. J exp Biol 64: 251-266.
Pflüger, H.J., 1980a. Central nervous projections of sternal trichoid sensilla in locusts. Naturwissenschaften 67: 316.
Pflüger, H.J., 1980b. The function of hair sensilla on the locust's leg: The role of tibial hairs. J exp Biol 87: 163-175.
Pflüger, H.J. and Burrows, M., 1987. A strand receptor with a central cell body synapses upon spiking local interneurones in the locust. J comp Physiol A 160: 295-304.
Pflüger, H.J., Bräunig, P. and Hustert, R., 1981. Distribution and specific central projections of mechanoreceptors in the thorax and proximal leg joint of locusts. II. The external mechanoreceptors: hair plates and tactile hairs. Cell Tissue REs 216: 79-96.
Pflüger, H.J., Bräunig, P. and Hustert, R., 1988. The organization of mechanosensory neuropiles in locust thoracic ganglia. Phil Trans Roy Soc Lond B 321: 1-26.
Siegler, M.V.S., 1985. Nonspiking Interneurons and Motor Control in Insects. Adv Insect Physiol 18: 249-304.
Siegler, M.V.S. and Burrows, M., 1979. The morphology of local nonspiking interneurones in the metathoracic ganglion of the locust. J comp Neurol 183: 121-148.
Siegler, M.V.S. and Burrows, M., 1983. Spiking local interneurones as primary integrators of mechanosensory information in the locust. J Neurophysiol 50: 1281-1295.
Siegler, M.V.S. and Burrows, M., 1984. The morphology of two groups of spiking local interneurones in the metathoracic ganglion of the locust. J comp Neurol 224: 463-482.
Watson, A.H.D. and Burrows, M., 1987. Immunocytochemical and Pharmacological Evidence for GABAergic Spiking Local Interneurons in the Locust. J Neurosci 7: 1741-1751.

NOCICEPTIVE MICROCIRCUITRY IN DORSAL HORN

V. Bijlani, S. Wadhwa and Tilat A. Rizvi

Department of Anatomy
All-India Institute of Medical Sciences
New Delhi-110029, India

ABSTRACT

The dorsal grey in spinal cord is the first site of interaction between primary afferent nociceptive fibres, intrinsic neurons and a variety of descending fibres. Observations were made on the dorsal grey of human fetal spinal cord ranging in age from 8 to 37 weeks of intrauterine life with a view to understand the development of neuronal circuits which modify/modulate the afferent input. Primary afferent fibres form a marginal plexus and extend along the lateral and medial margins of the grey matter of dorsal horn. Substance P immunoreactivity was seen in primary afferent fibres at 8 weeks which increased with gestational age. GABA immunopositive terminals were also seen at this age. The area of their distribution corresponded to the location of bulbospinal fibres described to have modulatory influences on the primary afferent fibres carrying pain sensation (substance P positive). Enkephalin immunoreactivity was seen at 12 weeks and serotonin at 16 weeks. GABA immunopositive neurons were first identified at 12 weeks. Several varieties of interneurons are present which play an important role in nociception and analgesia. Encased neuron of Lima and Coimbra (1983) can be identified with Golgi's method in the marginal zone at 13-14 weeks of intrauterine life. Electron microscopic study of the serial sections of this neuron revealed excitatory and inhibitory contacts on soma and dendrites. The characteristics of vesicles indicate the existence of substance P, acetylcholine, serotonin, enkephalin and GABA in the terminals (Rizvi et al., 1986). Substantia gelatinosa (lamina II and III) is densely populated with interneurons. The stalked and islet cells could be identified at 13-14 weeks of gestation. There is evidence of existence of a complex microcircuitry and neurochemical substrate by which the nociceptive input is modulated in the dorsal grey prior to projection to higher centre.

INTRODUCTION

The neurons in the dorsal grey of the spinal cord from the first relay station for the sensory input of primary afferent fibres from the receptors in the skin, subcutaneous tissue, muscle, fascia, joint capsules and viscera etc. The neurons of origin of the primary afferent

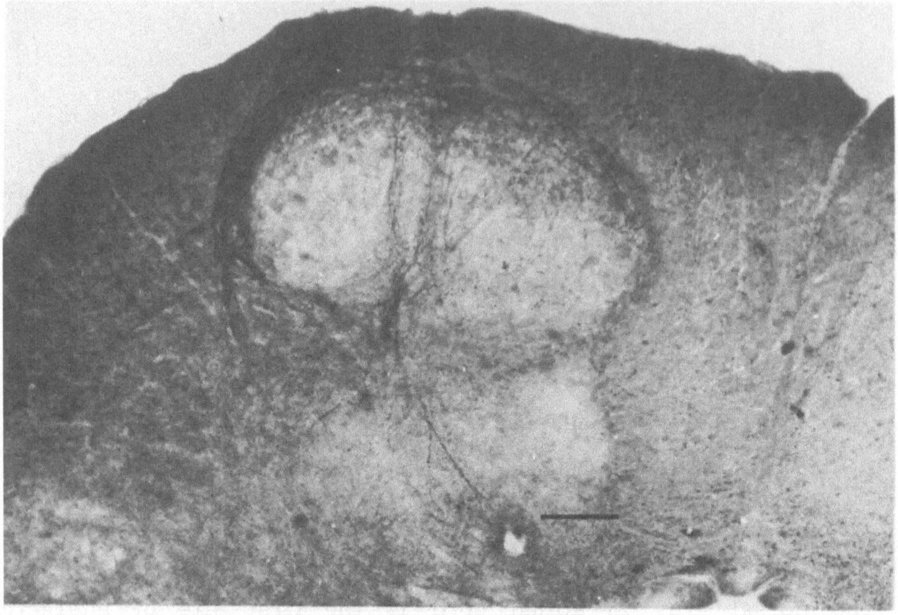

Fig. 1. Photomicrograph of dorsal horn of 16 weeks human spinal cord shows substance P immunoreactivity in the form of a black precipitate in the superficial marginal zone, substantia gelatinosa and in the medially and laterally curving fibres. Scale bar = 100 µm.

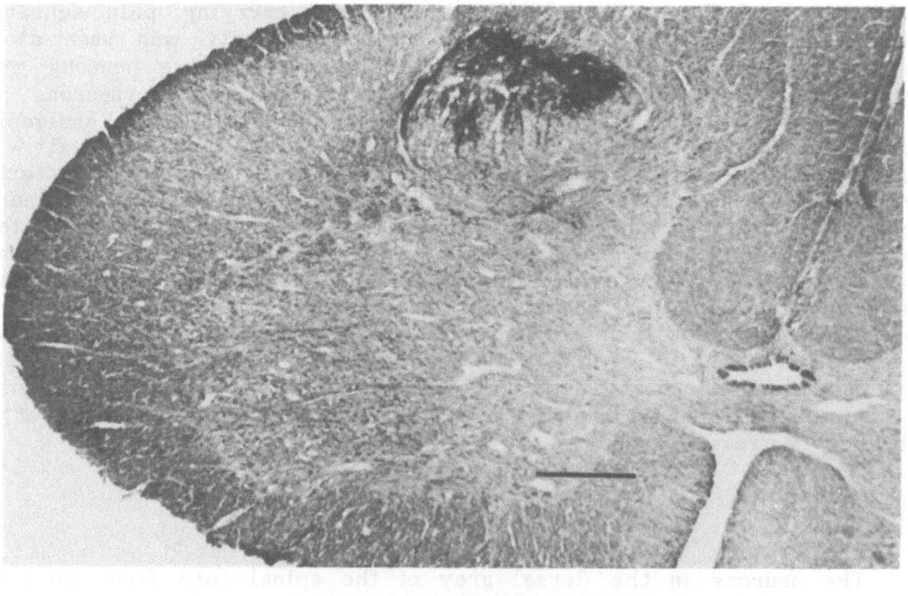

Fig. 2. Immunoreactivity for substance P in the dorsal horn of spinal cord at 37 weeks of gestation. Note the absence of immunoreactivity along the medial margin of dorsal grey. Scale bar = 500 µm.

fibres are in the dorsal root ganglia. The central processes of these neurons form the dorsal rootlets which are attached to the spinal cord. According to most descriptions, there is segregation of fibres depending on sensory modalities. Small diameter axons, called C and A delta fibres are said to carry noxious stimuli. Transmission at the spinal level is regulated in part by activity in the primary afferent neurons as well as by intrinsic neurons. Though increasing amount of information is being made available regarding the distribution of primary afferent fibres, but very little is known about the precise connectivity of intrinsic neurons. There is agreement that the sensory input to dorsal grey gets integrated and modulated before reaching the projection neurons. The dendrites and axons of the neurons of dorsal horn along with ramification of primary afferent fibres and collaterals of ascending and descending fibres form a complex neuropil. The precise wiring and connectivity of this neuropil is largely unknown.

In the present study, an attempt is made to study the development of this neuropil in human foetuses with a view to analyse the morphological substrate of nociception in adult spinal cord.

MATERIAL AND METHOD

The present investigation was carried out on human foetal material obtained on autopsy and hysterotomy performed for medical termination of pregnancy. Specimens ranged in age from 8 to 37 weeks of gestation. Cervical 4th to cervical 8th segment of the spinal cord was dissected, dorsal columns isolated and processed for light microscopy, electron microscopy and immunocytochemical procedures for localizing substance P, enkephalin, serotonin and GABA.

Development of neuronal somata was studied on the Nissl stained material. Golgi stained neurons were traced with camera lucida and their axonal and dendritic spread studied over the different age periods. Observations were made on the maturation of neuronal cytoplasm and developmental growth of dendritic tree. One interneuron from the marginal zone was processed for Golgi electron microscopy for studying the synaptic profiles. Immunocytochemical localization of substance P, enkephalin and serotonin was carried out by Avidin Biotin complex technique (Sternberger, 1974). GABA was localized by peroxidase antiperoxidase technique. Monoclonal antibodies were used except in case of GABA wherein polyclonal antibodies were used. Synaptic profiles in the different regions of the dorsal grey were observed under the electron-microscope with a view to elucidate the microcircuitry.

RESULTS

Golgi stain is picked up by the specimen obtained from 13-14 weeks old foetus. Earlier specimens do not pick up the stain. At this age dense plexus formed by thick and thin fibres is seen capping the dorsal grey. Some thin fibres can be seen coursing through the grey matter towards the deeper zones of grey. At 23-24 weeks, bundles of fibres are seen along the medial and lateral edges of the grey matter. Some of these fibres are visualized curving inwards towards the prospective laminae III and IV, wherein they are seen forming dense arbors (Bijlani et al., 1986).

Immunocytochemical localization of substance P was done to identify the nociceptive primary afferent fibres. Immunoreactivity is seen in the marginal zone at 8 weeks of intrauterine life. At 12 weeks,

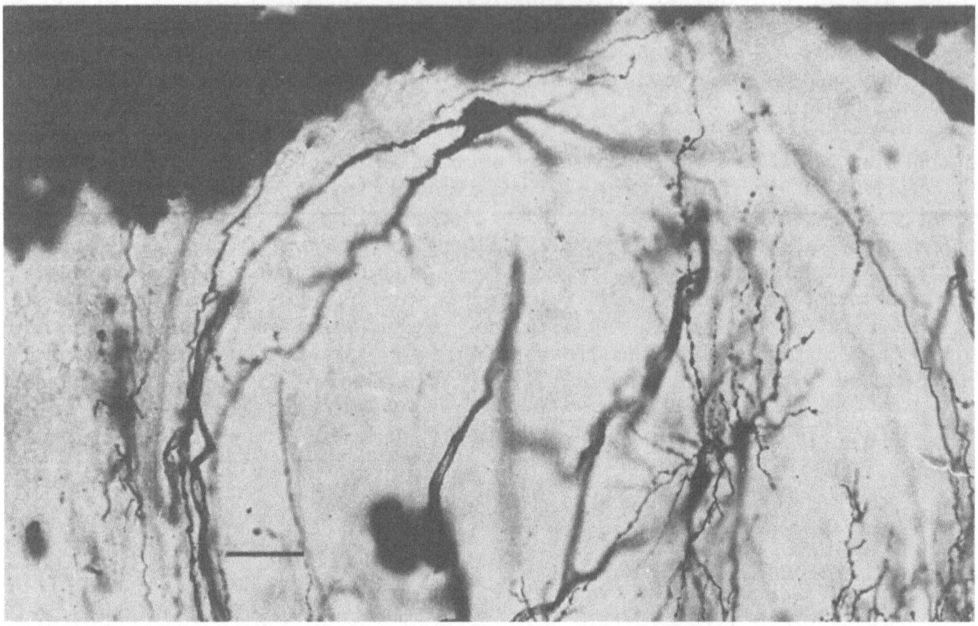

Fig. 3. Photomicrograph of Golgi stained projection neuron from the marginal zone of 20 weeks human fetal spinal cord with medially and laterally directed dendrites. Note the primary afferent fibres forming apparent contacts. Scale bar = 100 µm.

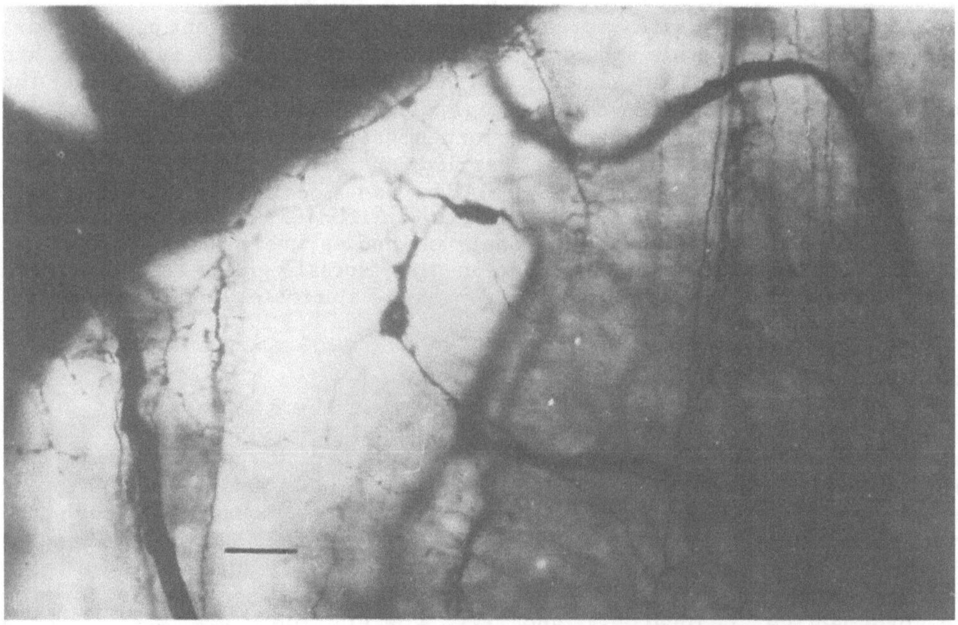

Fig. 4. Golgi stained substantia gelatinosa neurons from 13 weeks human fetal spinal cord. Note the islet cell with axon curving in the vicinity of the soma. Scale bar = 100 µm.

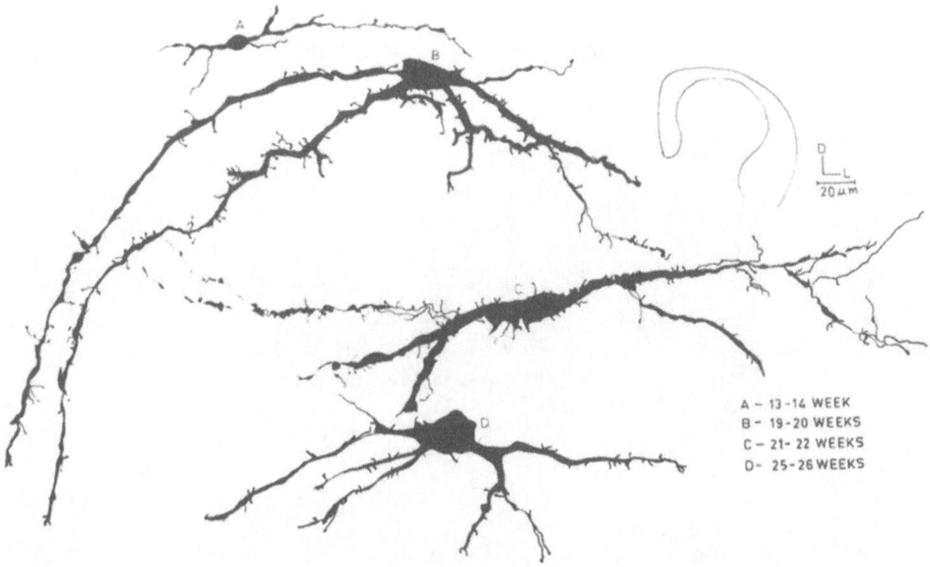

Fig. 5. Camera lucida drawings of Golgi stained projection neurons from the marginal zone showing developmental features at different age periods.

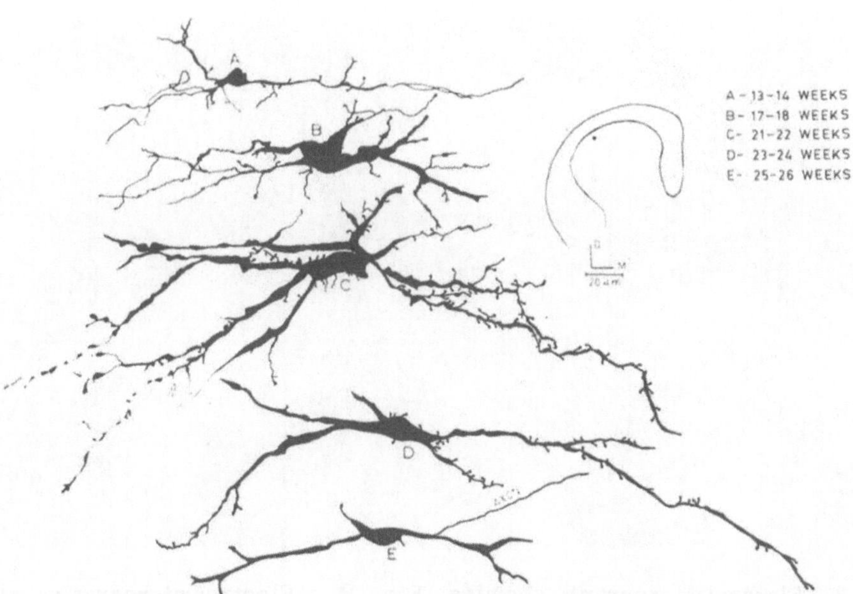

Fig. 6. Drawings of "encased" neuron at different gestational ages to show the developmental changes.

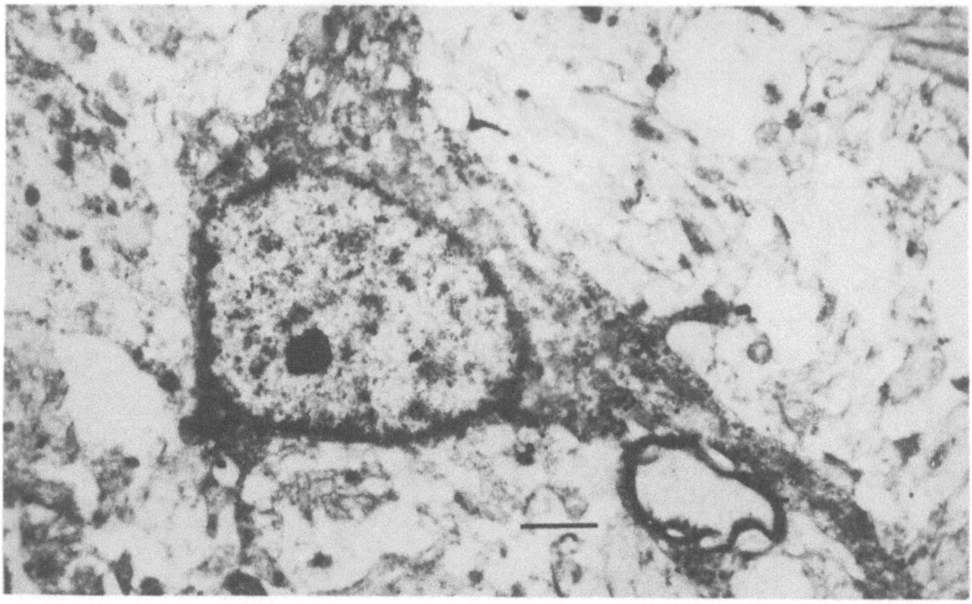

Fig. 7. Electronmicrograph through the deeper part of a gold toned "encased" neuron at 16-17 weeks of age. Scale bar = 1.0 µm.

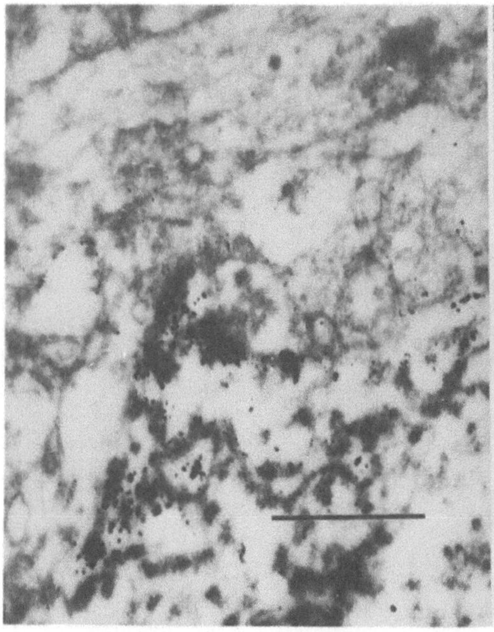

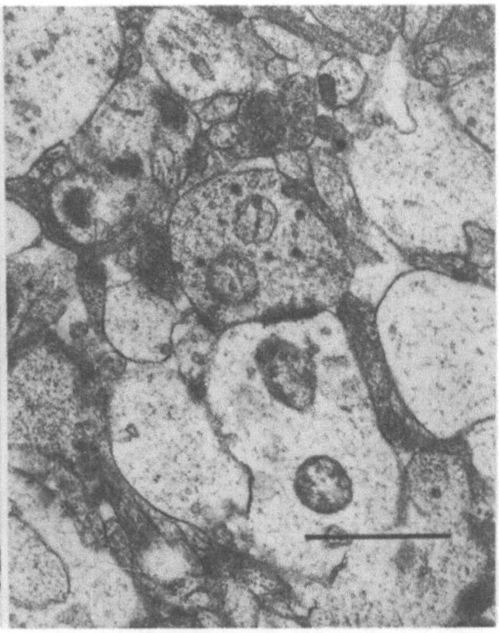

Fig. 8. Electronmicrograph showing symmetric synapses on the proximal part of dorsally directed dendrite of neuron in Fig. 7. Scale bar = 0.5 µm.

Fig. 9. Electronmicrograph shows an axodendritic asymmetric synapse. Axon terminal has dense core and clear spherical vesicles. Scale bar = 1.0 µm.

substance P positive immunoreactivity is seen in the fibres along the lateral edge. At this age substance P positive fibres are also seen coursing from the marginal zone ventrally through the grey matter. At 16 weeks of age immunopositivity is also present along the medial edge of dorsal grey (Fig. 1). Some fibres can be traced curving inwards and coursing towards the laminae III and IV. The density of immunoreactivity increases with age, the maximum being at 37 weeks age period. A notable feature is the absence of substance P positive fibres along the medial edge of dorsal grey after 16 weeks of intrauterine life (Fig. 2). Enkephalin immunoreactivity became apparent later than substance P. None is seen prior to 14 to 16 weeks. At this time a dense plexus in the marginal zone with some fibres coursing through the grey matter ventrally and incurving fibres along the lateral edge are seen. The density of immunoreactivity increases with age and the pattern of enkephalin positive fibres appears similar to substance P positive fibres. Appearance of serotonin-positivity is somewhat different. It appears at 12 weeks in the dorsal marginal area along its lateral edge. At 16 weeks the medial border also shows immunopositivity. The incurving fibres are also seen. No serotonin positive fibres are seen coursing from dorsal margin through the grey matter. The prospective laminae II, III and IV show punctate honeycomb lattice of immunopositive serotonin terminals. GABA immunoreactivity is seen in the terminals in the peripheral and dorsolateral part at an early period of 8 weeks of gestational age.

On the basis of soma size, axonal configuration as well as extent of dendritic arborization, both projection (Fig. 3) and interneurons can be identified in the marginal zone. Fig. 5 illustrates the development of projection neurons as observed in Golgi preparation. "Encased" neuron of Lima and Coimbra (1983) is also identified at 13-14 weeks. Fig. 6 illustrates the development of this interneuron. Development and dendritic remodelling of interneurons occurs somewhat later than the projection neurons (Figs. 5 & 6). At 13-14 weeks bipolar and multipolar neurons are seen. Fig. 4 shows "islet" cell with the axon coursing in the vicinity of cell soma. Stalked cells are seen at 13-14 weeks while stalked appendages appear at 20 weeks of age. The "encased" interneuron, in the present study, is processed for Golgi electron microscopy at 16-17 weeks of age period. Serial electronmicroscopic sections of this neuron (Fig. 7) reveal it to be surrounded by unmyelinated axonal profiles. Asymmetric synaptic contacts with clear spherical vesicles/mixture of clear spherical and dense core vesicles, are seen on the soma and dendrites. Symmetric synapses (Fig. 8) containing pleomorphic vesicles are seen on the dorsally directed dendrite and soma (Wadhwa et al., 1986). The synaptic contacts observed in the dorsal grey (Fig. 9) at different age periods are summarised in the Table I.

DISCUSSION

The pattern of neuronal connectivity in the dorsal grey of spinal cord has been the subject of many investigations since the classical description of Ramon y Cajal. In the recent past, Rexed (1952) provided for the first time a detailed account of the cytoarchitecture of the spinal grey and introduced the widely used nomenclature of lamination. Lamina I or marginal zone of Waldeyer is sparsely populated with neurons. Lamina II of Rexed contains densely packed columns of neurons and with lamina III corresponds to the substantia gelatinosa. The lamina IV, V and VI have also been described in detail by him. The dorsal horn is the first site of interaction between primary afferents and intrinsic neurons of the dorsal grey. There is also large evidence

TABLE I

SYNAPTOGENESIS IN DORSAL GRAY OF HUMAN FETAL SPINAL CORD

AGE(WKS)	MARGINAL ZONE	SUBSTANTIA GELATINOSA
8	i. Axodendritic with clear spherical vesicles (50-65 nm) ii. Axodendritic with clear spherical (40-60 nm) and few dense core vesicles (100-106 nm)	i. Axodendritic with clear spherical vesicles (55-70 nm) and dense core vesicles (100-105 nm). ii. Axodendritic with pleomorphic clear vesicles (40-50 nm). iii. Axodendritic with clear spherical vesicles (55-70 nm). iv. Dendrodendritic contacts
17-18	i. Axosomatic with pleomorphic clear vesicles (45-65 nm). ii. Axodendritic a. clear spherical vesicles (65-90 nm). b. Dense core vesicles (99-115 nm). iii. Dendrodendritic with clear spherical vesicles (55-73 nm) and dense core vesicles (100-120 nm).	i. Axodendritic with spherical vesicles (50-60 nm). ii. In addition to i. with dense core vesicles (95-100 nm). iii. Dendroaxonic with clear spherical vesicles (50-60 nm).
20	Two additional types seen i. Axosomatic symmetrical with pleomorphic vesicles (35-50 nm). ii. Axoaxonic with symmetrical with pleomorphic vesicles (37-50 nm).	
22		Several symmetrical i. axodendritic with pleomorphic vesicles (50-60 nm) on a single dendrite. ii. axoaxonic with pleomorphic vesicles (30-35 nm).
25	Axodendritic contacts on a single dendrite with spherical clear vesicles (47-73 nm) with dense core (112-135 nm).	i. Axosomatic contacts seen.

in literature about the existence of a variety of descending fibres taking part in the neuronal circuitry. The activities of dorsal horn play a fundamental role in nociception and other sensory modalities (Kerr, 1975). The small thinly myelinated and unmyelinated fibres of dorsal root establish synaptic connections with the neurons of laminae I, II and III. These small diameter fibres of the dorsal root are known to be nociceptive fibres. Neurons giving rise to the spinothalamic tracts are also described to be dispersed in lamina I, IV, V and VI.

Gobel (1978a) stated that all neurons in the marginal zone are projection neurons. Our studies are not in agreement with Gobel (1978a) as we have identified both projection and interneurons (Figs. 3,5,6). 'Encased neuron', an interneuron, in depth described by Lima and Coimbra (1983) has been studied. Synaptic profile on this encased neuron showed symmetric contacts with pleomorphic vesicles on the dendrite (Fig. 8) and soma which perhaps mediate serotoninergic, GABAergic or enkephalinergic inhibition. The asymmetric synapses on the soma and dendrites contained either mixed population of dense core and clear spherical or only clear spherical synaptic vesicles (Wadhwa et al., 1986). The asymmetric contacts are usually regarded as excitatory in function. The type of vesicles indicate presence of cholinergic, substance P, or enkephalinergic transmitters. The present study indicates the existence of both excitatory and inhibitory influences on encased neuron. This neuron appears to be contacted by descending supraspinal fibres and could thus be involved in modulation mechanism (Lima and Coimbra, 1983). We found this neuron situated in close proximity of supraspinal GABAergic terminals at this age period (Rizvi, 1988). Game and Lodge (1975) have described role of GABA in mediating pre and post-synaptic inhibition. Nishikawa et al. (1983) reported serotonin as a post-synaptic modulator. The present study indicates the presence of both these transmitters in addition to enkephalin and substance P. All of these seem to be involved in the microcircuitry around this interneuron. Cajal's central and limiting cells and Gobel's (1978b) stalked, islet, arboreal and lamina II-III border cells were identified in substantia gelatinosa. According to Trevino and Carsten (1975) the neurons of lamina II and III do not project for long distances. On the other hand, Light and Kavikjan (1988) have described long axonal projections. In our Golgi studies at 13-14 weeks we could find primitive stalked cells with their axons projecting to lamina I. According to Gobel (1978b) these neurons are equivalent to the limiting cells of Cajal and may be excitatory interneurons. According to Bennett et al. (1982) some stalked cells contain enkephalin - an inhibitory neurotransmitter. Csillik and Csillik (1981) have demonstrated enkephalin containing vesicles in the stalked appendages. We could identify islet cell with its axon in the vicinity of the lamina (Fig. 4). According to Gobel (1978b) these neurons may be inhibitory interneurons of Golgi type II and central cells of Cajal. Kerr (1975) has also suggested inhibitory neurons in substantia gelatinosa. The substantia gelatinosa has a dense population of opiate receptors which are located on the primary afferent terminals (Lamotte et al., 1976). Recent studies of Ditirro and Ho (1980) and Delanerolle and Lamotte (1980) have shown that the neuronal cell bodies located in the substantia gelatinosa contain enkephalin. Enkephalin has been localized immunohistochemically in the spinal cord of man (Lanerolle and Lamotte, 1982), monkey (Aronin et al., 1981) and in human fetuses and infant spinal cord (Charnay et al., 1984). Enkephalin immunoreactive fibres in these studies were predominant in the marginal layer and substantia gelatinosa (Sar et al., 1978) while enkephalin containing cell bodies were located in the substantia gelatinosa (Ditirro and Ho, 1980). Neuromodulatory role of enkephalin in the pain sensory system was supported by physiological (Basbaum and Fields, 1978) and anatomical (Glazer and

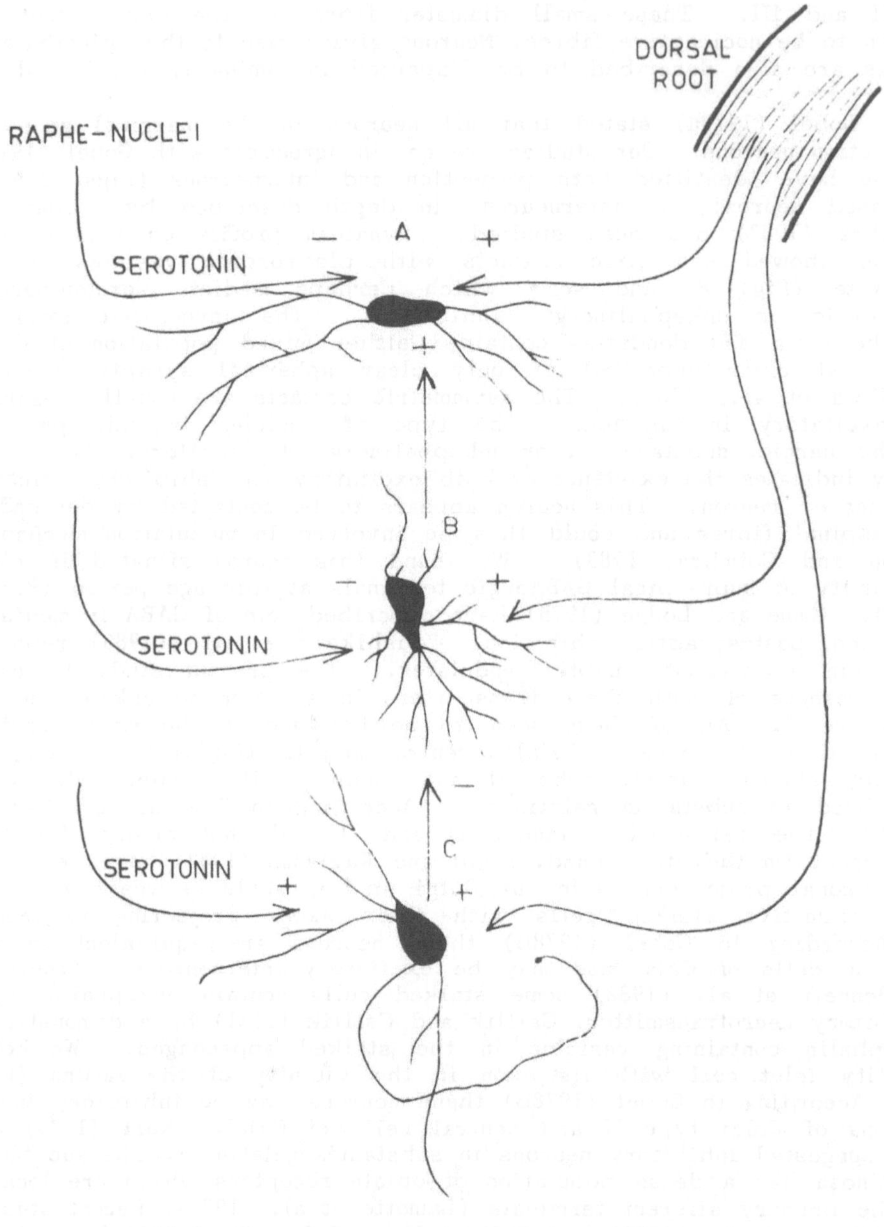

Fig. 10. Mode of action of descending serotoninergic fibres based on Ruda and Gobel's hypothesis.

A - projection neuron
B - stalked cell
C - islet cell

Basbaum, 1981) findings. Lamotte et al. (1976) and Atwah and Kuhar (1977) have illustrated the mechanism of presynaptic inhibition of primary afferent fibres by enkephalinergic neurons of substantia gelatinosa. In contrast physiological studies have shown that enkephalin acts at post-synaptic sites in dorsal horn. Ultrastructurally, enkephalin has been localized in axon terminals in the rat dorsal horn (Pelletier and Leclere, 1979) and found to form synapses with dendrites and soma (Hunt et al., 1980). Axodendritic synapses with clear spherical and dense core vesicles (Fig. 9) (Table I) seen in our study are probably enkephalinergic. According to Aronin et al. (1981) enkephalin binds mostly to opioid receptors contained on dendrites and soma within the spinal cord while direct synapses by enkephalinergic axons with primary afferent fibres form a secondary site of interaction. Therefore, earlier concept that opiate receptors are on primary afferent terminals and form axoaxonic synapses for presynaptic inhibition may not be strictly true. Serotoninergic terminals gave an appearance of honey comb network around unstained neurons specially in substantia gelatinosa. At ultrastructural level axosomatic and axodendritic synapses with pleomorphic vesicles correspond to serotoninergic terminals (Rizvi et al., 1986). Ruda and Gobel (1980) have described in detail the mode of action of descending serotoninergic fibres on islet, stalked and projections neurons and have stressed its modulatory role in nociception (Fig. 10). Oliveras et al. (1979) have correlated the stimulation of descending serotoninergic fibres to analgesia. Activation of serotoninergic pathways has been described to inhibit nociceptive responses of spinothalamic neurons (Messing and Lytle, 1977).

The presence of pleomorphic vesicles suggests the presence of GABA which may be used by the spinal cord interneurons to mediate both pre and post-synaptic inhibition (Game and Lodge, 1975). It appears that the enkephalinergic, GABAergic interneurons; supraspinal serotoninergic, enkephalinergic and GABAergic fibres form the intermediate microcircuitry in the nociceptive path in dorsal grey. The substance P positive nociceptive fibres are susceptible to the modulatory influences of these neuronal components thus indicating the "plasticity" in transmission of pain impulses. This may explain the varied pattern of pain perception, the role of acupuncture in relief of pain and the gate control theory of Melzack and Wall (1965).

REFERENCES

Aronin, N., Di Figlia, M., Liotta, A.S., and Martin, J.B. 1981. Ultrastructural localization and biochemical features of immunoreactive leu-enkephalin in monkey dorsal horn. J Neurosci., 1:561-577.

Atwah, S.F., and Kuhar, M.J. 1977. Autoradiographic localization of opiate receptors in rat brain. 1.Spinal cord and lower medulla. Brain Res., 124:53-67.

Basbaum, A.I., and Fields, H.L. 1978. Endogenous pain control mechanism: Review and hypothesis. Ann Neurol., 4:451-462.

Bennett, G.J., Ruda, M.A., Gobel, S., and Dubner, R. 1982. Enkephalin immunoreactive stalked cells and lamina IIb islet cells in cat substantia gelatinosa. Brain Res., 240:162-166.

Bijlani, V., Rizvi, T.A., Wadhwa, S., and Mehra, R.D. 1986. Development of neuronal circuitry in marginal zone of dorsal spinal grey in man, in: "Current trends in pain research and therapy, stimulus produced analgesia," K.N. Sharma and U. Nayyar, eds., Vol. II, Indian Society for Pain Research and Therapy, pp. 1-15.

Charnay, Y., Paulin, C., Dray, P., and Dubois, P.M. 1984. Distribution of enkephalin in human fetus and infant spinal cord: An immunofluorescence study. J Comp Neurol., 223:415-423.

Csillik, K.E., and Csillik, B. 1981. Selective labelling by transsynaptic degeneration of substantia gelatinosa cells: An attempt to decipher intrinsic wiring in the Rolando substance of primates. Neurosci Lett., 23:131-136.

Delanerolle, N., and Lamotte, C. 1980. Distribution of substance P and met enkephalin in human and monkey spinal cord. Anat Rec., 196:44A.

Ditirro, F.J., and Ho, R.H. 1980. Distribution of substance P (SP), somatostatin (SOM) and methionin enkephalin (ENK) in spinal cord of domestic Rabbit. Anat Rec., 196:47A.

Game, C.J.A., and Lodge, D. 1975. The pharmacology of the inhibition of dorsal horn neurons by impulses in myelinated cutaneous afferents in the cat. Exp Brain Res., 23:75-84.

Glazer, E.J., and Basbaum, A.I. 1981. Immunohistochemical localization of leucin enkephalin in spinal cord of cat: Enkephalin containing marginal neurons and pain modulation. J Comp Neurol., 196: 377-389.

Gobel, S. 1978a. Golgi studies of neurons in layer I of the dorsal horn of the medulla (Trigeminal nucleus caudalis). J Comp Neurol., 180:375-394.

Gobel, S. 1978b. Golgi studies of the neurons in layer II of the dorsal horn of the medulla (Trigeminal nucleus caudalis). J Comp Neurol., 180:395-414.

Hunt, S.P., Kelly, J.S., and Emson, P.C. 1980. The electron microscopic localization of methionin enkephalin within the superficial layers (I & II) of the spinal cord. Neuroscience 5:1871-1890.

Kerr, F.W.L. 1975. Neuroanatomical substrates of nociception in the spinal cord. Pain, 1:325-356.

Lamotte, C., Perl, C.B., and Snyder, S.H. 1976. Opiate receptor binding in primate spinal cord. Distribution and changes after dorsal root detection. Brain Res., 113:407-412.

Lanerolle, N.C., and Lamotte, C.C. 1982. The morphological relationships between substance P immunoreactive processes and ventral horn neurons in the human and monkey spinal cord. J Comp Neurol., 207:305-313.

Light, A.R., and Kavikjan, A.M. 1988. Morphology and ultrastructure of physiologically identified substantia gelatinosa neurons with axons that terminate in deeper dorsal horn laminae (III-V). J Comp Neurol., 267(2):172-189.

Lima, D., and Coimbra, A. 1983. The neuronal population of the marginal zone (Lamina I of the rat spinal cord). A study based on reconstructions of serially sectioned cells. Anat Embryol., 167:273-288.

Melzack, R., and Wall, P.D. 1965. Pain mechanisms. A new theory. Science, 150:971-979.

Messing, R.B., and Lytle, L.D. 1977. Serotonin containing neurons: their possible role in pain and analgesia. Pain, 4:1-21.

Nishikawa, N., Bennett, G.J., Ruda, M.A., Lu, G.W., and Dubner, R. 1983. Immunocytochemical evidence for a serotonergic innervation of dorsal column postsynaptic neurons in cat and monkey: Light and electron microscopic observations. Neuroscience, 10(4): 1333-1340.

Oliveras, J.L., Guilband, G., and Besson, J.M. 1979. A map of serotonergic structures involved in stimulation producing analgesia in unrestained treaty moving cats. Brain Res., 164:317-322.

Pelletier, G., and Leclere, R. 1979. Localization of leu-enkephalin in dense core vesicles of axon terminals. Neurosci Lett., 12: 159-163.

Rexed, B. 1952. The cytoarchitectonic organization of the spinal cord in the cat. J Comp Neurol., 96:415-495.

Rizvi, T.A., Wadhwa, S., Mehra, R.D., and Bijlani, V. 1986. Ultrastructure of marginal zone during prenatal development of human spinal cord. Exp Brain Res., 64:483-490.

Rizvi, T.A. 1988. A cytoarchitectural and immunohistochemical study on the developing dorsal horn of man. Thesis submitted for the award of Doctor of Philosophy.

Ruda, M.A., and Gobel, S. 1980. Ultrastructural characterization of axonal endings in the substantia gelatinosa which take up serotonin. Brain Res., 184:57-83.

Sar, M., Stumpf, W.E., Miller, R.J., Chang, K.J., and Cuatrecasas, P. 1978. Immunohistochemical localization of enkephalin in rat brain and spinal cord. J Comp Neurol., 182:17-38.

Sternberger, L.A. 1974. Immunocytochemistry. Prentice Hall, Inc. Englewood Cliffs, N.J., pp. 246.

Trevino, D.L., and Carsten, E. 1975. Confirmation of the location of spinothalamic neurons in the cat and monkey by retrograde transport of horseradish peroxidase. Brain Res., 98:177-182.

Wadhwa, S., Rizvi, T.A., and Bijlani, V. 1986. A combined Golgi electronmicroscopic study of "Encased neuron" in lamina I of dorsal horn of the developing human spinal cord. J Anat Soc India, 35(2):65-72.

DISTRIBUTION AND INNERVATION OF CUTICULAR SENSE ORGANS IN THE SCORPION, <u>HETEROMETRUS FULVIPES</u>

A.Kasaiah, V.Sekhar, G.Rajarami Reddy and K.Sasira Babu

Department of Zoology
S.V.University P.G.Centre
Kavali - 524 202, A.P., India

ABSTRACT

Five types of cuticular sense organs are found on the walking legs and pedipalps of the scorpion, <u>Heterometrus fulvipes</u>: long straight hairs, small white hairs, short straight bristles, trichobothria and slits. A large compound slit sensillum with 13 slits is present in the basitarsal segment of each walking leg. Just born animal possess only small white hairs on pedipalps and long straight hairs on walking legs, while the 1st instar animals possess all the five types of cuticular sensilla. Sensory hairs on pedipalps and walking legs increase from just born to adult scorpion except the wind sensitive trichobothria. The density of receptors increase from proximal to distal segments of walking legs and pedipalps. Cobalt back filling and silver intensification of leg and pedipalp nerves revealed single innervation to short straight bristles and multiple innervation to long straight hairs (7),small white hairs (20),trichobothria (6)and slit sensilla (2).

INTRODUCTION

Among arthropods, the sensory system of arachnids received less attention compared to insects and crustaceans. Gross morphology, fine structure, and physiology of cuticular sense organs in spiders were well studied (Barth, 1971, 1972a,b; Seyfarth and Pfluger, 1984; Seyfarth et al., 1985; Babu and Barth, 1988). In scorpions, the topography of slit sense organs (Barth and Wadepuhl, 1975; Barth and Stagl, 1976), anatomy of trichobothria (Venkateswara Rao, 1963) and distribution of hair sensilla (Foelix and Schabronath, 1983) were also described. However, a more detailed study of the cuticular sense organs on the pedipalps and walking legs of scorpion appeared desirable because of their principal involvement in predation and defensive behaviour (Brownell and Farley, 1979a, b). Hence, using the cobalt backfilling method, the innervation pattern of these cuticular sense organs in the ground scorpion, <u>Heterometrus</u> <u>fulvipes</u> is described, along with the topography and development of sense organs.

MATERIAL AND METHODS

The scorpion, H.fulvipes is abundantly available in nearby fields of the P.G.Centre, Distribution of cuticular receptors on pedipalps and walking legs was studied in detail selecting just born, first instar (1.8 cm), and adult animals (8 cm). The animals were brought from the field to the laboratory and were maintained in glass vivarium. Pregnant females collected in the months of may to june gave birth to young ones in the laboratory. Scorpions were fed with cockroaches and termites twice a week. The nomenclature used by Vachon (1952) for scorpion leg segments were adopted.

The innervation of cuticular sense organs of pedipalps and walking legs was studied by adopting the cobalt chloride and silver intensification method (Bacon and Altman, 1977). Pedipalpal and leg nerves were exposed in scorpion ringer (Padmanabha Naidu, 1967) and the nerve cut ends were immersed in 5% $CoCl_2$ for 8 h. Later the preparations were washed in 70% alcohol and sulphided for 20 minutes; fixed for 5 minutes in carnoy's fixative, silver intesified, dehydrated in ethanol and cleared in methyl salcylate.

RESULTS

Despite a large variety in size and shape, the cuticular hairs on the pedipalps and walking legs of Heterometrus fulvipes are classified into four different types: 1) short curved white hairs 2) long straight hairs 3) short straight bristles and 4) trichobothria. The hair sensillae are distributed over the entire body surface except the filiform trichobothria which are confined to pedipalps. The other characteristic cuticular mechano sensory structures, the slit sensilla are also found as single slits on pedipalps and compound slit sensilla on the walking legs.

Development and distribution of sensilla

In a new born H. fulvipes, small white hairs on pedipalps, long straight hairs on walking legs alone are present. In the first instar stage however, short curved hairs, long straight hairs and trichobothria occur on pedipalps whereas short curved hairs, long straight hairs and bristles are present on the walking legs. Interestingly, the number of trichobothria from first to abult stage remains constant whereas the other types of hair sensilla increase in their number (Tables 1 & 2).

The small white hairs in new born are 40 to 60 μm long while the long straight hairs measure 120 to 140 μm. In adults the length of both small white hairs and long straight hairs increase to 80-100 μm and 600-3800 μm respectively. These hairs are movably articulated within the cuticle. The hair shaft has a thin wall surrounding a central lumen. The bristles appear in the first instar with a length of 120 to 150 μm long and measure up to 400-650μm long in adults. The hair shaft has a rather thick wall surrounding a wide central lumen. The bristles have a narrow socket, consequently has a limitted range of movement. Trichobothria appear in the first instar stage with a length of 750 μm and measure upto 2000 μm in adults. All trichobothrial hairs are hollow with a central lumen. The hair base is suspended in a typical socket, characteristic of trichobothria. Inside the cup a longitudinal slit is present which limits the range of hair deflection.

Pedipalps

Most of the hair sensilla in general are arranged in longitudinal rows. The hair density increases from proximal to distal segments and highest number occur on terminal segment of pedipalps. A total of 54 small white hairs appear in just born with highest numbers (48) on the movable and fixed fingers of the hand. The number of small white hairs increases from 281 in first instar to 1047 in adults (Table 1). In adults, the small hairs are present on all faces of the four pedipalpal segments. However, the highest density of receptors occur on the movable and fixed fingers.

The long straight hairs also increase from the first instar (136) to adult (385) (Table 1). Similar to the small hairs, the highest density of long hairs occurs on the movable and fixed fingers of hand segment. In coxa-trochanter, maximum number occurs on dorsal and internal faces, while minimal number occur on the external face during post natal development. In other three segments, the highest number occur on the external face.

Table 1. Distribution of cuticular hair sensilla on the pedipalp of scorpion, *H. fulvipes* during post-natal development. LSH-Long straight hairs; SWH-Small white hairs; T-Trichobothria.

Animal stage		Coxa-trochanter	Femur	Tibia	Hand	Total
Just Born:						
	LSH	.	.	.	..	..
	SWH	4	1	1	48	54
	T					
	Total	4	1	1	48	54
1st instar:						
	LSH	18	17	14	87	136
	SWH	10	11	24	236	281
	T	..	3	19	26	48
	Total	28	31	57	349	465
Adult:						
	LSH	40	34	34	277	385
	SWH	75	71	101	800	1047
	T	..	3	19	26	48
	Total	115	108	154	1103	1480

The filiform trichobothrial hairs first appear in the first instar and continue through the other stages of development having a constant number (3-femur; 19-tibia; 26-hand), fixed location and orientation. In femur, one hair occurs on dorsal, external and internal faces and are absent on ventral face. On tibia, 13 are external, 3 each on ventral and dorsal, and absent on internal face. In the terminal hand segment, 15 are external, 5 dorsal, 4 ventral and 2 internal.

On the pedipalps of *Heterometrus*, only isolated single slits are present and the compound or lyriform organs are absent. Slit sense organs are located on all segments except femur. Most of these occur close to the distal or proximal part of a joint. Out of the 40 slits (20 µm or more) identified, 28 occur on hand segment. The location, number and to a large extent the size of slits are constant through post-natal development. The slits are not noticeable at the light microscopic level in the new born animals. Most of these cuticular receptors have rather conspicuous cuticular lips bordering the slits. In coxa-trochanter, the slits are present only on dorsal and ventral faces. In tibia, the slits are on ventral face, towards the distal region. Slits are absent on ventral face of the hand.

Walking legs

Hair sensilla in general are arranged in longitudinal rows along prominent ridges and transversely near proximal and distal regions of leg segments. During post-natal development, the three types of hairs increase in number from first to adult stage. Both long and short hairs occur on all leg segments, whereas shrot bristles are present only on the basitarsus and tarsus. There is a slight decrease in the number of hair sensilla from 1st to 4th walking leg (Table 2).

The small white hairs appear only after the first moult with a total number of about 78 and increasing upto 193 in adults. They are present on posterior and anterior faces of all leg segments except the anterior face of the coxa of first leg. At all developmental stages there is a general increase in receptor number from proximal to distal segments.

Table 2. Distribution of cuticular hair sensilla on the walking legs of the scorpion. H. fulvipes during post-natal development. LSH-Long straight hair; SWH-Small white hair; SB-Short straight bristles.

Animal stage		Walking leg I	Walking leg II	Walking leg III	Walking leg IV	Total
Just born:						
	LSH	41	39	36	36	152
	SWH	..	..	..	..	...
	SB	..	..	..	..	...
	Total	41	39	36	36	152
1st instar:						
	LSH	104	101	99	91	395
	SWH	78	78	74	68	298
	SB	14	14	14	14	56
	Total	196	193	187	173	749
Adult:						
	LSH	188	170	162	161	681
	SWH	193	166	158	143	660
	SB	16	17	17	17	67
	Total	397	353	337	321	1408

The long straight hairs are present on all walking legs from just born to adult stage. The hair sensilla occur both on anterior and posterior faces of walking legs and the number decrease from ls to 4th walking leg. A total of 41 sensilla appear on the 1st walking leg of just born while the adult has about 188 hairs. Similarly the total hairs on the 4th leg increase from 36 in just born to 161 in adult animals (Table 2).

The short bristles which appear from 1st instar stage are confined exclusively to basitarsus and tarsal segments. The 14 bristles in 1st instar stage increase to 16 on the first walking leg and 17 on the remaining legs in adult (Table 2).

Slit sensilla are not fully formed in just born animals. The slit size increases marginally from 1st to adult stage while their number remains constant. Large isolated slits and compound slit sensilla are absent on coxa. A total of 23 slits occur on trochanter alone constituting the highest number of slits of all the leg segments. A group of three slits (20-40 μm) are present on the anterior face of tibia while the posterior face has two slits (50-70 μm).

The largest compound sensillum consisting of thirteen slits is present on the anterior face of the basitarsus. The slits are arranged in close parallel rows and hence qualifies the term "lyriform organ", used for similar structures in spiders (Barth and Libera, 1976). The slit lenghts range from 30 to 225 μm long. The anterior face of the tarsus is free from slits while the posterior face has two slits of 15-30 μm long.

Innervation of cuticular sensilla

Cobalt fills of leg nerve and pedipalp nerve revealed that the various types of hairs and slit sensilla described earler are all innervated. Sensory cells are located beneath the hypodermis close to the base of sense organs. The dendritic terminals always end at the proximal side of the hair base.

The somata innervating the small white hairs (Fig. 1a) lie 50 μm deep to the hair base. The receptor cells measuring 10-15 μm long are arranged in a single mass consists of about twenty neurons. The dendrites run parallel to each other and terminate at the rim of the hair base. Axons arising from all these neurons form into a single bundle and join the main nerve (Fig.2).

The somata of long straight hairs occur 50 to 70 μm deep to the hair base (Fig. 1b). The cell mass consists of seven cells, some are globular and other are spindle shaped. The largest globular cell measures 40 μm and other cells measure 10 to 15 μm across. The dendrites arising from these cells run parallel to each other and terminate near the rim of the hair base. The largest dendrite close to the cell body measures 5 μm. The axons arising from these neurons also run parallel to each other as a single bundle and join the main nerve.

The short bristles are innervated by single sensory neurons located 60 μm deep to the hair base (Fig. 1c). The somata is spindle shaped and measure 20 μm across and 30 μm long. The diameter of dendrite close to the cell body is 3 μm and terminate at one side of the hair base.

Each trichobothrium is innervated by a group of six sensory cells, lying 50 μm deep to the hair base (Fig. 1d). Two of six somata are large

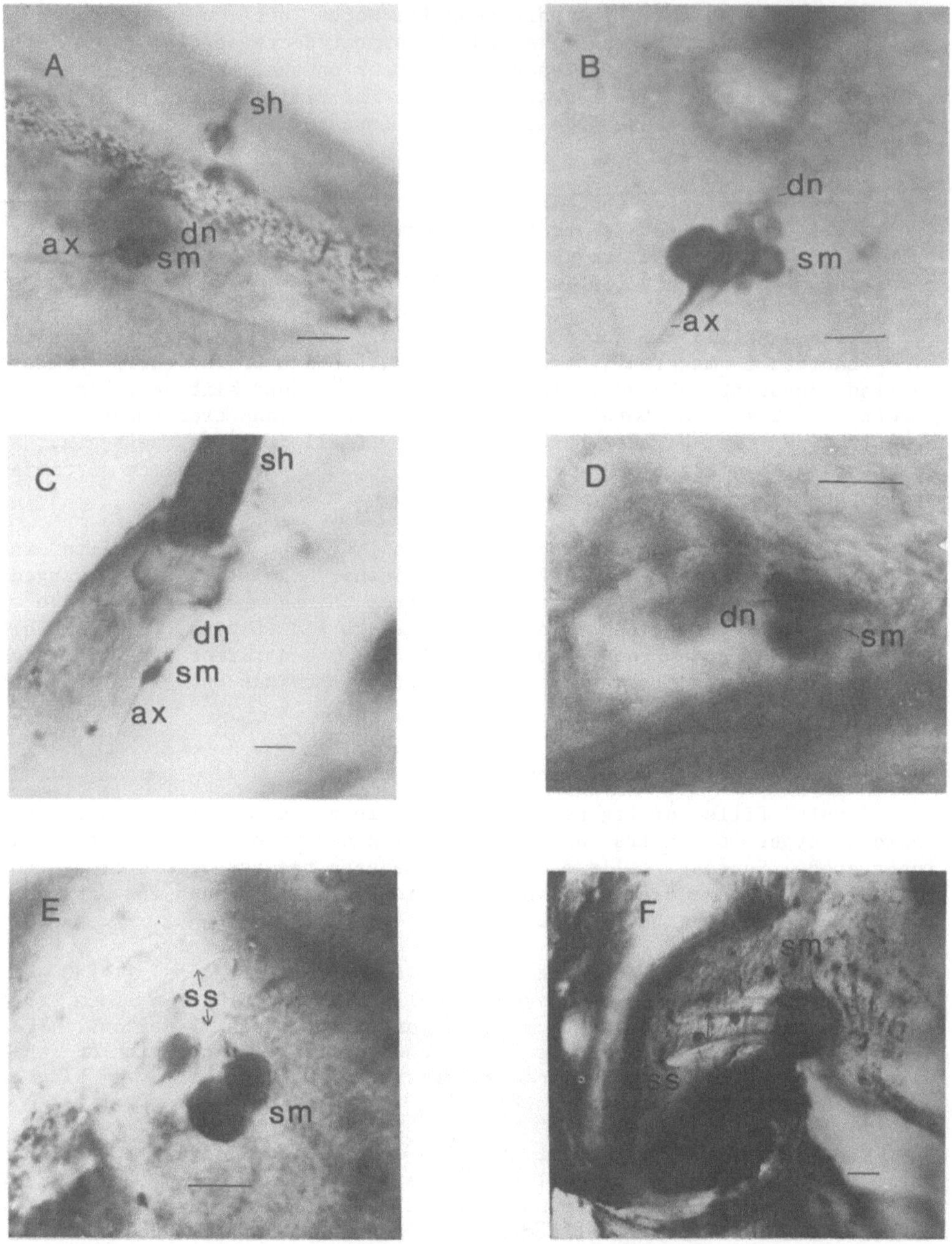

Fig. 1 Microphotographs showing innervation pattern of cuticular sensilla in H. fulvipes. A) Cluster of somata consisting of 20 sensory cells innervating a small white hair. B) Long hair innervated by 7 bipolar neurons of both globular and spindle shaped somata. C) Single bipolar cell of short straight bristle. D) Six sensory neurons of trichobothrium. Only three somata are seen and other three are not in focus. E) Isolated slit sensillum (arrows) innervated by two bipolar cells. F) Basitarsal Compound Slit Sensillum showing 13 individual slits (arrows) and their sensory cells. ax- axons; Sh- hair shaft; dn- dendrites; Sm- somata; SS- single slit. (Scales, 40μm).

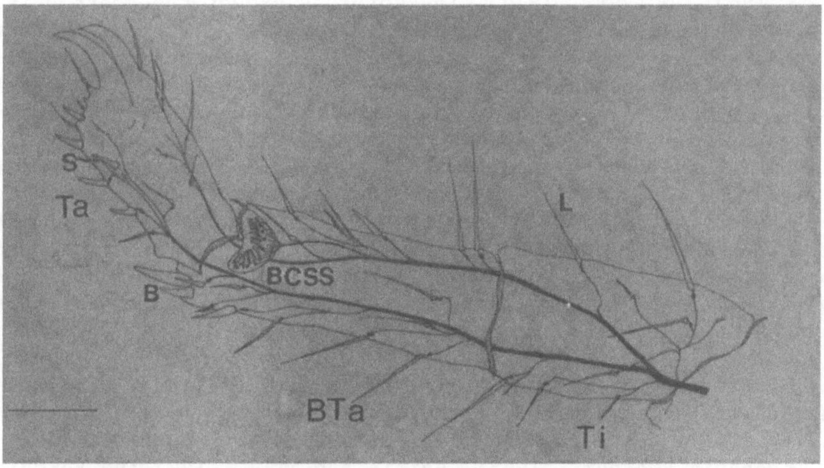

Fig. 2 Camera-lucida drawing as viewed from the anterior face of distal segments [tibia (Ti), basitarsus (BTa) and tarsus (Ta)] of walking leg. Axons arising from cuticular receptors join two major nerve bundles traversing the basitarsal and tarsal segments. A bundle of axons from the BCSS join one of these nerves. L- long straight hair; B- short bristle; S- small white hair; BCSS- Basitarsal Compound Slit Sensillum. (Scales, 2mm).

(12 μ m) and globular while the rest of them are spindle shaped and measure to 10-15 μm long. The dendrites of these somata run together as a bundle and terminate at one side of the hair base. The axons also travel as a bundle of fibres and join the main nerve proximally.

The slit sensilla whether isolated or in a compound slit sensillum are innervated by two somata (Fig. le & f). The two cell bodies are large (35-40 μm) close together and lies 30 to 35 μ m deep from the base of the slit. The two dendrites join together and terminate close to the broadened area of the slit. The axons travel parallel to each other and join the main nerve. The somata innervating the thirteen slits in the BCSS are however smaller and measure 10-20 μ m across and 20 to 30 μ m long. Axons from these cells converge and from a single bundle that join the main nerve proximally (Fig. 2).

DISCUSSION

The present cobalt studies have revealed that not all mechano receptors in scorpions are multiply innervated as has been hitherto reported. For the first time we have demonstrated that the short straight bristles found on basitarsus and tarsus segments of the walking legs, are innervated by single bipolar sensory cells. Whereas, the earlier reports of multiple innervation for other cuticular receptors are confirmed. A common characteristic feature of all these hair sensilla is that their dendrites terminate at the hair base unlike in insects.

In general, arthropod mechanoreceptors are reported to be singly innervated and chemoreceptors are multiply innervated (Bullock and

Horridge, 1965; McIver, 1975). In scorpions except the short bristles all other mechanoreceptors are multiply innervated. Earlier electron microscopic studies in A. australis (Foelix and Schabronath, 1983) and P. mesaensis (Brownell and Farley, 1979a) have revealed seven dendritic processes to each long hair on the walking legs. Anatomical (Venkateswara Rao, 1963) and electrophysiological (Sanjeeva Reddy, 1971) studies on H. fulvipes showed 4-5 sensory cells for each long straight hair. The present cobalt fill studies on these long hairs of H. fulvipes revealed 7 bipolar sensory neurons. On the basis of their structure (Foelix and Schabronath, 1983) and electrical responses (Babu and Sanjeeva Reddy, 1967; Sanjeeva Reddy, 1971; Brownell and Farley, 1979a,b; Kasaiah, 1988) these long straight hairs and short bristles in scorpions are now known to serve as tactile receptors.

The trichobothria (thread hairs) of insects are innervated by one mechanosensitive receptor cell (Schmidt and Gnatzy, 1971; Edwards and Palka,1974) while those in arachnids possess several such cells- 4 in spider Tegenaria (Gorner, 1965; Harris and Mill, 1977 and Reissland, 1978); 6-7 cells in Buthus (Ignatiev et al., 1976) and one cell in Euscorpius (Hoffmann, 1967). In the present study, the trichobothria of scorpion H. fulvipes are found to be innervated by 6 bipolar sensory cells. It is now clear that those trichobothria of scorpion respond predominently to weak air currents (Hoffmann, 1967; Ignatiev et al., 1976; Kasaiah, 1988).

Of all the cuticular receptors the small white hairs are innervated by the largest number of receptor cells. Venkateswara Rao (1963) reported 4-6 sensory cells in H. fulvipes while our cobalt studies of the same species stained a cluster of 20 sensory cell bodies. These small white hairs in Androctonus and Euscorpius (Foelix and Schabronath, 1983), in the cribellate spider Ciniflo (Harris and Mill, 1973) were also reported to possess more than 20 sensory cells. Structural, physiological and electron microscopic studies (Alexander and Ewer, 1957; Abushama, 1964; Rao, 1964; Foelix and Schabronath, 1983) reveal that these small hairs function as chemoreceptors.

Cos-impregnation studies on spider Cupiennius showed that each slit sense organ is innervated by two bipolar sensory cells (Seyfarth and Pfluger, 1984; Seyfarth et al., 1985; Babu and Barth, 1988). The EM studies of Foelix and Schabronath (1983) on the scorpion, Androctonus australis reported two neurons to each slit sensillum. The present investigation on the scorpion H. fulvipes also confirms the presence of the two bipolar cells innervating each slit sensillum be it a single slit or several slits as in basitarsal compound slit sensillum. These slit sense organs are uniquely arachnid cuticular mechanoreceptors. It was convincingly demonstrated by structural and physiological studies (Pringle, 1955; Barth and Wadepuhl, 1975) that these slit sense organs function as cuticular stress sensing elements and also capable of distinguishing substrate vibrations (Brownell and Farley, 1979b; Kasaiah, 1988).

The number of trichobothria in spiders increase gradually during ontogeny (Emerit, 1970; Haupt, 1986). The present study in Heterometrus shows that they are not found in just born animals but emerge only after the 1st moult. From 1st to adult stage however their number (96), position and grouping on pedipalps are constant. Even in classification of scorpions, the distribution of these trichobothria on pedipalps is used as an important criteria (Vachon, 1952). Since the new born scorpions mount on to the dorsum of the mother for nuture and protection, the wind sensitive mechanosensory trichobothria that aid in prey capture are not required. The present study also reports that just born animals possess only small white hairs on pedipalps and long straight hairs on walking legs. It is suggested that these receptors provide the just born scrpions with necessary chemical and tactile sense for climbing to the back of the mother and for leading a social life along with the other young ones. Only after the first moult, the young ones move down the mother in search of prey at which time the trichobothria and other types of sensory haris on walking legs and pedipals are developed.

Foelix and Schabronath (1983) in scorpions have reported that the density of hair sensilla is highest on first legs and lowest on the fourth leg. Moreover hair sensilla are found in larger number towards the distal pedipalpal and leg segments. A similar pattern of distribution was also reported in Thelyphonus (Moro and Geethabali, 1985; 1986; Geethabali and Moro, 1988). In present study on Heterometrus also hair sensilla are found in higher densities on distal segments of pedipalps and all walking legs. Since these appendages reach out farthest from the animals body, they perceive sensory information at a considerable distance from the animal. This distribution pattern of hair sensilla have survival significance to the animal because through these distal sensilla the animal can quickly detect a prey or predator and release an appropriate behavioural activity (Linsenmair, 1968; Brownell and Farley, 1979a,b; Kasaiah, 1988).

The present study also showed a higher number of hair sensilla on the 1st pair of legs and a gradual decrease in 2nd, 3rd and 4th pairs of walking legs. This is becuase that the first two pairs of walking legs, apart from locomotion are also used for certain other specialised activities (Kasaiah, 1988). For instance during burrow digging, these two paris of legs are used for loosening and collecting soil into a mound. During birth process the distal segments of these two paris of walking legs form a "birth basket" into which the emerging young ones drop (Williams, 1966). During these activities, the hair sensilla on these distal leg segments provide the necessary tactile sensation to the animal and hence a higher number of receptors are found on these anterior walking legs.

ACKNOWLEDGEMENTS

This study was supported by DST grant SP/YS/L19/86 to GRR and a UGC Research Fellowship to AK.

REFERENCES

Abushama, F.T., 1964, On the behaviour and sensory physiology of the scorpion Leiurus quinquestratus (H & E), Anim Behav., 12:140-153.

Alexander, A.J., and Ewer, D.W., 1957, On the origin of mating behaviour in spiders, The Amer Nat., 91: 311-317.

Babu, K.S., and Barth F.G., 1988, Central nervous projections of a lyriform organ in the spider Cupiennius salei keys, Tissue & Cell (In press).

Babu, K.S., and Sanjeeva Reddy P., 1967, Unit-hair receptor activity from the telson of the scorpion, Heterometrus fulvipes. Cur Sci., 36: 599-600.

Bacon, J.P., and Altman, J.S., 1977, A silver intensification method for cobalt-filled neurons in wholemount preparations, Brain Res., 138: 359-363.

Barth, F.G., 1971, Der sensorische apparat der spaltsinnesorgane

(Cupennius salei keys, Araneae), Z Zellforsch., 112: 212-246.
Barth, F.G., 1972a, Die physiologie der spaltesinnesorgane: I. Modellversuche zur Rolle des cuticularen spaltes beim Reiztransport, J Comp Physiol., 78: 315-336.
Barth, F.G., 1972b, Die physiologie der spaltesinnesorgans. II. Funktionelle morphologie eines mechanorezeptors, J Comp Physiol., 81: 159-186.
Barth, F.G., and Stagl, J., 1976, The slit sense organs of arachnids, Zoomorph., 86: 1-23.
Barth, F.G., and Wadepuhl, M., 1975, Slit sense organs on the scorpion leg (Androctonus australis L., Buthidae). J Morph., 145: 209-228.
Brownell, P., and Farley, R.D., 1979a, Detection of vibrations in sand by Tarsal sense organs of the nocturnal scorpion, Paruroctonus mesaensis. J Comp Physiol., 131: 23-30.
Brownell, P., and Farley, R.D., 1979b, Orientation to vibrations in sand by the nocturnal scorpion, Paruroctonus mesaensis: Mechanism of target localization, J Comp Physiol., 131: 31-38.
Bullock, T.H., and Horridge, G.A., 1965, "Structure and function in the nervous systems of invertebrates Vol.II," W H Freeman, San Francisco.
Edwards, J.S., and Palka, J., 1974, The cerci and abdominal giant fibres of the house cricket Acheta domestica. I. Anatomy and physiology of normal adults, Proc Roy Soc Lond B., 185: 83-103.
Emerit, M., 1970, Nouveaeu apports a la theorie de I'arthrogenese de I'appendice arachnidien. Bull Mus Hist Nat 2 Ser., 41: 1398-1402.
Foelix, R.F., and Schabronath, J., 1983, The fine structure of scorpion sensory organs. I. Tarsal sensilla, Bull Br Arachnol Soc., 6(2): 53-67.
Geethabali, and Moro, S.D., 1988, The disposition and external morphology of trichobothria in two arachnids, Acta Arachnol., 36: 11-23.
Gorner, P., 1965, A proposed transducing mechanism for a multiply-innervated mechanoreceptor (trichobothrium) in spiders, Cold Spring Harbor Symp Quant Biol., 30: 69-73.
Harris, D.J., and Mill, P.J., 1973, The ultrastructure of chemoreceptor senilla in Ciniflo (Araneida, Arachnida), Tissue & Cell., 5: 678-689.
Harris, D.J., and Mill, P.J., 1977, Observations on the leg receptors of Ciniflo (Araneidae: Dictynidae). I. External Mechanoreceptors, J Comp Physiol., 119: 37-54.
Haupt, J., 1986, Postembryonal development and trichobothriotoxie in Heptaethelidae: Possibilities and limits of a phylogenetic analysis, Actas X Congr Int Arachnol Jaca/Espana., 1: 349-354.
Haffmann, C., 1967, Bau and funktion der trichobothrien von Euscorpius carpathicus L, Z Vergl Physiol., 54: 290-352.
Ignatiev, A.M., Ivanov, V.P, and Balashov, Y.S., 1976, The fine structure and function of the trichobothria in the scorpion Buthus eupeus Koch (Scorpiones, Buthidae), Entomol Rev., 55: 12-18.
Kasaiah, A., 1988, Neuroethology of the scorpion, Heterometrus fulvipes : Habitat, cuticular receptors and behaviour. Ph.D. Thesis, Sri Venkateswara University, Tirupati.
Linsenmair, K.E., 1968, Anemotaktische orientierung bei skorpionen (Chelicerata, Scorpiones), Z Vergl Physiol., 60: 445-449.
McIver, S.B., 1975, Structure of cuticular mechanoreceptors of arthropods, Annu Rev Entomol., 20: 381-387.
Moro, S.D., and Geethabali.,1985, Distribution of cuticular sensory hairs on the legs and whip of Thelyphonus indicus stoliczka (Arachnida: Uropygi), Monitore Zool ital., 19: 207-218.
Moro, S.D., and Geethabali., 1986, The topography of slit sense organs in the whip scorpion, Thelyphonus indicus (Arachnida, Uropygida), Veh Naturwiss Ver Hamburg., 28: 91-105.

Padmanabha Naidu, B., 1967, A new perfusion fluid for the scorpion, Heterometrus fulvipes, Nature Lond., 213: 410.

Pringle, J.W.S., 1955, The function of the lyriform organs of archnids, J Exp Biol., 32: 270-278.

Rao, K.P., 1964, Neurophysiological studies on arachnid scorpion, Heterometrus fulvipes, J Anim Morph Physiol., 11: 133.

Reissland, A., 1978, Electrophysiology of trichobothria in orbweaving spiders (Agelenidae, Aranea), J Comp Physiol., 123: 71-84.

Sanjeeva Reddy, P., 1971, Function of the supernumerary sense cells and the relationship between modality of adquate stimulus and innervation pattern of the scorpion hair sensillum, J Exp Biol., 233-238.

Schmidt, K., and Gnatzy, W., 1971, Die feinstruktur der sinneshaare auf den cerci von Gryllus bimaculatus Deg (Saltatoria, Gryllidae). II. Die Hautung der fadenund keulenhaare, Z Zellforsch., 122: 210-226.

Seyfarth, E.A, and Pfluger, H.J., 1984, Proprioceptor distribution and control of a muscle reflex in the tibia of spider legs, J Neurobiol., 15: 365-374.

Seyfarth, E.A., Eckweiler, W., and Hammer, K., 1985, A survey of sense organs and sensory nerves in the legs of spiders, Zoomorph., 105: 190-196.

Vachon, M., 1952, "Etudes sur les scorpions", Inst Pasteur d'Algerie.

Venkateswara Rao, P., 1963, Studies on the peripheral nervous system of the scorpion Heterometrus fulvipes. Doctoral dissertation, Sri Venkateswara University, Tirupati.

Williams, S.C., 1966, Burrowing activities of the scorpion Anuroctonus phaeodactylus (Wood) (Scorpionida: Vaejovidae), Proc Calif Acad Sci., 34: 419-428.

NEURAL ORGANIZATION OF THE SENSORY APPENDAGES OF THE WHIP SCORPION, *THELYPHONUS INDICUS* STOLICZKA (ARACHNIDA, UROPYGI)

Rajashekhar K.P., Geethabali, and Y. Ramamohan *

Department of Zoology
Bangalore University
Bangalore - 560 056, India

*Electron Microscope Laboratory
Department of Neuropathology
National Institute of Mental Health and Neurosciences
Bangalore - 560 029, India

ABSTRACT

In whip scorpion, the multisegmented whiplike flagellum and the antenniform legs are the modified telson and the first pair of legs respectively. These two appendages bear numerous sensory hairs. Each segment of the flagellum also bears at its base an oval membranous region called 'Fenestra Ovalis'. The neural organisation of these two sensory appendages was examined with light and electron microscope and compared with the sensory appendages of other arthropods.

Each sensory hair of the antenniform leg has polyneural innervation. The afferent axons form tiny nerves. In contrast to the conventional arthropod neural organisation, an unusual finding is the formation of synapses by some of the axons in the periphery. Various types of synapses based on pre- and postsynaptic elements, and the nature of synaptic vesicles have been studied. The aggregation of the axons results in the formation of two large nerves which contain predominantly fine fibres and a few exceptionally large "Giant" fibres.

In the antenniform legs just below the cuticle, cells with striking microvillar formations similar to those found in rhabdoms of photoreceptor cells were observed. Associated with these structures were cells containing abundant osmiophilic granules.

The innervation of sensory hairs of the flagellum is similar to that of the antenniform legs. Fenestra Ovalis shows a thin and membranous cuticle below which sensory cells with broad, expanded dendrites containing elongated mitochondria were found.

These observations are discussed from a standpoint of neural organization in relation to information processing at the periphery and in comparison with the sensory appendages of arthropods.

INTRODUCTION

Sensory systems mediate the perception of external environment. The sensory appendages form an important part of sensory systems in arthropods. These appendages bear numerous and varied types of sensory structures. The associated neural tissue consisting of the receptor neurons and their processes comprise the sensory component of the peripheral nervous system. The architecture of the sensory structure and the response characteristics of the receptor neurons contribute partly to the analysis of the external environment. It is generally believed that the peripheral nervous system does not contribute much further in information analysis as the receptor axons conduct the information to the central nervous system without interacting among themselves and the second order neurons lie in the central nervous system.

Whip scorpions which belong to the order Uropygi possess antenniform legs which are modified forms of the first pair of legs (Fig. 1). They are sensory appendages and are not used for locomotion. The telson which is the last metasomatic segment is modified into a multisegmented flagellum which also is a sensory appendage. Previous studies by Foelix (1975) on whip spiders have revealed interesting and intriguing features in the organization of the peripheral nervous system of arachinds, which in many aspects, deviate from the general arthropod neural organization. These observations are suggestive of a novel role of the peripheral nervous system in information processing. Studies on the cuticular sensory hairs of whip scorpion were carried out by Moro and Geethabali (1985) at light microscopic level. Rajashekhar and Geethabali (1988) made preliminary observations on the peripheral nervous system of sensory appendages. The sensory appendages of whip scorpion remain largely unexplored. The present study at the light and electron microscopic level was undertaken to observe the sensory structures on the appendages and the associated neural tissue. These observations would contribute towards understanding the role of the peripheral nervous system in processing of information in arachnids.

MATERIAL AND METHODS

The surface of the sensory appendages of whip scorpion _Thelyphonus indicus_ stoliczka was observed using light microscope and scanning electron microscope. The tissue was fixed in 5% glutaraldehyde in 0.2 M cacodylate buffer (pH 7.4), washed in buffer in and dehydrated in grades of acetone. Following air drying the surface was coated with gold-palladium (10 nm) and scanned using JEOL JSM-35CF scanning electron microscope.

For transmission electron microscopy, the tissue was fixed in cacodylate buffered glutaraldehyde (5%) for 8 hours, washed in buffer and post fixed in osmium tetroxide following which it was dehydrated in grades of alcohol and embedded in araldite. Toluidine blue stained semithin sections were used for orientation and light microscopic observations. Ultrathin sections were cut using Reichert Ultracut E microtome and stained by the method of Reynolds (1963) and observed using JEOL JEM 100CX II electron microscope.

RESULTS

Sensory Structures

Antenniform leg: The segments of the antenniform leg are long and thin. The tarsus is made up of eight secondary segments which are densely covered with sensory hairs. About 300 hairs are found on each segment of the tarsus

(Fig. 2). The hair sensilla could be classified into 1. Short chemosensory hairs, 2. Long ornamented chemosensory and 3. Tactile hairs . The short chemosensory hairs measure 20 µm long and bear a pore at the tip (Fig. 3). The long chemosensory hairs measure 50 µm long and bear spiny ornamentations on their walls with a pore at the tip of the hair(Fig. 4). The pores measure 0.3 to 0.6 µm in diameter. The long mechanosensory hairs lack a pore at the tip and measure 250 µm in length. They are few per segment (3 or 4). Apart from sensory hairs no other sensory structures were found on the surface.

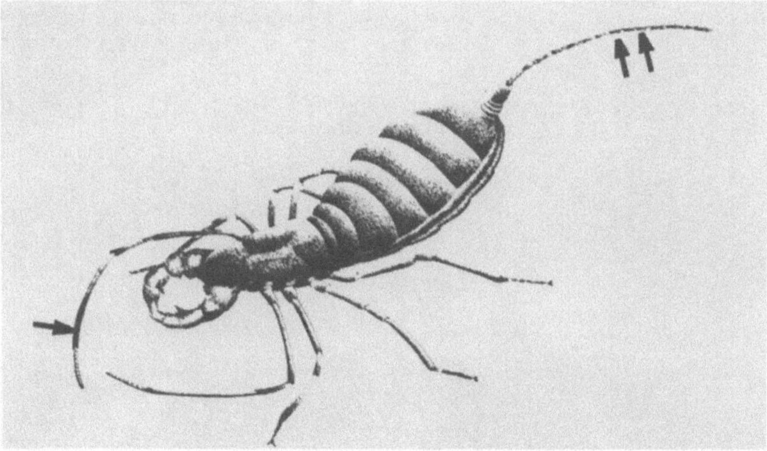

Fig. 1. Whip scorpion. The first pair of legs is modified into elongated antenniform leg (arrow). The telson is whip like and multisegmented and is called flagellum (double arrow).

Flagellum: The flagellum is long and whip like consisting of 20 to 30 segments covered with hairs. It is mounted on peg like post abdominal segments. The density of hairs is more on the distal segments than on the proximal segments (Fig. 5). These hair sensilla do not show any surface ornamentation. There are broadly two types of hairs. The long mechanosensory hairs (800 to 1000 µm) and the short chemosensory hairs with a pore at the tip (Fig. 6). Compared to the tarsal segment of the antenniform legs, the flagellum is not so densely covered with hairs. Each segment of the flagellum bears at its base an oval membranous structure called as "Fenestra Ovalis" (Yoshikura 1965) measuring about 0.2 mm long (Fig. 7). Under the scanning electron microscope, no surface specilisation was observed. However, this region can be observed using transmitted light.

Receptor Neurons and Sensory Nerves

The organization of receptor neurons, their processes as well as the sensory nerves is similar in the tarsal segments of the antenniform leg and the flagellum. The cuticle is thick and is indented at the base of the hair sensillum. Hypodermal cells are found below the cuticle. At the base of each hair sensillum are groups of receptor neurons (Fig. 8). All the

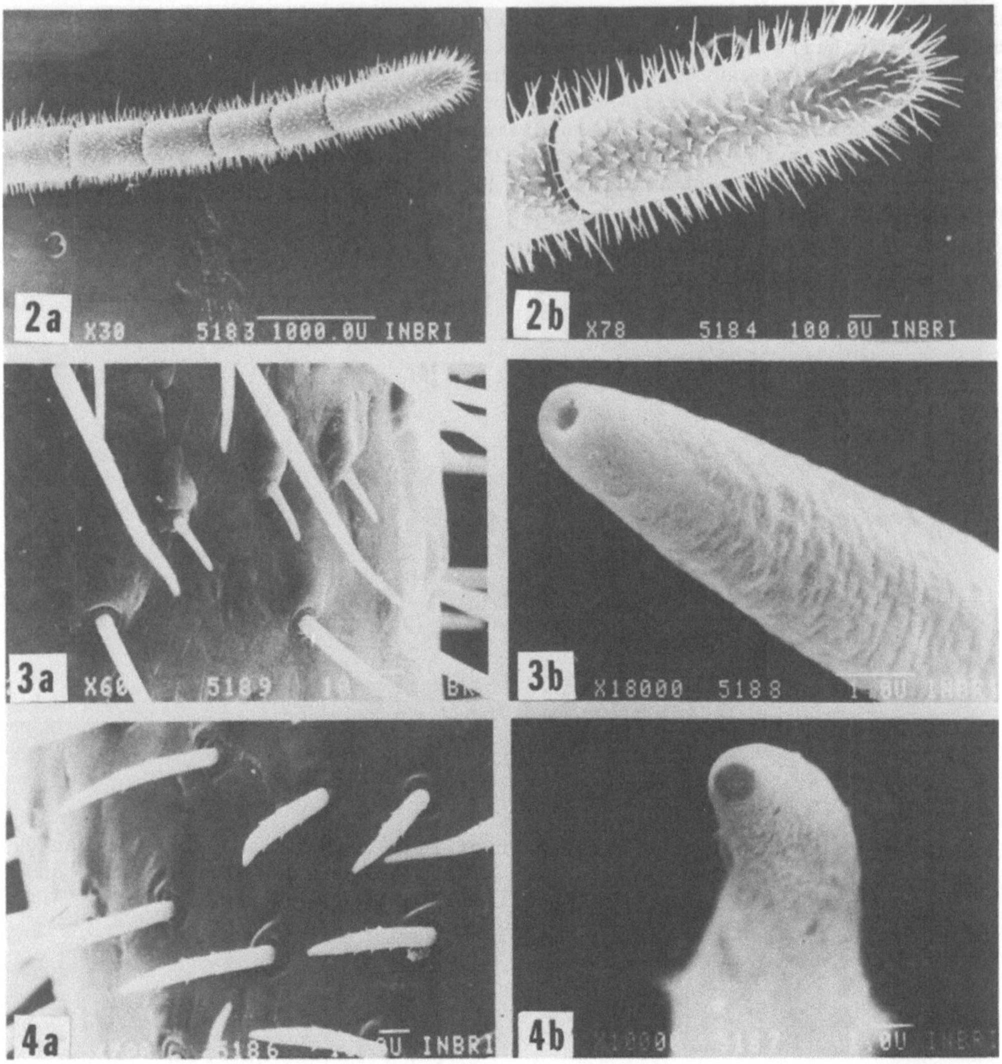

Fig. 2-4. Scanning electronmicrographs of the tarsal segments. Fig. 2.(a) The tarsal segments of the antenniform leg. (b) First segment of tarsus demonstrating the disposition of sensory hairs. Fig. 3.(a) The short chemosensory hairs on tarsal segments and (b) pore at the tip of the hair. Fig. 4.(a) Ornamented chemosensory hairs (b) pore at the tip of these hairs.

hair sensilla have polyneural innervation. The mechanosensory hairs have a smaller number of 3 to 5 neurons while the chemosensory hairs have as many as 30 neurons in the largest group observed. These groups of neurons are enclosed in thin, often multilayered glial sheaths. The nuclei of sensory neurons stain less intensely than the hypodermal nuclei. A cross section of the flagellum at the region of 'Fenestra Ovalis' shows a thin and membranous cuticle. Underlying this membranous cuticle are neurons which have a broad dendritic region. About 100 such neurons are found in each 'Fenetra Ovalis' (Fig. 9).

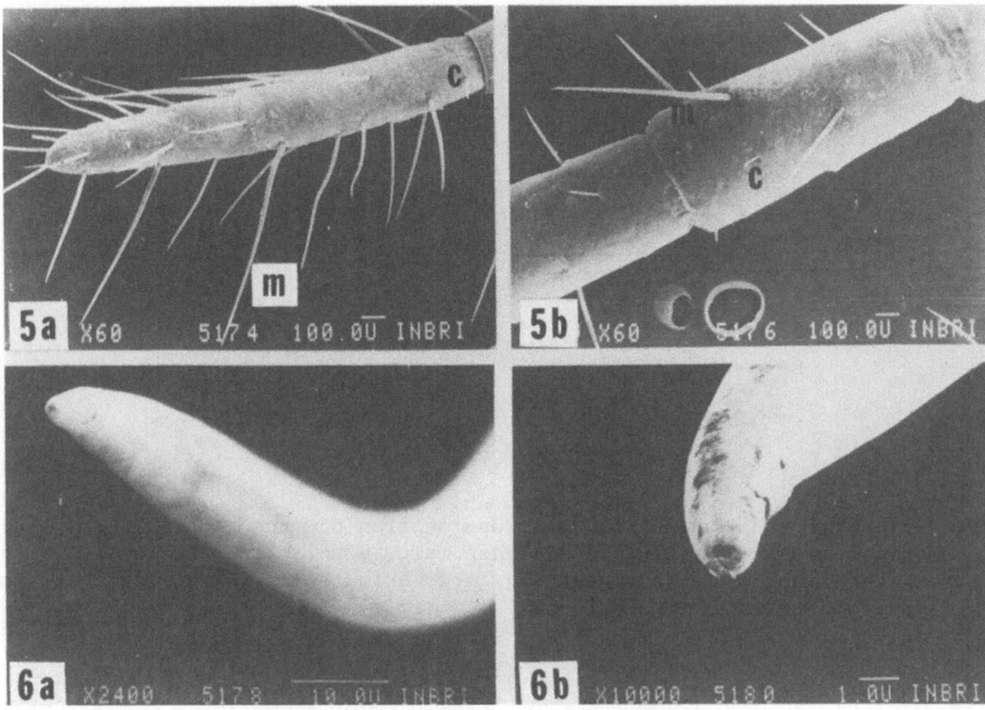

Fig. 5-6. Fig. 5 (a) Distribution of sensory hairs on the first segment of the flagellum. Long mechanosensory hairs (m) and chemosensory hairs (c) are seen. The segments proximal to the body show lesser number of hairs per segment. Pores seen at tip of chemosensory hairs are shown in (6a) and (6b).

The axons from groups of sensory cells assemble together forming small sensory nerves containing few axons. These groups of axons are covered by a glial sheath (Fig. 10). The smaller nerves may contain fascicles of axons enveloped by a glial sheath. Interstingly, the neurites of the small nerves form numerous synapses. These synapses are of different types based on the cell processes involved in synapse formation viz, Axo-axonal, axo-dendritic and axo-somatic. Often the synapses show synaptic bars with vesicles around them. The synaptic vesicles vary in shape and electron opacity and can be differentiated into clear, dense cored, and pleimorphic.

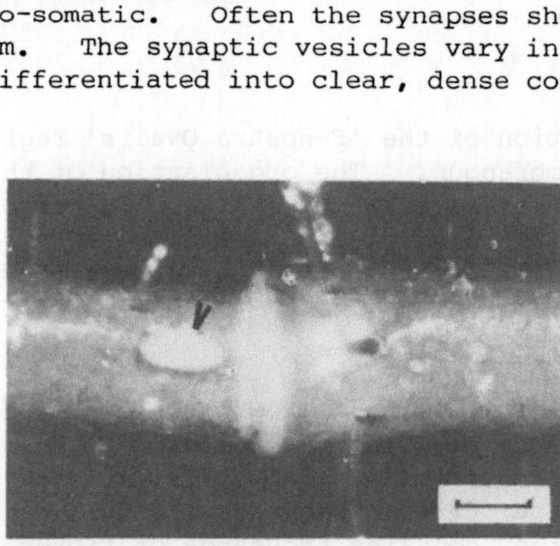

Fig. 7. 'Fenestra Ovalis' region at the base of a segment of the flagellum. Scale 0.2 mm.

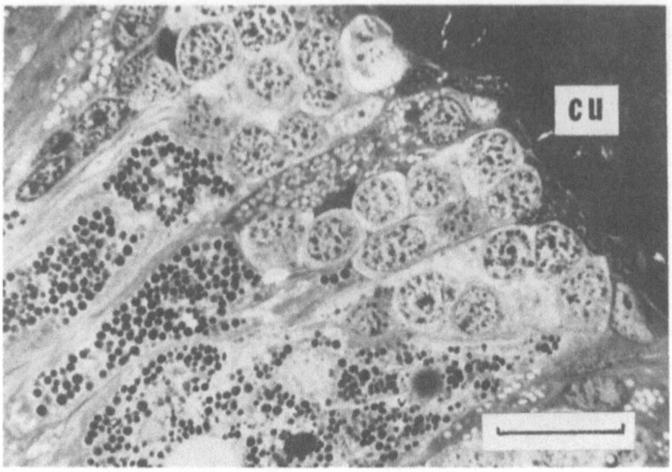

Fig. 8. Light micrograph of the cross section of the antenniform leg. Groups of receptor neurons are seen below a thick cuticle (cu). Note the presence of cells containing large granules below the level of the receptor neurons. Scale 25 µm.

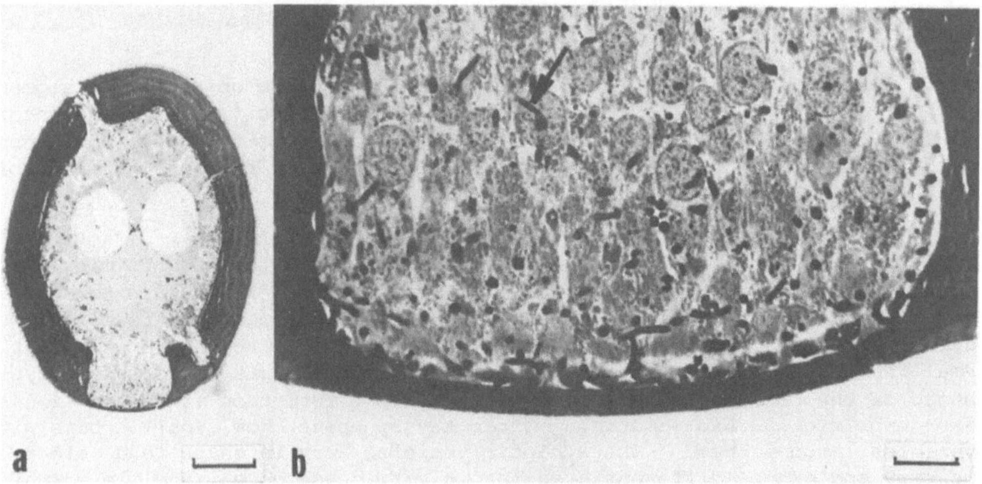

Fig. 9. (a) Cross section of the 'Fenestra Ovalis' region. The cuticle is thin and membranous. The organisation of the receptor neurons below the membrane is shown in (b). Receptor neurons with long broadened dendrites are seen. Chitinous thread like structures (arrow) form a mesh work inside the 'Fenestra Ovalis'. Scale: (a) 50.0 µm (b) 5.0 µm

Post synaptic density is lacking. Few convergent and diad synapses were also seen (Fig. 10). The small nerves assemble to form two large nerves in the lumen of the antenniform leg. The fibres in these two nerves are sensory nerves as the tarsal segments of the antenniform leg lacks musculature. Each nerve contains thousands of fibres and is covered by a perineurium (Fig. 11). Synapses which are abundant in the small nerves are rarely seen in these two large nerves. Fibres of a wide range of diameter are found in these nerves. They are predominantly fine in size.

However, the presence of few exceptionally large fibres is noteworthy. These fibres measure 6-12 μm in diameter and resemble the giant fibres found in the sensory nerves of whip spiders(Foelix and Troyer, 1980). They are wrapped individually in glial sheaths.

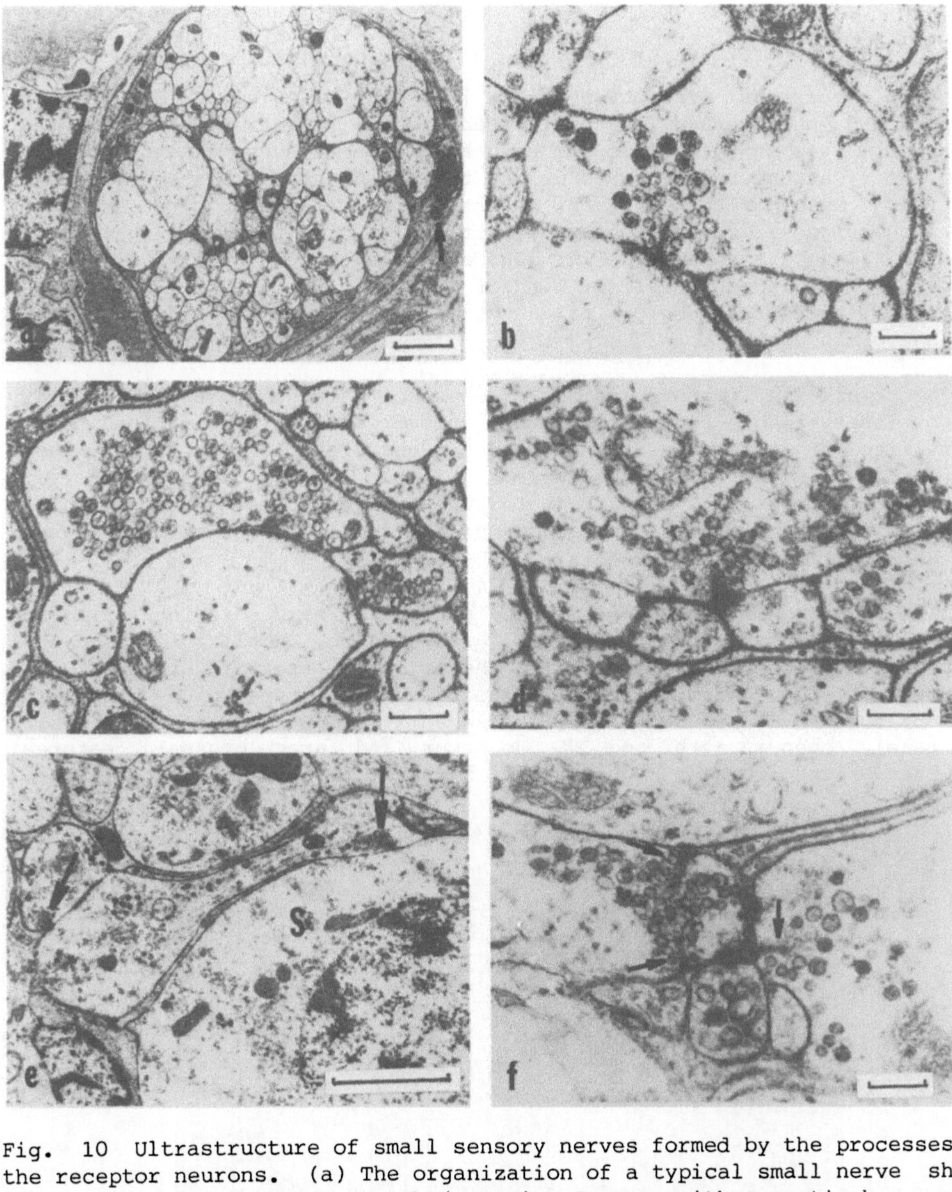

Fig. 10 Ultrastructure of small sensory nerves formed by the processes of the receptor neurons. (a) The organization of a typical small nerve shows fibres wrapped by a glial sheath (arrow). Synapse with synaptic bar around which the synaptic vesicles are clustered is shown in (b). (c). An example of convergent synapse involving two presynaptic fibres and a post synaptic fibre. (d) A diad synapse (e) Axo-somatic synapse. The fibre presynaptic to the soma is post synaptic to another fibre. Synapses are indicated by arrows. (f) Serial-like synapses are seen with the post synaptic fibre being presyanaptic to other fibres. Scale a. and e 5.0 μm, b,c,d,f 100 nm

A striking feature is the presence of cells containing large granules below the level of receptor neurons of the antenniform leg. They are large and osmiophilic measuring 0.5 µm (Fig. 8). Electron microscopic observations revealed few cells with striking microvillar formations of the cell membranes (Fig. 12). These cells also show desmosomes at specific regions. The organization of highly convoluted cell membranes in association with granule bearing cells resemble photosensory systems in their morphology.

DISCUSSION

Whip scorpions are nocturnal and like many other arachnids rely on mechano and chemosensory systems. Behavioural observations show that antenniform legs and flagellum are used as probes to sense the environment. The sensory structures observed during the present study may assist such a function. Most of the hairs on the antenniform legs bear pores at their tip suggesting a contact chemosensory role. Such receptors have been described among arachnids (Foelix, 1985 a). The sensory appendages of whip scorpions lack the specialised kind of sensory structures like sensilla basiconica, sensilla coelocnica and sensilla styloconica found on the antennae of insects.

A remarkable feature is the presence of a large number of sensory neurons innervating a sensillum. Their number is more in comparison to the innervation of insect sensillum. Insect mechanoreceptors have one neuron innervating each sensillum,whereas 3 to 5 neurons innervate mechanoreceptors/hairs of arachnids.The significance of multiple innervation is unclear as they do not confer any advantage and is presumed to be a primitive feature (Foelix, 1985b)

The occurrence of synapses in the peripheral nerves is intriguing and was first demonstrasted in various arachnids by Foelix (1975). Such peripheral synapses were also seen in the primitive arachnid <u>Limulus</u> (Hayes and Barber, 1982). Based on their morphology various types of synapses have been observed in the present study. Axo-somatic and serial synapses which are rarely seen in arthropod CNS do occur and are unusual features. The observations suggest interaction between receptor neurons. Thus a

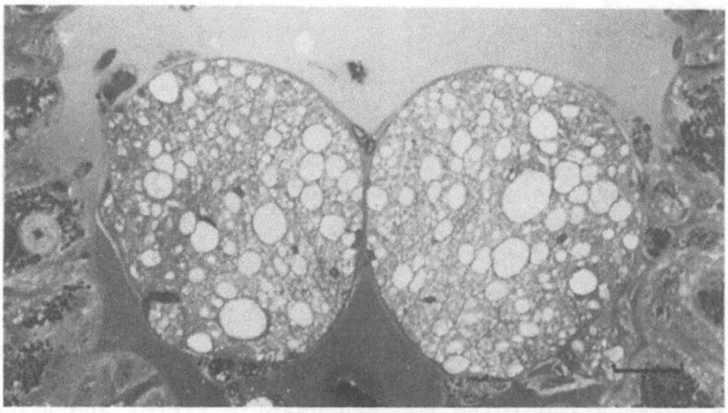

Fig. 11. Light micrograph showing the two large nerves in the lumen of the antenniform leg. The fibres are predominantly fine but a few of them are exceptionally large and can be termed "giants". Scale 10.0 µm

considerable degree of information processing may occur at the periphery. The functional significance of such a 'peripheral neuropil' is, at present, largely speculative. A functional analysis needs to be carried out to assess the role of peripheral nervous system of arachnids in integrating the sensory input. Foelix(1985 b) considers the occurence of synapses in the peripheral sensory nerves as a primitive feature. The peripheral synapses do not occur diffusely and are found in the small nerves formed by the processes of receptor neurons. Schurmann's study (1978) on Peripatus, which is regarded as a link between annelids and arthropods, demonstrated synapses to be confined to the ventral nerve cord and were not found in the leg nerves. In their ultrastructre peripheral synapses of whip scorpion and other arachnids resemble those of insects and vertebrates. The presence of synapses in the peripheral nerves suggesting processing of information in the periphery was, probably, an experiment of nature in the course of evolution.

A feature in the organization peripheral nerves which draws attention is the presence of exceptionally large fibres in the two large nerves of the antenniform leg. These fibres can be termed 'giants' by virtue of their size. They are comparable to the peripheral giant fibres in the sensory nerves of the whip spider demonstrated by Foelix and Troyer (1980). The cell bodies of giant fibres remains to be localised in whip scorpion and efforts are underway. The presence of giant fibres indicates fast conduction and rapid responses. While probing the surroundings whip scorpions tend to withdraw the antenniform legs rapidly when a noxious stimulus is encountered. Similar reactions were recorded in whip spider by

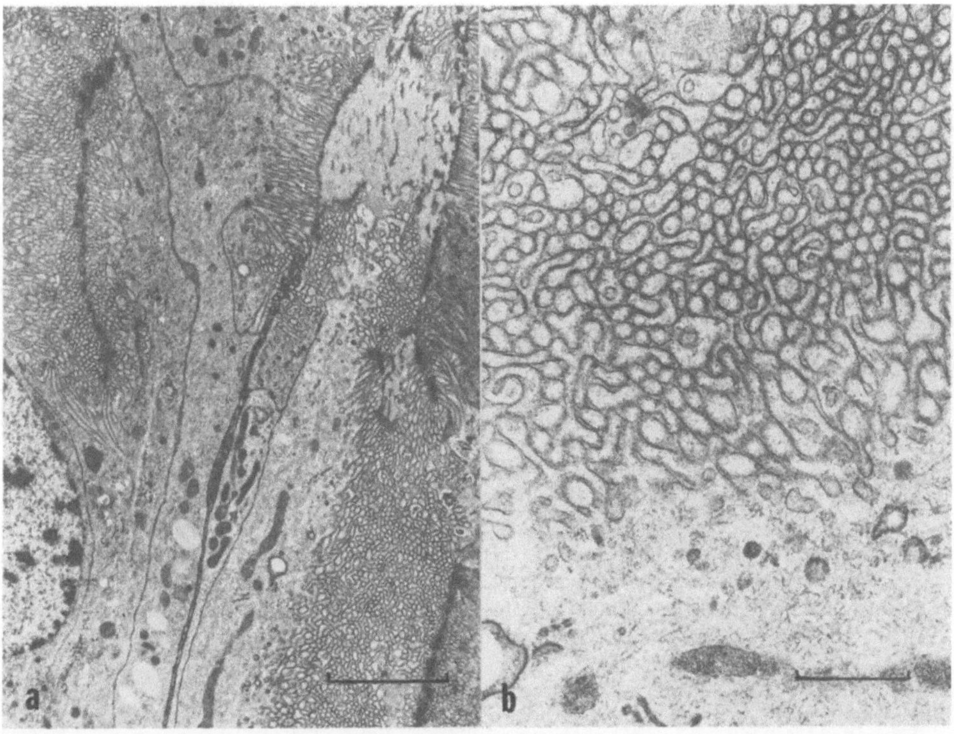

Fig. 12 (a). Cells with extensive microvillar membrane formations seen in the antenniform leg, below the cuticle. The microvillar arrangement is illustrated in (b). Scale: (a) 5.0 µm (b) 0.5 µm

Foelix and Troyer (1980). Peripheral giant fibres may confer an advantage by rapidly conducting the information to the CNS from the periphery. Peripheral sensory giant fibres appear unique to whip scorpion and whip spider as all known giant fibre systems in invertebrates are either interneurons or motor in nature. Whip scorpions thus appear to possess all the three types of giant fibre systems as giant interneurons in the ventral nerve cord (Rajashekhar and Geethabali, 1987) and giant efferent fibres (Rajashekhar and Geethabali, 1988) have also heen demonstrated in addition to the peripheral sensory giant fibres demonstrated in the present study.

The presence of cells with microvillar membrane formations associated with cells containing large osmiophilic granules tempts us to speculate a photosensory function of these strucutres in the light of demonstration of extraocular photoreception in whip scorpion by Patten (1919). Whether the strucutres are dermal photoreceptors needs to be examined.

ACKNOWLEDGEMENT

We thank Dr. Ramana Rao, IDL-Nitro Nobel Basic Research Institute, Bangalore for Scanning Electron Microscopy.

REFERENCES

Foelix, R.F. 1975. Occurence of synapses in the peripheral nerves of arachnids. Nature 254 : 146 - 148.

Foelix, R.F. 1985a. Mechano and Chemoreceptive Sensilla. In : F.G.Barth (ed) Neurobiology of Arachnids.

Foelix, R.F. 1985b. Sensory nerves and Peripheral Synapses In : F.G. Barth (ed) Neurobiology of Arachnids.

Foelix, R.F. and D.Troyer 1980. Giant neuorns and associated synapses in the peripheral nervous system of whipspiders. J.Neurocytol. 9 : 517 - 537.

Hayes, W.F. and S.B.Barber 1982. Peripheral synapses in Limulus Chemoreceptors. Comp. Biochem. Physiol 72A : 287 - 293.

Moro, S.D. and Geethabali 1985. Distribution of cuticular sensory hairs on the legs and whip of Thelyphonus indicus stoliczka (Arachnida Uropygi) Monitore Zool. Ital. 19 : 207 - 218.

Patten, B.M. 1919. Photoreception of partially blinded Whiptail Scorpions. J.Gen. Physiol 1 : 435 - 458.

Rajashekhar, K.P and Geethabali, 1987. "Giant" Fibres in the ventral nerve cord of whip scorpion Thelyphonus indicus Stoliczka. Curr. Sci, 56 : 1300 - 1301.

Rajashekhar, K.P. and Geethabali, 1988. Organization of the nervous system of whip scorpion Thelyphonus indicus Stoliczka. Rev. Arachnol (in Press).

Reyonolds, E.S. 1963. The use of citrate at high pH as an electron opaque stain in electron microscopy. J.Cell Biol. 17 : 208 - 212.

Schurmann F.W. 1978. A note on the structure of synapses in the ventral nerve cord of the Onycophoran Peripatoides leuckartii Cell Tissue Res. 186 : 527 - 534.

Slifer, E.H. 1970. The structure of arthropod Chemoreceptors. Ann. Rev. Entomol. 15 : 121 - 142.

Yoshikura, M. 1965 Post embryonic development of a whip scorpion Typopeltis stimpsonii (Wood). Mem. Fac. Gen. Educ. Kumamoto Univ. Ser. Nat. Sci. 1: 31-70.

PARTICIPANTS

Acharya, S.
Arunan
Balakrishnan, Rohini
Molecular Biology Unit
Tata Institute of Fundamental
  Research, Bombay-5, India.

Bijlani, V.
Bijlani (Mr.)
Department of Anatomy
All India Institute of
  Medical Sciences
Ansari Nagar, New Delhi-110 029
India.

Blest, A. David
Developmental Neurobiology Group
Research School of Biological
  Sciences
The Australian National University
P.O.Box 475
Canberra City, A.C.T. 2601
Australia.

Borst, Alexander W.
Max-Planck-Institut für
  Biologische Kybernetik
Spemannstrasse 38
D-7400 Tübingen 1
West Germany (F.R.G.).

Bult, R.
Laboratorium voor Algemene
  Natuurkunde
Rijksuniversiteit Groningen
Westersingel 34
9718 CM Groningen
The Netherlands.

Carlson, John
Department of Biology
Kline Biology Tower
Yale University
P.O.Box 6666
New Haven CT 06511-8112
U.S.A.

Chauhan, H.G.
Homi Bhabha Auditorium
Tata Institute of Fundamental Research
Bombay-5, India.

Chopra, Maninder S.
Centre for Cellular and Molecular
  Biology
Hyderabad-500 007.

Datta, Sumana
Department of Biology
Kline Biology Tower
Yale University
P.O.Box 6666
New Haven CT 06511-8112
U.S.A.

Deshpande, Seema
Molecular Biology Unit
Tata Institute of Fundamental Research
Bombay-5, India.

Elepfandt, Andreas
Faculty of Biology
University of Konstanz
D-7750 Konstanz
West Germany (F.R.G.).

Fernandes, Joyce
Molecular Biology Unit
Tata Institute of Fundamental Research
Bombay-5, India.

Fischbach, K.-F.
Institut für Biologie III
Schänzlestrasse 1
D-7800 Freiburg im Breisgau
West Germany (F.R.G.).

Gayatri, Archana
Molecular Biology Unit
Tata Institute of Fundamental Research
Bombay-5, India.

Geethabali
Department of Zoology
Bangalore University
Bangalore-560 056, India.

Gopinath, G.
Gopinath (Mr.)
Department of Anatomy
All India Institute of
Medical Sciences
Ansari Nagar, New Delhi-110 029
India.

Gopinathan, Asha
Department of Neurobiology
SUNY, Stonybrook
N.Y.11794-5230
U.S.A.

Götz, Karl G.
Max-Planck-Institut für biologische
Kybernetik,
Spemannstrasse 38
D-7400 Tübingen
West Germany (F.R.G.).

Gupta, Anil
Molecular Biology Unit
Tata Institute of Fundamental
Research, Bombay-5, India.

Hardie, R.C.
Department of Zoology
University of Cambridge
Downing Street
Cambridge CB2 3EJ
U.K.

Hausen, Klaus
Hausen, Marguerite
Zoologisches Institut der
Universität Koln
5000 Koln 41
West Germany (F.R.G.).

Hendrickson, Anita E.
Department of Biostructure &
Department of Ophthalmology
University of Washington
Seattle WA 98195
U.S.A.

Hensler, Klaus
Zoologisches Institut der
Universität Basel
Rheinsprung 9
CH-4051 Basel
Switzerland.

Horridge, G. Adrian
Centre for Visual Sciences
Research School of Biological
Sciences
The Australian National
University
P.O.Box 475, Canberra City, ACT 2601
Australia.

Järvilehto, H.
Department of Zoology &
Department of Physiology
University of Oulu
Linnanmaa
SF-90570 Oulu
Finland.

Jallon, J.M.
Laboratoire de Biologie et
Génétique Evolutives du CNRS
F-91190 Gif-sur-Yvette
France.

Joshi, Swati
Kakeri M.M.
Molecular Biology Unit
Tata Institute of Fundamental
Research, Bombay-5, India.

Kankel, Douglas R.
Department of Biology
Kline Biology Tower
Yale University
P.O.Box 6666
New Hacen CT 06511-8112
U.S.A.

Kasaiah, A.
Department of Zoology
Sri Venkateswara University
Post Graduate Centre
Kavali-524 202
District Nellore, India.

Kenkare, U.W.
National Tissue Culture Facility
Zoology Department
Poona University
Pune, India.

Kokila, S.K.
Department of Zoology
Bangalore University
Bangalore-560 056, India.

Krishna Prasadan, T.N.
Department of Biosciences
Sardar Patel University
Vallabh Vidyanagar 388 120
Gujarat, India.

Kucheria, K.
Department of Anatomy
All India Institute of
Medical Sciences
Ansari Nagar, New Delhi-110 029
India.

Lobo, Cheryl
Molecular Biology Unit
Tata Institute of Fundamental
Research, Bombay-5, India.

McKay, R.
Departments of Biology & Brain
and Cognitive Sciences
E 25-435 Massachusetts Institute
of Technology (MIT)
Cambridge MA 02139, U.S.A.

Mehra, Raj D.
Department of Anatomy
All India Institute of
Medical Sciences
Ansari Nagar, New Delhi-110 029
India.

Mistri, Rashid
Molecular Biology Unit
Tata Institute of Fundamental
Research, Bombay-5, India.

Mizunami, Makoto
Department of Biology
Faculty of Science 33
Kuushu University
Fukuoka 812
Japan.

Nässel, Dick R.
Department of Zoology
University of Stockholm
Svante Arrhenius v. 14-16
S-106 91 Stockholm
Sweden.

Nilsson, Dan -E.
Department of Zoology
University of Lund
Helgonavägen 3
S-223 62 Lund
Sweden.

Paranjpe, Jaishri
Molecular Biology Unit
Tata Institute of Fundamental
Research, Bombay-5, India.

Parelker, M.A.
Public Relations Officer
Tata Institute of Fundamental
Research, Bombay-5, India.

Pflüger, Hans -Joachim
Freie Universität Berlin
Institut für Neurobiologie
Königin-Luise-Strasse 28-30
D-1000 Berlin (F.R.G.).

Pinto, Ludwin
Molecular Biology Unit
Tata Institute of Fundamental
Research, Bombay-5, India.

Pollak, George D.
Department of Zoology
The University of Texas at Austin
Austin TX 78712-1064, U.S.A.

Ponder, Betty M.
School of Graduate Studies & Research
University of New Brunswick
P.O.Box 4400, Fredericton, N.B.
Canada E3B 5A3.

Premani, Chetan
Raghuram, V.
Molecular Biology Unit
Tata Institute of Fundamental
Research, Bombay-5, India.

Rajashekhar, K.P.
C/o. Prof. J.L.Wilkens
Department of Biological Sciences
The University of Calagary
2500 University Drive, N.W.
Calagary, Alberta T2N 1N4
Canada.

Ramamohan, Y.
Electron Microscope Laboratory
Department of Neuropathology
National Institute of Mental Health
and Neurosciences (NIMHANS)
Bangalore-560 029, India.

Ramaswami, Mani
Division of Biology
California Institute of Technology
Pasadena Ca 91125, U.S.A.

Redkar, V.D.
Rodrigues, Veronica
Molecular Biology Unit
Tata Institute of Fundamental
Research, Bombay-5, India.

Rospars, Jean Pierre
Laboratoire de Biométrie
Institut National de la Recherche
Agronomique
Route de Saint-Cyr
F-78000 Versailles, France.

Rudolph, A.
C/o. Dr. R.C.Hardie
Department of Zoology
University of Cambridge
Downing Street, Cambridge CB2 3EJ
U.K.

Subberwal, U.
Deapartment of Anatomy
All India Institute of
Medical Sciences
Ansari Nagar, New Delhi-110 029
India.

Sakai, Hiroko M.
National Institute of
Basic Biology
Okazaki 444 Japan.

Sakai, Masaki
Department of Biology
Faculty of Science
Okayama University
Tsushima-Naka 3-1-1
Okayama 700 Japan.

Sasira Babu, K.
Department of Zoology
Sri Venkateswara University
Post Graduate Centre
Kavali 524 202
District Nellore, India.

Seabrook, William D.
School of Graduate Studies
University of New Brunswick
P.O.Box 4400, Fredericton, N.B.
Canada E3B 5A3.

Shanbhag, Shubha R.
Molecular Biology Unit
Tata Institute of Fundamental
Research, Bombay-5, India.

Shikata, Ayako
Shikata, Yoshihiro
Department of Mathematics
Faculty of Science
Nagoya University
Chikusa-Ku, Nagoya 464 Japan.

Siddiqi, Ashya
Siddiqi, Obaid
Molecular Biology Unit
Tata Institute of Fundamental
Research, Bombay-5, India.

Siddiqui, Shahid S.
Laboratory of Molecular Biology
Toyohashi University of Technology
Tempaku-cho, Toyohashi 440 Japan.

Singh, Aalok R.
Singh, Kusum
Singh, R. Naresh
Molecular Biology Unit
Tata Institute of Fundamental
Research, Bombay-5, India.

Singh, Satpal
Department of Biology
The University of Iowa
Iowa City, Iowa 52242, U.S.A.

Sinha, Anindya
Molecular Biology Unit
Tata Institute of Fundamental
Research, Bombay-5, India.

Sivasubramanian, P.
Department of Biology
University of New Brunswick
Bag Service No. 45111
Fredericton, N.B.
Canada E3B 6E1.

Sood, Rashmi
Molecular Biology Unit
Tata Institute of Fundamental
Research, Bombay-5, India.

Srinivasan, M.V.
Centre for Visual Sciences
Research School of Biological Sciences
The Australian National University
P.O.Box 475, Canberra City A.C.T. 2601
Australia.

Strausfeld, Camilla
Strausfeld, N.J.
Division of Neurobiology
Arizona Research Laboratories
611 Gould-Simpson Science Building
The University of Arizona
Tucson AZ 85721, U.S.A.

Sur, Mriganka
Department of Brain and Cognitive
Sciences
Massachusetts Institute of Technology
Cambridge Ma 02139, U.S.A.

Szél, Ágoston
Semmelweis University of Medicine
Laboratory I of Electron Microscopy
H-1450 Budapest, IX.
Túzoltó Utca 58, Hungary.

Venard, Renée
Laboratoire de Biologie et Génétique
Evolutives du CNRS
F-91190 Gif-sur-Yvette, France.

Venkataramana, N.K.
Department of Neurosurgery
National Institute of Mental Health &
Neurosciences (NIMHANS)
Bangalore-560 029, India.

Vijay Sarthy, P.
Department of Ophthalmology,
Physiology and Biophysics
University of Washington
Seattle WA 98195
U.S.A.

Vishwas Sarangdhar
Molecular Biology Unit
Tata Institute of Fundamental
Research, Bombay-5, India.

Wadhwa, Shashi
Department of Anatomy
All India Institute of
Medical Sciences
Ansari Nagar, New Delhi-110 029
India.

White, Kalpana
Department of Biology
Brandeis University
Bassine 235
Waltham, Massachusetts 02254
U.S.A.

Zeil, Jochen
Lehrstuhl für Biokybernetik
Universität Tübingen
Auf der Morgenstelle 28
D-7400 Tübingen
West Germany (F.R.G.).

Fig. 1. Left to Right . (a) J.P.Rospars, S.R.Shanbhag, K.Singh, S.D. Deshpande. (b) A.G.Gayatri, S.N.Joshi, R.J.Mistri, M.C.Arunan, L.Pinto. (c) D.R.Nässel, D.-E.Nilsson. (d) J.Zeil. (e) R.N. Singh, M.A.Parelker. (f) S.S.Siddiqui, K.G.Götz, A.Elepfandt, Á.Szél. (g) V.Sarangdhar, M.Sakai, H.M.Sakai. (h) A.Siddiqi, G.D.Pollak, O.Siddiqi, C.Strausfeld. (i) M.Sur, C.Lobo, R.Sood, S.Acharya.

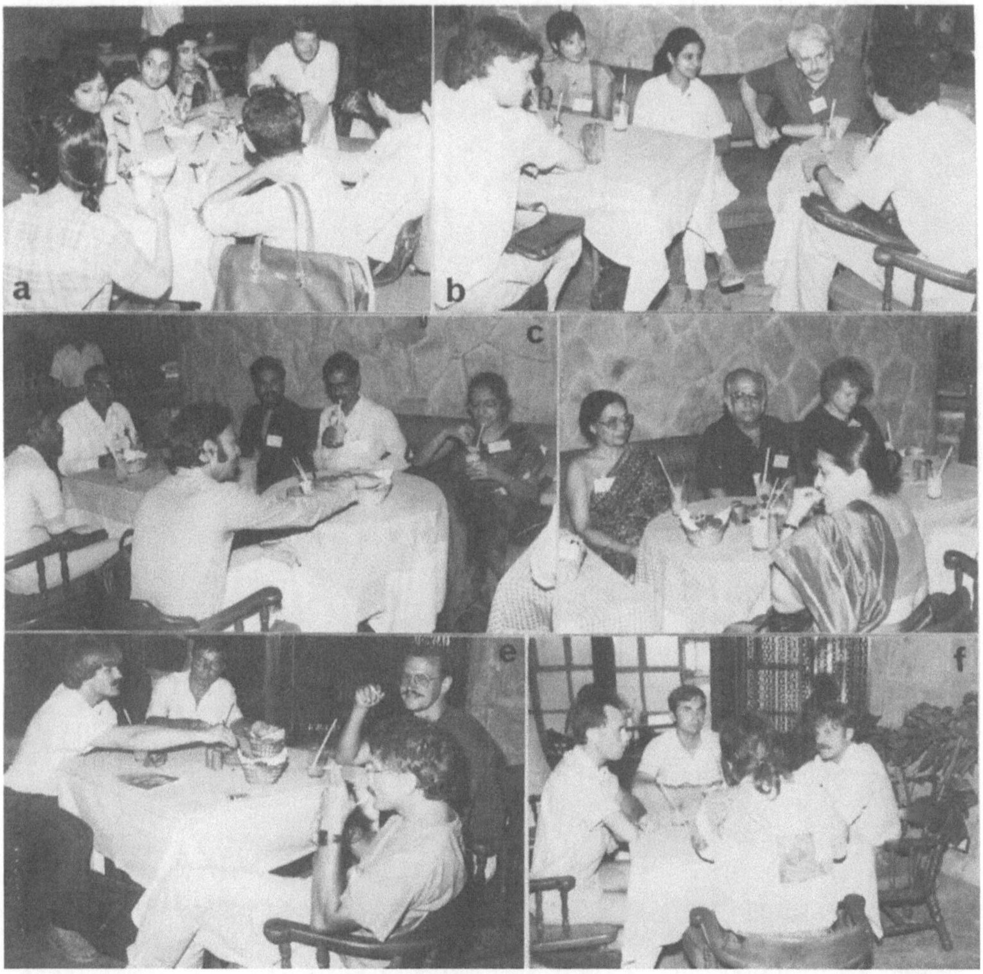

Fig. 2. Left to Right. (a) J.M.Paranjpe, A.Gayatri, L.Pinto, M.Järvilehto. (b) R.Bult, S.Datta, K.White, D.R.Kankel, J.Carlson. (c) A.Kasaiah, K.Sasira Babu, Y.Ramamohan, N.K.Venkataramana, K.P.Rajashekhar, Geethabali. (d) V.Bijlani, Mr. Bijlani, A.Hendrickson, V.Kucheria. (e) K.F.Fischbach, K.Hausen, A.Borst, J.Zeil. (f) D.E.Nilsson, A.Elepfandt, H.J.Pflüger.

Fig. 3. Left to Right. (a) S.D.Deshpande, G.A.Horridge, R.N.Singh. (b) G.A.Horridge, R.N.Singh. (c) 1st row : H.G.Chauhan, K.G.Götz, K. Hausen. 2nd row : J.Fernandes, R.Balakrishnan, A.Gayatri, J.M. Jallon, R.J.Mistri. 3rd row : Y.Raghuram, R.Sood, C.Lobo, J. Paranjpe. (d) 1st row : K.Hensler, Á.Szél. 2nd row : A.Borst, A. P.Gupta, M.M.Kakeri, V.D.Redkar. (e) S.D.Deshpande, S.R.Shanbhag, R.N.Singh. 2nd row : M.Järvilehto, K.Hausen, A.Elepfandt. (f) 1st row : G.D.Pollak. 2nd row : G.A.Horridge, G.Gopinath, S.Wadhwa. 3rd row : D.A.Blest, U.W.Kenkare, P.Vijay Sarthy, K.Singh. (g) P. Vijay Sarthy, R.Mehra, A.Hendrickson, S.N.Joshi, L.Pinto, C.Premani. (h) R.Venard, M.V.Srinivasan, S.S.Siddiqui, V.D.Redkar, G.A.Horridge, A. Elepfandt. (i) R.McKay, A.Gopinathan, T.N.Krishna Prasadan, S.Datta, M.C.Arunan.

Fig. 4. Left to Right. (a) G.A.Horridge, R.Mckay. (b) A.Borst, M. Mizunami. (c) R.C.Hardie, J.Zeil. (d) M.V.Srinivasan. (e) H.M.Sakai. (f) D.R.Kankel. (g) K.Hausen, R.N.Singh. (h) G.D. Pollak, J.M.Jallon. (i) M.Ramaswami. (j) K.Hensler.

Fig. 5. Left to Right. (a) M.Järvilehto. (b) M.Sur. (c) S.Wadhwa. (d) N.J.Strausfeld. (e) D.A.Blest. (f) K.-F.Fischbach. (g) J.Carlson. (h) V.Rodrigues. (i) K.P.Rajashekhar. (j) H.J. Pflüger.

Fig. 6. Left to Right. (a) A.Shikata, H.J.Pflüger, M.Sakai. (b) S.D. Deshpande, C.Lobo, S.R.Shanbhag. (c) M. Hausen, M.Sakai, A. Gayatri, S.R.Shanbhag, H.M.Sakai, C.Strausfeld. (d) A.Gayatri, S.Datta, C.Lobo, K.Singh, A.Shikata. (e) A.Rudolf, H.M.Sakai, R.Sood, K.Singh, U.Subberwal. (f) R.Sood, K.Hensler, A.Sinha, J.Carlson, T.N.Krishna Prasadan, S.Acharya, K.P.Rajashekhar, S.K.Kokila. (g) M.A.Parelker, D.A.Blest, Y.Shikata, A.Gayatri, A.Borst, R.C.Hardie, S.Singh, S.N.Joshi, S.D.Deshpande, P. Vijay Sarthy, K.Singh, R.N.Singh, M.Sur.

INDEX

MIX
Papier aus verantwortungsvollen Quellen
Paper from responsible sources
FSC® C105338

If you have any concerns about our products,
you can contact us on
ProductSafety@springernature.com

In case Publisher is established outside the EU,
the EU authorized representative is:
Springer Nature Customer Service Center GmbH
Europaplatz 3, 69115 Heidelberg, Germany

Printed by Libri Plureos GmbH
in Hamburg, Germany